Practical Fluorescence Spectroscopy

Practical Fluorescence Spectroscopy

Zygmunt (Karol) Gryczynski
Professor, Department of Physics and Astronomy, Texas Christian University
and University of North Texas Health Science Center

Ignacy Gryczynski
University of North Texas Health Science Center

CRC Press is an imprint of the
Taylor & Francis Group, an **informa** business

CRC Press
Taylor & Francis Group
6000 Broken Sound Parkway NW, Suite 300
Boca Raton, FL 33487-2742

First issued in paperback 2022

ISBN-13: 978-1-439-82169-5 (hbk)
ISBN-13: 978-1-03-233737-1 (pbk)
DOI: 10.1201/9781315374758

Publisher's Note

The publisher has gone to great lengths to ensure the quality of this reprint but points out that some imperfections in the original copies may be apparent.

Visit the Taylor & Francis Web site at
http://www.taylorandfrancis.com

and the CRC Press Web site at
http://www.crcpress.com

Contents

Preface

FLUORESCENCE SPECTROSCOPY HAS BECOME an indispensable tool in many scientific disciplines from biology and ecology to chemistry and physics. This book is meant to be a practical reference and guide for those who plan to use fluorescence in their respective field, or to those who are new to using fluorescence. The main focus is to describe how to perform experiments, from initial design and set up to data analysis and presentation. Advantages and disadvantages of various techniques are presented with an emphasis on the design of experiment (DOE) to suit the reader's specific needs. This book is designed to include sufficient theory and background to bridge the gap between introductory classroom textbooks and current journal articles as related to particular experiments. It is intended for students starting lab work and professionals from different fields venturing into the fluorescence spectroscopy. This is not a theoretical textbook with problem sets, but rather a tutorial for practical applications. Starting from basic experiments on transmission and absorption of light, simple design and interpretation of steady-state fluorescence experiments, to advanced experiments involving oriented systems, time-resolved fluorescence, fluorescence quenching, and Förster Resonance Energy Transfer (FRET).

Design and discussions of specific experiments are meant to teach less known nuances of experimental strategy, data fitting, and analyses, as well as potential problems and pitfalls. It is important to remember that a single fluorescence technique/experiment usually is applicable across many fields and sample types. Data from an absorption measurement can be used to determine how many fluorophores have labeled a protein (i.e., after using a commercial labeling kit) or the percent of incorporation of magnesium into a ZnMgO thin film on sapphire (i.e., after semiconductor growth). Thus, it is important to consider that while the exact experiment one wishes to perform may not be described within this text, an identical or at least similar experimental design is likely to be presented but applied to a different type of problem and/or sample.

The first two chapters introduce necessary theoretical concepts, experimental designs, and the basics of data analysis. Chapters 3 through 8 present sets of different experiments. Experiments are divided into two parts. Chapters 3 through 5 describe basic steady-state experiments starting from absorption/transmission measurements, basic concepts of steady-state fluorescence measurements, and simple examples of basic applications of steady-state fluorescence like fluorescence quenching. This part is meant to introduce the new experimentalist/user to the principals of proper fluorescence measurement. Chapters 6 through 8 cover more advanced fluorescence concepts from fluorescence anisotropy and time-resolved fluorescence to examples of more advanced experiments that will typically involve both steady-state and time-resolved approaches. This part demonstrates more powerful applications of fluorescence technology and potential experimental problems that an experimentalist may encounter.

Each chapter in the experimental section is preceded by an abbreviated introduction that serves as a refresher of basic principles applicable to a particular group of experiments. This acknowledges that the nomenclature and certain formalisms differ across disciplines. It aims to show that these are different ways of writing the same thing and that they facilitate cross-disciplinary collaboration. Each introduction presents and discusses experimental basics and instrumentation used in presented experiments. Guidelines are presented, as well as, a general rule of thumb for conducting experiments and selecting the most appropriate types of tools with some practical limitations. In addition, each experiment is preceded by a short description of a problem and the theoretical and experimental basis needed for this particular experiment. Each experiment ends with brief "*Conclusions.*" Also, to address some potential problems in the experiment description, a "*Note*" is indicated in the text addressing/emphasizing certain specific issues.

Chapter 3 presents a series of introductory experiments on how to properly measure sample transmittance/absorption and discuss problems and limitations of such experiments. Chapter 4 includes a series of basic fluorescence experiments that are independent of time (steady-state measurements), as well as interpretations applicable to those experiments. Chapter 5 follows the format of Chapter 4 but expands to some practical steady-state applications and examples. In Chapter 6, steady-state anisotropy experiments are discussed, Chapter 7 describes time-resolved experiments, and Chapter 8 is dedicated to examples of more advanced fluorescence experiments.

Readers may be somewhat surprised that this book contains only a few original references. This is because all experiments were done by us and our students from the beginning to the end.

For a broader perspective, we recommend these textbooks on fluorescence:

- Joseph R. Lakowicz: *Principles of Fluorescence Spectroscopy*, 2006, Springer Science + Business Media, LLC.
- Bernard Valeur and Mario N. Berberan-Santos: *Molecular Fluorescence. Principles and Applications*, 2012, Wiley-VCH Verlag GmbH & Co. KGaA.
- David M. Jameson: *Introduction to Fluorescence*, 2014, Taylor & Francis Group/CRC Press.

This book has been a lengthy project that took over 5 years to complete. During this time, many of our colleagues and friends became involved and contributed numerous valuable suggestions and insights. Initially, we were unsure of how to approach the problem of writing a book on the experimental aspects of fluorescence and Dr. Karol Gryczynski (son of Dr. Zygmunt (Karol) Gryczynski) was instrumental in keeping us running and patiently writing/rewriting initial chapters. We also need to thank many of our students who were involved in testing and reading the numerous experiments in this book. We would like to acknowledge Dr. Sangram Raut, Dr. Sunil Shah, M.Sc. Zhangatay Nurekeyev, Dr. Hung Doan, Dr. Joe Kimball, Dr. Rahul Chib, Dr. Sebastian Requena, Jose Chavez, Luca Ceresa, and Bobby Pendry. We also want to thank our long-term collaborators and friends for reading and correcting parts of the experiments—Dr. Anna Synak, Dr. Rafal Fudala, Dr. Rafal Luchowski, and Jack Gryczynski (son of Dr. Ignacy Gryczynski) for help with proofreading. We also thank the University of North Texas Health Science Center and Texas Christian University for providing the intellectual environment that supported this project.

We also would like to acknowledge the publishers, Taylor and Francis. With particular thanks to Mr. Lou Chosen for originally supporting the idea for writing the experimental handbook and Mr. Aoife McGrath, in the final phase, for his patience and understanding.

Finally, we have to sincerely acknowledge the patience of our families, especially our wives, for their long and tireless support over this time, as well as over 40 years supporting our adventure with fluorescence.

Authors

Dr. Zygmunt (Karol) Gryczynski, PhD, is "Tex" Moncrief Jr. chair and professor of Physics, Department of Physics and Astronomy, Texas Christian University, and director and professor, Center for Fluorescence Technologies and Nanomedicine (CFTN), Department of Microbiology, Immunology, and Genetics, University of North Texas, Health Science Center. Dr. Gryczynski received his MS in experimental physics in 1982 from the University of Gdansk, Poland, and PhD in spectroscopy in 1987, working on the basic spectroscopic studies of isotropic and oriented systems of organic molecules. In 1991 he became a research assistant professor in the Department of Biochemistry and Molecular Biology, University of Maryland and 1998–2004 he was an assistant director in the Center for Fluorescence Spectroscopy at the University of Maryland. From 2005 he is a professor of Molecular Biology and Immunology at the University of North Texas Health Science Center at Fort Worth, Texas. In 2006 with the support from the Emerging Technology Funds (ETF) of Texas together with his colleagues he established a Center for Commercialization of Fluorescence Technologies (CCFT) that in 2013 has been transformed to the Center for Fluorescence Technologies and Nanomedicine (CFTN). In 2010 he became the "Tex" Moncrief Jr. Chair and professor of Physic in the Department of Physics and Astronomy, Texas Christian University at Fort Worth.

He has authored over 300 peer-review publications, 12 book chapters, 10 patents, and 15 edited books. He is also a member of editorial boards of *Journal of Experimental Biology and Medicine* and *Methods and Applications in Fluorescence.*

Dr. Ignacy Gryczynski, PhD, is professor in the Department of Microbiology, Immunology and Genetics, University of North Texas Health Science Center. Dr. Gryczynski received his MS (1973) and PhD (1977) in Physics from the University of Gdansk, Poland. For 20 years, since 1985, he was working in the Center for Fluorescence Spectroscopy (at the University of Maryland School of Medicine in Baltimore) directed by Dr. Joseph R. Lakowicz. In 2005 he moved to UNTHSC in Fort Worth, Texas, where he directed Microscopy Core Facility and co-directed Center for Commercialization of Fluorescence Technologies at UNTHSC. His research interests cover basics of fluorescence spectroscopy and its applications in biochemistry and biology.

He has authored over 470 peer-reviewed publications and is a member of editorial boards of *Journal of Experimental Biology and Medicine* and *Journal of Photochemistry and Photobiology B: Biology.*

CHAPTER 1

Theory of Light and Light Interaction with Matter

Conventionally, we call "light" a small range of a broad spectrum of electromagnetic radiation that corresponds to the visible spectral range (400–700 nm) as shown in Figure 1.1. Light is electromagnetic radiation which presents properties of both a wave and a particle. A "particle" of light is called a photon. The dual nature of light is often a source of confusion, but experiments confirming both interpretations exist. For example, experiments involving single-photon double-slit interference and the photoelectric effect have shown both the wave and particle nature of light, respectively. A photon is a quanta or the smallest possible amount/part of an electromagnetic energy/wave. Since we almost always refer to multiple photons, a photon can be referred to as a quantum of electromagnetic radiation. Both definitions will be used throughout this text. For example, a single chromophore (molecule) absorbs or emits a photon of a certain energy. It is important to remember that when discussing photons or electromagnetic waves, we are talking about light; both representations are equivalent, but one is often much easier applied to a particular scenario. For example, it is more natural to discuss a metal interacting with an oscillating electric wave rather than a stream of "particles" (photons) that never physically collide with the metal surface.

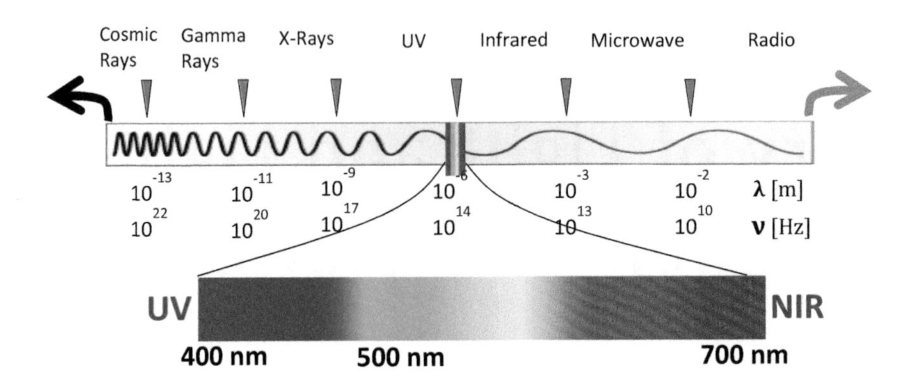

FIGURE 1.1 Electromagnetic radiation. The expanded range of 400–700 nm represents visible light.

1-1 BASICS OF LIGHT

This section is meant as a review of the basics of light as electromagnetic waves or photons. More importantly, it should demonstrate that there are many different ways of looking at light from a mathematical perspective, but they all lead to the same interpretation. Various uses of light are presented across many different fields, especially biology, chemistry, physics, and engineering. Within each field, and indeed specializations within each individual field, the nomenclature and properties of interest vary. For example, in biology, spectra are frequently given in a wavelength scale (nanometers) but in the semiconductor field, they are typically presented on a scale in terms of energy (electron volts). However, any spectrum may be presented in wavelength or energy without a loss of information. By the end of this chapter, the reader should have a fluid understanding of the relationships between different notations.

When looking at the light, one most readily thinks about its brightness and color. The brightness of the light is determined by the intensity (energy), passing through a certain surface area orthogonal to the direction of light propagation per unit of time (we also call it photon flux). The intensity of light is typically denoted as I. A simple way of quantifying the brightness of the light is to think about how many photons hit a detector (e.g., human retina, charge-coupled device, photodiode) per unit time. The more photons that hit a detector, the brighter (i.e., more intense) the light source is. From the wave point of view, the intensity of the electromagnetic radiation is proportional to the square of the electric field amplitude. Color is related to the period (duration of time of one cycle of a wave) of the electromagnetic radiation; equivalently, color (wavelength) is related to the frequency or energy of the light wave.

Figure 1-1.1 shows a typical schematic of a light wave. The electromagnetic field is depicted as orthogonal electric and magnetic fields $\boldsymbol{E}$ and $\boldsymbol{B}$, respectively. It should be noted that $\boldsymbol{E}$ and $\boldsymbol{B}$ are rarely drawn to scale as the amplitude of $\boldsymbol{B}$ is roughly $1/c$ the size of that of $\boldsymbol{E}$. In Figure 1-1.1, the wave is traveling in the $\hat{z}$-direction, the electric field points in the $\hat{x}$-direction, and the magnetic field points in the $\hat{y}$-direction. It is important to notice

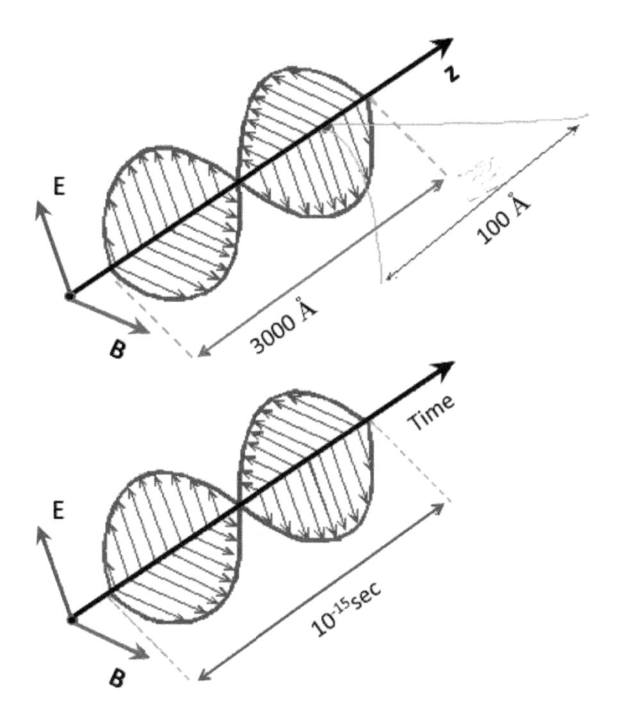

FIGURE 1-1.1 A light wave and its comparison with a typical chromophore.

that the electric field, magnetic field, and the direction of propagation are all orthogonal (akin to the "right-hand rule"). The electric field of light is given as:

$$\boldsymbol{E}(x,t) = E_0 \hat{\boldsymbol{e}} \sin(kx + \omega t + \varphi) \tag{1-1.1}$$

E_0 is the maximum intensity of the electric field or the amplitude of the electric vector and $\hat{e}$ is the directional unit vector. This directional unit vector indicated allows for the direction of the electric field to point in any arbitrary direction. In the example in Figure 1-1.1, $\hat{e} = \hat{x}$; one may also see $\boldsymbol{E}_0$, a constant vector, in place of $E_0\,\hat{x}$. In this case, $E_0 = |\boldsymbol{E}_0|$, and $\hat{e} = \frac{E_0}{|E_0|}$. Thus, Equation 1-1.1 becomes:

$$\boldsymbol{E}(x,t) = \boldsymbol{E}_0 \sin(kx + \omega t + \varphi) \tag{1-1.2}$$

Thus, the intensity of this light ray is:

$$I = \frac{cn\varepsilon_0}{2}|\boldsymbol{E}|^2 = \frac{cn\varepsilon_0}{2}\boldsymbol{E}_0^2 \tag{1-1.3}$$

where c is the speed of light, n is the refractive index of the medium the light is traveling in, and ε_0 is the permittivity of vacuum. Light sources such as lamps and lasers are often described by the power, P of the light they emit:

$$P = \int I dS \tag{1-1.4}$$

where S is the surface area of the light used, usually the focused area of the lamp or laser is called the spot size. Equation (1-1.4) often simplifies to the product of an intensity and an area $P = I \times S$. The area of the spot size is usually a circle ($S = 4\pi r^2$) or a rectangle (width × height). Power is given as a time average, for pulsed light sources this includes the time intervals when there is no light emitted. Thus, the instantaneous intensity of a pulsed laser, while hitting a sample, is greater than that of a non-pulsed (steady state) light source for the same given average power. This is an important consideration for sample damage. Power relates well to the number of photons emitted per second, $N_\gamma\, s^{-1}$:

$$\frac{N_\gamma}{s} = P\frac{\lambda}{h\,c} \tag{1-1.5}$$

where λ is the wavelength, a parameter corresponding to the color. Equation (1-1.5) relates the particle and wave natures of light. It gives the number of light particles (photons) as a function of the power of a light wave.

The electric field, in the example of Figure 1-1.1, is a sine wave on the XZ plane depicted temporally in Figure 1-1.2.

E_0 is the amplitude of the wave. The angular frequency ω (rad/s) is $\omega = 2\pi\nu$ where ν is the frequency of light in cycles per second (Hz). The distance from crest to crest of the wave is the wavelength, $\lambda = c/\nu$. An increase in ω decreases the distance from crest to crest or increases the frequency with which the wave repeats. The phase, φ, relates to a translation of the wave on the t-axis. In the example of Figure 1.1-2, the wave is a sine wave and φ is depicted as the distance from the origin to the first zero. The phase shift (translation on t-axis) is the same at any point of the wave.

Light is characterized in space by a wave vector $\boldsymbol{K}$:

$$\boldsymbol{K} = \frac{2\pi}{\lambda}\hat{r} = \frac{2\pi\nu}{c}\hat{r} = \frac{\omega}{c}\hat{r} \tag{1-1.6}$$

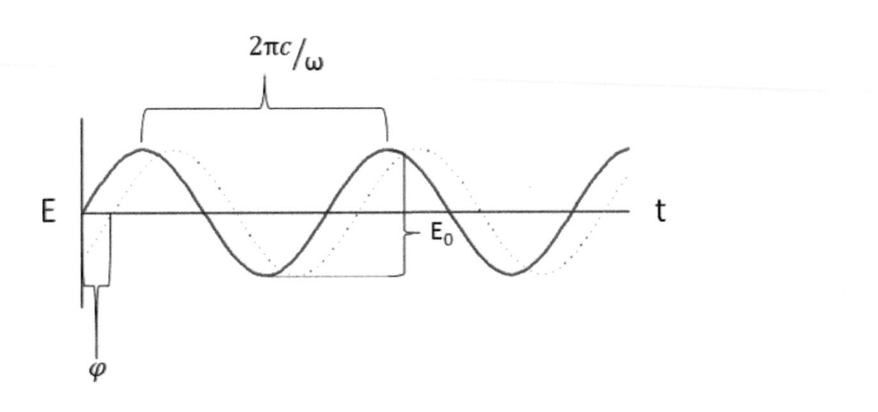

FIGURE 1-1.2 Electric field wave (solid) and electric field with a non-zero phase shift (dotted).

where $\hat{r}$ is the propagation direction of the wave in Figure 1-1.1, $\hat{r} = \hat{z}$ since the wave propagates in the $\hat{z}$-direction. The wavenumber k is defined as:

$$k = |\boldsymbol{K}| = \frac{2\pi}{\lambda} \tag{1-1.7}$$

It is important to note that not all disciplines include the factor of 2π in the definition of $\boldsymbol{K}$ and k, and thus other texts may use:

$$k = |\boldsymbol{K}| = \frac{1}{\lambda} \tag{1-1.8}$$

The factor of 2π will be somewhere else in the equations. The momentum of a photon is:

$$p = \boldsymbol{K}\hbar \tag{1-1.9}$$

where $\hbar$ is the Plank's constant, h, divided by 2π.

The color of light is determined by the period of the sine wave of the electromagnetic field or in other words frequency or wavelength. The electric and magnetic fields are always in phase. Depending on the application, the periodicity of the light wave may be given in wavelength, frequency, wavenumber, or energy. All of these properties of waves are related to each other in a one-to-one fashion. Table 1-1.1 is meant as a review on how to convert among these representations. These conversions are important because these will be the units of the independent axis in most absorption and fluorescence data. Any of these properties are equally applicable to waves and photons (particles), and it is correct to say that a photon has a wavelength.

TABLE 1-1.1 Different Units Used to Describe a Property of Electromagnetic Radiation (Light)

	Wavelength (λ)	Wavenumber (k)	Frequency (ω)	Energy ($\mathcal{E}$)
Commonly Encountered Units	Distance (nm, μm, or Å)	Inverse distance (nm^{-1} or cm^{-1} or kK)	Inverse time (Hz or MHz or GHz or s^{-1})	Energy (eV or J)
Conversion to Wavelength		$\lambda = \frac{1}{k}$	$\lambda = \frac{2\pi c}{\omega}$	$\lambda = \frac{hc}{\mathcal{E}}$
Conversion to Wavenumber	$k = \frac{1}{\lambda}$		$k = \frac{\omega}{2\pi c}$	$k = \frac{\mathcal{E}}{hc}$
Conversion to Frequency	$\omega = \frac{2\pi c}{\lambda}$	$\omega = 2\pi ck$		$\omega = \frac{2\pi\mathcal{E}}{h}$
Conversion to Energy	$\mathcal{E} = \frac{hc}{\lambda}$	$\mathcal{E} = hck$	$\mathcal{E} = \frac{h\omega}{2\pi}(= \hbar\omega)$	
Why we use it	Many instruments are set up to output wavelength. Easier to draw in the diagram.	Often used in equations. Relates well to the wave vector.	Often used with plasmonics. In natural units is same as energy.	Study transition/vibration energies.
Visible Region	400–700 nm	25,000–14,286 cm^{-1}	4.7×10^{15} Hz–2.7×10^{15} Hz	3.1–1.77 eV

It is important to consider that the relationship between energy and wavelength is an inverse one with energy and frequency being directly proportional. The inverse relationship between energy and wavelength means that the number of electron volt (eV) in a 50 nm window will differ for different wavelength ranges; in fact, over a large range of energies, the difference can be quite large. For example, the difference between 200 and 250 nm is 50 nm or 1.24 eV, while the difference between 800 and 850 nm is also 50 nm, but 0.9 eV. This is the reason why one may see a graph with an axis labeled in both wavelength and energy with different scales. This also means that spectral fitting in a broad range may differ depending on if a wavelength or energy scale is used and should be done with the proper conversion.

To gain some familiarity with the scales of light waves, we can consider that the "size" of the Bohr Atom is roughly 1 Å. For 1 nm = 10 Å, the wavelength of 500 nm (green) light is 5000 Å = 0.5 μm which is roughly 5×10^3 atoms long. Considering that a typical molecule (fluorescence dye) is only tens of atoms in size, the wavelength of typical light is much larger than the molecule. Even quantum dots are on the scale of a few to tens of nm in size, making their physical structure non-resolvable to visible light. The frequency of current microprocessors is about 3 GHz, with some liquid helium microprocessors that can operate up to about 8.5 GHz, a time scale of 10^9 Hz. In contrast, the frequency for 500 nm of light is about 3.8×10^{15} Hz or 3.8×10^6 GHz or 3.8 PHz (Petahertz). An electron volt is defined as the energy needed to move an electron through a potential difference of 1 V. A 500 nm wavelength of light has an energy about 2.5 eV that is roughly 100 times greater than the average thermal energy of a molecule at room temperature, and about 10 times greater than the energy of an average non-covalent bond. For comparison, it takes 13.6 eV (Rydberg constant) to ionize a single hydrogen atom with its electron in the ground state.

1-2 LIGHT POLARIZATION

Polarization of light occurs when the electric field vibrates in one direction (in one plane). The polarization of light is given by the direction of its electric field $\hat{e}$ from Equation (1-1.1). Figure 1-2.1 shows the single wave traveling in the direction z that has vertical polarization. Most light in everyday life is a composition of many single waves that have no constraints on the $\hat{e}$ direction in the plane orthogonal to the traveling direction. For example, the direction of the electric field of any given light wave from the Sun has an arbitrary random direction in the plane orthogonal to the direction of light propagation. Any two or more photons emanating from the Sun are very likely to have different electric field directions, and thus the light coming from the Sun has randomly distributed electric field vectors around the direction of light propagation. This type of light is non-polarized, and we call it isotropic light. A single wave (Figure 1-2.1) has only one direction for the electric field $\hat{e}$, and any discussion of linear polarization therein is trivial.

This section is mostly concerned with linearly polarized light. Light is polarized in a direction, $\hat{\boldsymbol{r}}$, when its electric field vector points in that direction, $\hat{r}$. When light passes through a polarizer, a device that only allows the light of a certain polarization to be transmitted (say $\hat{r}$), it may lose some or all of its intensity. If $\boldsymbol{E}$ points along $\hat{r}$ no intensity is lost as the light is already polarized. If $\boldsymbol{E}$ and $\hat{r}$ are orthogonal, all intensity is lost (no light travels

FIGURE 1-2.1 A single electromagnetic wave traveling in the direction z.

through the polarizer). If $\mathbf{E}$ and $\hat{r}$ are neither parallel nor orthogonal, the amount of light transmitted is determined by Malus' Law:

$$I = I_0 \cos^2\left(\frac{\mathbf{E} \bullet \hat{r}}{|\mathbf{E}||\hat{r}|}\right) = I_0 \cos^2\left(\hat{r} \bullet \hat{e}\right) = I_0 \cos^2(\alpha) \qquad (1\text{-}2.1)$$

where α is the angle between the polarization direction of the polarizer and the direction of $\mathbf{E}$ of the incident light (Figure 1-2.2a). Isotropic light loses half of its intensity traveling through a polarizer. It is important to remember that light has no memory, for lack of a better word, of its previous states, including information about its polarization. A common example of this is light passing through two orthogonal polarizers ($\theta = 90°$) where there is no light transmitted (Figure 1-2.2b). However, when there is a third polarizer inserted in-between the two orthonormal polarizers light is indeed transmitted (Figure 1-2.2c). To illustrate this mathematically, consider the light that has passed through the first polarizer and now enters the orthonormal one:

$$I_{\text{final}} = I_0 \cos^2(90) = 0 \qquad (1\text{-}2.2)$$

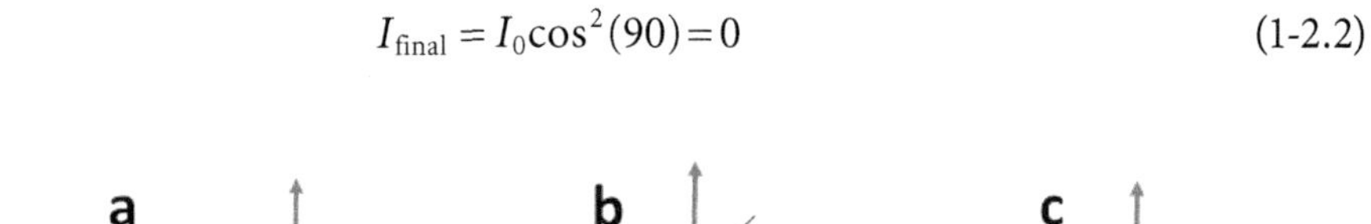
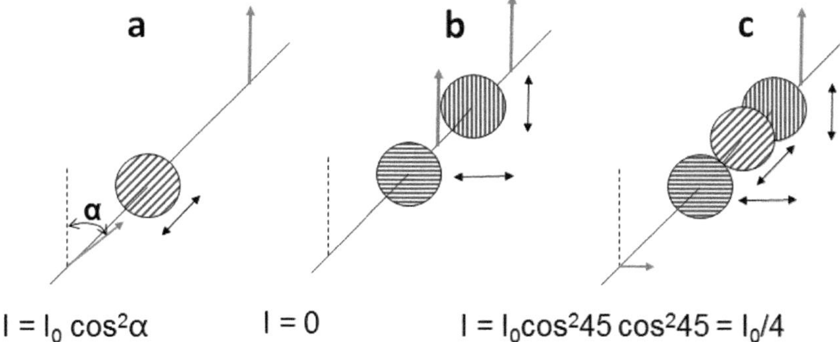

FIGURE 1-2.2 Light being transmitted through polarizers. Light is traveling out of the page and downward. a) Single polarizer; b) Two crossed polarizers; c) Inserted third polarizer enables transmission of the light through a crossed polarizers.

As expected, no light is transmitted. However, in the presence of a third intermediate, polarizer offset to 45° to both original polarizers, the final intensity becomes:

$$I_{\text{intermediate}} = I_0 \cos^2(45) = \tfrac{1}{2} I_0 \tag{1-2.3}$$

$$I_{\text{final}} = \tfrac{1}{2} I_0 \cos^2(45) = \tfrac{1}{4} I_0 \tag{1-2.4}$$

Light is indeed transmitted for any angle in-between 0° and 90°. Thus, light entered both systems in the same state and left through the same final polarizer, but its final intensity varied greatly due to the presence of a third polarizer in one of the systems. The reason for this is that after the light passes through the intermediate polarizer in the second example (Figure 1-2.2b); all information about its original state was completely lost.

Linearly polarized light is a collection (ensemble) of light waves (photons) that all have the same direction of the electric field, thus $\hat{e}_i = \hat{e}_j$. Here i and j are arbitrary iterators overall photons in the ensemble. If we have the light that is an ensemble of randomly oriented electric field vectors (isotropic light) entering an arbitrary linear polarizer, we consider the average angle of all vectors on the $\hat{e}$ direction in the ensemble, which turns out to be 45°. In other words, the isotropic light passing through a linear polarizer loses half its intensity.

Polarizers

The equations in the previous section deal with ideal polarizers and media. A polarizer which transmits through 50% of isotropic light is called a perfect polarizer since 50% is the theoretical maximum of its transmission. In reality, polarizers will reflect and absorb light due to material imperfections so transmitted light is typically no greater than 90–95% of the theoretical limit. Surfaces of many optical elements (polarizers, mirrors, etc.) are specially coated and optimized to work in a specific spectral range achieving efficiencies greater than 95%, but their efficiencies will significantly decrease with light outside the specified range. It is also important to remember that light does not have "memory" due to what process its intensity was lost or how it was polarized other than an apparent intensity decrease or polarization change.

The first modern study of polarizers was conducted by the Danish scientist Bartholin in 1667. He studied a rare transparent mineral called Icelandic spar. Light propagating through this stone is separated into two orthogonally polarized components via refraction. This results in light passing through the stone, creating two side-by-side images. This double image only hinted at polarization since the human eye is not capable of resolving such a phenomenon (this is the principle by which 3D movies operate, only they use circularly polarized light, and each eye sees the differently polarized light through the special polarizing glasses). Polarizers made from these types of minerals are still used in Glan-Thompson polarizers that also split incident light into two beams of different polarizations. In 1929, the Polaroid Corporation filed for a patent of a plastic sheet polarizer. The common sheet polarizers, we see today, were invented in 1938 by Edwin Land who worked for Polaroid Corporation. Polarizers consist of free moving charges on parallel conducting wires. Light with an electric field parallel to these wires will be absorbed; whereas light with a perpendicular electric field will pass through. Current polarizer sheets polarize visible light using stretched hydrocarbon polymer chains covered in conducting iodide ions. The spacing of these conduction wires must be less than the wavelength of light they polarize. The polymers are stretched to be as thin as possible, ideally to the width of a monomer. For visible light, the spacing of these wires is to be less than 400 nm. Today, we can also use wire grid polarizers where very thin metallic (aluminum) wires are layered (embedded) between fused silica or quartz. Such polarizers may have the working range from UV to NIR. Longer wavelength electromagnetic radiation such as radio waves may be polarized using conductors that are much sparser such as those found on the surface of a metal grill.

1-3 INTERACTION OF LIGHT AND CHROMOPHORES

A *chromophore* is a chemical molecule or group responsible for giving color to a substance through absorption of certain wavelengths of light. If the chromophore is capable of the emission of light (fluorescence) is called a *fluorophore*. Chromophores may also be referred to as "color radicals." For the purposes of this book, a chromophore will usually be an organic molecule or dye.

For organic molecules, the loosely bound electrons in the "outer shell" orbital(s) are shared between atoms forming "molecular orbitals" that are responsible for interacting with light. The energy of molecular orbitals in a molecule is quantized similar to an atom's energy. When the molecule does not have extra energy, the electrons are in a relaxed state called the *ground state*. When light is shone at such a molecule it may be absorbed and the energy from the light (photon) is transferred to the outer electrons leading to a transition (excitation) to a higher energy state. Such a molecule is said to be in an *excited state*. The excitation to the first excited state requires the smallest discrete amount of energy the molecule can absorb and each consequent excitation state requires larger discrete amounts of energy. It is important to remember that electronic states, including the ground state, are discrete, meaning the molecule is in one of these states and nowhere in-between. This means that molecules take on only a discrete amount of extra energy of the light they have absorbed. The energy of the electrons is also quantized corresponding to an eigenstate of a wave function ψ. The electronic energy of a molecule is $\mathcal{E}^n$ where the superscript n denotes the excited state of the molecule; $n = 0$ in the ground state. At higher excited states, there will be higher values of n, and the molecule will possess more energy.

The total energy of a real molecule is determined by its electronic energy state, vibrational energy state, and rotational energy state. When a molecule transition (decays/relaxes) from an excited state to a lower state (usually ground state) it may emit a photon. The total energy of a molecule at given state i, $\mathcal{E}_i$, is:

$$\mathcal{E}_i = \mathcal{E}_i^n + \mathcal{E}_i^v + \mathcal{E}_i^r \tag{1-3.1}$$

where $\mathcal{E}_i^n$, $\mathcal{E}_i^v$, and $\mathcal{E}_i^r$ are electronic, vibrational, and rotational energies for the state i, respectively. The wavelength of the emitted photon is thus proportional to the energy difference $\Delta\mathcal{E} = \mathcal{E}_i \rightarrow \mathcal{E}_0$:

$$\lambda = \frac{hc}{\Delta\mathcal{E}} \tag{1-3.2}$$

The wavelength of the emitted photon depends on the change of the total energy, not just the electronic transition. Experimentally, the number of photons with wavelengths λ emitted by the chromophore is counted for all wavelengths of interest. The vibrational energy of molecules is typically in a range of 0.004–0.4 eV. Since molecules are made of many bonded atoms they can vibrate in many different ways called modes. Each mode of vibration has its own set of energy levels, and combined energy levels are also observed. For example, benzene has 30 vibrational modes while water has 3. However, individual molecular species have a smaller range and only the change in vibrational energies, $\Delta\mathcal{E}_v$, not $\mathcal{E}_v$, contributes to the wavelength of emitted light. For most liquid and solid samples, rotational energy levels are broad and overlap, so that no rotational structure is distinguishable. The rotational energy is typically an order of magnitude smaller than the vibrational energy.

A single chromophore may absorb and emit light multiple times. However, every time a chromophore is in the excited state, the probability for degradation is significantly higher than in the ground state. The excited-state energy is significantly higher than typical thermal energy, and after the excitation cycle has been repeated many times, the chromophore will no longer emit light due to photochemical degradation. This process is called photobleaching. The probability for photobleaching of an individual fluorophore is relatively low and an average fluorophore can be excited hundreds of thousands to millions of times before it undergoes photochemical degradation. For a perfect emitter, millions of excitations will lead to millions of emitted photons. This may sound like a very big number, but one needs to remember that a typical laser pointer (5 mW) emits about 10^{16} photons per second. In a real case, it may take a single fluorophore a fraction of a second to photobleach. The photobleaching of dyes in solution, where we have a very large number of dye molecules, depends on the excitation light intensity and the volume of the sample; typically, photobleaching is not a fast process. However, under powerful laser excitation conditions, in many cases, a significant decrease of intensity for common fluorophore solutions can be easily detected in minutes. This can be a significant problem in microscopy, where the intensity of light in the focal point is very high and the number of dyes available in excitation volume is much smaller.

Absorption of Light

The three primary factors affecting light absorption by molecules are:

1. The energy of the incident photon corresponds to the energy difference of two states, $h\nu = \mathcal{E}_i - \mathcal{E}_0$.
2. The intensity of the excitation light source.
3. The relative orientation between the polarization of the excitation light and the molecular transition moment.

Liquid samples are typically placed in a transparent rectangular container called a cuvette. The molecules of interest are dispersed in a solvent. Different wavelengths of light are passed through the liquid and the light coming out on the other side of the cuvette is observed. The intensity of light is decreased (attenuated), as it passes through the sample. The attenuation of the light as it passes through a cuvette containing a sample (chromophore molecules) depends on:

- The absorption cross-section, σ, of the used molecule (chromophore); parameter that reflects the probability of the photon passing in the molecule proximity to be absorbed. The cross-section of a molecule depends on the wavelength/frequency of light.
- The concentration of molecules in the solution, C.
- The path, l, the light travels through the sample and cuvette.

The concentration of molecules in a solution, C, is given in moles per liter (mol/L). Sometimes, the symbol (M) is used and it means the same as mol/L. The number of molecules per cubic centimeter (number of molecules/cm^{-3}) is often used, in which case, the conversion factor from mol/L is $C \times N_A/1000$, where N_A is Avogadro's number (6.0225×10^{23} mol^{-1}). The absorption cross-section of the molecule at a given wavelength is σ and has units of (cm^2/mol). Absorption cross-section reflects probability for a photon of a given wavelength (energy) to be absorbed as it passes in chromophore proximity. It is instructive to calculate the number of chromophores in a typical sample. Most often concentrations of measured samples will be in a range of micromole per liter (μM).

How many chromophores are in a typical 2 mL sample at 1 μM concentration? 1M means that there are about 6×10^{23} molecule in 1 L of the solution. In 2 mL of this solution (1 μM), it will be about $6 \times 10^{23} \times 10^{-6} \times 2 \times 10^{-3} = 1.2 \times 10^{15}$ chromophores. One needs to remember that a typical molecule is much smaller than the wavelength of a given light and incoming/passing electromagnetic radiation produces a local field perturbation at the chromophore location. The chromophore in such a field has a certain probability to absorb energy that depends on the wavelength (frequency of the field). As light travels through the solution, it will get absorbed by individual chromophores and the number of absorbed photons will depend on the number of chromophores in the light path. In typical conditions, one chromophore may absorb one photon and the number of absorbed photons will be proportional to the length light travels through the solution, Δl. Thus, the change in intensity, ΔI of the light as it travels through the solution layer is:

$$\Delta I = I_0 n \Delta l \sigma \tag{1-3.3}$$

where I_0 is the intensity of the incoming light wave (number of photons per surface unit per second), Δl is the path length, and n is the number of chromophores per unit of volume. The intensity of the transmitted light, I for a sample of thickness l is:

$$I = I_0 e^{-\sigma n l} \tag{1-3.4}$$

Equation (1-3.4) represents the Beer–Lambert law. It is useful to rewrite the Beer–Lambert law to be a function of wavelength since the absorption cross-section is wavelength dependent:

$$I(\lambda) = I_0(\lambda)e^{-\sigma(\lambda)nl} \tag{1-3.5}$$

Absorption may be calculated using log base 10 or log base e, called decadic or natural (Napierian), respectively. In photochemistry and photobiology, the extinction coefficient ($\varepsilon(\lambda)$) is more frequently used. The extinction coefficient is a measure of how much a chromophore at a concentration of 1 mole in a 1 cm layer absorbs at a particular wavelength. The units for the molar extinction coefficient are (L mol^{-1} cm^{-1}), although the units used may vary by field (i.e., [m^2 mol^{-1}]). In this text, the decadic molar extinction coefficient, $\varepsilon(\lambda)$, is used. It relates to the absorption cross-section as.

$$\sigma(\lambda) = \frac{2.303\ \varepsilon(\lambda)}{N_A} = 3.823 \times 10^{-21}\varepsilon(\lambda) \tag{1-3.6}$$

Thus, as a general rule, when the absorption cross-section σ is used then the natural logarithm will also be used. When the extinction coefficient ε is used, the logarithm will be taken with base 10. The Beer–Lambert law then becomes:

$$I(\lambda) = I_0(\lambda)10^{-\varepsilon(\lambda)Cl} \tag{1-3.7}$$

where C is the molar concentration of the molecule and the exponent's base is 10 instead of e (this convention is not adopted by all texts). The absorbance or optical density, Ab (OD), is defined as:

$$Ab(\lambda) = \varepsilon(\lambda)Cl \tag{1-3.8}$$

The absorbance is a unitless quantity which is the argument of the exponent in the Beer–Lambert law. It is a quantity that is measured experimentally. The fraction of light absorbed by the sample can then be expressed:

$$\frac{I_0 - I}{I_0} = 1 - 10^{-\varepsilon(\lambda)Cl} = 1 - 10^{-Ab(\lambda)} \tag{1-3.9}$$

Samples with low absorbance ($Ab(\lambda) < 0.1$), we called optically thin and the fraction of light absorbed by the molecules is approximated by $(I_0 - I)/I_0 = 2.303Ab(\lambda)$. As discussed later, optically thin samples will be used for fluorescence measurements.

The extinction coefficient, $\varepsilon(\lambda)$, is the principle parameter characterizing spectroscopic properties of a chromophore. In order to experimentally evaluate $\varepsilon(\lambda)$, a sample of known chromophore concentration is placed in a cuvette of a well-calibrated thickness (typically 1 cm), and $\varepsilon(\lambda)$ is then evaluated from the measured absorbance according to Equation (1-3.8).

It is important to clarify the terminology we are using. By ***absorption***, we will understand the physical process of converting the wave energy to the matter. ***Absorbance*** is a measure of absorption expressed in the common (decadic) logarithm of the ratio of incident to transmitted light. We will use alternatively both terms for the description of our absorption measurements. ***Extinction*** is a total attenuation of the light due to absorption, scattering, and reflections expressed in the common logarithm of the ratio of incident to transmitted light. ***Optical density (OD)*** is the measure of the transmission of an optical medium for a given wavelength, also expressed in the common logarithmic scale. The higher the OD, the lower the transmittance and vice versa, e.g., optical density of 1 means 90% of incident light is extinct and 10% is transmitted. We will sometimes use this term to characterize our samples.

Emission of Light

Similar to atoms, in a molecular system the character of a given state depends on the spin of the excitable electron(s) (molecular orbital), and typically we are dealing with singlet states or much more rarely with triplet states. Both are physically allowed states, however, transitions between singlet and triplet states are forbidden. The ground state is typically a singlet state, and by absorbing a photon (energy), a molecule can be excited to any of its allowed singlet states. Even if all allowed states are quantum mechanically real states, the molecule in solution will very quickly relax to the lowest energy excited state, which is typically the most stable state. In fluorescence, the emissive state is typically the lowest excited state, a principle also called Kasha's rule. A molecule in an excited state $\mathcal{E}_i$ stays for a specific (finite) amount of time and then it transitions back to the ground state (or decays or relaxes to the ground state). If this transition involves emission of a photon it is a radiative transition and the wavelength of the emitted photon is determined by the energy difference between the two states. It is possible for the chromophore to lose the extra energy via a non-radiative process, one not involving the emission of light. Such non-radiative decay may be the result of a collision and heat dissipation. Table 1-1.1 may be used to determine other properties of this emitted photon. While the extra energy absorbed from a photon is in the molecule, portions of this extra energy may be used to alter rotational and vibrational states. This is why the energy of the photon absorbed by the molecule is not necessarily the same as the energy of the photon being emitted by the molecule. The energy difference between the absorbed and emitted photons is called the Stokes' shift. When the molecule transitions in one step from the excited state to the ground state, it releases the excess energy in the form of a photon (light). This photon is emitted in an arbitrary/random direction. It is conceptually wrong to think of the emitted photon in some way being related to the photon absorbed, it is merely the result of the molecule releasing the excess energy. In fact, the excess energy may not even come from absorbing a photon; it can come from a chemical reaction (chemiluminescence), electric (current/electroluminescence), or heat (thermoluminescence) interactions. The molecule has no recollection of how it got into the excited state. There are no phase dependences between absorption and emission processes.

Most fluorescence experiments are done on large collections or ensembles of molecules and involve many photons. However, even a single fluorophore when excited many times will stay in the excited state for a different time after every consecutive excitations. In other words the transition to the ground state is a statistical process and we can only talk about the probability for a given transition per unit of time. This is why we talk about averages, such as the average lifetime or the center of the emission peak. An individual molecule stays in an excited state for a random time t_i after each excitation. Another molecule in the experiment will stay in the excited state for a random time t_j. In general, t_i does not equal t_j but an average time for a large number of events for each molecule will be the same. So, to characterize the ensemble of relaxing molecules the average lifetime, τ, is used. The average lifetime, τ represents the time after which the population of excited molecules decreases $1/e$ of the initial number of excited molecules. In fact, only a few molecules may decay at $t = \tau$ but most excited molecules in the ensemble (~63%) will decay during the period of time from $t = 0$ to $t = \tau$. In general, the number of molecules in the excited state, $N(t)$ can be described:

$$N(t) = N_0 e^{-t/\tau} \tag{1-3.10}$$

At time $t = \tau$, the number of molecules remaining in the excited state will be:

$$N = N_0 e^{-1} = 0.368\ N_0 \text{ or } 36.8\% \text{ of } N_0 \tag{1-3.11}$$

It is important to notice that in this discussion once N_0 molecules are excited, no more molecules in the ensemble become excited. Since the decay process is a statistical process and because there is no relationship between a molecule's excitation source and how long it is in the excited state, after the next time interval:

$$N' = Ne^{-1} = 0.368\ N \text{ or } 0.135\ N_0 \tag{1-3.12}$$

The number of molecules in the excited state decreases exponentially.

The decay of excited molecules depends on both radiative relaxation (radiative decay) and non-radiative process occurring in the excited state. Oscillations, rotations, and collisions of molecules may induce transitions to the ground state (without emission of the photons) with the dissipation of the excitation energy into heat. The photo-processes are described with the rates, which are related to probabilities. The deactivation of the excited state can be described with the decay rate k, which is the sum of the radiative rate (Γ) and non-radiative rate k_{nr}:

$$k = \Gamma + k_{nr} \tag{1-3.13}$$

A cumulative rate k reflects how long in average the molecule stays in the excited state before it relaxes via one of the specific channels. The number of molecules in the excited state as a function of time is described as follows:

$$N(t) = N_0 e^{-kt} = N_0 e^{-(\Gamma + k_{nr})t} \tag{1-3.14}$$

This process may be generalized to account for more decay channels.

The average time fluorophores spend in the excited state, called *lifetime* is:

$$\tau = 1/(\Gamma + k_{\text{nr}}) \tag{1-3.15}$$

The probability that a molecule will emit a photon is the ratio between the radiative rate and the sum of all rates, and describes the efficiency of the radiative process:

$$Q_Y = \frac{\Gamma}{\Gamma + k_{\text{nr}}} \tag{1-3.16}$$

The efficiency of the radiative process, Q_Y, is called quantum efficiency (quantum yield) and reflects the number of emitted photons as compared to the total number of absorbed photons. Importantly, in fluorescence we do not use energy efficiency since the wavelength of emission is typically longer than the wavelength of excitation light (Stoke's shift), a fact that already reflects energy being lost. A $Q_Y = 1$ typically does not mean energy efficiency is equal to 1, it only means that each absorbed photon leads to the emission of a photon typically of lower energy.

If the non-radiative rates are equal to zero the fluorescence lifetime $\tau_n = 1/\Gamma$ is called *natural* fluorescence lifetime.

Jablonski diagram: Figure 1-3.1 shows a classic energetic representation of molecular energy levels called a Perrin–Jablonski diagram (historically called a Jablonski diagram). Each horizontal line represents an energy state, $\mathcal{E}_i$. The solid lines are electronic states and the dashed lines are sums of electronic and vibrational states. Under normal conditions,

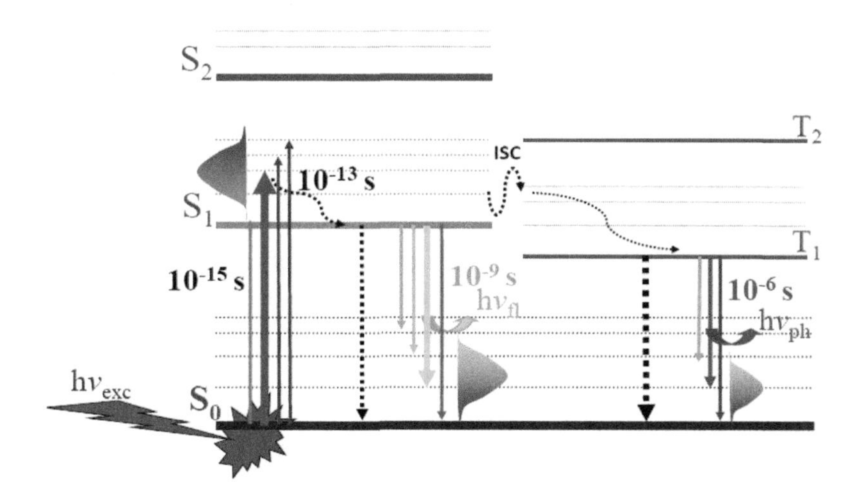

FIGURE 1-3.1 Jablonski diagram describing molecular photo processes. Solid horizontal lines indicate electronic levels and dotted horizontal lines indicate vibrational levels. The left side shows the singlet states and the right side shows the triplet states. Arrows depict transitions labeled with a typical time of that transition (1 ns = 10^{-9} s).

molecules are in the lowest energetic state (ground state, $\boldsymbol{S}_0$ that has energy $\mathcal{E}_0$ or the black line on the bottom of the diagram). Incoming excitation light causes mixing of the ground state with the excited states, leading to the transition to the excited state (light absorption). The absorbed energy of light ($E = h\nu$) brings the molecule from the ground state energy $\mathcal{E}_0$ (typically at room temperature the lowest vibrational state of the state $\mathcal{E}_0$) to a new energy state for which quantized total energy is $\mathcal{E}_i^j$. Absorption is a very fast process happening on a 10^{-15} s time scale. Depending on the energy of the light (wavelength), the transition is to one of a number of vibrational energy levels, j:

$$\Delta\,\mathcal{E} = \mathcal{E}_i^j - \mathcal{E}_0 \tag{1-3.17}$$

After absorption, the molecule quickly relaxes to the lowest vibrational level of the first excited state $\mathcal{E}_1$. This relaxation process depends on the molecule and molecular environment and is in the order of 10^{-14} to 10^{-12}. Emission is typically a transition from the lowest vibrational level of the first excited state, $\mathcal{E}_1$ to one of the vibrational levels of the ground state, $\mathcal{E}_0^j$. The energy difference $\Delta\mathcal{E}'$ can be expressed as:

$$\Delta\mathcal{E}' = \mathcal{E}_i - \mathcal{E}_0^j \tag{1-3.18}$$

$\Delta\mathcal{E}'$ is not the same as the energy difference for absorption, $\Delta\mathcal{E} \geq \Delta\mathcal{E}'$. This difference is due to molecular relaxation in the excited state and the fact that the emission transition could be to higher vibrational levels of the ground state. In real absorption and emission measurements where many molecules contribute to the measured spectrum, we will observe a distribution of all $\Delta\mathcal{E}$ due to the distribution of vibrational and rotational states. Typically, the distribution of vibrational energy levels in combination with the probability for the transition has a so-called normal distribution (also called a Gaussian or bell curve) centered at specific transition energies. Such a normal distribution in a spectrum is referred to as a *peak*. The peak is centered at the mean value of the normal distribution called the maximum. The wavelength value of the maximum of the peak is normally given as absorption and emission positions of the dye. For absorption transitions that have more than one electronic state involved in the same experiment, their energies are typically far enough apart to result in multiple distinguishable peaks in the spectrum. Contrary to this, the emission is typically one electronic transition with vibrational and rotational states contributing to the width of the distribution, resulting in a bell-shaped emission. For some rigid molecules like anthracene, the emission spectrum is structured and vibrational states can be clearly seen as separated peaks in the spectrum.

The schematic representation of the bell-shaped absorption and emission spectra are shown in Figure 1-3.1 drawn horizontally. Correctly, the intensity should be drawn as a function of wavenumber that directly corresponds to energy. As we discussed above, the distribution in the spectrum corresponds to the energy levels of vibrational and rotational states. Drawing the intensity as a function of energy (wavenumber or eV) is a proper way of doing so for many applications that we will discuss later. However, for many historical reasons, absorption and emission spectra are frequently represented on a wavelength scale. The apparent shapes, for wavenumber and wavelength representations, are different and any data operation involving surface area calculations under the emission curve should be done carefully.

The Jablonski diagram is often called an energy diagram because the horizontal lines represent energy levels. Compare transitions in absorption $S_0 \rightarrow S_1$ with transitions in fluorescence $S_1 \rightarrow S_0$. The energy differences for absorption transitions are larger than for fluorescence transitions (lengths of the arrows in Figure 1-3.1 pointing up (absorption) are longer than arrows pointing down (fluorescence)). So, the energy of the emitted photon will typically be lower than the energy of the absorbed one. In effect, the wavelengths for absorption spectrum are shorter than for emission spectrum. It could be rephrased that the fluorescence emission is shifted toward longer wavelengths as compared to the absorption. First, who realized this was Sir Stokes and this phenomenon is now called a *Stoke's shift*.

1-4 INTERACTIONS AMONG CHROMOPHORES

One of the most important properties of fluorescence molecules is the fact that the absorption and emission process are separated in time. Typically excited molecules will stay in the excited state for a short time in the range of a nanosecond. But this time is comparable to the timescale of molecular processes and interactions. So, such excited molecules may interact during their fluorescence lifetime with other molecules. Due to such interactions, the fluorescence signal may significantly change bringing fundamental information about interacting entities. Such interactions can occur by direct contact (collisions due to the diffusion of molecules during the finite time when they are in the excited state). Or the interaction can be through space (due to physical distances larger than the molecule size) when there is no physical contact between interacting molecules.

We need to realize that in typical fluorescence experiments we will deal with many chromophores. Depending on the fluorophore concentration, the number of fluorophores in 1 cm^3 is very large (for very dilute 1 nM solution it is about 6×10^{11}). As a consequence of this high number, the separations between chromophores are small, comparable to the average distance molecules may diffuse during their fluorescence lifetime. Also, even if in most fluorescence experiments we excite only a small percentage of molecules, a chromophore in the excited state may easily find a molecule of solvent, some other charged atom/molecule, an unexcited chromophore, or even another excited chromophore. In many cases, interactions with atoms or molecules like iodide, oxygen, or acrylamide lead to a non-radiative deactivation of the fluorophore's excited state. Such collisions with other molecules or among chromophores itself result in a non-radiative deactivation with an extra non-radiative deactivation rate, leading to a decrease in fluorescence intensity and fluorescence lifetime. A different type of quenching is fluorescence (Förster) resonance energy transfer (FRET). In this case, molecules that have suitable spectroscopic properties can interact through space without direct contact. Excitation energy from one molecule (called donor) can be transferred to another suitable molecule (called acceptor) through a significant distance (up to 10 nm). In this case, the donor fluorescence is quenched and fluorescence lifetime of the donor is shortened.

Fluorescence Quenching

In molecular fluorescence, the number of emitted photons is usually lower than the number of absorbed photons, which is reflected in quantum yields lower than one. This is because of intrinsic non-radiative processes such as non-radiative transition or intersystem crossing to a triplet state. Atoms and functional groups incorporated in molecular structures of fluorophores can accelerate these processes. In this case, we observe an *intramolecular quenching*. Examples of such quenchings include heavy atoms within molecular structures or covalently attached quenchers, like in molecular beacons. If the fluorescence emission is decreased by species not linked to fluorophores, we refer to this as *external quenching*.

It should be noted that fluorescence depends on temperature (higher temperatures result usually in lower quantum yields); often this is a dramatic dependence, especially in the case of indoles.

Therefore, quenching experiments should be conducted with rigorous temperature control.

External quenching falls into two categories: static and dynamic (the last is also called collisional).

Static quenching: If quenching species (atoms or molecules) form non-fluorescent complexes with fluorophore molecules in a solution, then the observed fluorescence will be decreased. For example, when half of the fluorophore molecules form non-fluorescent complexes with quenchers, the observed fluorescence will be reduced to 50% compared to a non-quenched solution. Static quenching can be described with the association constant K_a.

Assuming that the solution has a concentration of $[N_0]$ fluorophores and part of these fluorophores, $[N_c]$, formed complexes with quenchers, the concentration of unquenched fluorophores will be reduced to $[N]$. The concentration of complexes, N_c, is proportional to the number of free fluorophores and to the number of quenchers, $[Q]$, and the complex formation depends on the product $[N] \times [Q]$:

$$\boldsymbol{N_c = N_0 - N = K_a \cdot N \cdot [Q]} \tag{1-4.1}$$

This leads to:

$$\frac{N_0}{N} = 1 + K_a \cdot [Q] \tag{1-4.2}$$

The fluorescence intensity is proportional to the number (concentration) of free fluorophores, therefore:

$$\frac{I_0}{I} = 1 + K_a \cdot [Q] \tag{1-4.3}$$

The ratio I_0/I depends linearly on the quencher concentration.

In the case of static quenching, unaffected (free) fluorophores behave exactly as they would in the absence of the quencher, and other fluorescence properties such as lifetime and polarization are not changed. The solution behaves as if it has a lower concentration of fluorophores.

The complex formation depends on the temperature; at higher temperatures, static quenching is less effective because there are fewer complexes in the solution.

If external species do not perturb fluorophores in the ground state (do not form non-fluorescent complexes) they can still affect fluorophores while they are in the excited state. However, this requires direct contact of the excited fluorophores and quenchers (as in static quenching). In solutions, diffusion of quenchers and excited fluorophores allow this direct contact. Such quenching is called *dynamic or collisional.*

Dynamic quenching: Dynamic (collisional) quenching depends on the effectiveness of the quenching process described by a bimolecular constant k_q and on the quencher concentration [Q] which result in a quenching rate equal $k_q \times [Q]$.

It is convenient to illustrate molecular processes on the Jablonski diagram. In particular, the dynamic quenching is represented in Figure 1-4.1 by a non-radiative transition $S_1 \rightarrow S_0$ with the constant rate $k_q \times [Q]$. Other transitions are: absorption, Abs; fluorescence (radiative transition) with the constant rate Γ (k_f is sometimes used instead of Γ); non-radiative deactivation of the excited state (independent on the quencher), k_{nr}; and non-radiative transition to the triplet state (intersystem crossing), k_{isc}.

For consistency, we will use Γ in the following equations.

The constant rates describe the probability of deactivation processes and are expressed in units of 1/s. The time needed for the given process k_i is equal to $1/k_i$. In the case of fluorescence, $\tau_n = 1/\Gamma$ is called a natural lifetime. The average time of fluorophores in the excited state will depend on all deactivation processes and is given by:

$$\tau = \frac{1}{\Gamma + \sum k_i} \tag{1-4.4}$$

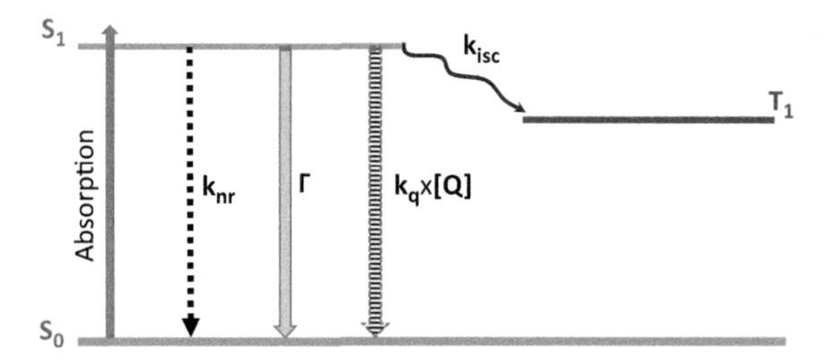

FIGURE 1-4.1 Jablonski diagram describing fluorescence processes in the presence of the quencher.

In the absence of the quencher:

$$\tau_0 = \frac{1}{\Gamma + k_{nr} + k_{isc}} \tag{1-4.5}$$

And in the presence of the quencher:

$$\tau = \frac{1}{\Gamma + k_{nr} + k_{isc} + k_q \cdot [Q]} \tag{1-4.6}$$

The efficiency of the given process is equal to the ratio of a given rate constant to the sum of all rates in the processes:

$$E = \frac{k_i}{\Gamma + \sum k_i} \tag{1-4.7}$$

The efficiency of the fluorescence (fluorescence intensity) in the absence of the quencher is:

$$E_0 = I_0 = \frac{\Gamma}{\Gamma + k_{nr} + k_{isc}} \tag{1-4.8}$$

And in the presence of the quencher:

$$E_f = I_f = \frac{\Gamma}{\Gamma + k_{nr} + k_{isc} + k_q \times [Q]} \tag{1-4.9}$$

Equations 1-4.5 through 1-4.9 give:

$$\frac{\tau_0}{\tau} = \frac{I_0}{I_f} = \frac{\Gamma + k_{nr} + k_{isc} + k_q \times [Q]}{\Gamma + k_{nr} + k_{isc}} = 1 + \tau_0 \times k_q \cdot [Q] = 1 + K_{sv} \cdot [Q] \tag{1-4.10}$$

This relationship is known as the *Stern–Volmer* equation, and K_{sv} is a Stern–Volmer constant. The graphical illustration of this dependence is called a Stern–Volmer plot.

In the case of dynamic quenching, a simultaneous decrease of the lifetime and fluorescence intensity is expected, whereas in the case of static quenching only intensity is decreased. Another difference between static and dynamic quenching is the temperature dependence—dynamic quenching intensifies with temperature because diffusion increases, whereas static quenching decreases with increased temperatures.

Often, the quencher acts in a dual role as a static and dynamic quencher, simultaneously. In such a case, the change in the fluorescence intensity will be given by the product of both processes:

$$\frac{I_0}{I} = \left(1 + K_a \cdot [Q]\right) \cdot \left(1 + K_{sv} \cdot [Q]\right) \quad (1\text{-}4.11)$$

The lifetime is dependent only on the dynamic process. The time-resolved (lifetime) measurements will directly reveal K_{sv} for dynamic quenching. The expression (1-4.11) is square dependent on the quencher concentration which results in an upper-curvature of the Stern–Volmer plots.

There are a number of approaches to the theoretical description of quenching mechanisms. These include a model of a sphere of efficient quenching, diffusion dependence with a transient effect, and Collins–Kimball's theory with distance-dependent quenching. Detailed descriptions of these models are provided in the recommended books (see Preface). In practice, most quenching experiments can be interpreted with the Stern–Volmer approach.

What information is provided by fluorescence quenching measurements?

There are two major applications of fluorescence quenching. First is the accessibility of quenchers to fluorophores—quenching experiments can directly provide information on the location of fluorophores. For example, a tryptophan moiety located on the surface of a protein will be quenched more effectively than if it was buried deep in the protein. Fluorescence quenching is applicable to macromolecules, proteins, membranes, and DNA. Second, the dependence of the intensity and lifetime of fluorescence on the quencher can be used for fluorescence-based sensing. For example, oxygen is an efficient dynamic quencher and can be detected with lifetime measurements of many fluorophores; of course, the quenching effect will be stronger with long-lived fluorophores which allow sufficient diffusion of oxygen.

Fluorescence (Förster) Resonance Energy Transfer (FRET)

Resonance energy transfer (FRET) is a process in which one chromophore transfers its excitation energy to another chromophore in a non-radiative way (without emission of the photon). The concept of radiationless resonance energy transfer originated almost 100 years ago with reports on fluorescence self-depolarization in solutions in the 1920s. The first approximation for the radiationless interaction of two oscillating dipoles based on classical physics was developed by Perrins in 1925. However, a classical model or its subsequent quantum mechanical expansion by Perrins in the early 1930s still was unable to give a quantitative account of the experimental data measured for the fluorophores at high concentrations. The exact theory of fluorescence resonance energy transfer (FRET) was correctly explained by Förster almost 20 years later. Recently, the radiationless energy transfer is also called Förster resonance energy transfer (FRET) to recognize Förster's contribution.

FRET is the most widely applied fluorescence-based technology today that allows researchers to study molecular processes with sub-nanometer resolution. The enormous power of FRET lies in the fact that it brings not only static information but also may give insight into the internal mobility/flexibility of a biomolecular system.

The concept of FRET for two interacting chromophores is presented in Figure 1-4.2. The chromophore which gives up excitation energy is called the donor and the chromophore which takes the energy is called the acceptor. In Figure 1-4.2, we also presented the schematic representation for electronic levels for acceptor and donor molecules. The energy level for an excited state of an acceptor is typically lower than the energy of the excited state of the donor. In practice, this means that the donor emission energetically overlaps with the acceptor absorption. When the donor and the acceptor are two different molecular species (typically the case), we call it hetero-FRET to distinguish from the FRET between the same molecular species that we call homo-FRET or energy migration. Often the ratio of donors to acceptors is not one to one especially in free solutions where typically the acceptor concentration is much higher than that of the donor. The efficiency of FRET is determined by multiple factors:

1. The separation between the donor and acceptor, *r*.
2. The overlap integral of the donor's emission and acceptors absorption.
3. The relative orientation of the donor and acceptor transition moments.
4. Donor fluorescence lifetime and quantum yield in the absence of acceptor.
5. Refractive index of the medium between the donor and acceptor.

The exact theory for RET is rather complex and we will present only the final results. Readers interested in the physical basis and mathematical derivation of FRET are referred to the original papers. For a weak coupling between two dipole moments, where the energy of interaction between the donor and acceptor is small compared to the vibrational splitting of the donor energy levels, for a single donor and a single acceptor separated by the

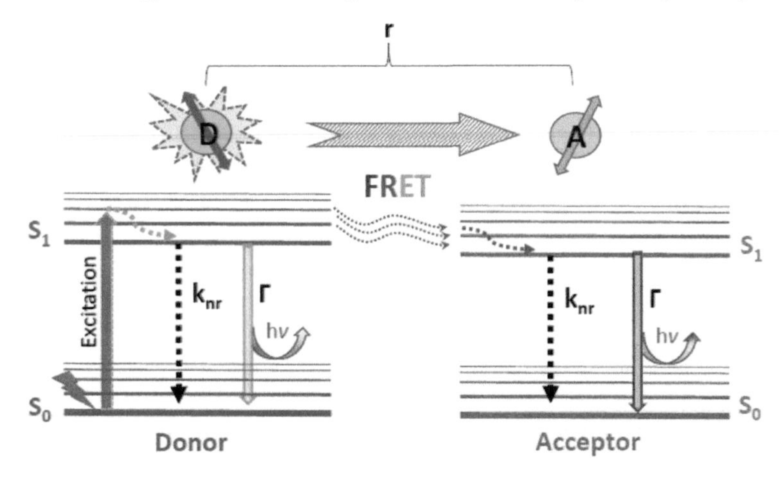

FIGURE 1-4.2 Schematics for electronic levels of an interacting donor-acceptor system.

distance, r one may calculate the rate of transfer (probability of transfer of energy quantum from donor to acceptor per unit time):

$$k_T = \frac{Q_D \kappa^2}{\tau_D r^6} \left(\frac{9000 ln10}{128 \pi N n^4} \right) \int_0^\infty F_D(\lambda) \varepsilon(\lambda) \lambda^4 d\lambda \tag{1-4.12}$$

where Q_D is the quantum yield of the donor in the absence of the acceptor; τ_D is the lifetime of the donor in the absence of the acceptor; n is the refractive index of the medium; N is Avogadro's number; F_D is the normalized fluorescence intensity of the donor (emission spectrum normalized to unity of surface area); $\varepsilon_A(\lambda)$ is the extinction coefficient of the acceptor at wavelength, λ; and κ^2 is the orientational factor describing the relative orientation of the transition moment of the donor and acceptor. The integral in Equation (1-4.12) is referred to as an overlap integral, $J(\lambda)$ expresses the spectral overlap between the donor emission and acceptor absorption and is given by:

$$J(\lambda) = \frac{\int_0^\infty F_D(\lambda) \varepsilon(\lambda) \lambda^4 d\lambda}{\int_0^\infty F_D(\lambda) d\lambda} \tag{1-4.13}$$

$F_D(\lambda)$ is dimensionless. If the extinction coefficient $\varepsilon(\lambda)$ is expressed in units of M^{-1} cm^{-1} and λ in nanometers, then $J(\lambda)$ is in units of M^{-1} cm^{-1} nm^4. The overlap integral has been defined in several ways with different units. This sometimes causes confusion if one tries to calculate the so-called R_0 value for a specific donor-acceptor (D-A) system. We usually recommend the units of nanometers or centimeters for the wavelength and M^{-1} cm^{-1} for the extinction coefficient.

R_0 is a characteristic Förster distance and its meaning can be understood from the frequently used form of Equation (1-4.12) where the right-hand side can be written in the form:

$$k_T = \frac{1}{\tau_D} \left(\frac{R_0}{r} \right)^6 \tag{1-4.14}$$

where $R_0 = 8.79 \times \left[Q_D \kappa^2 n^{-4} J(\lambda) \right]^{1/6}$

In this expression, R_0 is given in Å and the overlap integral is given in M^{-1} cm^{-1}. It is important to realize that the energy transfer process competes with the spontaneous decay of the donor, which proceeds with the rate constant $1/\tau_D$. The probability p that the donor will not lose its energy at a time t after excitation is given by:

$$-\frac{1}{p} \frac{dp}{dt} = \frac{1}{\tau_D} + \frac{1}{\tau_D} \left(\frac{R_0}{r} \right)^6 \tag{1-4.15}$$

From that relationship, one may understand the phenomenological meaning of R_0 that it is the distance between the donor and acceptor at which a probability for the excited donor to transfer the excitation energy to the acceptor is equal to the probability of losing excitation energy by all other processes. In other words, half of the excitation energy of the donor is transferred to the acceptor, while the other half is dissipated by all other processes, including emission.

If the transfer rate is much faster than the decay rate ($1/\tau_D$), then the energy transfer process will be efficient. If the transfer rate is slower than the decay rate, then the probability for a transfer to occur during the excited-state lifetime is low.

The efficiency of energy transfer (E) can be defined by rate constants for all processes involved in excitation energy losses. E represents the fraction of photons absorbed by the donor that is transferred to the acceptor and is given by:

$$E = \frac{k_T}{k_T + k_{nr} + \Gamma} = \frac{k_T}{k_T + 1/\tau_D} \tag{1-4.16}$$

which is the ratio of the transfer rate to the total decay rate (sum of all deactivation rates) of the donor. Taking into account Equation 1-4.14, we can describe the transfer efficiency as:

$$E = \frac{R_0^6}{R_0^6 + r^6} = \frac{1}{1 + \left(r / R_0\right)^6} \tag{1-4.17}$$

This equation shows that the energy transfer efficiency is strongly dependent on distance when the D-A separation is near R_0 as shown in Figure 1-4.3. The efficiency quickly increases to 1.0 as the D-A separation decreases below R_0. For instance, if $r = 0.5$, R_0 the energy transfer

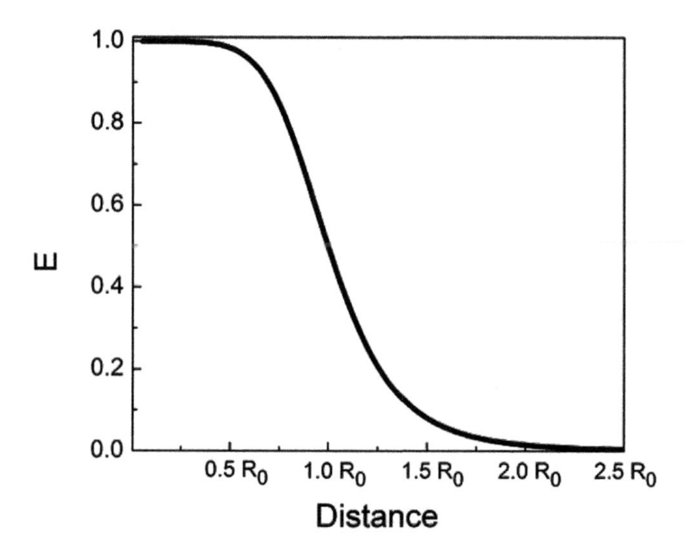

FIGURE 1-4.3 Dependence of energy transfer efficiency (E) as a function of donor-acceptor distance.

efficiency is 98.5% and if $r = 2R_0$ the energy transfer efficiency is only 1.5%. This outstanding dependence on $1/r^6$ of FRET has been extensively tested experimentally and leads to very precise distance estimation between donor and acceptor molecules. Because of the FRET capabilities to measure separation very precisely, it is frequently called a molecular ruler.

1-5 SCATTERING

When measuring absorption and emission, we may have an additional process affecting our result. In absorption and emission, an incident photon is absorbed by a chromophore which enters an excited state as a result of the interaction. When the chromophore relaxes from the excited state, it may release a photon. Interaction of electromagnetic radiation with the matter also leads to scattering in an outward direction of field energy at various directions. This process is much more effective for non-continuous media, such as solutions of biomolecules (proteins or DNA) and especially suspensions of large objects like cells. Scattering is an interaction between light and matter in which the light is not completely absorbed. The matter which scatters light may be referred to as a scatterer. During scattering, the traveling direction of a photon is altered and sometimes also is its energy. When the scattered photon does not change its energy (color), the scattering is called *elastic*; when the photon changes in energy, the scattering is called *inelastic*. Elastic and inelastic often refer to collisions, and in a way scattering of photons can be described as collisions between photons and matter particles. This description is not accurate, especially when the scatterer is small compared to the wavelength of light. A photon incident on a molecule will alter its direction and possibly its energy as a result of passing near a molecule, the energy of the molecule will also be altered (equal and opposite). The details of this mechanism are quite complex. The common way in which scattering is described is that a molecule absorbs a photon into a "forbidden" energy level during which time the molecule is in a "virtual excited" state. The energy, the molecule has in a virtual state is forbidden by quantum mechanics (the value of the energy lays inside the band-gap). This leads to a messy situation wherein it becomes acceptable to violate the laws of physics, but only if it is for a short enough time, something that should typically be avoided. All virtual states are extremely short-lived. A virtual state is not a real state but its effects are observable through scattering. Similarly, the centrifugal force is not a real force, but its effects may be observed by driving a car around a curve very quickly. Figure 1-5.1 shows a Jablonski diagram where the solid lines are electronic states and dotted lines are vibrational states (rotational states not shown). In elastic scattering, the molecule goes back to its initial state (left) and in inelastic scattering, it alters its rotational state (right). Thus the energy lost/gained by the photon results in the gain/loss of vibrations. The vibrations of molecules are quantum oscillators which have non-zero ground energy. This sometimes results in Jablonski diagrams being drawn with parabolic shapes with discrete lines in them where one parabola corresponds to an electronic state and the horizontal lines within it correspond to quantized vibrational states in the electronic state. However, it is often the case that the excited energy of a vibrational state is given as the difference between its energy and the vibrational ground energy. This is the quantity which is experimentally observed, $\Delta \mathcal{E}_v$.

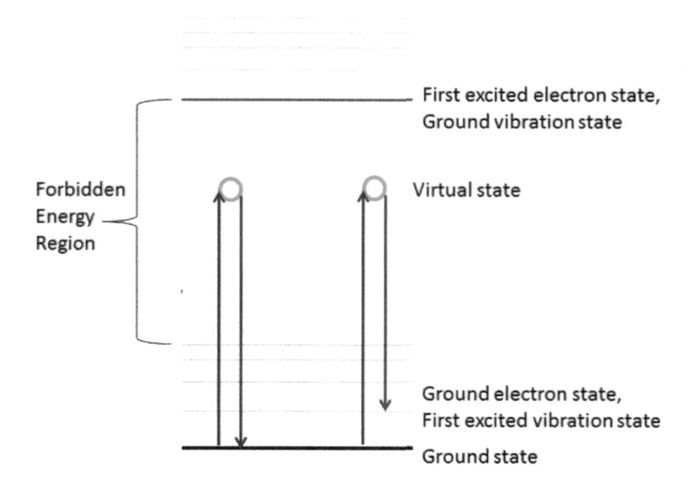

FIGURE 1-5.1 Jablonski diagram with virtual states depicting elastic (left) and inelastic (right) scatterings.

Elastic Scattering

Rayleigh scattering is a type of elastic scattering which appears in many experiments. It only alters the path of a photon and does not result in a change of the vibrational state of the scatterer. Rayleigh scattering is commonly observed when powerful lasers are shone through an air medium. For example, consider a green laser pointer, which would appear as a line that streaks through the air. If all photons emitted from the laser traveled in a straight line, they would not reach the observer's eye (unless the laser is shined directly into the point of view of the observer). All photons emitted from a laser travel in a straight line, but they also interact with air via Rayleigh scattering, causing some of the photons to change direction and be visible to an observer not looking directly into the laser (about 0.0001% of a laser beam's intensity is lost per meter while traveling in air under 1 atmosphere of pressure). In a 90° fluorescence measurement set up, the excitation light is scattered and reaches the detector which is perpendicular to the light's emission direction. Usually, it is scattered by aqueous samples. Rayleigh scattering is proportional to λ^{-4}, thus shorter wavelengths are scattered much more. The cross-section of Rayleigh scattering is given by:

$$\sigma_{\text{Rayleigh}} = \frac{1}{\lambda^4}\frac{2\pi^5 d^6}{3}\left(\frac{n^2-1}{n^2+1}\right)^2 \tag{1-5.1}$$

where d is the diameter of the particles. The λ^{-4} dependence is important because it is observed in the spectra. Rayleigh scattering is applicable to scattering particles that are much smaller than the wavelength of light they scatter. A strong dependence on wavelength results in a much higher scattering at shorter wavelengths making it frequently a concern for measuring protein absorption in the UV spectral range. This phenomenon is also responsible for the blue color of the sky where tiny particles scatter much more blue/UV light than other colors. Mie scattering theory is a full solution to Maxwell's equations with a scattering sphere, whereas Rayleigh scattering is an approximation for Mie scattering for small scatterers.

Raman Scattering

Raman scattering is an inelastic form of scattering, also referred to as the *Raman Effect*. In Raman scattering, some of the energy of the incident photon is absorbed by the molecule and the scattered photon loses energy or the molecule absorbs some of the photons energy into its vibrational state. When the photon loses energy (increased observed wavelength), the change is called a Stokes shift; when it gains energy (decreased observed wavelength) this is called an anti-Stokes shift. The shift refers to the wavelength shift of the photon. In Raman scattering, the photon interacts with the vibrational energy states of the molecule, $\mathcal{E}_v$. Thus, Raman spectroscopy can be used to probe the vibrational states of molecules, proteins, crystal lattices, etc. Vibrations are quantized, thus it is possible to identify a species by its rotational energy transitions. A quantum of vibration is called a phonon; this term is often used when talking about vibration in solid-state lattices. The change in incident photon's energy $\Delta\mathcal{E}$ is equal and opposite to the change of the molecule's vibrational energy. The quantity which is experimentally observed is a light with a wavelength $\lambda + \Delta\lambda$ where $\Delta\lambda$ corresponds to $\Delta\mathcal{E}_v$, the molecule's vibrational energy where λ is the excitation wavelength. If Raman scattered light redshifts, the vibrational energy of the molecule increases, and if the scattered light blueshifts, the vibrational energy of the molecule decreases. Many instruments set up to collect Raman show $\Delta\lambda$ in their output window. Raman scattering may also be observed while performing fluorescence emission measurements. In this case, the Raman signal will look like an emission peak. Since $\Delta\lambda$ depends on the excitation wavelength while $\Delta\mathcal{E}_v$ is constant, a Raman peak will move with an inverse dependence to the excitation energy. An emission peak will stay in the same place or vanish, with varying excitation energy. Normally, a Raman signal is much weaker (many orders of magnitude) than an emission signal. Typically, the Raman signal of a solvent becomes a concern for fluorescence measurements made with very low fluorophore concentrations as an abundance of solvent compared to molecules of dye may result in a significant Raman contribution. For example, a nanomolar concentration of a UV fluorophore in water will give an easily measurable signal, but the signal can be weaker than the water Raman signal due to an overwhelmingly higher concentration of water (over 55.5 moles).

Water (similar to other transparent media) scatters light mainly via two processes, Rayleigh scattering and Raman scattering. Rayleigh scattering is due to any small particles (impurity or very small air bubbles) present in the water and Raman scattering is the scattering process on water molecules. Raman scattering is an inelastic scattering meaning that the scattered photon changes its energy (wavelength). In Raman scattering, a photon is "absorbed" by the water molecule but to a very short-lived virtual state. A virtual state is a forbidden energy state and a distinct form of an excited state. A molecule's decay forms the virtual state practically immediately but as it relaxes some of the energy of the photon is transferred into vibrational energy of the water molecule. Thus, the vibration energy, $\mathcal{E}_v$, of the molecule is increased and the energy of the photon is decreased. The photon transfers part of its energy to the molecule. Transitions of vibrational states of water do not emit light, what is observed in this experiment is the excitation light which lost some of its energy due to the Raman interaction.

The vibrations of water can be modeled as a collection of harmonic quantum oscillators. The ground (not excited) state of water, $\mathcal{E}_0$, is not 0 but 0.574 eV, no quantum harmonic oscillator has a zero ground energy. Everyday liquid water is in the vibrational ground state. Since H_2O is a compound molecule it can undergo several types of vibrations, specifically, it has three vibrational modes. These modes are illustrated in Table 1-5.1. The vibrations of a water molecule may be symmetric, anti-symmetric, or bend. These types of vibrations are called modes and have distinct frequencies. Each mode has its own first excited state vibrational energy of 1.03, 0.772, and 1.104 eV respectively. What is observed in the spectra is the energy differences between the ground state energy and the first excited state energy of each of these vibrations subtracted from the energy of the excitation beam.

$$\mathcal{E}_{\text{observed}} = \mathcal{E}_{\text{excitation}} - \mathcal{E}_{\text{vibration}}$$

$$\lambda_{\text{observed}} = \lambda_{\text{excitation}} - \lambda_{\text{vibration}}$$

The electronic state does not change in Raman scattering. It is important to remember that in the above equation in wavelength $\lambda_{\text{vibration}}$ varies as a function of excitation wavelength but $\mathcal{E}_{\text{vibration}}$ does not. Table 1-5.1 lists the vibrational energy states and other useful information for the Raman effect in the water.

Each vibrational state of water has higher excited states and linear combinations of different modes, not listed in Table 1-5.1. What is observed in the water Raman spectra are

TABLE 1-5.1 Vibrational Modes of Water

	Symmetric	Anti-Symmetric	Bend
Motion Red: oxygen Blue: hydrogen			
$\mathcal{E}_0$ (eV)	0.574	0.574	0.574
$\mathcal{E}_1$ (eV)	$\mathcal{E}_1^{\text{symmetric}} = 1.04$	$\mathcal{E}_1^{\text{anti-symmetric}} = 1.03$	$\mathcal{E}_1^{\text{bend}} = 0.772$
$\mathcal{E}_2$ (eV)	$\mathcal{E}_2^{\text{symmetric}} = 1.5$	$\mathcal{E}_2^{\text{anti-symmetric}} = 1.47$	$\mathcal{E}_2^{\text{bend}} = 0.965$
$\Delta\mathcal{E}_{0-1}$ (eV)	0.466	0.456	0.198
Δk_{0-1} (cm^{-1})	3758	3678	1597
Average Δk (cm^{-1})		3300	

peaks with three components, one for each of the vibrational modes of water. In most experiments, only the ground to a first excited state transition is observed. These peaks have their width form rotational states and since the relative intensities of these peaks change the center of the visible spectral shape appears to change. However, $\Delta\mathcal{E}_{\text{vibration}}$ is always constant and is not wavelength dependent per each vibrational mode. This also explains the change in the spectral shapes as we change the excitation wavelength. The 3300 cm^{-1} wavenumber is commonly used to describe this shift, it is a linear combination of the three peaks. The observed spectral shape is always shifted by the same amount of energy thus the change in wavelength increases as the excitation wavelength increases.

Raman scattering is a useful tool for studying molecular and protein systems. Molecules which do not have fluorescence may have Raman, as is the case for water. This allows for their detections spectroscopically. A Raman signal requires many more Raman active molecules than an emission measurement requires fluorophores.

Vibrations which occur in a solid lattice such as semiconductors also have a Raman signal.

Another situation when virtual states are used to explain a phenomenon is two-photon excitation. In this process, two photons of the same energy, $\mathcal{E}_\gamma$, excite a chromophore with $\Delta\mathcal{E} = 2\mathcal{E}_\gamma$. The first photon absorbed excites the chromophore to an excited virtual state, and if the chromophore absorbs another photon of the same energy while in this virtual state, it will excite to a real excited state. This is the reason why high intensity is needed for two-photon excitation because the virtual state is very short-lived and the second photon needs to be absorbed very quickly after the first. This idea also works for three-photon excitation where $\Delta\mathcal{E} = 3\mathcal{E}_\gamma$ and so on. When a chromophore is excited by absorbing multiple photons (two or more) the process is called multi-photon excitation.

SUMMARY

This section is meant to remind the reader of which quantities are important to consider while conducting experiments. Each of the quantities is measured as a function of wavelength.

- $I(\lambda)$ – The intensity is measured as a function of the instrument and setup parameters. It measures how many photons at a certain wavelength reach the detector, useful to determine spectral shapes. This observation alone is not always enough to make a convincing case in support of a hypothesis since it strongly depends on the equipment used. For these reasons, most emission spectra use arbitrary units (a.u.) or are normalized. Relative intensities in the same spectrum (one intensity measurement/spectrum) can make a compelling case, however.

- $I(\lambda)/I_0(\lambda)$ – The ratio of intensities is a common way to communicate data because it is reproducible among different instruments. This ratio (normalization) is also often calculated by instruments.

- $P(\lambda)$ – The power of emitted light measured in Watts (W, mW). Usually relates to a light source. The power of a light wave relates to the number of photons per second as $\frac{N_\gamma}{s} = P\frac{\lambda}{hc}$.

- $N(t,\lambda)$ – The quantity measured is the number of photons emitted per time interval and directly corresponds to the number of chromophores in the excited state. For steady-state experiments, $N(t,\lambda) = N(\lambda)$ and is a constant in time. Lifetimes and relaxations rates are calculated from this quantity thus a lifetime measurement will result in $N(t,\lambda)$ data which is then fitted to find the lifetime, τ.
- $Ab(\lambda)$ – The absorbance is a quantity often measured to characterize chromophores and provides information on many types of other samples. Although this quantity is a difference and a ratio of intensities it is important enough to consider separately.

CHAPTER 2

Experimental Basics

MOST OF THE DATA in this book will be acquired as a function of wavelength, meaning there will be multiple data points, one for each wavelength. This wavelength scale may be converted to other scales such as wavenumber or frequency (see Table 1-1.1). Plotted optical data such as emission intensities or absorption versus wavelengths as the independent axis is called a spectrum. Sometimes, instead, the wavelengths wavenumbers are used. Both presentations are equivalent and often both scales are used simultaneously.

Real instruments have random errors resulting from a dark current of the photodetector and other limitations such as threshold current, which affects the reading. Perhaps one of the most dissatisfying limitations is the difficulty of measuring light intensity on a meaningful unit scale. This is the reason why many equations in Chapter 1 use ratios of intensities. Fluorescence intensities are usually presented in arbitrary units (a.u.).

The term *fluorescence* was originally introduced by Stokes and referred to Sir John Herschel's experiment reporting the celestial blue color glowing from the solution of quinine sulfate. Now, the term fluorescence describes the process of light emitted when a molecule, called a fluorophore, transitions from its excited singlet electronic state to its electronic ground state. Usually, fluorophores achieve the excited state via stimulation by light (absorption). Therefore, fluorescence belongs to a family of photoluminescence, together with phosphorescence which is an effect of the transition from an excited triplet state to the ground state. However, experimentalists from various fields seldom limit their tools by strict definitions because instruments will report any light they see no matter of its origin.

In this book, we will use term "fluorescence" strictly referring to light emitted from a molecule (chromophore) when it transitions from the singlet excited state to a singlet ground state after a finite duration of time subsequent to light absorption. The light emitted when transiting from a triplet state to a singlet ground state will be called phosphorescence.

In this chapter, we will introduce the most fundamental concepts that are frequently needed when running and describing specific experiments. The term "Experiment" will refer to physical measurements of light absorbed or emitted by the sample, basic data analysis, and results in interpretation. To familiarize a new researcher, not yet familiar with fluorescence technology, how absorption and fluorescence experiments should be done correctly we will introduce basic concepts and terms ("fluorescence slang"). Starting from the description of optical cuvettes, optical filters, principles of absorption/transmission and emission measurements, most crucial corrections, signal-to-noise ratio, and statistical data analysis.

2-1 TYPES OF CUVETTES

A cuvette is a vessel used for spectroscopic measurements. Cuvettes come in different shapes, volumes, and are made of different materials, and should be transparent for the wavelengths of interest. An ideal cuvette would only hold samples (most commonly liquid) and does not interact with the light used in the experiment. This is often an acceptable approximation and it is often worthwhile to obtain a cuvette that satisfies this approximation. Real cuvettes transmit light of a limited spectral range, present dielectric mismatch of refractive index (different refractive index than air and solvent), and can have wear and tear damage such as scratches (some may be small and go unnoticed). All of these factors can adversely affect the outcome of an experiment.

The material a cuvette is made of determines which wavelengths of light it is suitable for. The cuvette should have a sufficient transmission such that light attenuation on the cuvette walls will not affect the outcome of an experiment. Typical optical cuvettes do not have a uniform transmission for all wavelengths, and most often the transmission in the UV or IR is the limiting factor. There is no universal agreement upon minimum transmission to determine the suitability of a cuvette for use at a specified wavelength, and manufacturers tend to use different standards (between 10% and 90%). Table 2-1.1 Transmission Ranges of Different lists some materials cuvettes are made from and their theoretical (usable) transmission range.

Table 2-1.1 should not be taken literally nor be used as a guide. It is presented only to make the reader aware that the transmission ranges of cuvettes may vary greatly. It is important to be sure that the cuvette is adequate for the wavelengths of interest in your experiment. There are plenty of plastic cuvettes that work with wavelengths lower than 380 nm and there are hundreds of different transparent plastics not suitable for fluorescence or absorption measurements. In reality, the fabrication process (quality of surface and the purity of the cuvette walls) contribute to light attenuation (its absorption). Thus, cuvette performance may vary across different brands. Furthermore, many manufacturers

TABLE 2-1.1 Usable Transmission Ranges of Different Cuvettes

Material	Theoretical Transmission (nm)
Far UV quartz	170–2700
Near IR quartz	220–3800
UV silica	220–2500
Optical glass	320–2500
Plastic	380–850
UV plastic	290–900

use their own proprietary materials and surface coating to increase the transmission range and/or decrease the price of their cuvettes. A high transmission may be required to yield meaningful results, thus working at the boundaries of cuvette transmission can cause problems. For a protocol for finding the transmissions of cuvettes, see Experiment 3-3. The only way to truly know the quality of your cuvette is to measure it in your laboratory as described in Experiments 3-3 and 3-4.

The most common absorption and fluorescence cuvettes have an external base of 12.5 mm × 12.5 mm and a height of 45 mm and internal dimensions 10 mm × 10 mm. This is a standard size for which most spectrophotometers and spectrofluorometers holders are designed. There are also many types of larger and smaller cuvettes. Some vendors sell adaptors which allow placing smaller cuvettes into the standard holder that accept 12.5 mm × 12.5 mm square cuvettes. Unless otherwise stated, it should be assumed that a cuvette used in the experiment has external dimensions 12.5 mm × 12.5 mm and internal dimensions 10 mm × 10 mm. These are the so-called standard 1 cm × 1 cm cuvettes (Figure 2.1-1a). Cuvette sizes often refer to the volume of liquid they are designed to hold. A 1 cm cuvette holds 1 mL of solution per 1 cm of height. So, a 45 mm tall cuvette may hold up of 4.5 mL of solution. For many biological experiments, a few milliliters of the volume is difficult to achieve. The volume needed to reach a certain height of solutions can be changed by making two or four of the cuvette walls thicker inward. For example, a "0.4 cm cuvette" ("4 mm cuvette") is a cuvette with external dimensions of 12.5 mm × 12.5 mm and internal dimensions of 4 mm × 10 mm. Such a 4 mm cuvette will require a solution volume of 0.4 mL per ~10 mm of height; similarly, a 1 mm cuvette corresponds to a solution volume of 0.1 mL per ~10 mm of solution height. It is important to remember that due to the surface tension and the formation of a meniscus, the height

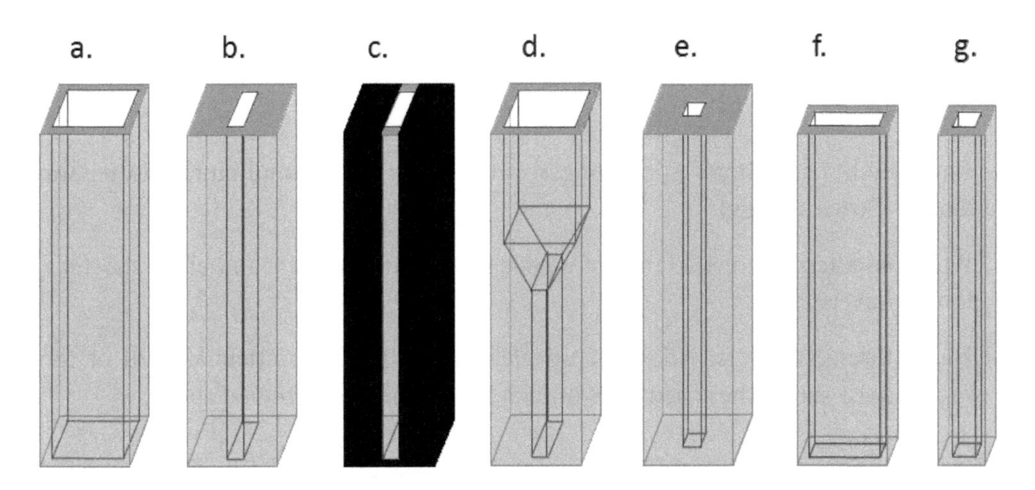

FIGURE 2-1.1 Schematics of cuvettes, a–e are standard size. Red lines inside represent internal volume holding the liquid sample. All cuvettes are drawn to be 12.5 mm × 12.5 mm × 400 mm external dimensions. a: 10 mm × 10 mm fluorescence cuvette; b: 2 mm × 10 mm fluorescence cuvette; c: 10 mm × 2 mm absorption cuvette; d: 2 mm × 10 mm cuvette; e: 2 mm × 2 mm fluorescence cuvette. Cuvettes f and g are examples of not standard dimension cuvettes.

is only approximate. Also, cuvettes with different exterior base dimensions are used, for example, 4 mm × 12.5 mm. Such cuvettes are more difficult to mount in standard spectrophotometers without a proper adapter. Figure 2-1.1 shows the schematics of several different cuvettes.

The path length of the cuvette is the length light travels through the sample inside the cuvette. Cuvettes a and c in Figure 2-1.1 have 1 cm path lengths. Cuvettes b, c, and f have the same dimensions (2 mm × 10 mm) and cuvettes e and g have the same internal dimensions (2 mm × 2 mm) but different external dimensions. Cuvettes f and g have non-standard exterior dimensions and are not referred to in this text unless specifically described. However, those who are comfortable with non-standard sizes cuvettes should certainly use them, though with the appropriate adaptors or specially designed sample holders.

Cuvette c is a small volume absorbance cuvette. It has two dark (black) sides that do not transmit any light. This is useful because a 1 cm path length may be used with a smaller volume and any light not traveling through the sample will be blocked from reaching the detector. Importantly, using cuvette b instead of c in an absorbance measurement will result in getting an incorrect result even when the background has been measured with this cuvette (b). For example, see Experiments 3-11 and 3-12. Cuvette c, however, cannot be used for an emission measurement, which is typically done in a different geometry. Fluorescence cuvettes are cuvettes that are transparent on all four sides (some specialized cuvettes may have three transparent sides or only two orthogonal sides).

2-2 TYPES OF FILTERS

A filter is an optical element that selectively allows a specific range of wavelengths of light to pass through while blocking or attenuating others. Filters are typically labeled for the wavelengths they work for. Filters can be categorized in the following ways:

- Short (low) pass filters: Allow shorter wavelengths to pass while blocking longer wavelengths.
- Long (high) pass filters: Allow longer wavelengths to pass and block shorter wavelengths (lower energy).
- Bandpass filters: Allows a range of wavelengths (band) to pass and block shorter and longer wavelengths.
- Neutral density filters: Allow all wavelengths to pass but attenuates all of them. Figure 2-2.1 shows the schematics of each type of filter mentioned above.

A λ long pass filter allows wavelengths above λ to pass and blocks shorter wavelengths than λ. For example, a 540 long-wave pass filter means that wavelengths above 540 nm are transmitted and below 540 nm are blocked. Similar, a λ short pass filter allows wavelengths below λ to pass and blocks wavelengths longer than λ. A bandpass filter will typically have defined central wavelength for the light transmittance and a bandwidth of the transmitted band. So, a λ bandpass filter allows wavelengths centered at around λ to pass, blocking

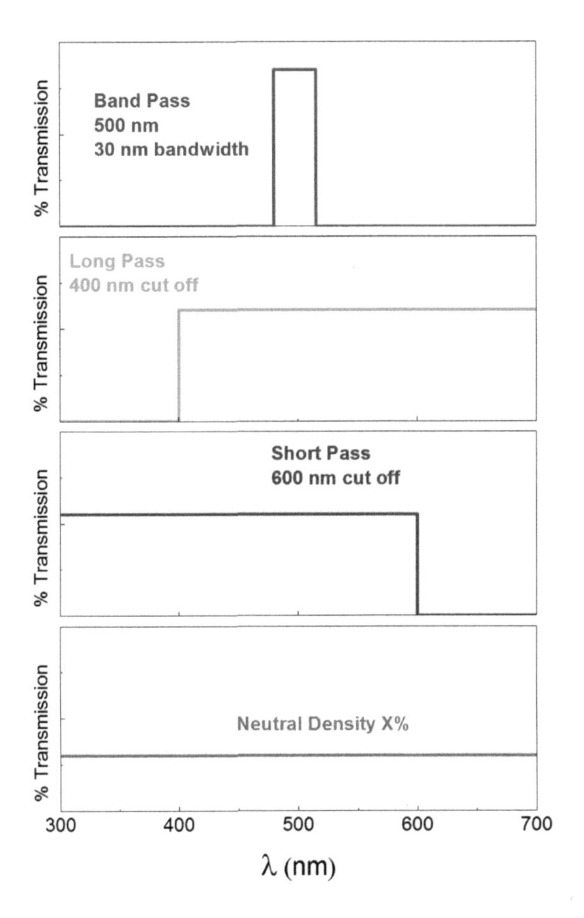

FIGURE 2-2.1 Schematics of idealized filters.

wavelengths longer than $\lambda + \Delta\lambda/2$ and shorter than $\lambda - \Delta\lambda/2$. The filter passes a band of width $\Delta\lambda$ thus wavelengths passed are in the range $\lambda - \Delta\lambda/2$ to $\lambda + \Delta\lambda/2$. A neutral density filter is a filter that in the broad range of wavelengths attenuates the transmitted light by a specified amount. For example, a neutral density filter 1 will have absorption 1 and will attenuate transmitted beam about 10-fold (will transmit 10% of light).

Various filters can be made from different types of materials and work differently. Most general types are dichroic filters and absorptive filters. A dichroic filter is typically an interference thin film deposited on glass/quartz which reflects certain wavelengths and allows other to pass, and an absorptive filter is typically dyed glass or plastic which absorbs certain wavelengths and allows others (outside dye's absorption) to pass. Some absorptive filters can also be based on liquid solutions of chromophores.

Either type of filter is effective for most fluorescence experiments; however, there are some differences that may prove relevant for certain experiments. Since dichroic filters reflect light instead of absorbing it they are less prone to be damaged via heating. This is only a concern when using high-intensity laser light or high power xenon lamps. Absorptive filters frequently have a residual fluorescence that can perturb measurements. When using such filters (called sometimes colored glass), we need to be alert that filter intrinsic

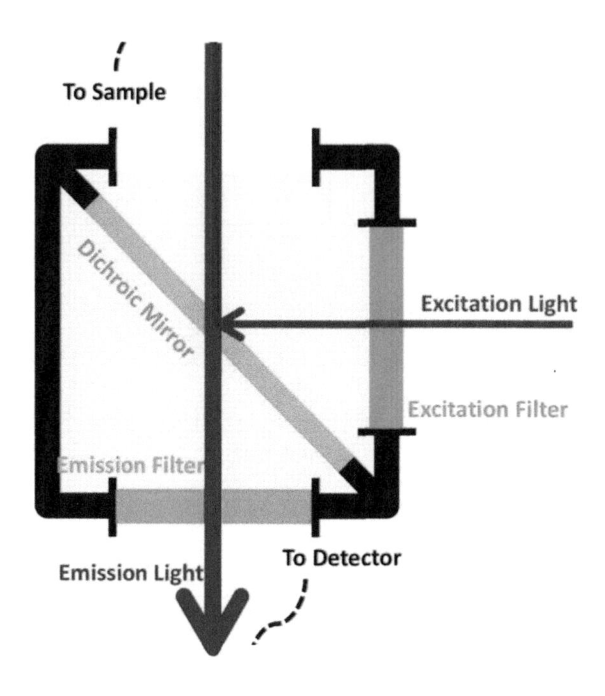

FIGURE 2-2.2 Microscope cube with two filters and a reflective dichroic filter (mirror).

fluorescence may corrupt our measurement. So, for some experiments, it can be important to what filters we will use, and sometimes also the filters sequence can be important. For example, to eliminate short-wavelength scattering (scattered excitation light) when measuring a weak fluorescence, we prefer to put first an interference filter and the absorptive filter next. The interference filter that strongly reflects unwanted light will attenuate short-wavelength scattering limiting intrinsic emission of the absorptive filter. The opposite sequence will first expose to the scattering absorptive filter that would emit a longer wavelength that is not attenuated by interference filter (it overlaps with emission we want to detect).

One common microscopy component that uses filters is called a cube. A schematic of a cube is presented in Figure 2-2.2. There are excitation and emission filters (typically dichroic) at right angles to each other and a third dichroic filter (mirror) at a 45° angle to both of them. The dichroic filter at 45° acts as a mirror that reflects a range of wavelengths and very efficiently transmits the other wavelength. In a standard microscope, the emission light from the sample travels the same path as the excitation light between the sample and the dichromic mirror only in the opposite direction. The longer wavelength emission light that is transmitted by the dichroic mirror and leaves the cube at the bottom.

2-3 MEASURING TRANSMITTANCE/ABSORBANCE

Before doing a fluorescence experiment, the first step should typically be a measurement of sample absorption/transmittance. This will let us know the spectral range where the sample absorbs and what is the optical density of the sample. It is important to realize that the absorbance and transmittance are related and it is easy to convert from the sample absorbance to sample transmittance.

An instrument for measuring sample transmittance or absorbance (sometimes called also transmission or absorption) spectra in the UV/VIS range is called a *spectrophotometer.* A spectrophotometer typically consists of a light source, an optical system including a spectral apparatus (monochromator), a sample compartment/holder, a radiation (light) detector, and a system for data acquisition and data processing (connection to a computer). Figure 2-3.1 shows schematics of two concepts for spectrophotometers, a dual-beam spectrophotometer, and a single beam spectrophotometer. The dual-beam spectrophotometer is shown in Figure 2-3.1a. In the dual-beam spectrophotometer, a special pair of cuvettes with identical (or very similar) optical properties should be used. The dual-beam spectrophotometer divides the light beam into two parts that travel through two parallel trajectories into both cuvettes. One cuvette holds the sample while the other is the reference (or blank) cuvette containing the solvent or a reference solution. The light passes simultaneously through both cuvettes, and the detection system compares the intensities of the two beams, converting them into transmission or absorption readings. The second type of spectrophotometer, more commonly used today, is a single beam spectrophotometer (Figure 2-3.1b). In the single-beam instruments, the cuvettes are moved sequentially into a single beam of light. First, the cuvette containing a solvent (reference) is used for collecting the baseline, which is stored and remembered by the

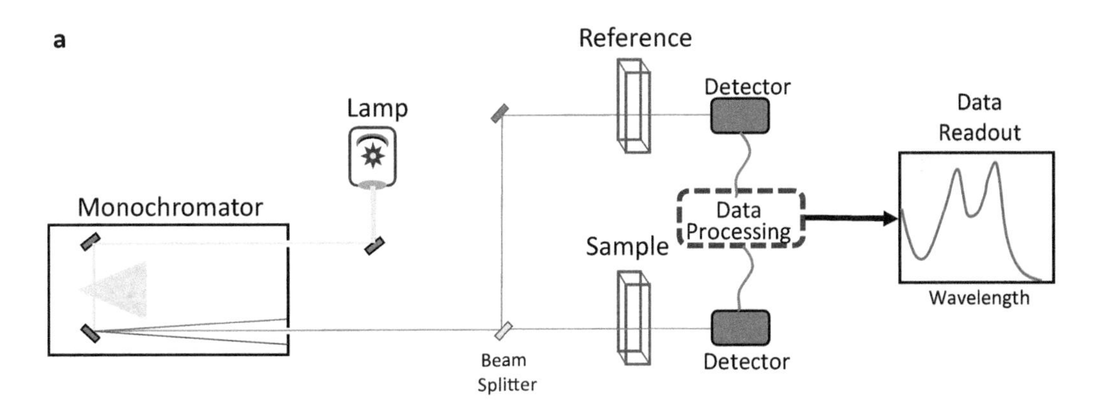

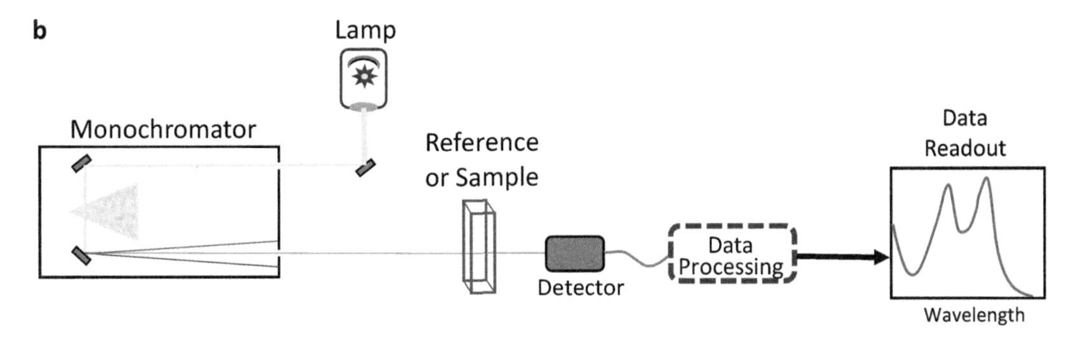

FIGURE 2-3.1 Schematics of dual-beam (a) and single beam (b) spectrophotometers.

operating computer. Next, the baseline cuvette is replaced with an identical cuvette filled with the sample of interest. Then, after setting the measurement software from baseline to sample, readings for each consecutive sample can be taken and will be corrected with the previously measured baseline.

As the light travels through the sample (in cuvette), photons are absorbed, scattered, and reflected, decreasing the intensity of the transmitted light. If the intensity of the initial beam is I_0, its intensity when it reaches the detector will be I, which will, of course, be lower than I_0. The fraction of incident light that passes through the sample is called transmitted light—*light transmittance*. The light transmittance, T is defined as a percentage of the light that has passed through the sample and is:

$$T = \frac{I}{I_0} \times 100\,\% \tag{2-3.1}$$

Or, as a wavelength-dependent transmittance in terms of absorbance:

$$T(\lambda) = \frac{I(\lambda)}{I_0(\lambda)} = \frac{1}{10^{Ab(\lambda)}} \times 100\% \tag{2-3.2}$$

For the definition of absorbance, please revisit Section 1-3 "Interaction of Light and Chromophores" (p. 9).

Light transmittance is measured as the percentage of light from the light source that was ultimately received by the detector. Typical samples used in fluorescence spectroscopy are liquid solutions, but occasionally solid samples (polymer films or glasses) are used. It is important to realize that there are other ways in which light intensity may be lost apart from being absorbed by the sample. For example, light passing through any interface of two media (such as air and glass wall) will be partially reflected. In practice, even the cleanest, highest-quality optical glass will still reflect a few percents (typically about 1–3%) on each interface (total on two cuvette walls 5–10%). A typical cuvette has two glass walls and the traveling light beam crosses four interfaces (air/glass, glass/solvent, solvent/glass, and glass/air), significantly attenuating the beam intensity due to the multiple reflections. These reflections are due to a dielectric mismatch (refractive index mismatch) and are the reason why an empty cuvette attenuates light more than a cuvette filled with water (water has a higher refraction index than air). The refractive index of air is practically 1 and the refractive index of normal glass is about 1.5. Since the refractive index of water is 1.33, the reflection on glass/water and water/glass interfaces are significantly smaller than on-air/glass interfaces (for more details on this, see Experiments 3-1 and 3-2). When measuring the baseline, one must remember that the solvent itself may absorb light independently of the chromophore dissolved in it (this is why a baseline measurement is necessary). Also, some samples containing larger particles like proteins may scatter light via Rayleigh scattering. In order to evaluate the true absorption by the chromophore, the measurement has to be corrected for light reflections, light scattering, and solvent absorption. This is typically done using matched (or the same) cuvettes filled with solvent only.

Special Cases of Measuring Absorption

On many occasions, we must deal with samples that would not be considered typical. A common case when working with modified and recombinant proteins is that the sample volume is limited. Frequently, we also may have to measure solid samples that have a limited size or atypical shape. A good example would be filters that are used for laser beam "cleaning" which can be as small as 4 mm in diameter. In this case, where the sample is small, we have to remember a few important factors such as beam diameter and beam position (how high, in relation to the cuvette bottom, the beam passes through the sample) to properly position the sample.

Measurements of Absorption of Low Volume Samples

When preparing samples of concentrated proteins, DNA, or other biomolecules we will be forced to measure absorptions using a very low volume of samples. In these cases, we will typically have less than 1 mL of total solution volume. Putting half a milliliter of the sample into a standard (1 cm × 1 cm) cuvette will give us a sample layer only 0.5 cm high, and due to the meniscus, the layer thickness in the center will be even less than 0.5 cm. For a typical spectrophotometer, direct measurements of the absorption of such a sample will be very difficult or impossible for the following reason. The light beam size (diameter) could be larger than 5 mm and part of the beam will miss the sample. A practical solution is using a smaller cuvette, for example, a cuvette that is 0.2 cm × 1 cm (as shown in Figure 2-1.1). A variety of such cuvettes are available today with different thicknesses ranging from 0.5 to 5 mm. Half of a milliliter of the sample in the cuvette that has a thickness of 2 mm will have a height of 2.5 cm, which would be very easy to measure in any spectrophotometer.

Another problem is when the small volume sample has low absorbance. To increase the measured absorption, we can use the 1 cm path of a 0.2 cm × 1 cm cuvette, but to do this, we have to consider the beam shape/diameter. In most spectrophotometers, the beam diameter (beam cross-section) is much larger than 2 mm, and using the 1 cm path will create significant problems. Typically, these problems can be solved by using a cuvette made for absorption measurements that has two black walls (not transparent to light) as shown in Figure 2-1.1c. One must measure the baseline with such a cuvette first, and it is always necessary to use identical (matched) cuvettes (or the same one, cleaned of course). Also, when using such absorption cuvettes, it is important to always place the cuvette in the holder in exactly the same way—a small displacement on one of the sides may strongly change the reading. This error is manifested by an artificially high or low (negative) absorption reading outside the absorption spectrum (wavelength where there is no absorption), as will be shown in Experiments 3-5 and 3-11. Of course, such black-walled cuvettes cannot be used for fluorescence measurements. The sample will need to be transferred to a different type of cuvette (like in Figure 2-1.1b) that does not sacrifice transparency on any of its walls. What happens if someone, trying to speed-up the experiment, will use this cuvette (2-1.1b) for the absorption? This will create a significant error in the measured absorption. Making a blank out of such a cuvette filled with a buffer will **not** protect against error and the absorption reading will be distorted (see Experiment 3-11). To understand this type of situation, one could consider a sample that has very high absorption. In this case, practically no light should reach the detector, with absorption readings higher than 3 (limits of most typical spectrophotometers) corresponding to light intensity attenuation 10^3-fold or the transmission of $T = 0.1\%$. However, a significant part of the light traveling through the glass side of the cuvette will always reach the detector. If 50% of the beam is traveling through the glass and is not absorbed the apparent total beam attenuation detected by the spectrophotometer will be about 50% of what corresponds to absorption only 0.3, which is obviously incorrect. We can solve this problem by limiting the beam size below 2 mm or adopting a fluorescence cuvette for absorption measurements as shown in Experiment 3-12. For a quick fix, one may use a pinhole or vertical slit to limit the beam cross-section. In this case, a significant part of the light will be cut off, thus sacrificing the overall spectrophotometer sensitivity.

Measurement of Absorption of Small (Solid) Samples

There are a few examples where we will deal with small samples. One is a sample in the form of a small piece of glass or plastic. This for example could be an optical filter used to "clean" laser beams. Other small sample types are microstructures deposited on large glass supports. Some other common examples of small samples are cells deposited/grown on a microscopic slide, surface deposited proteins (as seen in an immunoassay), or dyes deposited on a small surface.

Typically, such surface depositions are not uniform and only a small fragment of the area is sufficiently uniform, usually less than 1–2 square mm. In such cases, limiting the beam size with a pinhole can be problematic since it may dramatically reduce the beam intensity; especially for array type spectrophotometers where the beam size could be 4–5 mm diameter. For such small samples, the spectrophotometer beam can be collimated/focused to a small spot. In Figure 2-3.2, we present the concept for focusing the beam to a small spot using two lenses and a pinhole. We and others used such a set-up for measuring absorption from plasmonics nanostructures (500 μm × 500 μm) deposited on glass surfaces. The first lens focuses the beam and the second re-focuses the beam to fit it into the spectrophotometer detection system. In this case, the beam intensity is not significantly attenuated and spectrophotometer sensitivity is not compromised, but the correct baseline and appropriate reference must be made with the lenses and the pinhole/slit in place.

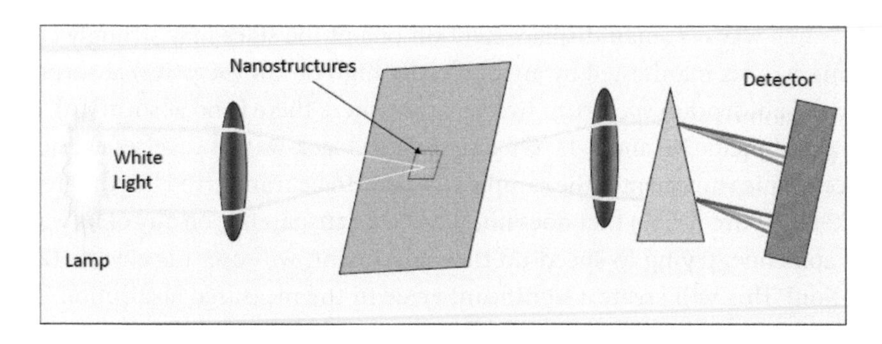

FIGURE 2-3.2 Concept for collimating/focusing the light beam to measure absorption from a small sample deposited on transparent background (glass).

2-4 FACTORS AFFECTING THE PRECISION OF ABSORPTION MEASUREMENT

There are many vendors of spectrophotometers and each vendor will typically offer a few different models. However, there are a few rules of thumb in place to help maximize precision. Typically, the highest precision is obtained for absorbance measurement, Ab, when in the range of 0.1–1.0. The theoretical minimum error of an absorbance measurement depends on the type of detector used. Detectors that are thermal noise-limited (i.e., phototube/photodiode) are the most precise for $Ab = 0.43$, while for detectors that are shot noise-limited (i.e., photomultiplier), maximum precision is achieved at $Ab = 0.86$. This corresponds to 2.7- and 7.2-fold light attenuation, respectively, for thermal noise and shot noise-limited detectors.

In typical experiments, the absorption may vary from 0.02 to 2 and can be measured with good precision. For small absorptions, where light attenuation is minimal, we are trying to extract a small difference from a high light intensity. For high absorptions, the attenuation is very strong and the intensity of light reaching the detector is very low, which will disturb the precision. In this case, the actual precision is limited by the random fluctuations of the detection system. Stray light, or light not from the spectrophotometer's light source, can also be a limiting factor in absorption measurements. Typical detectors are broadband detectors (detect light in a broad range like 200–900 nm) and respond to all light reaching them. For high concentrations, when the absorption is very high, the contribution of stray light can be significant, leading to the instrument reporting an incorrectly lower absorbance. It is a good practice to test the empirical detection limit of the spectrophotometer. This can be done by a calibrated increase/decrease of the sample concentration. When absorption is so high that it reaches a point where an increase in sample concentration will not result in linear (proportional) increase in the reported absorbance, the detector is not detecting enough light or/and is simply responding to stray light. In practice, the concentration of the sample or the optical path length must be adjusted to place the unknown absorbance within a range that is valid for the instrument. Most recent spectrophotometers use modulated/flashlight sources with a "lock-in" type detection mode and are not significantly affected by the stray light, but keeping a spectrophotometer in a dark place is a good practice that helps to limit stray light. Additionally, one must be aware that a computer monitor may be an unintended source of stray light perturbing absorbance measurements. This is a much more significant problem for fluorescence measurements where very weak light signals are detected.

Besides the above limitations, there are a few other factors that can decrease the accuracy and precision of absorption measurements. The most common are: (1) sample scattering, (2) optical beam distortion, (3) sample photostability, (4) defected or inadequate cuvette, and (5) sample fluorescence.

1. *Sample scattering.* Light scattering by the sample is due to the presence of particulate matter, emulsions, air bubbles, micelles, or large proteins that may cause scattering. Extensive sample scattering will result in the attenuation of the transmitted beam and the measured absorbance will be higher than expected. Typical scattering by small particles (Rayleigh scattering) strongly depends on the wavelength (λ^{-4}) and

quickly increases at shorter wavelengths. Knowledge of how much light is scattered will help determine a more accurate absorbance reading. This is a typical problem for measuring proteins solutions, cells, or even quantum dots. When the size of scattering particles is much smaller than the wavelength of light, the correction for Rayleigh scattering can be determined (calculated) directly from the measurement. In cases of larger objects like micelles or cells, the correction can only be made by measuring the proper baseline, which is in most cases very difficult or just impossible (Experiments 3-17 through 3-19).

2. *Optical beam distortion.* There are many factors that may contribute to beam distortion. A simple reflection from the cuvette walls depends on the angle of incidence. For an incident beam orthogonal to the cuvette walls, the reflection is minimal. As the angle of incidence increases (angle measured from the normal to the cuvette wall surface), the reflection increases. The reflection strongly depends on light polarization. Incidentally, a cuvette held not orthogonal to the beam during measurement will have different reflections than a properly positioned cuvette. In addition, a twisted cuvette will displace the beam in a way that may significantly distort intensity readings by the detector. See Experiments 3-1 and 3-2.

3. *Sample photostability.* To measure absorption it is necessary to expose the sample to the light of different wavelengths. This may lead to chromophore photodegradation. Byproducts of degradation may have different absorptions. This is a minor problem for absorption measurements, but it will be much more serious for fluorescence experiments. In general, the absorption/fluorescence measurements should be done with the smallest possible illumination.

4. *Defected or inadequate cuvette.* Using a cuvette that has two parallel flat walls is very important. Not parallel walls and/or the cuvette position not orthogonal to the light direction will result in light banding and redirecting the beam from its original trajectory. The other obvious problems are cracked or dirty cuvettes that are easy to see and should be avoided.

5. *Sample fluorescence.* Chromophores with high fluorescence quantum yields will emit strong fluorescence light. A small part of this fluorescence emission will reach the spectrophotometer detector, which may distort the absorption reading at the longer wavelengths. This is a minor problem, but sometimes may lead to surprising results, manifesting as negative or decreased absorption. This effect is detectable for spectrophotometers that illuminate the sample with white light (all wavelength) and the monochromator selects the wavelength later just before the detector. Typically, diode-array spectrophotometers separate the wavelength before the array detector.

2-5 EFFECT OF SAMPLE SCATTERING ON MEASURED TRANSMISSION AND ABSORBANCE

As we mentioned before, typical samples will have some residual scattering that in many cases can be neglected. However, working with biological samples (proteins, DNA, cells), we frequently will see significant scattering that may considerably alter results. A typical suspension of cells will have high scattering that is manifested by whitish/yellowish apparent color of the suspension. Even working with protein solutions that visually appear perfectly clear in reality are not free of scattering. We need to remember that proteins are in fact particles, and may have (especially in the UV spectral region) significant scattering contributions to transmitted light attenuation. Such scattering is often difficult to measure or even to estimate.

Precise scattering measurements require specialized systems. In Figure 2-5.1, we present systems we have been using to monitor sample scattering. In this case, a solution is placed in the center of the sample holder in-between two hemi cylinders and the intensity of scattered light is collected to the fiber optics mounted on a movable arm. In many instances, scattering will be a directional phenomenon and it is necessary to measure the relative intensity of scattered light as a function of the direction of observation (in reference into the direction of incoming light) and polarization of incoming light.

For the purpose of our discussion of absorption and emission, we do not need to study the details of the special distribution of scattered light. In the case of absorption, it is important to realize what contributes to observed light attenuation—increase of apparent absorption. In this case, we need to realize that light that is scattered out will not reach the spectrophotometer detector, and will be counted as beam attenuation. For large objects (comparable to or larger than the wavelength of light) like very large protein aggregates, cells, etc., the effect is significant but has limited wavelength dependence in the optical range and will result in overall baseline increase. For objects much smaller than the wavelength of light, the effect is rather small but quickly increases toward shorter wavelengths. For particles smaller than 20–30 nm (practically all proteins), scattering should behave as Rayleigh scattering with the overall dependence as λ^{-4}. So, for a dye-labeled protein that

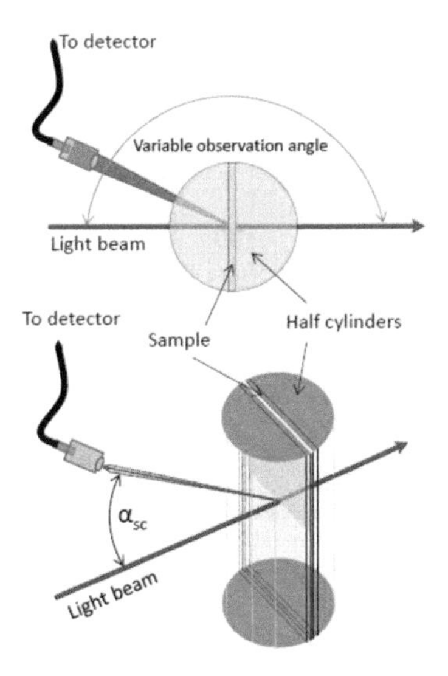

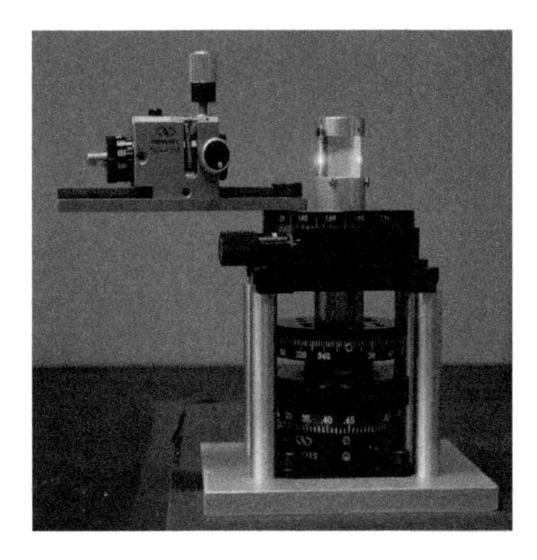

FIGURE 2-5.1 Design for the variable angle scattering detection. Left: concept for the design; Right: real photograph of the device.

absorbs above 400 nm, the effect can be negligible, but moving to a range near 300 nm, the scattering contribution becomes significant. See Experiments 3-17 through 3-19.

It is important to realize how scattering can affect absorption measurement. As an example, we will consider silica particles (called Ludox) that are 20–30 nm in size. Such particles are frequently used as scatterers (references) for fluorescence lifetime measurements. Silica has no significant absorption above 250 nm and when suspended in water it is visually clear. For high concentrations, a bluish opalescence can be seen. The Rayleigh scattering cross-section is given by Equation 1-5.1. We can approximate that the intensity of out-scattered light will depend on the dielectric constants of solvent and particles, particle size, and most importantly, wavelength. Simplifying the Rayleigh Equation 1-5.1, we can approximate scattered light intensity:

$$I_s(\lambda) \cong \frac{B}{\lambda^4} \tag{2-5.1}$$

where B substitutes for all constant parameters in Equation 1-5.1. The observed attenuated intensity due to sample scattering will be $I(\lambda) = I_0(\lambda) - I_s(\lambda)$. We can recalculate this to measured transmittance T and absorbance for thin samples as:

$$T(\lambda) = \frac{I_0(\lambda) - I_s(\lambda)}{I_0(\lambda)} 100\% \quad \text{or} \quad Ab(\lambda) = -\log \frac{I_0(\lambda) - I_s(\lambda)}{I_0(\lambda)} \tag{2-5.2}$$

Figure 2-5.2 shows the expected wavelength-dependent transmittances (solid lines) and absorbance (dashed lines) as a function of wavelength for two different values of B corresponding to 20 nm (thick lines) and 30 nm (thin lines) particle sizes. In Figure 2-5.2, we can see the characteristic behavior of measured transmittance or absorbance due to

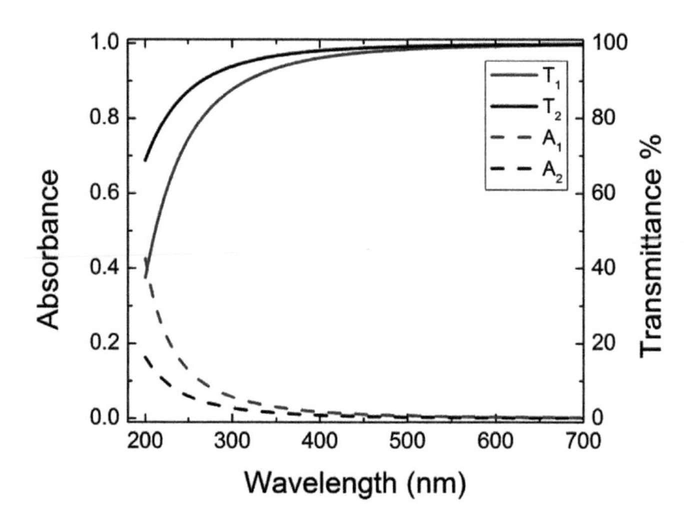

FIGURE 2-5.2 Calculated absorbance and transmittance of silica particles (Ludox) suspended in water. The thicker dark lines are calculated for 20 nm particles; the thinner light lines represent larger, 30 nm particles.

light scattering by small particles. It is important to stress that such sample behavior is only expected if the scattering particles are much smaller than the wavelength of light. In the case of particles absorbing in the UV spectral range, we can estimate scattering contribution by measuring absorbance/transmittance at longer wavelengths. A few examples are presented in Experiments 3-17 through 3-19, where we measure absorption spectra in the presence of small silica particles or absorbance of large proteins.

2-6 MEASURING EMISSION

In contrast to absorbance/transmittance measurements, measurements of emission are more complicated and frequently more prone to errors due to improper configurations and sample preparation. Figure 2-6.1 presents a schematic configuration of a typical instrument for fluorescence measurements—*spectrofluorometer.* This is a T-format instrument configuration where the emission excitation line and emission detection line form a 90° angle (so-called *square geometry*). There are two emission detection channels (so-called left and right detection lines), but in most common instruments only one detection line (left or right) is present. Single line detection configurations are called L-format configurations.

A full-spectrum light from the excitation lamp goes to the excitation monochromator (equipped with a diffraction grating) where an appropriate wavelength can be selected.

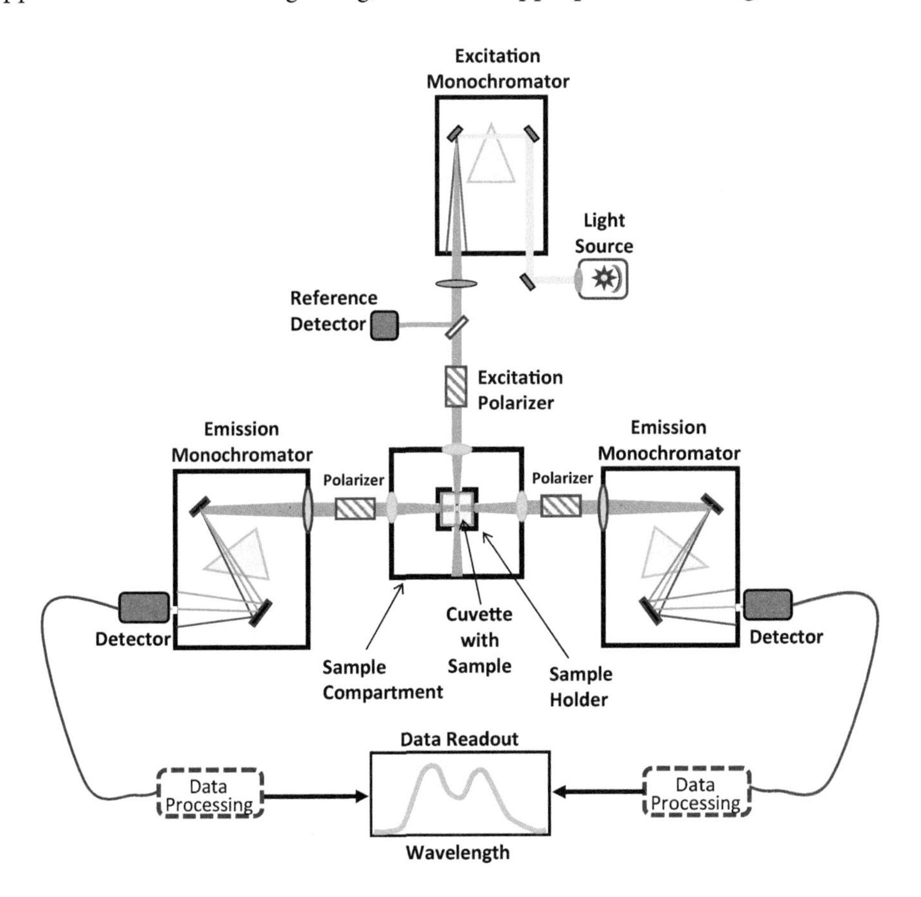

FIGURE 2-6.1 Schematic of a T-format spectrofluorometer (square geometry).

One needs to remember that selecting a given wavelength from a monochromator really means selecting a bandwidth of wavelengths at each step. Depending on the selected slits (or slit width settings), the bandwidth can typically vary from 1 nm to 20 nm. Next, the excitation beam passes through the polarizer and is focused on a sample. The fluorescence is collected through the lens, passes the emission polarizer, and is focused on a diffraction grating in the emission monochromator. The monochromator scans the selected wavelength range through the emission spectrum. Again, depending on the slit openings, a bandwidth of wavelengths is selected at each step. Most diffraction grating-based monochromators select the bandwidth in nanometers, which is *kept constant* through the entire scan. A prism-based monochromator may select the bandwidth in an energy scale (e.g. 500 kK) that is kept constant through the scan. The light from the monochromator is focused on the detector, typically a photomultiplier (PMT). The signal from the PMT is measured at each wavelength point and graphed as an emission spectrum by the data collection electronic system. The measurements are done by scanning the emission monochromator with a fixed excitation wavelength within the absorption of the sample (emission spectra collection). Another possibility is to scan wavelengths on an excitation monochromator with a fixed observation wavelength (within the fluorescence of the sample) on the emission monochromator (excitation spectra collection). Sometimes, we can also use a synchronous scan where both monochromators are scanned with a fixed wavelength separation (excitation at λ and emission at $\lambda + \Delta\lambda$). In Figure 2-6.2, we present

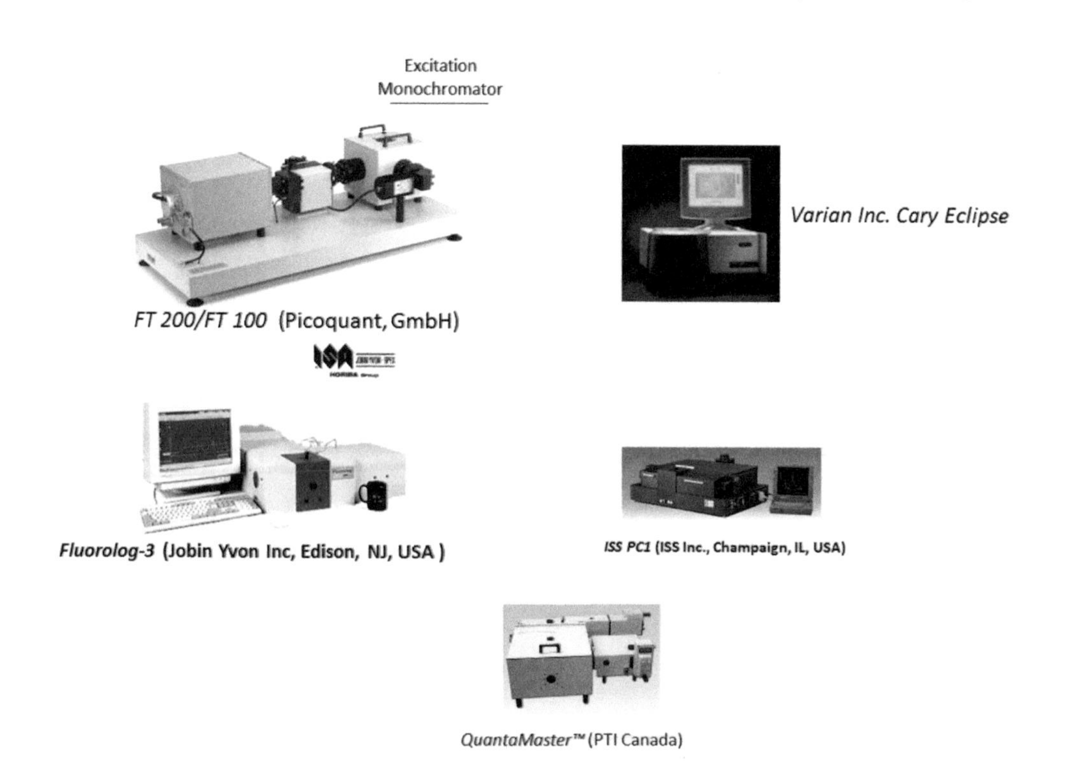

FIGURE 2-6.2 Photographs of spectrofluorometers that have been used by us. FT200/FT100 instruments use laser/LED excitation sources.

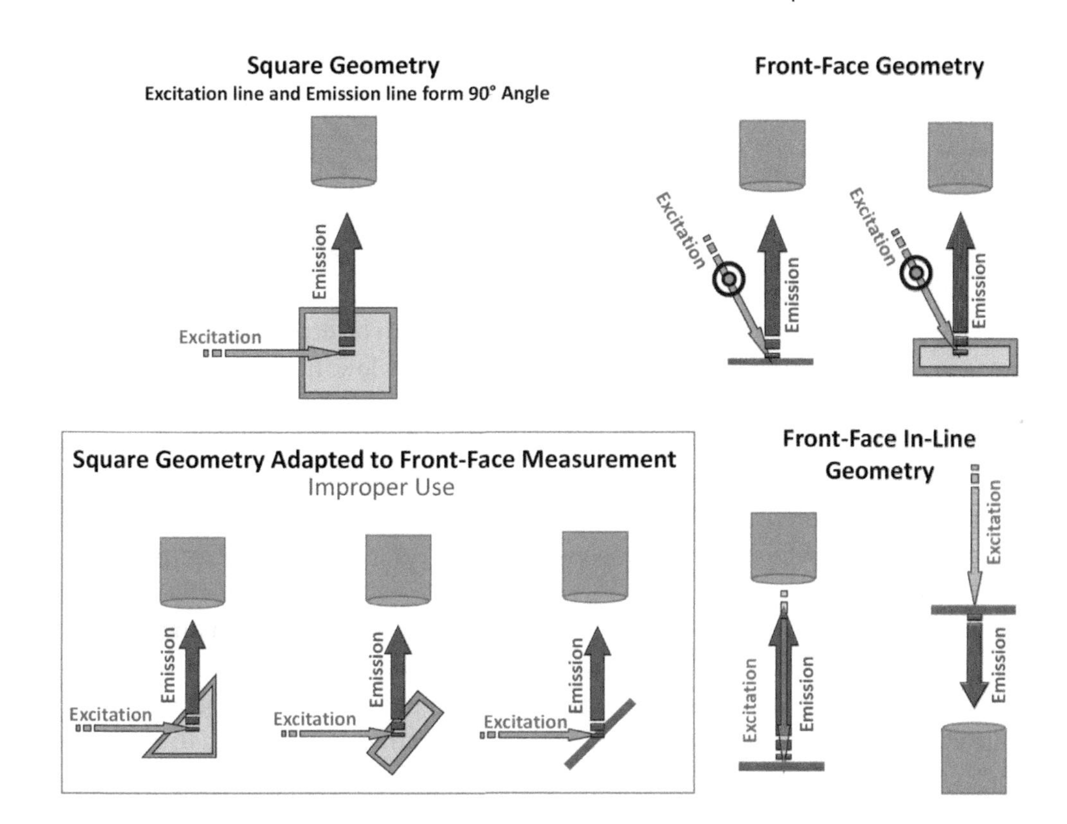

FIGURE 2-6.3 Most common emission geometries used for fluorescence measurements (in the box are improper geometries).

photographs of a few fluorescence instruments we have been using. These are only a few examples and many other instruments are on the market, but all of them fall to these few basic categories.

Fluorescence measurements can be done in many different configurations that are defined by relative orientations of excitation and emission lines as well as sample orientation. Figure 2-6.3 presents three commonly used configurations for fluorescence measurements. Most common is the *square geometry* configuration, where the excitation line and emission line form a 90° angle (Figure 2-6.3 (top)). In this case, the excitation beam passes through the sample while the emission is observed in the orthogonal direction. Since a free-space emission is equally distributed in all directions it does not matter which way, left or right, we measure the 90° angle. The main advantages of the 90° angle are: (1) an easy way to obtain the G-factor when measuring polarization; (2) contribution from the scattered excitation light is minimal. In addition, square cuvettes are typically used, and in the right-angle configuration, we do not have uneven perturbations due to reflections from the cuvette walls. If the walls of the cuvette form an angle other than 90°, reflections on the two interfaces (solvent/glass and glass/air) will strongly depend on light polarization and transmitted light will have not only lower-intensity but more importantly will have a distorted polarization.

It is important to keep a proper orientation of excitation polarization and observation polarization, to fulfill the so-called magic angle (MA) condition. A fluorescence signal will depend on the dynamics of chromophores and the polarizer orientation can play an important role. The MA conditions will be discussed in detail in the "Introduction" of Chapter 4.

In some experiments, the researcher may need to work with highly concentrated solutions (in other words, solutions that have very high optical density—high absorption at the excitation wavelength) and using square geometry will not permit such measurements. The excitation beam will simply not penetrate the solution and the emission can only be observed from the surface. Also, sometimes the sample can be deposited on a surface and square geometry cannot be used even if absorption is very low. In some cases, the manufacturer will suggest the configuration as marked in Figure 2-6.3 (bottom-left) as a potential solution. This configuration uses a triangular cuvette or twisted thin 2 mm cuvette. The excitation happens only on the surface (excitation light does not penetrate the solution) and emission emerges only from the surface. This approach may give an approximate value for overall fluorescence response, but it is completely wrong for polarization measurements.

To understand why this is a problem, let us consider a schematic of a triangular cuvette in Figure 2-6.4. The excitation comes from the left and is vertically polarized (a light electric vector is perpendicular to the page surface). Light approaches the glass wall of the cuvette where it is partially reflected and the beam is refracted according to Snell's Law. Then, it enters the sample where it is again partially reflected and refracted. For marked light polarization (orthogonal to the plane), penetrating light changes the direction of propagation but does not change the light polarization. However, for light polarized in the figure plane, both direction of propagation and the direction of polarization will change. Excited sample emits fluorescence and emission needs to exit the cuvette. During this process, the light again encounters two interfaces (liquid-glass and glass-air) where it is

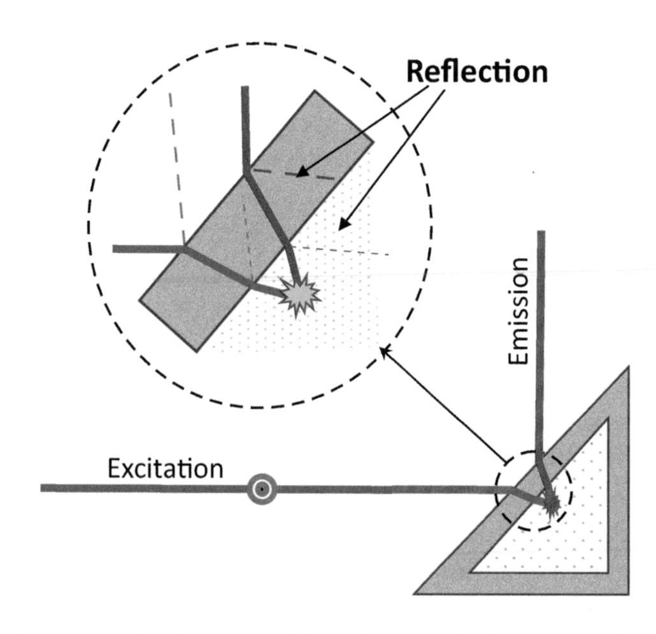

FIGURE 2-6.4 Reflections and refractions of the light on a tilted cuvette wall.

partially reflected and refracted (as marked in the inset of Figure 2-6.4). We will lose some intensity of the transmitted light, but more importantly, these two reflections strongly depend on the light polarization. Since the wall of the cuvette forms an angle of about 45° that is close to the Brewster angle for a glass–air interface, the polarization distortion of transmitted light can be very significant. Certainly, such a configuration should not be used for any polarization measurements and may also generate problems for lifetime measurements.

The proper front-face configuration is shown in Figure 2-6.3 (top-right) where the excitation line and emission line are in the plane orthogonal to the light polarization and form an angle smaller than 90°. *The important requirement is that the direction for emission observation is orthogonal to the front surface of the sample.* Just twisting the cuvette in the square geometry or using the triangle cuvette as shown at the top in Figure 2-6.3 for reasons mentioned above, may lead to significant errors. In a proper front-face configuration, the magic angle conditions are as in the square geometry (vertically polarized excitation and observation polarizer oriented at 54.7°). A significant limitation for a front-face configuration is the fact that we cannot evaluate the G-factor by just rotating the excitation polarization to horizontal.

The front-face configuration will typically not be supported by the instrument manufacturer, but sample compartments can be easily adopted for front-face measurements. In Figure 2-6.5, we present schematics of such an adaptor made for SLM and ISS sample compartments with the picture of the sample compartment. It is important to remember that in the front-face configuration we use only vertical excitation, with the sample surface orthogonal to the emission line. Only in such conditions, the polarization can be properly measured.

The third geometry that we present in Figure 2.6-3 (bottom) is the in-line geometry as used in microscopy and high-throughput screening (HTS). In-line is the front-face

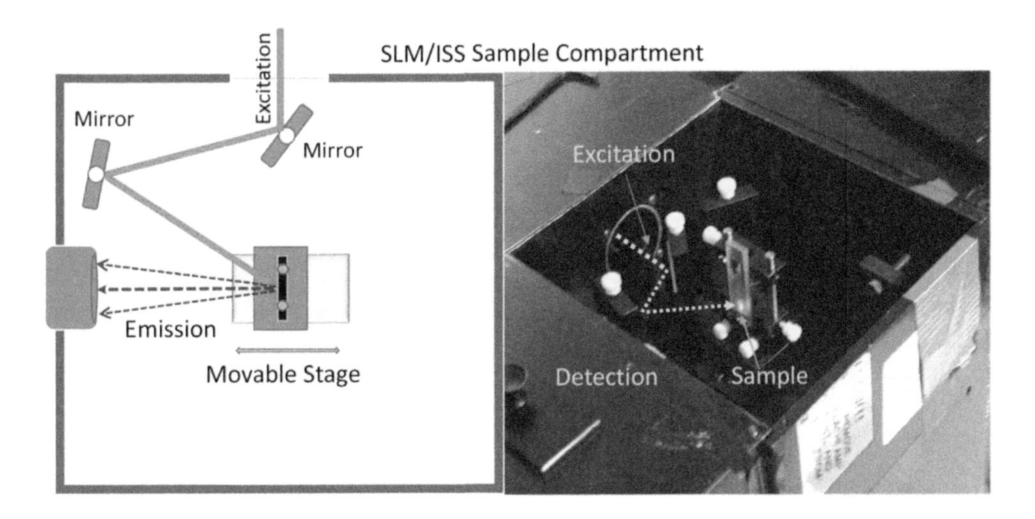

FIGURE 2-6.5 Front-face configuration: left: schematic representation; right: photograph of an open sample compartment with the front-face adapter.

geometry where the angle between excitation and emission is 0°. This geometry is proper for intensity and lifetime measurements (it is possible to achieve magic angle condition), however, this configuration requires special precautions in order to avoid a straight excitation light leak to the detector.

2-7 EFFICIENCY OF COLLECTING EMISSION: MEASURING QUANTUM YIELD

When comparing the emission of fluorophores in different solvents, we need to consider solvents refractive indexes. A typical observation system detects fluorescence intensity from one well-defined direction. The only light coming through a collecting lens will reach the detector. The amount of collected light depends on the solvent refraction index. As schematically presented in Figure 2-7.1, light exiting the cuvette undergoes refraction that affects the collection angle that the lens can see. This directly affects the detected number of photons and may distort measurements of quantum yield that rely on a direct comparison of two different samples that may have a different index of refraction. Below, we discuss how this influences quantum yield calculations.

Quantum Yield Calculation

A common example where we should be careful how to analyze the results can be calculating the quantum yield (QY) of unknown fluorophores using a known standard (reference). This is the most common way of evaluating the quantum yield for unknown dyes, and we will discuss how it should be done properly. Equation 1-3.16 gives a general definition for quantum yield as a ratio of emitted photons to absorbed photons by the sample or equivalently defines QY through the radiative and non-radiative rates. Using these definitions, there is no simple way to evaluate/calculate the quantum yield of an unknown fluorophore. The reason for that is that we will need to have a perfectly calibrated excitation lamp to know exactly the number of photons emitted by the lamp per second and a perfectly calibrated detection system that will be able to exactly evaluate the total number of emitted photons (photons emitted in all directions). For evaluating the total number of emitting photons, sometimes integrated spheres are used that does not always give satisfactory

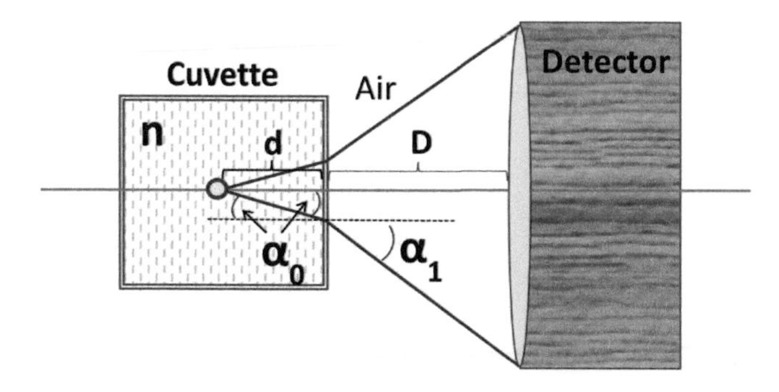

FIGURE 2-7.1 Refraction of light exiting a cuvette. At a higher refractive index, the light exits the cuvette at a bigger angle and less light is collected by the lens.

results. But the most convenient and efficient approach for quantum yield evaluations is to use a reference standard of known quantum yield. In general, emission of the fluorophore (number of emitted photons) is proportional to the number of absorbed photons and the fluorophore quantum yield. Photons are emitted in all directions (and only a small part reaches the detector), and additionally, photons are spread through the emission spectrum. Using a known standard/reference in exactly the same configuration as our unknown sample, we can safely assume that the portion of photons emitted toward the detector from our sample is the same. *Caution! Remember to use MA detection to avoid distortion due to sample polarization.* So, in principle, it is sufficient to measure absorptions for the reference and sample and then emission spectra in identical conditions (if the spectral range is different we need to properly recalculate the number of detected photons). The number of absorbed photons is equal to the number of excited molecules per unit of time and is proportional to the change in the intensity of light going through the sample. According to the Beer–Lambert law (1-3.7), the number of absorbed photons (excited molecules) of the standard and sample will be:

$$N_R \sim \Delta I_R = I_0\left(1-10^{-OD_R}\right) \text{ and for sample } N_S \sim \Delta I_S = I_0\left(1-10^{-OD_S}\right) \quad (2\text{-}7.1)$$

where OD_R and OD_S are optical densities (absorbances) for the reference and sample, respectively. The number of emitted photons for the reference and sample will be:

$$N_R^{Ph} = N_R\, QY_R \quad \text{and } N_S^{Ph} = N_S\, QY_S \quad (2\text{-}7.2)$$

A typical system does not collect all the photons, but for the same detection configuration (same system and the same/identical cuvette), the fractions of collected photons for the reference and sample should be identical if the samples have the same refractive indexes. As it is relatively easy to keep the detection geometry constant, quite frequently our sample and reference can be in different solvents that may have different refractive indices. To discuss how this may affect our signal and how to correct for this effect, let us consider the configuration presented in Figure 2-7.1. The reference is in the solution that has a refractive index, n_R, and the sample refractive index n_S. The emitting point in the center of the cuvette is located at a distance d from the cuvette wall and emits uniformly in all directions. The detector is at a distance D ($D>>d$) from the cuvette. For a fixed geometry, the system will always see the light entering into the detector direction (collecting lens) under the angle α_1. Due to refraction, this will correspond to the angle α_0 that will depend on the refractive index, n, of the solvent in the cuvette. Only the light emitted from a point source into the cubical angle α_0 will reach the detector. At any given distance, the intensity I is distributed on a surface $4\pi d^2$ and the portion of energy flowing to the detector is proportional to the surface area covered by the angle α_0, $s = \pi r^2 = \pi(d^*\sin\alpha_0)^2$, and using the notation as in Figure 2-7.1 for sample and reference, we will get:

$$S_0^R = \pi D^2 \text{tg}^2 \alpha_0^R \quad \text{and} \quad S_0^S = \pi D^2 \text{tg}^2 \alpha_0^S \quad (2\text{-}7.3)$$

We assumed $D \approx D+d$. Snell's law lets us calculate angles for reference $\sin\alpha_0^R = n_R \sin\alpha_1$ and for a sample $\sin\alpha_0^S = n_S \sin\alpha_1$. We can now write a number of photons emitted toward the detector at a given wavenumber (wavelength) by reference and sample respectively:

$$N_R^{Ph}(\nu) = \frac{N_R\, QY_R}{4\pi R^2}\pi D^2 n_R^2 \mathrm{tg}^2\alpha_1 \text{ and } N_S^{Ph}(\nu) = N_S\, QY_S \pi D^2 n_S^2 \mathrm{tg}^2\alpha_1 \tag{2-7.4}$$

We can use the wavelength or wavenumber scale to calculate the total number of photons (energy) emitted by the sample. The total number of emitted photons is proportional to the surface area under the emission spectrum. We can enter it to Equation 2-7.4 and by comparing a signal measured for reference and sample we can calculate the quantum yield of the sample:

$$QY_S = QY_R \frac{\left(1-10^{-OD_R}\right) n_S^2 \int F_S(\nu)d\nu}{\left(1-10^{-OD_S}\right) n_R^2 \int F_R(\nu)d\nu} \tag{2-7.5}$$

where $F_S(\nu)$ and $F_R(\nu)$ are integrated intensities (areas under the emission spectra profile) for sample and reference respectively. The spectrum profile will generally include a detector sensitivity factor (correction for different sensitivity to a different wavelength—later discussed in "Measuring Emission Using Different Detectors—Detector Sensitivity"). Equation 2-7.5 lets us calculate the quantum yield of a sample if we know the quantum yield of reference. In this general representation, we do not have to worry about differences in sample and reference absorptions.

For most recent monochromators (with gratings as dispersing elements), the linear step will be in wavelength and we will use the wavelength version of Equation 2-7.5:

$$QY_S = QY_R \frac{\left(1-10^{-OD_R}\right) n_S^2 \int F_S(\lambda)d\lambda}{\left(1-10^{-OD_S}\right) n_R^2 \int F_R(\lambda)d\lambda} \tag{2-7.6}$$

We need to remember that when converting spectra between wavelength and wavenumber we need to use the proper conversion.

How to Convert from Wavenumbers to Wavelengths?

Let assume that we measure the intensity of $F(\lambda)$ with the bandpass of $\Delta\lambda$. The signal will be proportional to the number of photons in the $\lambda-\Delta\lambda$ wavelength range, $F(\lambda)\Delta\lambda$. In the wavenumber representation, this will be $F(\nu)\Delta\nu = F(\nu)\left(1/\lambda - 1/\lambda+\Delta\lambda\right)$. Of course, both representations must provide the same number of photons, therefore $F(\lambda) = F(\nu)/\lambda(\lambda+\Delta\lambda)$. For $\Delta\lambda$ much smaller than λ, this dependence can be written as $F(\lambda)\lambda^2 = F(\nu)$.

Sometimes to evaluate QY, a simplified version of the Equation 2-7.6 is used:

$$QY_S = QY_R \frac{OD_R n_S^2 \int F_S(\lambda) d\lambda}{OD_S n_R^2 \int F_R(\lambda) d\lambda} \tag{2-7.7}$$

A significant error can be made when the absorption of the sample and reference are a different and even larger error when the emission spectra of reference and sample are in different spectral ranges and not using the proper conversion. We will briefly analyze how big of an error it can be made when the simplified equation is used.

Absorption

Generally, we would recommend the absorption for a sample and reference to be low (below 0.1) and they should be adjusted (if possible) to be as close as possible. This is not always possible, so we would like to know how much error we can make by using the simplified Equation 2-7.7. The variation will be in the difference of an exact ratio $R_e = (1-10^{-OD_R})/(1-10^{-OD_S})$ versus approximate ratio $R_a = OD_R/OD_S$. For discussion, we will define variance in both values as:

$$V = \frac{R_e - R_a}{R_e} \tag{2-7.8}$$

In Figure 2-7.2, we are plotting the dependence for variance V as a function of OD_R for a few values of OD_S. For graphing purposes, we consider the absorption range from 0 to 0.5. We do not recommend measurements when absorptions are greater than 0.1 due to significant excitation beam attenuation and limited sample penetration that may alter the geometry. The deviation (difference between exact and approximate value Equation 2-7.6)

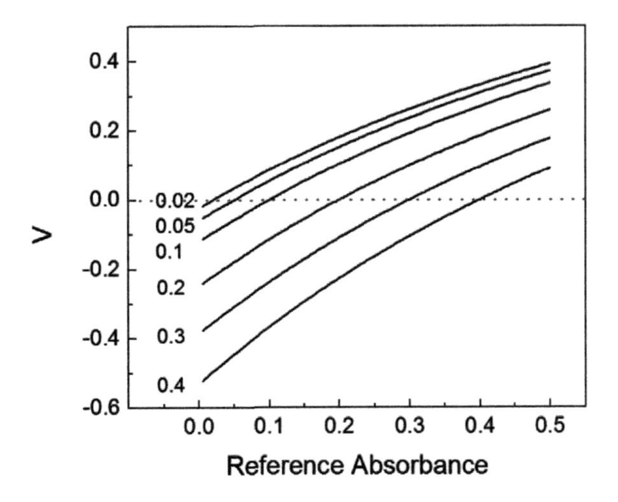

FIGURE 2-7.2 Dependence for variance in experimental error as a function of reference absorption.

depends on the absorption difference between the sample and reference in an almost linear fashion (curves in Figure 2-7.2 are almost parallel). So, if the differences in absorption between the sample and reference are less than 0.1 the expected error is below 0.1 (10%). For example, when sample absorption is 0.2 and reference absorption is 0.3 the error will be within 0.1 (10%). Obviously, for absorptions below 0.1, the difference between the sample and reference is much smaller than 0.1 and the error will be well below 10%. But still, for the absorption of reference 0.1 and sample 0.05, we will make 5% error.

Area under the Emission Spectra

As we discussed earlier, to calculate quantum yield we need to evaluate the area under the emission curves (total number of photons) for reference and the sample, and the integration can be done for the wavelength or wavenumber scale $\left(\int F(v)dv \text{ or} \int F(\lambda)d\lambda\right)$. Depending on the type of the used monochromator, the proper integration will be with wavelength or wavenumber. Since most, if not all, modern spectrofluorometers use gratings, we will use wavelength representation. In principle, if a proper conversion between wavelength and wavenumber has been used ($1/\lambda^2$), the result will be correct. In many cases, the emission spectrum of the reference and sample are shifted, sometimes significantly and we should remember about the proper conversion. Let us consider a simple example of two emission bands as shown in Figure 2.7-3. One centered at 25 kK (400 nm) and the other at 20 kK (500 nm) with a width of 3 kK (note: 1 kK = 1000 cm^{-1}, K stands for Kajzer). As a 5 kK (100 nm) shift may sound large, just looking on spectra in the wavelength scale

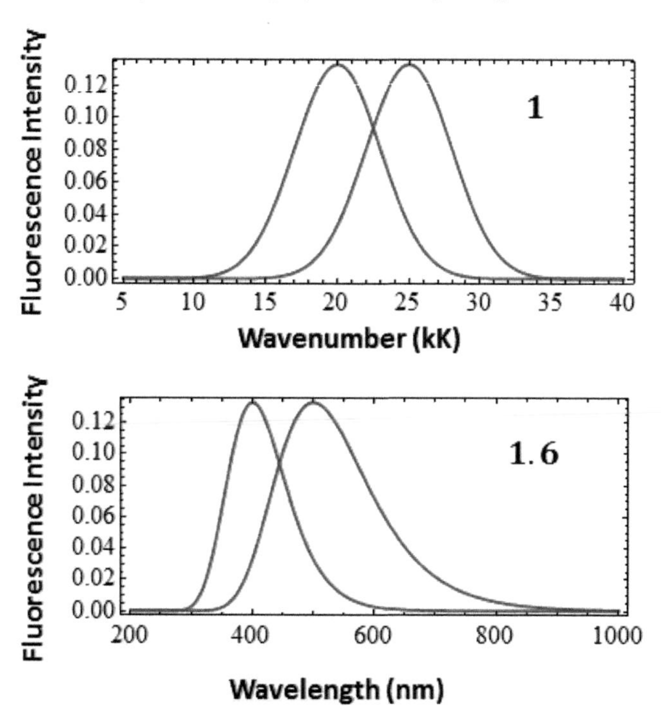

FIGURE 2-7.3 Two emission spectra presented in a wavenumber scale (top) and wavelength scale (bottom).

in Figure 2-7.3, one would realize it is not that uncommon. An evaluation of the area under the emission curve using the wavenumber scale gives the same values for the reference and sample, and the ratio between areas of the sample and reference is 1. However, measuring the same spectra in the wavelength scale and evaluating areas in the wavelength scale (**no** $1/\lambda^2$ correction factor) the areas are very different, giving the ratio between the area of sample and reference of 1.6. This means a colossal error in evaluating the quantum yield of 60%. It is clear that even a small spectral shift between the reference and sample may lead to a significant mistake in quantum yield evaluation. Obviously, when integrating the area under the curves when using $1/\lambda^2$ correction factor the ratio is 1, as it should be.

There are more examples where the surface area under the spectrum is used and improper integration may result in significant mistakes. To get a perspective of how large the effect is, let us consider a symmetrical bell-shaped spectra of fixed width in the energy scale (wavenumber) with precisely defined peaks as presented in Table 2-7.1. All bell-shaped spectra have identical shapes (width and peak intensity) and they are shifted by multiples of 5 kK (kilo Kaisers). These spectra are illustrated in Figure 2-7.4 in wavenumber (top) and wavelength (bottom). In the wavenumber, they are all equally spaced and the area under the peaks are equal. In the wavelength representation, the peaks are not evenly spaced and the area under each curve grows as it moves to longer wavelengths. Furthermore, the order of peaks is reversed when plotted in wavenumber/wavelength; the peak with the lowest wavenumber has the longest wavelength.

Thus, in the wavenumber representation, it may be concluded that all peaks represent emissions of the same intensity (area under the curve), that they are symmetric, and that they are equally spaced. The above conclusions would not be reached using the wavelength representation.

The top and bottom spectra of Figure 2-7.4 present exactly the same data, computer-generated normal distributions as shown in Table 2-7.1. They look quite different because the independent axis of each plot is an inverse transformation of the other as shown in Table 2-7.1. In reality, when generating the wavelength data, each point (x, y) was

TABLE 2-7.1 Sample Peaks and Some Properties Relative to the First Peak Computed in Wavenumber and Wavelength

Peak (kK)	Ratio of Area under Peak to Area under First Peak in Wavenumber	Ratio of Area under Peak to Area under First Peak in Wavelength	Shift from First Peak (kK)	Shift from First Peak (nm)
$\frac{1}{2\sqrt{2\pi}}e^{-\frac{(\lambda-18)^2}{2*2^2}}$	1	1	0	0
$\frac{1}{2\sqrt{2\pi}}e^{-\frac{(\lambda-22)^2}{2*2^2}}$	1	1.5	4	−101
$\frac{1}{2\sqrt{2\pi}}e^{-\frac{(\lambda-26)^2}{2*2^2}}$	1	2.1	8	−170.9
$\frac{1}{2\sqrt{2\pi}}e^{-\frac{(\lambda-30)^2}{2*2^2}}$	1	2.8	12	−222.2
$\frac{1}{2\sqrt{2\pi}}e^{-\frac{(\lambda-34)^2}{2*2^2}}$	1	3.7	16	−261.4

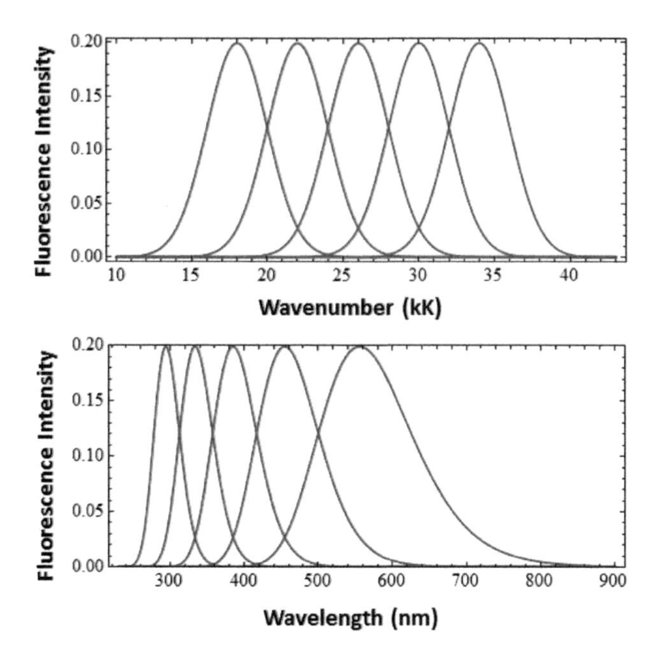

FIGURE 2-7.4 Sample peaks presented in wavenumber (top) and wavelength (bottom).

transformed to (10,000/x, y). Notice that the dependent axis value (intensity) is not transformed at all. Surface areas under each curve in Figure 2-7.4 (top) are identical, but comparing the surface area under the first curve (short wavelength) in Figure 2-7.4 (bottom) to the last one (long wavelength), we can see that the change is over 3-fold (over 300%).

We presented this example as a warning on how big errors can be made when improperly converting from wavenumber to wavelength representation.

2-8 WAVELENGTH CALIBRATION

A spectrum (transmission or emission) is an intensity reading as a function of wavelength. We typically consider error associated with the measured intensity, but we should be aware of possible errors in the wavelength reading. Most error analysis will not consider error for an independent variable and it is assumed that this error is statistically negligible. However, even if statistical error for reading each particular wavelength is very small it is possible (quite frequent in fact) that systematic error could be a big problem. Typically mechanical use of the monochromator or the entire system may lead to small mirror or diffraction grading displacement that shifts all wavelength reading toward smaller or higher values. In the case of wavelength, this is a problem of monochromator calibration. Monochromators are used to separate a beam of light into its wavelength components using refraction (prism) or diffraction (grating). The detector then performs a reading on each separate wavelength. Most monochromators use movable diffraction gratings to separate different wavelengths of light. We often rely on the manufacturer for service to calibrate the instrument and almost always such calibrations are reliable. However, an instrument may lose calibration over time or due to inadvertent misuse (e.g., sending the monochromator to an out of range wavelength). In some gratings, this may happen when the stepping motor slips or somebody selects an impossible to reach wavelength (i.e., −1 nm or 100,000 nm). A slight change in entering beam alignment may cause a prism-based monochromator to lose calibration.

There are a few methods that can be used to check the calibration. All of these methods rely on spectral features that are known to the experimentalist. Most frequently today, we depend on widely available lasers or other discreet line sources such as mercury, sodium, or xenon lamps. Laser light is typically monochromatic and has a well-defined wavelength. Widely available laser pointers are good sources of monochromatic light at 405 nm (violet laser diode), 532 nm (green laser diode), and 635 nm (red laser diode). Using these laser diodes, we can easily calibrate the monochromator in the visible spectral range. But we encourage the experimentalist to verify independently that the nominal wavelength specified for the laser is correct. Some more sophisticated spectroscopy labs will also have lasers with many other wavelengths, such as 224 nm (HeAg), and 325 nm (HeCd).

However, there are two problems with this. As it is easy to calibrate a monochromator externally, typically it is not easy to incorporate a laser beam into a spectrophotometer. The best way of doing this is to use scattering in the place of a sample. Second, UV lasers are not always easily accessible. In this case, we recommend using known spectral features of some fluorescent compounds. For example, benzene is a common solvent available in most chemistry labs. A drop of benzene closed in a quartz cuvette evaporates quickly, and gaseous benzene has very well defined and structured absorption/emission spectra that can be very easily used for calibration. It may be a surprise to some less-experienced researchers, but the absorption and emission of such vapors are quite significant. In Experiment 3-14, we present a simple routine to perform calibrations of the spectrophotometer and spectrofluorometer using this approach.

A small calibration error may be readily fixed with a single wavelength. However, when the calibration is off by a large amount in an unknown direction, a single wavelength is not enough. Oftentimes line sources such as lamps, which have multiple distinct emission lines, are used. Identifying these lines is most often done by finding the distance between them or their relative heights. Figure 2-8.1 shows the spectra of two commonly used lamps with mercury and xenon sources.

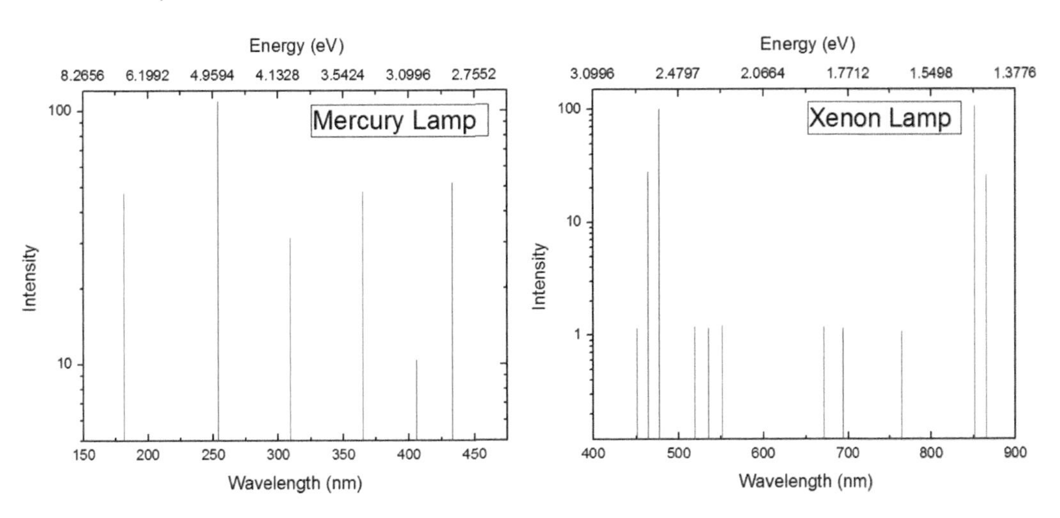

FIGURE 2-8.1 Spectra of lamps used for alignment.

Comparing the wavelength scales (bottom) and energy scales (top), it can be noticed that if the separation of two spectral lines in wavelength is 10 nm, and their separation in energy will differ based on the location of the two spectral lines. Thus, if the wavelength scale is shifted by 20 nm the distances between these spectral lines will be the same in wavelength but not in energy. The energy-to-wavelength conversion can also be calculated as $\mathcal{E} = 1239.8/\lambda$. Often while calibrating an instrument, the operational software asks for three-wavelength lines. These lamps provide more than enough peaks of known location. If the calibration is correct, the spectral lines plotted in an evenly spaced wavelength and evenly spaced energy will be mirror images of one other.

2-9 INSTRUMENT RESPONSE

Real instruments have photodetectors that have different sensitivities to different wavelengths; they have light sources that emit different intensities at different wavelengths, and they have mirrors that have different reflectivity at different wavelengths. A spectrophotometer or spectrofluorometer corrects for these differences in its software when they are set to do so (usually by default). However not all data acquisition modes do these corrections automatically, and the internal calibration has to be redone. As an example, Figure 2-9.1 presents two typical relative response functions of two different photodiodes. The dependent axis has been normalized and shows how much less sensitive the detector is at different wavelengths. The 1 on the scale corresponds to the maximum efficiency of the detector which is usually around or less than 80%, or it will detect four (or less) out of five photons. It is common for a certain detector to be five times as sensitive to blue light than to red light. Thus, the measured intensity of 500 a.u. in blue wavelengths and 100 a.u. in red wavelength would mean that there were the same amount blue and red wavelength photons emitted/detected. This conversion is typically done in the detector unit itself and data corrected for the relative response is a readout. Alternatively, software from the detector manufacturer may correct the data in the time of measurement.

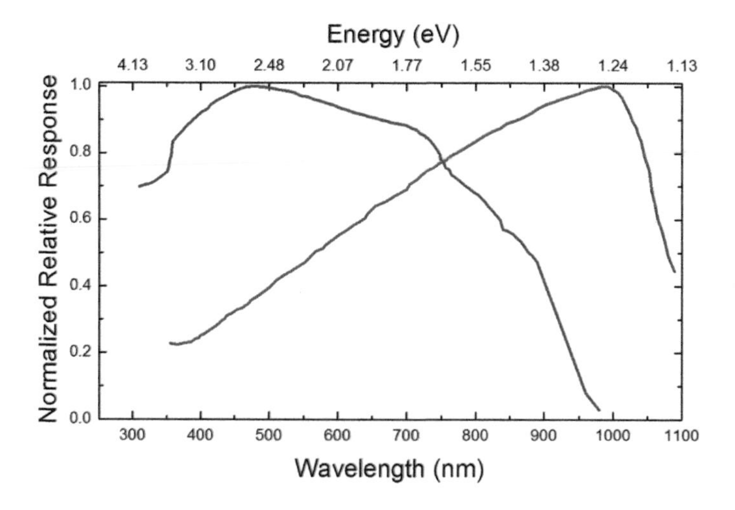

FIGURE 2-9.1 Relative responses of two photodiodes.

The intensities of excitation lamps per wavelength are just as varied as the detector sensitivity/response. When measuring excitation spectra, some instruments may automatically correct for the excitation light intensity. However, we recommend regularly checking such calibrations since the lamp profile typically changes with time. Such procedures will be discussed in Experiment 5-4.

2-10 LAMP PROFILE

Most instruments use various types of lamps (light sources) to generate light that can be used for measuring absorbance/transmittance or for excitation in fluorescence experiments. In the case of absorption that is a relative measurement in light intensity change, the only limiting factor is sufficient light intensity. So, there is no need to calibrate a lamp since we are always measuring the baseline. Also, there is no universal lamp that will cover the entire spectral range with sufficiently high light intensity and will have constant or known intensity. The spectrophotometers use two lamps, one to cover UV (usually a deuterium lamp) and one to cover the visible spectral range (usually a halogen lamp). The instrument will automatically use and change appropriately the lamp for the requested range.

A bigger problem is the emission measurement because the fluorescence signal directly depends on the intensity of the excitation lamp, and in addition, all lamps will have different spectral profiles. Also, the apparent lamp profile may depend on the monochromators and optics being used. A xenon lamp has good intensity in UV but in many lamp constructions, the glass used for the lamp production limits the UV range to only ~300 nm. One additional problem is the fact that the lamp profile depends on many factors, like sufficient warming time for the lamp and age of the lamp. In this case, comparing fluorescence signals of two chromophores with different excitation wavelengths can be a real challenge, especially when measuring and calibrating the lamp profile. We will discuss some examples of determining the lamp profile in the experimental part.

2-11 PEAKS AND SPECTRAL SHAPES

Typically observed absorption and emission spectra are a composition of multiple peaks. A peak is a normal distribution or a spectral profile (envelope). Often, what is called a peak is really a sum of a few normal distributions. In this section, some basic ideas of recognizing peaks are presented. Most commonly assumed spectral profile for a single peak is a Gaussian distribution. Since absorption and fluorescence spectra represent distribution in an energy scale, to properly represent the Gaussian profile we should use an energy-equivalent scale (eV or wavenumber, ν). The peak can be represented with a function P, given by:

$$P(\nu)=\frac{A_0}{\sigma\sqrt{2\pi}}e^{-\frac{(\nu-\nu_0)^2}{2\sigma^2}} \tag{2-11.1}$$

TABLE 2-11.1 Description of Properties of Peaks

Parameter	Mathematical Description	Spectroscopic Description
P "peak"	A normal distribution. Also called bell curve or Gaussian. A function.	Spectral shape in a spectrum. Often what is called a peak may be a group of peaks or a near Gaussian distribution.
ν "wave number"	The parameter of the function P and the independent axes in a graph of P.	The wavelength of light, an intrinsic property of a photon. It is often converted to wavenumber, frequency, etc.
A_0 "amplitude"	Vertical translation factor of each point in the function P.	Related to the height of the peak. It is often useful to work with A_0/σ since both parameters affect the height.
σ "standard deviation"	Parameter describing the deviation from the mean value. 68.2% of the area under P is located in the interval $[-\sigma, \sigma]$.	Related to the width of the peak (spectral width). This parameter is often discussed as the full-width half max, $FWHM = 2\sqrt{2Ln(2)}\ \sigma = 2.35\ \sigma$.
ν_0 "mean" or "center"	The mean value of the distribution.	The peak center. The wavelength at which there is the greatest emission/scattering/optical response. This gets messy when a spectral shape is made up of multiple normal distributions and is called a peak. For example, the peak center may not be the mathematical center of any normal distribution.
$\int P(\nu)d\nu$ "area"	The area under the curve of P equals A_0. It happens that $\int_{-\infty}^{\infty} \frac{1}{\sigma\sqrt{2\pi}} e^{-\frac{(\nu-\nu_0)^2}{2\sigma^2}} d\mu = 1$	Directly proportional to the intensity of observed light and the number of emitted photons. Brightness.
$\int P_1(\nu) \cdot P_2(\nu)d\nu$ "overlap integral"	Area which is present under both functions P_1 and P_2. If P_1 is to the left of P_2 the quantity is found by integrating P_2 from its lowest left value to the intersection point of P_1 and P_2 and then adding the integral of P_1 from the intersection point to its lowest right point. If P_1 and P_2 do not intersect the overlap integral is 0.	Often used with absorbance and emission spectra. Useful in calculating resonant energy transfer emission of the donor and absorption of the acceptor.
$\frac{d^4}{d\lambda^4} P(\lambda)$ "fourth derivative"	The fourth derivative of the function P.	Used as a means of finding peak centers, in a spectrum consisting of multiple difficult to distinguish peaks. The zeros of the fourth derivatives correspond to peak maxima.

The parameters of Equation (2-11.1) and common uses are listed in Table 2-11.1.

What is observed in the fluorescence spectra is a peak or a sum of multiple peaks. In this sub-section, simulated normal distribution peaks with three different means and three different standard deviations will be presented. This is done to demonstrate the line shape of the sum of peaks, and to show how altering parameters alters a spectrum. The peaks in this section are centered at 490, 500, and 520 and are labeled P_1, P_2, and P_3, the standard deviations will be mentioned in the text and kept the same for individual spectra of summed peaks for simplicity. Also $A_0 = 1$ for all presented peaks. Units are left out of both axes to concentrate on the shape of the curves and connect sample cases with results encountered in the lab.

TABLE 2-11.2 Equations Used for Simulations

$P_1(\lambda)=\frac{1}{3\sqrt{2\pi}}e^{-\frac{(\lambda-490)^2}{2*3^2}}$	$P_1(\lambda)=\frac{1}{10\sqrt{2\pi}}e^{-\frac{(\lambda-490)^2}{2*10^2}}$	$P_1(\lambda)=\frac{1}{30\sqrt{2\pi}}e^{-\frac{(\lambda-490)^2}{2*30^2}}$
$P_2(\lambda)=\frac{1}{3\sqrt{2\pi}}e^{-\frac{(\lambda-500)^2}{2*3^2}}$	$P_2(\lambda)=\frac{1}{10\sqrt{2\pi}}e^{-\frac{(\lambda-500)^2}{2*10^2}}$	$P_2(\lambda)=\frac{1}{30\sqrt{2\pi}}e^{-\frac{(\lambda-500)^2}{2*30^2}}$
$P_3(\lambda)=\frac{1}{3\sqrt{2\pi}}e^{-\frac{(\lambda-520)^2}{2*3^2}}$	$P_3(\lambda)=\frac{1}{10\sqrt{2\pi}}e^{-\frac{(\lambda-520)^2}{2*10^2}}$	$P_3(\lambda)=\frac{1}{30\sqrt{2\pi}}e^{-\frac{(\lambda-520)^2}{2*30^2}}$

Table 2-11.2 presents equations used to simulate the spectra presented in this section. Each curve will be one of those functions or a sum of thee of them in the same column.

2-12 STANDARD DEVIATION AND HEIGHT

Figure 2-12.1 presents peaks centered at 500 (P_2) with standard deviations (bandwidth) of 3, 10, and 30. The assignments on dependent and independent axes are arbitrary.

The area under all curves in Figure 2-12-1a has been fixed to 1 for all peaks and the center of all peaks is 500. The tallest thinnest peak has $\sigma = 3$, the middle peak has $\sigma = 10$, and the shortest widest peak has $\sigma = 30$. Thus, as the standard deviation increases, the peak height decreases and the width increases in such a way as to keep the area under the curve the same.

Figure 2-12.1b presents all peaks scaled to be the same height. P_2 is unaltered and P_2' and P_2'' are multiplied by 10/3 and 10, respectively, or by $\sigma 2/\sigma 1$ and $\sigma 3/\sigma 1$, respectively. The number by which P_2' and P_2'' are multiplied is A_0 of Equation (2-11.1). This did not alter the center or the shape of the peak, but it did increase the area under curves P_2' and P_2'' to 10/3 and 10, respectively. Thus, if these peaks are interpreted as spectra, the same intensity of light (total number of photons) would be emitted in Figure 2-12.1a. But for peaks in Figure 2-12.1b, the wider peaks would contribute higher intensity (a larger number of photons will be emitted by the wider peak).

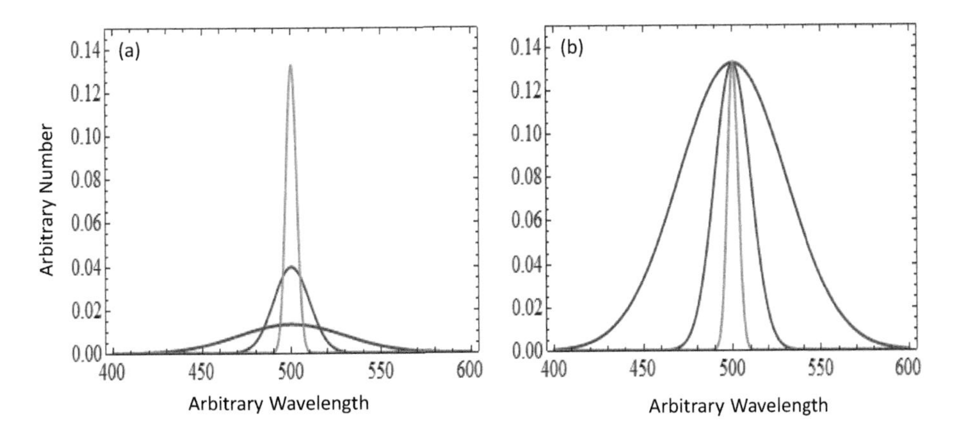

FIGURE 2-12.1 Examples of different width peaks with a maximum position at the same wavelength. a) Peaks normalized to the area; b) Peaks normalized to the height.

In real experiments, the amplitude A_0 is controlled by factors such as chromophore concentration, excitation intensity, and detector sensitivity. When A_0 is too high, problems such as saturation can occur.

2-13 SUMS OF PEAKS

Often a spectrum has multiple peaks that are close together. Figure 2-13.1 presents the sums of three normal $P_1 + P_2 + P_3$ (top), with varying standard deviations of 3, 10, and 30, and their component peaks (bottom). Every individual graph uses the same σ and $A_0 = 1$, as will be done for the rest of the section. Every spectrum in a figure has the same independent axes scale and interval.

The leftmost graph has $\sigma = 3$ nm and each center is clearly distinguishable. Two of the peaks are joined and the third is clearly distinguishable, usually, it would be said that there are three peaks in this spectrum. The center plot has $\sigma = 10$ nm and only one maximum is visible. None of the peaks are clearly visible, however, there is a spectral feature at 530 nm which is a bump sticking out called a shoulder. A shoulder results from a distribution that is close enough to another to blend as one shape but not far enough to have a distinguishable asymmetric shape. Also, the height of the combined feature is greater than the height of any one of its components (note the scale changed). This feature may be referred to as a single peak to two non-resolved peaks. The rightmost spectrum has all three peaks with $\sigma = 30$ nm and at a glance, it looks like a single normal distribution centered at 503 or one peak. Upon closer examination, the peak is asymmetric suggesting more than one normal distribution is present in the spectral shape.

There are some important observations to make comparing a spectral shape (top) to their components (bottom). In the left side, the maximum heights are the same on top and bottom. On the right-hand side, the height of the one spectral shape (top) is almost triple the height of one of its components (bottom)—note the scales on graphs. The height is the sum of all component normal distributions at a particular value of the independent axes.

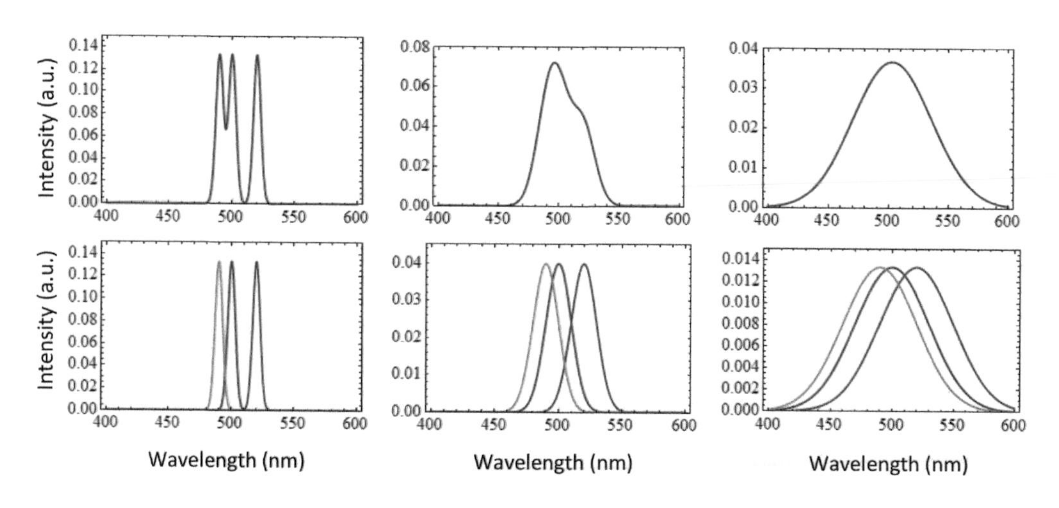

FIGURE 2-13.1 Sums of peaks of different half-width (top) and their individual components (bottom).

Also note that as σ increases, the overlap of the peaks increases. The bottom images are perfect fits to the top images. That can only be said because the top images were generated by a computer as the sum of the three peaks in the bottom graphs and there is no noise present.

2-14 NOISE—SIGNAL-TO-NOISE RATIO

Any real-world instrument for collecting absorption or fluorescence data will have noise or a nonzero reading where, theoretically, one should not exist. A signal is useful data that is theoretically expected, such as an absorbance peak or intensity reading. Noise is always present and a capable experimentalist will collect enough signal so that a measured signal will be statistically significant despite the noise—akin to listening to a single conversation in a loud, crowded place. Experimental noise will arise from a dark current in detectors, stray light, or from part of the experimental setup (i.e., vibrations or random perturbations).

Instrumental noise is statistically random, thus if a measurement is taken, and then taken again one second later, it will not be identical. However, both signals will have the same interpretation and statistical significance. Figure 2-14.1 illustrates this point with two absorbance baseline spectra. Both spectra are straight lines roughly corresponding to 99.993% transmission (absorbance below 0.0001) and amount to random noise. However, the red and black spectra do not align point by point.

The peaks in the previous Figure 2-13.1 are all perfect in the sense that they have no noise and they can be fitted with 0% error. A fit result in a function, P, which does not have noise such as the functions in Figure 2-13.1. To demonstrate the effects of noise, noise is added to the spectral shape from the center of Figure 2-13.1 center (the sum of three peaks centered at 490, 500, and 520 with $\sigma = 10$ and $A_0 = 1$ in all three). Figure 2-14.2 shows the results with noise intensities equal to H, $H/5$, $H/10$, $H/20$, $H/100$, and $H/1000$ where $H = 0.724$ is the maximum height of the spectral shape. For noise equal to H, the spectrum has no useful information. In the $H/1000$ case, the signal-to-noise ratio is very large and the spectral shape looks just like that of Figure 2-13.1 center. It is important to realize that the lines of the spectral shapes in Figure 2-14.2 are all the same thickness, generated and plotted in the same software; any perceived variation of line thickness is

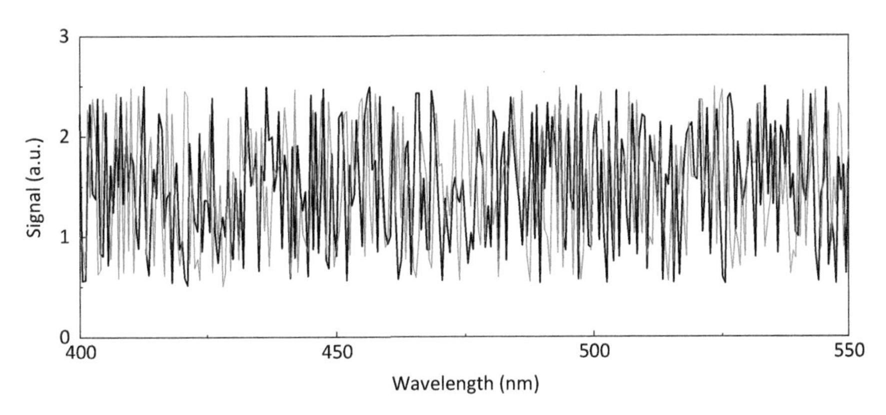

FIGURE 2-14.1 Example of random noise.

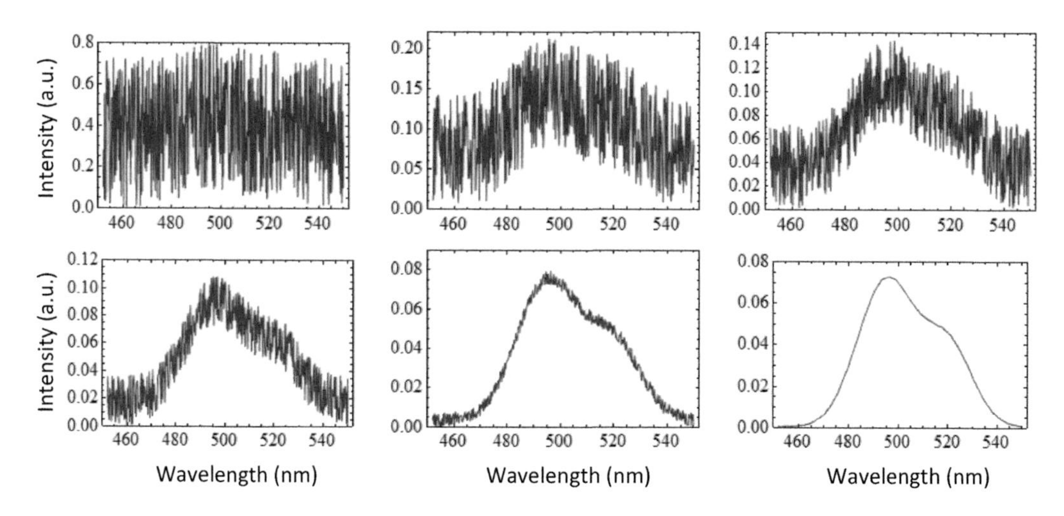

FIGURE 2-14.2 Effect of experimental noise on the measured spectrum.

due to noise. Also, the horizontal scale (wavelength) in these plots is smaller than in those of Figure 2-14.1, this results in the spectral shapes appearing wider (broader). Although the spectra in Figure 2-14.2 are generated as the sum of three peaks with added random noise and the fit will now be characterized with an error. This error decreases as the signal to noise ratio increases but never reaches 0%. The first figure would not produce any kind of reasonable fit since the entire signal is washed out by the noise. In any experiment, it is desirable to have the highest signal to noise ratio possible. In reality, the noise of a well-aligned instrument will be constant and the maximum height of the curve, H, will increase due to an increase in A_0. This is why the noise is given as a ratio of H.

In a real experiment, the question can be asked: *In Figure 2-14.1, is there a real absorbance peak between 400 nm and 550 nm?* Both the black and red curve seem to have multiple higher peaks in this range, perhaps 1% above the noise level. The answer is no. Even if there is a chromophore absorbing light in that range and there really is a 1% absorbance increase, the data above does not prove it, as it is too noisy to do so. If you were told that there is a chromophore absorbing light in this situation, could you prove me wrong? For a spectrum to be used as evidence of absorbance the signal intensity should be at least twice as high as the noise, or a signal to noise ratio of 2. This ratio is an arbitrary number and various experimentalists (or researchers) may have different preferences. Such a signal to noise ratio is usually a bare minimum requirement, and still shows a lot of noise in the baseline and in the signal peak itself. In general, the signal to noise ratio should be as high as possible. A common way to tell that the signal-to-noise ratio is acceptable is if the spectra profile is clearly visible over random fluctuations such as those in Figure 2-14.3.

The signal to noise ratio is typically increased by increasing the exposure time (collecting photons for a longer time) or summing multiple repeated runs of the same experiment. In both cases, the random noise will add up and decrease average noise signal variation, while a real signal will remain approximately constant leading to a better signal-to-noise ratio. As an example, let us consider the emission spectrum of anthracene when the signal is low as compared to the noise. Figure 2-14.3a shows the emission spectrum of anthracene as

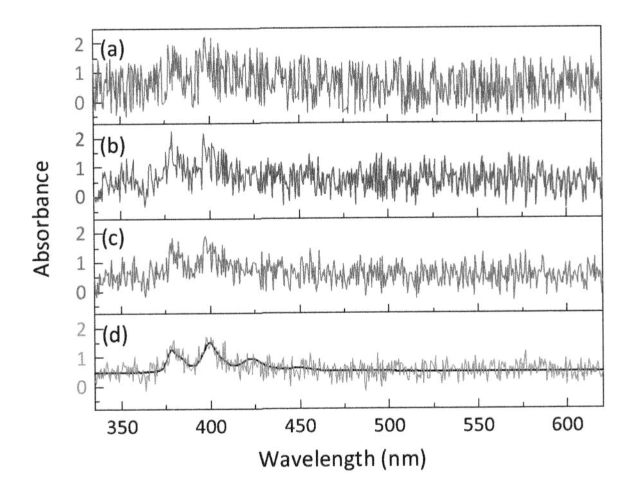

FIGURE 2-14.3 Example of spectra with different signal/noise ratios.

measured with a single scan. A small increase of around 400 nm is barely visible. Repeating measurement twice and averaging (Figure 2-14.3b) shows some improvement, and with four (Figure 2-14.3A-2c) and six (Figure 2-14.3d) scans, averaging the scans, the anthracene emission pattern is clearly revealed. Included in the background of Figure 2-14.3d is the actual anthracene spectrum, which shows a high level of agreement with the average of six scans. In practice, it is recommended to take multiple measurements and averaging them after. This gives a better result than would be obtained by merely increasing the exposure time (averaging time). In general, commercial chromophores are designed to be very bright, and usually, a low signal-to-noise ratio means the concentration of chromophores is too low or the instrument is misaligned.

2-15 SAMPLING RATE

In practical experiments, the instrument measures intensity of light as a function of the wavelength (or wavenumber or eV)—more precisely at any given wavelength the instrument measures the average intensity. The distance between each point on the wavelength scale (independent variable) is called step size. Each point corresponds to a number of photons of a specific wavelength that reaches the detector. In reality, it is a small band of wavelengths that reaches the detector $\lambda \pm \Delta\lambda$, and the bandwidth, $2\Delta\lambda$ is constant across the wavelength scale for most grating-type monochromators. Thus, a spectrum is a graph of (λ, I) points where I is intensity (number of photons) getting to the detector within the bandwidth $\lambda \pm \Delta\lambda$. The density of these points is determined by the operator by setting the step size for the scan. However, the lower limit is dictated by the instrument (i.e., the number of grooves in a grating—more grooves means smaller steps between measured wavelengths are possible). Typically, the user may select settings and alter the sampling interval. This is illustrated in Figure 2-15.1. Each panel displays the graph from Figure 2-13.1 ($\sigma = 3$) with added noise level $H/100$. The sum of three peaks has been plotted, the only difference is how many data points (coordinate pairs) are used to represent the spectral shape. The measured points are marked with dots that are connected, resulting in curves. The six spectra correspond to measuring every 0.2, 1, 4, 6, 8, and 10 units on the independent axes in a range of 450–550. The separation between successive measurements is called the step or sampling rate. If the units in Figure 2-15.1 were nanometers, the step would be 0.2 nm, 1 nm, 4 nm, 6 nm, 8 nm, and 10 nm for spectra a, b, c, d, e, and f, respectively. The first spectrum has (550−450)/0.2 = 500 points and the last has 100/10 = 10 points. Spectra a and b replicate the expected result well and look smooth. Spectrum c looks as though it is made up of straight lines, and it is questionable if the height of all three peaks is the same. Spectra d, e, and f lose resolution to the point where the peaks at 490 and 500 appear as one peak. Spectra d and f have the merged 490 and 500 peaks appear as a symmetric peak that would suggest fitting it as one peak. Spectrum e is asymmetric, suggesting two peaks or a shoulder. Information about a shoulder is more easily lost than information about distinct peaks. Spectrum d has more data points then spectrum e, however, because of the locations of the sampling points; spectrum e shows a more accurate resolution. The sampling rate profoundly affects curve fitting and resolution. Spectra d and f without extra knowledge of the system would most likely result in the use of two normal distributions. The system here literally is the sum of three normal distributions plotted by a computer (with insignificant noise level). It is always a good idea to use other knowledge while fitting data; spectrum d can be well fitted with three normal distributions. Of course, with more sampling points a fit will be more accurate.

Oftentimes sparse sampling will cause problems for the spectral features that are less than a nanometer apart, and this is usually a limitation of the instrument. The scale of the independent axis in this example may as well be one one-thousandth of a nanometer. In real experiments, $\Delta\mathcal{E}$ may be less than the resolution of the instruments. This is almost always the case for the rotational, often for vibrational, and scarcely for electronic transitions. It is important to use fitting parameters based on what is known about a spectrum. For example, there is no reason not to fit spectrum f of Figure 2-15.1 with three normal

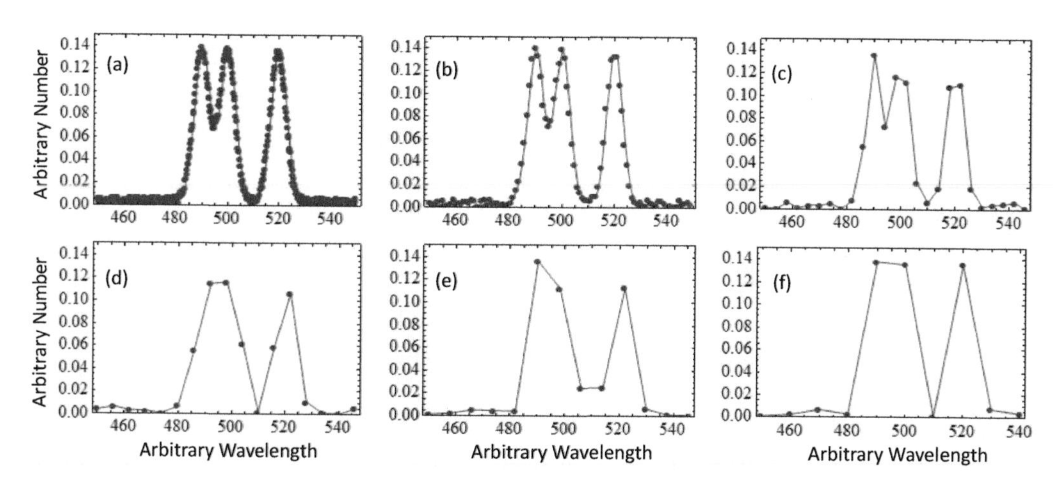

FIGURE 2-15.1 Six figures of the same curve with different sampling rates: (a) 0.2 (b) 1 (c) 4 (d) 6 (e) 8 (f) 10 units on the independent axis. Measured values are depicted as dots.

distributions given the knowledge of how it was generated (written in the previous paragraph). However, with no other knowledge, fitting with two distributions would produce just as good a fit, possibly even better, than using three. However, the interpretation of the spectrum would be altered. Spectrum in Figure 2-15.1f is $P_1 + P_2 + P_3$ with a sampling every 10 units. However, if given only spectrum f, it is just as reasonable to conclude that it is made up of two peaks, one larger and one smaller, instead of three peaks with the same A_0 and σ.

In real instruments, the smallest step sampling rate is determined by hardware, such as the grating, but it can be changed in the software to larger step size. An experiment with a higher sampling rate (smaller step) will take a longer time to run. It is not always possible to have a sampling rate high enough to distinguish all the peaks that make up a spectral shape and the same problem as with spectrum f of Figure 2-15.1 will be present. Thus, if a chromophore is known to have three emission peaks and the best-fit results in two, it is most likely a problem with the sampling rate or bandwidth setting on your instrument.

2-16 BANDWIDTH

Typically any instrument today will also have a setting for bandwidth (often called "slit") at the excitation and observation channels. Size of the slit is directly related to the wavelength bandwidth. For most instruments today, this means that a certain band of wavelengths will reach the detector $\lambda \pm \Delta\lambda$ for each step during the scanning. It is routine that for step 1 nm we will use the bandwidth of 2 nm or even 5 nm. But we have to remember that bandwidth will have its effect on the measured spectra profile. A larger bandwidth gives more light but subtle structural features could be lost. Typical grating-based monochromators will transmit bandwidth $\lambda \pm \Delta\lambda$ with the probability distribution around the central wavelength approximated by a Lorentzian shape:

$$P(\lambda_0,\lambda,\Gamma)=\frac{1}{\pi}\frac{\Delta\lambda}{(\lambda-\lambda_0)^2+\Delta\lambda^2} \tag{2-16.1}$$

where $\Delta\lambda$ is the set bandwidth. If the intensity is measured with a set slit an observed intensity will be:

$$I(\lambda,\Delta\lambda)=P(\lambda_0,\lambda,\Delta\lambda)I(\lambda) \tag{2-16.2}$$

As an example, let's consider the emission spectrum of anthracene in cyclohexane. This is a well-structured emission spectrum when it is measured with a very small bandwidth (0.2 nm in our case). Figure 2.16-1 shows the original emission spectrum of anthracene and normalized spectra calculated according to the Equation (2-16.2) for different bandwidth, $\Delta\lambda$. As the bandwidth increases, clearly, the structure of the original spectrum is disappearing. For a large bandwidth, the structured features can be completely lost. The experimentally measured spectra are presented in Experiment 4-6.

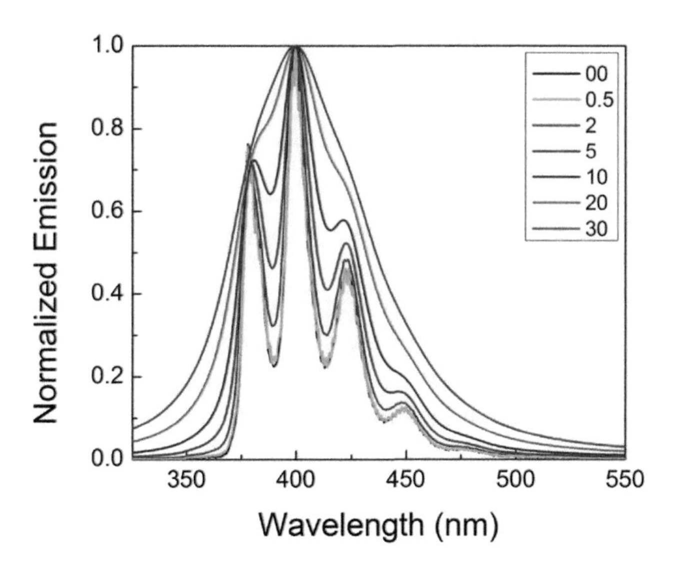

FIGURE 2-16.1 Theoretically predicted (Equation 2-16.2) of normalized emission spectra profiles as a function of slit width (as seen in the legend).

2-17 ARTIFACTS AND ERRORS

Peaks observed in spectra are made/measured by instruments in the real world. Although most of the time these instruments work well and stray light is properly controlled, spectra with errors may be encountered. Figure 2-17.1 shows some possible errors using an emission peak centered at 500 with a standard deviation of 10. Sometimes, measuring emission spectra, various artifacts and errors may corrupt the detected spectrum. These errors, their possible causes, and possible solutions are listed below with letters corresponding to those of Figure 2-17.1. In the experimental part, we are also presenting how these errors affect real measurements and how to identify certain errors and eliminate them.

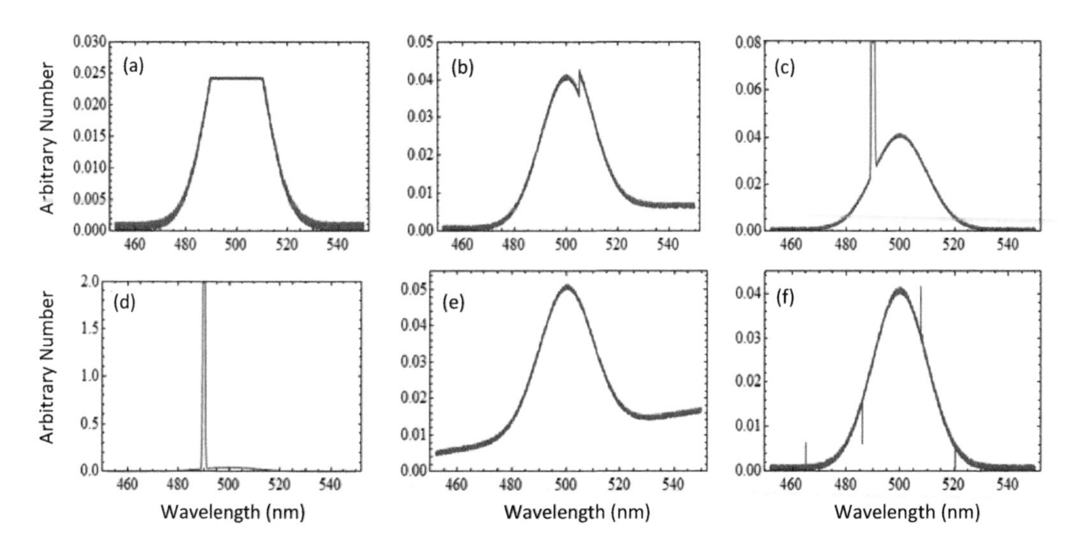

FIGURE 2-17.1 (a–f) Examples of potential errors when measuring emission spectra.

a. **The top of the peak is a straight horizontal line (or close to it) without or only with partial noise (Figure 2-17.1a).** This is most likely saturation of the detector. The straight line corresponds to the maximum value accepted by the detector, and most counts in the region are actually greater than the detector can accept. To fix this error, the intensity of the excitation light may be attenuated, the concentration of a used chromophore may be reduced (if applicable), or the parameters of the detection path may be altered (decrease voltage, slit width, and/or integration time).

b. **After a certain point all values in the spectrum are shifted vertically (either up or down) (Figure 2-17.1b).** This is a systematic error caused by a change in parameters while the measurement was running. Possible causes include a misaligned grating—as the grating moves, it also tilts in the wrong direction raising or lowering the beam path. This effect can show up at points where the instrument changes the lamp from UV to visible region. Another possibility is the introduction of stray light during the experiment (an opened cover or door). It is best to check for stray light first, as it is easier to fix this way.

c. **A tall thin peak appears in the spectrum (Figure 2-17.1c).** This is usually an artifact of the excitation source, such as a harmonic of the excitation (present at $2\lambda_{excitation}$, $3\lambda_{excitation}$, and so forth). To check if the problem is due to a harmonic from the excitation, one may shift the excitation wavelength a few nanometers. If the observed peak moves proportionally, it originates from the excitation source. Similar peaks (or sometimes just a bump) on the emission spectrum may indicate Raman scattering from the solvent (typically water). Raman scattering is typically not very strong and its position is correlated with the excitation wavelength. For water, it is shifted by about 3.3 kK. Raman shift is related to energy and it has a constant shift in energy scale. A water Raman peak for 290 nm excitation will be at about 320 nm while for 480 nm will be at about 560 nm. So, a trick with changing the excitation wavelength will show a proportional change in the observed peak.

   Some lasers, such as HeCd or HeNe, have multiple laser lines. A similar effect could be due to a leaking (not completely cleaned) laser line. For a problem arising from excitation light, it is best to use an extra filter on the excitation line that transmits excitation but does not transmit in the emission wavelength range. If the peak does not depend on excitation wavelength one should check for possible leaks of stray light since many fluorescent bulbs (used to light the room) have distinct peaks also.

d. **A very tall thin peak appears and the spectrum is practically not visible (Figure 2-17.1d).** This happens for the same reason as c, but the sample emission is weak, not close to saturating the detector. Such a strong peak may also appear when accidentally scanning through the excitation wavelength. The light intensity of the excitation wavelength is much stronger than the emission signal. This can be resolved in the same way as c; increasing chromophore concentration or altering detector parameters are, most likely, not necessary.

e. **The entire spectrum appears tilted (Figure 2-17.1e).** This is similar to the case of scattering. It may frequently happen for absorption measurement when scattering by small particles increases toward the shorter wavelength. In such a case, the tilt will be opposite to that shown in Figure 2-19.1e (baseline will be raising up more at a shorter wavelength). A random noise should be a horizontal line and Figure 2-17.1e shows a systematic error since the background is changing. It is possible that there is stray light or an electrical problem with the instrument.

f. **The spectrum has vertical lines sticking out of it (Figure 2-17.1f).** This normally results from a current spike or an under-responsive pixel in the detector array. Extra current will produce a reading at one coordinate spike up, and an under-responsive pixel will make it spike down. These artifacts usually do not reach the range of the detector (zero or saturation). This is a random error in the electronics of the setup and should disappear or change when the measurement is repeated with the exact same parameters. This often happens when the detector is too hot (some detectors need to be cooled), or if the detector is nearing the end of its lifetime.

Another possible error is a translation of the dependent axes, and this results in the spectrum being shifted to the left or right but otherwise, everything else is identical. This type of error is especially visible when studying the distances between spectral shapes or curve fitting. In this case, alignment is necessary and the later experimental section describes the process; although calling a technician is also a good option.

Often the fluorescence spectrum is overlapping the Raman spectrum of solvent used. In this case, the total observed spectrum is a combination of the fluorescence spectrum and Raman peak. This can be corrected by the measurement of the solvent only, followed by the subtraction of the "clean" Raman peak from the total spectrum.

Spectral shapes and errors will be visible in any coordinate system on any scale. No conversion of the dependent axes will change the interpretation of a spectrum or create a peak out of a shoulder. Thus, the horizontal and vertical axes scales in the above examples are not important. The horizontal axes are labeled to allow visualizations of peak centers and how they translate when multiple peaks are merged into one. The vertical scales show how the height of peak changes with σ and additions of peaks.

2-18 SPECTRAL OVERLAP

Figure 2.18-1 schematically shows absorption spectrum, $Ab(\nu)$ and emission spectrum, $F(\nu)$. The area shaded green is called the spectral overlap, as the area colored in green under both curves. Two closely spaced spectra will present a significant range where both spectra are covering the same wavelength range. The spectral overlap is typically defined between absorption and emission of the single dye. Most fluorophores will show a significant overlap between the absorption spectrum and emission spectrum. However, there are also fluorophores with very little spectral overlap and a very large Stoke's shift. More pronounced overlap between the emission spectrum of one chromophore and absorption spectrum of a different chromophore can be observed for donor-acceptor systems like in FRET pairs. In this case, the pair of dyes is typically selected to present the largest overlap. Also, it is important to remember that for a proper calculation of

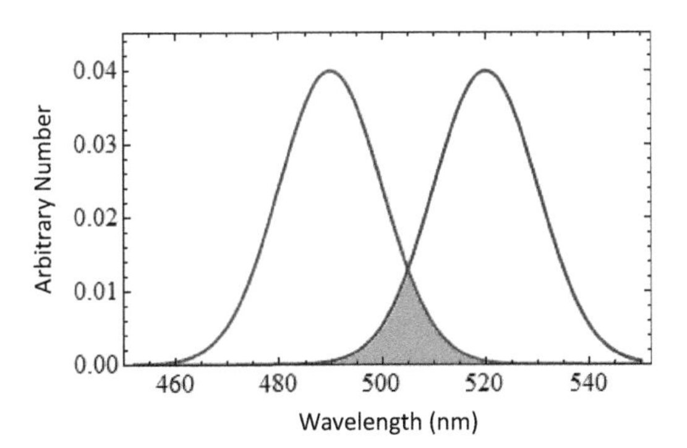

FIGURE 2-18.1 Two overlapping spectra. Overlapping region is marked in green.

the overlap integral it is important to properly represent the spectra (in wavelength or a wavenumber scale) and properly normalize the spectra.

Since a normal distribution never reaches zero, it is necessary to estimate the points where a peak ends. Normally, the point at which the peak reaches noise level is used. In Figure 2-18.1, if the left peak represents the absorption spectrum (extinction coefficient, $\varepsilon(\lambda)$) and the right peak represents the emission spectrum (fluorescence intensity $I(\lambda)$) the overlap integral can be represented by:

$$J = \int_{0}^{\infty} \frac{I(\upsilon)\varepsilon(\upsilon)}{\upsilon^{4}} d\upsilon \qquad (2\text{-}18.1)$$

The overlap integral is a rather complicated product between absorption and emission. In practice, it is best to use numeric integration within a reasonable range where both spectra have meaningful values. Like in this case, integration between 480 and 530 should be sufficient. In practice, it is common to use a Riemann sum instead of an integral in order to find the area under the overlapping spectra. As it will be shown later it is also important to properly normalize spectra before integration.

2-19 PEAK NORMALIZATION

It is often informative to compare multiple spectral shapes that may be in different scales. Experiments performed on different instruments (excitation/detector/cuvettes/completely different labs on other continents) often have different intensity scales. Sometimes, it might be useful to compare your data to published data, in which case a data digitizer is often used to convert a graph (image) into "data points." Moreover, the dependent (intensity) axis is usually expressed in arbitrary units (a.u.), and little meaningful comparisons about emission intensity may be inferred from such data. As an example, consider each panel in Figure 2-19.1. The solid curve is always the same peak.

Now, how do the dotted curves compare to the solid curves? An immediate answer may be that the sample represented by the dotted line emitted less light than the sample represented by the solid curve. This is not necessarily true. What if the dotted line came from data published in the last decade, but with 20× lower intensity of excitation light? Further

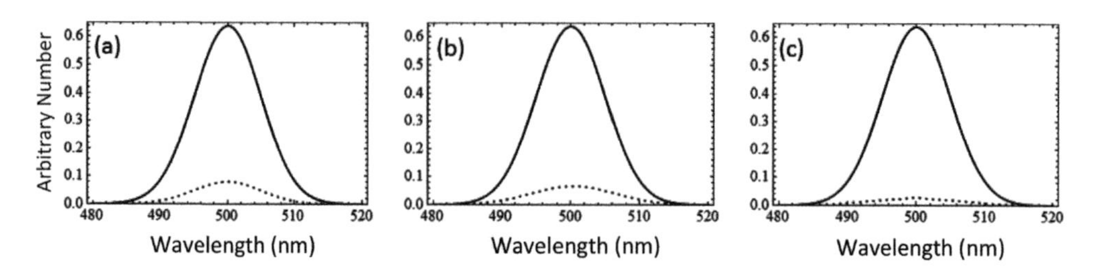

FIGURE 2-19.1 Each panel shows two spectral shapes (solid line and dotted line).

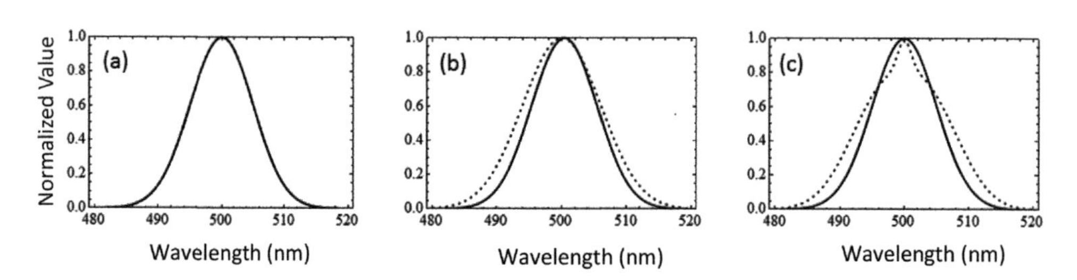

FIGURE 2-19.2 Normalized spectral shapes from Figure 2-19.1.

still, perhaps different cuvettes were used for each data set. Perhaps a collaborator measured the same sample on a different instrument with a detector that had different sensitivity. It is well known that the intensity of the emission is the least reliable parameter for comparison between two different experiments. In the case of spectra measurements, the spectral shape should not depend on the instrument (assuming the spectra are properly corrected for instrument wavelength-dependent sensitivity). When comparing spectral shapes with arbitrary or an unknown dependent (y) axis, it is customary to *normalize* the data. This normalization process involves dividing each data point by the maximum intensity in that one spectral shape, thus setting the emission maximum to one. This is different from a probability normalization, which involves setting the area under the curve to one. Figure 2-19.2 shows the same spectral shapes as Figure 2-19.1, only normalized.

Now the maximum intensity of each peak is the same, 1 arbitrary units. However, comparing the emission centers becomes much easier. In panel (a), the two spectral shapes are the same. In panel (b), the dotted spectrum is broader and shifted. In panel (c), the dotted spectrum is not a single peak and would originate from multiple emission centers. Differences that are now obvious to us in Figure 2-19.2 were not so readily observable in Figure 2-19.1.

Of course, there are experiments designed to compare the brightness of multiple centers and normalization is often detrimental to the presentation of such results. If the intensities are very different it is also helpful to plot the intensity axis in a logarithmic scale, which allows for easier observation of the spectral features of the less bright emitter.

2-20 RESIDUALS

The word residual has several meanings across multiple disciplines. It is usually associated with a small leftover quantity of what is being discussed. In this context, a residue is a distance between a data point and a point on its fit function on the dependent axes

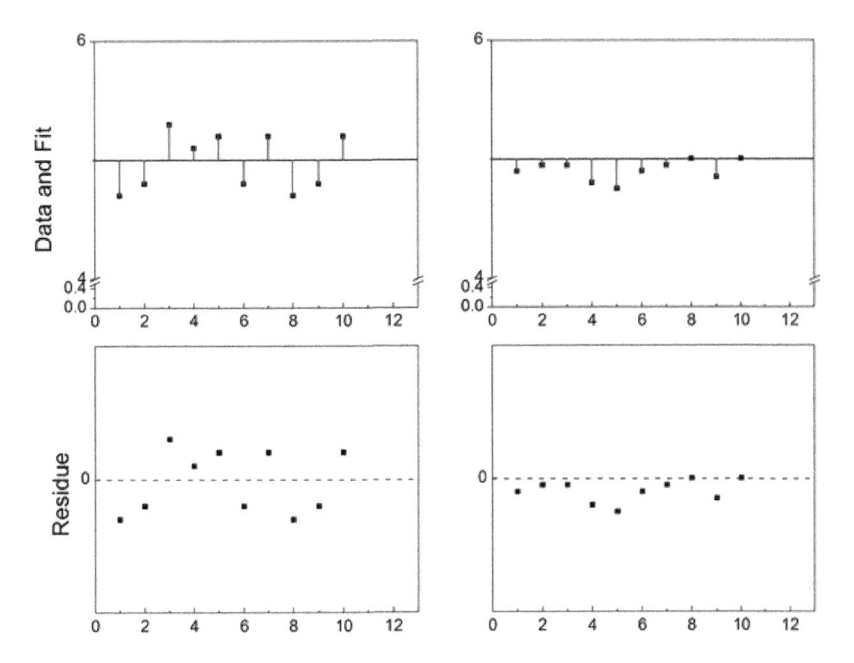

FIGURE 2-20.1 Two examples of residues distribution.

(measured along dependent axes). Residues are very useful in observing systematic errors and are often plotted purposefully. They are related to an error, but they are not the same. As it will be demonstrated in Figure 2-20.1, it may be necessary to discard a fit with a smaller least square difference (see Equation 2-20.1) due to its graphs of residues.

Here, we will present a residue analysis of four sets of data fitted to two arbitrary functions:

Horizontal Line	:	$y=5$
Exponential Decay	:	$y=e^{-x/3}$

This section is meant to introduce the visualization of residues, and the interpretation of the data and scales are not important. Figure 2-20.1 presents two sets of data points (black squares) fitted with the line $y = 5$ (black solid line). All graphs have the same independent axis, with the two top graphs having an identical-dependent axes and the two bottom graphs having their identical dependent axes. The residues are vertical lines between each individual data point and the fit function with the same value of the dependent axes (red). The bottom plots of Figure 2-20.1 present the graphs of the residues for each fit, the line $y = 0$ is drawn dashed. The residue's value on the independent axis is the same as the value at which it was calculated at, and its value on the dependent axes is the distance between a data point and its fit on the dependent axes, red lines in the top plots. The residues on the left side are scattered about zero, about half the points above zero and a half below it, this is what is expected for random noise. All the residues on the right plots are negative, meaning that the fit function is above all the data points. This is a systematic error, as the function $y = 5$ overestimates the value of the data. If the data for the example were intensity as a function of time, the fit would predict too high intensity. It is interesting to observe that in this example; least square analysis would favor the systemic error. The least-square analysis is used to calculate the error by minimizing the sum of the square of the difference between the fit function and data points for every data point:

$$\mathrm{Err} \sim \sum \left(y(x) - \mathrm{Data}_x \right)^2 \quad \text{(2-20.1)}$$

where $y(x)$ is the value of the fit function at point x and Data_x is a data point at point x. Least square errors cannot check for systematic errors, but plots of residuals can. Even though the fit on the right of Figure 2-20.1 more closely matches the data, it is not a good description of a real-world process because the data is below the fit; this is a systematic error (the line $y = 4.9$ would be a much more acceptable fit regardless of the least square analysis).

Figure 2-20.2 shows a fit and residuals of an exponential decay function, $y = e^{-x/3}$. The figure is set up as Figure 2-20.1. The right side of Figure 2-20.2 has a systematic error in the first part of the data and fit. This can be seen because the data points for $y < 5$ are all above the fit. The fitting presented here is similar to fitting in lifetime measurements.

So far only sparse data have been presented to visualize the procedure of taking residues. In reality, a residue plot will be as dense as the actual data and trends that show systematic error will consist of many tens if not hundreds of points. Figure 2-20.3 shows residues from fitting experimental data (fluorescence intensity decay). The top (corresponding to the right plot at Figure 2-20.2) is a systematic error because of the residual takes on values below and above the line $y = 0$. The residual on the bottom is

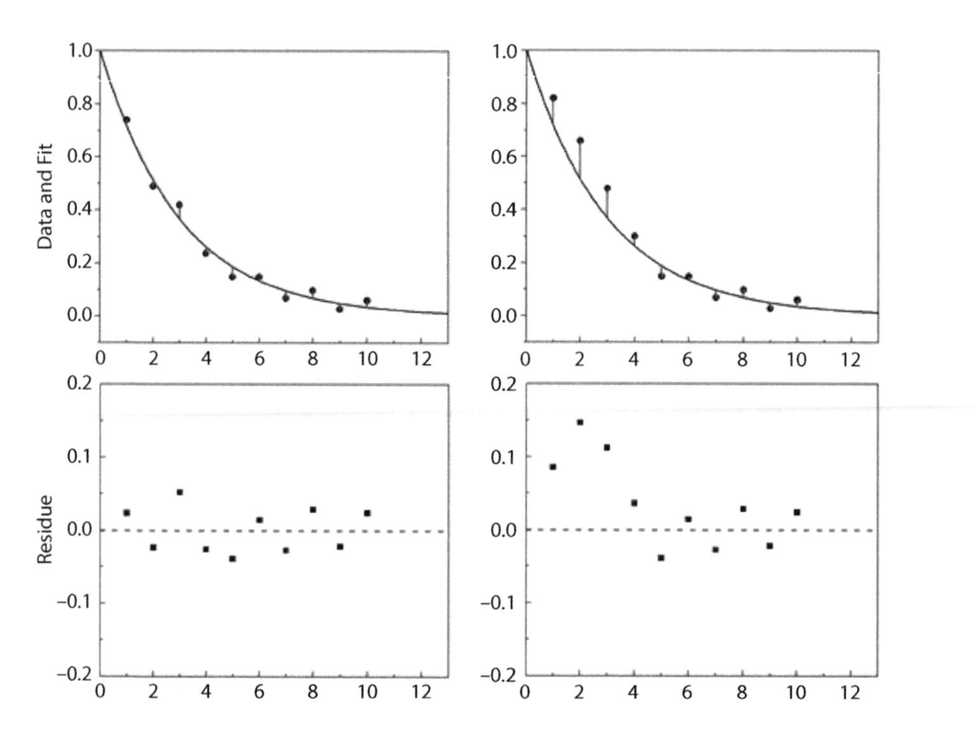

FIGURE 2-20.2 Residues in an exponential decay function (left - random error; right - systematic error).

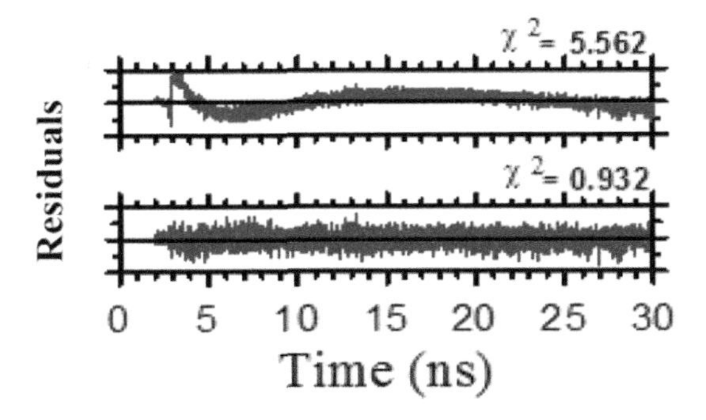

FIGURE 2-20.3 Experimentally determined residues for exponential decay. Top: Insufficient fit. Bottom: Good fit.

for a correct fit because all points are randomly distributed about $y = 0$ and there are no regions which are always above or below.

2-21 INTERPOLATION

Something that often comes up during data analysis is an interpolation. Interpolation is not fitting, and it does not reveal information about the nature of physical processes nor assumes a model for analysis. Interpolation is used to estimate a value at a particular point on the independent axes if no data is presented at that point. It is necessary when comparing two data sets that do not align well and it is best to avoid interpolation when possible by taking data on the same scale and offsets. Interpolation may be used for any kind of data. Common reasons to use interpolation include

- Two instruments are used in an experiment, and it is not possible or practical to obtain data from both on the same independent axis.
  - One instrument must have the same distance between successive data points and another does not produce data in constant intervals—see the example, below in this section.
  - The two instruments must use different fundamental quantities for the independent axis of data, and data from one is presented as a function of the data from the other.
- One piece of data is obtained non-experimentally.
  - Data from the work of others that is digitized (data extracted from a graph or picture).
  - Data from a common source which comes as a list of points—for example, the index of refraction of water (Experiment 3-4).
  - Data obtained from iterative solutions of non-analytical equations.

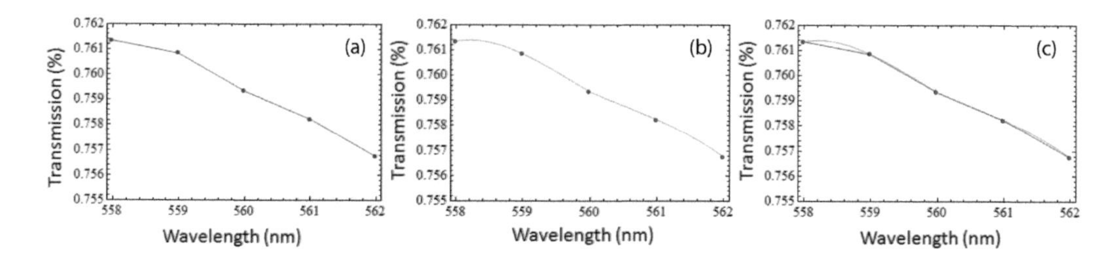

FIGURE 2-21.1 (a–c) Various methods for data interpolations.

TABLE 2-21.1 Experimental Data Points

Wavelength (nm)	Transmission (%)	Wavelength Offset from Two Data Sets (nm)	Wavelength (nm)	Normalized Emission Intensity (a.u.)
558	0.76134	0.06	557.94	0.50533
559	0.76083	0.03	558.97	0.49408
560	0.75933	0	560	0.47683
561	0.75822	0.04	561.04	0.46383
562	0.75673	0.06	561.94	0.43696

It is always best to use software for interpolation; most commercial softwares have dedicated tools called interpolation or interpolate. As an example, consider correcting emission data with the transmission data of a filter, perhaps measuring transmission/absorption on a spectrophotometer like a diode array (Agilent) that gives a reading in equal 1 nm steps and measures emission on a Varian Eclipse spectrofluorometer that does not have an equal number steps. Below is a list of the data points for this example as they might be acquired.

In the above example, the wavelength values are almost the same; they are even the same for some wavelengths. How would a point-by-point comparison for the same wavelengths be done? Is it ok to assume that 558 = 557.94? Will the scales always line up for 200 data points? What if the offsets are greater? Will the software allow rounding? Figure 2-21.1 shows the five transmission data points from Table 2-21.1 plotted and interpolated by connecting them with: straight lines (a) splines (curves) (b), and a comparison of both (c). The five data points are identical in each plot, and only their interpolation differs. The benefit of having these interpolated functions is that now it is possible to get a transmission or emission value anywhere on the dependent axes. The interpolation results in a continuous function. It is always good practice to check an interpolation function visually by graphing, especially when using splines. Interpolation should be avoided for sparse data points and for regions where there are no data points.

A commonly overlooked place where interpolation is present is in spectra. Making a line graph instead of a scatter graph results in interpolation or connecting individual data points. This makes the spectra look continuous when in reality measurements are taken at discrete wavelengths and it is possible to find a scale on which the data looks discrete. Interpolated data should not be used for peak fitting.

CHAPTER 3

Steady State Experiments—Transmission/Absorption

INTRODUCTION

The term "steady-state experiments," implies studies/measurements of phenomena that are constant in time. In practice, these represent measurable parameters (e.g., absorption, emission, or polarization/anisotropy) averaged over time, for which the outcome is constant in time.

In this chapter, we will present the basics of transmission/absorption measurements that we qualify as steady-state experiments. The first few experiments here are very basic and are used to introduce common procedures, as well as warn the reader of common errors or oversights; they include more descriptions and connections with theories presented in Chapters 1 and 2. For groups of experiments (and, occasionally, for individual ones), we will provide a brief introduction to the problem and some basic concepts to familiarize the reader with a more general problem. An understanding of earlier experiments is assumed in later experiments for the sake of clarity and briefness. The types of experiments found in this chapter include measurements of the following:

- Transmittance
- Absorbance
- Scattering

Note: By *absorption*, we will refer to the physical process of converting the wave energy to the matter. *Absorbance* is a measure of absorption expressed in the common (decadic) logarithm of the ratio of incident to transmitted light. We will use alternatively both terms for the description of our absorption measurements. Similarly, transmission and transmittance are used alternatively in describing the transmission of the light. However, it should be noted that the correct dependent axes labels should be transmittance and absorbance.

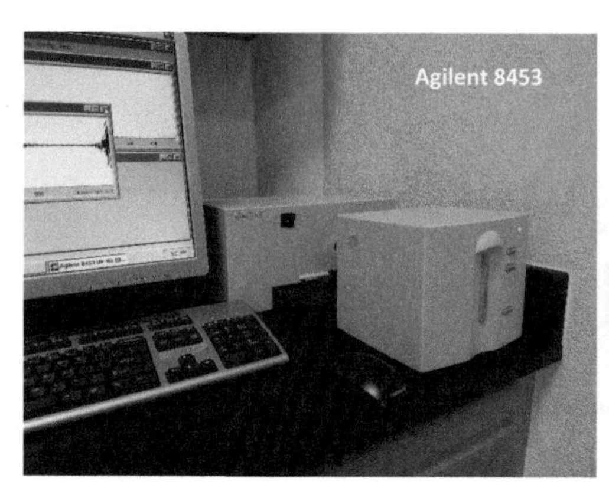

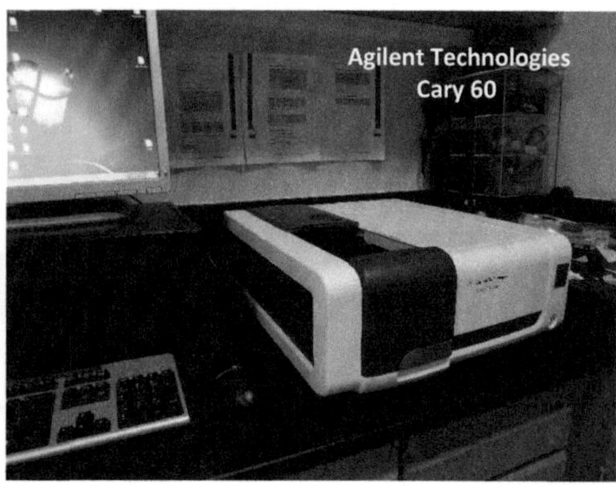

FIGURE 3.1 Examples of spectrophotometers used in experiments.

Experiments presented in this chapter can be performed with any spectrophotometer typically available in the lab. Throughout our own careers, we have used various spectrophotometers ranging from single beam diode arrays like Agilent 8453 or flash lamp Varian Cary 50 and Agilent Technologies Cary 60 UV-Vis to double beam spectrophotometers like Aviv 14DS (earlier Cary 14 Model). Figure 3.1 shows examples of some modern spectrophotometers. Experiments are presented as goals, thus an experiment will not be limited to a single technique. More advanced experiments will have analyses, interpretation of data, data fitting presented as needed, and will be ended with general conclusions. Many sections contain tips on how the data may be visualized, along with comparisons of different representations, as it is often useful to be able to think of data in other representations for different experiments and for multidisciplinary work. This is typically indicated in the text as "*Note.*" Discussions of various potential problems and how errors in the presented experiments may manifest in other experiments are presented at the end of each Experiment as "*Conclusions.*" It is advised that the reader uses the methods they are most comfortable with during experimentation to prepare samples and obtain expected results. There are several "stepping stone" experiments meant as introductions to different types of equipment, which give detailed instructions and guidelines for operation while presenting potential problems/pitfalls. These detailed instructions are not repeated in every experiment to limit repetition.

The "stepping stone" experiments of this section are as follows:

Transmission of transparent materials. The purpose of these experiments is to demonstrate that all transparent materials will attenuate light intensity in some way due to reflection and scattering. These experiments also demonstrate how transmittance and/or absorbance measurements should be done and why spectrophotometric cuvettes are rectangular with parallel walls and why they must be placed for measurement orthogonally to the direction of the beam of light.

- ***Transmission of glass and optical elements.*** The purpose of these initial experiments is to familiarize the reader how optical elements affect the transmitted beam and how this will demonstrate in (affect) transmittance/absorption measurement.

- ***Transmission of cuvettes—An introduction to absorbance.*** The purpose of these experiments is to demonstrate how common cuvettes affect light attenuation, and that their placement is important; the reader should understand the basic principles by which cuvettes that are apparently identical and transparent may, in fact, attenuate light differently.

- ***Measuring the absorbance of chromophores in a solution.*** The goal of these experiments is to familiarize the reader with how to measure absorption using liquid samples in cuvettes while also showing how different cuvettes (among other factors) may influence the outcome.

Measuring scattering. Scattering for most experiments is considered negligible. In some cases, however, such as in measuring the absorption of beads, cells, or quantum dots, scattering must be taken into account. Sometimes scattering of proteins should also be taken into consideration. The goal of these experiments is to demonstrate what scattering is and when (and why) an experimenter should be concerned with it.

- ***Measuring absorbance in the presence of scattering.*** Here, we present two approaches: (1) how to account for scattering when it can or must be measured independently within the experimental design; and (2) when scattering particles are small and one can use the Rayleigh approximation to account for it.

Measuring absorbance in thin films/layer. These types of measurements are typically difficult. The overall sample absorption is low and reflectivity is comparable to the regular cuvette wall. Therefore the relative contribution of reflected light to the overall attenuation of the transmitted beam can be significant. Special care must be taken to properly correct for such sample reflectivity. A useful example would be poly(vinyl) alcohol (PVA) films. These are common polymer films that can make very thin layers with spin-coating (>10 nm) to a film that are close to a 1 mm thickness. Also, PVA transparency may vary with its molecular weight, but in general, these films are transparent in UV down to 250 nm. We will present methods for preparing PVA films, as well as how to measure absorbance therein. The polymer film is just an example of thin-layer samples and similar approaches can be used for measuring different films or layers deposited on transparent background.

How to prepare isotropic and oriented polymer films. This part presents a brief discussion on methods for preparing PVA films. The general method is simple and easily achievable in a chemistry/physics laboratory. Such films offer many practical applications for studying dyes embedded and immobilized in a polymer matrix.

Introduction to Transmittance/Absorbance

The first step before measuring emission spectra (or fluorescence lifetime) of a given sample is to measure its absorption spectrum. This is crucial and is the simplest way of identifying a sample and establishing its optimal excitation conditions.

First, it is important to remember that the spectrophotometer measures just the intensity of light reaching the detector. In the typical measurement of transmittance/absorbance we just compare the light intensity reaching the detector with the sample present to the light intensity detected without the sample present. It is assumed that transparent materials (like glass, quartz, or plastic) do not interact with light and thus should not contribute to light attenuation. Indeed, in broad spectral ranges, these materials do not absorb light. However, it is not only absorption that contributes to the total attenuation of light (extinction) that is going through the transparent media/materials. Here, the term "extinction" is used to emphasize that there are a few different physical processes that may contribute to the attenuation of transmitted light through transparent materials. These processes include absorption, scattering, and light reflection at the interfaces of material and the surrounding air; they are schematically shown in Figure 3.2. A detector that is placed after the sample sees only light that passed through and has no information about how it was attenuated.

In most cases, only light extinction that takes place due to material absorption is considered, in such a case it is simply referred to as "absorption." However, it is important to remember that reflection, and to a smaller extent scattering also contribute to the overall attenuation of light when taking measurements, and these processes must be taken into account during experiments. Many visually transparent materials will have significant absorption in UV and/or NIR spectral ranges that are not discernable to the human eye and therefore are not immediately apparent.

A distortion of the beam path (e.g., shifting the beam or diffracting the beam) by the sample is another factor that can contribute to the apparent extinction of light. The sample compartment in a spectrophotometer will hold a cuvette in a way that the walls through which light will travel are orthogonal to the direction of the beam. There are two important reasons for that. First, reflection on a flat, dielectric interface depends on the incident angle and is minimal when that angle is 90° (0° to the normal to the cuvette surface). Second, a typical cuvette that is about 12 mm thick (10 mm internal path and two cuvette

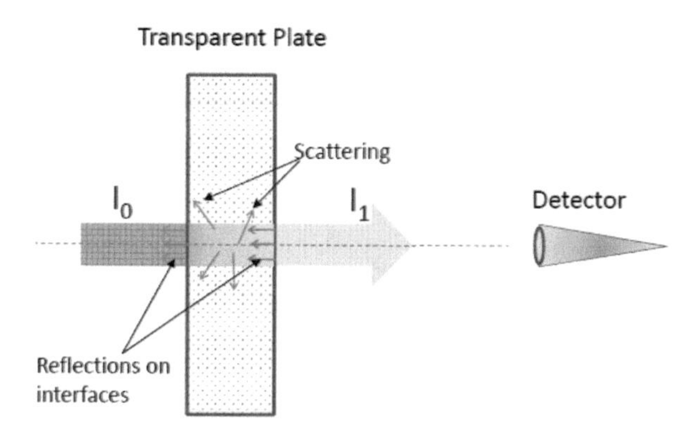

FIGURE 3.2 Processes that contribute to light attenuation when light travels through a transparent plate.

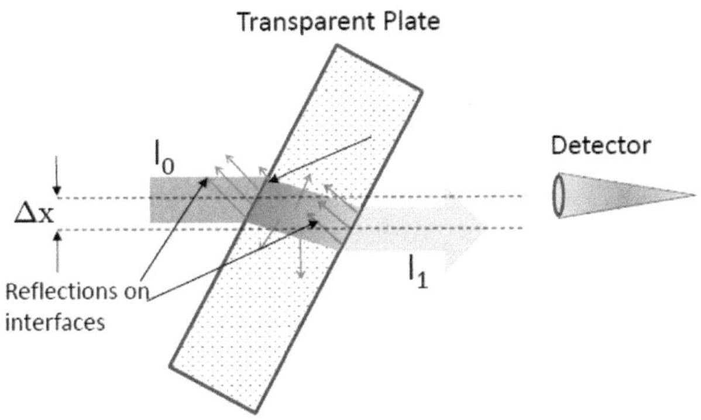

FIGURE 3.3 Tilted block of glass and path of the light beam.

walls that are about 1 mm thick each) will shift the beam of light when inserted at an angle different than 90°. Figure 3.3 schematically illustrates the propagation of light through the tilted glass block (or cuvette) by an angle α, as compared to the proper orthogonal orientation.

In this case, the effective parallel shift (Δx) of the incident beam (in any direction, left, right, up, or down that will depend on the tilt angle (α) and the direction of the tilt) will displace the beam that in effect will not hit the center of the detection system. So, even if the attenuation due to reflection and scattering does not change much, the apparent attenuation can be dramatic. Typically, detection systems are equipped with a limiting slit and the beam will not be completely transmitted through and/or will hit the detector at a different spot. Also, measuring the transmittance/absorbance of a sample that is not flat will be problematic. For example, when trying to measure the absorbance of a convex or concave entity such as a lens, the light beam will be diffracted and only a small central component of the beam will reach the detector.

The introductory Experiments 1 and 2 aim to demonstrate the following:

- How the transmittance of a sample depends on its position in the sample compartment of a spectrophotometer.
- How to conduct tests to compare the transmittance of transparent plates made from different materials.
- How tilting the sample can affect the measured transmittance by altering beam trajectory and artificially lowering measured transmittance.
- How the shape of a sample (e.g., the curvature of a surface) can affect transmittance and therefore alter measurements.

These demonstrations serve to show the experimenter why only square cuvettes must be used, and why absorption cannot be measured directly in a beaker, vial, test tube, or any other type of non-uniform container.

Introduction to Absorption Measurements for Chromophores in Solutions

Note: As chromophore, we refer to the molecule capable of absorbing light and we will alternatively use terms chromophore and dye.

In most cases, the absorption of a specific chromophore is measured in a well-defined buffer and under well-defined conditions (cuvette size, type, temperature, etc.). The baseline or a blank measured for these kinds of samples should be an identical cuvette filled with an identical solution without the chromophore (just a buffer or solvent). The following sample measurement will be identical (the same) cuvette filled with chromophore suspended in the solvent or buffer. In this way, any light extinction due to the cuvette (e.g., reflection, scattering, etc.) and buffer can be subtracted (corrected) from the sample measurement and only the absorption of the chromophore will be measured. For a two-channel spectrophotometer, the measurements are parallel and for a single channel spectrophotometer, the background is stored in the computer memory. Typical concentrations of chromophores are in the micromolar range that has little effect on the optical properties of the solvent/buffer. However, if for some reason the needed chromophore concentration is very high, that may change the optical properties of the buffer (e.g., refractive index) we should account for that.

Most of the presented experiments use low chromophore concentrations and a measurement is taken for both the baseline (e.g., cuvette + buffer (solvent)) and the sample (cuvette + buffer (solvent) + chromophore) simultaneously in the dual-beam spectrophotometer or sequentially for a single beam spectrophotometer. In this case, the measurement is just a comparison of light intensity reaching the detector with and without the sample (chromophore) present. Correcting the sample measurement for the baseline contribution means subtracting the amount of light that has been attenuated by the buffer and cuvette only from the amount of light attenuated by the sample (cuvette + buffer (solvent) + chromophore of interest). To calculate actual chromophore transmission (more precisely light attenuation by the chromophore absorption), it is necessary to correct the intensity of transmitted light by the sample for intensity attenuation due to the baseline only at each wavelength. Using the definition of transmission, we can describe transmissions for a buffer, T_B and the sample in a buffer, T_{B+S} as:

$$T_B(\lambda) = \frac{I_B(\lambda)}{I_0(\lambda)} * 100\% \text{ and } T_{B+S}(\lambda) = \frac{I_{B+S}(\lambda)}{I_0(\lambda)} * 100\% \tag{3.1}$$

where $I_0(\lambda)$ is the intensity of light detected with nothing in the optical path, $I_B(\lambda)$ and $I_{B+S}(\lambda)$ are intensities detected with buffer only and sample in buffer. Taking the ratio of the transmission for buffer plus sample to the transmission of just the buffer alone, we have:

$$T_S(\lambda) = \frac{T_{B+S}(\lambda)}{T_B(\lambda)} = \frac{I_{B+S}(\lambda)}{I_B(\lambda)} * 100\% \tag{3.2}$$

That gives the transmission of the sample in reference to the buffer and other factors like the cuvette. The important thing to remember is that the transmission of the sample is the ratio of the transmission of the buffer + sample to that of the buffer, both measurements made in identical cuvettes and conditions.

To find the dependence for measured absorptions for sample plus buffer (solvent) and buffer (solvent) consider Equations 3.1 and 3.2 and write the absorption of buffer plus sample and buffer only as follows:

$$\log\frac{I_{B+S}(\lambda)}{I_0(\lambda)}=-\varepsilon_{B+S}(\lambda)Cl \text{ and } \log\frac{I_B(\lambda)}{I_0(\lambda)}=-\varepsilon_B(\lambda)Cl \tag{3.3}$$

For sample transmission, T_S we can calculate sample absorption A_S:

$$A_s(\lambda)=\varepsilon_{B+S}(\lambda)Cl-\varepsilon_B(\lambda)Cl=A_{B+S}(\lambda)-A_B(\lambda) \tag{3.4}$$

When using absorption correction for a buffer, it becomes a simple difference between absorption of sample plus buffer (solvent) and absorption of buffer (solvent) only. We say that the absorption is an *additive property*. So, for multiple absorbing species, a measured cumulative absorption is simply, the sum of all the absorbing components.

Measuring the absorption spectrum is the first step in dye evaluation. Typically, we are interested in finding the spectral profile (shape of the spectrum and position of the peaks) and the extinction coefficient. It is easy to compare extinction coefficients measured in two experiments since these are just values. So, measuring extinction coefficients of a specific dye (chromophore) at two different concentrations should give the same values (very close). When measuring the same dye in a different solvent, the extinction coefficient will typically be different and it will be clearly evident from measured and calculated values.

The same dye at two different concentrations that differ several folds (e.g., 1 μM and 100 nM) in the same solvent should show an identical spectrum that is proportionally lower (10-fold when the concentration differs 10-fold). Comparing normalized spectra (spectral profile) can then be more useful. The differences that result from different sample concentrations indicate the formation of new species such as aggregates. So, just comparing absorption spectra measured for samples that have different concentrations will bring valuable information on the formation of potential aggregates. As described in Section 2-19 "Peak Normalization" (p. 71), the spectra normalization procedure is just adjusting all absorption values across all measured wavelengths range to the single value (typically maximum). This is done by dividing all values across the absorption spectrum by the maximum value of measured absorption. In this case, the measured absorption at each wavelength will be divided by the maximum value of measured absorption (typically the peak value). So, a maximum value of 1 will be at the peak and all other points will be proportionally smaller. Normalized spectra for two different concentrations should perfectly overlay unless there are some processes (or errors/artifacts) that lead to an artificial change in the shape of the absorption spectrum.

Measuring Absorption of Thin Samples (Films)

In some fluorescence experiments, the sample can be a thin film or surface. It can be a membrane, a glass with chromophore inside, a slide with deposited dye on the surface, or a polymer film in which chromophore has been embedded. We will discuss several simple examples for measuring absorption spectra for samples that are thin. In such samples the absorption is typically low and other secondary effects like light reflection or scattering may significantly contribute to attenuation of transmitted light. One common example of thin samples is polymer films. Various polymer films are used, like poly(vinyl) alcohol (PVA), polyacrylamide, and polyethylene just to name a few. PVA film will serve as an example to discuss in detail how to work with thin films. Any other thin films can be treated in a similar way. Polymer films are very attractive since in many instances they represent a solid (rigid) system. In contrast to the solution, such films immobilize dyes, significantly restricting their mobility. PVA is probably the simplest to make and form a matrix that is closest to true rigid systems. There are a few advantages of using PVA films. Besides easy preparation of PVA films, PVA has low permeability to oxygen that makes fluorophores much more stable and frequently allows studying phosphorescence. Films also have high plasticity that allows easy stretching and orienting ghost fluorophores to study absorption and emission in oriented systems.

Most polymer films are visually transparent but present significant absorption in the short wavelength range (blue-UV) and in many cases, the contribution of the film absorption to overall measured absorption in this range should be carefully accounted for.

The following experiments show how the polymer film (PVA film) can be prepared. Beginning with a simple approach on how to determine the absorption of the films, and when measuring the dye absorptions how to correct for the film contribution. An important aspect to remember in the case of a thin layer sample is that the biggest problem is the inaccurate estimation of the film thickness. Even if we can measure the physical thickness of the film using a very precise caliper, we have to remember that the apparent thickness of such films is sensitive to many factors like external humidity or temperature.

EXPERIMENT 3-1 TRANSMITTANCE OF A TRANSPARENT PLATE

Keywords: transmittance, light reflection, scattering, glass slide, sapphire slide, rectangular glass block.

A single absorption/transmission measurement requires two measurements: (1) a reference measurement and (2) a sample measurement. The reference measurement establishes the amount of light reaching the detector when there is no sample in the light path (typically referred to as the baseline measurement); the sample measurement then records the amount of light reaching the detector when the sample is present. The difference between these two readings allows us to calculate the amount of light being extinct (attenuated) by the sample (i.e., light absorbed, reflected, and scattered by the sample). In other words, the baseline measures the light intensity reaching the detector after traveling through all optical elements of the spectrophotometer and all other elements needed to hold the sample (e.g., cuvette and solvent). For the purpose of this experiment, the baseline will be determined/measured by putting nothing in the beam path (empty spectrophotometer—no cuvette or cuvette with solvent). In the following step, we insert the sample (in this case, a transparent plate) and repeating the measurement procedure. In this case, the light extinction/attenuation by the plate is represented by a simple ratio of light intensity reaching the detector with the sample in the path as compared to the light intensity when the sample is not present (baseline).

(Continued)

EXPERIMENT 3-1 (CONTINUED) TRANSMITTANCE OF A TRANSPARENT PLATE

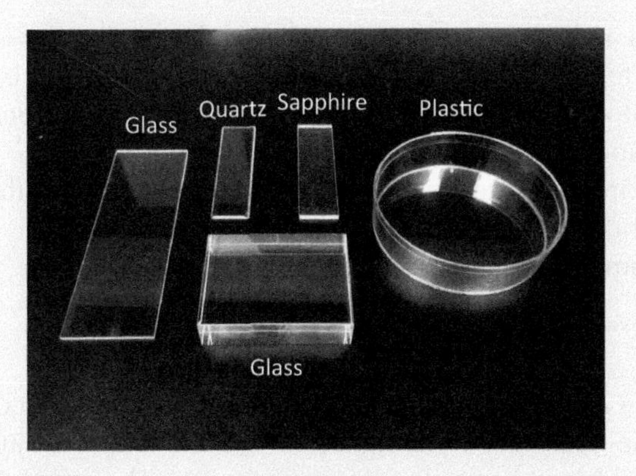

FIGURE 3-1.1 Photograph of used transparent materials.

This protocol describes measurements of light transmission through various transparent materials that are typically used in an experimental setting (quartz, glass, sapphire, and plastic). The goal is to demonstrate that any transparent material will impact the intensity of transmitted light to some extent, and underscore the importance of taking an appropriate baseline measurement to account for this. Figure 3-1.1 shows a photograph of selected transparent elements. In the case of a Petri dish, we will measure transmission through the bottom of the dish (this is typically a clear flat surface).

MATERIALS

- Glass plate
- Quartz plate
- Sapphire plate
- Glass microscopy slide (1 mm thick)
- Petri dish (plastic)

EQUIPMENT

- Spectrophotometer

METHODS

1. Turn on the spectrophotometer.
2. Turn on the "lamp" in the software (if needed).
   a. Wait a few minutes for the lamp to warm up.
3. Take a baseline measurement by clicking the baseline button in the software or on the spectrophotometer.
   a. Make sure there is nothing in the beam path.
4. Take a measurement of the sample but do NOT insert the sample yet. We want to make sure the lamp has warmed-up. This is not a necessary step when you already have good experience with the instrument but we recommend this as a check if the lamp

(Continued)

EXPERIMENT 3-1 (CONTINUED) TRANSMITTANCE OF A TRANSPARENT PLATE

has warmed up and if it is stable. If the measured transmittance is very close to 100% (absorbance 0) for the whole range then the spectrophotometer is ready to be used.

5. Insert a plate into the beam path. Make sure the plate is orthogonal to the direction of the spectrophotometer beam propagation.
6. Take the sample measurement by clicking sample in software or on the spectrophotometer.
7. Change plate and repeat from step 5 for each element.
   a. Keep the same baseline for all plates.

RESULTS

Figure 3-1.2 presents a photograph of the sample compartment of a spectrophotometer overlaid with labels in the beam path. The light path (yellow line) is emitted from the source, collimated to a parallel beam, and enters the detector side. The sample/cuvette holder has a hole through which the light travels (the holder does not affect the light). As a first step, we will measure the baseline with nothing in the light path. This represents the intensity of the beam leaving the source of light and reaching the detector after going through all optical elements. The detected intensity as a result of this is called I_0. A typical instrument's software will repeat this measurement a few times and store average values for each wavelength, and will only display signal variations as seen in Figure 3-1.3. Figure 3-1.3 shows a computer screenshot of a baseline measurement (more precisely, the baseline signal variation represented in percentage of transmission) with nothing in the sample compartment. Although it appears to be noisy, the scale limits are very small and the signal variation is mostly caused by random noise in the detector (unavoidable). The maximum variation range will depend on the instrument but it should not be more than 0.2% and for most instruments will be much less. If the baseline variation is much larger, it might mean that the detector is not detecting

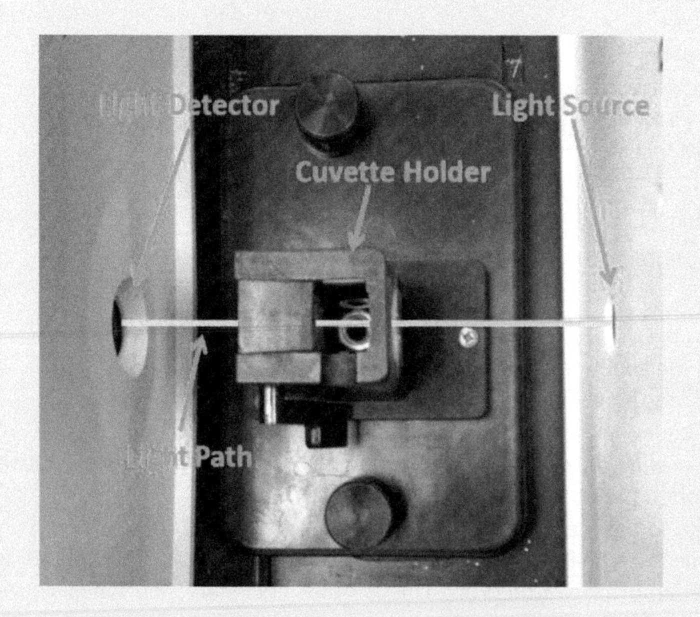

FIGURE 3-1.2 Picture of a typical sample compartment within a spectrophotometer.

(Continued)

EXPERIMENT 3-1 (CONTINUED) TRANSMITTANCE OF A TRANSPARENT PLATE

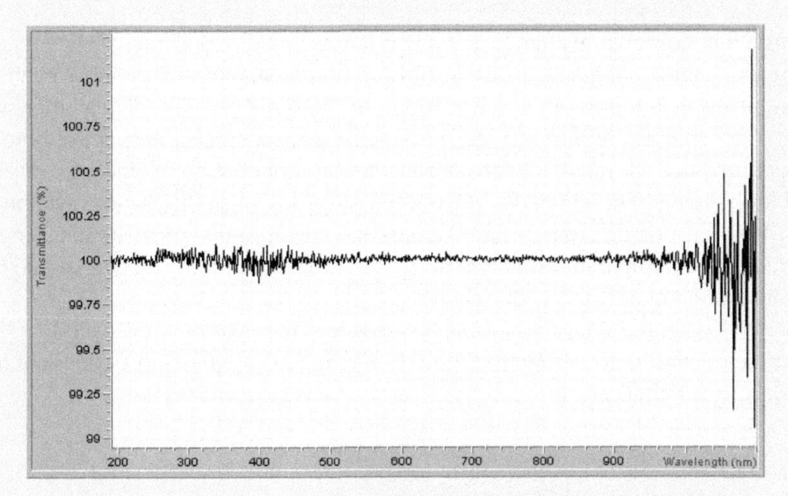

FIGURE 3-1.3 Spectrum of the baseline: screen capture of the operating software.

a sufficient amount of light or that the light source is not working properly. This may happen when you do not switch on the lamp, something not transparent is in the path of the light (something is blocking light), or your detector is broken. In Figure 3-1.3, the signal variation that can be seen at wavelengths above 950 nm indicates that lamp efficiency in this range is low. It should also be noted that significant (but smaller) baseline fluctuations can happen when the lamp is not properly warmed up. Figure 3-1.4 shows examples for the baseline

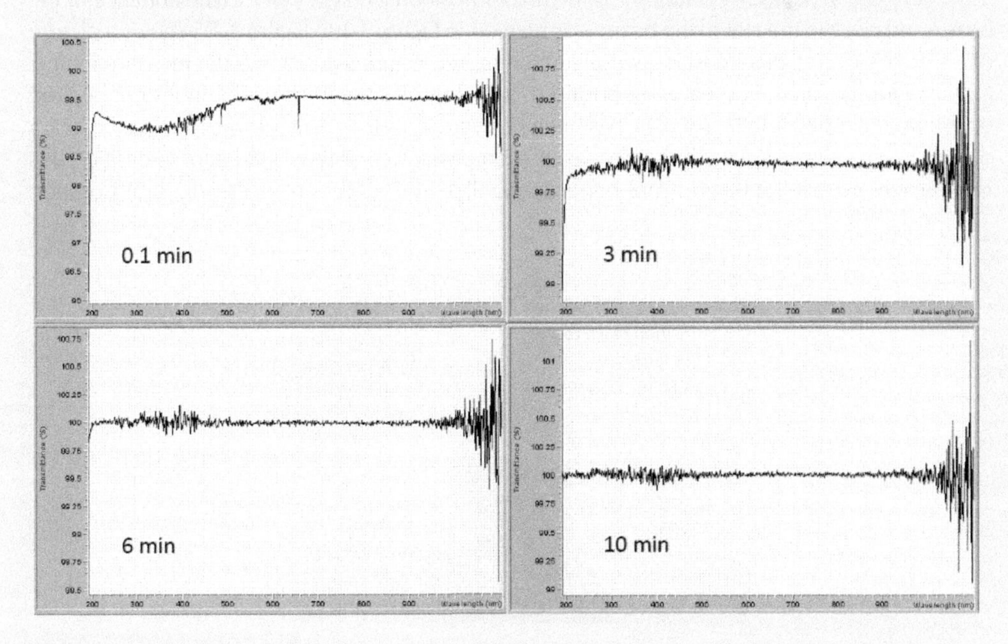

FIGURE 3-1.4 Baseline variation measured as a function of time after the lamps were turned on.

(Continued)

EXPERIMENT 3-1 (CONTINUED) TRANSMITTANCE OF A TRANSPARENT PLATE

variation as a function of time after switching on lamps. In this case, the baseline reading will have larger fluctuations (larger than 0.2% and a different shape) and possibly even features resembling peaks. It is a good practice to measure the baseline immediately after switching it on, wait a few minutes and repeat the baseline measurement. If the spectrophotometer is properly warmed up, the range for the baseline should not change significantly (the random noise will look slightly different each time but the range for noise fluctuations will be roughly constant). Also, it is possible to measure the baseline and after that measure the sample without introducing the sample into the light path. The measured transmittance should be close to 100% (absorption close to 0) with only noise.

The intensity detected with the sample inserted into the light beam is called I. Obviously both intensities (I_0 and I) are a function of wavelength, and wavelength-dependent transmittance, $T(\lambda)$, is determined as:

$$T(\lambda) = \frac{I(\lambda)}{I_0(\lambda)} * 100\%$$

In an experiment that covers an 800 nm range (e.g., from 200 nm to 1000 nm), when taking a measurement every 0.5 nm, the transmission is acquired/calculated 1600 times for each scan.

Transmittances of the plates, ($T(\lambda)$) at each wavelength are the quantities of interest for this exercise. First, we will test if the measured transmittance depends on the position of the plate. We set the spectrophotometer to a 250–800 nm range. In the case of a diode array spectrophotometer, the software will only display data in this range even if it detects a broader range. In the case of a wavelength scanning spectrophotometer, the software will only scan the user-selected range and display transmittance in this range. We then insert a quartz plate orthogonally into the beam path at different positions—at the holder position (center), close to the entrance of the beam, and close to the exit of the beam—as marked in Figure 3-1.5. Figure 3-1.5 shows a picture of a system with the plate positioned/held in the center. Figure 3-1.6 shows the measured transmittance for the three different plate positions. It is clear that the transmission does not depend on the plate position where our plate has been positioned, as long as it is orthogonal to the beam direction (we keep the plate the same way at each position). For low light beam intensities, this is expected, even if the beam is not perfectly parallel. The reflection, scattering, or absorption

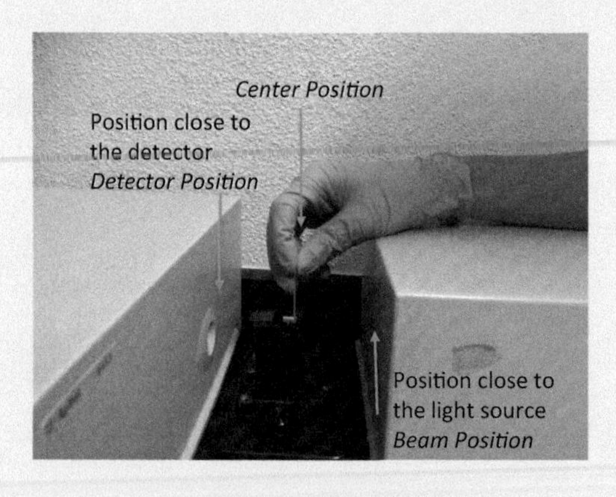

FIGURE 3-1.5 Picture of a quartz plate held in the light path within a spectrophotometer.

(*Continued*)

EXPERIMENT 3-1 (CONTINUED) TRANSMITTANCE OF A TRANSPARENT PLATE

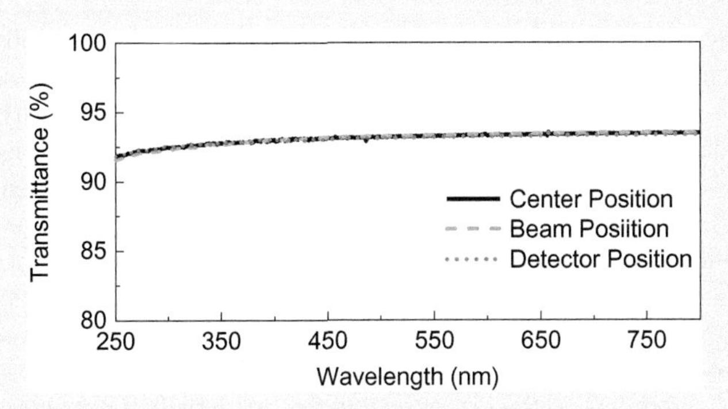

FIGURE 3-1.6 Measured transmittances for a quartz plate positioned at different places in the sample compartment.

of light should not depend on local light intensity (intensity/photon flux). This can be not true for very high intensities that can be obtained only with high-energy laser beams (which are required in order to observe non-linear processes or when doing saturation experiments).

Next, we will test the transmission of light through different materials, all positioned in the center of the chamber when the plates are orthogonal to the beam. Figure 3-1.7 presents measured transmittances for the four different materials that were shown in Figure 3-1.1. Interestingly, each plate looks very similar upon visual inspection (their visual transparencies are very close); through this quick investigation, however, we can now see that each of them has significantly different transmittances.

It is clear that glass and plastic lose transmittance property (do not transmit light) starting at the UV spectral range (below 400 nm). It is also interesting to note that in the visible range (400–750 nm, wherein all plates look identical) we measure significant differences. As we can see, the best transmittance is through quartz (fused silica), whereas the worst (on average) is through sapphire. In the range shown (250–800 nm), these differences are due to differences between the materials refractive indexes, surface quality, and other potential minor impurities. A higher

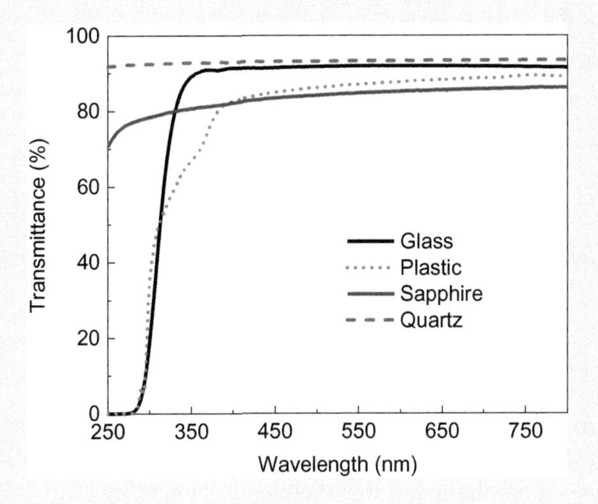

FIGURE 3-1.7 Measured transmittances for different materials (different types of slides).

(Continued)

EXPERIMENT 3-1 (CONTINUED) TRANSMITTANCE OF A TRANSPARENT PLATE

refractive index leads to a higher reflectance (see next experiment). Also, it is important to realize that differences of a few percentages easily measured are difficult to be spotted visually.

Finally, we will demonstrate how significant differences in transmittance can be induced by positioning the medium within the path of the beam in a way that is *not* orthogonal—simply put, we want to find out what happens when we tilt the plate diagonally. To demonstrate this, we used a high refractive index glass block approximately 10 mm thick (Figure 3-1.1) used in total internal reflection (TIRF) experiments. First, we positioned the glass block in a proper orthogonal position to the beam direction in the center of the sample compartment and measured the transmittance. Then, we tilted the block about 10° and took another measurement; after that, we tilt the same direction but more to about 20°. Figure 3-1.8 shows the transmittances for these measurements. In the orthogonal to the beam path orientation, the block has a transmittance very similar to what we previously measured for the Sapphire plate. This is a high-quality glass that has a higher refractive index than ordinary glass, so its reflectivity is larger than ordinary glass. We can see that even a small 10° tilt from the orthogonal position results in a significant drop in transmittance; this drop is even larger at 20°. The main reason behind this drop in transmittance is beam shifting and part of the beam missing an entry slit to the detector (see Figure 3.2); also contributing to the drop is the increase in reflectivity that increases when the incident angle is less than 90°.

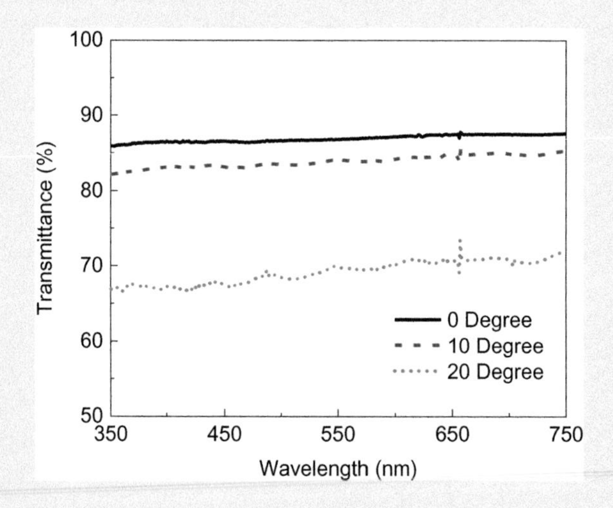

FIGURE 3-1.8 Measured transmittance for a glass block (used for TIRF) when tilted from the orthogonal position.

CONCLUSIONS

The transmittance of light depends on the type of transparent material being used, as different materials will have different wavelength ranges through which light will efficiently permeate. The biggest factor contributing to the attenuation of light by clear transparent materials is a reflection on the interface. For materials with a higher refractive index, the amount of

(Continued)

EXPERIMENT 3-1 (CONTINUED) TRANSMITTANCE OF A TRANSPARENT PLATE

reflected light is higher, and thus the transmittance will be lower. The sapphire slide in this experiment had the largest refractive index of any other material used, and as a result, had the lowest apparent transmittance. These properties may seem trivial but can be important experimentally if, for example, a researcher is comparing two slides that happen to be made from different materials. It's important to realize that the extent of light attenuation by transparent materials mostly depends on the type of material (refractive index). So, two microscopy slide made from the same material, one 1 mm thick and second only about 100 μm thick (#1 slide) will present a similar attenuation.

A typical cuvette is a rectangular element (cuvette walls and solution) and when using regular cuvettes, the measured attenuation of light will depend on the orientation. A tilted cuvette will parallel shift the transmitted beam (similar to the tilted block in Figure 3-1.8) and thus change the intensity of light detected by the detector beside an attenuation expected from the sample absorption. For a proper sample shape (rectangular) and proper cuvette positioning (two cuvette walls orthogonal to the beam), the measured transmittance should not depend on where the sample is positioned in the spectrophotometer sample compartment (see Figures 3-1.5 and 3-1.6). Since the spectrophotometer beams are not always perfectly parallel, we recommend a central position where the sample holder is typically positioned.

For our experiment, we used just simple optical elements. Attenuation of the beam due to reflection at the interfaces can be reduced by an optical coating. The anti-reflection coatings are offered by most vendors providing optical hardware.

Sometimes experimentalists may notice that an improperly positioned cuvette (tilted cuvette) shows over 100% transmittance. This is obviously wrong. If the baseline was measured properly, it typically means the beam alignment in the spectrophotometer is wrong and needs to be corrected. Slight shifting of the beam by a tilted cuvette accidentally aligns the beam that enters the detector. An opposite tilt (other direction) will significantly drop the transmission much below 100%. This may happen for old instruments or instruments for which the frame was slightly banded (for example when moving the instrument or during transportation of the instrument).

EXPERIMENT 3-2 TRANSMITTANCE THROUGH IRREGULAR (NOT SQUARE) OBJECTS

Keywords: transmittance, light reflection, scattering, lens, non-uniform glass elements.

The previous experiment established that even small deviations from perfect orthogonal positioning within a spectrophotometer can lead to significant changes in the apparent transparency of a material. Tilting a glass slide just a few degrees is all it takes to significantly alter measured transmittance. This is mostly due to a small parallel shift/displacement (see Figure 3.3 in the "Introduction") of the beam being propagated through the sample. As a result of this shift, part of the transmitted beam does not hit the detector. The effect of bending a beam (changing the direction of light propagation) while refracting it (as would happen within the two nonparallel walls of a prism or simple a lens) could be dramatic as well. As a result, transmittance measurements of solid transparent samples confined within an irregular entity, like a bottle (Figure 3-2.1) or a small vial are practically impossible. Taking these types of measurements requires advanced

(*Continued*)

EXPERIMENT 3-2 (CONTINUED) TRANSMITTANCE THROUGH IRREGULAR (NOT SQUARE) OBJECTS

FIGURE 3-2.1 Optical elements with curved surfaces.

techniques which will be explained in detail later. Nevertheless, herein we will demonstrate the effects of taking measurements in or through such objects (e.g., vials). These experiments are intended to demonstrate why we do not measure transmittance/absorbance directly in vial or bottle where we prepare our sample but always will transfer the sample to a square cuvette.

MATERIALS

- Different lenses
- Glass vial
- Water

EQUIPMENT

- Spectrophotometer

METHODS

1. Turn on the spectrophotometer.
2. Turn on the "lamp" in the software.
   a. Wait a few minutes for the lamp to warm up.
3. Take a baseline measurement (with nothing in the light path).
4. Position a lens in the beam path. Take transmittance measurements while holding the lens at three positions: close to the light source, centrally, and close to the detector. Ensure that the beam path is going roughly through the center of the lens.
5. Repeat step 4 using an empty bottle followed by a bottle filled with water.

(*Continued*)

EXPERIMENT 3-2 (CONTINUED) TRANSMITTANCE THROUGH IRREGULAR (NOT SQUARE) OBJECTS

RESULTS

When measuring transmittances of the optical elements shown in Figure 3-2.1, we positioned them close to the light source, near the center (where the cuvette mount would otherwise be), and close to the detector. For each measurement, we made sure the lens was orthogonal to the beam, and that the beam traveled through the center of the lens. Similarly, when measuring transmittance through a bottle we ensured that the beam would hit it while it was standing vertically, at all three positions. Transmission measurements for two different lenses ($F = 50$ mm and $F = 300$ mm) at three different positions are presented in Figure 3-2.2.

It is clear that both lenses do not transmit below 350 nm. For a 50 mm lens, the transmittance is lower than 100% and strongly depends on its position. A lens with a short focal length produces a highly divergent beam when light passes through it, and almost no light will reach the detector. Lower apparent transparencies are the more expected result in this case. Much more surprising results are obtained when using a lens with a longer focal distance (300 mm). The measured transmittance through this type of lens, when held near the light source, reaches almost 140%, which is obviously an anomaly. This occurs because the lens slightly focuses the beam in the short distance between the lens and the detector; as a result, some light that was blocked by the slits when the baseline measurement was taken now reaches the detector, giving us a reading above 100% (which would otherwise be an impossible result). This phenomenon is rather unusual and typically depends on the spectrophotometer being used as well as the focal distance of the lens.

Next, we tested the transmittance of light through an empty glass bottle (Figure 3-2.3) while empty (left) and filled with water (right), placed in three positions in the sample compartment.

It is clear that the glass bottle only transmits light above 300 nm in either case. The measured apparent transmittances strongly depend on the position of the bottle, and the measured light attenuation is larger for a curved bottle than for a flat glass slide. More surprising are the measured transmittances of the bottle filled with water. In theory, water should not contribute to absorption itself in the range where the bottle is transparent. However, our results clearly indicate changes not only in the transmission profiles (transmission spectra) at

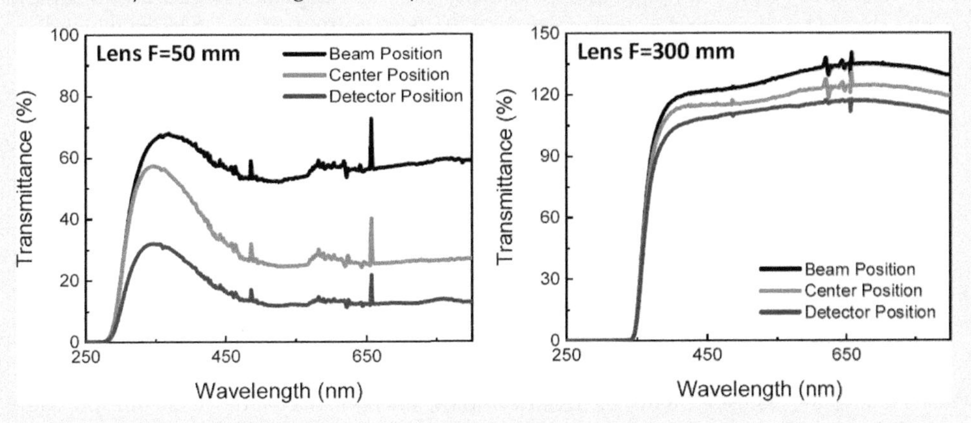

FIGURE 3-2.2 Measured transmissions for two lenses that have different focal points (left $F = 50$ mm and right $F = 300$ mm).

(Continued)

EXPERIMENT 3-2 (CONTINUED) TRANSMITTANCE THROUGH IRREGULAR (NOT SQUARE) OBJECTS

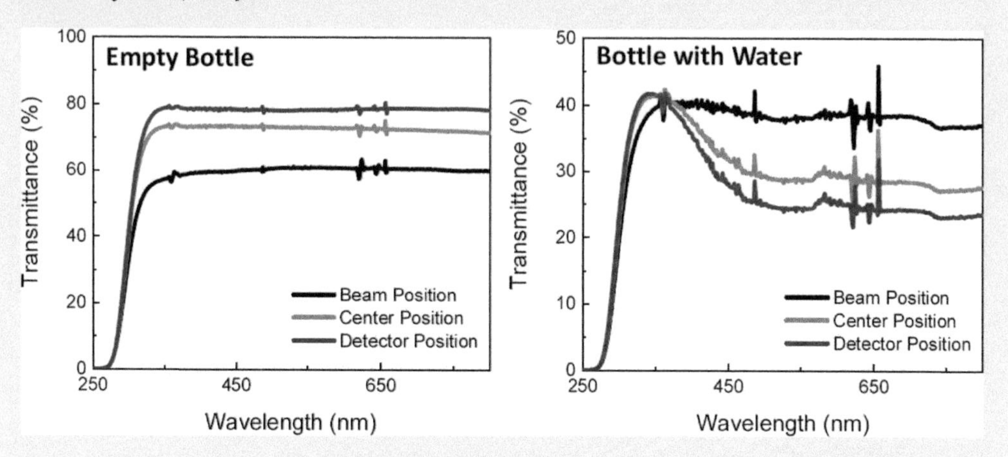

FIGURE 3-2.3 Measured transmittances for an empty bottle (left) and a bottle filled with water (right).

the three positions but also in the total intensity (pay attention to the scale). This behavior is a result of light refraction—an empty bottle will refract light negligibly, based only on the thickness of its walls; a bottle filled with water works akin to a cylindrical lens, refracting light to a much more significant extent.

CONCLUSIONS

The measurements presented in these two experiments clearly indicate that if we want to measure light attenuation (transmittance), a sample should have two parallel walls through which a beam of light can pass without refracting. Any irregularity in these surfaces (either being curved or simply nonparallel) will lead to a significant distortion of the measured signal. When measuring liquid samples, one should use only rectangular cuvettes and a sample cuvette should be positioned in such a way that the flat cuvette surfaces through which light passes are orthogonal to the beam propagation.

It is also clear that we cannot measure absorption directly in the flask or vial in which the sample was made.

EXPERIMENT 3-3 TRANSMISSION OF DIFFERENT CUVETTES

Keywords: transmittance, absorbance, cuvette.

When taking the measurements of liquid samples, an optical cuvette is an indispensable part of the experiment. It is important to realize that each cuvette can and will optically contribute to measured light attenuation/extinction (absorption and reflection), similar to a quartz or a glass slide (as demonstrated in Experiments 3-1 and 3-2). This being the case, it

(*Continued*)

EXPERIMENT 3-3 (CONTINUED) TRANSMISSION OF DIFFERENT CUVETTES

is important to understand how a cuvette contributes to the attenuation of transmitted light. This experiment will present transmission/absorption measurements for different cuvettes (cuvettes made from different materials). This experiment may also be useful for selecting cuvettes for further experiments. As mentioned at the introduction of this chapter, a single absorption/transmission measurement requires two measurements, the reference (baseline) and the sample. The baseline reading in a typical single-beam spectrophotometer is meant to correct for the properties of the instrument, as well as any optical elements in the beam path which are not being studied (including the solvent and walls of the cuvette). For the purpose of this experiment, however, the baseline will be measured with nothing in the beam path, as was done in the first two experiments. The second measurement will include everything in the baseline measurement plus the sample being analyzed—in this case, the "sample" will actually be a cuvette. The useful data acquired from an absorbance measurement is the difference between these two measurements, or how much light intensity is attenuated due to the presence of the sample. Typically these attenuations are very low, but in some optical regions, or using specific materials, they can become quite substantial. Such attenuation is corrected by reference measurements, but if the intensity of the light drops too much due to the cuvette (or other optical elements in the light path), it can result in significant noise and error. It is important to realize that while these measurements are traditionally referred to as "absorbance," they can provide valuable insight for us to consider this phenomenon in terms of transmittance instead. For example, a measured absorbance of 0.7 may sound reasonable, but this actually means only 20% of light is transmitted by the cuvette, and measuring a sample with an additional absorbance of 1 will only allow 2% of the lamp's original light intensity to reach the detector. As one might imagine, this can cause an error/noise in the reading.

The following protocol describes how to measure the transmission of cuvettes made out of various materials. The goal here is to emphasize that choosing the right cuvette or pair of cuvettes is crucial, as it can impact measurements.

MATERIALS

- Plastic cuvette
- Glass cuvette
- Quartz cuvette

EQUIPMENT

- Spectrophotometer

METHODS

1. Turn on the spectrophotometer.
2. Turn on the "lamp" in the software.
   a. Wait a few minutes for a lamp to warm up.
3. Take a baseline measurement by clicking the "baseline" in the software or on the spectrophotometer.
   a. Make sure there is nothing in the beam path.
4. Insert a cuvette into the cuvette holder.

(Continued)

EXPERIMENT 3-3 (CONTINUED) TRANSMISSION OF DIFFERENT CUVETTES

5. Take sample measurement by clicking sample in software or on the spectrophotometer.
6. Change cuvette and repeat from step 5.
   a. Keep the same baseline for all cuvettes.

RESULTS

For this experiment, we used an *Agilent*™ 8453 diode array spectrophotometer presented in Figure 3.1. The light path (yellow line) is emitted from the source and enters the detector. The cuvette holder has a hole through which the light travels (the holder does not affect the light). Figure 3-1.2 shows the configuration we used for the baseline measurement and Figure 3-3.1 shows the baseline measured in this experiment. For comparison, the baseline is now measured in absorption units. We want to remind the reader that, as with any experiment, one should allow sufficient time for the lamp to be warmed up to limit fluctuations when taking the baseline measurement.

As in Experiment 3-1, single measurement displays the ratio of the intensity of the beam entering the detector, I, to the intensity of the beam leaving the source, I_0, at predetermined wavelengths. The detector does not account for how the intensity, if any, has been lost between the source and detector. The transmission is defined as:

$$T(\lambda)=\frac{I(\lambda)}{I_0(\lambda)}*100\%$$

Using this, the wavelength-dependent absorbance, $A(\lambda)$, can then be calculated:

$$A(\lambda)=Log_{10}\left(\frac{1}{T(\lambda)}\right)=-1*Log_{10}(T(\lambda))$$

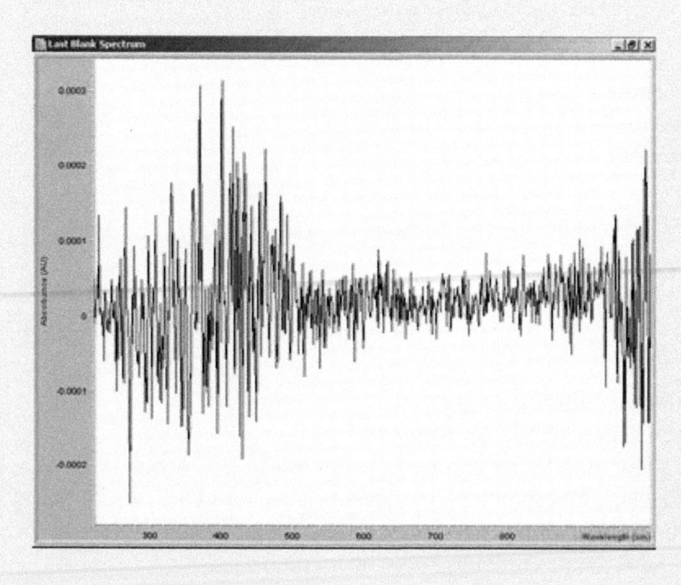

FIGURE 3-3.1 The spectrum of the baseline: a screen capture of the operating software.

(Continued)

EXPERIMENT 3-3 (CONTINUED) TRANSMISSION OF DIFFERENT CUVETTES

$A(\lambda)$ is the parameter of the dependent (y) axis in Figure 3-3.1. Going from absorbance to transmission:

$$T(\lambda)=\frac{1}{10^{Ab(\lambda)}}*100\%$$

The transmission or absorption of the cuvettes ($T(\lambda)$ or $A(\lambda)$), respectively, are the quantities of interest in this experiment. Contrary to our previous two experiments, here the cuvette can be mounted in the holder to ensure proper cuvette positioning and high reproducibility when inserting and removing samples (as opposed to doing so by hand as we did before). Figure 3-3.2 presents the absorbance (A) and transmission (T) spectra for glass, plastic, and quartz cuvettes. We purposely present results as absorption and transmission to familiarize the reader with the fact that both representations are equivalent and acceptable in the scientific literature; both top and bottom plots present the same measured data in different representations. It may come as a surprise that the plastic cuvette in this experiment has less absorbance (especially below 350 nm) than the glass one. This is not a general result and may differ based on the source of the cuvette, as plastic cuvettes are optimized by manufacturers for use in the UV spectral range (<350 nm). A glass cuvette is typically better, but sometimes maybe less optimal (strictly in terms of absorbance/transmittance) than a plastic cuvette, especially if the glass cuvette is older and has been used many times before. The only way to know this for certain though is to perform this measurement/experiment. It should also be noted that there are other general problems that may arise when using plastic cuvettes, which we will discuss in later experiments. In general, since a

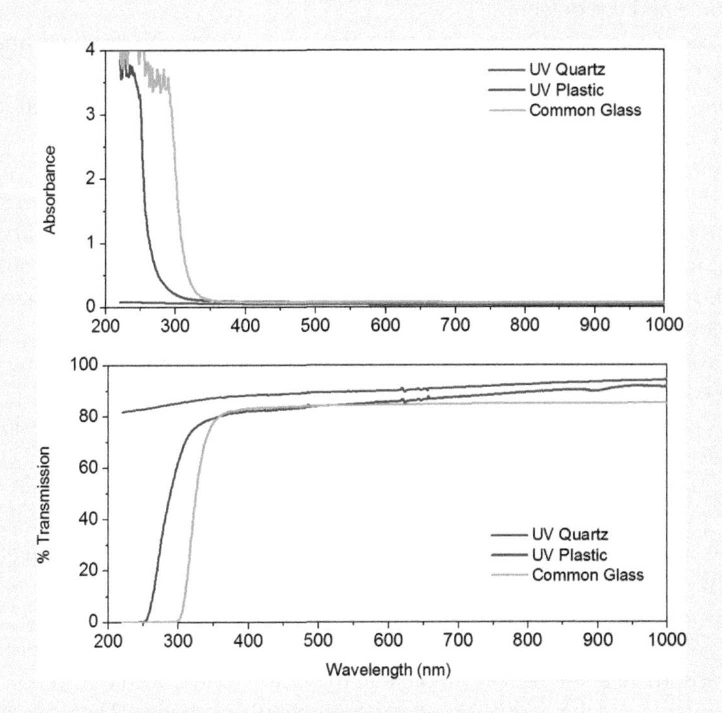

FIGURE 3-3.2 Absorbance and transmission of various cuvettes.

(Continued)

EXPERIMENT 3-3 (CONTINUED) TRANSMISSION OF DIFFERENT CUVETTES

cuvette is meant to hold a sample and not interfere with the measurement, it should absorb as little light as possible (i.e., transmit as much light as possible) at the range of interest.

According to the results shown in Figure 3-3.2, a common glass cuvette would be unusable below ~340 nm; a common plastic cuvette would be unusable below 300 nm, and common quartz (or fused silica) cuvette can be used at wavelengths no lower than 200 nm. For measurements below 220 nm, it should be noted that oxygen absorption becomes a problematic factor in addition to the cuvette interference.

There are multiple potential errors that can lead to an incorrect reading. Any errors leading to sample transmission reading greater than 100% (or, alternatively, a negative absorption reading) are easy to spot, typically being an extra source of light (the sample compartment is not fully closed) or an angular distortion of light through the medium (as demonstrated in the previous two experiments). Apart from these and other common errors (e.g., the lamp not being sufficiently warmed up), mixing different types of cuvettes (cuvettes mismatch) within the same set of measurements is one of the most common mistakes, especially being that glass and quartz cuvettes look almost identical to the naked eye.

In contemporary spectrophotometers, all of the mathematics is done by the software; all a user needs to do tell the machine what measurement is being taken (either baseline or sample) and then command the machine when to take it. The software should also allow the user to easily toggle between viewing the data in terms of absorbance or transmission. This being the case, one might ask—Why are any calculations even worth discussing?

- It is crucial to understand that the acquired data is the ratio of the beam's intensity at the source and the detector.
  - Factors such as scattering and reflection may change this reading; understanding this can be invaluable in the troubleshooting process.
- Knowing the equations helps us remember that a baseline is always subtracted from a sample measurement.
  - An incorrect baseline will corrupt results.
  - Remember what you are measuring the absorbance of. A common error is to measure the absorbance of a chromophore in some solvent or buffer, without correcting for the absorbance of the solvent/buffer alone. Some buffer components or even high-purity solvents may significantly contribute to absorbance, especially below 300 nm, and thus should always be accounted for.
- The equations give us the ability to compare data from the literature in one representation to data in another.
  - We can use the equations to convert between absorbance/transmission in a software such as Excel.
    - =1/10^(A1)*100 (A1 ∈ [0,4])
    - =–LOG10 (A1) (A1 ∈ [0,1])
  - Absorbance and transmission have a logarithmic relationship; knowing this, we can further manipulate the original equations when we have data.

Figure 3-3.3 presents the data for only the glass cuvette: the left and right *y*-axes are for transmission and absorbance, respectively (note that the scales differ), with the green curve representing transmission and the dark blue curve representing absorbance. The *x*-axis corresponds

(Continued)

EXPERIMENT 3-3 (CONTINUED) TRANSMISSION OF DIFFERENT CUVETTES

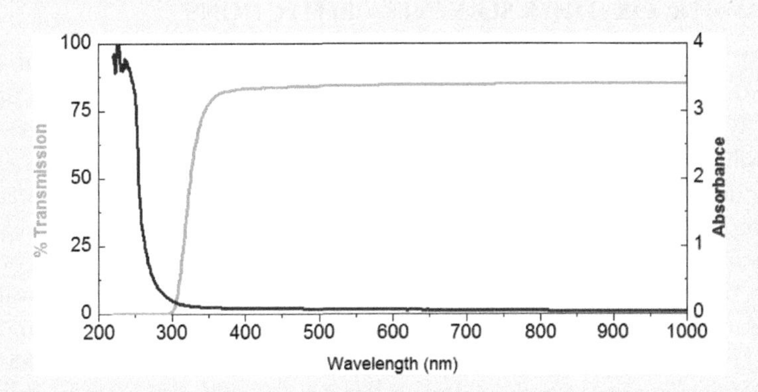

FIGURE 3-3.3 Transmission and absorbance of glass cuvette.

to wavelength and is the same for both curves. The position of the intersection of the two curves shows the logarithmic relation between transmission and absorbance. This should not be confused with a common reflection (*R*)/transmission (*T*) where $1 = R + T$ (in which case, the two curves would intersect at 50% and be presented on the same scale, percent).

CONCLUSIONS

Similar to quartz or glass slides, cuvettes can significantly attenuate the intensity of transmitted light. The attenuation depends on the material from which the cuvette has been made. Furthermore, transmission depends on the quality of the material.

EXPERIMENT 3-4 TRANSMITTANCE OF CUVETTE FILLED WITH AIR, WATER, OR OTHER SOLVENTS—REFLECTIONS

Keywords: absorbance, water, dielectric, reflection.

For most measurements and chromophores studies, we prefer to use the wavelength range for which a given solvent is well transparent. Common solvents like water, ethanol, glycerol, or air are considered to be transparent (do not absorb light) at a UV/visible/NIR spectral range. Since some solvents are not easy to handle; they can be very viscous (e.g., glycerol) or smelly and even harmful (e.g., benzene); extra effort may be needed to utilize and manage a given solvent properly to perform reference measurements. For example, in the 300–800 nm range, air, water, propylene glycol, and quartz do not absorb light (the absorption is negligible). So, why not use an empty cuvette as a reference when the sample is in glycerol or another transparent solvent? Instead, it is always necessary to use the same solvent and the same type of cuvette for reference measurements. Also, we would not recommend using as reference a different transparent solvent as that we use for the sample. For example, when measuring a sample in glycerol, an identical cuvette with water is inappropriate to use as a reference measurement, as is ethanol, air, or any other solvent, even if they all are equally transparent in the given wavelength range.

(*Continued*)

EXPERIMENT 3-4 (CONTINUED) TRANSMITTANCE OF CUVETTE FILLED WITH AIR, WATER, OR OTHER SOLVENTS—REFLECTIONS

In this experiment, we will demonstrate some less recognized processes that significantly influence absorption measurements. When using a cuvette, the light needs to pass through two cuvette walls—multiple (four) interfaces (air/glass; glass/solvent; solvent/glass; glass/air) at which it will be reflected (see Experiment 3-1). The refractive indices of glass and solvents are typically different and depend on the wavelength of light in question. Because a spectrophotometer only detects the intensity of transmitted light, the attenuation of light intensity due to reflections will impact any apparent absorbance reading. It may not be apparent, but the reflections that occur for light traveling through interfaces of cuvette walls may significantly perturb the overall intensity of transmitted light and measurements of absorption. These reflections are well predictable, and an effective perturbation can be predicted using the known refractive indexes of the materials. Even if the reflection from a single glass surface is fairly small, we do not remember that every cuvette presents four reflective surfaces for transmitted light. So, the cumulative effect becomes significant. The reader will realize that using a reference cuvette that is identical in shape but made from a different transparent material (e.g., glass and plastic), or alternatively using a different solvent in a reference measurement (despite using an identical cuvette), can yield a significant error, especially within high precision measurements.

In this protocol, we will measure the absorbance using cuvette with air (empty cuvette) and cuvette filled with water.

MATERIALS

- Cuvette
- Water

EQUIPMENT

- Spectrophotometer

METHODS

1. Start the spectrophotometer.
2. Take a baseline measurement with nothing in the beam path.
3. Insert a cuvette.
4. Measure the absorbance/transmittance of the cuvette.
5. Fill the cuvette with water.
6. Measure the absorbance/transmittance of cuvette filled with water.

RESULTS

The transmittance spectra of the empty cuvette and the cuvette filled with water are presented in Figure 3-4.1.

The presented spectra at first may appear quite surprising. Apparently, an empty cuvette transmits less light than a cuvette filled with water. Any researcher, even a novice in the field, should question this result. And rightly so—if it is correct, we need to know what the basis is

(*Continued*)

EXPERIMENT 3-4 (CONTINUED) TRANSMITTANCE OF CUVETTE FILLED WITH AIR, WATER, OR OTHER SOLVENTS—REFLECTIONS

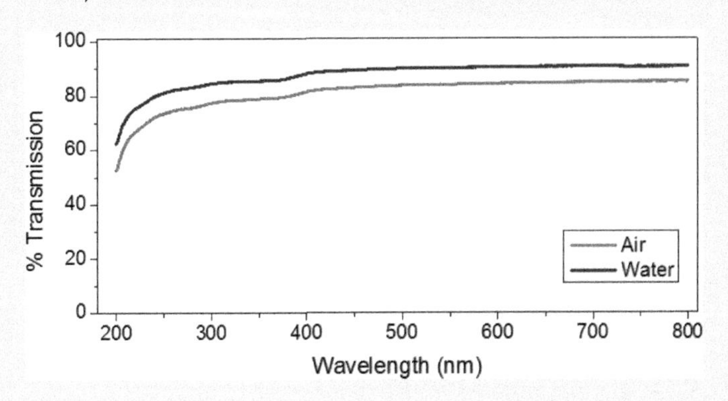

FIGURE 3-4.1 Transmission spectra of empty cuvette and cuvette filled with water.

for such behavior. And how could this result be obtained if the experimental procedure was followed and the experiment was done correctly?

The reader may also notice the difference in transmission between empty cuvettes presented in previous experiments and the transmission presented here. This is not an anomaly; in this experiment, we intentionally used a cuvette from a different vendor. We did this on purpose so that the reader will realize that there are many different cuvettes available and it is difficult to distinguish the difference between them visually. In a typical lab, there will be many cuvettes acquired at different times and from different sources. Thus, consistency in the use of cuvettes during an experiment is critical if we are trying to get a very precise measurement.

In this case, the reflections of light at the boundaries of the cuvette (cuvette walls) is the biggest contributor to light attenuation. Light reflection when going through the interface between two media of different dielectric constants contributes more to the attenuation of light than the absorbance of air or water (which, in most cases is so low that cannot be even accurately measured). At the boundary of two dielectrics, a light beam with normal incidence into the interface surface has a transmittance, T, of:

$$T=\frac{4n_1n_2}{\left(n_1+n_2\right)^2} \tag{3-4.1}$$

where n_1 and n_2 are indices of refraction for two media on opposite sides of an interface. Since the cuvette, air, and water in the optical frequency range are not magnetic ($\mu = 1$), the index of refraction is the square root of the dielectric constant. The index of refraction is dependent on the wavelength, and the experimentally obtained values are depicted in Figure 3-4.2 for air (n_a), water (n_w), and glass (n_g). The absorption of water is more complex at longer wavelengths, and these equations are not shown. The index of refraction of air is not constant, although it may appear that way since a constant scale was chosen to depict all three materials in a graph. To perform calculations at each wavelength, it is important to remember that the index of refraction is not constant, but rather a function of wavelength,

(Continued)

EXPERIMENT 3-4 (CONTINUED) TRANSMITTANCE OF CUVETTE FILLED WITH AIR, WATER, OR OTHER SOLVENTS—REFLECTIONS

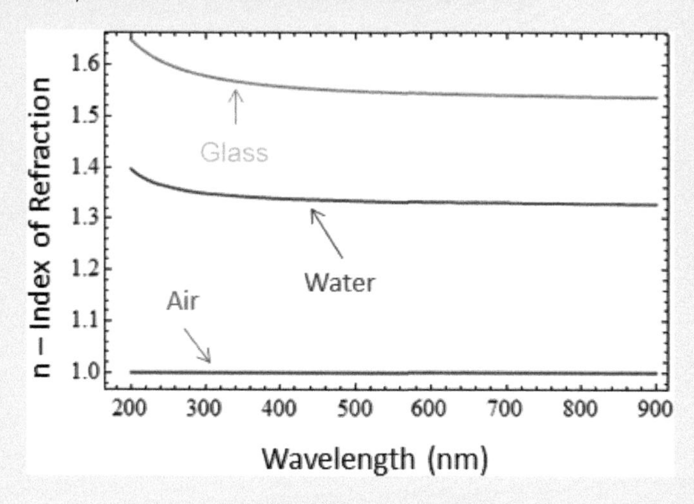

FIGURE 3-4.2 Indices of refraction for materials presented.

n(λ). We could use digitalized numbers from Figure 3-4.2 or we can use interpolation by approximated function such as:

$$n_a(\lambda) = 1 + \frac{5792105 * 10^{-8}}{238.0185\ \lambda^{-2}} + \frac{167917 * 10^{-8}}{57.352 - \lambda^{-2}} \tag{3-4.2}$$

$$n_q(\lambda) = \sqrt{1.286 + \frac{1.070 * \lambda^2}{\lambda^2 - 1.006 * 10^{-2}} + \frac{1.102 * \lambda^2}{\lambda^2 - 100}} \tag{3-4.3}$$

Figure 3-4.3 presents an approximation of what is happening. At every change of the refractive index, part of the beam is reflected and part is transmitted, as in the equation above. There are four such interfaces depicted in Figure 3-4.3, they are circled. The large yellow beam (arrow) represents the light arriving at the detector while the small red arrows depict the reflected beams that do not reach the detector. It is possible to do this calculation considering the superposition of the reflected beams, but that level of mathematics is not necessary to understand the experimental result.

To ascertain the effect these reflections have on the measurement, it is necessary to determine how much of a change in the light beam these reflections are responsible for. This must be done on a per wavelength basis (the calculation must be performed multiple times, once per wavelength). To accomplish this, it is necessary to find the difference of the beam transmission only due to reflections, as follows:

1. Assume the initial beam intensity is I_0.
2. Calculate the attenuation of the beam going through an empty cuvette.
   a. Initial beam intensity: I_0.
   b. After the first interface: $(I_0) * \dfrac{4 n_c n_a}{(n_c + n_a)^2}$.

(Continued)

EXPERIMENT 3-4 (CONTINUED) TRANSMITTANCE OF CUVETTE FILLED WITH AIR, WATER, OR OTHER SOLVENTS—REFLECTIONS

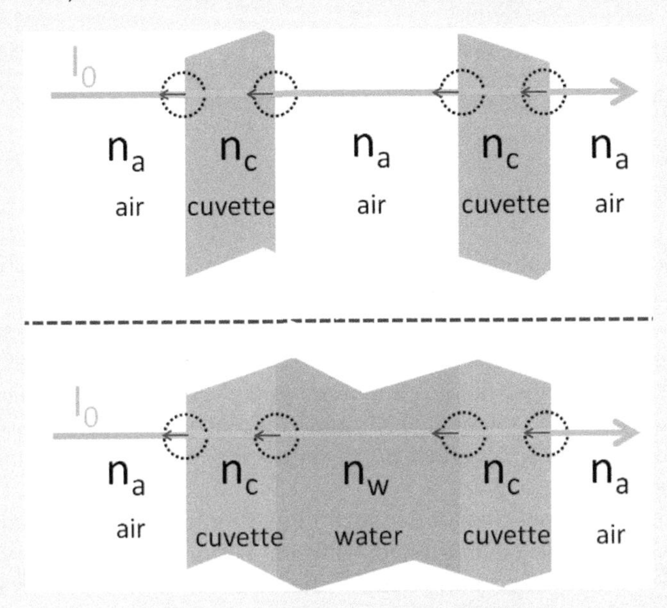

FIGURE 3-4.3 Schematic representations of light reflections through cuvette walls filled with air (top) and water (bottom).

c. After the second interface: $\left(\frac{4n_c n_a}{(n_c+n_a)^2} * I_0\right) * \frac{4n_c n_a}{(n_c+n_a)^2}$.

d. After the third interface: $\left(\frac{4n_c n_a}{(n_c+n_a)^2} * \frac{4n_c n_a}{(n_c+n_a)^2} * I_0\right) * \frac{4n_c n_a}{(n_c+n_a)^2}$.

e. And after the fourth interface $\left(\frac{4n_c n_a}{(n_c+n_a)^2} * \frac{4n_c n_a}{(n_c+n_a)^2} * \frac{4n_c n_a}{(n_c+n_a)^2} * I_0\right) * \frac{4n_c n_a}{(n_c+n_a)^2}$.

3. Calculate the attenuation of the beam going through the cuvette filled with water.
   a. Initial beam intensity: I_0.

   b. After the first interface: $(I_0) * \frac{4n_c n_a}{(n_c+n_a)^2}$.

   c. After the second interface: $\left(\frac{4n_c n_a}{(n_c+n_a)^2} * I_0\right) * \frac{4n_c n_w}{(n_c+n_w)^2}$.

   d. After the third interface: $\left(\frac{4n_c n_a}{(n_c+n_a)^2} * \frac{4n_c n_w}{(n_c+n_w)^2} * I_0\right) * \frac{4n_c n_w}{(n_c+n_w)^2}$.

(Continued)

EXPERIMENT 3-4 (CONTINUED) TRANSMITTANCE OF CUVETTE FILLED WITH AIR, WATER, OR OTHER SOLVENTS—REFLECTIONS

e. And after the fourth interface: $\left(\frac{4n_c n_a}{(n_c+n_a)^2} * \frac{4n_c n_w}{(n_c+n_w)^2} * \frac{4n_c n_w}{(n_c+n_w)^2} * I_0\right) * \frac{4n_c n_a}{(n_c+n_a)^2}$.

4. Subtract the difference in of intensity loss due to reflections for both setups.

a. $$\left(\frac{4n_c n_w}{(n_c+n_w)^2}\right)^2 * \left(\frac{4n_c n_a}{(n_c+n_a)^2}\right)^2 - \left(\frac{4n_c n_a}{(n_c+n_a)^2}\right)^4.$$

Figure 3-4.4 shows the resulting (calculated) transmissions of both the empty cuvette and the cuvette filled with water. The bottom line is the difference between the two top lines or, the total attenuation due to reflections when the cuvette is empty versus when it is filled with water. The difference, in this case, varies between 7% (longer wavelengths) and 9% (shorter wavelengths). But it is important to understand that transmission through the empty cuvette can be as low as 80%, and when filled with water this improves to almost 90%. This is just a small effect of the difference in the refractive index between glass, water, and air.

After the theoretical consideration, an extra intensity loss due to reflections is added to the transmission of the cuvette filled with air. The transmissions of water and air cannot be distinguished when corrected for the reflections, as shown in Figure 3-4.5. This result shows that experimental data can be very well approximated by theoretical predictions confirming that measured attenuation is due to reflection processes on dielectric interfaces.

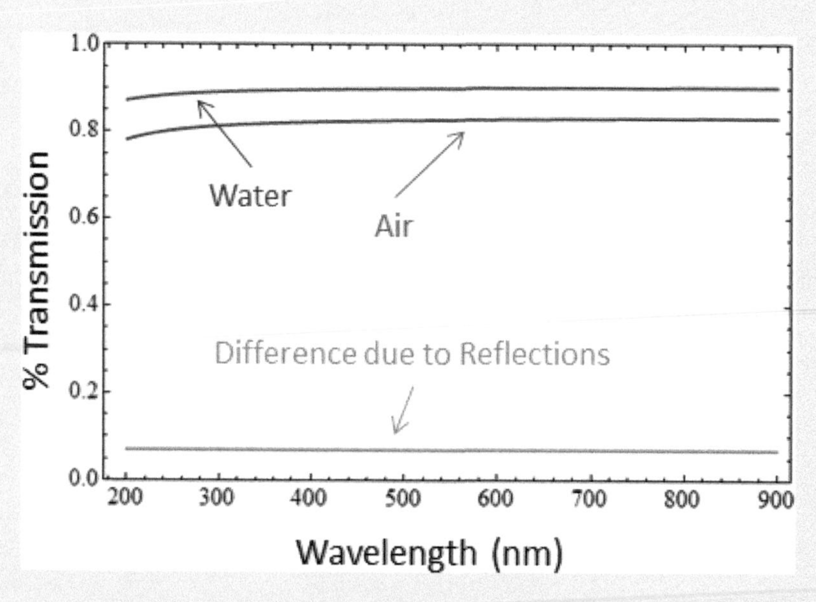

FIGURE 3-4.4 Calculated theoretical transmissions for an empty cuvette and a cuvette filled with water, taking into account only reflections due to the dielectric mismatch.

(Continued)

EXPERIMENT 3-4 (CONTINUED) TRANSMITTANCE OF CUVETTE FILLED WITH AIR, WATER, OR OTHER SOLVENTS—REFLECTIONS

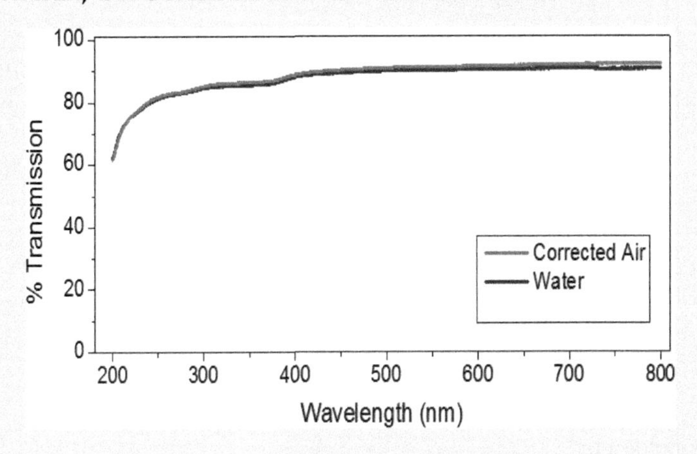

FIGURE 3-4.5 Experimental result corrected for reflections (two air/glass and two glass/water interfaces) and overlaid with theoretically calculated transmission.

CONCLUSIONS

Effects due to reflections at dielectric boundaries are most often accounted for by taking the proper baseline measurement (using an identical cuvette filled with the same solvent as used for the sample), and thus need not be treated mathematically as shown earlier. It is best to avoid having to calculate these theoretical corrections as much as possible. Many buffers or suspensions contain salts and other components and would have an index of refraction different from that of pure water. Thus it is important to use the identical buffer for baseline measurement as that used for the sample even if it is known that it does not absorb/scatter in the regions of interest. One possible pitfall from these reflections is that they may vary significantly over a large range of wavelengths (due to the dependence of the refractive index on wavelength, measured transmittance presents a changing slope—it is not a flat line as seen in Figure 3-4.5), and the relative heights of two peaks in the absorption spectrum positioned at different wavelengths (typically one in UV and one in visible) can be affected. The effect will depend on how large the sample absorption is; most notably, for low absorptions (e.g., low solute concentrations) the effect can be significant.

It may appear that this effect is only important for absorption measurements. But we need to remember that to excite fluorophore in the cuvette, the excitation beam must enter the cuvette (pathing two reflective interfaces) and emission must exit the cuvette (again pathing two reflective interfaces). When comparing fluorescence of a dye from two different solvents (like water and glycerol) for which refractive indexes are different the effect can be significant.

EXPERIMENT 3-5 MEASURING ABSORBANCE OF CHROMOPHORES IN SOLUTION

Keywords: absorbance, absorption spectra, organic dyes, spectrophotometer, Rhodamine 6G.

In the previous Experiment 3-4, the transmittance of the empty cuvette versus cuvette filled with water were compared in detail. As discussed in the introduction to this and following

(Continued)

EXPERIMENT 3-5 (CONTINUED) MEASURING ABSORBANCE OF CHROMOPHORES IN SOLUTION

experiments, absorption of the chromophore is simply the difference between the absorption of solvent with chromophore and the absorption of the solvent only (Introduction, Equation 3.2.4). In this experiment, we will just test and demonstrate this simple fact.

MATERIALS

- Water
- Solution of Rhodamine 6G (R6G) in water
- Glass vials
- 1 cm × 1 cm cuvettes

EQUIPMENT

- Spectrophotometer

METHODS

1. Make an aqueous solution (first sample) of R6G (~3 μM) using your favorite procedure.
   a. Make sure the compound is completely dissolved.
   b. Make the second solution of half concentration (by diluting the previous solution with water).
2. Set the spectrophotometer into absorption mode.
3. Take a reference (measure baseline) with nothing in the sample compartment.
4. Place the empty cuvette in the sample holder of the spectrophotometer and measure absorption.
   a. Make sure that the transparent cuvette wall is facing the light source and plane of the cuvette wall is orthogonal to the path of light. Alignment is important to have the correct line path inside the cuvette. As discussed in previous experiments when tilted under an angle the cuvette will reflect light differently and cause different diffraction/shift of the light.
5. Place a cuvette filled with water into the cuvette holder of the spectrophotometer in an identical way as in point 3 and measure absorption.
6. Fill the cuvette with the first solution and measure absorption (remember to wash the cuvette after measurement).
   Results of all four measurements (air, empty cuvette, cuvette filled with water, and cuvette filled with first R6G solution) are shown on a Figure 3-5.2.
7. Fill the cuvette with the second solution (diluted R6G) and repeat absorption measurement.
8. Wash carefully the cuvette and fill it with water. Set it in spectrophotometer and measure baseline.
9. Repeat the measurements of absorption with the second solution as in point 6 and 7 (now the difference is the baseline).
10. Empty and dry cuvette. Place the empty cuvette in the sample holder and measure baseline.
11. Fill the cuvette with a second R6G solution and measure absorption.

(*Continued*)

EXPERIMENT 3-5 (CONTINUED) MEASURING ABSORBANCE OF CHROMOPHORES IN SOLUTION

RESULTS

In these experiments, we are measuring the absorbance of R6G at each wavelength $Ab(\lambda)$ according to the equation:

$$Ab(\lambda) = \varepsilon(\lambda)Cl$$

C, the concentration, is determined while making the solution, l, the beam path (sample/cuvette thickness), and $\varepsilon(\lambda)$ is an extinction coefficient that is a property of the dye, see Chapter 1, Eq. 1-3.5).

Figure 3-5.1 shows the photograph of three 1 cm × 1 cm cuvettes; one empty, second filled with water, and third filled with R6G solution. A cuvette on the right filled with a higher concentration of R6G solution shows color characteristic for the chromophore. We used higher concentration to better expose the color of the solution in the photograph.

Figure 3-5.2 shows measured absorptions for air (nothing in the light path), empty cuvette, cuvette filled with water, and cuvette filled with the first solution (higher concentration of R6G) for which the blank was just empty chamber (air). As expected, the measurement of air shows zero line (no absorption). The empty cuvette (orange line) shows significant absorption that drops almost to half when the cuvette is filled with water (blue line). Measurement of the R6G solution shows a distinct absorption spectrum with a maximum of about 535 nm. It is important to notice that at a longer wavelength (above 580 nm) R6G absorption and water absorption merge showing almost no difference. Indeed over 600 nm, R6G does not absorb and the only contributions to light attenuation (apparent absorption) are reflections from the cuvette walls.

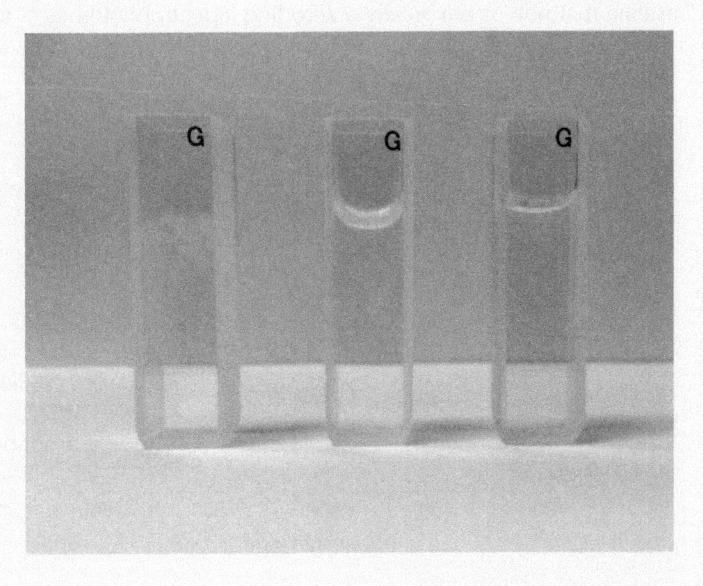

FIGURE 3-5.1 Photographs of three absorption cuvettes (empty, filled with water, and filled with R6G solution).

(Continued)

EXPERIMENT 3-5 (CONTINUED) MEASURING ABSORBANCE OF CHROMOPHORES IN SOLUTION

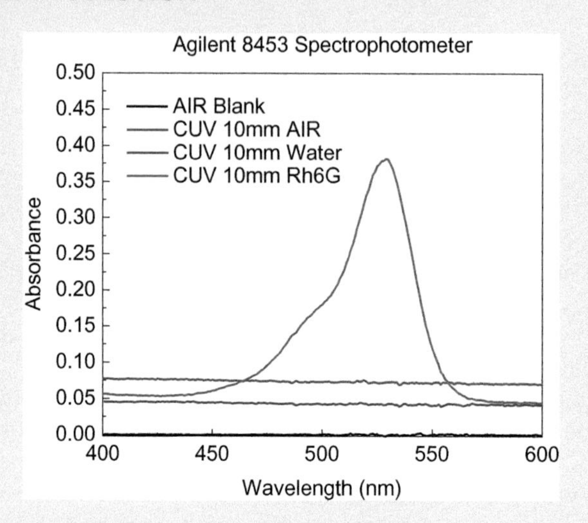

FIGURE 3-5.2 Measured absorptions for air (baseline), empty cuvette, cuvette filled with water, and cuvette filled with RhG solution.

Figure 3-5.3 shows the absorption of lower concentration of R6G (blue) against air baseline (as specified in number 7 of the "Method"). For comparison, we also included the apparent absorption of water (green). Similar to for higher concentration in Figure 3-5.2, the absorption of Rh6G merges with water for longer wavelength. Next, we used the cuvette with water and measured baseline. In Figure 3-5.3, at the lower part presents measured water absorption against water baseline that now is practically a zero line. Spectrophotometers automatically correct for baseline and since the baseline and water (sample) are the same solutions the

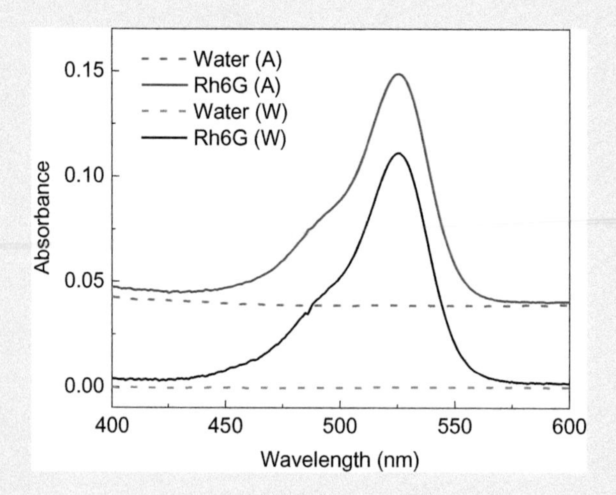

FIGURE 3-5.3 Absorption spectra measured for water and R6G solution against air baseline and against water baseline.

(*Continued*)

EXPERIMENT 3-5 (CONTINUED) MEASURING ABSORBANCE OF CHROMOPHORES IN SOLUTION

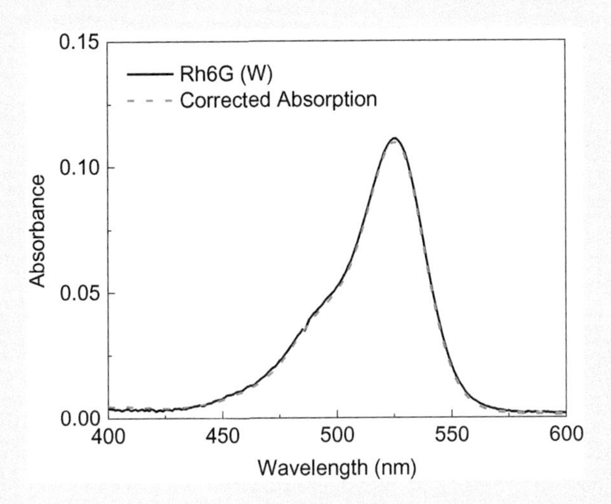

FIGURE 3-5.4 Corrected absorption spectra. Black solid line: absorption spectrum of R6G measured with water baseline (correction by the instrument); orange dashed line: absorption spectrum recovered from measurements against air baseline of R6G in solution and water.

output is a zero line with only instrumental noise. The black curve in Figure 3-5.3 shows a measured R6G second solution against the water baseline. This is a proper spectrum of R6G that at long wavelength starts with values close to zero.

For comparison in Figure 3-5.4, we present measurement of R6G against air baseline (blue line in Figure 3-5.3) from which we subtracted water absorption measured with air baseline also (green line in Figure 3-5.3). Such a subtraction can be easily done with any software including Excel. The calculated absorption spectrum of R6G is presented by the orange dashed line that is overlaid with directly measured R6G absorption with proper water baseline. It can be clearly seen that both absorption spectra are practically identical. This was the same R6G solution and as long as we correct properly the result should be the same spectrum.

Sometimes, we do not have an easily accessible exact buffer solution that can be used for a reference. If we want to use a different solution that looks similar and have identical absorption (usually no absorption) in the spectral range of interest (like water, alcohol, glycerol, or even air that does not absorb in visible spectral range) we have to be careful. Different solvents (or even water buffer that has significantly different salts concentration) may have different refractive indices like for example water, glycerol, or propylene glycol, that can impact measured absorption. As an example, we are presenting a measurement of the same R6G solution measured against the background from an empty cuvette (refractive index of air is 1 as compared to water 1.33). Figure 3-5.5 shows the measured absorption of R6G against reference from the empty cuvette. The shape of the absorption spectrum is not much perturbed but the absorption is significantly shifted down and for the large range we can read negative absorptions. This is obviously wrong and in this case, when only reading maximum absorption at the peak we will make a large error. We want to stress that this

(Continued)

EXPERIMENT 3-5 (CONTINUED) MEASURING ABSORBANCE OF CHROMOPHORES IN SOLUTION

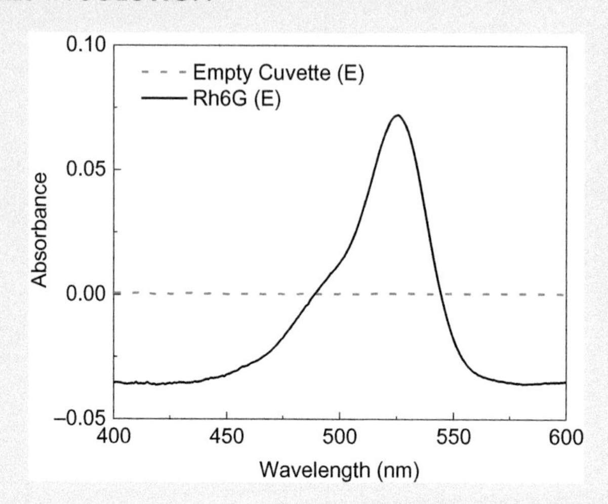

FIGURE 3-5.5 The measured absorption spectrum of the Rh6G solution against baseline taken with empty cuvette.

effect will depend on total sample absorption. For much larger concentrations, for example, when absorption is 1 or more the baseline shift is constant and its relative contribution will be much lower (in the range of 2–4% in this case).

CONCLUSIONS

The important conclusion from these measurements is to remember to use a proper baseline. The baseline correction can be done by the instrument when using appropriate reference solution or done by subsequent measurements against air and simply by subtracting solvent contribution as in Figures 3-5.3 and 3-5.4. It is also important to stress that two spectrally transparent solvents that do not absorb light in the region of interest typically should not be substituted for the baseline. For example, water and glycerol do not absorb in visible spectral range. But glycerol has a much higher refractive index and light reflection will be different. Also, glycerol is very vicious and cleaning the cuvette after taking a baseline reading can be a pain. While water is also transparent and can be clean easily. When having the chromophore in glycerol to save time and effort it could be tempting to use water for measuring the baseline. The refractive index difference between water and glycerol is almost comparable to the difference between air and water. So, we would expect a similar shift as presented in Figure 3-5.5. Purifying proteins by HPLC, we collect solutions that may contain significant and different concentrations of salts. If the concentration of a protein is very low (frequently a case), it is dangerous to measure the absorption of the collected solution against water since we can make a significant error.

EXPERIMENT 3-6 MEASURING ABSORBANCE OF CHROMOPHORES IN SOLUTIONS: ABSORBANCE PRESENTED IN DIFFERENT SCALES

Keywords: absorbance, absorption spectra, organic dyes, spectrophotometer, acridine orange, ruthenium bipyridyl, sulforhodamine 101.

The quantity being measured is absorbance at each wavelength $A(\lambda)$ according to the equation:

$$A(\lambda) = \varepsilon(\lambda)Cl$$

C, the concentration, is determined while making the solution, l, the beam path (sample thickness), is determined by the cuvette used, and $\varepsilon(\lambda)$ is an extinction coefficient that is a property of the dye.

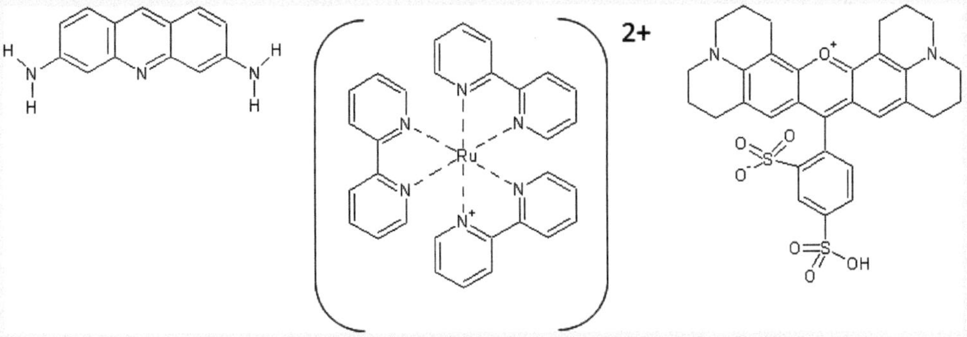

This protocol describes measurements of absorption spectra of solutions of organic dyes (acridine orange, ruthenium bipyridyl $[Ru(bpy)_3]^{2+}$, and sulforhodamine 101).

Traditionally, in most biological and biochemical applications of spectroscopy, spectra are recorded and presented as a function of wavelength (nm). However, in some cases, the spectrum needs to be presented as a function of wavenumber (cm^{-1}) (physicochemical applications) or electron volts (eV) (spectroscopy of solid-state materials). In some fields, the units used for wavenumbers are called Kaisers. 1 Kaiser (K) is defined as 1 cm^{-1} and 1000 cm^{-1} is called a kilo Kaiser (1 kK).

MATERIALS

- Acridine orange (AO)
- Ruthenium bipyridyl $[Ru(bpy)_3]^{2+}$
- Sulforhodamine 101 (S101)
- Water
- Glass vials
- 1 cm × 1 cm cuvettes

EQUIPMENT

- Spectrophotometer

(*Continued*)

EXPERIMENT 3-6 (CONTINUED) MEASURING ABSORBANCE OF CHROMOPHORES IN SOLUTIONS: ABSORBANCE PRESENTED IN DIFFERENT SCALES

METHODS

1. Make a stock aqueous solutions of dyes using your favorite procedure.
   a. Make sure all compounds are completely dissolved.
   b. Prepare samples of the dyes (few milliliters) using water and small amounts of stock solutions.
2. Place a cuvette filled with water (reference) into the cuvette holder of the spectrophotometer.
   b. Make sure that the cuvette wall is facing the light source and the detector (plane of the cuvette wall is orthogonal to the path of light). Alignment is important to have the correct line path inside the cuvette, a straight line from the source to the detector. Also, an angle will cause a diffraction of the light.
3. Set the spectrophotometer in absorption mode.
4. Run the baseline measurement.
5. Transfer dye solution into a clean cuvette.
6. Replace the reference with the sample (solution in a cuvette).
7. Measure spectrum—this acquires the absorbance spectrum.
8. To measure another sample(s) repeat steps 5–7. Ensure that water is the correct baseline for all samples.

RESULTS

The absorption spectra of acridine orange, ruthenium bipyridyl $[Ru(bpy)_3]^{2+}$, and sulforhodamine 101 are presented in Figure 3-6.1. The measured absorptions are plotted as functions of wavelength in nm and wavenumbers in kK for comparison.

Figure 3-6.1 presents single, correct spectra of all three compounds in water represented in two unit systems. The corresponding top and bottom spectra in Figure 3-6.1 for each

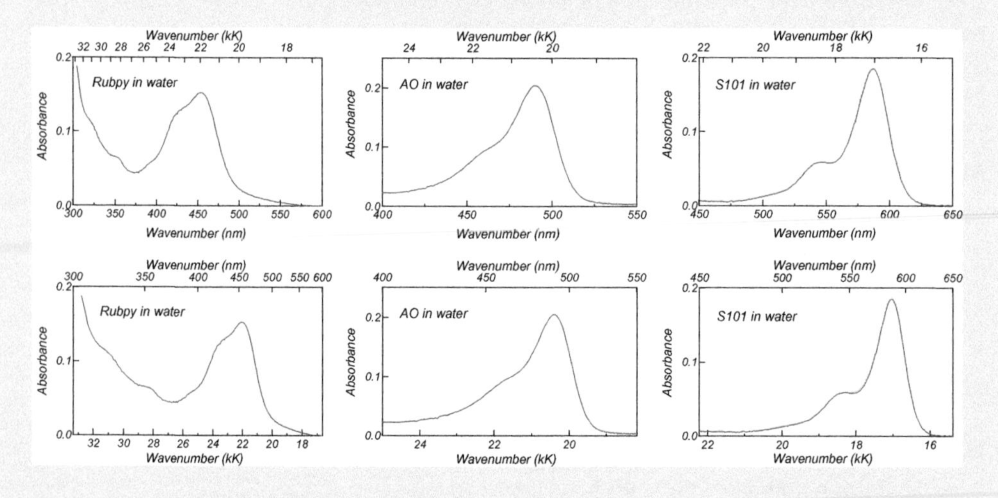

FIGURE 3-6.1 Absorption spectra of $Ru(bpy)_3^{2+}$ in water; acridine orange (AO) in water; and rhodamine S101 in water. Top: presented in wavelength scale and bottom: presented in wavenumber.

(Continued)

EXPERIMENT 3-6 (CONTINUED) MEASURING ABSORBANCE OF CHROMOPHORES IN SOLUTIONS: ABSORBANCE PRESENTED IN DIFFERENT SCALES

compound appear shifted when presented in wavelength (top) versus wavenumber (bottom) (even if the range for top and bottom is the same). This shift is a result of the fact that two-unit systems (nm and kK) are inversely related to one another. Thus a distance of *x* units is always the same in one representation but it differs in another. This is illustrated in Figure 3-6.1; the distance between any set wavelength of tick marks (50 nm) is always the same at the top of the figure, but the distance between the wavenumber tick marks (2 kK) increases as the wavenumber decreases, and vice versa in the bottom of Figure 3-6.1. It should also be noticed that in the figures above the wavelength is always increasing from left to right while wavenumber is decreasing from left to right.

Note: Always prepare samples for absorption/fluorescence measurements using stock solutions. Be sure that the dye in the stock solution is completely dissolved. This can be achieved through sonication, shaking, or gentle heating.

EXPERIMENT 3-7 ABSORBANCE OF RHODAMINE AND ANTHRACENE—NORMALIZED ABSORPTION SPECTRA

Keywords: absorbance, cuvette, spectrophotometer, Rhodamine 6G, anthracene.

The extinction coefficient of the dye (chromophore) and spectral profile (absorption spectra) are intrinsic characteristics of the compound/chromophore and should be independent of the type of used cuvette, spectrophotometer, and other experimental factors. Spectra (shape) measured in different conditions can be compared easily by normalizing them. This procedure becomes very useful when we want to compare change spectral shape due to changing conditions (solvent, concentration, temperature, pH, etc.).

The goal of this experiment is to show that measured spectra are independent of the cuvette or spectrophotometers used. This is also important since any comparison can be easily done between measurements made in different labs.

MATERIALS

- Water and ethanol (EtOH)
- Solution of Rhodamine 6G in water (R6G)
- Solution of anthracene in ethanol
- Glass vials
- Cuvettes 10 mm × 10 mm, 4 mm × 10 mm, 2 mm × 10 mm, and 1 mm × 10 mm (see Figure 3-7.1)

EQUIPMENT

- Spectrophotometer

(Continued)

EXPERIMENT 3-7 (CONTINUED) ABSORBANCE OF RHODAMINE AND ANTHRACENE—NORMALIZED ABSORPTION SPECTRA

METHODS

1. Make a stock aqueous solution of R6G (~10 μM) in water and anthracene in EtOH using your favorite procedure. *Note:* Anthracene is not soluble in water.
2. Make sure the compounds are completely dissolved.
3. Set the spectrophotometer into absorption mode.
4. Take a reference (measure baseline) with the reference solution (water) using a 10 mm × 10 mm cuvette.
5. Place a cuvette in the sample holder (10 mm × 10 mm cuvette) of the spectrophotometer and measure absorptions for R6G solution.
6. Repeat steps 4 and 5 using 4 mm × 10 mm, 2 mm × 10 mm, and 1 mm × 10 mm cuvettes.
7. Take a reference (measure baseline) with a reference solution (EtOH) in the sample compartment using the 10 mm × 10 mm cuvette.
8. Repeat steps 4–5 using anthracene in different cuvettes and EtOH as a reference.

RESULTS

First, we want to check if the absorption spectrum of a given compound will depend on how big is the maximum value of absorption. For that, we will use for the R6G concentration of about 0.4 μM and for anthracene about 12 μM. We arbitrarily choose these concentrations to cover a large range of measured absorptions. For our measurements, we will use four different path cuvettes (10, 4, 2, and 1 mm). Figure 3-7.1 shows the photography of used cuvettes. The first, 10 mm cuvette is the typical absorption cuvette. Next three are fluorescence cuvettes (4 × 10, 2 × 10, and 1 × 10) that we will use as absorption cuvettes with the absorption path along the shorter distance (marked by arrows). This is a proper use of these cuvettes (in Experiment 3-11, we are discussing why this is an appropriate use). Using these cuvettes along the 10 mm path can be done but it has to be done with very special care (see next Experiment 3-7.12). The absorptions in 1 cm (10 mm) cuvette for R6G and anthracene solutions are about 0.43 and 1.3, respectively. And in the 1 mm cuvette should be 10 times less. In this way, we will cover the absorption range from about 0.04 to 1.3.

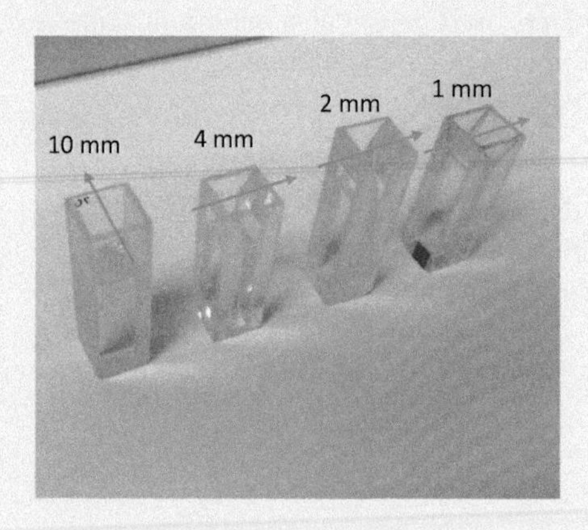

FIGURE 3-7.1 Cuvettes used in the experiment.

(*Continued*)

EXPERIMENT 3-7 (CONTINUED) ABSORBANCE OF RHODAMINE AND ANTHRACENE—NORMALIZED ABSORPTION SPECTRA

Figure 3-7.2 shows measured absorptions for both solutions (remember about appropriate blanks/references) in Agilent 8453 diode array spectrophotometer (top) and in Agilent Cary 60 wavelength scanning spectrophotometer (bottom). Just visually, we can see that the spectra measured in both spectrophotometers (top and bottom) are practically identical in values and spectral shape.

Figure 3-7.3 shows spectra from Figure 3-7.2 that have been normalized (each value in the spectrum has been divided by the maximum value of the absorption in this spectrum). This procedure (spectra normalization) will adjust each point (value) of the spectrum so the maximum value for the spectrum will be 1. It is clear that the spectral profiles are practically identical for each spectrophotometer. A very small deviation at the short wavelength part seen in normalized spectra is just due to small noise. Just for comparison in Figure 3-7.4, we are presenting normalized absorption spectra in the 4 mm cuvette in Agilent Cary 60 (solid line) and Agilent 8453 (dashed line) for both used compounds. The agreement between spectra is practically perfect.

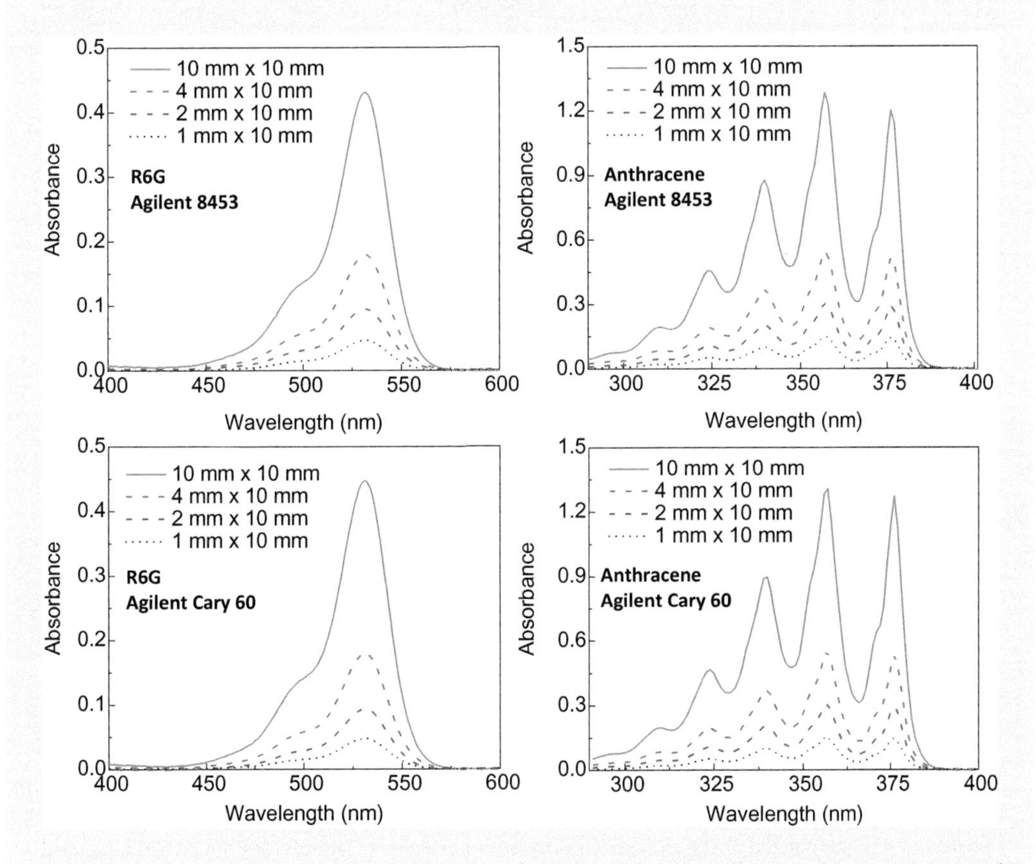

FIGURE 3-7.2 Measured absorptions for R6G (left) and anthracene (right) in Agilent 8453 (top) and Agilent Cary 60 (bottom) spectrophotometers.

(Continued)

EXPERIMENT 3-7 (CONTINUED) ABSORBANCE OF RHODAMINE AND ANTHRACENE—NORMALIZED ABSORPTION SPECTRA

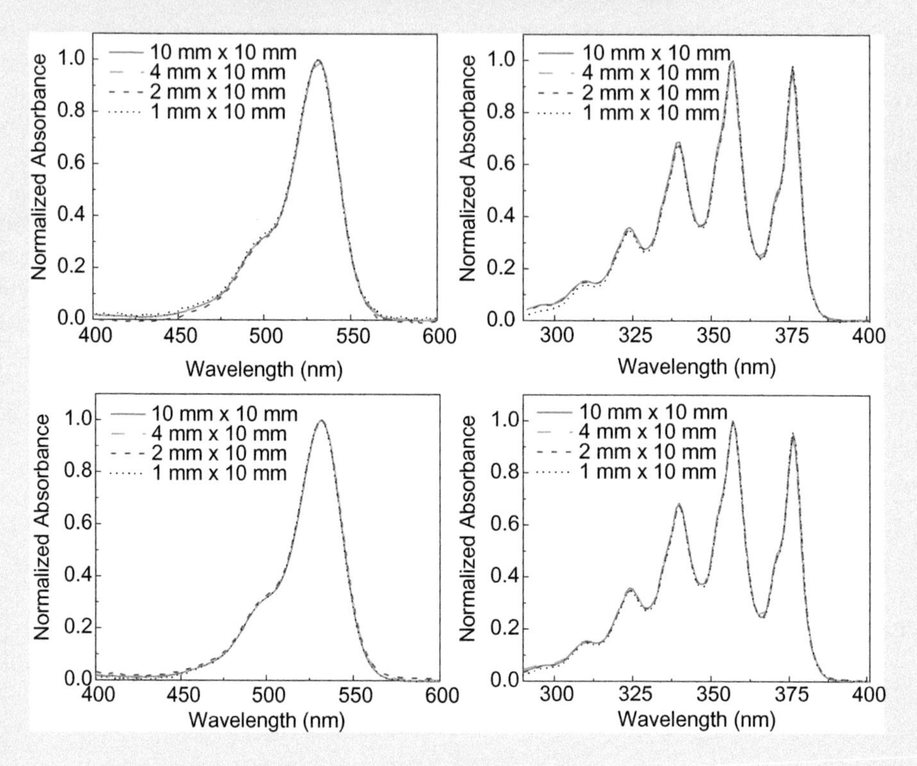

FIGURE 3-7.3 Normalized absorption spectra for R6G (right) and anthracene (left) measured in Agilent 8453 (top) and Agilent Cary 60 (bottom).

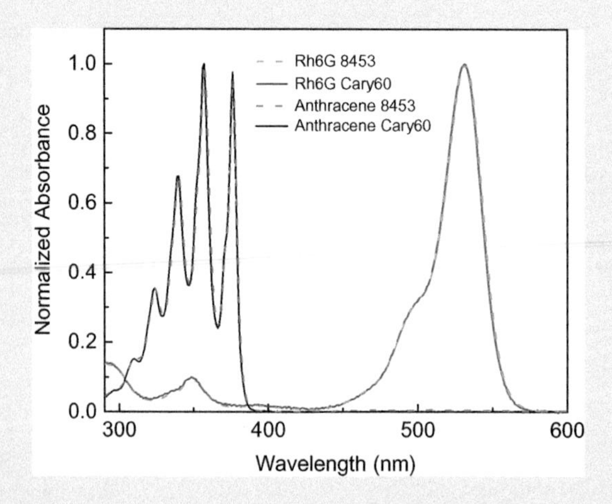

FIGURE 3-7.4 Normalized absorption spectra measured in Agilent Cary 60 for Rh6G (solid blue) and anthracene (solid black) and absorption spectra measured in Agilent 8453 for Rh6G (dashed orange) and anthracene (dashed green).

(Continued)

EXPERIMENT 3-7 (CONTINUED) ABSORBANCE OF RHODAMINE AND ANTHRACENE—NORMALIZED ABSORPTION SPECTRA

Comparison of measured spectra in Figures 3-7.3 and 3-7.4 shows that in measured range of absorption 0.04–1.3 the absorption spectrum profile does not depend on absorption value (cuvette thickness) or used instrument.

CONCLUSIONS

It is clear from these two experiments that a common spectrophotometer will work well in the absorption range 0.04–1 and from our experience, we know that this range is typically much larger (0.01–1.5). In the following Experiments 3-8 and 3-9, we will specifically test this range. We want to bring to the reader attention that the absorption 0.01 corresponds to ~98% transmittance and absorption 1.5 corresponds to just ~3% transmittance. This is a relatively large range for excellent linearity. In Chapter 1, we discussed the optimization of absorption measurement and we show that the highest sensitivity for measured absorption is about 0.4 or 0.8 depending on the type of used detector. The problem arises when measured absorption is too low or too high. For too low absorption, the light intensity change is so minimal that detecting the difference can be within the instrument noise. For too high absorption, the amount of light absorbed by the sample is so high that the amount of light reaching the detector (transmitted light) is so low that the detector cannot distinguish signal change from dark current. Typically, the problem can be solved by using the appropriate cuvette or diluting the sample.

EXPERIMENT 3-8 ABSORBANCE OF RHODAMINE 800—EFFECT OF OUT OF RANGE ABSORPTION

Keywords: absorbance, cuvette, spectrophotometer, Rhodamine 800.

The shape of the absorption spectrum should be independent of the sample concentration or from the cuvette used for measurement. As discussed in Chapter 1 and Experiment 3-7, in a large range of absorption the measured spectrum, when normalized, should be and is always the same. Unless we will have some additional processes/interactions that lead to inherent absorption spectrum change which should not be overlooked (e.g., aggregation of chromophores). We now want to check how the absorption spectrum may be changed when the concentration reaches absorptions significantly over 2. For this experiment, a highly concentrated solution of Rhodamine 800 (R800) is used, that presents an absorption of much over 3. Absorption 3 means a 1000-fold light intensity attenuation, and less than 0.1% of original light intensity reaches the detector. This represents the limit of detection for most common spectrophotometers and typical spectrophotometers are generally unable to reliably distinguish differences for absorptions above 3.

The goal of this experiment is to demonstrate how too high absorption (too high concentration) can alter the measurement and how to recognize and overcome the problem.

There are two ways of changing solution absorption. We can use one standard cuvette 10 mm × 10 mm and dilute solution, or use one solution with cuvettes of different thickness. To demonstrate the effect cuvettes of varying thickness will be used with one fixed

(Continued)

EXPERIMENT 3-8 (CONTINUED) ABSORBANCE OF RHODAMINE 800—EFFECT OF OUT OF RANGE ABSORPTION

concentration of R800. There are two reasons for that. First, this will ensure that diluting the solution does not change any potential aggregation state (aggregation for the used concentration is not something to expect, but necessary to mention to avoid any ambiguity). Second, commercial cuvettes typically have a very well calibrated thickness. So, precision should be very high compared to a simple dilution, which would have quite a few potential errors, especially considering the small volumes in question.

MATERIALS

- Rhodamine 800 (R800)
- Water
- 0.05 cm × 1 cm absorbance cuvette
- 0.1 cm × 1 cm absorbance cuvette
- 0.2 cm × 1 cm absorbance cuvette
- 0.4 cm × 1 cm absorbance cuvette
- 1 cm × 1 cm absorbance cuvette

EQUIPMENT

- Spectrophotometer

METHODS

1. Prepare a solution of Rhodamine 800 (R800) in water with a concentration ~100 μM (we start with high absorbance and we want to adjust concentration for absorption in the peak to be ~10 in 1 cm path cuvette in the maximum; we should measure absorption of stock solution in 1 mm cuvette and it should be ~1).
2. Use water as the baseline. Be sure to measure baseline with each different cuvette. Baseline from different cuvettes could be different (see Experiments 3-3 and 3-4).
3. Measure baseline with the 1 cm × 1 cm cuvette filled with water.
4. Measure absorbance of R800 with above set up.
5. Measure baseline with the 0.4 cm × 1 cm cuvette filled with water using the 0.4 cm path.
6. Measure the absorbance of the same rhodamine solution with 0.4 cm × 1 cm cuvette using the 0.4 cm path.
7. Measure baseline with the 0.2 cm × 1 cm cuvette filled with water using the 0.2 cm path.
8. Measure absorbance of the same as above R800 solution with above cuvette.
9. Measure baseline with the 0.1 cm × 1 cm cuvette filled with water using the 0.1 cm path.
10. Measure absorbance of the same as above R800 solution with above cuvette.
11. Measure baseline with the 0.05 cm × 1 cm cuvette filled with water using the 0.1 cm path.
12. Measure absorbance of the same as above R800 solution with above cuvette.

RESULTS

The above procedure should result in four spectra. These spectra are plotted in Figure 3-8.1. The acceptable absorptions are spectra collected for 1 mm and 2 mm path cuvettes. This is easy to check by normalizing the spectra (see Experiment 3-7). Absorptions in 1 mm and 2 mm will perfectly overlap. For 1 cm × 1 cm cuvette, the absorption spectrum in the visible range

(*Continued*)

EXPERIMENT 3-8 (CONTINUED) ABSORBANCE OF RHODAMINE 800—EFFECT OF OUT OF RANGE ABSORPTION

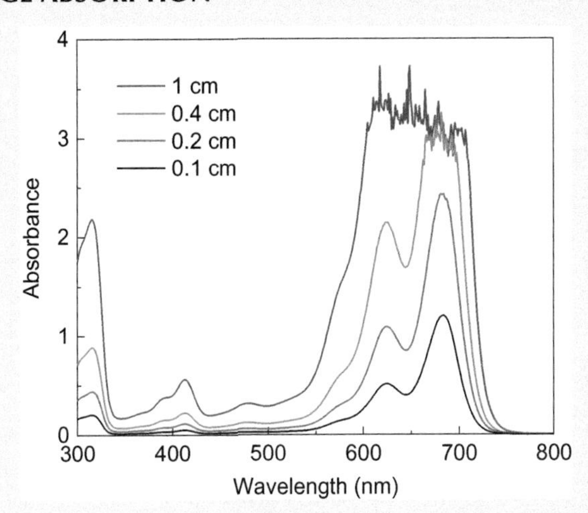

FIGURE 3-8.1 Absorbance spectra of R800 measured in different path cuvettes.

looks flat with a maximum in the range of 3–4. We want to stress that absorbance 3 is a 1000-fold transmitted light intensity attenuation (about 0.1% of light reaches the detector). For most spectrophotometers, this is a limit for measured absorbance. The spectrum measured with 0.4 cm × 1 cm cuvette looks better, but still, the maximum absorbance at the peak is too high.

How to decide which spectra (or spectral range) are correct?

One way we discussed in Experiment 3-7 is to normalize the spectra and see if they overlap. This is a good approach if we know which spectrum is correct. But in some conditions, this procedure may not be easy to do or may not work for various reasons (e.g., when measured spectra are mixtures of two or more components). If the absorption is in the correct range its dependence on the cuvette path (or concentration) should be linear and independent of how many components (chromophores) contribute to the absorption. This could be seen by observing the absorption as a function of cuvette thickness (path) at different observation wavelengths. We will select wavelengths that correspond to significantly different extinction coefficients. Importantly, when measuring absorption as a function of wavelength, each wavelength is measured individually (independently) and lower absorptions should be within the linear range of the spectrophotometer. The example of this is presented in Figure 3-8.2.

At the wavelength range 650–690, the absorption is the highest and quickly saturates for 4 mm and 10 mm cuvettes. The dependence becomes already better for observation at ~625 nm (shoulder) and already quite linear for 600 nm and below.

The lessons of Figure 3-8.2 are:

1. If the absorption is too high measured values can be not correct. This typically may happen if absorption is larger than 2.5. So, the measured absorptions should be below 2 for most spectrophotometers.
2. Measured absorptions at different wavelengths range, where the extinction coefficient is lower, are still correct. Even if measured absorption in the maximum is much over 3 (or over spectrophotometer range) than we can properly predict the value at the absorption maximum.

(Continued)

EXPERIMENT 3-8 (CONTINUED) ABSORBANCE OF RHODAMINE 800—EFFECT OF OUT OF RANGE ABSORPTION

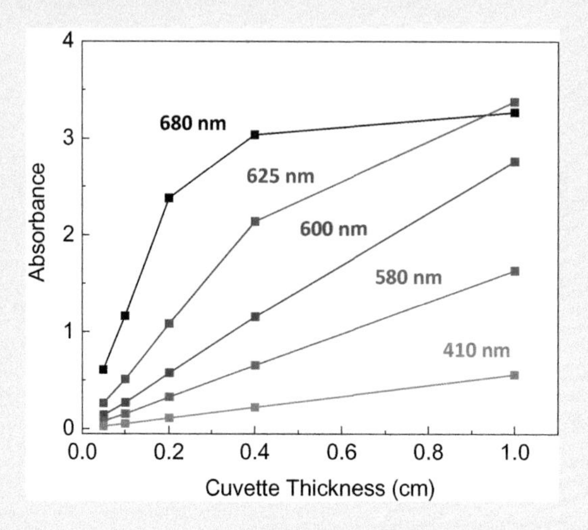

FIGURE 3-8.2 Dependence of measured absorptions for different cuvette thickness for different wavelength.

Can we estimate absorption value at the maximum absorption-based only on the measurement in 10 mm cuvette?

There are a few examples where this could be useful. When we have a series of increasing concentrations of the sample. For example, when purifying dye (or labeled protein) via HPLC, we are collecting multiple fractions. When approaching the peak for the elution of the purified compound the concentration increases and absorption in the maximum can reach values exceeding the detectability range of the detector. Since the absorption spectra in the linear spectral range of the spectrophotometer when normalized overlap very well, we can use the measurement at a different wavelength where the extinction coefficient is much smaller to calculate true absorbance at the peak. As shown in Figures 3-8.1 and 3-8.2, absorption at the maximum (680 nm) does not show proper values for the two highest concentrations. But absorptions for the observation wavelength of 600 nm and below are perfectly linear. Let's consider observation wavelengths 680 nm and 580 nm. For the 1 mm cuvette, the measured absorptions are at the linear range. If we take the ratio of absorption at 680 nm to absorption at 580 nm as measured in 1 mm path, we will have $r = (Ab(680)/Ab(580)) = 1.18/0.163 = 7.24$. This factor should be constant for each cuvette (or concentration). Absorbance at 2 mm for 580 observation is 0.352 and we calculated absorbance at 680 nm observation should be $0.352 \times 7.24 = 2.55$. The measured value is 2.43 which is slightly off. Similarly, the absorption value measured in 10 mm for 580 nm observation is 1.64 and calculated absorption value at 680 nm is 11.87 that is drastically different from 3.21. Since we are using calibrated cuvettes, 11.87 value is expected (it should be 10 times bigger than reading in 1 mm cuvette). This routine is not needed when using calibrated cuvettes but becomes very handy when dealing with different not well-defined concentrations. This justifies why when dealing with high concentrations of samples eluted during HPLC purification, we can set observation of the maximum absorption peak. This is valid when at different (off-peak) observation wavelength, we do not have some other component (or solvent) absorbing at that wavelength.

EXPERIMENT 3-9 ABSORBANCE OF ANTHRACENE—LOW ABSORPTION RANGE

Keywords: absorbance, cuvette, spectrophotometer, anthracene.

In the previous experiment, the effect of a high absorption over the range of the spectrophotometer was discussed. In this case, the intensity of the light beam reaching the detector was too low to be distinguished from the dark current. In the case of very low absorption, the intensity of the light beam is affected very little and the difference between the intensity of the beam traveling through the reference (baseline measurement) and the sample is very small. The intensity change of the beam traveling through the sample is therefore too small to be distinguished from the instrumental/experimental noise.

If the sample absorption is too low we have two options. The simplest option is to increase the concentration of the sample. But this may be impossible under certain circumstances, for example, low sample solubility. In this case, in place of a standard cuvette, one may use a 20 mm or 50 mm path cuvette (the path is referred to the length through which the beam travels). Occasionally, a longer cuvette (100 mm or more) can be used. These cuvettes would require special holders not always available with standard spectrophotometers. It is important to remember that the same (or equivalent) cuvette should be used for reference.

The goal of this experiment is to test the lower detection limit of a spectrophotometer. Refer to Chapter 2 for the discussion on the signal-to-noise ratio.

MATERIALS

- Ethanol (EtOH)
- Solution of anthracene in ethanol
- Glass vials
- Cuvettes 10 mm × 10 mm, 4 mm × 10 mm, 2 mm × 10 mm, and 1 mm × 10 mm

EQUIPMENT

- Spectrophotometer

METHODS

1. Prepare a stock solution of anthracene in EtOH. Using the stock solution and EtOH, prepare a sample with the absorbance at maximum in 10 mm cuvette of about 0.025 (~2.5 μM). From previous Experiment 3-7, we know that 0.025 absorption will be well within the spectrophotometer range.
2. Use EtOH as the baseline. Measure baseline with the 1 cm × 1 cm cuvette filled with EtOH.
3. Measure absorbance of anthracene solution.
4. Measure the baseline with the 0.4 cm × 1 cm cuvette filled with EtOH using the 0.4 cm path.
5. Measure the absorbance of the same anthracene solution with 0.4 cm × 1 cm cuvette using the 0.4 cm path.
6. Repeat procedure 5 and 6 with 2 mm and 1 mm path length cuvettes.

RESULTS

Figure 3-9.1 shows measured absorptions for the anthracene solution in EtOH in four different thicknesses cuvettes. On the left are presented measured absorption spectra and on the

(Continued)

EXPERIMENT 3-9 (CONTINUED) ABSORBANCE OF ANTHRACENE—LOW ABSORPTION RANGE

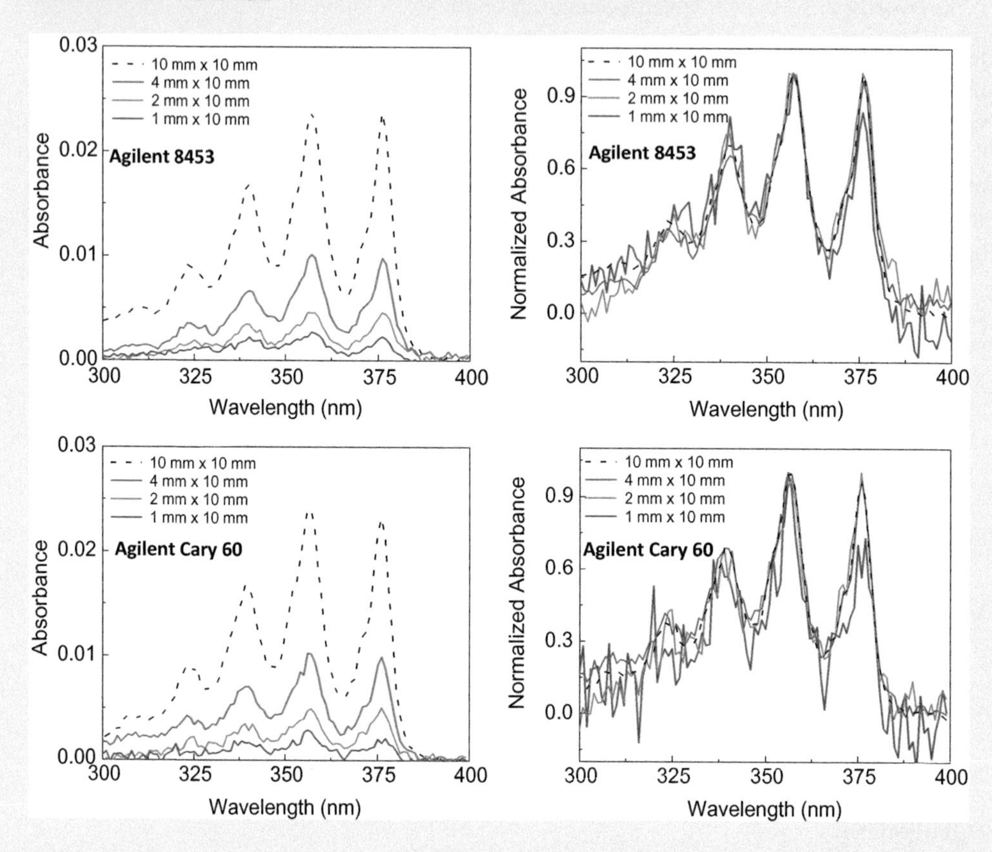

FIGURE 3-9.1 Measured absorption spectra of anthracene in EtOH in different thickness cuvettes in Agilent 8453 (top) and Agilent Cary 60 (bottom).

right normalized spectra in two spectrophotometers (Agilent 8453: top and Agilent Cary 60: bottom). The measured absorption spectra for 10 mm path (thickness) is in a good spectrophotometer range and it very well agrees with the spectra measured in previous Experiment 3-7. For the smaller thickness of cuvettes in normalized spectra, we observe increasing noise contribution. For the lowest 1 mm thickness, the noise is already significant and we observe some spectra deformation. But for 2 mm path cuvette where measured absorption is 0.0005, we are still measuring good spectra. For lower absorptions, the noise becomes a significant factor.

CONCLUSIONS

It is clear that absorptions greater than 0.005 can be measured with significant precision. But in the case of lower absorptions, we have to be very careful with the baseline measurement. Like in this case, the baseline should be from the same cuvette carefully cleaned. Otherwise, the baseline fluctuation will corrupt the result. For example, using a different type of quartz (fused silica) cuvette that has slightly different refractive index may introduce an error of more than 0.001 in absorption. We also want to point the reader to one more factor affecting the

(Continued)

EXPERIMENT 3-9 (CONTINUED) ABSORBANCE OF ANTHRACENE—LOW ABSORPTION RANGE

precision of the measurement. For low absorptions, where beam intensities are high, the limiting factor is the instrument noise. This noise can be reduced by increasing the averaging time. For the next experiment, we will use 2 mm and 1 mm cuvettes and we will change averaging time from 0.1 s to 5 s.

It is trivial but an important observation that for absorptions values in the working range of the spectrophotometer (typically between 0.01 and 2), the measured absorption should be a linear function of cuvette thickness or (sample concentration) and normalized spectra should well overlap.

Note: In the case of very low absorbance measurements, it is very important to clean the cuvette surfaces. When not precisely cleaned and wiped surfaces it can significantly elevate/distort the measured signal.

EXPERIMENT 3-10 LOW ABSORBANCE—HOW TO IMPROVE QUALITY OF MEASURED SPECTRUM

Keywords: absorbance, cuvette, spectrophotometer, rhodamine 800.

The previous experiment demonstrated that if absorption drops below 0.002 (transmitted light >99.5%) the experimental noise can be a significant factor disturbing measured absorption. As mentioned before, the quality of measured spectra can be partially improved by using a longer optical path cuvette, or using a much more expensive spectrophotometer that will have lower noise. But still, the improvement will be limited. The following experiment shows how to improve the signal-to-noise ratio by increasing the averaging time or/and repeating the measurement multiple times and averaging them. Most of the spectrophotometers will have the option to change integration time for the measured spectrum (for some, it will be the scanning speed). What this effectively does is improve the signal-to-background ratio. When doing so, the experimentalist needs to remember that the change should be done for both baseline and sample measurements.

The goal of this experiment is to test the lower detection limit of a spectrophotometer. Again, the reader is reminded of the discussion in Chapter 2 on signal-to-noise ratio.

MATERIALS

- Ethanol (EtOH)
- Solution of rhodamine 800 (R800) in ethanol
- Glass vials
- 10 mm × 10 mm cuvette

EQUIPMENT

- Spectrophotometer

(Continued)

EXPERIMENT 3-10 (CONTINUED) LOW ABSORBANCE—HOW TO IMPROVE QUALITY OF MEASURED SPECTRUM

METHODS

1. Prepare a solution of R800 in EtOH with a low concentration, so measured absorbance at maximum in 10 mm cuvette will be 0.05–0.1.
2. Use EtOH as the baseline.
3. Set the standard averaging time in a spectrophotometer (typically 0.1 s for scanning spectrophotometer and 1 s for diode array spectrophotometer) and measure baseline with the 1 cm × 1 cm cuvette filled with EtOH.
4. Measure absorbance of R800 solution using the same averaging time.
5. Dilute R800 solution 50–100-fold, so the maximum absorption will about 0.001.
6. Measure baseline using longer averaging time (2 s in our case) and measure the absorbance of the diluted R800 solution.
7. Repeat procedure 5 and 6 for shorter averaging time (0.2 s in our experiment).

RESULTS

Figure 3-10.1 shows measured absorption of R800 solution that has absorption well within a good experimental range (in our case about 0.1). In this range, the spectrum looks very good and noise is negligible. In the figure, we also included measurement of a solvent (EtOH) that represent a baseline (zero absorption). Figure 3-10.2 shows the measured absorptions for a diluted solution (80-fold in our case) with 2 s and 0.2 s averaging time in a diode array spectrophotometer. In both cases, the maximum peak in the visible range is about 0.001. It is clearly a visible significant noise contribution that for the short wavelength range becomes dominant (lamp provides less light intensity in the shorter wavelength range). The difference between 0.2 s and 2 s averaging time in short wavelength range is clear. Also, we need to notice that measured absorption in the 350–450 nm range for short averaging time is negative. This could be due to a small baseline shift that is associated with positioning the cuvette or not precisely cleaned and wiped the cuvette walls. For very small absorptions, this will

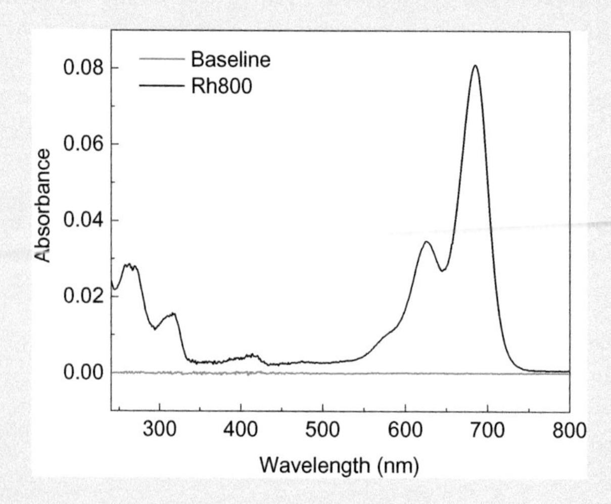

FIGURE 3-10.1 The measured absorption spectrum of R800 higher concentration with standard averaging time. The orange line show baseline comparison.

(Continued)

EXPERIMENT 3-10 (CONTINUED) LOW ABSORBANCE—HOW TO IMPROVE QUALITY OF MEASURED SPECTRUM

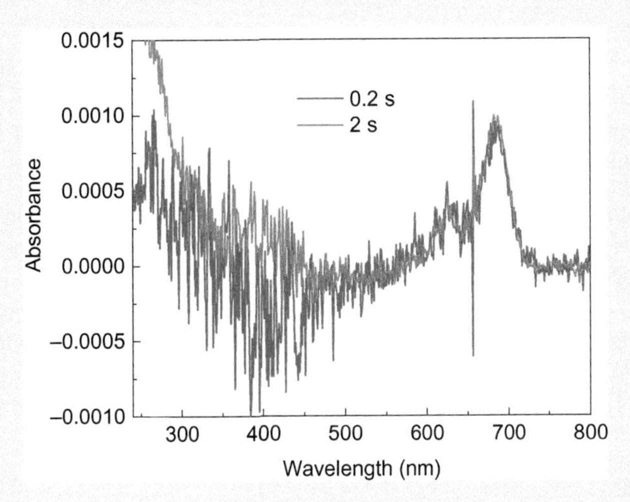

FIGURE 3-10.2 Measured absorption spectra for 80 fold diluted solution of R800 with 0.2 s (green) and 2 s (orange) averaging times.

happen (lower or higher than expected readings) due to minimal changes in cuvette positioning or some very small impurity (or imperfection) on the cuvette wall. *Note: This problem can be minimized when we do not remove cuvette for solution exchange but do it gently with the cuvette mounted in the holder using a pipette.* Figure 3-10.3 shows normalized to the absorption peak (~700 nm) spectra from all three measurements. The spectrum measured for higher R800 concentration is shown as a black line and can be used as a standard (this should be a reference spectral profile). The spectrum measured with short 0.2 s

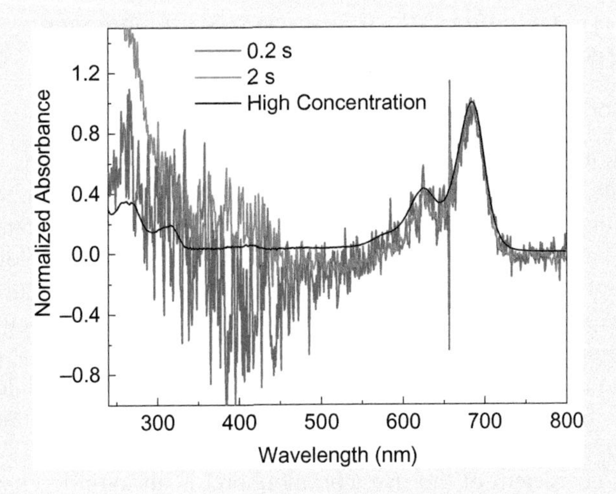

FIGURE 3-10.3 Normalized absorption spectra of high concentration and diluted Rh800 solutions (0.2 s and 2 s averaging time).

(Continued)

EXPERIMENT 3-10 (CONTINUED) LOW ABSORBANCE—HOW TO IMPROVE QUALITY OF MEASURED SPECTRUM

averaging time shows much higher noise below 450 nm and significant deviation from the expected profile. A 2 s averaging improves noise and down to 350 nm, the spectral profile is closer to our standard spectrum. Below 350 nm, the deviation from the standard spectrum increases. The deviation in shorter wavelength is typically much higher for low absorptions due to much smaller lamp power and higher lamp fluctuations.

Note: We can improve the signal-to-noise ratio for measured spectrum by increasing the averaging time. However, we need to pay special attention to the wavelength range where the lamp intensity is typically lower (UV and NIR) where spectral deviation can be significant. When measuring very low absorptions some variations are also due to imperfect cuvette positioning and some impurities/imperfections in/on cuvette walls. Also, in the UV spectral range (below 300 nm), we could observe elevated noise and artificially high absorbance as compared to the expected values. This is due to low lamp intensity and some impurities in the solvent.

CONCLUSIONS

For these measurements, we used a relatively old diode-array spectrophotometer. Even then, measured precision in the visible range is very good down to 0.001 with longer averaging time. Such low absorbance corresponds to less than 0.3% light attenuation. This can be improved by repeating the measurement of the sample multiple times and averaging the results. Also, a new spectrophotometer can do a better job and a specialized spectrophotometers can measure the order of magnitude better. But for lower absorptions, the stability of the baseline and solvent purity becomes a bigger problem. In conclusion, we can safely assume that we can have good confidence for absorption measurements above 0.001.

EXPERIMENT 3-11 ABSORBANCE OF RHODAMINE B—PROPER USE OF CUVETTES

Keywords: absorbance, cuvette, spectrophotometer, rhodamine B.

For many samples (typically biological samples like reconstituted proteins), we do not have a sufficient quantity (volume) of the sample to fill a standard 1 cm (10 mm × 10 mm) cuvette (>2 mL). Also, dilution of the sample may not be an option and measurements have to be done in a much smaller volume. There are a few cuvette types that may allow measurements for sample volumes of 100 μL or even less. Examples of some such cuvettes are shown in the previous experiment (Figure 3-7.1). Typically, these kinds of cuvettes are made for fluorescence measurements and using a longer path for absorption is not possible without special modification. Using such cuvette for absorption measurement will lead to significant error. There are some specialized absorption cuvettes available for measuring small volume samples discussed in this and the next experiment.

The extinction coefficient of the dye (chromophore) is an intrinsic characteristic of the compound and should be independent of the type of cuvette used, spectrophotometer, and other experimental factors. So, data obtained from an absorbance measurement

(*Continued*)

EXPERIMENT 3-11 (CONTINUED) ABSORBANCE OF RHODAMINE B—PROPER USE OF CUVETTES

should be independent of the cuvette, as long as the cuvette is used properly. The initial Experiments 3-1 and 3-2 already discussed some aspects on how to properly measure transmittance/absorbance. But there are a few additional factors to be considered when using non-standard cuvettes. A cuvette may be misused when it is inserted into the cuvette holder at a wrong (not intended by manufacturer) angle (90° offset), not inserted completely, or just filled partially so that only part of the beam travels through the solution (meniscus in the beam path).

The goal of this exercise is to present some common errors made when using non-standard cuvettes. We will demonstrate the effects of misusing (using cuvette in not intended way).

The term "standard cuvettes" we use meaning cuvettes with exterior dimensions of 1.25 cm × 1.25 cm, and interior dimensions of 1 cm × 1 cm (10 mm × 10 mm). Most cuvettes used for absorption and fluorescence measurements will have standard external dimensions while having different internal sizes that refer to the size of the sample holding cavity. This is done on purpose to make sure that the cuvette will properly fit into the spectrophotometer's (fluorimeter's) standard sample holder and will be properly positioned in the light beam path. When using a non-standard size cuvette with external dimensions different from those of a standard cuvette, extra care should be taken to always position the cuvette properly in the beam path. These kinds of cuvettes typically use an "adapter" that is a metal or plastic element with external dimensions of the standard cuvette (12.5 mm × 12.5 mm) and internal dimensions to perfectly fit the cuvette and position it in the center.

Note: When talking about the use of a cuvette, the length through which the beam travels is written first.

MATERIALS

- Rhodamine B
- Water
- 0.2 cm × 1 cm absorbance cuvette
- 0.2 cm × 1 cm frosted cuvette
- 0.2 cm × 1 cm clear cuvette
- 1 cm × 1 cm clear cuvette

EQUIPMENT

- Spectrophotometer

METHODS

1. Prepare a solution of Rhodamine B (RhB) in water with a concentration of about 10 μM.
2. Use water as the baseline.
3. Measure baseline with the 1 cm × 1 cm cuvette filled with water.
4. Measure absorbance of rhodamine with above set up.
5. Measure baseline with the clear 0.2 cm × 1 cm cuvette filled with water using the 0.2 cm path.
6. Measure absorbance of rhodamine with above set up.
7. Measure baseline with the clear 0.2 cm × 1 cm cuvette filled with water using the 1 cm path.

(Continued)

EXPERIMENT 3-11 (CONTINUED) ABSORBANCE OF RHODAMINE B—PROPER USE OF CUVETTES

8. Measure absorbance of rhodamine with above set up and cuvette positioning.
9. Measure baseline with the absorbance cuvette filled with water and 1 cm path.
   a. There is no 0.2 cm beam path since the black walls block all light.
10. Measure the absorbance of rhodamine with above set up, but filling the cuvette about halfway so that the entire beam does not pass through the solution.
11. Measure absorbance of rhodamine with above set up filling the cuvette all the way.
12. Measure baseline with the 0.2 cm × 1 cm frosted cuvette using the 1 cm path.
13. Measure absorbance of rhodamine with above set up.

RESULTS

The above procedure should result in six spectra, three correct and three incorrect. The spectra are plotted in Figure 3-11.1. Short dashes are used for the clear 1 cm × 1 cm cuvette, dashes are used for the 0.2 cm × 1 cm cuvette, a solid line for the absorbance cuvette, and a dotted line for the frosted 0.2 cm × 1 cm cuvette.

In principle, all measurements along the 10 mm path should give similar (identical) results and in the 2 mm path measured absorption should be five times lower (see previous Experiment 3-7). When looking at Figure 3-11.1, this is definitely not the case. A 10 mm × 10 mm cuvette is a standard cuvette that most probably gives proper result. A 2 mm × 10 mm cuvette should give 1/5 of the previous reading and the result is very close. A 10 mm × 2 mm absorption cuvette (cuvette with two black walls) should also give a very similar result. But looking at Figure 3-11.1, we notice both results are very close but differ by a few percents. We want to stress that with the 10 mm × 2 mm absorption cuvette, it is possible to get a result very close to the regular cuvette. We present average spectra from three randomized trials that we believe closely represents the average difference. The reason for the small deviation is the spectrophotometer we have been using for these measurements. Agilent 8463 spectrophotometers that have a relatively large beam size, so as explained later a significant part of the beam is cut off

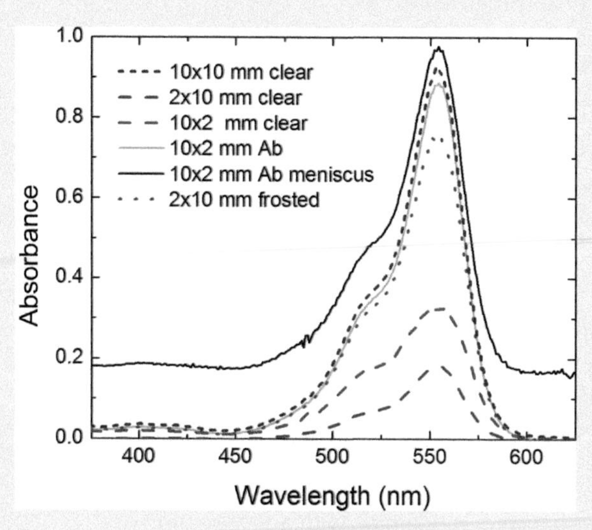

FIGURE 3-11.1 Measured absorption spectra for Rhodamine B solution in different cuvettes.

(Continued)

EXPERIMENT 3-11 (CONTINUED) ABSORBANCE OF RHODAMINE B—PROPER USE OF CUVETTES

by the black cuvette walls. Since the sample holder is a little loose (typically intended for the sample to be easily mounted), putting the cuvette for a baseline measurement and taking it out and putting it back with the sample, the position slightly differs that leads to a different beam attenuation by the black cuvette walls. Similar measurement made in the Agilent Cary 60 that has a much smaller beam size will show a much smaller difference (less than 1%).

The rest of the measurements are definitely unacceptable. The 10 mm × 2 mm clear cuvette (this is fluorescence cuvette) shows artificially low absorption. The 10 mm × 2 mm frosted cuvette (this is fluorescence cuvette that has frosted edges) shows better results but is still far from what we would expect. Finally, the 10 mm × 2 mm absorption cuvette (black sidewalls cuvette) when we filled it not sufficiently and a small part of the beam hits the meniscus and travels over the sample solution which gives a strange result. A moderately good reading at the maximum absorption (~1) but completely off at low absorption reading (elevated baseline). We present this result since underfilling the cuvette is quite common when dealing with a small amount of the sample.

The lessons of Figure 3-11.1, we would like to take home are:

1. The same extinction coefficient should always be calculated if the absorbance is measured properly.
2. If the beam spot size is larger than the width of the facing sample cuvette side the measured absorbance will typically be too low.

The absorbance coefficient and optical density are related as

$$A(\lambda) = \varepsilon(\lambda)Cl$$

A is the quantity measured experimentally and presented in Figure 3-11.1, l is the distance the beam travels through the sample, 1 cm or 0.2 cm in this experiment, C is the concentration of the sample, and $\varepsilon(\lambda)$ is the extinction coefficient. When the experiment is done correctly, the calculated extinction coefficient should always be the same regardless of the path length and used cuvette. In the table, we present calculated extinction coefficients at maximum absorption from the three closest measurements.

1 cm × 1 cm clear	$Ab(554\text{ nm}) = 0.93$	$\varepsilon(554\text{ nm}) = \frac{0.93}{10\,\mu\text{M} \times 1\text{cm}} = \frac{0.93}{10\,\mu\text{M cm}} = 93{,}000\ \mu\text{M}^{-1}\text{cm}^{-1}$
0.2 cm × 1 cm clear	$Ab(554\text{ nm}) = 0.184$	$\varepsilon(554\text{ nm}) = \frac{0.184}{10\,\mu\text{M} \times 0.2\text{cm}} = \frac{0.92}{10\,\mu\text{M cm}} = 92{,}000\ \mu\text{M}^{-1}\text{cm}^{-1}$
0.2 cm × 1 cm absorbance	$Ab(554\text{ nm}) = 0.89$	$\varepsilon(554\text{ nm}) = \frac{0.089}{10\,\mu\text{M} \times 1\text{ cm}} = \frac{0.089}{10\,\mu\text{M cm}} = 89{,}000\,\mu\text{M}^{-1}\text{cm}^{-1}$

These values are relatively close, thus the calculated extinction coefficients are almost independent of cuvette used. As a comparison, the calculated extinction coefficient from the 1 cm × 0.2 cm cuvette measurement is 36,000 μM^{-1} cm^{-1}. Much different from the expected value of about 93,000 μM^{-1} cm^{-1}.

(*Continued*)

EXPERIMENT 3-11 (CONTINUED) ABSORBANCE OF RHODAMINE B—PROPER USE OF CUVETTES

The goal of the following considerations is to explain what may lead to significant errors in absorption measurements when using a non-standard cuvette. Understanding this will allow prevention of such mistakes in different experiments.

The most common error in using a cuvette comes from allowing parts of the beam to reach the detector without traveling through the sample solution. The baseline measurement even if it is taken with the exact same set up will not prevent the problem. Two common causes of this are using a thin clear cuvette with a thin side facing the beam or not filling the cuvette high enough for the whole beam to travel through the sample. If the level is so low that the beam does not hit the sample at all, the resulting measurement would be near zero absorbance with no clear absorption profile. This would be easy to spot as an incorrect result since that would mean there are no chromophores in the sample (at least in the wavelength range selected). In that case, it is best to test over the entire accessible wavelength range of the spectrophotometer.

In Figure 3-11.2, we are showing diagrams of three types of cuvettes that most of the readers will be familiar with. The first is a regular square cuvette 10 mm × 10 mm. Second is a fluorescence cuvette 2 mm × 10 mm but for absorption, we are using it as a 10 mm × 2 mm cuvette (the light is traveling along 10 mm path), and the third one is a 10 mm × 2 mm absorption cuvette. This cuvette has two black (nontransparent) walls and the transparent path is only along the 10 mm length. This cuvette cannot be used for fluorescence. On a side of each cuvette, we schematically have drawn the profile of the transmitted beam. For the 10 mm × 10 mm standard cuvette, the beam profile is unchanged only the intensity of the beam decreased proportionally to the sample absorption. Measurement of absorption will be correct. For the second 10 mm × 2 mm fluorescence cuvette, the beam size is larger than 2 mm width and part of the beam travels through cuvette glass. On the right is the beam profile after pathing through the cuvette. The part of the beam that travels through the sample will be adequately attenuated but parts of the beam traveling through the glass will be practically unchanged. In this case, the spectrophotometer will see artificially higher intensity (not the whole beam was attenuated) leading to artificially lower absorption reading (see red dashed line in Figure 3-11.1). This will happen even if the baseline (blank) measurement was done using the same cuvette filled with buffer. For the third cuvette, the only part of the beam that has been transmitted is the open 2 mm space. The beam profile shows

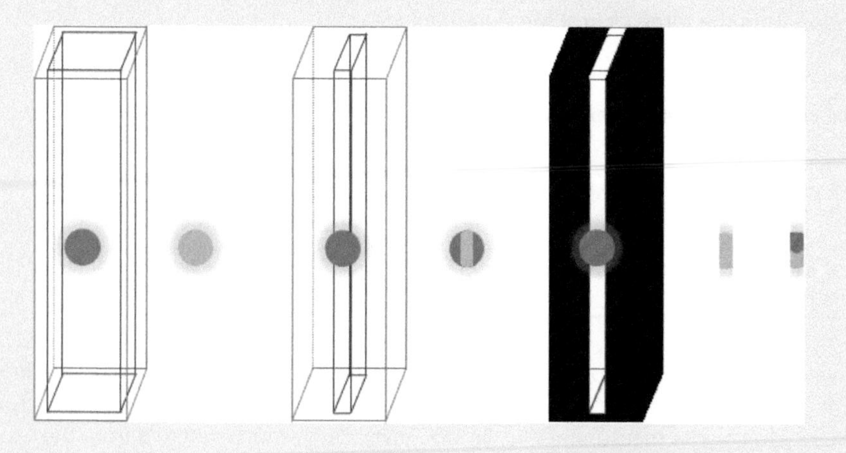

FIGURE 3-11.2 Cuvettes and schematic of beam spot before and after traveling through the cuvette.

(Continued)

EXPERIMENT 3-11 (CONTINUED) ABSORBANCE OF RHODAMINE B—PROPER USE OF CUVETTES

that the intensity is adequately attenuated and absorption measurement will give a proper result. *Important*: The baseline must be taken with the same cuvette filled with buffer and the cuvette should be positioned in exactly the same way for the blank and sample measurement. Typically the beam profile in the spectrophotometer is not uniform and attenuation will depend on the cuvette position and shifting the cuvette slightly left or right will significantly affect the reading. In such a case, it is typically very difficult to position the cuvette exactly the same way for baseline and sample measurement. A small fraction of the millimeter displacement will affect the reading. For this reason, in Figure 3-11.1, we can see a small (2–3%) difference between the expected value and measured one in an absorbance cuvette (two black walls).

CONCLUSIONS

Any absorption/transmission measurement will be corrupted if cuvettes are not used properly. It is also important to make sure that if different cuvettes are used for the sample and the reference it should be matched cuvettes (identical cuvettes). This can be checked by measuring the transmission of both cuvettes as described in Experiment 3-4. Transmission of Cuvettes—An Introduction to Absorbance.

EXPERIMENT 3-12 HOW TO ADOPT A FLUORESCENCE CUVETTE FOR ABSORBANCE MEASUREMENT

Keywords: absorbance, cuvette, spectrophotometer, rhodamine 101.

In the previous Experiment 3-11, we conducted absorption measurements using 2 mm × 10 mm fluorescence cuvettes and showed that when using the 10 mm path to measure absorption, we are making a significant error. Having small volume samples requires the use of special absorption cuvettes that have two black walls. Examples of different 4 mm × 10 mm cuvettes are shown in Figure 3-12.1. These cuvettes are available but typically expensive and not always easy to find in a typical lab. *The question becomes whether the small volume fluorescence cuvette can be used to properly measure absorption?* This protocol will show how to adopt the 4 mm × 10 mm and 2 mm × 10 mm fluorescence cuvettes to the absorbance measurements along 10 mm path. This is a generic approach and can be used with any cuvette.

MATERIALS

- Water
- Rhodamine 101 (R101) solutions
- Two 4 mm × 10 mm fluorescence cuvettes
- 2 mm × 10 mm fluorescence cuvettes

EQUIPMENT

- Spectrophotometer

(Continued)

EXPERIMENT 3-12 (CONTINUED) HOW TO ADOPT A FLUORESCENCE CUVETTE FOR ABSORBANCE MEASUREMENT

METHODS

1. Start the spectrophotometer.
2. Fill the 2 mm × 10 mm cuvette with water and measure blank using 2 mm path.
3. Fill the cuvette with R101 solution and measure absorption using 2 mm path.
4. Empty and wash cuvette. Filled with water and measure blank using a 10 mm path.
5. Fill the cuvette with R101 and measure absorption using a 10 mm path.
6. Use the same or identical cuvette and mask sides of the cuvette with black tape (stick the black tape to edges of the cuvette). See Figure 3-12.2.
7. Fill the cuvette with water and measure blank using 10 mm path.
8. Fill the cuvette with R101 solution, place the cuvette an identical way in the holder and measure absorption.
9. Compare measured absorptions in # 3, #5, and #8.

RESULTS

Perhaps most fluorescence labs will be equipped with various fluorescence cuvettes but much less with specialized absorption cuvettes. These cuvettes (two black walls) are relatively expensive and can only be used for absorption. In Figure 3-12.1, we are presenting an example of three different (4 mm × 10 mm) cuvettes filled with R101 solution.

While the left and right cuvettes are relatively easy to find in the fluorescence lab, the middle one is more difficult. But we can relatively easy convert any fluorescence cuvette to an absorption cuvette. It is important to realize that the sidewalls for absorption measurement should not be transparent or even partially transparent as in the frosted cuvette. To ensure the sidewall will not transmit any light, it is sufficient to block them with strips of black electric tape (we can use any other black materials). Figure 3-12.2 shows the picture of 4 mm × 10 mm and 2 mm × 10 mm cuvettes made in this way. On the left is presented unmasked cuvette and on the right black tape masked cuvette. The 2 mm cuvette is transparent why the 4 mm has frosted edges. It takes a little bit of patience but it is relatively quick

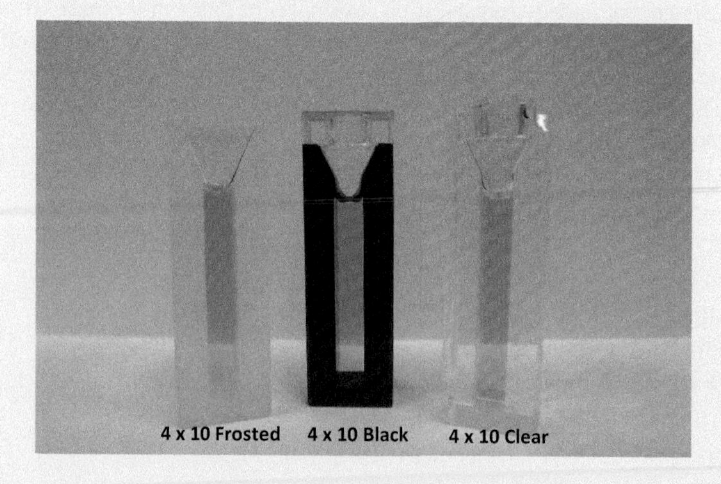

FIGURE 3-12.1 Examples of three types of 4 mm × 10 mm cuvettes filled with R101 solution.

(*Continued*)

EXPERIMENT 3-12 (CONTINUED) HOW TO ADOPT A FLUORESCENCE CUVETTE FOR ABSORBANCE MEASUREMENT

FIGURE 3-12.2 Fluorescence cuvettes adapted for absorption measurement along a 10 mm path. Top. 4 mm × 10 mm frosted fluorescence cuvette unmasked (left) and masked (right). Bottom. 2 mm × 10 mm clear fluorescence cuvette unmasked (left) and masked (right).

and an inexpensive approach. *Important*: Since the spectrophotometer beam is not perfectly parallel it is crucial to put black strips also on the opposite side (not well visible on the photography). The reason for that is that parts of the beam may enter the cuvette glass after passing through the front wall and without the black strips on the backside will exit unaffected leading to a slightly lower absorbance reading. We want to stress that this procedure does not damage the cuvette. The strips can be easily removed and the cuvette can be washed.

When taking an absorption measurement with the masked cuvette (blank and sample measurement), we have to make sure the cuvette is well-positioned. So, we recommend that after taking the blank with water to take out cuvette (with water) and put it back and re-measure absorption. The result should be a flat line zero line (see in the earlier experiment Figure 3-10.1). The deviations should be within the instrument noise. Any more significant deviation means that the cuvette was not positioned correctly (differently as it was positioned for the blank measurement). In this case, we recommend to try to position the cuvette again. As an example, we present R101 absorption measurements using 2 mm × 10 mm fluorescence cuvette (bottom cuvette in Figure 3-12.2). Figure 3-12.3 shows measured absorption in 2 mm path (black line). This is a correct measurement but absorption is relatively low. To increase measured absorption we wanted to measure along the 10 mm path. First, we filled cuvette with water (reference) and measured blank (baseline). Resulting baseline does not show any problems. Then we empty and filled the cuvette with R101 solution (without

(*Continued*)

EXPERIMENT 3-12 (CONTINUED) HOW TO ADOPT A FLUORESCENCE CUVETTE FOR ABSORBANCE MEASUREMENT

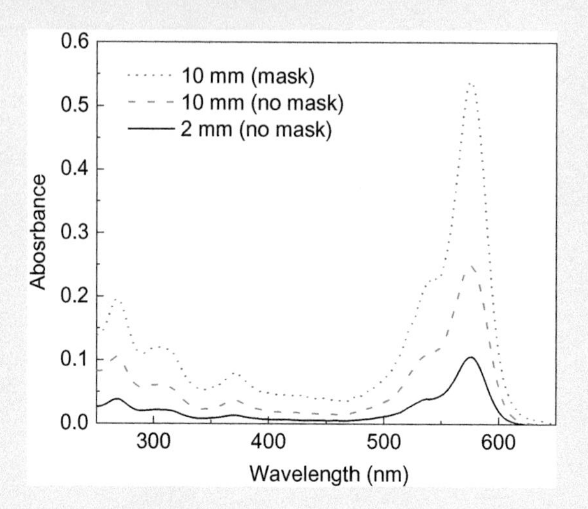

FIGURE 3-12.3 Measured absorption spectra of R101 in 2 mm × 10 mm clear fluorescence cuvette along 2 mm path (black), 10 mm path and no mask (dashed orange), and 10 mm path with masked edges (dotted blue).

taking the cuvette out of the holder). Measured absorption is presented in a dashed orange color. The maximum of absorption is about twice of that in the 2 mm path. The spectral profile is correct but the value of measured absorption is lower than expected. The third measurement we made using the masked cuvette shown in Figure 3-12.2 (bottom right). First, we measured the blank and without removing the cuvette from the holder we emptied the cuvette and filled with R101 solution. Measured absorption is shown as a blue dotted line. The absorption maximum is exactly five times of that measured in the 2 mm path.

Why the measurement along the 10 mm path with no mask gave a wrong absorption value even if we did measure a proper blank?

We used the diode-array spectrophotometer for which the beam diameter is about 5 mm. So, as shown in Figure 3-11.2 in the previous experiment, a significant part of the beam traveled through the glass when the 10 mm path has been used. In such a case, almost half of the light was pathing in the glass without interacting with the absorber and measured absorption was smaller than expected. However, the shape of the measured spectrum has been conserved since the reference measurement accounted for any potential artifacts (measurements taken with blank measured along 2 mm path will show significant spectra distortion). Masking the cuvette, we eliminated the part of light traveling through the glass. Overall we lost over 50% of total beam intensity but as it will be shown later in Experiment 3-16 this does not affect strongly precision of the measurement (it will effect if the loss would be significantly more than 90%).

CONCLUSIONS

It is possible to measure absorption along the longer path of the fluorescence cuvette but the cuvette has to be adopted by masking edges with the black tape. As shown in Figure 3-12.3, without masking the cuvette, the error can be significant. Another option is to limit the beam size by putting the slit in the optical path. As long as the beam width is smaller than the

(*Continued*)

EXPERIMENT 3-12 (CONTINUED) HOW TO ADOPT A FLUORESCENCE CUVETTE FOR ABSORBANCE MEASUREMENT

cuvette width the measurement will be correct. Also, many spectrophotometers use a much smaller beam size (about 2 mm) and using 4 mm width cuvette is typically not a problem. However, measurements with 2 mm or less cuvette are always problematic.

Note: To avoid problems with the cuvette positioning, we recommend to not remove the cuvette from the sample holder. If possible, empty the cuvette using a pipette (with the long tip) and file it with the sample.

EXPERIMENT 3-13 DETERMINING THE CONCENTRATION OF CHROMOPHORES IN A SOLUTION FROM KNOWN OPTICAL DENSITY (ABSORPTION)

Keywords: absorbance, concentration, spectrophotometer, fluorescein.

The extinction coefficient is a characteristic parameter for each fluorophore/chromophore in a specific solvent. It is important to remember that an extinction coefficient is specific to the solvent typically at room temperature (temperature dependence is relatively small). So, many experiments can take advantage of it and evaluate the concentration of a given substance (chromophore) just from a measurement of absorption. This becomes very useful for many experiments where losses due to the biological/physiological treatment are very difficult to control but which need to evaluate a final dye concentration in the product. Also, most companies providing dyes specify the extinction coefficient at the absorption maximum, so when making a solution of the dye it is easier to measure absorption and calculate its concentration than precisely weighing a very small amount and making the solution. The concentration of chromophores in a solution is given as:

$$C = \frac{A(\lambda)}{\varepsilon(\lambda)l}$$

Because C and l are constants, the shape of the optical density graph will look like the absorbance spectrum, identical shape with different units on the dependent variable scale. The absorbance, A, is measured experimentally. The extinction coefficient, ε, can be taken from literature or calculated using an absorbance spectrum of a solution with a known chromophore concentration. In this protocol data from Seybold, P.G., M. Gouterman and J. Callis (1969b) *Calorimetric, photometric and lifetime determinations of fluorescence yields of fluorescein dyes*. Photochem. Photobiol. 9, 229–242 is used for the extinction coefficient of fluorescein shown below.

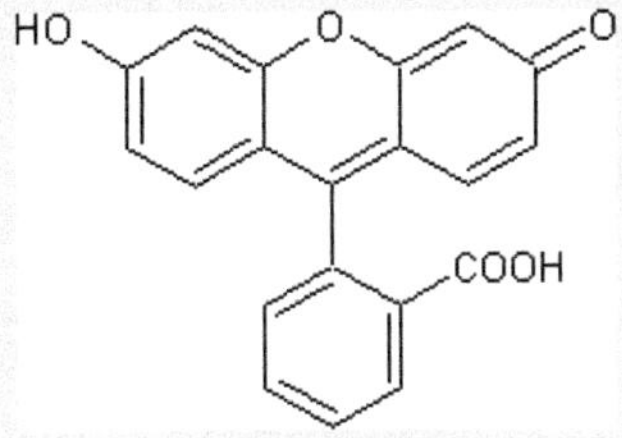

This protocol describes the way for obtaining the concentration of a chromophore in solution from an absorbance spectrum.

(Continued)

EXPERIMENT 3-13 (CONTINUED) DETERMINING THE CONCENTRATION OF CHROMOPHORES IN A SOLUTION FROM KNOWN OPTICAL DENSITY (ABSORPTION)

MATERIALS

- 1 cm × 1 cm cuvette
- Fluorescein
- Ethanol (EtOH)

EQUIPMENT

- Spectrophotometer

METHODS

1. Prepare a stock solution of fluorescein in ethanol with an arbitrary concentration.
2. Use ethanol to measure the blank with a 1 cm path length cuvette (path is referred to the length through which the beam travels).
3. Measure the absorbance of the solution in the same (or matched) 1 cm cuvette. We expect for the fluorescein solution absorption in the good range of the spectrophotometer (0.1–1). If the measured absorption is too high dilute the solution with ethanol.

RESULTS

The Seybold et al. are reporting standard extinction coefficient spectrum of fluorescein and in this experiment, we will use data from that report.

Figure 3-13.1 shows the measured absorbance of fluorescein in EtOH. As expected, the absorbance line shape in Figure 3-13.1 correspond to the shape reported by Seybold et al. The absorbance is plotted on the same independent axis and the measured absorbance data

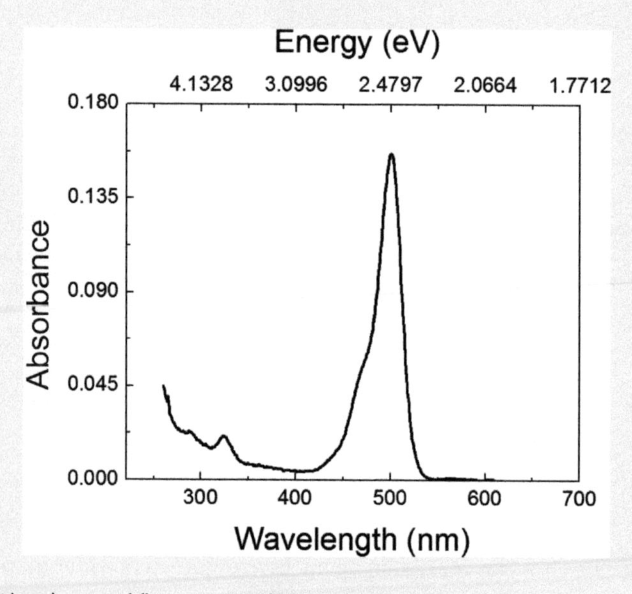

FIGURE 3-13.1 Absorbance of fluorescein solution of unknown concentration.

(Continued)

EXPERIMENT 3-13 (CONTINUED) DETERMINING THE CONCENTRATION OF CHROMOPHORES IN A SOLUTION FROM KNOWN OPTICAL DENSITY (ABSORPTION)

does not have data for all the wavelengths presented in the literature. This is not a problem, the wavelengths of interest vary among applications and typically a single data point at the absorbance maximum is sufficient for the concentration calculation.

It is enough to do the calculation presented in the introduction of this experiment only once. It is usually best to use the data point with the maximum absorbance, which should correspond to the maximum value of the molar extinction. *Caution: It is important to make sure that the maximum absorbance is in a linear range of the spectrophotometer and is not too high (see Experiment 3-8).* The percent error in high-intensity data is usually lower than at other data points. Table 3-13.1 presents the found extinction, measured absorbance, and calculation to find the concentration at four different wavelengths. The wavelength of the maximum absorbance, 499 nm, is highlighted.

This calculation is a function of experimental data from two different experiments, done decades apart (our experiment and experiment done by Seybold et al.). Thus, it is not surprising that the calculation at every wavelength is not exactly the same. The presented discrepancy is within 0.1 μM (below 10%).

It should be noted that it is possible to find the concentration for every point in an absorbance spectrum. In theory, the ideal experiment would produce a horizontal line on vertical concentration axes. Figure 3-13.2 shows a graph of calculated concentration at every wavelength for the absorbance spectrum in this section, along with the absorbance spectrum. There are three regions on the graph: peaks below 345 nm, the group of peaks between 430 nm and 545 nm and the gap in-between. The calculated concentration is almost a horizontal line in the regions where there are high peaks. The dashed line in Figure 3-13.2 is the calculated concentration from this section; it fits the line in the high wavelength absorbance peak well. For the wavelength range, 340–480 nm calculations are noise and are significantly off from the expected value. There are a number of reasons for that. Absorption in this range is very small and contributions from any impurities or dye degradation products will have the biggest effect. Also, some differences in solvent (buffer) purity may affect calculated optical density. We purposely used relatively low

TABLE 3-13.1 Calculated Fluorescein Concentrations at Different Wavelengths

Wavelength (nm)	Extinction ($cm^{-1}M^{-1}$) from literature	Absorbance (a.u.) from experiment	Concentration (M) from calculation
515	32,300	0.06257	$\frac{0.06257}{32{,}300\ cm^{-1}M^{-1} \times 1cm} = 1.9\ \mu M$
499	91,829	0.15574	$\frac{0.15574}{91{,}829\ cm^{-1}M^{-1} \times 1cm} = 1.7\ \mu M$
470	30,506	0.05652	$\frac{0.06257}{30{,}506\ cm^{-1}M^{-1} \times 1cm} = 1.8\ \mu M$
280	10,846	0.02029	$\frac{0.06257}{32{,}300\ cm^{-1}M^{-1} \times 1cm} = 1.8\ \mu M$

(Continued)

EXPERIMENT 3-13 (CONTINUED) DETERMINING THE CONCENTRATION OF CHROMOPHORES IN A SOLUTION FROM KNOWN OPTICAL DENSITY (ABSORPTION)

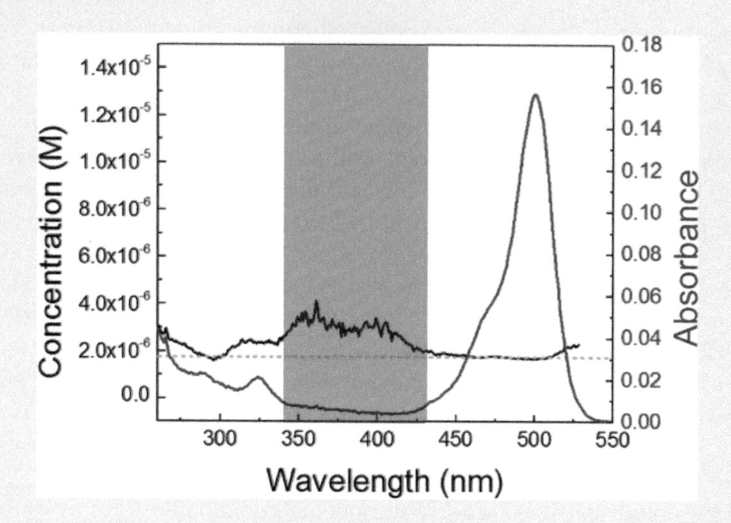

FIGURE 3-13.2 Absorption spectrum of fluorescein with calculated value of concentration.

concentration so the absorption maximum is below 0.2 and the absorption readings in the 340–480 nm range (area in-between the two peaks shaded in red) are below 0.01 what may contribute to the noise and error. Since fluorescein does not absorb in that region the noise level in the current experiment and the noise level from Seybold et al.'s experiment from 1969 is a factor. For this reason, fitting a concentration value from the entire spectrum is not useful. Increasing the concentration 5-fold would greatly improve noise in our experiment but we cannot change error in the original data. For these reasons, it is always best to compare higher signals (larger absorptions) because the impact of experimental noise is minimal there.

The ratio of molar extinction at different wavelengths may be used to estimate the height of absorbance at one wavelength given the absorbance of at another wavelength. See "Degree of Labeling of Avidin Labeled with Fluorescein, Experiment 3-22."

CONCLUSIONS

This should not be a surprise that estimated concentration is best approximated at the spectral range where absorptions are larger. In the spectral range 425–525 nm, the concentration plot is a flat line. Below 425 nm, the error is significant. The extinction coefficient of fluorescein is small and any potential impurity will have a significant impact. Also at a longer wavelength above 525 nm, we can see some discrepancies. This is where the absorbance has a very steep slope and reading of absorption values can be less precise.

Note: The extinction coefficient can be estimated independently from the known molecular weight of the dye and weighting an accurate amount of dye with a precise balance.

EXPERIMENT 3-14 ABSORBANCE OF BENZENE—RESOLVABLE VIBRIONIC STATES

Keywords: benzene, spectrophotometer, benzene, heptane.

Benzene is a common hydrocarbon with a well-studied absorption spectrum which is highly structured similarly as typical aromatic compounds. Positions of individual absorption peaks are well defined in common solvents. Also, benzene evaporates very quickly at room temperature, filling the cuvette with benzene vapor. Absorption and emission of benzene vapor are well defined and can be quite significant and easy to measure. This can be a useful reference for checking the calibration and the spectral bandwidth of a spectrophotometer. Heptane is a common non-polar solvent that works well for benzene. Their chemical structures are presented below with benzene on the left and heptane on the right.

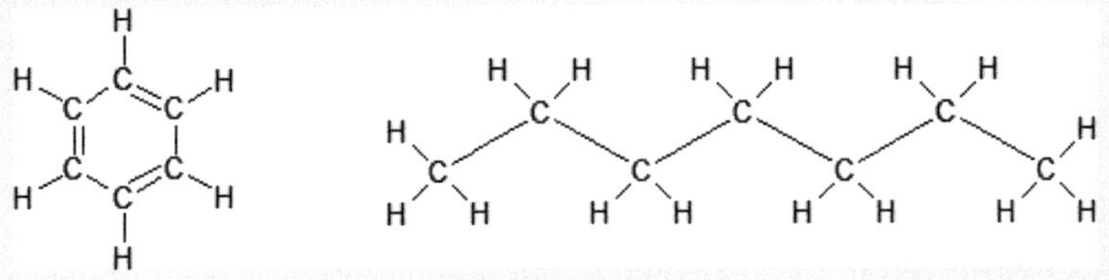

This protocol describes measuring the absorbance of benzene vapor in air and benzene dissolved in heptane.

It may come as a surprise for even experienced researchers that a small drop of benzene in the closed cuvette will quickly evaporate and present an excellent absorption spectrum. This is a very well-structured spectrum with multiple vibrational bands. Such well-defined narrow bands can sometimes be used to calibrate/check the wavelength calibration of the spectrophotometer.

MATERIALS

- Benzene
- Heptane
- Standard 1 cm × 1 cm quartz cuvette with Teflon cap.

EQUIPMENT

- Spectrophotometer

METHODS

1. Make a baseline measurement with an empty cuvette.
2. Put a drop of benzene into the cuvette and close tightly with the cap (*be careful and do not breath benzene vapors since this is a toxic substance*). Wait for the drop to evaporate (if it is cold in the room, hold the cuvette in your hand to slightly warm it up).
3. Insert the cuvette with benzene and measure absorption.
4. Add a small amount of benzene solution to heptane.
5. Measure the baseline with the cuvette filled with heptane.
6. Add the solution of benzene in heptane into the cuvette and measure absorption.

(Continued)

EXPERIMENT 3-14 (CONTINUED) ABSORBANCE OF BENZENE—RESOLVABLE VIBRIONIC STATES

RESULTS

The absorbance of benzene vapor in the air is presented in Figure 3-14.1. The features of interest lie between 230 nm and 270 nm and below 218 nm (but we will not consider the far-UV part).

First, the absorption spectrum from a small drop of benzene should be quite good. The measured absorption in the long-wavelength band (230–270 nm) should be about 0.5 (but it could be less or more—as long as it is between 0.2 and 0.8, we would not recommend any adjustments). It could be seen that at short wavelength range ~200 nm is over the range. *Note: The absorption value is below 3 and we already observe a significant peak deformation (compare to Experiments 3-7 and 3-8). In the deep UV efficiency of the UV lamp is much lower and relative beam intensity is much lower compromising the spectrophotometer linear range to below 2.* The absorbance spectrum of benzene shows two groups of peaks, there is one distinct spectral shape at 200 nm (that we will not consider) and a group of about seven distinct peaks centered at 250 nm. This group of peaks coincides with different vibrational energy transitions and one electronic transition, $\Delta\mathcal{E} = \mathcal{E}n + \mathcal{E}v$. Not all 30 vibrational modes of benzene are observed in this experiment. But it is rare to have such distinct vibrational modes; this means that the spacing between vibrational levels, $\Delta\mathcal{E}n$ is large enough to be resolved by an absorbance measurement at room temperature.

Figure 3-14.2 represents an energy diagram for the ground and first excited state (Jablonski diagram within Franck-Condon quantum oscillator potentials) that relates the observed absorbance spectrum to the energy states of benzene. The measured quantity, $\Delta\mathcal{E}$, is represented by the length of the blue arrows. The electronic energy states are represented as quantum oscillator potentials in Figure 3-14.2. The electronic ground state is the green (bottom) potential and the first excited state is the red (top), the vibrational energies are represented as horizontal lines inside the oscillator potentials. The region outside the potentials cannot support states. What is measured is the difference in energy between the initial and final states; Figure 3-14.2 shows only transitions from the ground state which does not have

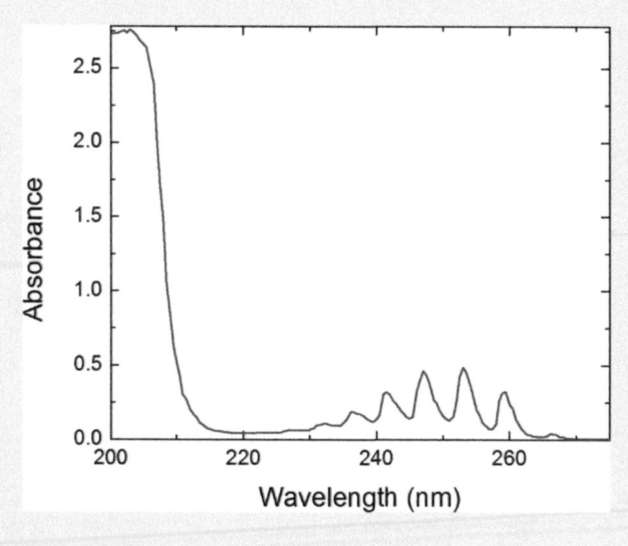

FIGURE 3-14.1 The absorption spectrum of benzene vapor.

(Continued)

EXPERIMENT 3-14 (CONTINUED) ABSORBANCE OF BENZENE—RESOLVABLE VIBRIONIC STATES

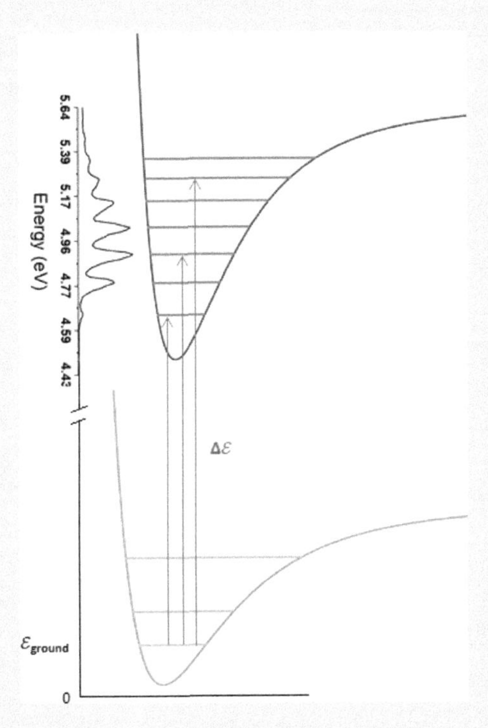

FIGURE 3-14.2 Energy diagram for absorption transitions of benzene.

zero energy. Details and labels of states of benzene transitions may be found elsewhere. The absorbance spectrum in Figure 3-14.2 (vertical) is the same as that of Figure 3-14.1 but rotated, scaled, and units converted to electron volts. The distance between the peaks in the absorbance spectrum corresponds to the separation energy of the vibrational states of benzene, both are shown on the same scale and aligned. Often the separation of the vibrational states is too small to be resolved in experiments. The peak at 200 nm in Figure 3-14.1 is the transition to the second excited electronic state, not shown in the Jablonski diagram in this section. It does not have readily resolvable vibrational states. This Jablonski diagram looks somewhat different than the diagram in Figure 1.3.1 but the measured quantity and scales are reading the same. If the quantum oscillator potentials were removed and the horizontal line lengths would be made all equal length the diagrams Figure 3-14.2 would look like a classic Jablonski diagram. The measured quantity, the vertical distance between two energy states, is unaltered by the representation of the energy states.

The energy of vibrations for a simple harmonic oscillator is often a good enough approximation for molecular vibrations, it is given by:

$$\mathcal{E}_v = \hbar\omega\left(n + \frac{1}{2}\right) = h\nu\left(n + \frac{1}{2}\right)$$

where ω is the angular frequency of vibrations and ν is the frequency of the vibration. This relation is valid per vibrational mode. The diagram in Figure 3-14.2 does not show

(Continued)

EXPERIMENT 3-14 (CONTINUED) ABSORBANCE OF BENZENE—RESOLVABLE VIBRIONIC STATES

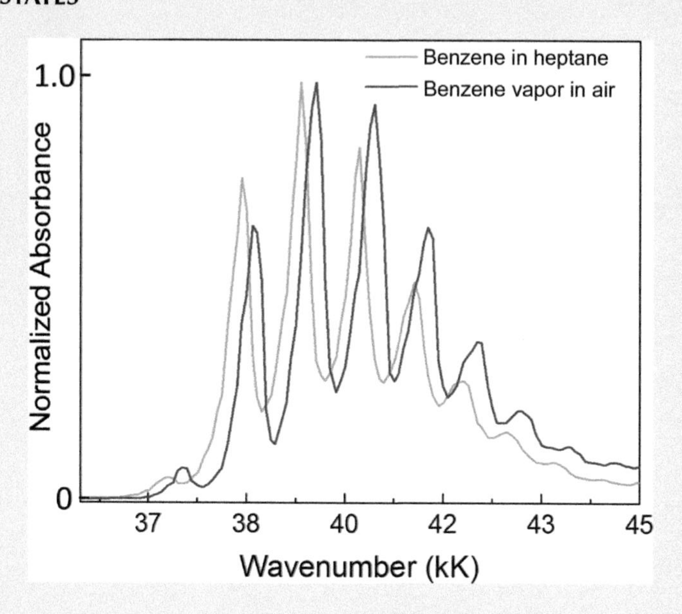

FIGURE 3-14.3 Measured absorption of benzene in vapor and heptane.

different potentials for different modes. The frequency of vibrations depends on both the state of the molecule and the medium it's in. However, the normal modes of oscillation remain the same.

Figure 3-14.3 shows the absorbance spectra of benzene in a gas phase in air and benzene dissolved in heptane. The same vibrational peaks with the same separations are presented in both spectra, but the groups of the peaks are shifted relative to each other.

In Figures 3-14.1 through 3-14.3, we purposely present absorption data on three different scales of the dependent axes, wavelength (nm), energy (eV), and wavenumber (kK). It is useful to have the same scale in the spectrum and Jablonski diagram because it allows for a direct comparison of the process responsible for the experimental result. Different fields also have different preferences for units and benzene is a very simple, important, and often used the substance. It is a common mistake to assume that the benzene in air spectra of Figures 3-14.1 and 3-14.3 are different. But it is important to remember that the dependent axes in the two spectra are inversely proportional to each other and the exact same data points are presented in both graphs (Figure 3-14.1 covers larger spectral range). Table 1.1 shows the appropriate conversions for the independent axes units.

CONCLUSIONS

A simple use of well-structured emission of benzene vapor can be used to check and calibrate your spectrophotometer monochromator. Measured positions of main peaks will depend on the wavelength resolution of the spectrophotometer, temperature, and other factors. The approximate positions of the three strongest absorption peaks are about 260, 254, and 248 nm.

EXPERIMENT 3-15 MEASURING THE TRANSMISSION OF FILTERS

Keywords: filter, band-pass filter, interference filter, spectrophotometer.

Filters are very important in many fluorescence experiments. Often they are used to block excitation light from reaching the detector or isolating/selecting a range of wavelengths which can reach a detector. If there is no grating or prism monochromator in the beam path, filters may be the only means of controlling which wavelengths of light reach the detector, a situation common in microscopy. But even if we have a monochromator on the detection line, filters are important supporting elements for cleaning the signal. Today, the most commonly used monochromators are grating (often double grating) monochromators that will transmit a significant signal at harmonics (second harmonics) and will also transmit a small amount of light off the selected wavelength. This is normally not a problem since second harmonics is well separated (double of wavelength, so setting a monochromator to 300 nm will transmit a significant amount of 600 nm light) and the intensity of the leaked light is typically only a few percents. But when measuring very weak emissions that uses strong excitations this could be a problem.

Also, it is important to realize that real filters have imperfections which allow some unwanted light to be transmitted or distort the spectral shape of emission from a sample. Since filters are frequently used to select emission wavelengths range (band path) for time-resolved (lifetime), measurements and such an unexpected marginal transmittance in different regions can be a big problem. Therefore it is very important to know the exact transmittances of individual filters and filter's compositions. This has to be tested experimentally.

As an example, a few different filters were selected, as shown in Figure 3-15.1. This protocol describes testing the transmission of various filters and how to design filter composition to eliminate various unwanted transmittance ranges.

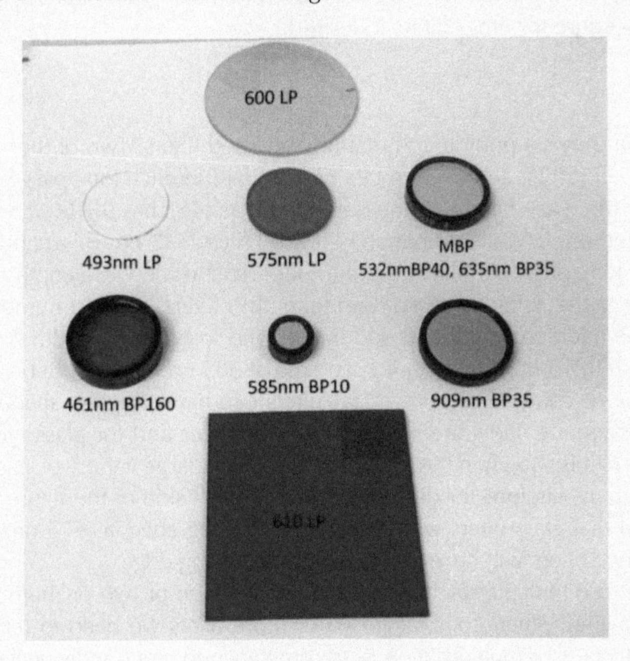

FIGURE 3-15.1 Set of selected filters.

(Continued)

EXPERIMENT 3-15 (CONTINUED) MEASURING THE TRANSMISSION OF FILTERS

MATERIALS

- Optical filters

EQUIPMENT

- Spectrophotometer

METHODS

1. Measure air as a reference baseline.
2. Place a filter in the beam path (see Experiment 3-1).
   a. Make sure the plane of the filter is normal to the beam path. Tilting an interference filter will lead to a significant shift of transmittance. This could be occasionally useful when we want to shift a transmittance band.
   b. If the spectrophotometer does not have a mount for solid samples it is best to use a filter holder which attaches to an optics post. Most filters will not fit in a cuvette holder.
   c. Using double-sided sticky tape on the cuvette holder to attach a filter is convenient, but this often leaves adhesive residue on a filter if it is taped directly. This residue not only scatters and absorbs light but it also emits light.
3. When the filter is mounted measure the transmission.
4. For a quick filter test in many spectrophotometers that are not affected by the external light, it is convenient just to hold the filter in the optical path while measuring. We recommend holding filters by the edges to avoid making the filter surface dirty.
5. Repeat procedure for other filters if desired.

RESULTS

Figure 3-15.1 here shows a photograph of a few selected filters. Two of these filters are glass long pass filters (575 nm LP and 610 nm LP), two are interference long pass filters (493 nm LP and 600 nm LP), three are bandpass interference filters (461 nm BP160, 585 nm BP10, and 909 nm BP35), and one is multiple bandpass (MBP) filter (532 nm BP40 and 635 nm BP35). Typically, for the long pass (LP) filters, the number refers to the wavelength of 50% transmittance of the filter at the rising edge. For bandpass (BP) filters, the first number relates to the wavelength in the middle of the band and the second to half-bandwidth of the transmitted band (e.g., 585 nm BP10 is the filter band centered at 585 nm and 10 nm broad as measured at half maximum). As shown in Figure 3-15.1 filters can have different shape and size. Most common for interference filters are round 1-inch diameter and for glass type filters square 2 × 2 inches. But as it is shown in the figure, we can find a large variety of sizes. Figure 3-15.2 shows measured transmissions for different filters. The left side of the figure show long-pass filters. It is typical that glass filters will have a slower rising edge (as 575 nm LP and 610 nm LP) and interference filters will have the sharp quickly rising edge.

It is possible to use multiple stacked filters (a combination of two or more filters) to cut off unwanted wavelengths. When combining two or more filters we need to remember that the resulting transmittance is a multiplication of transmittance for each individual filter. For example, putting together two filters with transmissions at a given wavelength 80% and 90% the

(Continued)

EXPERIMENT 3-15 (CONTINUED) MEASURING THE TRANSMISSION OF FILTERS

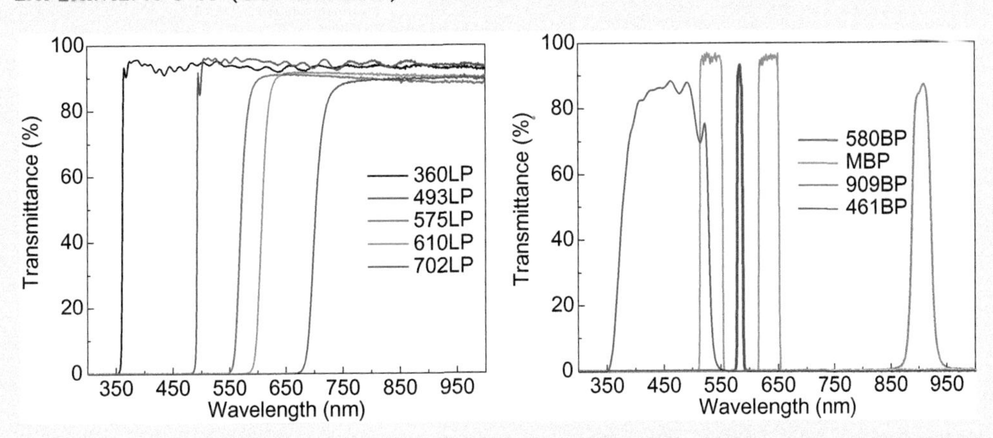

FIGURE 3-15.2 Measured transmissions for a long pass (LP) filters (left) and various bandpass (BP) filters (right).

final transmittance will be multiplication (0.8 × 0.9 = 0.72) yielding about 72% transmittance. Figure 3-15.3 shows two different combinations of filters. Left, MBP filter (red dashed line) and 575 LP filter (green dashed line) and resulting transmittance of this combination is shown in black. A shorter wavelength band 532 nm is not transmitted by a used 575 LP filter and we can clearly see only a single 635 nm band. The right side is a combination of 461 BP filter with 493 LP filter. The resulting transmittance (in black) is a much narrower band centered at about 515 nm.

There are several important aspects to observe in Figures 3-15.2 and 3-15.3 which should always be considered when using filters.

- No filter will have 100% transmission. This should come as no surprise since a quartz plate does not have 100% transmission due to reflections (see Experiment 3-1). As presented in Experiment 3-1 just a reflection from a dielectric surface reduces the

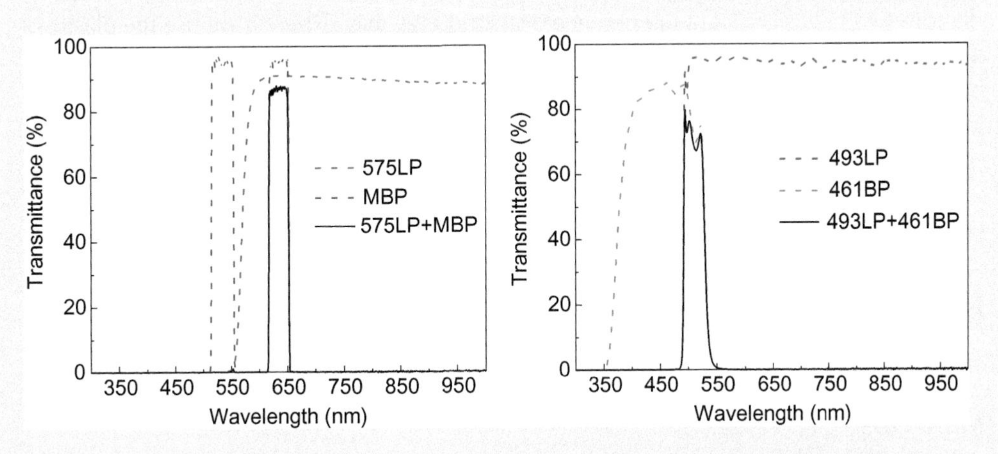

FIGURE 3-15.3 Combinations of two different filters and resulting transmittance. Left. Multiband (540 and 640 nm) crossed with 575 nm LP filter. Resulting transmittance shown in black. Right. Broad bandpass (375–545 nm) filter crossed with 493 nm LP filter. Resulting transmittance shown in black.

(Continued)

EXPERIMENT 3-15 (CONTINUED) MEASURING THE TRANSMISSION OF FILTERS

transparency of about 5–10%. Filters reflect light in the same way and in the case of absorbing filters (glass filters doped with dyes) the maximum transmittance is not much above 90%. The dichroic filters with an antireflection coating for which the mechanism for light reflection is different may present at very specific wavelength range with transmittance close to 100% (98–99%).

- Filter' regions which transmit light do not have the constant transmission as a function of wavelength; oscillations of varying transmittance can be present for interference filters.
- Filters may pass multiple bands and have effectively zero transmission in-between them.
- Combining filters are used to allow more selection over which wavelengths of light are passed.

Note: Special optical filters with antireflection coating can transmit almost 100% of light in the desired range.

EXPERIMENT 3-16 MEASURING UV ABSORPTION IN DIFFERENT CUVETTES

Keywords: absorption, absorbance, glass cuvette, quartz cuvette, plastic cuvette.

Many absorption measurements in biology/biochemistry are done in the UV range. Absorption of amino acids or DNA is in the range below 300 nm. We would always recommend using a quartz cuvette (the most transparent in this range) not glass or plastic. But the quartz cuvette is always much more expensive and not always accessible. Also, frequently we can use a glass cuvette by mistake since glass and QUARTZ cuvettes are very difficult to differentiate visually. Can a reliable measurement of absorption of tryptophan or tryptophan analog (*N*-acetyl-L-tryptophan amide (NATA)) be made if one does not have a quartz cuvette (or uses the glass cuvette by mistake)? This is a rare case, but in this experiment, we want to demonstrate how significant interference/distraction we may expect if we use the plastic or glass cuvette for an absorption measurement in the UV spectral range.

MATERIALS

- *N*-Acetyl-L-tryptophan amide (NATA)
- Water
- 1 cm × 1 cm cuvettes— quartz (fused silica), plastic, and glass cuvette

EQUIPMENT

- Spectrophotometer

METHODS

1. Prepare water solutions of NATA with an optical density of about 0.3.
2. Measure blank with an empty chamber.
3. Measure the transmittance of each cuvette (quartz, glass, and plastic) filled with water

(*Continued*)

EXPERIMENT 3-16 (CONTINUED) MEASURING UV ABSORPTION IN DIFFERENT CUVETTES

4. Next, measure the blank with a quartz cuvette filled with water.
5. Measure the absorbance of NATA solution.
6. Repeat 2 and 3 steps for glass and plastic cuvettes

RESULTS

Figure 3-16.1 shows the photograph of NATA solution in quartz, glass, and plastic cuvettes. Visually, in all cases the solution and cuvette are transparent and it is difficult to see the difference. In practice, telling the difference between quartz and glass cuvette is difficult even for the very experienced researcher.

Figure 3-16.2 shows measured transmittances for each cuvette. Above 350 nm, all cuvettes are well transparent and below 350 nm there are significant differences. While the transmittance of glass cuvette starts significantly dropping already below 300 nm the residual transmittance extends to 240 nm. In the case of plastic, the transmittance between 250 nm

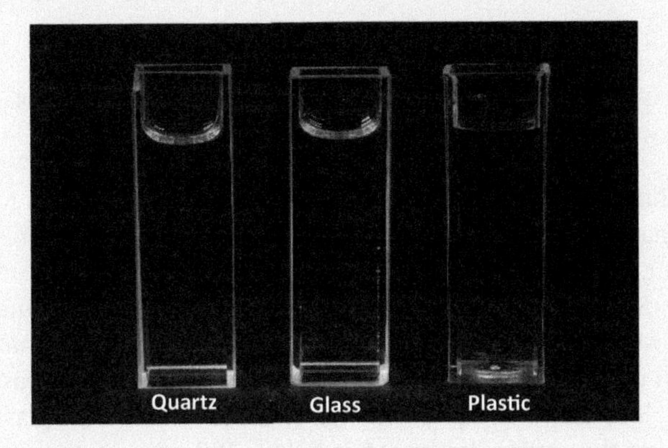

FIGURE 3-16.1 Photograph of quartz, glass, and plastic cuvette filled with NATA solution.

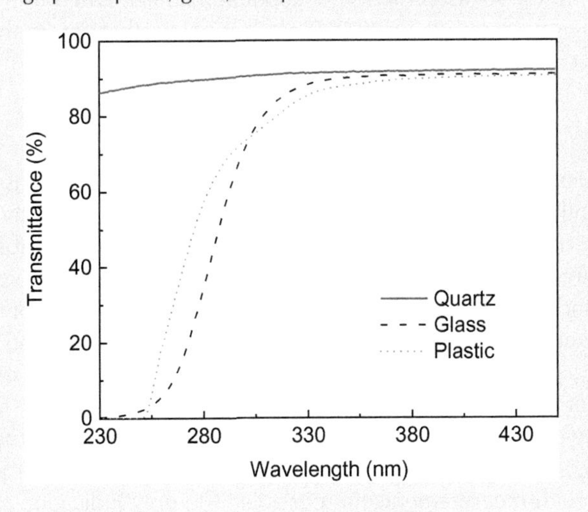

FIGURE 3-16.2 Transmittances measured for quartz, glass, and plastic cuvette.

(*Continued*)

EXPERIMENT 3-16 (CONTINUED) MEASURING UV ABSORPTION IN DIFFERENT CUVETTES

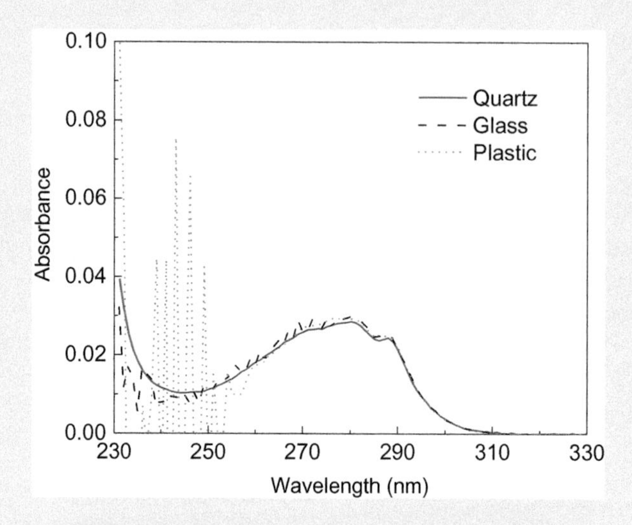

FIGURE 3-16.3 Measured absorptions of NATA solutions in quartz, glass, and plastic cuvette.

and 300 nm is higher than for glass but below 250 nm it abruptly drops to practically zero. We want to stress that the absorption range of NATA falls in the range where the transmittance of glass and plastic cuvette quickly decreases.

Figure 3-16.3 shows measured absorptions in all three cuvettes. The measurement in the quartz cuvette shows a clear absorption spectrum with only very minimal noise. The measured absorption in quartz we will consider as the standard for next measurements in glass and plastic cuvette. It could be a surprise but measured absorption in a glass cuvette matches well that measured in the quartz. In the range below 280 nm, we can observe some noise but down to 240 nm, the measurement is acceptable. Similar, when using a plastic cuvette the measurement is very good down to 260 nm but below that where the transmittance is practically zero and measured absorption becomes very noisy.

CONCLUSIONS

The presented results demonstrate that within the reasonable range measured absorption is proper, independently on used cuvette, even if in the range of wavelengths where the cuvette transmission drops dramatically. It is important to remember that an absorption measurement is a comparative measurement between the reference (cuvette and solvent) and sample (cuvette, solvent, and chromophore of interest). As long as the amount of transmitted light is sufficient (light reaching the detector is much above noise), the absorption measurement will be correct. In this experiment, we used low sample absorption and the main limiting factor was transmittance of the cuvette. So, for the plastic cuvette below 260 nm range, the intensity of light reaching the detector is low what is manifested by high noise in the measurement.

We do NOT recommend to measure absorption in UV using glass or plastic cuvette but want to show that such measurements are possible. The only indication of using the wrong type of the cuvette will be the increased noise as shown in Figure 3-16.3 but the value will be more or less correct.

EXPERIMENT 3-17 MEASURING SCATTERING OF SILICA MICROSPHERES—APPARENT ABSORPTION DUE TO OUT-SCATTERING OF LIGHT

Keywords: scattering, absorbance, silica particles, Ludox.

Occasionally, some samples present significant absorbance, sometimes even glowing (opalescence) similar to fluorescence. But the observed absorbance or emission does not indicate absorbance or fluorescence peaks. It happens when small nanometer-sized particles are in the solution. It could be small silica beads (~20–30 nm) called Ludox, often used as a scatterer for time-resolved fluorescence experiments. But it also may happen for solutions of proteins or some other nanoparticles. Typically accounting for sample scattering (scattering that originates from the sample e.g., cells, protein aggregates, etc.) can be very difficult since getting an appropriate reference is practically impossible. One needs to remember that scattering (more precisely out-scattering) of light contributes to transmitted light extinction and shows as an apparent sample absorption. For particles that are much smaller than the wavelength of light, the scattering can be purely Rayleigh-type scattering. Accounting/correcting for that is possible and can be done without prior knowledge of the exact sample absorbance. However, if the contributing scattering particles are larger (diameter > $\lambda/10$) and this is not a Rayleigh-type scattering, then separating measured light extinction to absorption and scattering component can be very difficult, or even impossible. This is especially true with a large distribution of particle sizes predicting the scattering profile (spectrum). Since the extinction of light is a convoluted effect of sample absorption and sample scattering, separating the two components without prior knowledge of one is impossible.

In the next few experiments, two examples will be presented on how to account/correct for scattering contribution. (1) when one can predict the scattering profile from Rayleigh law and (2) when one cannot assume that particles are small enough and it is necessary to use a particle reference measurement to correct for particle contribution to measured absorption. However, in the last case, it will only be an approximation since in most cases finding an appropriate reference (exactly the same sample that only scatters but does not absorb) is very difficult.

It is important to stress that such effects are more significant for short-wavelength absorbers (below 350 nm) like proteins.

MATERIALS

- Silica beads (Ludox)—suspension of 20–30 nm silica nanoparticles, 30% w/w
- Water
- 1 cm × 1 cm cuvettes
- Pipettes

EQUIPMENT

- Spectrophotometer

METHODS

1. Prepare a high concentration of Ludox solutions in water. Different lower concentrations can be prepared from a stock solution. It is important to make sure that there are no aggregates (silica particles have a tendency to aggregate during storage and

(*Continued*)

EXPERIMENT 3-17 (CONTINUED) MEASURING SCATTERING OF SILICA MICROSPHERES—APPARENT ABSORPTION DUE TO OUT-SCATTERING OF LIGHT

extensive sonication can be needed). It is difficult to judge the concentration of Ludox particles but typically higher concentration will have a bluish opalescent color. A high concentration will have an apparent absorbance of about 0.1 at 300 nm. It is important to remember that in this case, apparent absorbance is not really typical absorbance. Ludox particles do not absorb light in this wavelength range but only out-scatter the light what results in attenuation of the transmitted light beam (light extinction).

2. File the cuvette with high purity deionized water and collect blank.
3. File the cuvette with a Ludox solution and measure absorbance.

RESULTS

Possible results for couple different Ludox concentrations are presented in Figure 3-17.1. Measured spectra will not have any special features with increasing absorbance at a shorter wavelength.

It is important to be able to decide if the light extinction is only due to Rayleigh scattering. We selected one absorption curve from Figure 3-17.1 for data analysis. According to the previously discussed effect of scattering on absorbance and transmission in Chapter 1, the intensity of scattered light is proportional to $1/\lambda^4$. According to the Equation 1-9a.1 in Chapter 1, the intensity of out-scattered light, I can be described:

$$I \sim \frac{1}{\lambda^4} * \frac{2\pi^5 d^6}{3}\left(\frac{n^2-1}{n^2+1}\right)^2 = \frac{B}{\lambda^4} \tag{3-17.1}$$

where B is a constant that include all parameters in the Rayleigh equation (index of refraction, particle size, etc.). Apparent absorbance due to Rayleigh scattering can be expressed:

$$Ab(\lambda) = -\log\left(\frac{I_0 - \left(B/\lambda^4\right)}{I_0}\right) \tag{3-17.2}$$

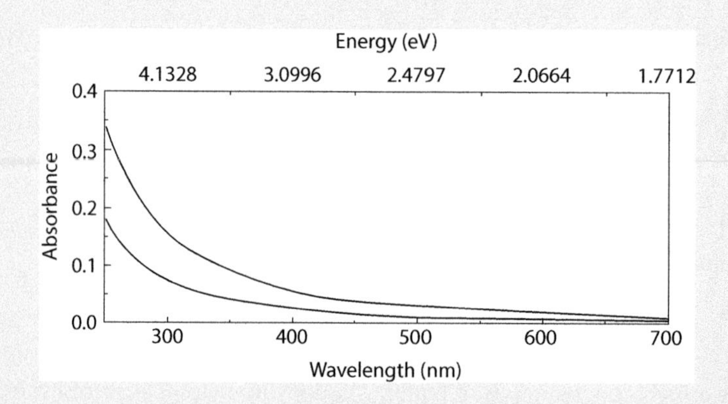

FIGURE 3-17.1 Experimental measurements of absorbance (extinction) for two different concentrations of Ludox solution.

(Continued)

EXPERIMENT 3-17 (CONTINUED) MEASURING SCATTERING OF SILICA MICROSPHERES—APPARENT ABSORPTION DUE TO OUT-SCATTERING OF LIGHT

To fit experimental data, we prefer to use a single constant, B that will be a fitting parameter. This parameter will include size distribution and typically undefined index of refraction.

We will fit experimentally measured values of absorbance (Ab_{exp}) with calculated absorbance according to Equation (3-17.1) by minimizing the error: $\sum_{\lambda}(Ab_{exp}(\lambda)-Ab(\lambda)+c)^2$. The arbitrary constant, c is to account for small baseline shifts. Similar fit can be made to measured transmittance.

For our analysis, we consider the highest concentration of Ludox particle. When trying to fit all points in the range 240–700 nm (Figure 3-17.2a), one will notice a very high standard deviation and unacceptable error distribution. The same data points fitted only within the range 280–700 nm gives a much better result, but still a significant standard error and systematic error distribution (Figure 3-17.2b). When skipping the first 120 points at UV and fitting the range 360–700 nm, the standard error is very good (below 1.1) and error distribution in the range 400–700 nm is excellent. However, in the 350–370 nm range, we observe a small bump and in the range below 300 nm, the data are completely out of range (Figure 3-15.2c).

There are two possibilities for such behavior. (1) Our particles are too large and the Rayleigh model is not valid in the shorter wavelength range. However, from our experience in such a case, the error will behave differently. (2) Our Ludox is slightly contaminated and when used in high concentrations we see residual absorption contribution from silica itself and from some organic contaminations. The small bump in the 350 nm is clearly visible in the residuals corresponds to the wavelength where the grading changes and measured points do not match perfectly.

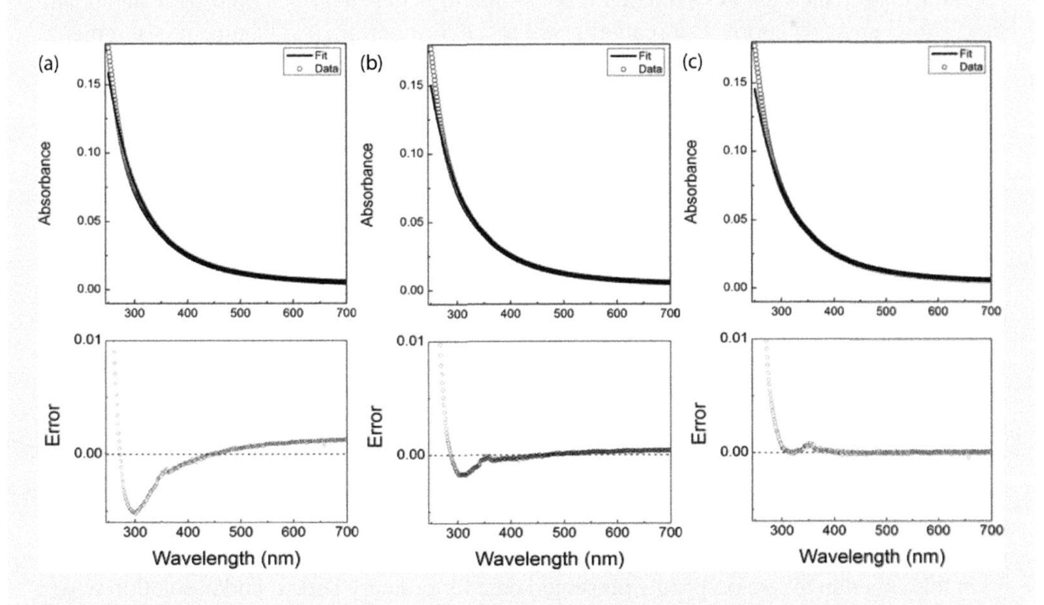

FIGURE 3-17.2 Fits to a measured absorbance when considering all 240–700 nm range (a); reducing the range to 280–700 nm (b); and reducing the fitting range to 360–700 nm (c). Lower panel presents error distribution.

(Continued)

EXPERIMENT 3-17 (CONTINUED) MEASURING SCATTERING OF SILICA MICROSPHERES—APPARENT ABSORPTION DUE TO OUT-SCATTERING OF LIGHT

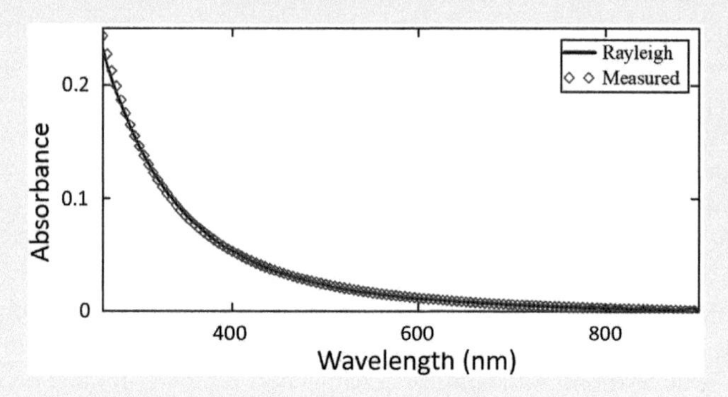

FIGURE 3-17.3 Example of fresh extra pure Ludox absorbance (open point) and fitted Rayleigh function (black line).

The fit of experimental results is excellent in the range of above 370 nm which suggests that such contamination is the main problem. Too large particles size will manifest itself with a declining contribution also in longer wavelengths. In our case, the error just explodes below 300 nm where silica and the main protein components (tryptophan, tyrosine, etc.) start contributing.

Warning. We observed many times for old solutions of Ludox that some bacteria (organic component) may grow and after some time the solution may become cloudy. For this experiment, we used our good Ludox stock, but we suspect the high increase in the absorption below 300 nm is a result of residual silica absorbance and some residual organic contamination which are very difficult (impossible) to eliminate. This could be a significant problem when using Ludox as a scattering reference for fluorescence lifetime measurement. When exciting with UV we can observe small residual fluorescence that may corrupt weak fluorescence of some compounds or a pulse response function.

In Figure 3-17.3, we are presenting absorption for a freshly prepared new Ludox solution and a fit to Rayleigh function in the 300–700 nm. A very fresh solution that has been sonicated for a few hours yields in this range an excellent fit. The behavior is very good down to 280 nm. Below 280 nm, we expect the silica particles start also absorbing light and the Rayleigh approximation will not work.

CONCLUSIONS

Presented results demonstrate that scattering can significantly alter the transmitted light intensity and in some experiments should be taken into account. We presented the experiment using ordinary Ludox that behaves well down to 350 nm. An important conclusion is that when scattering particles are small it is easy to account for their contribution just by measuring apparent absorption at the extended range where the compound of interest does not absorb and fitting the extended range with the Rayleigh function that then can be extended down to the absorption range.

In this experiment, we purposely presented data for ordinary Ludox. Ludox solution when not kept properly can very easily get contaminated that will present significant problems when using Ludox as a scatterer in lifetime measurements. The last absorbance spectrum (Figure 3-17.3) shows best Ludox we could find (very fresh) for which experimental data can be well fitted with Rayleigh function down to 280 nm.

EXPERIMENT 3-18 MEASURING ABSORPTION IN THE PRESENCE OF KNOWN SCATTERING

Keywords: Absorbance, scattering, baseline.

It is recommended that before starting this experiment, the reader is familiar with the previous experiments on measuring transmission and absorbance. In this experiment, the absorbance of the sample in a solution will be determined taking into account that it also scatters light.

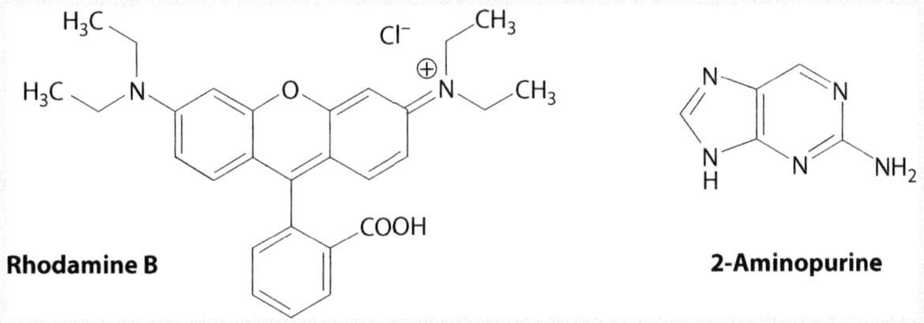

Rhodamine B

2-Aminopurine

A spectrophotometer only reports on the light that reaches its detector (measure light extinction), it does not know how or why it was attenuated. Previous experiments discussed how to measure the absorption of the chromophore in pure solvents that does not scatter. More precisely solvent scattering is negligible or can be corrected through the baseline measurement. Light attenuation due to scattering appears as an apparent absorption even if the sample does not absorb but only out-scatters the light (Experiment 3-1). In this section, we recognize that the sample or buffer scatters and use this to correctly find the true absorbance of the chromophore, $Ab(\lambda)$. Herein model experiments are presented using solutions of 2-aminopurine (UV absorber) with Ludox and Rhodamine B (yellow/orange) with microspheres.

MATERIALS

- 2-Aminopurine (2AP)
- Rhodamine B (RhB)
- Microspheres—suspension of 0.4 μm nonfluorescent beads
- Silica beads (Ludox)—suspension of 30 nm silica nanoparticles, 30% w/w
- Water
- 1 cm × 1 cm cuvettes
- Pipettes

EQUIPMENT

- Spectrophotometer

METHODS

1. Prepare solutions for this experiment using your favorite procedure(s). Ensure that the concentration of the dye is the same in solutions with and without the scatter. The best way to ensuring identical dye concentrations is to prepare a solution of the dye and split

(Continued)

EXPERIMENT 3-18 (CONTINUED) MEASURING ABSORPTION IN THE PRESENCE OF KNOWN SCATTERING

into two identical parts. Then, add the calibrated amount of concentrated solution of Ludox or beads to one part and exactly the same amount of water (buffer) to another.

a. Dye and scatterer suspended in water.
    i. 2-Aminopurine with Ludox.
    ii. Rhodamine B with microspheres.
b. Scatterer suspended in water.
    i. Ludox.
    ii. Microspheres.

2. Take a baseline using a cuvette filled with only water.
3. Measure the absorption of the dye and scatterer.
4. Measure the absorption of the scatterer alone.
5. Subtract the absorbance of the scatter alone from the absorbance of the dye and scatter in the same solution. The result is the absorbance of the dye.

RESULTS

For small particle scattering, it is wavelength dependent ($\sim\lambda^{-4}$) and plays a more significant role in the shorter wavelength region (UV) (we will discuss this in the next Experiment 3-19). In the case of biological macromolecules such as proteins, DNA, or membranes scattering may strongly affect the absorption and fluorescence measurements. In many cases, a properly measured scattering background can be subtracted from the absorption spectrum.

Figure 3-18.1 shows the absorption spectrum of 2AP and Ludox in solution when the water has been used as a baseline. The absorbance seems to be increasing in the UV and there appears to be a shoulder at about 310 nm.

Figure 3-18.2 presents the measured absorption spectrum of Ludox. It is known that Ludox in the presented range does not absorb light but only out scatters and that light does not reach the detector. In fact, Ludox is frequently used as a light scatterer and measured absorbance is only apparent absorption that is a result of light attenuation due to out-scattered light intensity. Since light absorption is additive to correct for the scattering it is possible to subtract the Ludox absorption spectrum from the 2AP with Ludox spectrum. The result is presented in Figure 3-18.3.

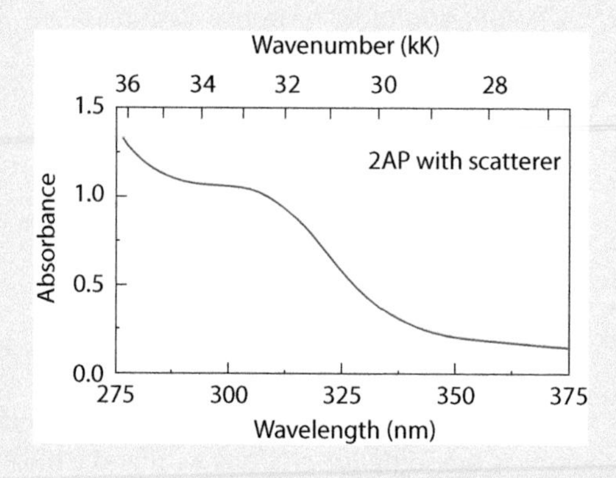

FIGURE 3-18.1 2AP absorption spectrum with Ludox, water used for a baseline.

(Continued)

EXPERIMENT 3-18 (CONTINUED) MEASURING ABSORPTION IN THE PRESENCE OF KNOWN SCATTERING

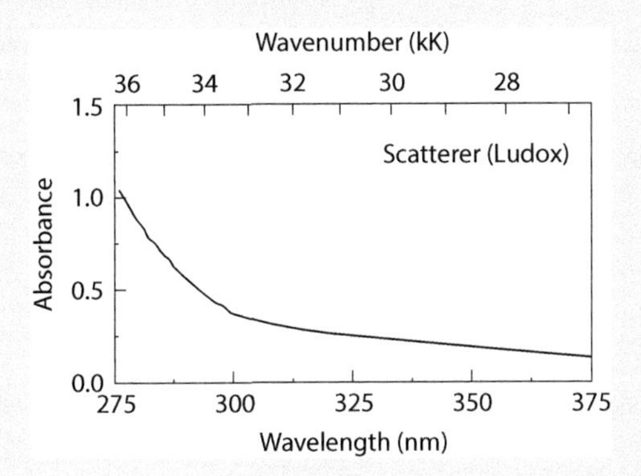

FIGURE 3-18.2 Absorption spectrum of Ludox with water used as a background.

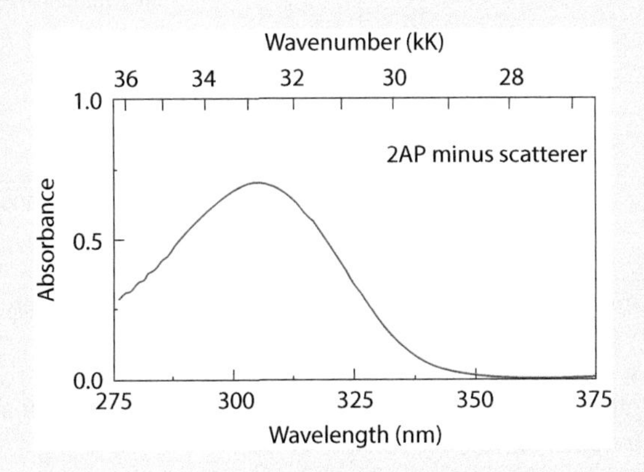

FIGURE 3-18.3 The difference of absorption spectrum and 2AP and Ludox and Ludox only absorption spectrum.

Subtracting is done point-by-point, for each value of wavelength the Ludox only value is subtracted from 2AP + Ludox value, the resulting point is plotted at the same wavelength. This is exactly the same procedure used to correct for the baseline. After correction, it becomes clear that there is one dominate absorption peak at 310 nm and the spectrum drops off as it goes into the UV.

Below the same procedure is demonstrated for rhodamine B with microspheres (Figure 3-18.4).

In the case of not strongly scattering solutions, the scatterer can be directly used as a reference for the baseline, but this is a rare case. Typically we do not have an adequate scattering reference solution as in this experiment. For example, measuring a solution of a protein that is very large (or contains aggregates) or dye embedded into beads that are commercially available and we do not have a proper reference from dye-free beads. If a proper scatterer is not available one can be mimicked with Ludox or beads. In the case of small particles (much smaller than the wavelength of light), scattering can also be described and predicted theoretically. We recommend this only as a last resort. Using an improper baseline may cause serious problems in

(Continued)

EXPERIMENT 3-18 (CONTINUED) MEASURING ABSORPTION IN THE PRESENCE OF KNOWN SCATTERING

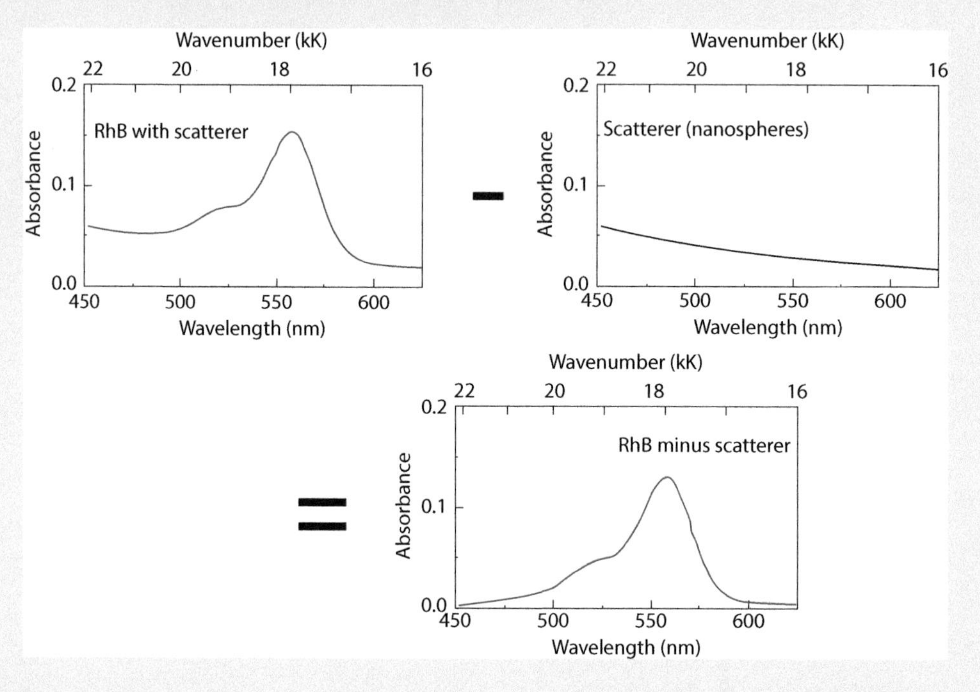

FIGURE 3-18.4 The absorbance spectrum of rhodamine B with microspheres corrected with the absorbance spectrum of only microspheres.

experimental results interpretation. If the experimenter has a knowledge that the particles diameter, d, is much smaller than a wavelength of light ($d < \lambda/10$) the scattering can be predicted using Rayleigh law and measuring absorption in the range that extends significantly over the range of chromophore absorbance as presented in following experiments. See, next Experiment 3-19.

EXPERIMENT 3-19 MEASURING ABSORPTION IN THE PRESENCE OF SMALL SCATTERING PARTICLES

Keywords: absorbance, scattering, baseline, rhodamine 110, quinine sulfate.

In Experiment 3-17, measurements of light extinction due to small silica particles scattering were discussed. For wavelengths over 300 nm, the measured absorption closely followed the Rayleigh scattering law. For shorter wavelengths, the small impurities and silica absorbance started to contribute leading to significant deviations from theoretical predictions as presented in Experiment 3-17. Also, Ludox particles are about 20–30 nm size, and below 300 nm a small divergence from the simple $1/\lambda^4$ dependence would be expected. This experiment will test how to recover the true absorption spectra without prior knowledge of the sample scattering profile. The only assumption is that scattering particles are much smaller than the wavelength of light. Two dyes will be used; rhodamine 110 (R110) and quinine

(Continued)

EXPERIMENT 3-19 (CONTINUED) MEASURING ABSORPTION IN THE PRESENCE OF SMALL SCATTERING PARTICLES

sulfate (QS). The R110 absorbs in the visible range while QS absorption is in the UV range. This experiment tests whether true (or approximate) absorption of a dye can be obtained in the presence of scattering particles when the only assumption is that particles are small. The following experiments will discuss this approach and how it could be useful when measuring the absorption of proteins or noble metal clusters that are relatively small particles but may significantly scatter light as compare to its absorption.

MATERIALS

- Rhodamine 110 (R110)
- Quinine sulfate (QS)
- Silica beads ~30 nm diameter (Ludox)—suspension in water
- 0.4 cm × 1 cm cuvettes
- Pipettes
- Sonicator and water

EQUIPMENT

- Spectrophotometer

METHODS

1. Prepare rhodamine 110 (R110) and quinine sulfate (QS) solutions in water (about 5 mL each) for this experiment using your favorite procedure(s). *For QS solution, we will not use usual acidic conditions but just water what results in a small short-wavelength shift of absorption.* Ensure that the concentrations of dyes are small so the absorptions in the peak are about 0.2 for each of them.
2. Prepare the Ludox solution (about 10 mL). Make sure that the concentration of Ludox is high enough to result in a measured absorption (light extinction) at 300 nm of about 0.1. Sonicate the solution for 60 min. Good sonication is important since particles tend to aggregate.
3. Place about 2 mL of Ludox solution in three separate vials.
4. Add to the first vile 1 mL of R110 solution, to the second 1 mL of QS solution, and to the third 1 mL of water.
5. Take a baseline using a cuvette filled with water only.
6. Measure the absorption of each dye solution (only in water). These will be our reference spectra. We will use spectral range 280–700 nm. The reader will notice that even for long-wavelength absorbing dye, this range is excessive. Absorption of R110 ends at 550 nm and the QS absorb in short (UV) wavelength range.
7. Measure the absorption of three solutions prepared in step 4 (solution of each dye-containing scatterer and only Ludox diluted with water) in the same 280–700 nm range.
8. To subtract the contribution of scatterer we will use two approaches
   a. First, we will use the same approach as in the previous Experiment 3-18. From the measured absorption of the dye in the presence of Ludox (mixture), we will subtract Ludox absorption as measured for a reference solution (Ludox solution proportionally diluted). The only difference is that this time we are using lower dye concentrations (absorption in the range of 0.1).

(*Continued*)

EXPERIMENT 3-19 (CONTINUED) MEASURING ABSORPTION IN THE PRESENCE OF SMALL SCATTERING PARTICLES

b. Second, we will test if we can extract true (approximate) absorption of a dye from measured absorption of dye-scatterer mixture only. For that, we will use the absorption measured in the long wavelengths range where dye does not absorb and light extinction (apparent absorption) is only due to light out scatted by particles. In this case, we do not know true scattering profile but we expect the particles are much smaller than the wavelength of light and we can approximate scattering pattern with the Rayleigh equation. Using Rayleigh parameters obtained from the fit at long wavelengths range, we extend the scattering profile spectrum to shorter wavelengths and subtract the theoretically predicted apparent absorbance due to the scattering from the absorbance of the dye-scatter mixture.

In many real experiments, we will not be able to measure background from scattering like measuring the absorption of proteins scattering of protein molecules can be significant but we do not know its real contribution—it will depend on overall protein size and amino acid composition. But if we can assume that our system should only absorb in a shorter wavelength, we can measure absorption extending the range to longer wavelengths and utilize a non-absorbing range to estimate the contribution of scattering to apparent light extinction. The following results show how much we can trust recovered spectra when using this simple approach.

RESULTS

To correct for transmitted light attenuation due to scattering, we are using a procedure discussed in Experiment 3-15. The intensity of scattered light, I in the range where chromophore does not absorb is proportional to $1/\lambda^4$ and can be presented in a simplified form:

$$I \sim \frac{B}{\lambda^4}$$

where B is a constant that include all parameters in Rayleigh equation (index of refraction, particle size, etc.). When particles scatter the light, the out-scattered photons do not reach the detector and apparent absorbance due to Rayleigh scattering can be expressed:

$$Ab_{RS}(\lambda) = -\log\left(\frac{I_0 - I}{I_0}\right)$$

We can fit experimentally measured values of absorbance (Ab_{exp}) in the range where chromophore does not absorb into a calculated absorbance by minimizing the error: $\sum_\lambda (Ab_{exp}(\lambda) - Ab_{RS}(\lambda) + c)^2$. The arbitrary constant, c is to account for small baseline shifts (small imperfection of background correction or defect/dirt on the cuvette wall). We fit experimental data to a single constant, B and baseline shift c. The B parameter will account for possible size distribution and typically undefined index of refraction of particles. By fitting experimentally measured absorbance at long wavelength (range where fluorophore does not absorb), we can estimate constants B and c. Using these constants, we can calculate apparent absorbance (light attenuation) due to the scattering for the shorter wavelength range. We will subtract predicted absorbance of Ludox particle scattering from the measured absorbance of dye solution

(Continued)

EXPERIMENT 3-19 (CONTINUED) MEASURING ABSORPTION IN THE PRESENCE OF SMALL SCATTERING PARTICLES

containing Ludox particle (mix). If the calculated/predicted absorbance due to scattering is correct the recovered absorption of a known compound (in our case R110 and QS) should be very close to its absorption measured in just water. In this case, we assume the chromophore is not interacting with silica particles. *Note:* Any adhesion of the dye to the particle surface may lead to spectral change and obviously, this procedure would not work, or rather the procedure will work but will yield different spectrum as affected by chromophores interactions with particles. To avoid this problem, we will compare recovered values using the Rayleigh equation with recovered spectra when subtracting (corresponding) absorption of the equivalent solution of Ludox particles. In this case, we will use Ludox solution diluted by adding the calibrated amount of water to the stock solution of Ludox (see # 3 and 4).

Figure 3-19.1 presents normalized absorption spectra of R110 and QS in water (remember, this is not water stabilized with H_2SO_4 as typically used for the standard). We will use these spectra as a reference spectra to compare to the recovered spectra. The R110 has a strong absorption band in the visible range (500 nm) with some weak bands below 400 nm and QS absorbs below 350 nm.

Figure 3-19.2 shows the measured absorption spectrum of R110 mixed with Ludox (mix). The measured absorption has clear features of the absorption spectrum of R110 in water but it is significantly different at a short wavelength range. The spectrum is significantly lifted up and in shorter wavelengths shows much higher absorption that one would expect from the reference spectrum shown in Figure 3-19.1. In Figure 3-19.2, we also added measured apparent Ludox absorption (blue dashed line on the left figure) and predicted absorption of Ludox solution in short wavelength range after fitting absorption in the long wavelengths range (560–650 nm where R110 does not absorb light) to the Rayleigh scattering function (blue dashed line at the right figure). In Figure 3-19.2, we also show recovered absorption of R110 after subtracting absorption of the measured corresponding Ludox solution (left) and recovered absorption of R110 after subtracting absorption of calculated (fitted) assuming the Rayleigh scattering.

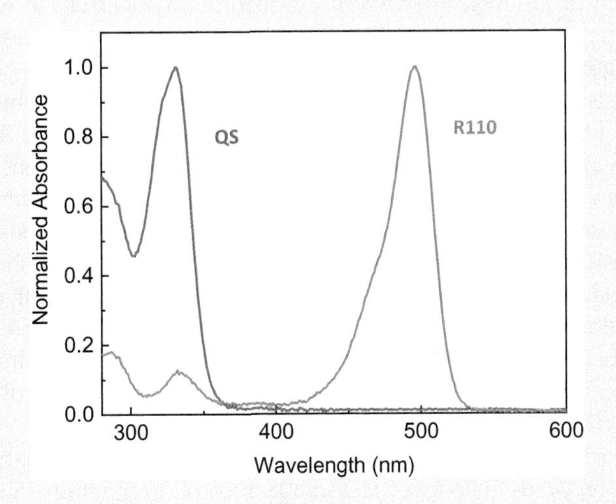

FIGURE 3-19.1 Normalized absorption spectra of quinine sulfate (QS) and rhodamine 110 (R110) in water.

(Continued)

EXPERIMENT 3-19 (CONTINUED) MEASURING ABSORPTION IN THE PRESENCE OF SMALL SCATTERING PARTICLES

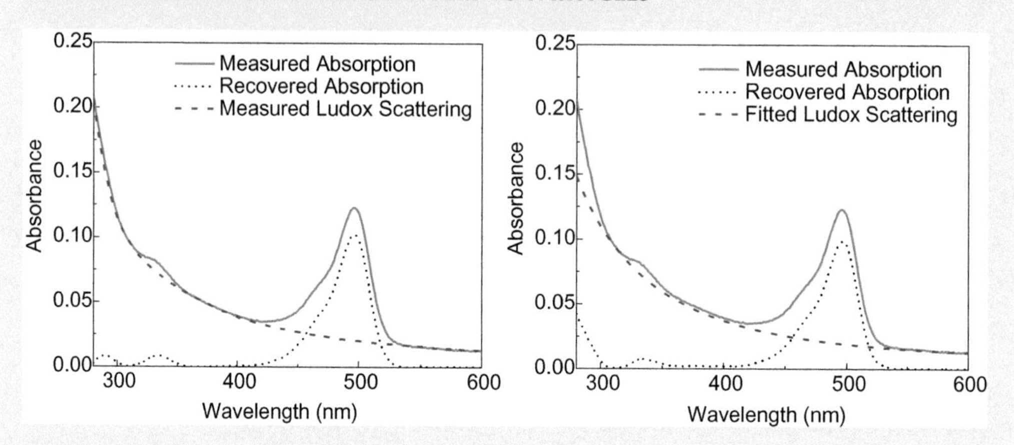

FIGURE 3-19.2 Measured absorption of R110 mixed with Ludox particles (orange). Left: Measured absorption of the corresponding (reference) Ludox solution (dashed blue) and recovered absorption of R110 (black dotted line). Right: Calculated apparent absorption of Ludox solution after fitting to Rayleigh function in the 560–650 nm range (dashed blue) and recovered absorption of R110 (black dotted line).

In both cases, simple subtraction (in one case measured Ludox reference and in the other calculated Ludox contribution) gives very good results leading to absorption very similar to that presented in Figure 3-19.1 for R110. To better compare recovered spectra, we present in Figure 3-19.3 normalized absorption spectra for R110 in water (reference spectrum of R110). Next, recovered absorption of R110 after subtracting absorption of the measured corresponding Ludox solution, and recovered absorption spectrum of R110 after subtracting predicted absorption of Ludox that was fitted in the 560–650 nm range. Comparison of the measured reference absorption spectrum in water with recovered absorption spectra in solutions containing silica particle shows a perfect agreement in the long-wavelength range.

There are increasing discrepancies at the shorter wavelength range. Subtraction of experimentally measured Ludox absorption gives really good results down to 300 nm. The recovered absorption by predicting the Rayleigh absorption pattern is in excellent agreement down to 350 nm and then differences increase and below 300 nm are significant.

Figure 3-19.4 shows measured absorption of QS mixed with the Ludox particle. Similar to Figure 3-19.2, we also included measured Ludox absorption (blue dashed line in the left figure) and predicted Ludox absorbance (blue dashed line on the right figure). The directly measured absorption (green line) is very different from that of the reference spectrum in Figure 3-19.1 especially at shorter wavelengths. This is expected since the contribution of scattering at shorter wavelengths is much greater. The recovered (corrected) absorption spectra are presented in black dotted lines.

Subtracting measured apparent Ludox absorbance (blue dashed line on the left figure) from measured absorption yields quite a satisfactory QS absorption spectrum. Since absorption of QS ends at about 360 nm, we can safely use 380–460 nm range to fit the Rayleigh function and recover constants B and c. Subtracting estimated Ludox scattering (blue dashed line in

(*Continued*)

EXPERIMENT 3-19 (CONTINUED) MEASURING ABSORPTION IN THE PRESENCE OF SMALL SCATTERING PARTICLES

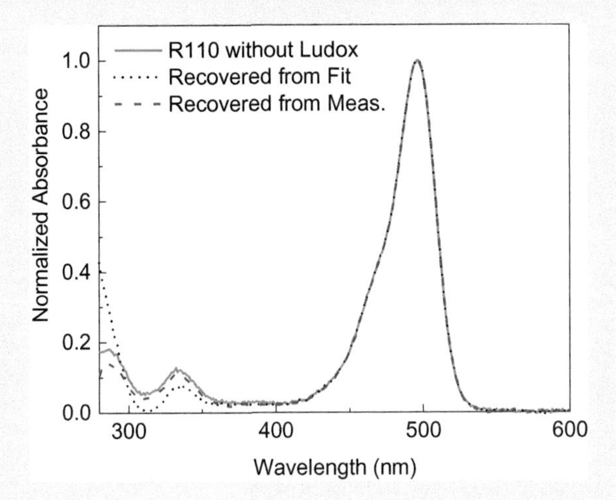

FIGURE 3-19.3 Normalized absorption spectra of R110 in water and recovered from a mixture with Ludox particles.

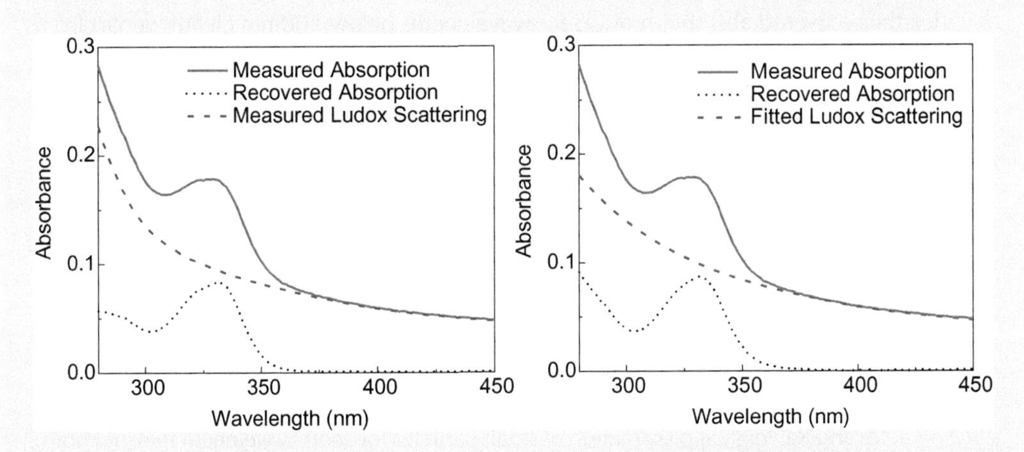

FIGURE 3-19.4 Left: Measured absorption of QS mixed with Ludox particles (green). Measured absorption of the corresponding (reference) Ludox solution (blue dashed line) and recovered absorption of QS (black dotted line). Right: Calculated apparent absorption of Ludox solution after fitting to the Rayleigh function in the 380–460 nm range (blue dashed line) and recovered absorption of QS (black dotted line).

the right figure) in the short wavelength range yields a reasonable absorption spectrum of QS. For easy comparison, we present three normalized spectra of QS in Figure 3-19.5. In the longer wavelength range, we observe a very good agreement. We can notice a slight shift of recovered spectra toward the longer wavelength. But this is a minor effect. The spectrum recovered by subtracting measured absorption of corresponding Ludox solution gives very good results also in short-wavelength below 300 nm. Contrary, using predicted absorption

(*Continued*)

EXPERIMENT 3-19 (CONTINUED) MEASURING ABSORPTION IN THE PRESENCE OF SMALL SCATTERING PARTICLES

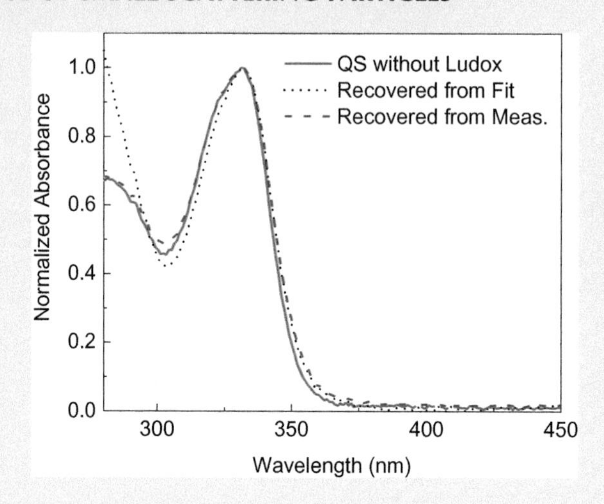

FIGURE 3-19.5 Normalized absorption spectra of QS in water and recovered from the mixture with Ludox particles.

of Ludox the recovered absorption of QS for wavelengths below 300 nm clearly depart from expected profile similar to the results for R110 (Figure 3-19.3).

There are a few observations we want to point out. The absorption spectra of QS recovered from the mixture by both approaches shows an almost identical small shift toward the longer wavelength. This could be the result of a small fraction of QS bind to silica resulting in the long-wavelength shift. The recovered spectrum by direct subtraction of Ludox absorption shows a good agreement for the whole spectral range while the procedure using the Rayleigh function clearly gives worse results in shorter (UV) wavelength range. There are two main reasons for this. One, silica particles start to absorb in UV and its absorption contribution becomes significant below 300 nm. This will strongly affect recovered spectra in the short wavelength range. Since Rayleigh scattering does not account for any absorption of particles, the measured spectra present higher absorption than expected from predicted by scattering and absorption of the dye (adding Ludox, we are also adding absorber/impurity). Second, the Ludox particles are 20–30 nm silica particles and Rayleigh approximation of small particles for short-wavelength may not hold. The true scattering will be very difficult to predict since it will depend on many factors.

CONCLUSIONS

Presented spectra demonstrate that correct (approximate) absorption profile of chromophore can be obtained by using Rayleigh approximation when scattering particles are significantly smaller from the wavelength. In the case of large Ludox particles in the range above 300 nm, the results using Rayleigh approximation are comparable to directly measured and corrected for particles scattering. In a shorter wavelength, the difference is larger and recovered spectra are higher than expected. This is probably the result of potential absorption of light by particles. Also, some contamination by organic impurities will have a significant effect. These results clearly indicate that we can calculate the pure absorption component of a chromophore when embedded in a highly scattering system. It is interesting to note that calculated

(Continued)

EXPERIMENT 3-19 (CONTINUED) MEASURING ABSORPTION IN THE PRESENCE OF SMALL SCATTERING PARTICLES

from Rayleigh scattering correction for R110 did not work well in the range below 400 nm while for QS worked pretty well down to 325 nm. Also, calculated parameters (*B* and *c*) for R110 and QS are slightly different indicating that we may have some small contribution of larger particles and in the shorter wavelength range scattering does not obey Rayleigh rule completely. In this case, an approach with fitting a Rayleigh function works well in the initial part (long wavelengths absorption side) and the error increases toward shorter wavelengths that are far apart from the fitted region. Because of this, we recommend fitting Rayleigh parameters from the baseline as close as possible to the first absorption transition.

A practical example would be measuring the absorption of large proteins where scattering by globular protein may have a significant contribution to measured light extinction (apparent absorption) by the solution. An example of the protein is presented in next Experiment 3-20.

Warning: To perform these model experiments we want to stress that Ludox solution needs to be of high purity and it has to be sonicated for a long time to make sure there are no aggregates.

EXPERIMENT 3-20 MEASURING ABSORBANCE OF TRYPTOPHAN AND TRYPTOPHAN IN PROTEINS

Keywords: absorbance, tryptophan, spectrophotometer, Rayleigh scattering, *N*-acetyl-L-tryptophanamide, human serum albumin.

Tryptophan is an amino acid which is found in proteins (like human serum albumin (HSA)). Tryptophan and other aromatic amino acids including tyrosine absorb UV light below 300 nm. The absorbance spectra of amino acids vary based on their environment and bonds they form. The absorption of a protein will depend on amino acids composition, but in the 280 nm range typically presents a clear peak that can be used to analyze/characterize proteins. One common application is to use absorption in the 280 nm range to estimate protein concentration. Obviously, any globular protein will have significant physical size. Tetrameric hemoglobin, for example, will have a diameter of about 3.5 nm, or human serum albumin is an elongated (cigar-shaped) protein with a diameter of about 3 nm and a length of about 7 nm. Also, antibodies and enzymes will typically have a larger physical size but most known proteins have an average size significantly below 20 nm. When measuring the absorption of such systems, a significant incline of the baseline becomes apparent in the range 300–400 nm. For many ratiometric readings (ratio of absorptions at two different wavelengths) and especially for a single point 280 nm reading, such a slope may have a significant impact. Although a reference (protein without aromatic amino acids) to directly measure apparent absorption due to light scattering is not available, it was shown in the previous Experiment 9-19 that by measuring absorption in an extended range above 300 nm and fitting it to the Rayleigh model, such a contribution can be estimated with relative accuracy. In this protocol, the absorbance of tryptophan and proteins containing tryptophan are measured and corrected for protein scattering. Tryptophan is an amino acid that absorbs light at the longest wavelength and the red edge of protein absorption is dominated by tryptophan absorption. The absorption spectrum at the 280 nm wavelength range is dominated by tyrosines especially that typical protein will contain many more tyrosine residues than tryptophan residues.

(*Continued*)

EXPERIMENT 3-20 (Continued) MEASURING ABSORBANCE OF TRYPTOPHAN AND TRYPTOPHAN IN PROTEINS

MATERIALS

- Tryptophan
- *N*-Acetyl-L-tryptophanamide (NATA)
- Human serum albumin
- Water
- 1 cm × 1 cm quartz cuvettes

EQUIPMENT

- Spectrophotometer

METHODS

1. Measure the blank with a cuvette filled with water.
2. Prepare stock solutions.
   a. Tryptophan.
   b. *N*-Acetyl-L-tryptophanamide.
   c. Human serum albumin
3. Measure the absorbance of each solution.

RESULTS

Measured absorption spectra of tryptophan, NATA (tryptophan derivative with a carboxyl side chain), and HSA protein are presented in Figure 3-20.1. For tryptophan and NATA, absorption ends completely at about 310 nm. On the contrary, the absorbance of HSA protein (Figure 3-20.1, right) does not go down to zero as quickly as the absorbance of tryptophan alone. Since we purposely measured absorption up to 500 nm the increasing (inclining) baseline toward shorter wavelength can clearly be seen. This is caused by Rayleigh scattering. Frequently, we would measure absorption only up to 350 nm and ignore the small baseline shift/incline.

HSA is a small protein and we can safely assume that the Rayleigh model should apply. We will use the same approach as in the previous Experiment 3-19. To correct for transmitted light attenuation due to scattering we are using the procedure discussed in Experiment 3-17. The intensity of scattered light, I by the protein in the range where tryptophan does not absorb is proportional to $1/\lambda^4$ and apparent absorbance due to Rayleigh scattering can be expressed as:

$$Ab_{RS}(\lambda) = -\log\left(\frac{I_0 - I}{I_0}\right)$$

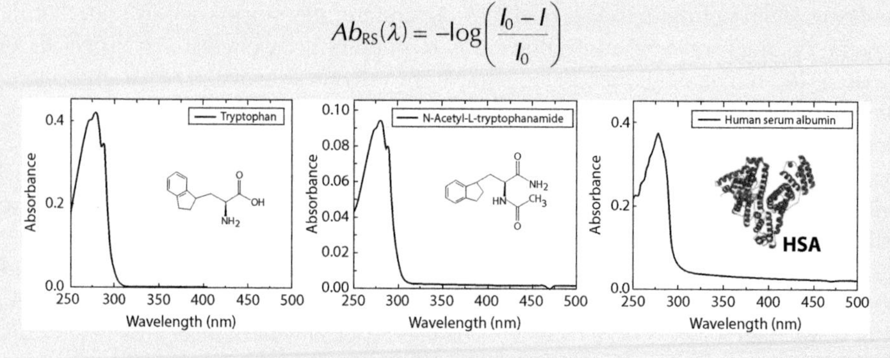

FIGURE 3-20.1 Absorbance of tryptophan in different environments.

(*Continued*)

EXPERIMENT 3-20 (CONTINUED) MEASURING ABSORBANCE OF TRYPTOPHAN AND TRYPTOPHAN IN PROTEINS

We can fit experimentally measured values of absorbance (Ab_{exp}) in the range where tryptophan does not absorb (320–500 nm) to a calculated absorbance by minimizing the error: $\sum_{\lambda}(Ab_{exp}(\lambda) - Ab_{RS}(\lambda) + c)^2$. As previously discussed, the arbitrary constant, c is to account for small baseline shifts (small imperfection of background correction or defect/dirt on the cuvette wall). We fitted experimental measured apparent absorption in the 330–450 nm range to a single constant, B and baseline shift c (see details in Experiment 3-19). Figure 3-20.2 (left) shows the fit in the considered range and on the right figure we present measured absorption in the full range with apparent absorption due to scattering extended to remaining wavelengths. The fit aligns very well with the measured spectrum for wavelengths over about 320 nm. As discussed in the previous Experiment 3-19, the apparent absorption due to light scatting (extinction of light due to out-scattering) can be subtracted from the measured absorption spectrum to obtain only the absorbance due to chromophores. The result is shown in Figure 3-20.3. The absorbance at higher wavelengths goes to zero, as is the case with tryptophan. The absorbance peak is centered at 280 nm and has a slightly different shape from that expected for tryptophan or NATA since multiple tyrosine residues significantly contribute to the absorption spectrum.

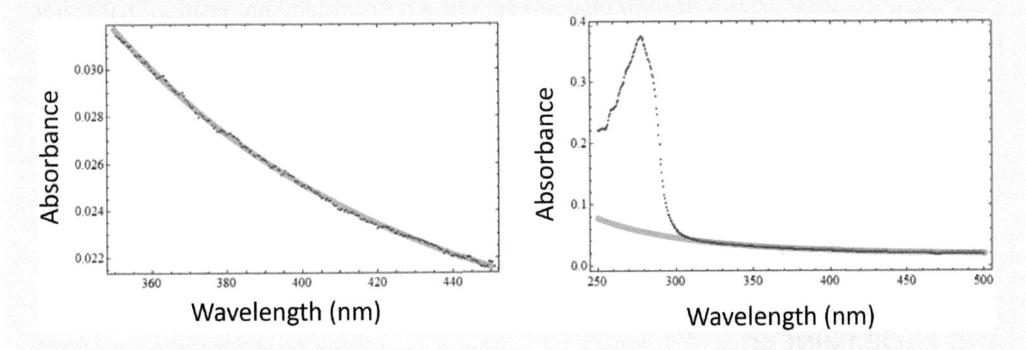

FIGURE 3-20.2 Scattering of human serum albumin fitted as Rayleigh scattering (line) and raw absorbance data (dotted).

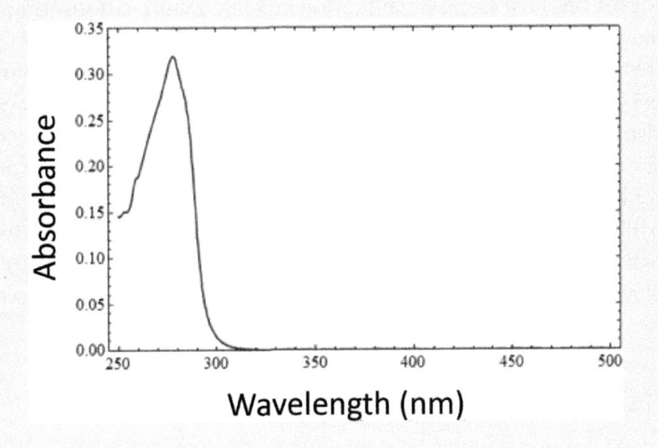

FIGURE 3-20.3 The spectrum of human serum albumin-corrected for scattering.

(Continued)

EXPERIMENT 3-20 (CONTINUED) MEASURING ABSORBANCE OF TRYPTOPHAN AND TRYPTOPHAN IN PROTEINS

CONCLUSIONS

Unlike in Experiments 3-18 and 3-19, in the present experiment, the scatterer and the absorber cannot be separated. This is why fitting is necessary here. As we noticed in Experiment 3-19 for silica beads at the short wavelength range, we observe a deviation from the true dye spectrum. *Would this be a case in this experiment?* If we used well-purified components the presented spectrum in Figure 3-20.3 should be very close to real protein absorption. The protein size is much smaller than the wavelength and the only absorbing components are amino acids. At shorter wavelengths below 250 nm absorption of the buffer, components will play the role. In conclusion, it is always possible to perform a scattering background fit in the range where there is no absorption and subtract it from the measured spectrum. However, it is often quicker and easier to use the scatter as the baseline when measuring background absorbance if possible. *Warning*: If the fit of Rayleigh model does not give good results in the range where we do not have absorption (like in this case, we would not be able to get a satisfactory fit in the range above 330 nm it would indicate that light attenuation is also due to some other factors. These are typically impurities that absorb in the range above 330 nm or protein aggregation that leads to much larger particle sizes.

Interestingly, the absorption spectrum of the protein after Rayleigh correction ends at about 310 nm as it would be expected for tryptophan absorption.

EXPERIMENT 3-21 pH DEPENDENCE OF ABSORBANCE OF FLUORESCEIN AND RHODAMINE 6G

Keywords: absorbance, scattering, baseline, pH dependence.

Typical fluorophores (in most cases organic molecules) contain various ionizing groups. So, it is expected they will have different chemical forms depending on the pH of the environment. This will result in different absorption and emission spectra. Typical biological systems have different pH or even protein surfaces that expose different amino acids will have locally different pH. Also, sometimes an experiment calls for local pH sensing/detection, such as the pH inside or outside a cell membrane. In many cases when using water and/or buffers, it is vital to control the pH throughout the experiment. Below two examples of very common fluorophores exhibiting very different dependences of absorption spectra (extinction coefficient) as a function of pH are presented.

This protocol measures the response of fluorescein and rhodamine 6G to changing pH.

MATERIALS

- Fluorescein and rhodamine 6G (R6G)
- Water
- 1 cm × 1 cm cuvettes

(*Continued*)

EXPERIMENT 3-21 (CONTINUED) PH DEPENDENCE OF ABSORBANCE OF FLUORESCEIN AND RHODAMINE 6G

EQUIPMENT

- Spectrophotometer

METHODS

1. Prepare buffer solutions of different pH and measure equal volumes of each (about 5 mL).
   a. 5.45
   b. 5.9
   c. 6.1
   d. 6.6
   e. 6.9
   f. 7.3
   g. 8.1
   h. 8.5

   In principle, you can use any buffer (or set of buffers) of your choice. In our experiment, we used a 0.1M phosphate buffer and adjusted pH for each solution using a calibrated pH meter.
2. Prepare highly concentrated solutions of fluorescein and R6G in water.
3. Mix fluorescein and R6G into each buffer separately.
   a. There should be 16 different solutions
4. Measure the absorbance of each solution.

RESULTS

Figure 3-21.1 shows the resulting spectra. The absorbance of fluorescein change dramatically with changing pH, the absorbance of R6G remains largely unchanged (any intensity changes are most likely due to small concentration differences).

The pH dependence of fluorescein is due to its carboxyl group (COOH). Rhodamine shows no pH dependence (very small) in the presented pH range and all the spectra overlap almost exactly. A change in absorbance may have consequences in applications such

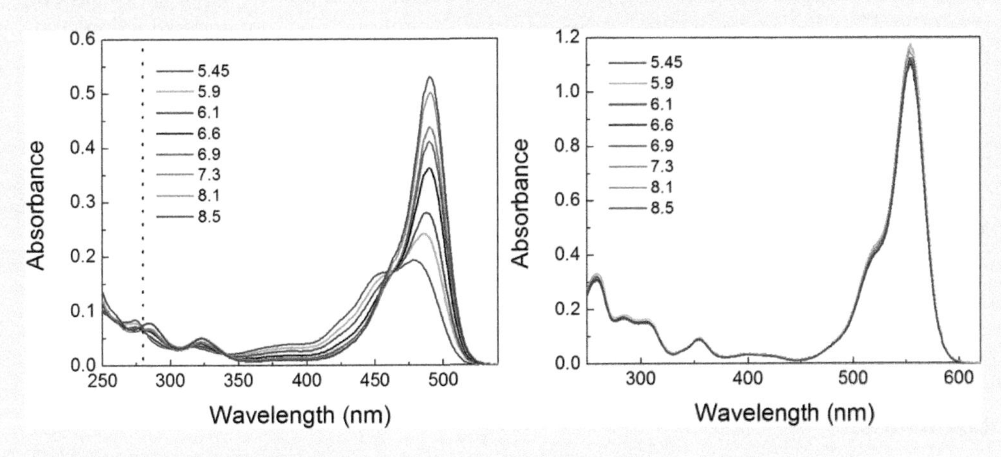

FIGURE 3-21.1 The pH-dependent absorbance of fluorescein (left) and rhodamine (right).

(Continued)

EXPERIMENT 3-21 (CONTINUED) PH DEPENDENCE OF ABSORBANCE OF FLUORESCEIN AND RHODAMINE 6G

as protein labeling. The correction factor in labeling is a ratio of the peak absorbance intensity to the absorbance at 280 nm (see in "Degree of Labeling of Avidin Labeled with Fluorescein—Experiment 3-22"). Importantly, the intensity of the absorbance of fluorescein at 280 nm (vertical dotted line) changes very little as a function of pH while the intensity at the absorbance maximum changes 3-fold. Thus the correction factor of fluorescein is pH-dependent and this should be accounted for when fluorescein is used as a label in systems that present different pH.

EXPERIMENT 3-22 DEGREE OF LABELING OF AVIDIN LABELED WITH FLUORESCEIN

Keywords: absorbance, degree of labeling, spectrophotometer.

Fluorescent labeling is frequently the first step in observing and tracking proteins both in-vitro and in-vivo. An important consideration in using fluorescent labels for proteins is how many chromophores are attached to the average protein in the used solution (ensemble). One would expect more chromophores would produce more light, however binding too many chromophores may cause the opposite effect, yielding a much lower than expected fluorescence. Also, often besides chromophore quenching, too many labels may cause problems like the formation of aggregates and a reduction in the protein's specificity/activity. In effect, quite often a protein labeled with 4–5 fluorophores will yield an overall better signal than the same protein labeled with 10–15 fluorophores. Thus, there is an optimal number of how many chromophores should be bound to a given protein for the best experimental results; it is definitely not as many chromophores as possible. The average number of chromophores bound to a protein is called the *degree of labeling* which may be obtained using an absorbance spectrum of the chromophore/protein conjugate and some simple calculations. There are a few reasons why using absorption is better than fluorescence. First, as mentioned above, due to a variety of possible fluorescence quenching processes, the overall fluorescence signal is not directly proportional to the degree of labeling. Second, the absorption (spectrum and extinction coefficient) is much less sensitive to the environment than fluorescence. This is because perturbing the ground state of a molecule is significantly more difficult than perturbing the excited state.

This protocol determines the degree of labeling of avidin with fluorescein using an absorbance spectrum.

MATERIALS

- Avidin/fluorescein conjugate
- Water
- 1 cm × 1 cm cuvette

EQUIPMENT

- Spectrophotometer

(*Continued*)

EXPERIMENT 3-22 (CONTINUED) DEGREE OF LABELING OF AVIDIN LABELED WITH FLUORESCEIN

METHODS

1. Obtain avidin labeled with fluorescein.
   a. This may be bought pre-made.
   b. Labeling kits are also available.
   c. Follow the directions which came with the specific labeling kit.
2. Measure the baseline using the buffer of the conjugate (i.e., water).
3. Transfer conjugate to a cuvette.
4. Measure the absorbance of the conjugate.
5. This measurement does not alter the conjugate in any way, the sample used for the absorbance measurement may be used for further investigation.

RESULTS

The resulting spectrum is presented in Figure 3-22.1. It is also possible to find the absorbance of avidin and fluorescein individually but this is not necessary for this protocol. Although fluorescein absorbs in the same UV wavelength range as avidin, correcting the absorbance of the conjugate is unnecessary and a correction factor is instead introduced into the calculation. This is the case for all protein label conjugates. The absorbance maximum of avidin is at 280 nm (this is true for most proteins, a few other proteins have the absorbance maximum at about 260 nm). It is important to use the value of the measured absorbance reading at 280 nm, A_{280}, even if other more obvious peaks are present in the vicinity. For example, Figure 3-22.1 has a discernible peak at about 325 nm, it is not relevant in this protocol (it actually comes from fluorescein—compare figures from previous Experiments 3-13.2 or 3-20.1). The spectrum contains an uncertain number of peaks below 300 nm, this again is not important. The absorbance value of 280 nm is 0.079. The large absorbance peaks centered at about 494–500 nm belongs to fluorescein and the absorbance value at that wavelength is 0.174.

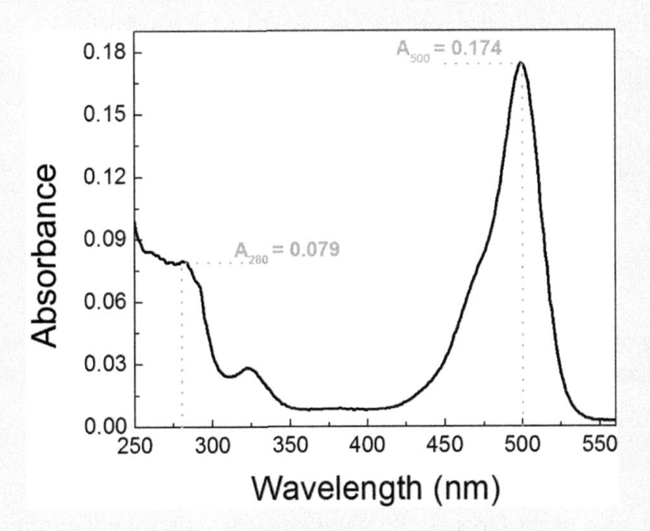

FIGURE 3-22.1 The absorption spectrum of avidin labeled with fluorescein.

(Continued)

EXPERIMENT 3-22 (CONTINUED) DEGREE OF LABELING OF AVIDIN LABELED WITH FLUORESCEIN

The exact center of the label (fluorescein) may vary by a few nanometers due to factors such as concentration and pH and this value is often given as $A\lambda$.

The degree of labeling (DOL) is given as:

$$\text{DOL} = \frac{A_{496} \times \varepsilon_{\text{avidin}}(280)}{\left(A_{280} - A_{496} \times \text{CF}_{280}\right) * \varepsilon_{\text{fluorescein}}(496)}$$

The variable for the degree of labeling are:

Variable	Description	Value
A_{500}	The maximum value of absorbance of the label (fluorescein), in this case, it happened to be at 496 nm. See Figure 3-22.1. The wavelength is dye dependent and often given as A_λ.	0.174
A_{280}	The maximum value of absorbance of the protein (avidin), occurs at 280 nm (in some cases 260 nm). Value is obtained from Figure 3-22.1.	0.079
CF_{280}	The correction factor of the label (fluorescein) at the wavelength of maximum absorbance of the protein. This is necessary because dyes absorb at 280 nm, and also this procedure does not involve fitting to find the absorbance of only the protein. This value is found in literature or by the label provider. This may change between different proteins a little but on average it is close to 0.19. But more strongly depends on the label (dye).	0.19
$\varepsilon_{\text{avidin}}(280)$	The molar extinction coefficient of the protein at the wavelength where its emission is maximal or 280 nm. Avidin contains a significant number of tryptophanes residues and its extinction coefficient we estimated to about 95,000 $\text{Mol}^{-1}\,\text{cm}^{-1}$	95,000 $\text{Mol}^{-1}\,\text{cm}^{-1}$
$\varepsilon_{\text{fluorescein}}(496)$	The molar extinction coefficient of the label at the wavelength where its emission is maximal. This can be determined from an absorbance measurement of fluorescein. Note that this is an assumed value and it is slightly lower than that reported in Experiment 3-13 for just fluorescein. Typically, a dye can be quenched in contact with the protein surface.	75,000 $\text{Mol}^{-1}\,\text{cm}^{-1}$

$$\text{DOL} = \frac{0.174 \times \varepsilon_{\text{avidin}}(280)}{\left(0.079 - 0.174 \times 0.19\right) * \varepsilon_{\text{fluorescein}}(496)}$$

The absorbance of the protein at 280 nm is given by (A_{280}–A_{496} × CF). The correction factor gives the absorbance of fluorescein at 280 nm as a fraction of its absorbance at 498 nm.

$$\text{CF} = \frac{\varepsilon_{280}}{\varepsilon_{496}}$$

Thus, the absorbance of the label at 280 nm is calculated as a fraction of its absorbance at its maximum. This allows for a simple approximation and not fitting the absorbance of the

(Continued)

EXPERIMENT 3-22 (CONTINUED) DEGREE OF LABELING OF AVIDIN LABELED WITH FLUORESCEIN

conjugate with the absorbance of the label, which is significantly more calculation intensive. The correction factor is an intrinsic property of the label and is provided with the label documentation.

Using numbers as presented in the presented outline we can calculate the degree of labeling DOL = 4.5. This means that on average, we have between 4 and 5 fluorescein molecules per a single molecule of avidin.

EXPERIMENT 3-23 ABSORPTION OF MIXTURES OF DIFFERENT CHROMOPHORES: ADDITIVITY OF ABSORPTION

Keywords: absorbance, absorption spectra, baseline, mixture of chromophores.

Earlier experiments demonstrated that absorbance is an additive property and the contributions from different components should be a simple sum. The additive nature of absorption assumes that absorbing components do not interact what would perturb their spectral properties. This is typically true when subtracting the contribution of the background, reflections on cuvette walls, or scattering. Now, it is time to test if the absorption of mixtures of different chromophores is also additive. In many biochemical processes, we observe mixtures of different chromophores (components). In many cases, various individual components are well characterized (known) but the resulting absorption of the mixture may not reflect a simple sum. *When working with the mixture, could we decide if these absorbing species form new entities, or behave just as independent components?* From basic absorption principles, it might be thought that individual components will be additive and any perturbation/interaction will (or may) result in a spectral change. The examples presented below include various chromophores that do not interact significantly between themselves, and also some which interact forming new (unpredictable) absorbing species.

MATERIALS

- Rhodamine B (RhB), rhodamin 6G (R6G), rhodamine 110 (R110), rhodamine 700 (R700), acridine orange (AO), uranin, coumarin 153 (C153), quinine sulfate (QS).
- EtOH, water
- 1 cm × 1 cm cuvettes
- Pipettes

EQUIPMENT

- Spectrophotometer

METHODS

1. Prepare solutions (~5 mL) of RhB, R6G, and R700 in EtOH using your favorite procedure(s). Ensure that absorptions of each dye are the same at maximum intensity (very similar).
2. Measure a baseline using a cuvette filled with only water.

(Continued)

EXPERIMENT 3-23 (CONTINUED) ABSORPTION OF MIXTURES OF DIFFERENT CHROMOPHORES: ADDITIVITY OF ABSORPTION

3. Measure absorptions of each dye. See normalized spectra in Figure 3-23.1.
4. Mix RhB and R6G at roughly 1:1. Measure absorption of the mixture, see Figure 3-23.2.
5. Add to the RhB:R6G mixture solution of R700 (about 1/3 volume of RhB:R6G mixture). Measure absorption of the mixture of three dyes, see Figure 3-23.3.
6. Using a linear minimization routine (Eq. 3-23.1) reconstruct measured absorptions of both mixtures using measured absorptions of individual components as standard spectra.
7. Prepare a solution of RhB and ErB in water and mix them in a 1:1 ratio. Measure absorption of all three samples (each individual component and 1:1 mixture), see Figure 3-23.5.
8. Prepare a solution of AO and ErB in water. Mix them in a 1:1 ratio. Measure absorptions of all three components.
9. Repeat procedure with AO and uranin.
10. Repeat procedure using QS and C153.

RESULTS

When preparing solutions of all individual components make sure their absorptions are closed (does not have to be perfectly identical) and should be below 0.2. Figure 3-23.1 shows normalized absorption spectra for R6G, RhB, and R700 in EtOH. Rhodamine 6G and B have very closely positioned absorption bands while R700 is significantly shifted toward the red.

For the first experiment, we made just 1:1 R6G:RhB mixture and measured its absorption spectrum. If mixed dyes do not interact with the measured absorption it should be a simple sum of two components. To analyze the measured absorption spectrum of the mixture, we will use a simple least-square analysis that allows to deconvolute measured spectrum of the mixture to individual standard components (absorptions of two mixed chromophores). If two (or more) components contributing to the absorption spectrum that are independent the measured absorption should be a simple weighted sum of all components. Meaning $A = \sum_i f_i A_i$

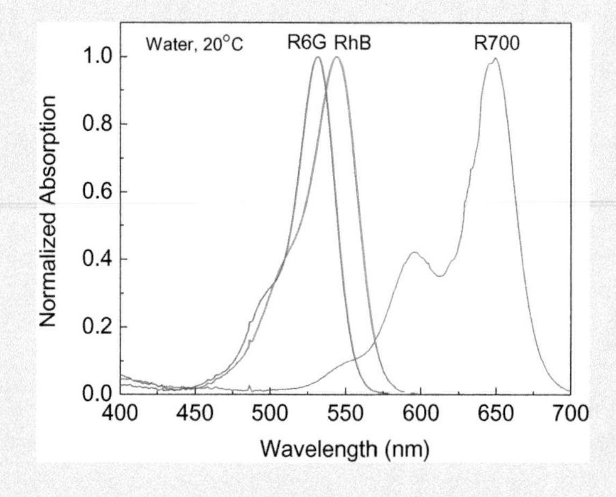

FIGURE 3-23.1 Normalized absorption spectra for R6G, RhB, and R700 in water.

(*Continued*)

EXPERIMENT 3-23 (CONTINUED) ABSORPTION OF MIXTURES OF DIFFERENT CHROMOPHORES: ADDITIVITY OF ABSORPTION

where f_i is the fraction of A_i absorbing component. So, if measured absorbance for the mixture is A^m we can minimize the error (see equation 2-20.1):

$$\mathrm{Err}(f_1,f_2,\ldots,f_n)=\sum_{\lambda_j}\left[A^m(\lambda_j)-\left(f_1A_1(\lambda_j)+f_2A_2(\lambda_j)+\cdots+f_nA_n(\lambda_j)\right)\right]^2 \quad (3\text{-}23.1)$$

where $A_1, A_2, \ldots, A_n$ are the absorption of standards (individual absorption of each component) and recovered fractions f_i for each absorption component A_i. In Figure 3-23.2, we are showing measured absorption of the mixture (circles), recovered absorptions for RhB (dashed green), R6G (dashed orange), and fitted absorption (solid blue line through the circle). The recovered fractions, f_1 and f_2 are also shown in the figure. The recovered fractions are very close to a 50:50 mixture and the fitted line is visually perfectly going through the measured values through the full spectral range. Also recovered fractions are very close to expected (0.49 and 0.51) in spite of strong spectral overlapping of individual absorptions (R6G and RhB). This is a clear indication that both chromophores are not interacting and are acting independently.

Next, to the RhB:R6G mixture, we added the third component R700 in EtOH. The added volume was about one-third of the volume of the mixture. The measured absorption of the new mixture of three components is shown in Figure 3-23.3. In the figure, we also plotted recovered by spectral deconvolution absorption spectra of RhB, R6G, and R700. Recovered fractions are shown in the figure. As expected the respective fractions are 0.36, 0.38, and 0.26 are very close to that we were mixed. Again the fitted line (solid blue) very well corresponds to measured points.

For the next two examples, we selected two more dyes that have similar structures (rhodamine 110 (R110) and erythrosine B (ErB)). In Figure 3-23.4, we are showing measured absorption spectra in water and structures of R110 and RhB. In Figure 3-23.5, we present an absorption spectrum for the 1:1 mixture (circles) and fitted decomposition to two components (R110—dashed green and RhB—dashed orange). The fitted line is shown in solid blue.

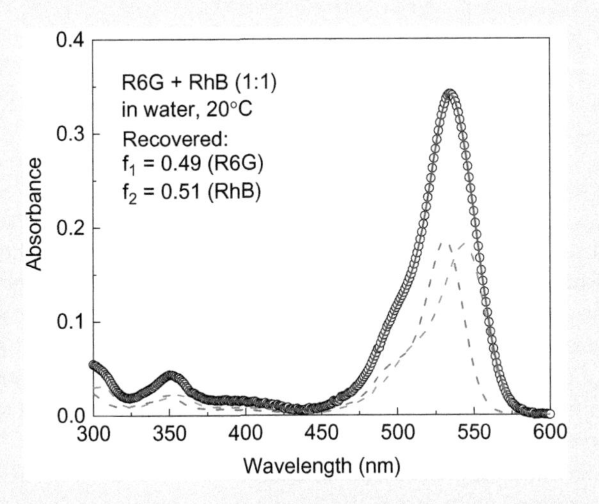

FIGURE 3-23.2 Measured absorption of R6G and RhB mixture (circles) and fitted two components (R6G—dashed orange and RhB—dashed green). A fitted line (solid blue) goes through the experimental points.

(Continued)

EXPERIMENT 3-23 (CONTINUED) ABSORPTION OF MIXTURES OF DIFFERENT CHROMOPHORES: ADDITIVITY OF ABSORPTION

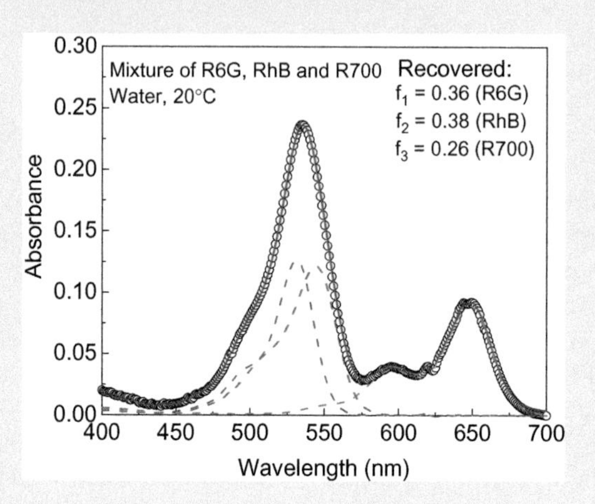

FIGURE 3-23.3 Measured absorption of R6G, RhB, and R700 mixture (circles) and recovered components of R6G (dashed pink), RhB (dashed green), and R700 (dashed orange).

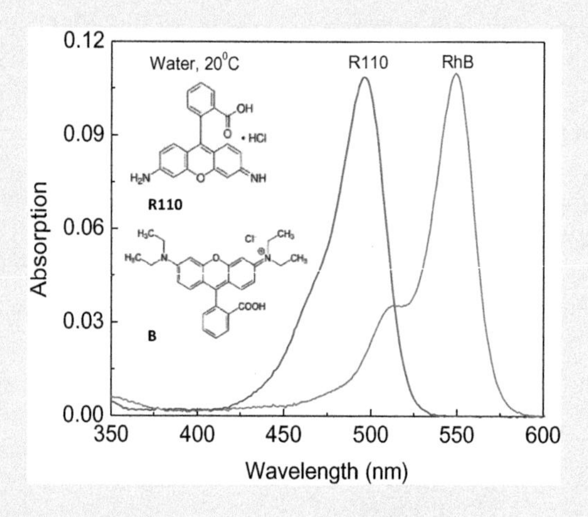

FIGURE 3-23.4 Measured absorption spectra of R110 and RhB in water.

Similar absorption spectra for ErB and RhB are presented in Figure 3-23.6 and the 1:1 mixtures of ErB and RhB is shown in Figure 3-23.7. In both cases, we observe a good fit and recovered fractions are appropriate. However, in both cases, at a shorter wavelength range, we can notice a small departure of the fitted line from the experimental points. This could be an indication of contribution from complexes that formed in the solution. However, the amount of formed interacting species is minimal and can only be detected at the spectral range where absorption is very low.

Next, we used AO and ErB. Measured individual spectra and structures are presented in Figure 3-23.8. Absorption spectrum for the 1:1 mixture is shown in Figure 3-23.9. In the figure, we present recovered spectra of AO and ErB together with the fitted line. This time the fit is not perfect and a significant deviation between the fitted line and measured points can

(*Continued*)

EXPERIMENT 3-23 (CONTINUED) ABSORPTION OF MIXTURES OF DIFFERENT CHROMOPHORES: ADDITIVITY OF ABSORPTION

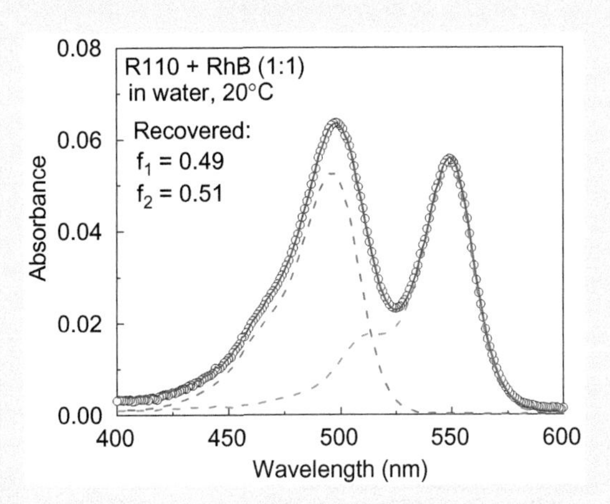

FIGURE 3-23.5 Measured absorption of R110 and RhB mixture (circles) and fitted two components (R110—dashed green and RhB—dashed orange). A fitted line (solid blue) goes through experimental points.

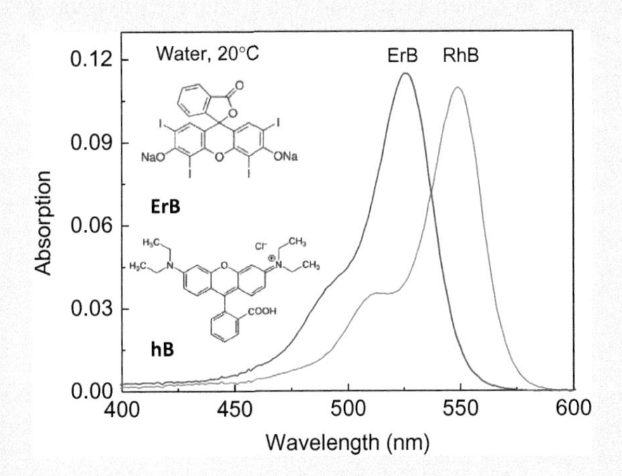

FIGURE 3-23.6 Measured absorption spectra of ErB and RhB in water.

be seen in the 550–600 nm range and near the 500 nm peak. Also recovered parameters are very different from that expected out of the 1:1 mix, much above experimental error that we could make when mixing the 1:1 solution. Just for comparison at Figure 3-23.10, we present expected absorption (dashed line) from 1:1 mixture and measured absorption (solid black line). Expected absorption is a simple sum of 50% absorption of AO plus 50% absorption of ErB. Here, the difference is much larger which is definitely indicating a problem with the mixture. Interestingly, spectral deconvolution shows the significant error for the fit but the fitting procedure was able to adjust relative parameters to minimize the error. But recovered individual fractions are not close.

Even more surprising is the mixture of acridine orange (AO) and uranin. In Figure 3-23.11, we present measured absorption spectra and structures for AO and uranin. The absorption

(Continued)

EXPERIMENT 3-23 (CONTINUED) ABSORPTION OF MIXTURES OF DIFFERENT CHROMOPHORES: ADDITIVITY OF ABSORPTION

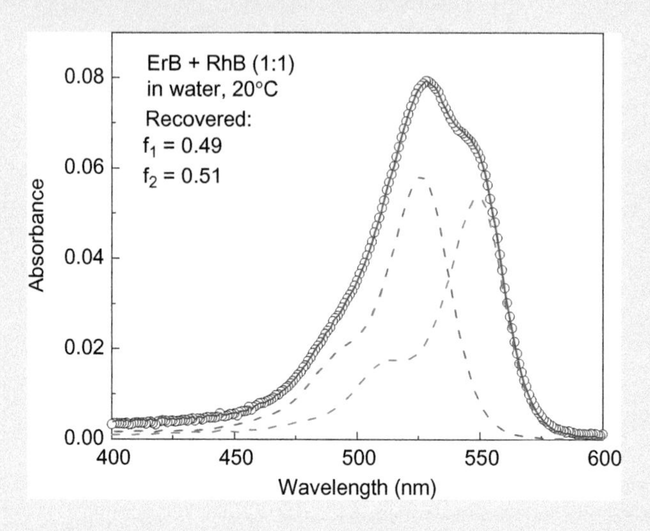

FIGURE 3-23.7 Measured absorption of ErB and RhB mixture (circles) and fitted two components (ErB—dashed green and RhB—dashed orange). A fitted line (solid blue) goes through experimental points.

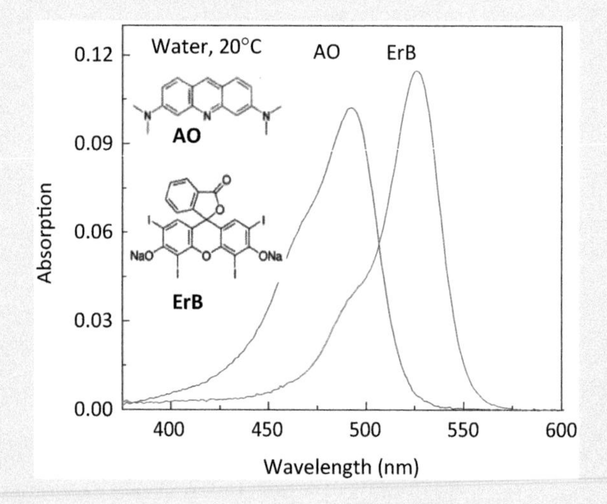

FIGURE 3-23.8 Measured absorption spectra of AO and ErB in water.

spectrum of the 1:1 mixture is presented in Figure 3-23.12 (green line) together with the calculated (expected) spectrum constructed using 50% absorption of AO and 50% absorption of uranin (dashed orange). The difference is dramatic. In Figure 3-23.13, we present the fitted spectra obtained from standard components (absorption spectra of AO and uranin). The fit is not good and one recovered fractional component is negative. This clearly indicates significant intermolecular interactions that lead to the formation of new absorbing species.

(Continued)

EXPERIMENT 3-23 (CONTINUED) ABSORPTION OF MIXTURES OF DIFFERENT CHROMOPHORES: ADDITIVITY OF ABSORPTION

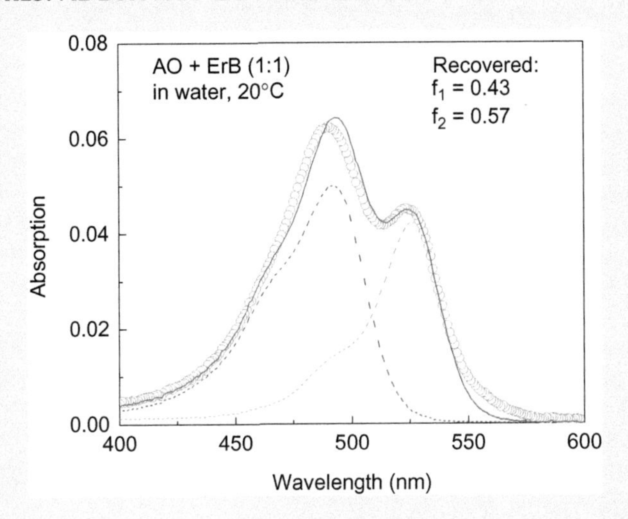

FIGURE 3-23.9 Measured absorption of 1:1 mixture of AO and ErB (circles) and fitted two components (AO—dashed blue and ErB—dashed green). A fitted line is shown in solid blue.

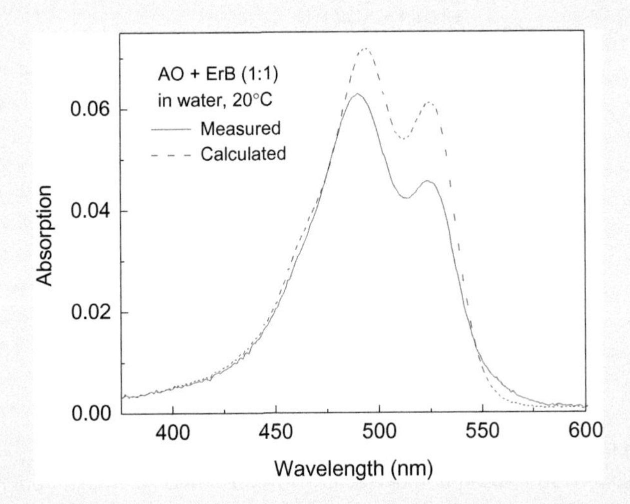

FIGURE 3-23.10 Measured absorption of 1:1 mixture of AO and ErB (solid green) and expected absorption (dashed red) from a sum of 50% absorption of AO and 50% absorption of ErB.

For the final experiment, we used QS and C153. The absorption spectra for both components are shown in Figure 3-23.14. In Figure 3-23.15, we present a measured spectrum of 1:1 mixture with the expected (calculated) spectrum for the 1:1 mixture (dashed red line). Again, the difference is dramatic. However, the spectral deconvolution procedure recovers a reasonable fit (Figure 3-21.16) with a slightly negative C153 fraction. Without prior knowledge, we would accept this fit as just absorption of coumarin with no QS added.

(Continued)

EXPERIMENT 3-23 (CONTINUED) ABSORPTION OF MIXTURES OF DIFFERENT CHROMOPHORES: ADDITIVITY OF ABSORPTION

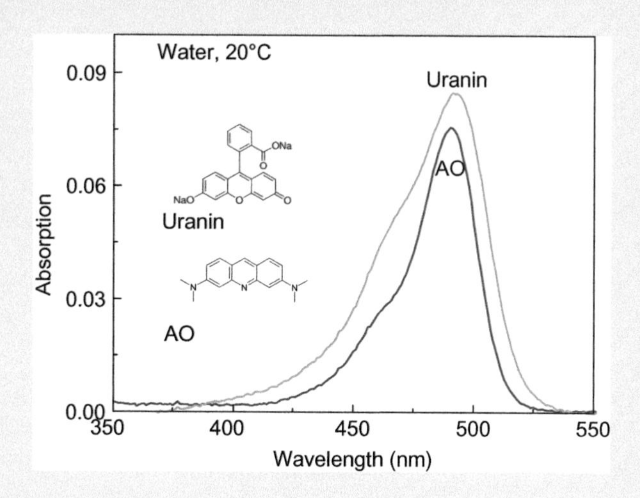

FIGURE 3-23.11 Measured absorption spectra of AO and uranin in water.

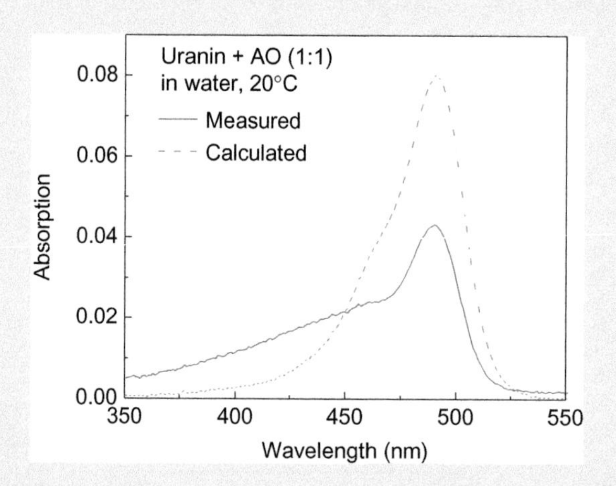

FIGURE 3-23.12 Measured absorption of 1:1 mixture of AO and uranin (solid green) and expected absorption from a sum of 50% absorption of AO and 50% absorption of uranin from Figure 3-23.11.

CONCLUSIONS

In this experiment, we presented different examples for mixtures of fluorophores. When mixing different chromophores we typically assume that they will act completely independent. However, this is rather a rare case even in relatively low concentrations. Quite frequently dyes will interact and form various unspecified complexes. The extent of formed complexes depends on many factors and chromophores with similar structures (look on chemical structures purposely shown in all graphs) will interact differently. When the mixed chromophores do not interact it is possible to recover individual components from measured absorption or predict the absorption spectrum from known components. But if the chromophores interact

(Continued)

EXPERIMENT 3-23 (CONTINUED) ABSORPTION OF MIXTURES OF DIFFERENT CHROMOPHORES: ADDITIVITY OF ABSORPTION

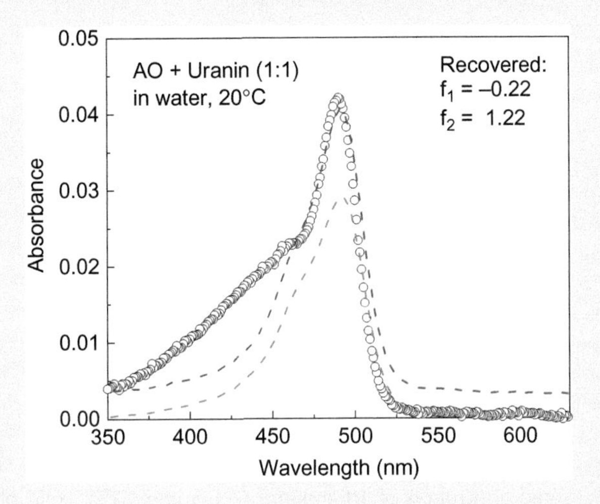

FIGURE 3-23.13 Measured absorption of 1:1 mixture of AO and uranin (circles) and fitted two components (AO—dashed orange and uranin—negative fraction is not shown). A fitted line is shown as dashed blue.

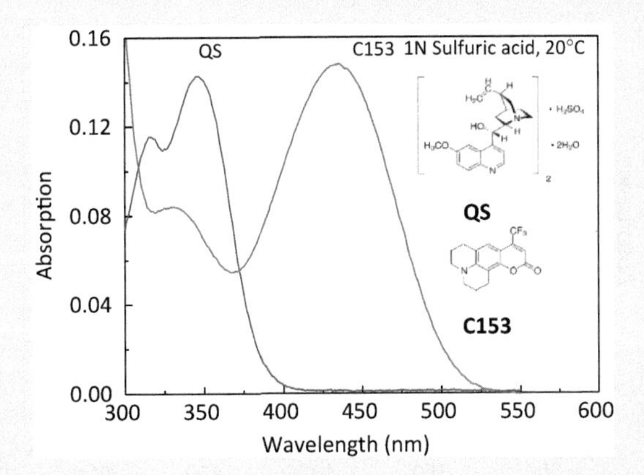

FIGURE 3-23.14 Measured absorption spectra of QS and C153 in 1 N sulfuric acid.

it is not true and spectral decomposition will not give correct results and sometimes even a misleading result can be obtained (see the mixture of QS and C153 in Figure 3-23.16). *So, how we can decide that the fit to the experimental data gives satisfactory results?* In our approach, we minimized the error (sum of differences between experimental points and fitted values). The error in an ideal case should be random. Unfortunately, the value of the error as defined by Equation 3-23.1 is relative to a total measured signal. So, for such absorption measurements, there are no general rules to what is a good value of the error that it is typically defined in a reference to a standard assumed value. The standard value will depend on many factors like the experimental noise, measured values, and other variables. In fact, when we compare the cumulative error (3-23.1) estimated for a solution of a given dye mixture that has a low

(*Continued*)

EXPERIMENT 3-23 (CONTINUED) ABSORPTION OF MIXTURES OF DIFFERENT CHROMOPHORES: ADDITIVITY OF ABSORPTION

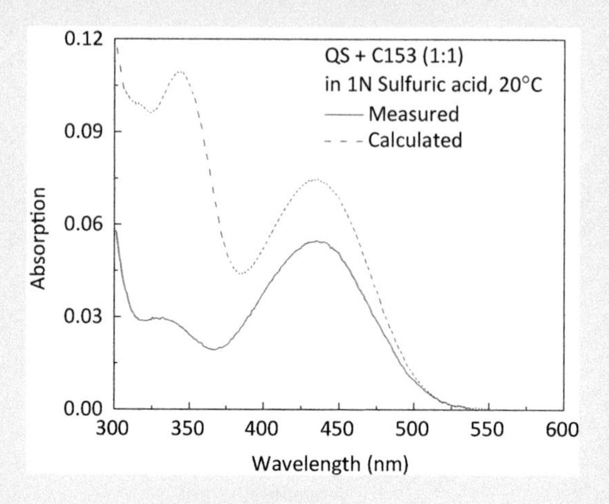

FIGURE 3-23.15 Measured absorption of 1:1 mixture of QS and C153 (solid blue) in 1 N sulfuric acid and expected absorption from a sum of 50% absorption of QS and 50% absorption of C153 from Figure 3-23.14.

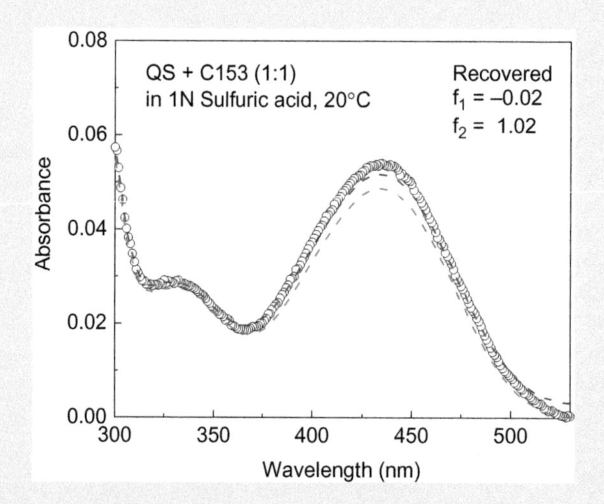

FIGURE 3-23.16 Measured absorption of 1:1 mixture of QS and C153 (circles) and fitted with two components (QS and C153 spectra) line (dashed blue). Individual components C153—dotted orange and QS—negative fraction not shown.

concentration (low absorption e.g., below 0.1) and high concentration (high absorption e.g., about 1) the error will typically be in favor of the low concentration (smaller individual point errors). Even if the fit for high concentration visually looks much better. The limiting factor in absorption measurements is typically overall signal, not an instrumental noise. So, to compare

(*Continued*)

EXPERIMENT 3-23 (CONTINUED) ABSORPTION OF MIXTURES OF DIFFERENT CHROMOPHORES: ADDITIVITY OF ABSORPTION

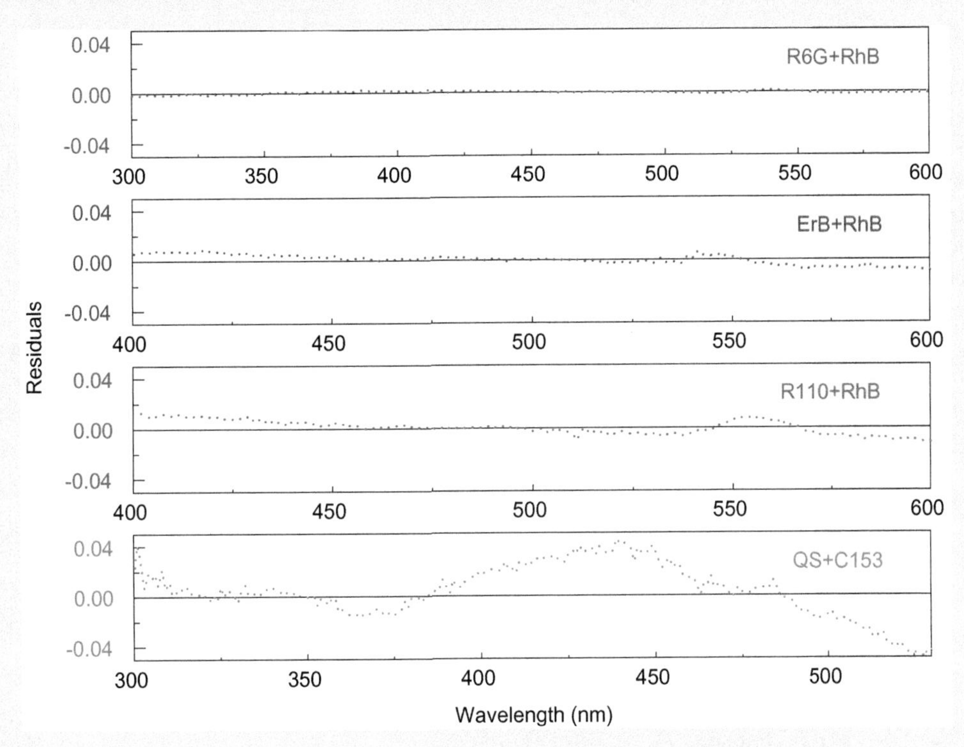

FIGURE 3-23.17 Distribution of residuals for four different mixtures.

fits between two absorption experiments we will need to normalize them. In our case of measuring absorption, we can conveniently divide the error by the maximum absorption.

Warning: When preparing a mixture of dyes for FRET or quenching experiments (or any other application), always measure absorption spectra to be sure that absorptions are additive.

This will normalize the error distribution (residuals) so we can compare the different fits. In Figure 3-23.17, we present residual distribution for the RhB/R6G, ErB/RB, RhB/R110, and QS/C153 mixtures. We used the same error range for all four plots. The first fit is the best and each of the following mixtures yields worse and worse fit. Already, the second fit shows a distinct trend in residuals distribution and the following two plots show also a large error variation. It could be a surprise but ErB/RhB mix and QS/C153 mix do not differ very much. However, while the ErB/RhB mix the fit corresponds to our expectation (1:1) the QS/C153 is completely out of range (−0.02:1.02). Without prior knowledge, we will not be able to decide which one is wrong. The other fits present a much larger error and are much easier to classify as not a simple mixture.

Note: We want to stress that in practice there are not many fluorophores that will not interact (affect each other) when mixed together for absorption measurement. And for those that behave properly, we typically have to use very low concentrations (this is one of the reasons why we did not use absorption higher than 0.2 and sometimes lower than 0.1).

EXPERIMENT 3-24 PREPARING ISOTROPIC PVA FILMS AND PVA FILMS DOPED WITH CHROMOPHORES

Keywords: absorbance, transmittance, light reflection, scattering, thin films, PVA.

Measurements of samples in thin polymer films offer various advantages like a rigid environment with very limited diffusion and very limited mobility of doped molecules (chromophores). In addition, polymers like PVA or polyethylene are easily stretchable, allowing orientation of chromophores. The possibility of orienting molecules opens new ways to study the intrinsic properties of molecular parameters like electronic transition moment's orientation. The spectroscopic methods discussed in this experiment and the following experiments are applicable to any thin film sample, including membranes, chromophores, glass slides, or molecules deposited on solid matrix-like silica.

In this experiment, a detailed description on how to prepare PVA films is provided. There are many chromophores which may be suspended in PVA. A liquid solution of PVA with a chromophore can be used to prepare films that are from a few microns to a millimeter thick, or films that are spin-coated on a solid substrate that can be very thin (10–1000 nm).

MATERIALS

- Polyvinyl alcohol (PVA) (e.g., Aldrich Mowiol—Mw 130,000; but other sizes over 50,000 are acceptable)
- Water
- Flask
- Small bottles
- Petri dishes
- 1 cm × 1 cm cuvettes

EQUIPMENT

- Heater with stirrer

METHODS

1. Preparing polymer solution in a flask:
   a. Add calibrated amount (weight) of PVA powder to water. We do not need to precisely weigh PVA since we want to achieve only an approximate concentration of 6–15% w/w. Of course, if you want to know precisely the PVA concentration you need to use a balance, usually, we prepare 10% w/w (after the preparation, this solution will have a honey-like consistency).

      Stir the solution for a short time (a magnetic stirrer is very useful). When stirring stops the PVA will form a thick layer at the flask bottom.

      i. The concentration will determine the viscosity of the solution and later the thickness of the film, the lower concentration will result in a thinner film. For a spin coating and obtaining very thin films, we will use very low PVA concentrations (1% or less).

   b. Warm the solution to 80–95°C continuously stirring for a few hours (the necessary temperature may depend on PVA molecular mass). The solution may slowly/gently boil for a short time. But be careful since overheating will result in the slight

(*Continued*)

EXPERIMENT 3-24 (CONTINUED) PREPARING ISOTROPIC PVA FILMS AND PVA FILMS DOPED WITH CHROMOPHORES

FIGURE 3-24.1 PVA powder/granules in water.

yellowish color of the solution. To avoid overheating, we usually put the flask inside a beaker filled with water.

i. Originally, the powder will form a solid layer on the bottom of the flask similar to undissolved sugar (Figure 3-24.1).
ii. After 2–4 hours at an elevated temperature (80–95°C), the PVA should be polymerized and form a viscous perfectly transparent solution; similar to propylene glycol or glycerin.

2. Introducing the chromophore into PVA solution in small bottles.
   a. Dissolve a high concentration of the selected chromophore in a small amount (~1 mL) of water or ethyl alcohol.
   b. Pour PVA solution into smaller bottles (10–20 mL) (this will be the solution used to make the film).
   c. Pour a few microliters of chromophores (in water or EtOH) into the solution and mix at a warm temperature. You may check absorbance or judge by the color to decide if the concentration is in an expected range. If it is too low, add more dye. The PVA solutions with chromophores will look like these in Figure 3-24.2. There is no need to precisely control dye concentration since during the drying process concentration will increase due to the polymer volume decrease 6–15-fold

(Continued)

EXPERIMENT 3-24 (CONTINUED) PREPARING ISOTROPIC PVA FILMS AND PVA FILMS DOPED WITH CHROMOPHORES

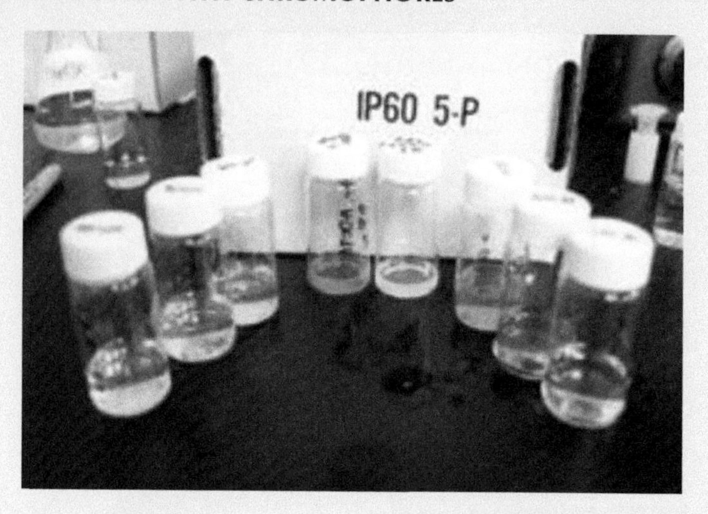

FIGURE 3-24.2 Various dyes in PVA solution.

depending on the initial PVA concentration. The 10% PVA solution will shrink its volume about 12-fold upon drying.

d. It is advisable to create multiple PVA + chromophore solutions with different concentrations and/or using a different amount of PVA solution to get films of needed/adequate spectroscopic properties.

3. Preparing polymer films using a Petri dish.
   a. We will prepare PVA only as a reference, and PVA + chromophore as a sample. Pour PVA and PVA + chromophore solutions into two leveled Petri dishes. The solution will form a uniform layer except at the edges where it will be thicker (due to meniscus).
   b. Leave the Petri dishes in a dark dust-free place and allow them to dry for at least 24–48 hours (Figure 3-24.3).
      i. Drying the solution too quickly will result in the film having uneven thickness (wavy film surface).
   c. When the film is dried it becomes solid and it can be gently removed from the Petri dish using a sharp edge (e.g., tip of the knife) starting from the edges.

Figure 3-24.4 shows examples of multiple films with various dyes embedded. If the film was prepared correctly it should be clear and transparent. When doped with visible and red dyes they will typically present colors.

Notes: There are many potential mistakes when preparing PVA films. Quick-drying may result in a not-even surface (a wavy surface). Too high concentration of lower solubility dye may result in dye crystallization showing small crystals or in some cases film may become visually cloudy/milky. For an example of good dried film see Figure 3-24.4. Do not use buffers; salts will crystallize and make films cloudy.

(*Continued*)

EXPERIMENT 3-24 (CONTINUED) PREPARING ISOTROPIC PVA FILMS AND PVA FILMS DOPED WITH CHROMOPHORES

FIGURE 3-24.3 PVA solutions in Petri dishes.

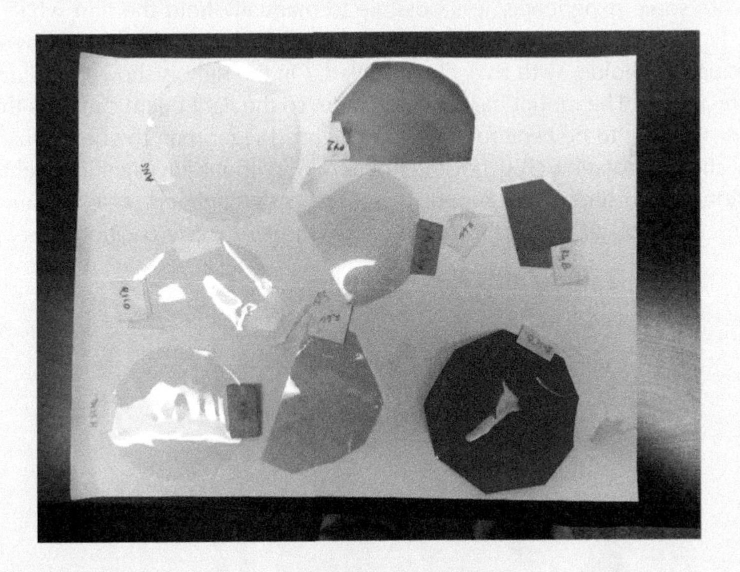

FIGURE 3-24.4 Dried PVA films containing various chromophores.

If the chromophore is not soluble in water we may try to dissolve in ethyl alcohol (EtOH). Many chromophores will dissolve much easier. But sometimes it could be difficult to prepare sufficient concentration of non-polar chromophores even in EtOH.

If the chromophore is dissolved in EtOH add the small amount of EtOH solution to the PVA solution. The part of PVA solution in-contact with EtOH may become cloudy. Do not worry, warm-up the PVA solution to about 70–80°C and mix vigorously. When it is

(Continued)

EXPERIMENT 3-24 (CONTINUED) PREPARING ISOTROPIC PVA FILMS AND PVA FILMS DOPED WITH CHROMOPHORES

clear add more EtOH solution of chromophore while stirring the hot solution. But remember, there will be a limit for maximum concentration of the dye in PVA so, when it is too high the dye will crystallize or precipitate during drying. The final solution should be transparent. If solution contains air bubbles keep warm the solution to let bubbles float to the surface. Sonicating solution for a short time will help.

The authors had success with chromophores such as anthracene, acridones, DPH, etc. when introducing them through EtOH solution into liquid PVA solution.

EXPERIMENT 3-25 MEASURING ABSORPTION OF PVA FILMS

Keywords: absorbance, transmittance, light reflection, scattering, thin films, PVA.

In this experiment, blank PVA films (films without the chromophore) prepared according to the procedure in the previous Experiment 3-24 will be used. Since the films are thin and flexible, a special mount will need to be prepared to measure absorption, which will allow for a gentle tautening of the film to make it straight when mounting them in the spectrophotometer chamber. With some experience, it is possible to manually hold the film while measuring absorption. Figure 3-25.1 shows the photograph of one of the mounts used by us. On the left is the actual film holder with PVA film installed. On the right is shown the stand with the mount for the holder. The mount has a central hole, so the light beam can pass through. It is important for the hole to be centered and only slightly bigger than the beam size (diameter) allowing for the free rotation (360°) of the holder/film without any significant change to the beam position on the film. If the mount is not perfectly centered, a significant change in transmission can be observed when rotating the film. *A properly positioned central hole is*

FIGURE 3-25.1 Rotary assembly for measuring films absorption.

(*Continued*)

EXPERIMENT 3-25 (CONTINUED) MEASURING ABSORPTION OF PVA FILMS

crucial when measuring oriented (stretched) films in later experiments. A small opening also ensures that we will monitor the same part of the film for any position. A large opening will obviously ensure that the beam will always completely pass through, but when rotating the film the beam will pass through the different parts of the film.

The goal of this experiment is to evaluate the extinction of light by the PVA film. The light extinction is due to reflection on two interfaces, small scattering, and absorption of PVA.

MATERIALS

- Polyvinyl alcohol (PVA) films made according to the procedure in Experiment 3-24.
- Film holder (custom made); see Figure 3-25.1

EQUIPMENT

- Spectrophotometer with custom-made film holder adapter; see Figure 3-25.2

METHODS

1. Measure the baseline with an empty sample compartment (not stand mounted).
2. Install the holder and measure absorption or transmission with only the film holder installed.

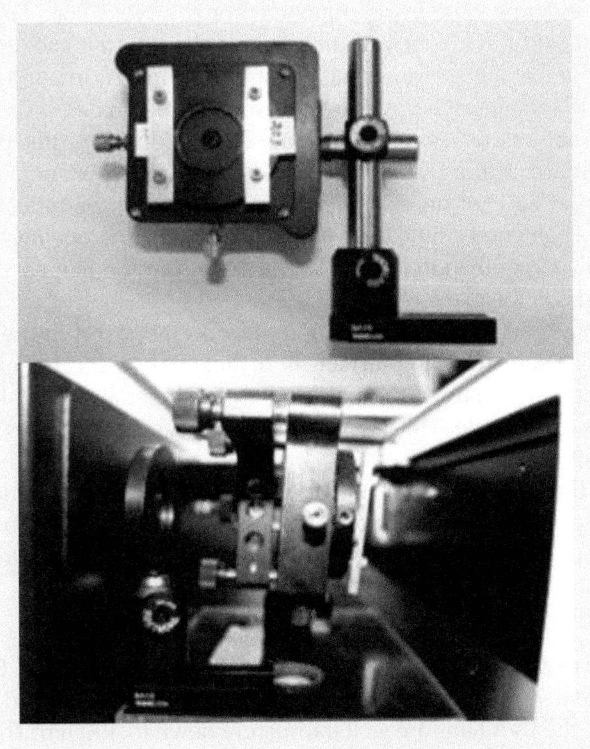

FIGURE 3-25.2 Assembly with mounted film (Top) and a mounted assembly in Cary 50 spectrophotometer for absorption measurement.

(Continued)

EXPERIMENT 3-25 (CONTINUED) MEASURING ABSORPTION OF PVA FILMS

a. Install the stand in the spectrophotometer chamber and attach the mount holder with no film mounted first. Measure absorption (transmittance) for a few positions of the holder when rotating holder 360°. Measured absorption should be minimal (transmission close to 100%) for any position around 360° rotation (most importantly for 0°, 90°, 180°, 270° the reading should be the same for each position that we will later use for oriented films measurements). *Note:* For some spectrophotometers, the beam diameter is large and it should be reduced by the large pinhole to ensure that the beam diameter will be smaller than a typically narrow strip of the film. In this case, the hole will reduce intensity showing a large apparent absorption (reduced transmittance). It is important that the absorption reading is the same for each holder orientation. The best way for ensuring this is to install a fixed pinhole before holder that will reduce the beam diameter to a smaller size than the opening in the holder (remember—baseline measurement in #1 should be done with pinhole only—not film holder installed).
b. Measure the baseline with the holder installed (no film). Then measure transmittance or absorbance for a few holder orientations. The measured absorbance should be very close to 0 (transmittance ~100%) for each holder orientation. Measuring non-zero absorption for some positions indicates that part of the beam does not go through (typically the center of rotation is not in the center of the hole opening) and the holder needs to be repositioned properly. *Note:* For a large beam diameter when installed the holder with the pinhole the transmittance can be lower than 100% if not the entire beam is pathing through. But it is important that attenuation will be identical in each position when rotating the holder.
c. Take the holder out and attach (install) the isotropic (un-stretched) film on a flat holder surface with a hole in the middle using the screw and aluminum clamp(s), see Figure 3-25.1 for an example of the holder with the mounted film. The film should be tightened a little to ensure a flat surface. The opening in the film mount is typically larger than in the holder, so make sure the film covers the entire hole and it is flat in the plane of the holder.
d. Mount the holder with the isotropic film on the stand in the spectrophotometer. Make sure the film surface is orthogonal to the beam of the spectrophotometer. The part of the holder with the hole should be in the beam path and only the sample will be exposed to the beam (not edges of the hole). See Figure 3-25.2.
e. Measure the absorbance with the PVA film installed. It is recommended to measure absorbance for a few holder orientations. Measured absorption (transmission) for the isotropic film should not depend on holder orientation (angle of rotation).

RESULTS

The measurements of polymer film do not require the use of the cuvette. Each film can and will have a slightly different (thickness) and accounting for the blank should be done in different ways. Typically, the PVA films are well transparent down to 300 nm. The refractive index of PVA will depend on molecular weight and how well the film has been dried. In practice, it may vary between 1.45 and 1.52 and is relatively close to quartz, UV silica, or just glass.

(*Continued*)

EXPERIMENT 3-25 (CONTINUED) MEASURING ABSORPTION OF PVA FILMS

For measurements, we used films mounted in the holder as shown in Figure 3-25.1. Two aluminum pieces can be screwed to push and hold PVA film that is slightly tensioned to be straight against the surface of the holder assembly (make sure the film is not stretched). The stand for the sample holder has an opening (hole) that is only slightly larger than the spectrophotometer beam size.

We recommend measuring light extinction (apparent absorption) for a few films of different thickness. The thickness of each film should be precisely measured (for measurement of film thickness we can use micrometer). As an example, we selected three films with different thickness. We measured apparent absorption for each film. Since a perfectly central mounting is not possible for practicality we recommend measuring absorption in four positions 0°, 90°, 180°, 270° by rotating the holder and averaging the results. It is important that the maximum difference for measured apparent absorption will not be more than 5%. If the deviation is significant, we need to realign the central hole and repeat the process.

Figure 3-25.3 shows measured average absorptions for three films of different thickness. The absorption spectra show distinct features for wavelengths below 400 nm and in the range over 450 nm is practically flat but significantly higher than 0 (as expected—see the Experiment 3-1). The absorption of the film starts about 450 nm and increases into shorter wavelengths. Also, at longer wavelengths independently of the film thickness, the light extinction that is mostly due to reflectance is very close. The reflection of light at boundaries of different dielectrics (air/PVA and PVA/air) is a dominant factor for light attenuation at wavelengths larger than 450 nm. As we discussed in earlier experiments (Experiment 3-1 and 3-3) at a boundary of two dielectrics, a light beam with normal incidence has a transmission, *T*, of:

$$T = \frac{4n_1n_2}{(n_1+n_2)^2} \tag{3-25.1}$$

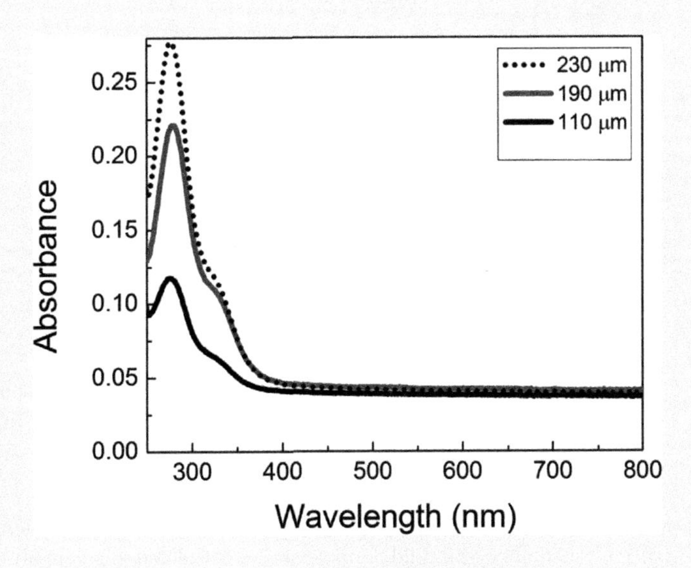

FIGURE 3-25.3 The absorbance of three PVA films of different thickness.

(Continued)

EXPERIMENT 3-25 (CONTINUED) MEASURING ABSORPTION OF PVA FILMS

where n_1 and n_2 are indices of refraction for air and PVA, respectively. We can safely assume that refractive index for air $n_1 = 1$ and refractive index for PVA is wavelength dependent and can be approximated by:

$$n_{PVA} = \sqrt{1.286 + \frac{1.070 * \lambda^2}{\lambda^2 - 1.006 * 10^{-2}} + \frac{1.102 * \lambda^2}{\lambda^2 - 100}} \tag{3-25.2}$$

It is important to remember that for different films (differently prepared) the parameters in Eq. 3-25.2 can vary slightly. The transmittance of PVA film due to reflectance on two dielectric interfaces is (see Experiment 3-3 for details):

$$T_{PVA} = \frac{4n_{PVA}}{(1+n_{PVA})^2} * \frac{4n_{PVA}}{(1+n_{PVA})^2} \tag{3-25.3}$$

And the corresponding apparent absorbance due to film reflectivity will be:

$$A_{PVA}^r = \log \frac{1}{T_{PVA}} \tag{3-25.4}$$

To calculate true absorption of the PVA film, we selected one film (230 μm thick) and the details for calculation are presented in Figure 3-24.4. In Figure 3-24.4, we present measured absorption of the PVA film (dotted line) in 240–800 nm range.

For comparison, we drawn calculated apparent absorption due to the film reflectivity as calculated from Equation 3-25.4 (dashed line). In the wavelength range over 400 nm, the

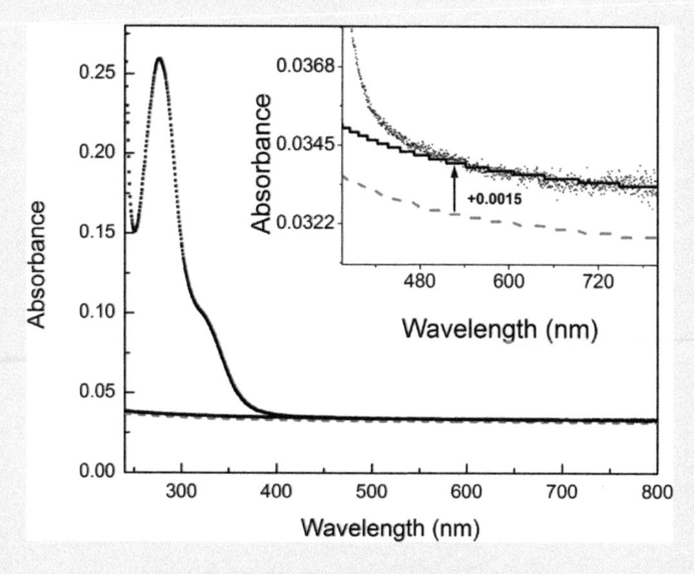

FIGURE 3-25.4 Measured absorption of 230 μm PVA film (dotted line) and calculated light attenuation due to reflectance (dashed line). The solid thin line shows calculated apparent absorption due to reflectance corrected for constant factor 0.0015 as presented in the insert.

(Continued)

EXPERIMENT 3-25 (CONTINUED) MEASURING ABSORPTION OF PVA FILMS

calculated apparent absorption is very close to the measured one. The insert in Figure 3-25.4 shows an expanded scale for long-wavelength range. We can notice that measured values are slightly shifted as compared to calculated values. The shift is shown in the insert and is 0.0015. It is impossible to make two perfectly identical films or even find two identical places on a single film. The film surface will typically have some small imperfections formed during drying or film can be contaminated with some dust particles. So, the measured beam attenuation in the non-absorbing range can be slightly higher than predicted from reflectivity only (*Note:* The shift of 0.0015 is close to the spectrophotometer resolution). This is mostly due to scattering on the surface and inside the film. The scattering particles and surface defects are relatively large (comparable to the wavelength) and predicting the pattern is very difficult. After many measurements, we can assume with good confidence that typical differences are within 10% of theoretically predicted attenuation due to reflectance. A higher value of the shift indicates major defects in the film and such film should not be used. To calculate true PVA film absorption, we need to subtract apparent absorption resulting from reflectance (that includes a small shift due to the film imperfection) from measured total absorbance. Two factors that we are correcting for (reflectance and small attenuation due to marginal scattering) are constant factors that are not affected by the film thickness. Corrected true absorption of the film will be proportional to the film thickness. Measured absorption of any PVA film can be represented as the sum of apparent absorption resulting from reflectance plus a small shift due to the film imperfection and absorption just corrected for the film thickness (similar to measured absorptions of dye solution in different thickness cuvettes). But reflective and scattering components are assumed to be independent of the film thickness.

Using the above approach, we calculated true absorbance for all three films presented in Figure 3-25.3. The thickness of PVA film can be precisely measured with a micrometer caliper, but we need to remember that this is an only approximation and the film thickness will depend on various external conditions (humidity, temperature, etc.). Out of these three measurements, we can calculate average PVA absorption per given thickness (100 μm for example). Such average absorbance of 100 μm thick PVA film is presented in Figure 3-25.5. In Figure 3-25.5, we also included an apparent absorption due to reflectivity as calculated from the wavelength-dependent refractive index of PVA. It is important to realize that in the visible range reflectivity of the film is a dominant contributor to light attenuation due to the PVA film.

In many simple experiments when ghost molecules (chromophores) absorb in the visible range (out of PVA absorption), we can use a reference (blank) PVA film. In single-beam spectrophotometer, we measure blank with just the reference PVA film. When we do not have good reference film, it is also ok in the visible range to use glass/silica plate/sheet (it can be a single cuvette wall or thin (#1 or #2) microscopy slide) for reference measurement. In the visible range, the background (reference) from glass/silica slide is close to PVA. We will not recommend using an empty cuvette since it has two walls and four reflective surfaces—see Experiment 3-3. When the refractive index is well matched the measured absorptions at a wavelength where chromophore and PVA are not absorbing should be 0 (transmittance 100%).

Warning: We want to bring to the reader's attention that measuring the thickness of PVA is not very reliable. For example, the measured thickness may vary with humidity, etc. So, typically one would use good judgment and common sense to approximate the absorbance of

(*Continued*)

EXPERIMENT 3-25 (CONTINUED) MEASURING ABSORPTION OF PVA FILMS

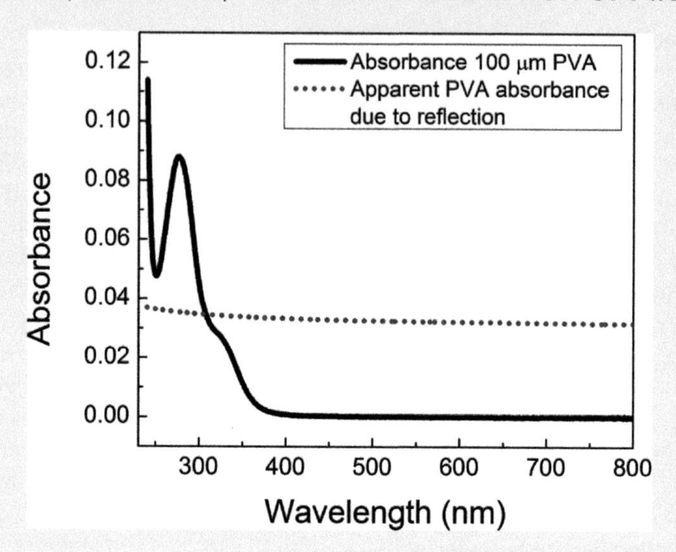

FIGURE 3-25.5 PVA absorption calculated for standard 100 μm PVA thickness (solid). Dotted line represent apparent absorption as calculated from the reflectivity.

PVA. After proper subtraction of PVA baseline, the absorption of embedded chromophore should reassemble expected absorption of this chromophore. Typically, we would recommend using PVA in the range where the contribution from PVA absorbance is below 20% and typically above 250 nm (below 250 nm PVA matrix absorbs too much and is unusable). The spectral range above 350 nm is rather safe and correction for PVA absorbance is straightforward.

EXPERIMENT 3-26 MEASURING ABSORPTION OF CHROMOPHORES IN THIN PVA FILM

Keywords: absorbance, transmittance, light reflection, scattering, thin PVA films, rhodamine 6G.

In this experiment, a step-by-step procedure for measuring chromophore absorption (rhodamine 6G) in PVA film will be presented. Such films containing a chromophore can be prepared according to the procedure described in Experiment 3-24. The procedure for measuring absorption of blank PVA film is discussed in the previous experiment.

MATERIALS

- Polyvinyl alcohol (e.g., Aldrich Mowiol—MW 130,000 but other MWs over 50,000 are ok) film prepared according to the procedure in Experiment 3-24.
- Polyvinyl alcohol film containing rhodamine 6G (R6G) prepared according to the procedure in Experiment 3-24.
- Film holder and mounting stand adapter to the spectrophotometer.

(*Continued*)

EXPERIMENT 3-26 (CONTINUED) MEASURING ABSORPTION OF CHROMOPHORES IN THIN PVA FILM

EQUIPMENT

- Spectrophotometer

METHODS

1. Prepare PVA film containing R6G (Figure 3-26.1), select a clear uniform part and cut a small piece of the film. Typically, a piece 0.5 × 1 inch is more than sufficient.
2. Mount an empty holder (no mounted film) in spectrophotometer and measure baseline (as in Experiment 3-24).
3. Mount the PVA film containing the chromophore and measure absorption.

RESULTS

In this experiment, we will present a measurement of PVA film containing R6G. To account for blank PVA absorption, we will use results from the previous Experiment 3-25 where we calculated film reflectivity and calibrated to 100 μm film absorbance.

Figure 3-26.2 shows measured total apparent absorption (dashed line) of the PVA film containing rhodamine 6G (R6G). Remember measured light attenuation (extinction) is due to dye absorption, PVA film absorption, reflection, and small scattering. In Figure 3-26.2, a distinct absorption band with maximum absorption in the range of 530 nm dominates the absorption spectrum but a number of small absorption bands in UV can be clearly identified. The absorption of the R6G film decreases toward long wavelengths and in the range above 600 nm light extinction is solely due to film reflectivity (similar to the previous Experiment 3-25 with just PVA film where absorption ended at 450 nm). The total apparent absorption of the film can be resolved to three main components: (1) reflectivity that can be fitted to points above 600 nm, (2) absorption of PVA that contribute to absorption spectrum below 450 nm, and (3) absorption of Rh6G. To obtain a true R6G spectrum in PVA matrix,

FIGURE 3-26.1 Photograph of PVA film containing rhodamine 6G.

(Continued)

EXPERIMENT 3-26 (CONTINUED) MEASURING ABSORPTION OF CHROMOPHORES IN THIN PVA FILM

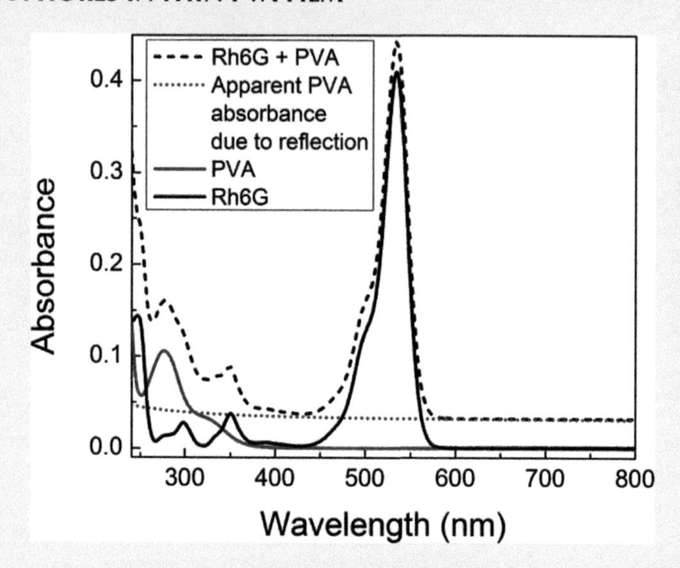

FIGURE 3-26.2 Measured absorption of R6G embedded into PVA film (dashed line), calculated from PVA thickness PVA absorption (solid gray line), and calculated apparent absorption due to film reflectivity (dotted line). Solid black line presents recovered absorption of R6G in PVA matrix.

we need to subtract from the total apparent absorption contribution of film reflectivity and PVA film absorption. *Note: For high chromophore concentration when chromophore absorption is dominant we can just use as reference a blank PVA film of similar (close) thickness made in identical conditions as the sample.* But in this example, we will present a detailed general procedure. In Figure 3-26.2, the dotted line represents calculated reflectivity that well matches measured values in the spectral range above 600 nm. Similar to PVA in Experiment 3-25, a small constant value of −0.0008 needs to be added (subtracted) to correct for film imperfection. As in this case, the correction value can be negative (the reflectivity of the Rhodamine film was lower than the predicted value). Perhaps our reference reflectivity value that has been calculated in a previous experiment based on an average of few films has been slightly overestimated. *Note: This is rather common and if the difference is really small like in this case (below 5%), we believe this is not a problem.* The solid gray line shows the PVA absorbance that was calculated from average PVA absorbance of 100 μm thick film calibrated for the thickness of the R6G film. In this case, the Rhodamine film was 190 μm thick and multiplication factor was 1.9. In Figure 3-26.2, the solid line represents a spectrum of R6G corrected for PVA absorption and reflectivity of the PVA matrix.

Figure 3-26.3 shows the corrected absorption spectrum of R6G (solid line) and spectrum with an expanded absorption scale (dashed line—multiplied by 10) to visualize UV spectral range. Also in the range above 600 nm, we observe a small variation from zero value for expanded absorption range. This indicates that used approximation for the refractive index of PVA was not exact or the film had some not-defined imperfection redirecting or scattering the beam. Upon drying, the film surface is not perfectly flat and some dust particles can

(*Continued*)

EXPERIMENT 3-26 (CONTINUED) MEASURING ABSORPTION OF CHROMOPHORES IN THIN PVA FILM

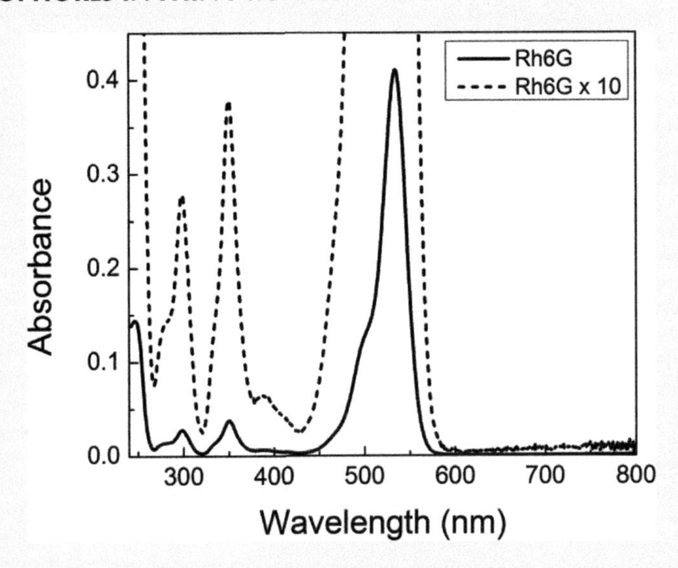

FIGURE 3-26.3 Expanded spectrum of R6G to visualize UV range.

randomly be captured in the film. Typically, we could select a better film to minimize these effects but such small deviations are quite common when measuring polymer films.

Note: We want to stress that in practice measured thickness of the film is only an approximation. The PVA is very hydroscopic and its thickness may depend on the humidity and other factors. So, when subtracting the PVA absorption we may need to vary PVA thickness by a few percent to obtain the best absorption spectrum of the host molecules. In practice, we can measure absorption in a much simpler way by just using as reference blank PVA film. For the visible part of absorption (above 400 nm), this is the easiest approach. But for absorptions below 350 nm, we will need good reference film. If the absorption of the ghost chromophore is small in UV range this could be a problem.

CHAPTER 4

Fluorescence—Steady-State Phenomena

INTRODUCTION

Jameson in his book "*Introduction to Fluorescence*" defined fluorescence in a simple sentence: "When the light of a particular color shines on certain types of materials they give off the light of a different color after a very, very short time interval." This simple definition says all of the most important attributes of the phenomenon called fluorescence. A higher energy of light is needed to stimulate the fluorophore to give a lower energy (different color) light after a short period of time. The important facts to remember are: (a) the color of emitted light in most cases is independent of the color of excitation light that is always a shorter wavelength (higher energy) than the emission (with the only exemption of multi-photon excitation). (b) The short time interval between excitation and emission is called the fluorescence lifetime, which is also independent of the wavelength of the excitation light. Figure 4-1 shows various fluorophores illuminated with UV light (left) and with room light (right). All fluorophores are illuminated with the same UV light but each one gives out a different and unique color presenting the beautiful rainbow of colors. Each of these colors reflects the characteristic emission spectrum of each fluorophore. The emission spectrum is one of the most important and most evident characteristics of fluorescence.

In this chapter, a set of steady-state experiments are described that should prepare the reader for a task on how to properly measure and interpret emission spectra. Emission spectra represent the distribution of energy given out (also called emitted) by a chromophore returning from the lowest excited state to the ground state as a function of wavenumber or wavelength. The emission spectrum is the basic characteristic of the chromophore, and in studying any system it is typically measured first after getting the absorption spectrum. It may seem trivial, but to properly measure an emission spectrum that could be compared between different instruments and different labs can sometimes be a difficult task. Detected fluorescence signals depend on many unstandardized factors that are difficult

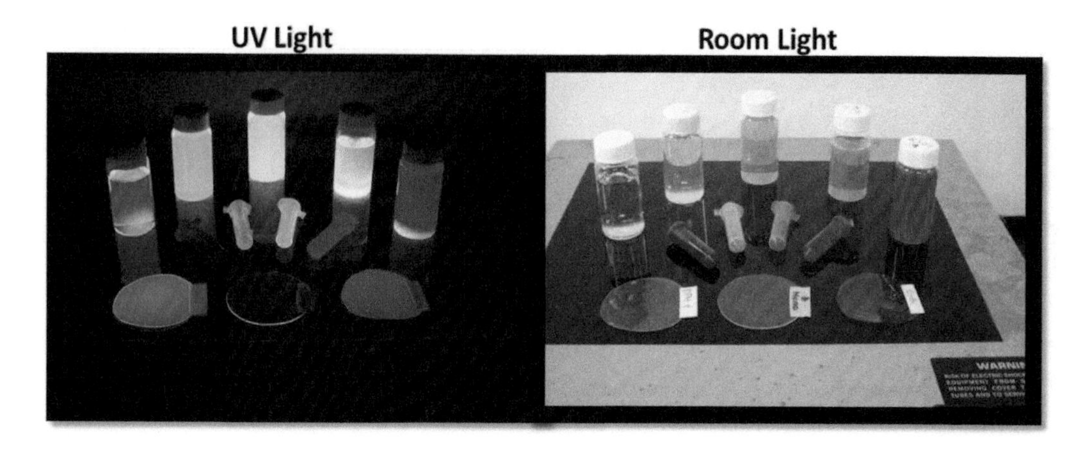

FIGURE 4-1 (Left) Examples of fluorescence from solution and solid materials. (Right) Photograph of elements shown in left in room light (no UV illumination).

to control. Among them is the intensity of an excitation light that is different for every source, geometry, and quality of the optical system that is impossible to standardize, and the efficiency of the detection system. These difficulties were not a problem in absorption measurements, discussed previously in Chapter 3 because absorption measurement always used a baseline reference. In the measurements of the ratio of baseline/sample intensities of transmitted light, experimental factors are canceled. In that sense, absorption measurements are intrinsically ratiometric. In contrast, fluorescence is measured on a dark background without the possibility of canceling experimental factors. Often, however, we will look at the relative change of fluorescence using the same conditions and instrument. But even a comparison of the relative changes in intensity within a single experiment should be done with caution with regard to experimental conditions, like magic angle (MA) for example.

Magic Angle (MA) in Steady-State Measurements

To avoid problems associated with photo selection by the excitation light and consequential emission polarization, the emission spectra should (must) be measured under MA conditions. These conditions can be different for various system geometries, but the overwhelming majority of fluorescence experiments are made in square or front-face geometry with cylindrical symmetry. In this case, it can be calculated that for a vertical polarization of excitation light the magic angle corresponds to an angle of 54.7° to the vertical on the observation line (excitation light polarization). In other words, the excitation polarizer must be set vertically (at 0°) and emission (observation) polarizer must be rotated to 54.7° (MA). Physically, MA observation means that the vertical (V) and horizontal (H) components equally contribute to the observed total emission intensity and intensity observed at this angle will not be affected by chromophore rotation. This also means that for the perfect instrument, where no corrections are needed, one should measure intensity at the magic angle equal to $I_{\mathrm{MA}} = \left(I_{\mathrm{V}} + 2I_{\mathrm{H}}\right)/3$. Where I_{V} and I_{H} are emission intensities measured for vertical and horizontal observations respectively when excitation light polarization is vertical. This concept will be tested later using Rhodamine 6G (R6G) that emits in 560–620 nm range when mixed in a solvent of different viscosities, see Experiment 4.3.

First, it is important to learn why and how emission spectra depend on observation polarization. For that one needs to realize that almost all fluorescence systems will have polarizers on excitation and on emission. Even without polarizers, the excitation beam may have some residual polarization; and yet a perfectly isotropic beam has electric vectors oriented/polarized in the plane orthogonal to the direction of propagation yielding significant polarization of the fluorescence. It is important to keep in mind that isotropic beams (light intensity distributed equally around the direction of propagation) are very rare (difficult to obtain) in spectrofluorometers or spectrophotometers. For these reasons, polarized excitation has always been used and the fluorescence signal will typically have well-defined emission polarization. To realize the potential error to be made, let's consider a few simple cases of samples that present identical total fluorescence but have different anisotropies. For a sample with the anisotropy of 0.4 (maximum for isotropic solutions), the ratio between parallel (vertical) and perpendicular (horizontal) components is 3 ($I_{\|}/I_{\perp} = 3$; or $I_V/I_H = 3$). In Table 4-1, the relative values are shown of expected intensity components seen for parallel, vertical, and MA (54.7°) observations respectively for samples with different inherent polarizations. The total emission intensity ($I_{\|} + 2I_{\perp}$) is fixed to be 1.

It is clear that both, parallel and perpendicular components change significantly, but the intensity observed under the MA is constant and completely independent from sample polarization. This is a very important property of MA observation for which observed intensity is equal to the average value of parallel and perpendicular (two perpendicular) intensity components. When comparing emission intensities between different samples one would make a significant error by using improper polarization (polarizer orientation) on observation or no polarizers at all. This may easily happen if someone wants to evaluate a quantum yield of an unknown fluorophore by comparing to a known standard, but the standard and sample are likely to present different polarizations. It is worth noting that $I_{iso} = 1/3\ (I_{\|} + 2I_{\perp})$ is called isotropic intensity for samples with cylindrical symmetry and is equivalent to intensity measured with MA observation.

Note: The term "polarization" is used to describe the incoming light, while the term "anisotropy" is frequently used to describe fluorescence emission. Both terms are related as will be explained in detail in Chapter 6.

TABLE 4-1 Anisotropy Values and Corresponding Relative Polarized Intensities of Fluorescence

Sample Anisotropy	$I_{\|}$	$I_{\perp}$	I_{MA}
0.4	0.6	0.2	0.333
0.25	0.5	0.25	0.333
0.1	0.4	0.3	0.333
0	0.333	0.333	0.333
−0.1	0.267	0.367	0.333

Spectral Overlap and Mirror Symmetry

In most cases, absorption and fluorescence spectra partially overlap each other. Examples of a relatively large spectral overlaps are presented in Figure 4-2. With such strong spectral overlaps, higher optical densities (absorptions) for a significant part of emission light will deform the observed fluorescence spectra due to a reabsorption process. Also, these fluorophores will be prompt to self-quenching and depolarization. The use of very low concentration (absorptions) is recommended. When the use of higher concentrations (absorptions) is necessary, the observed spectra should be properly corrected.

There are also many fluorophores with medium and small spectral overlaps. Figure 4-3 presents fluorophores with extremely small spectral overlaps.

Three of the four spectra presented below show relatively good mirror symmetry with a vertical axis going through cross-section point of the absorption and fluorescence spectra (called sometimes 0-0 transition). And, in general, most fluorophores show such mirror symmetry. However, there are some fluorophores with clearly asymmetrical spectra. Below, in addition to $[Ru (bpy)_3]^{2+}$ are two examples presented which lack mirror symmetry (Figure 4-4). The first, quinine sulfate in 1N H_2SO_4, has a deformed absorption spectrum. The second, p-terphenyl in cyclohexane shows vibrational (oscillation) structure in

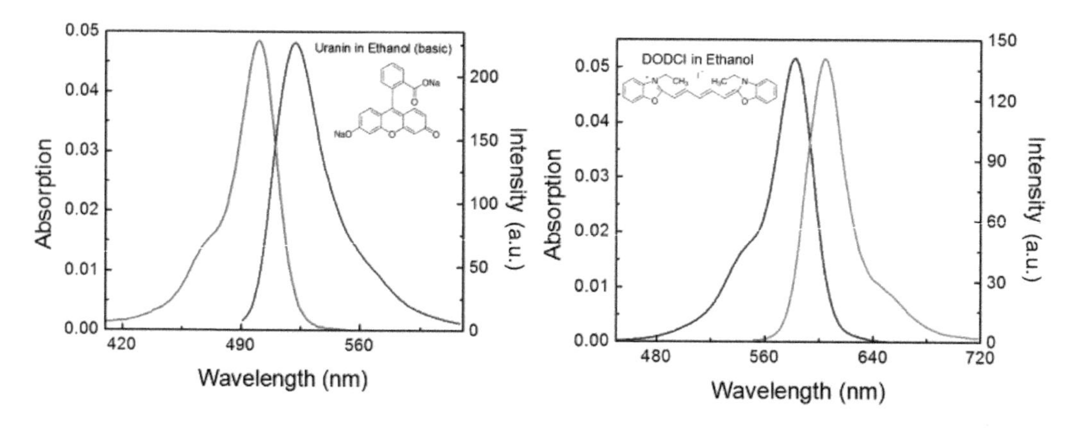

FIGURE 4-2 Absorption and fluorescence spectra of fluorophores with large spectral overlaps.

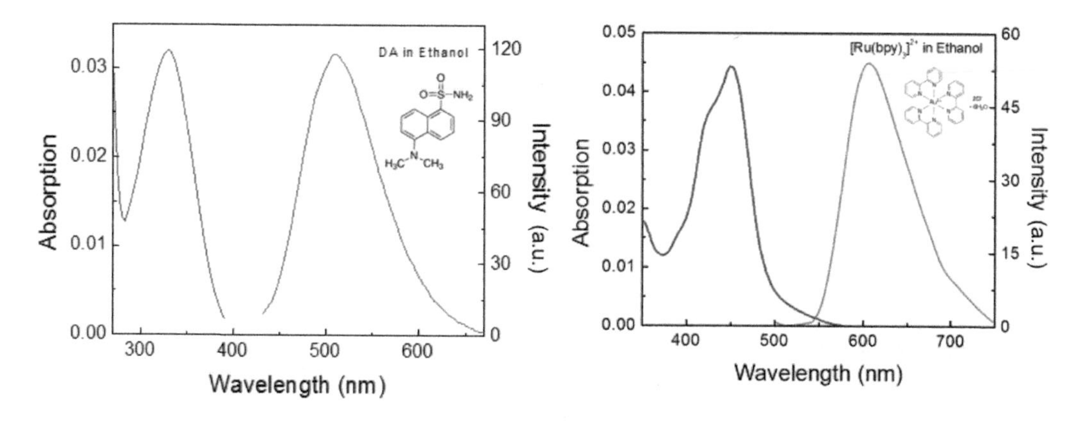

FIGURE 4-3 Absorption and fluorescence spectra of fluorophores with small spectral overlaps.

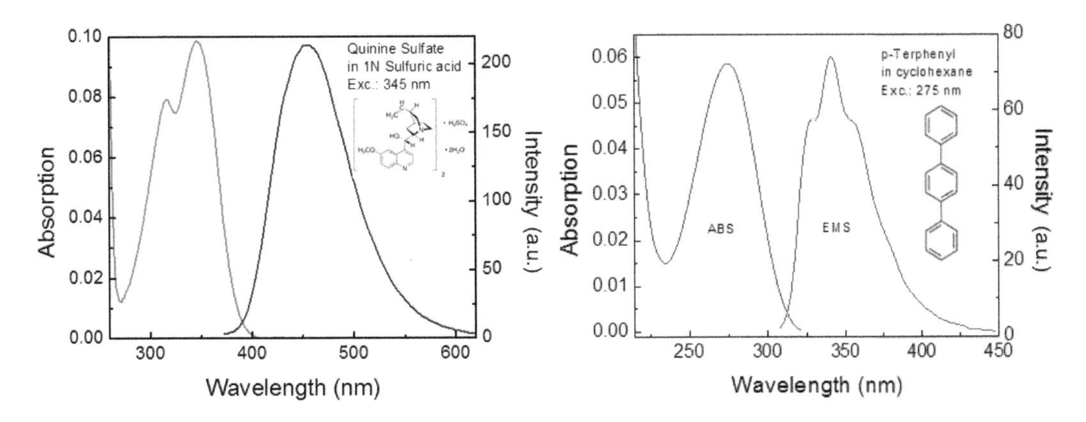

FIGURE 4-4 Absorption and fluorescence spectra of fluorophores with a broken mirror-symmetry. Left: quinine sulfate in 1N H_2SO_4. Right: p-terphenyl in cyclohexane.

the fluorescence spectrum. *Why do these fluorophores not show the mirror image symmetry?* In the case of quinine sulfate, the second electronic transition is very close to the first and manifests itself in the short-wavelength shoulder in the absorption spectrum. The second case (p-terphenyl) is different. In the ground state, the phenyl rings have full freedom of rotation around carbon bonds. While in the excited state, this freedom is hindered and the phenyl rings cannot rotate. The lack of freedom for phenyl ring rotations (molecule becomes rigid in the excited state) exposes oscillation states which are now visible in the fluorescence spectrum.

In Chapter 2, we discussed typical configurations for spectrofluorometers. Herein, in Figure 4-5, we present the photograph of various spectrofluorometers we have been using

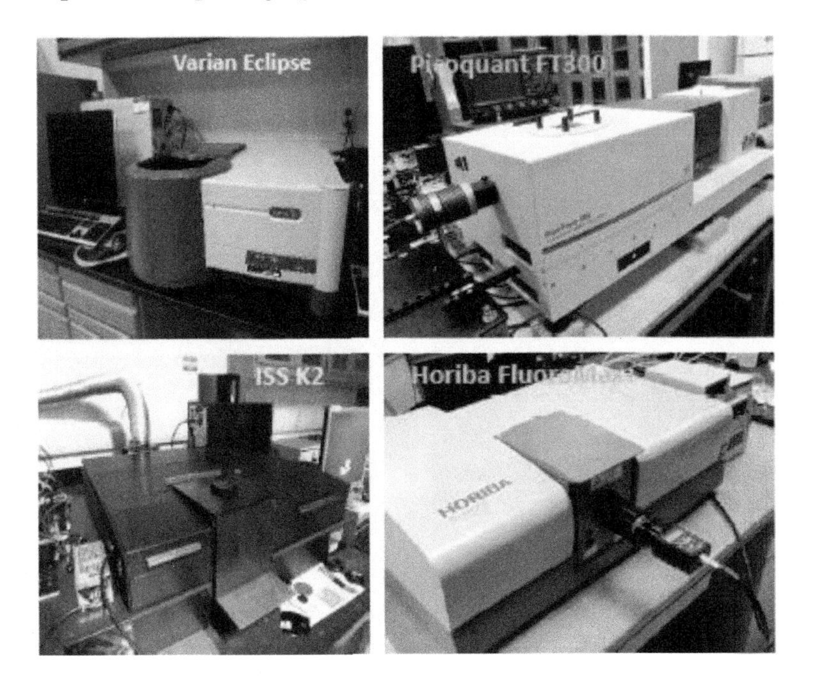

FIGURE 4-5 Example of various spectrofluorometers used in experiments.

for experiments. These four different spectrofluorometers are representative of general design and any other instrument (including homemade) will have a similar configuration.

Usually, one does not have easy access to electronics (this is now software controlled), excitation and detection lines. However, a sample chamber can be easily open and we can check multiple features or incorporate some modifications. For example, introducing or checking polarizers incorporate cuvette adaptors or changing a square geometry to a front-face geometry. Typically, the sample chamber is the central part of a spectrofluorometer. Figure 4-6 shows sample compartments (sample chambers) for four spectrofluorometers with marked excitations and emission lines. Conversely, Figure 4-7, shows just sample holders. The sample holder is the element that holds the cuvette (typically a standard 1 cm × 1 cm cuvette) that has three or four openings (for comparison, a sample holder in a spectrophotometer for absorption measurement has two openings only for light to go through). These openings are typically smaller than the cuvette size, depending on the vendor, to avoid any scatterings on cuvette edges/walls. As seen in Figure 4-7, the designs are slightly different. The type of sample holder used varies per application. For example, the Varian Eclipse sample holder has three openings and the experimentalist cannot observe a signal from the fourth direction. But the sample holder is easy to remove to externally check the cuvette positioning. The ISS sample holder allows for easy automatic change between two cuvettes (typically sample and reference). The FT300 or FluoroMax+ sample compartment allows access from each of the four sides.

FIGURE 4-6 Open sample compartments of Varian Eclipse, Picoquant FT300, ISS K2, and Horiba FluoroMax + spectrofluorometers.

FIGURE 4-7 Sample holders of the four used fluorometers.

For some special experiments when the sample volume is limited, we can use a special cuvette, like 2 mm × 10 mm with external dimensions identical to the standard cuvette. In some cases, cuvettes that are 3 mm × 3 mm (or smaller) can be used with a special adapter to fit it to a regular sample compartment.

Figure 4-8 shows such cuvettes, an adapter, cuvette in the adapter, and cuvette installed in the standard sample holder. When using such cuvettes, special attention should be taken to make sure the excitation illuminates the center. Typically, such a small cuvette can easily be used with laser excitation, but it is more difficult with regular lamp excitation. The advantage is that the sample volume can be really small and for a 2 mm × 2 mm cuvette 50 μL is sufficient (about 25-fold less than a 10 mm × 10 mm cuvette).

Note: Changing cuvette to smaller one or changing the configuration of the spectrofluorometer will always affect the fluorescence intensity signal. We need to remember that fluorescence signal is relative and does not have absolute meaning like absorbance measurements that is the relative measurement of light intensity change for a sample versus a reference/blank. Because of that different instrument will use a different way for representing the emission spectra and present value on the dependent axis (number related to the intensity of the signal) does not have absolute meaning. For example, an instrument may report intensity or just a number of counts. Even in a single instrument given number will change as we change the setting of the voltage or excitation/emission slits. So, through the presented experiments we will use various representation when presenting not normalized emission/excitation spectra.

FIGURE 4-8 Top: Non-standard cuvettes and cuvette adapters. Bottom: Eclipse sample holder with the cuvette in.

This chapter will discuss a number of experiments that will show how to properly measure the emission spectra, represent spectra in visual form, correct for detector sensitivity, what kind of errors/mistakes and artifacts one can expect to encounter, and how to avoid them. Among these are less known factors like Raman scattering or second harmonics that will be transmitted by typical grating type monochromators commonly used today.

EXPERIMENT 4-1 SAMPLE PREPARATION FOR FLUORESCENCE MEASUREMENTS IN SOLUTIONS

Keywords: emission, fluorescence, spectrofluorometer, solvent viscosity, rhodamine 6G.

Proper preparation of the sample is the required first step for reliable fluorescence measurements. A common mistake is an incomplete dissolution/dilution of the dye, especially in high viscosity solvents (glycerol, propylene glycol, etc.). Using a sample that is not well mixed may lead to unpredictable mistakes. Another very common mistake is underestimating the sample absorption, which may significantly change fluorescence emission intensity. If the sample absorption is too high, the measured fluorescence signal may decrease due to attenuation of the excitation beam intensity by the sample absorption (see Experiments 4-16 through 4-18) or absorption of the emission light by the sample (inner-filter effect). In the previous chapter, which discussed absorption measurements, showed that good results are obtained when the

(*Continued*)

EXPERIMENT 4-1 (CONTINUED) SAMPLE PREPARATION FOR FLUORESCENCE MEASUREMENTS IN SOLUTION

absorption reading is higher than 0.001 and lower than 3. This is a very broad range of sample concentrations. However, most fluorescence measurements are made with samples for which absorptions are at the very low end of this range. Measurements of absorption made with too high or too low absorptions are easy to spot. This is not the case for fluorescence where a sample with an especially high absorption may yield a visually appealing spectrum, which is in fact heavily distorted. Most new students use too much fluorescent compound while preparing solutions for measurements. They also tend to overlook proper dissolution of the studied fluorophores. We will visualize both issues in this and next experiments.

MATERIALS

- Rhodamine 6G (R6G)
- Ethanol (EtOH), water, glycerol
- Glass vials
- 1 cm x 1 cm cuvettes
- 495 nm long-wave pass filter

EQUIPMENT

- Spectrophotometer
- Spectrofluorometer
- UV illuminator

METHODS

1. Make stock solutions of R6G in EtOH (use a glass vial).
2. Fill the glass vial with glycerol and position it on UV illumination plate.
3. Add a small amount of R6G stock solution to glycerol in the vial. Observe and photograph diluting of R6G in glycerol under UV illumination. See Figure 4-1.1.
4. Prepare samples of R6G in water (using EtOH stock solution) with optical densities of about 0.2, 0.6, and 3. Measure fluorescence spectra of these solutions.
5. Make photographs of observed fluorescence spots in a spectrofluorometer (under the excitation). Alternatively, you can use a green laser pointer and make photographs outside the instrument.

RESULTS

Do not dissolve studied dyes directly. It is always easier to introduce fluorophores into a solution using a stock solution. In the stock solution, the dye is highly concentrated and should be completely dissolved.

Figure 4-1.1 shows the chemical structure of rhodamine 6G and the progressive dilution of R6G in glycerol. Glycerol in room temperature has high viscosity and it is not easy to mix with another solvent even if the solution is gently shaken (we do not recommend strong and vigorous shaking of viscous solutions since it will lead to air bubbles forming that are difficult to eliminate). First, in the left is a photograph after the addition of R6G and a few shakes. Next is the bottle after a few minutes of gentle shaking, then after 10 minutes, and finally

(Continued)

EXPERIMENT 4-1 (CONTINUED) SAMPLE PREPARATION FOR FLUORESCENCE MEASUREMENTS IN SOLUTION

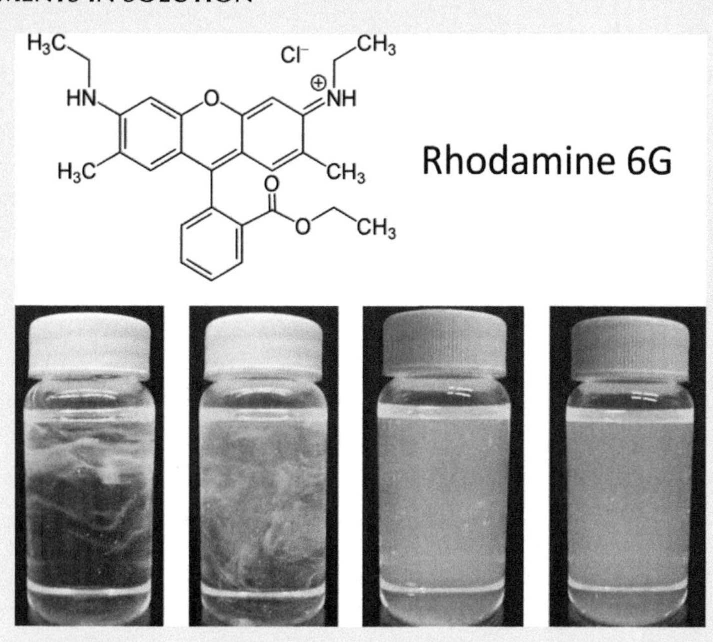

FIGURE 4-1.1 Structure of R6G (top) and progressive dilution of the R6G in glycerol. Although visually the solution looked homogeneous, the UV illumination reveals uncompleted mixing of R6G and glycerol.

after half-hour. Even after 10 minutes, the solution is not uniform and only after a significant amount of time and shaking finally becomes uniform. Only the final solution should be used for measurements. It is important to stress that just visual examination (not UV illumination) already for the first bottle seems ok and after a few minutes (second bottle) looks visually quite homogeneous, a common mistake.

As discussed in Chapter 2, the typical detection system in a spectrofluorometer detects light from the central part of the cuvette. A typical cuvette holder is made in such a way that about 1–2 mm on each side of the cuvette are not visible to the detector. This is not only to hold cuvettes firmly but also to eliminate any light that can be scattered on surfaces/walls of the cuvette and cuvette edges. The photomultiplier detector in the spectrofluorometer sees the fluorescence spot created by the excitation light. When the concentration of the dye is low (absorption is significantly below 0.2), the excitation light will not be significantly attenuated while passing the cuvette and the fluorescence spot will be homogeneous. In fact, light when passing through a sample that has an absorption of 1 will be attenuated 90%; absorption of 0.5 will be attenuated ~68%; absorption of 0.2 will be attenuated ~37%; and absorption of 0.1 only ~20%. In principle, this means that the intensity of fluorescence at the entry side of the cuvette is higher than at the exit side of the cuvette. This effect can be dramatic for higher concentrations. So, if the concentration of the fluorophore is higher, a few effects should be considered. First, the excitation light propagating through the cuvette will be absorbed, which will result in the change of the fluorescence intensity across the excitation beam. Second, the fluorophore may reabsorb fluorescence, which will then result in a change of the apparent emission color toward the red (we will demonstrate this effect later).

(Continued)

EXPERIMENT 4-1 (CONTINUED) SAMPLE PREPARATION FOR FLUORESCENCE MEASUREMENTS IN SOLUTION

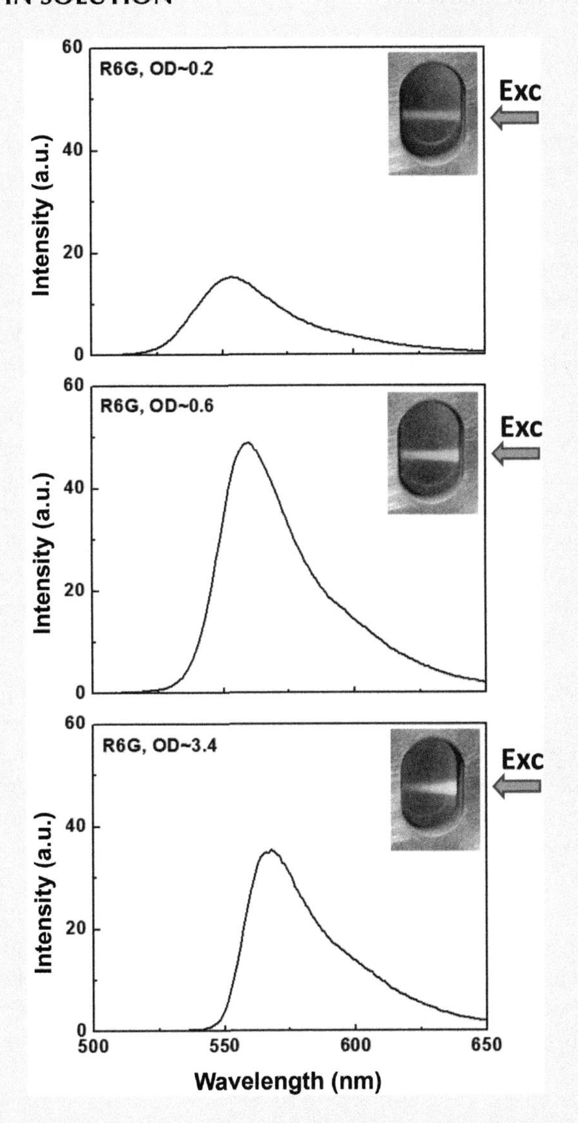

FIGURE 4-1.2 Measured fluorescence spectra of R6G solutions excited with 485 nm light for increasing R6G concentrations. Inserts show photographs of the emission spots as seen in the fluorometer holder and arrow marks the direction for excitation. With increasing concentration (increasing optical density—OD), the fluorescence spectra and observed spots change.

And third, the change of the fluorescence spot shape will result in a geometrical imperfection by a deviation from the orthogonal observation direction. This may affect not only the overall signal but the position and shape of the measured spectrum as well.

Figure 4-1.2 shows the measured emission spectra for three R6G solutions in water with different absorption (0.2, 0.6, and 3.4) using 485 nm excitation. The inserts show photographs

(*Continued*)

EXPERIMENT 4-1 (CONTINUED) SAMPLE PREPARATION FOR FLUORESCENCE MEASUREMENTS IN SOLUTION

of excitation profile. For the low concentration (top), the excited volume looks roughly uniform and as the concentration increases, we can see significant deviations. Note also an apparent change in the color of the emission. As we increase the concentration, we see a progressive change from green to yellow and orange. This is due to the reabsorption of the short-wavelength part of emitted fluorescence. Corresponding emission spectra are measured with the same settings so they can be directly compared. We can clearly see three effects: (1) The increase in intensity with the first concentration and then a significant intensity decrease. This is unexpected and clearly indicates problems. (2) The pointed shape of the excitation spot for higher concentrations. In the place where fluorescence intensity is high, the beam appears larger and as the signal decreases the beam becomes smaller. (3) With the increased concentration, the measured emission spectrum shifts toward red, yielding an apparent color change.

Note: Extra attention should be made to achieve complete dissolution of the fluorophore (a uniform solution). Proper dissolution of the dye for fluorescence measurements is important not only in the case of viscous solvents like glycerol but also in the case of water (buffer) solutions. In the case of water solutions, a common mistake is the preparation of stock solutions also in water, not in alcohol or DMF. If the compound is not highly soluble in water (even rhodamine for example), then the sample prepared from the water stock solution may contain small microcrystals of undissolved dye. Such a solution may be not stable and change the OD and fluorescence signal in time, even after shaking and ultrasonication.

One of the first assignments for students in fluorescence classes or lab rotations we give is the task of the sample preparation of a well soluble and well-known fluorophore. Quite frequently, the student reports that rhodamine is not much fluorescent. Then, we ask the student to progressively dilute the sample, and fluorescence appears. Doing progressive dilutions, the student can also measure absorption. Comparison of measurable absorption and fluorescence signal may give a good indication of the sensitivity difference between absorption and fluorescence. Typically for a highly concentrated sample that shows absorption about three or more emission is barely measurable. A 100-fold dilution is at the limit for absorption measurement but the fluorescence signal becomes excellent. Further 100-fold dilution for a total of 10,000-fold dilution is still easily measurable by fluorescence but it is just a baseline in absorption.

Also, in the case of fluorophores with a large absorption-emission spectral overlap, a high concentration will result in an artificial spectral shift (change of color of emission).

EXPERIMENT 4-2 EMISSION OF CHROMOPHORES IN SOLUTION—SPECTRA REPRESENTED IN WAVELENGTH AND ENERGY SCALE

Keywords: emission, fluorescence, fluorometer, rhodamine 6G

In the previous experiment, preparation/dilution of samples and their effects on results were discussed. This experiment will familiarize the reader with different data representations that are commonly used among the different fields. Biologists and frequently chemists would prefer a wavelength scale representation for the independent axis, while physicists and other chemists would prefer energy in electronvolts (eV) or wavenumber (cm^{-1}).

(*Continued*)

EXPERIMENT 4-2 (CONTINUED) EMISSION OF CHROMOPHORES IN SOLUTION—SPECTRA REPRESENTED IN WAVELENGTH AND ENERGY SCALE

In either case, proper magic angle conditions are necessary to be used if we want to compare emissions as discussed in the following Experiment 4-3 or measure/calculate quantum yields. As discussed in Chapter 3, evaluating the absorption properties of the chromophore (absorption spectra) is necessary to gain knowledge about where the chromophore can/should be excited to detect optimal fluorescence signal. Measuring the emission spectrum of a chromophore is often a necessary first step for utilizing a dye as a marker or sensor for applications in fluorescence-based detection and imaging techniques. Emission spectra can be used to obtain information about sample purity, chromophore concentration, and macromolecular interactions. To do it properly, an experimentalist should realize many factors that affect results. The next experiment will discuss why keeping magic angle conditions for emission spectra measurements is important. In this experiment, we only want to show how to present measured spectra and how this affects visual data presentation. As discussed in Chapter 2 spectral data is usually represented in one of two ways; fluorescence intensity as a function of wavelength and fluorescence intensity as a function of energy (wavenumber or eV).

This protocol describes measurements of emission spectra of R6G in a water solution.

MATERIALS

- Rhodamine 6G (R6G)
- Water
- 1 cm × 1 cm cuvettes

EQUIPMENT

- Spectrofluorometer
- Spectrophotometer

METHODS

1. Turn on the spectrofluorometer and spectrophotometer.
2. Turn on the excitation lamp of the fluorometer (if needed) and wait for it to warm up.
3. Pour a solution of 0.5–1.0 μM R6G into a cuvette.
4. Measure sample absorption (make sure to take proper background as discussed in Chapter 3 (absorption below 0.1).
5. Place cuvette into cuvette holder of the fluorometer.
   a. The cuvette holder is inside a dark chamber; make sure to close this chamber when the cuvette is safely positioned in the holder.
6. Select excitation wavelength to 490 nm
7. Select a scan range for emission to 500–700 nm.
8. Run the emission scan. If the signal goes over the scale adjust emission or excitation slit to a smaller opening (narrower band). For most systems, this R6G concentration should give a good signal. If the signal is low check if: (1) the sample is properly positioned; (2) the lamp is on.

(Continued)

EXPERIMENT 4-2 (CONTINUED) EMISSION OF CHROMOPHORES IN SOLUTION—SPECTRA REPRESENTED IN WAVELENGTH AND ENERGY SCALE

RESULTS

Figure 4-2.1 presents a labeled photograph of the inside of a spectrofluorometer chamber. For this photograph, we removed polarizers to clearly show windows (compare to Figure 4.3 left). During measurements, this chamber should be closed to reduce stray light entering the detector. The excitation light from the light source is traveling alongside the detector window (blue line), thus it can only reach the detector via scattering. The chromophores are emitting light in all directions and only light traveling perpendicular to the excitation beam can reach the detector. Thus, the only a small fraction of emitted light reaches the detector.

Figure 4-2.2 shows the normalized absorption and emission spectra of R6G in ascending wavelength scale and ascending energy scale (wavenumber and eV). Presenting absorption and emission on one figure (scale) gives an indication of how significant spectral overlapping is. All spectra in Figure 4-2.2 are made from the same data and present identical information. There are several visual differences due to using two different scales to describe the distribution of absorbed and emitted photons—wavelength (nanometers, nm) and energy (electron volts, eV, or wavenumber, cm^{-1}). First, each emission spectrum is a mirror image (top to bottom) of the corresponding absorption. Comparison of absorption and emission spectra in wavelength and energy scale (wavenumber or eV) shows that the spectrum in energy scale is a mirror reflection of the spectrum in wavelength scale. This results from the inverse relationship between wavelength and energy. Second, the line shape is a bit wider/narrower. If one of the axes was made descending instead of ascending, then the mirror image effect would go away but the line shape change would remain.

The independent axes may be converted to any representation as is shown in Table 1-1.1 in Chapter 1. The data of the axes is a list and multiple software can be used for conversion, for example, in Excel, the equation "= $(1.23984\cdot10^{-6})/(A1\cdot10^{-9})$ (hc/λ)" can

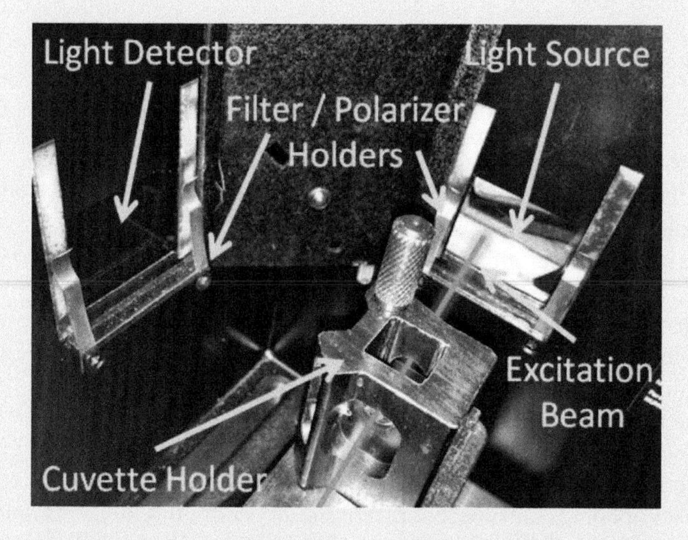

FIGURE 4-2.1 Photograph of a spectrofluorometer cuvette chamber with labeled parts. The light detector, cuvette holder, and emission source form a right angle (polarizers are not shown).

(Continued)

EXPERIMENT 4-2 (CONTINUED) EMISSION OF CHROMOPHORES IN SOLUTION—SPECTRA REPRESENTED IN WAVELENGTH AND ENERGY SCALE

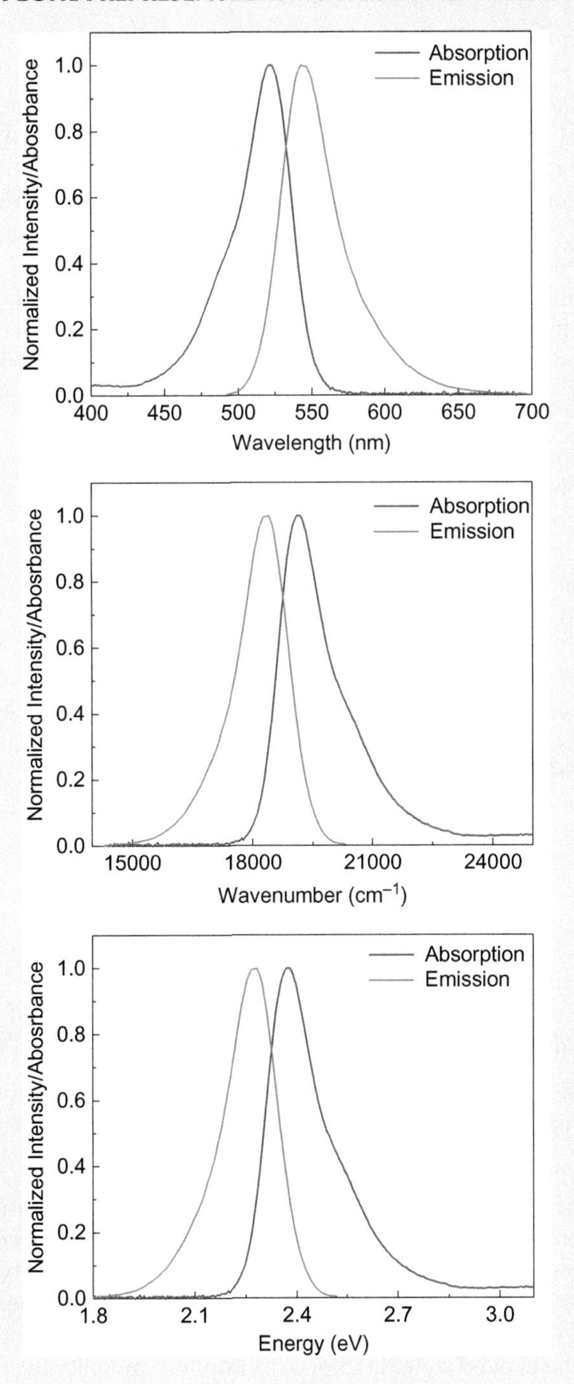

FIGURE 4-2.2 Normalized absorption and emission spectra of R6G in different representations of the independent axis: wavelength (nm), wavenumber (cm^{-1}), and electron volt (eV).

(Continued)

EXPERIMENT 4-2 (CONTINUED) EMISSION OF CHROMOPHORES IN SOLUTION—SPECTRA REPRESENTED IN WAVELENGTH AND ENERGY SCALE

be used to convert between nanometers and electron volts (works only for those two units). The dependent axis for absorption spectra typically represents light attenuation (see Chapter 3) but in this case, the values were normalized to the absorption peak. The dependent axis for emission spectra represents the intensity distribution of observed light that is usually in arbitrary units (a.u.) and as shown in Figure 4-2.2 are presented as normalized to the maximum of emission value. The intensity does not correspond to a particular standard value but rather represents characteristic parameters of the instrument used (thus it's called arbitrary).

The sensitivity and other parameters differ from one instrument to another and absolute reading values for emission intensity may have very different values. But properties that remain constant among different emission spectra from different spectrofluorometers should include:

- Peak center position.
- Intensity distribution in function of the independent variable.
  - Ratios of intensities of multiple peaks in the same spectrum.
    - I.e. peak 1 is 20% higher than peak 2.
  - Line shape.
    - For the same independent axes.
    - Up to random noise.

As the absolute values measured on the different instrument may vary significantly, the normalized spectra should be identical (after necessary corrections for instrument response that we will discuss later).

In practice, there are many factors which contribute to a successful fluorescence emission measurement and should be considered before beginning an experiment. We will discuss all factors in the following experiments.

EXPERIMENT 4-3 EMISSION OF RHODAMINE 6G IN ETHANOL/GLYCEROL MIXTURES: MAGIC ANGLE CONDITIONS

Keywords: emission, fluorescence, spectrofluorometer, solvent viscosity, polarization, rhodamine 6G, magic angle.

The majority of fluorescence experiments are done in liquids where fluorophores have significant freedom of rotational mobility. Imagine a molecule that absorbs a photon and after a finite time emits a fluorescence photon. During the time between absorption and emission, the molecule randomly rotates and the emitted photon will typically have a different plane of polarization. Since observations are made from one specific direction, the observed emission intensity will depend on how fast the molecule rotates as compared to the time it is in the excited state. In effect, fluorophore mobility may have a significant impact on the observed fluorescence intensity. So, to properly measure emission spectra (and later fluorescence lifetimes), it is necessary to find the precise conditions in which

(*Continued*)

EXPERIMENT 4-3 (CONTINUED) EMISSION OF RHODAMINE 6G IN ETHANOL/GLYCEROL MIXTURES: MAGIC ANGLE CONDITIONS

observed fluorescence intensity is independent of fluorophore mobility. In Chapter 2 and in the Introduction to Chapter 4, we already defined such conditions as magic angle conditions. When comparing the fluorescence intensities of two chromophores, it is critical to maintain magic angle conditions. Only spectra measured under MA condition of different solutions or the same solution at different conditions (e.g., different temperatures) can be properly compared. For example, when determining the quantum yield of the unknown fluorophore by comparing it fluorescence signal of the sample to the standard, it is imperative to make the measurement of the sample (fluorophore) and the standard under MA conditions.

The goal of this experiment is to demonstrate how fluorophore mobility impacts the observed emission intensity. The concept of MA is more advanced and difficult to explain without prior advanced knowledge of fluorescence polarization. However, it is critical to introduce it in an earlier experimental stage to be able to perform proper fluorescence measurements. It was not easy to design a simple approach that would demonstrate that MA is the only proper way for detecting emission spectra for chromophores in solutions. One possibility is to use a mixture of two solvents with very different viscosities. In this case, ethanol (EtOH) that has a very low viscosity (fluorophore will rotate fast) and glycerol that in contrast has very high viscosity (fluorophore will rotate very slowly). Both solvents mix very well, yielding solutions of different viscosities that depend on the EtOH/glycerol ratio. Dyes like R6G will rotate very quickly in EtOH thus yielding a very low emission polarization (dipole moments of excited molecules quickly randomize before emission). However, in glycerol R6G will tumble much slower, yielding a high emission polarization (dipole moments will not change their orientation much before emitting the photons).

MATERIALS

- Rhodamine 6G (R6G)
- Water
- Ethanol (EtOH)
- Glycerol
- Glass vials
- 1 cm × 1 cm cuvettes
- 495 nm long-wave pass filter

EQUIPMENT

- Spectrophotometer
- Spectrofluorometer

METHODS

1. Make a stock solution of R6G in EtOH (use a small glass vial and use very high concentration of R6G).
2. Prepare samples of R6G in glycerol (using a minimal amount of EtOH stock solution) with optical densities of below 0.1. The sample volume should be about 10–20 mL.

(Continued)

EXPERIMENT 4-3 (CONTINUED) EMISSION OF RHODAMINE 6G IN ETHANOL/GLYCEROL MIXTURES: MAGIC ANGLE CONDITIONS

3. Set the MA conditions on your fluorometer (in square geometry polarizers are set to vertical excitation and 54.7° angle (to the vertical) on observation) and measure the fluorescence spectra of the R6G solution in glycerol.
4. Prepare samples of R6G in EtOH (using EtOH stock solution) with an optical density close to that of R6G in Glycerol. The sample volume being 10–20 mL.
5. Using MA conditions, measure fluorescence spectrum of the EtOH solution of R6G. Adjust the concentration of R6G in EtOH to exactly match the emission intensity of R6G in glycerol (Note that the emission spectrum in EtOH will be slightly shifted toward shorter wavelengths. We match the intensities in the peak of emissions at the used excitation wavelength—not the solutions absorptions). We can add R6G stock solution or just EtOH.
6. Mix the R6G solutions in EtOH and glycerol in ratios 70/30, 50/50, and 30/70. This will produce mixtures of solvents with different viscosities.

Note: For solvents that mix in a perfect way, the emission signal should be unchanged. However, in our case the emission spectra are slightly shifted and mixing EtOH with glycerol may have an effect on total volume (molecular packing) and spectroscopic properties of the dye (solvatochromic effect). R6G is not very solvent sensitive and we expect that total effect will be less than 5%. But the reader should be aware that for some fluorophores this can be a significant effect and a problem for the experiment. Also, the reader should remember that we matched fluorescence signals, not absorptions or concentrations of R6G since quantum yields could be different.

RESULTS

Figure 4-3.1 A presents emission spectra measured at MA conditions for various mixtures. The emission spectra in EtOH (red) and glycerol (black) have similar intensities (as adjusted during sample preparation). As the ratio EtOH/glycerol changes, we first observe a very slight decrease in fluorescence intensity for the mixture and for a higher percentage of glycerol the maximum at the emission peak increases. Overall, this is a very small change and we can safely assume that the fluorescence intensity almost did not change when we were mixing our solutions.

However, it is known that the viscosity of such solutions increase as we increase glycerol contributions. Figures 4-3.1b–f show fluorescence intensities measured for vertical (VV—dotted black line), horizontal (VH dash red line), and MA (solid blue line) polarizer orientations for an increased concentration of glycerol starting from 100% EtOH (B) to 100% glycerol (F). As we mentioned in the introduction to this experiment, in EtOH we expect molecules to quickly randomize before emitting a photon, so the signal observed with any polarizer orientation should be similar. For the more advanced reader, the steady-state fluorescence anisotropy (we will discuss polarization experiments later) of R6G in EtOH is low, only about 0.024 as indicated in the figure. This means the vertical and horizontal components differ by only about 8%. The first thing to notice already in Figure 4-3b is that both (VV and VH) intensities are very different already for EtOH. These are uncorrected spectra (we did not take into account the G-factor correction for the spectrofluorometer—see Experiments 6-2 and 6-3). After correcting for the G-factor, the difference between VV and VH will be small and all (VV, VH, and MA)

(*Continued*)

EXPERIMENT 4-3 (CONTINUED) EMISSION OF RHODAMINE 6G IN ETHANOL/GLYCEROL MIXTURES: MAGIC ANGLE CONDITIONS

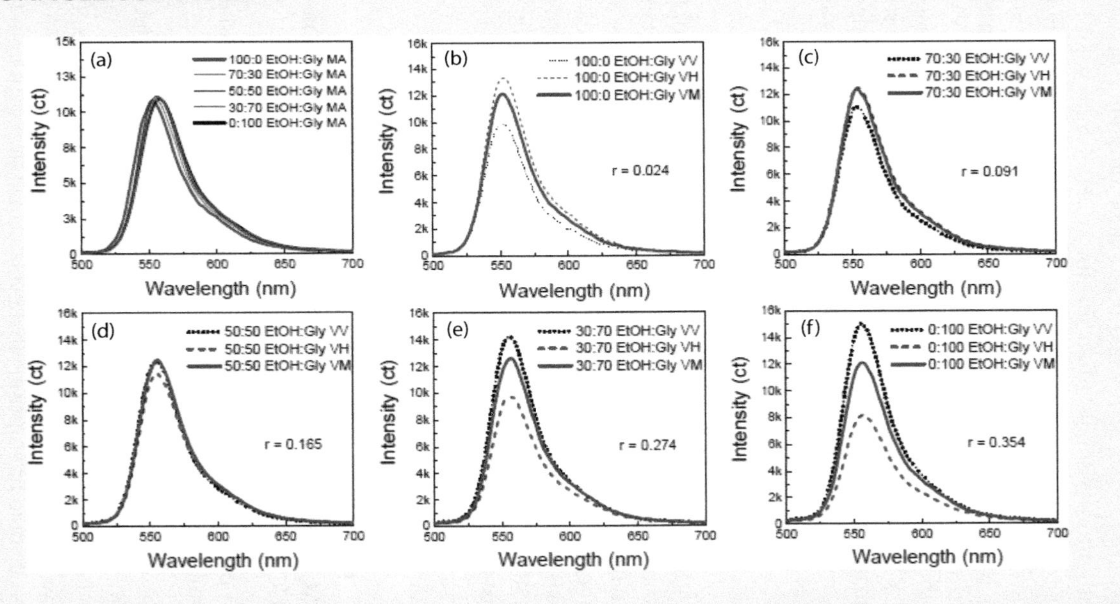

FIGURE 4-3.1 Fluorescence intensity measured for R6G dye in different mixtures solvents (EtOH/glycerol). (a) Magic angle conditions. (b–f) MA, VV, and VH measured emission intensities in increasing concentration of glycerol.

(Continued)

EXPERIMENT 4-3 (CONTINUED) EMISSION OF RHODAMINE 6G IN ETHANOL/GLYCEROL MIXTURES: MAGIC ANGLE CONDITIONS

intensities will be within 10% change. Figure 4-3.1c shows the spectra measured for 30% glycerol where viscosity is slightly higher. As expected, we observe a slight increase in steady-state anisotropy to 0.09. The spectra measured under MA conditions practically did not change (blue solid line) while the relative difference between VV and VH changed significantly (now is smaller than just for EtOH). Figure 4-3.1d is a 50/50 mixture that already has significant viscosity that is confirmed by relatively high anisotropy (0.165) and the relation between VV and VH progressively changed (now VH is lower from VV, opposite to the Figure 4-1.1b and c). But the intensity measured with MA is relatively constant. In Figure 4-3.1e, the difference between VH and VV increases and for 100% glycerol the difference becomes substantial. Just comparing Figure 4-3.1b with f when measuring VV or VH one would make a very large error in estimating relative signals while for measurements at MA the signal practically did not change, as it could be expected when we mix two solutions that initially have the same signals and components do not react/interact.

CONCLUSIONS

The ratio of VV to VH intensities changes significantly with the change of the solution viscosity, but at MA conditions the observed signal is constant. It is important to realize that to avoid mistakes when measuring emission spectra we should always use magic angle conditions. When measuring emission spectra the mistake can be significant, up to 50% of expected intensity. Using improper conditions (not MA) when measuring spectra may lead to significant mistakes in measured intensity but it will typically not affect the spectral profile. The problem is more dramatic for time-resolved intensity measurements.

EXPERIMENT 4-4 FLUORESCENCE SIGNAL DEPENDENCE ON GEOMETRICAL FACTORS: CENTER OF DETECTION AND EFFECT OF POSITION OF FLUORESCENCE SPOT ON DETECTED SIGNAL

Keywords: emission, fluorescence, spectrofluorometer, solvent viscosity, rhodamine 6G (R6G), rhodamine 123 (R123).

Experiment 4-1 presents examples of how sample preparation may affect the measured signal. Fluorescence measurements often compare signals from different samples, and it is important to realize how various factors may influence the result of the experiment. A commercial instrument would typically be optimized to have maximum sensitivity in the center of the sample holder. It could be surprising that even a small deviation from the center of the detection will cost significant signal modification. This is typically called inner filter effect type I where attenuation of the excitation light due to the sample absorption leads to a lower intensity reading than expected. Here we want to demonstrate how geometrical factors (e.g., position of the emission spot) affect measured signals. In most of the commercial systems with a fixed square geometry, the user will not be able to change/alter the position of the cuvette in the sample compartment. This is perfectly fine when diluted solutions are used

(*Continued*)

EXPERIMENT 4-4 (CONTINUED) FLUORESCENCE SIGNAL DEPENDENCE ON GEOMETRICAL FACTORS: CENTER OF DETECTION AND EFFECT OF POSITION OF FLUORESCENCE SPOT ON DETECTED SIGNAL

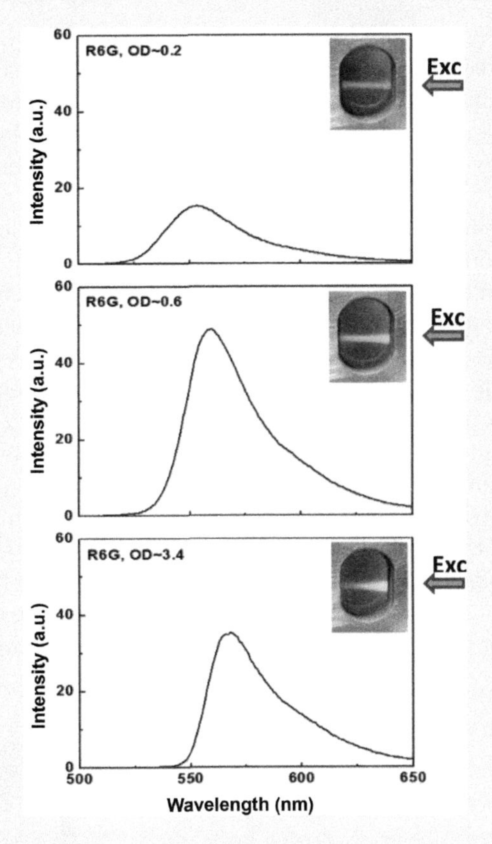

FIGURE 4-4.1 Measured fluorescence spectra of R6G solutions excited with 485 nm light for increasing R6G concentrations. Inserts show photographs of emission spots as seen in the fluorometer holder. With the increasing concentration (increasing optical density—OD), the fluorescence spectra and observed spots change.

and fluorescence spots are uniform across the entire length of the cuvette. Unfortunately, this may be an issue for more concentrated solutions. For many experimentalists, this could be a surprise, but unintended small changes in the fluorescence spot may cost a significant problem and misinterpretation of the results. Let's consider the situation in Figure 4-4.1 (taken from Experiment 4-1) which presents the emission trace and measured emission spectra of different concentrations of R6G in water. Each photograph in the insert shows an actual trace of the beam (emission) as it is seen through the holder opening.

Going from the first (top) panel to the second the concentration increases three times. The measured intensity increases slightly less than three times and since we see some spectral shift we can blame the slight loss to reabsorption effect. The photograph insert also shows a significant intensity increase in the center of the cuvette and in overall (total) intensity. However, moving to the third (bottom) panel the concentration increased again

(Continued)

EXPERIMENT 4-4 (CONTINUED) FLUORESCENCE SIGNAL DEPENDENCE ON GEOMETRICAL FACTORS: CENTER OF DETECTION AND EFFECT OF POSITION OF FLUORESCENCE SPOT ON DETECTED SIGNAL

almost 6-fold while the measured emission spectrum decreased quite substantially as compared to the middle panel. Even stranger, the photo shows a significant change in the apparent beam profile. The intensity at the entry to the cuvette is now very high and quickly decreases toward the exit from the cuvette. Overall, the total signal is significantly brighter than that in the middle or top panel. So, *why does our detector detect a smaller signal then?*

The one clearly visible effect is the change of the intensity distribution in the observed emission spot. In the top panel, the intensity is equally distributed across the spot (across the excitation beam). In the middle one, we can notice that the intensity at the beam entry is higher and lower at the end. And in the bottom panel, the intensity is mostly on the right side of the cuvette (excitation beam entry) and almost completely not visible on the left side. So, the "center of gravity" for the emission is completely on the right side of the cuvette while the top panel is in the center of the cuvette. Since this is a 1 cm cuvette, the shift is really small and cannot be more than half of the cuvette size (5 mm). This is a really small change in the geometry as compared to the size of the optical system (a typical sample compartment will have dimensions over 20 cm × 20 cm or more). *Would this relatively small shift of the fluorescence spot be able to alter the detected intensity? And if so, how large spot size the detection system is really observing?* In this experiment, we want to test how significantly the signal changes when we alter the emission spot position a few millimeters left or right. The concept of the experiment is shown in Figure 4-4.2. We

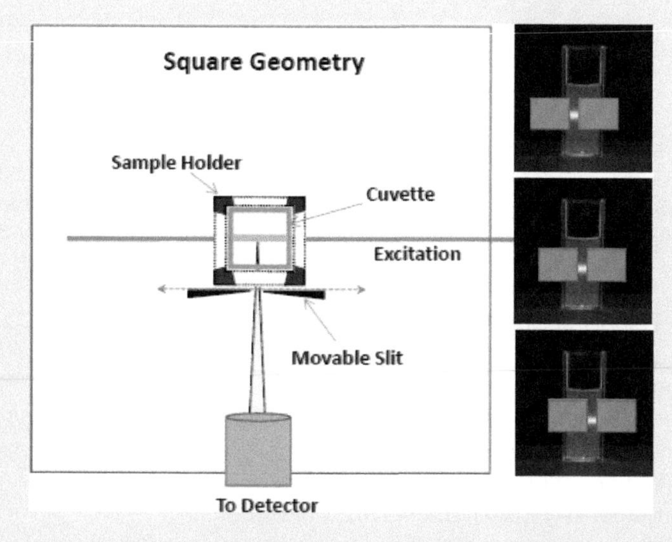

FIGURE 4-4.2 Sample compartment of square geometry system. The excitation beam comes from the left and observation is at the bottom. On the observation side of the cuvette (close to the sample holder), a slit has been installed. On the right are photographs of the cuvette with three slit positions (left, center, and right).

(Continued)

EXPERIMENT 4-4 (CONTINUED) FLUORESCENCE SIGNAL DEPENDENCE ON GEOMETRICAL FACTORS: CENTER OF DETECTION AND EFFECT OF POSITION OF FLUORESCENCE SPOT ON DETECTED SIGNAL

used low R123 concentration, so the fluorescence intensity along the excitation beam in the cuvette is uniform. We will observe fluorescence signal through a vertical slit of about 2 mm wide that we will shift from left to center, and to the right side of the cuvette.

MATERIALS

- Rhodamine 123 (R123)
- Water
- Glass vials
- 1 cm × 1 cm cuvettes
- 2 mm slit (or a microscope slide and black scotch tape)
- 495 nm long-wave pass filter

EQUIPMENT

- Spectrophotometer
- Spectrofluorometer (Varian Eclipse)
- Positioner mount
- Blue laser pointer
- Camera or cellphone

METHODS

1. Make a solution of R123 in water with absorption below 0.1.
2. Make a slit and mount it in front of the cuvette in the direction toward the detector. Not every lab will have easy access to various slits. But it is easy to make one. Take a microscope slide (in our case 1 × 2 inches) and stick two parallel strips of black tape (see Figure 4-4.3). It may take a few trials before you will succeed with a uniform slit. *Note:* It does not have to be exact, any slit below 2 mm will work.
3. Install the slit in three positions (center of the cuvette, 2–3 mm left and 2–3 mm right) in the front of the detector as schematically shown in the Figures 4-4.2 and 4-4.3.
4. Measure emission spectra of R123 for each slit position.
5. If possible, take photographs of observed fluorescence spots in a spectrofluorometer (under the excitation). Alternatively, you can make photographs outside the sample compartment using a blue laser pointer as shown in Figure 4-4.3.

RESULTS

Figure 4-4.3 shows three photographs of the cuvette with R123 with a slit in the front that is illuminated (excited) with 470 nm light beam from the left side. The concentration of R123 is low with absorption below 0.1 so, the intensity of fluorescence is uniform across the excitation beam. The slit is shifted 2 mm left or right as compared to the central position. It is visually clear that the visual intensity seen in all three slit positions are the same.

(Continued)

EXPERIMENT 4-4 (CONTINUED) FLUORESCENCE SIGNAL DEPENDENCE ON GEOMETRICAL FACTORS: CENTER OF DETECTION AND EFFECT OF POSITION OF FLUORESCENCE SPOT ON DETECTED SIGNAL

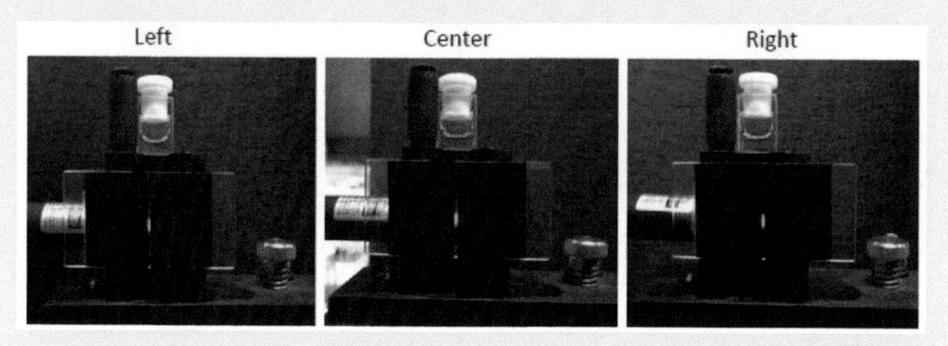

FIGURE 4-4.3 Photographs of a cuvette with R123 solution as seen through the slit positioned at three positions. These are 2 mm left of the center, center of the cuvette, and 2 mm right of the center. The brightness of the fluorescent spot is the same in each position.

Next, we measure fluorescence spectra at each slit position, see Figure 4-4.4.

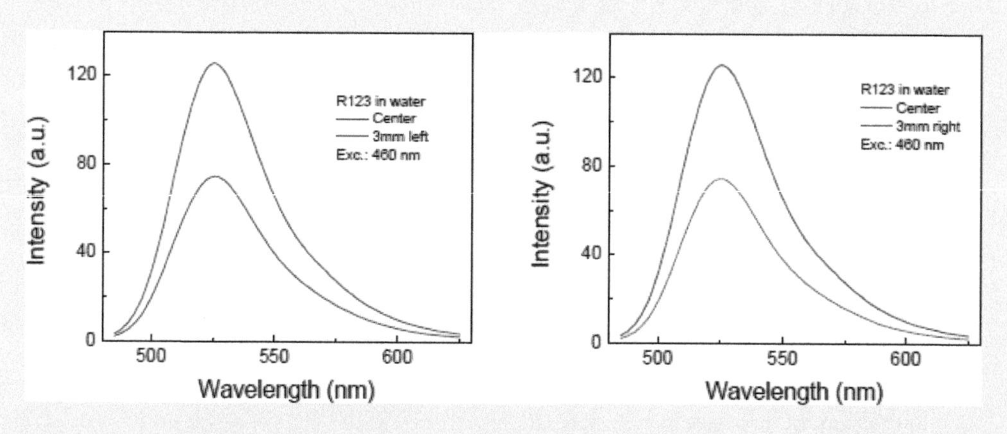

FIGURE 4-4.4 Left: Measured intensities through the slit in the central position (blue) and left position (red). Right: Measured intensities through the slit in the central position (blue) and left position (green).

In Figure 4-4.4, we present measured spectra in the left and right positions in comparison to the slit in the central position. Both in left and right positions the signal is significantly lower (~40%) what it should be (the emission is uniform and we expect intensity seen through the slit in each position to be identical). Observed change is very significant as for such small position modification (about 2 mm into the left and right).

It is important to mention that the measured signal also depends on how far from the cuvette the slit has been mounted. Typically, we would try to install a slit as close as possible that in practice may vary between a millimeter to about 1 cm separation from the cuvette.

(*Continued*)

EXPERIMENT 4-4 (CONTINUED) FLUORESCENCE SIGNAL DEPENDENCE ON GEOMETRICAL FACTORS: CENTER OF DETECTION AND EFFECT OF POSITION OF FLUORESCENCE SPOT ON DETECTED SIGNAL

CONCLUSIONS

We mentioned multiple times that a typical fluorescence system will have a detection system aligned to the center of the sample holder. But it is clear from this simple experiment that even a slight position modification/shift will have a significant effect. The consequences of this (inner filter effect type I) are very important and not everybody realizes this. For example:

1. When comparing samples that have significant (above 0.1) and different absorptions, we can make a trivial mistake by underestimating the emission of the sample with higher absorption.
2. When trying to make a measurement using a smaller cuvette (or small solid sample) but using a standard holder it is easy to slightly shift the position that will significantly modify the signal.
3. Biological samples are frequently highly scattering and absorptive and thus may contribute to significant excitation beam attenuation that will shift the apparent fluorescent spot position toward the cuvette wall.
4. In the next chapter, we will present quenching experiments. Some quenchers may have their own absorption at the excitation wavelength and contribute to the excitation beam attenuation.
5. In Chapter 6, we will discuss polarization. The shift of position is not only changing apparent intensity but will also influence a correction factor (G-factor).

EXPERIMENT 4-5 FLUORESCENCE SIGNAL DEPENDENCE ON GEOMETRICAL FACTORS: EMISSION SPOT POSITION AND DETECTED FLUORESCENCE INTENSITY FOR DIFFERENT SPECTROFLUOROMETERS

Keywords: emission, fluorescence, organic dyes, spectrophotometer, spectrofluorometer, rhodamine 6G.

Because geometrical factors play such an important role in fluorescence measurements this topic is covered in more detail to test the effects more precisely using different instruments. Typically this effect depends on observation monochromator slit orientation.

The previous experiment, showed how a different displacement of the slit position (how far the slit has been from the center of the cuvette) causes a significantly different signal from the uniformly illuminated sample (the absorption of the sample was significantly below 0.1 that gives the uniform distribution of fluorescence intensity along the excitation beam). A displacement as minuscule as just a few millimeters leads to a dramatic signal change. As has been shown in Figure 4-4.1 for highly absorbing samples, the emission signal strongly changes as a function of the excitation beam penetration depth. Already for samples with the absorption of 0.3, the fluorescence intensity of fluorescence at the exit side of the cuvette is only half of that at the entry side, the difference being apparent. Such uneven fluorescence intensity distribution will have a significant effect on the signal detected by the optical system focused on the sample center. For higher sample absorptions, the "center of gravity" of the

(*Continued*)

EXPERIMENT 4-5 (CONTINUED) FLUORESCENCE SIGNAL DEPENDENCE ON GEOMETRICAL FACTORS: EMISSION SPOT POSITION AND DETECTED FLUORESCENCE INTENSITY FOR DIFFERENT SPECTROFLUOROMETERS

fluorescence signal will be shifted toward the side where the excitation enters the cuvette, resulting in a stronger emission on this side of the cuvette. At first, the displacement of couple millimeters seems rather small and should not affect the signal from the entire cuvette. But measurement through the slit in the previous experiment demonstrated that the effect can be significant. In fact, most spectrofluorometers are precisely aligned to detect signals from the center of the sample compartment (cuvette) and in practice, all fluorescence systems will be biased toward such displacement. To experimentally test the extent of this effect, we will perform an experiment using 1 mm path cuvette. In this case, the position of the emitting spot in the horizontal plane along the excitation beam can be precisely monitored.

MATERIALS

- Rhodamine 6G (R6G)
- Water
- 1 cm × 1 cm cuvettes
- 0.1 cm × 1 cm cuvettes

EQUIPMENT

- Spectrophotometer
- Spectrofluorometer

METHODS

1. Prepare a stock solution of R6G with a concentration of about 0.1 mM. Dilute the stock solution to two samples with absorptions below 0.1 in 1 mm path and 10 mm path cuvette respectively. Make sure the absorbances in both cuvettes are equal (close).
2. Measure emission from the sample in 1 mm and the sample in 10 mm cuvette.
3. Remove the sample holder from the instrument chamber. Depending on the instrument, you may need to remove the entire holding plate or just only cuvette holder.
4. Install in the place of the regular holder a movable cuvette holder. In Figure 4-5.1, we are presenting the example. But in reality, it could be just a plate with the double-stick scotch tape to hold the cuvette at the bottom. As shown in Figures 4-5.1, the fluorescence is observed from 1 mm path and cuvette can be displaced with the micrometric screw along the excitation line. In the shown case (ISS chamber), we used 440 nm laser excitation (no lens on excitation line).
5. Place the 1 mm cuvette in the center of the observation lens (center of the original sample holder) and shifted back from the observation so the excitation beam will be passing close to the cuvette wall which is closer to the observation (see top schematic in Figure 4-5.1).
6. Remember to switch-off external light before adjusting. The signal from the R6G solution will be strong and you will need to use low voltage and a small observation slit. Align the position of the cuvette to get maximum signal. In practice, you can visually align the cuvette to be close to the center of observation lens. Then with external light off, you can move the cuvette with the micrometer screw when monitoring a signal at a single wavelength close to maximum. *Note:* The excitation beam has to be at the height

(Continued)

EXPERIMENT 4-5 (CONTINUED) FLUORESCENCE SIGNAL DEPENDENCE ON GEOMETRICAL FACTORS: EMISSION SPOT POSITION AND DETECTED FLUORESCENCE INTENSITY FOR DIFFERENT SPECTROFLUOROMETERS

corresponding to the vertical center of the observation and the cuvette displacement/shift should be parallel to the excitation beam direction.

7. Measure the emission spectrum at the maximum emission signal (central cuvette position). Shift the cuvette about 0.1–0.2 mm and repeat the measurement. Continue shifting the cuvette up to the point when measured intensity signal will be less than 10% of the maximum original signal. Then come back to the central position and repeat the process going in the opposite direction.
8. Repeat emission measurement using 1 cm × 1 cm cuvette where the 10 mm path is used for the excitation.

RESULTS

For the first measurement, we will use the regular cuvette holder and measure the signal emission spectra as observed from 1 mm and 10 mm cuvettes using the same excitation wavelength. Since the absorptions are identical (close), we could expect identical (similar) total fluorescence signal (the R6G concentration in 1 mm cuvette is about 10 times larger). *Note, the signal from 1 cm cuvette should be about 20% smaller since the edges of the holder block about 1 mm from each cuvette side.* But practically, all systems will show the fluorescence intensity measured in 1 mm cuvette significantly greater. Depending on the system, the difference could be almost 5–10-fold. These results are not shown.

In Figure 4-5.1, we present a schematic for the concept of the experimental configuration (top) and a picture of the experimental setup in the ISS K2 instrument. Insert in the

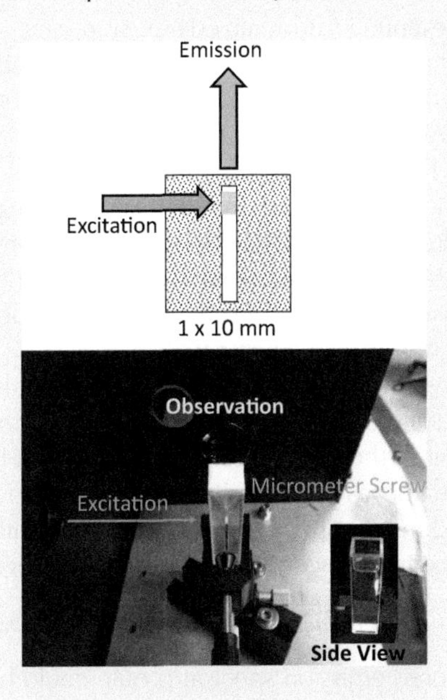

FIGURE 4-5.1 Top: Schematic of the top view of the cuvette and excitation beam. Bottom: Photograph of the experimental setup in the ISS K2 chamber (insert shows photography of the side view of the cuvette).

(Continued)

EXPERIMENT 4-5 (CONTINUED) FLUORESCENCE SIGNAL DEPENDENCE ON GEOMETRICAL FACTORS: EMISSION SPOT POSITION AND DETECTED FLUORESCENCE INTENSITY FOR DIFFERENT SPECTROFLUOROMETERS

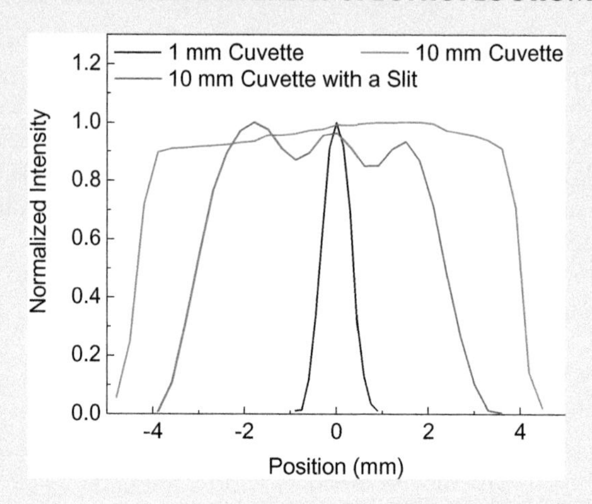

FIGURE 4-5.2 Measured fluorescence intensity profile for 1 mm (black) and 10 mm (orange) cuvettes as a function of displacement from the central position (maximum signal). The green line is for the measurement with using 1 cm cuvette and a 1 mm slit moved in front of the cuvette.

photograph clearly shows the position of the emission spot. The excitation and observation are fixed and we can shift (left–right) cuvette with the precision of about 10 μm. We perform an experiment with 1 mm and 10 mm cuvettes.

In Figure 4-5.2, we are presenting a normalized emission signal (measured at the maximum emission ~555 nm) as a function of the cuvette shift. We measured the emission signals when shifting the cuvette of about 0.2 mm in each step. The black line represents the measured signal from 1 mm cuvette. We expected sensitivity to the position of the emitting point but the decrease of about 15-fold when shifting about 1 mm came to us as a surprise. We repeated the experiment using a 10 mm cuvette. It is clear that for positions between ± 4 mm, the signal is relatively stable and drops quickly when approaching 5 mm shift. Just for comparison, we also included the measured normalized signal when using 1 cm cuvette path and a 1 mm slit scanned across in the observation as we used in previous Experiment 4-4 (green). The slit has been installed about 1 cm from the cuvette wall.

Such significant sensitivity to the spot location indicates that the collection lens, slit position, monochromator, and detectors are very precisely aligned to gain maximum sensitivity for the detection at the center of the sample holder. We checked how this effect will depend on the emission monochromator resolution using different observation slits and did find that for slits 0.5 mm, 1 mm, and 2 mm (2 nm, 4 nm, and 10 nm resolution) we do not observe any significant modification of the intensity dependence shown in Figure 4-5.2.

Note: Profiles on Figure 4-5.2 are intensity traces, not a fluorescence spectra.

Next, we tested if the measured emission spectral profile could be affected by the emission spot displacement. In Figure 4-5.3, we are presenting normalized emission spectra measured for R6G for different positions (center, and 1.2 mm and –1.2 mm shifted). The spectra measured with 1.2 mm shift are much lower (about 20-fold weaker) and to minimize the noise,

(Continued)

EXPERIMENT 4-5 (CONTINUED) FLUORESCENCE SIGNAL DEPENDENCE ON GEOMETRICAL FACTORS: EMISSION SPOT POSITION AND DETECTED FLUORESCENCE INTENSITY FOR DIFFERENT SPECTROFLUOROMETERS

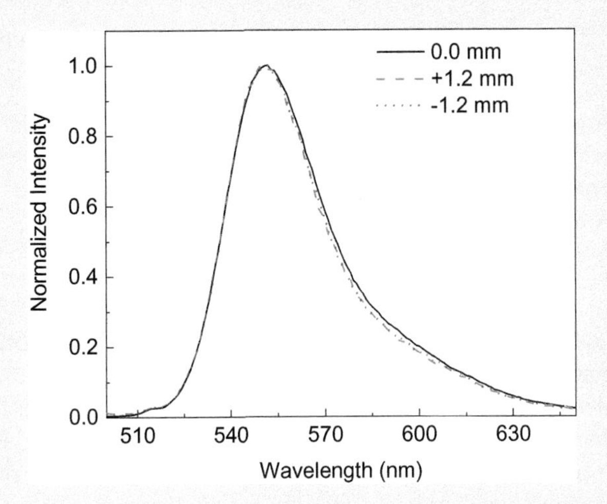

FIGURE 4-5.3 Normalized emission spectral profiles for central, shifted +1.2 mm and −1.2 mm position of the 1 mm cuvette.

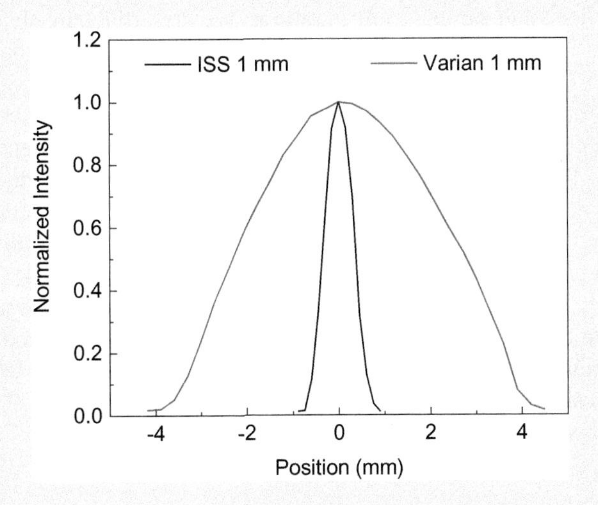

FIGURE 4-5.4 Normalized emission intensity profiles as a function of 1 mm cuvette displacement measured for Varian Eclipse spectrofluorometer (blue) and ISS K2 spectrofluorometer (black).

we averaged 10 scans. Presented spectra are almost perfectly overlapping which indicates that we do not observe any significant wavelength shift due to small displacement.

The sensitivity to the emission spot position will be characteristic for a given instrument. This effect depends on the monochromator slit orientation. Most of the commercial instruments have a vertical slit. Only Varian Eclipse has a horizontal slit. For comparison, we tested a Varian Eclipse spectrofluorometer. In Figure 4-5.4, we present an intensity profile as a function of 1 mm cuvette displacement as measured for Varian Eclipse (violet). For comparison in the graph, we

(Continued)

EXPERIMENT 4-5 (CONTINUED) FLUORESCENCE SIGNAL DEPENDENCE ON GEOMETRICAL FACTORS: EMISSION SPOT POSITION AND DETECTED FLUORESCENCE INTENSITY FOR DIFFERENT SPECTROFLUOROMETERS

also included intensity profile as we measured before for the ISS K2 spectrofluorometer. It is clear that Varian presents a much broader profile (less sensitivity to the cuvette displacement).

Note: Varian spectrofluorometer has horizontal monochromator slit orientation and shows much lower sensitivity to the position in horizontal displacement but shows strong sensitivity in the position orientation in the vertical direction. Most other system has a vertical slit and are less sensitive to vertical spot displacement.

CONCLUSIONS

We demonstrated that the detected signal strongly depends on sample position (the emission spot position). The sensitivity to the emission spot displacement depends on the particular instrument. Typically, precisely collimated systems will have higher sensitivity but will also present a larger vulnerability to small emission spot displacements. Using highly concentrated solutions with high optical densities result in an apparent displacement of the fluorescence spots. This effect (frequently called inner filter effect type I) may have a few important implications for other measurements:

1. When using too high sample concentrations, we are artificially shifting the emission gravity center into one side of the cuvette. This changes geometry that may lead to significant signal lowering. The simple solution is to use thin (1 mm) cuvette that will lower sample absorption and bring emission toward the spectrofluorometer detection center.
2. This effect becomes extremely important when measuring fluorescence in the presence of absorbers. Typical examples can be some quenching experiment (e.g., as discussed in Chapter 5 quenching of tryptophan (free or in the protein) with acrylamide (common quencher) and using shorter excitation wavelength (below 290 nm where acrylamide absorbs) will typically lead to artificial overestimation of quenching). Another example can be energy transfer (FRET) experiments where we increase the concentration of the acceptor. The acceptor will almost always absorb at the donor excitation wavelength and thus attenuate the excitation beam as it penetrates the sample. We will observe an artificial intensity drop of the donor that is not due to FRET.

EXPERIMENT 4-6 STRUCTURED AND UNSTRUCTURED EMISSION SPECTRA

Keywords: emission, fluorescence, fluorometer, anthracene, anthranilic acid, 1,6-diphenyl hexatriene, coumarin 106, 1,8-diphenyl octatriene, 4,4′dimethylamino-cyano-stilbene.

Profiles of the emission and excitation spectra for a fluorophore are important characteristics of the emitter. Typically, emission and excitation spectra are bell-shaped but sometimes may have a very characteristic structure. Visually, it is very difficult (practically impossible) to distinguish if the emission is smooth and bell-shaped or composition of multiple emission lines. A simple example can be a fluorescence bulb that presents distinct lines but is perceived by

(*Continued*)

EXPERIMENT 4-6 (CONTINUED) STRUCTURED AND UNSTRUCTURED EMISSION SPECTRA

the eye as white color light. This experiment presents a few examples of spectra that have very similar colors but very different spectral profiles.

MATERIALS

- Anthracene, anthranilic acid (AA), 1,6-diphenyl hexatriene (DPH), coumarin 106 (C106), 1,8-diphenyl octatrene (DPO), 4,4′dimethylamino-cyano-stilbene (DCS)
- Ethanol (EtOH), DMF, *n*-heptane
- Glass vials
- 1 cm × 1 cm cuvettes

EQUIPMENT

- Spectrophotometer
- Spectrofluorometer
- Cell phone or camera
- UV illuminator

METHODS

1. Make stock solutions of all six fluorophores in DMF (use glass vials).
2. Prepare samples in pairs: AA in EtOH and anthracene in *n*-heptane; DPH and C106 in DMF; DPO and DCS in EtOH. Measure absorption and adjust concentrations to optical densities of below 0.1.
3. Make photographs of the observed fluorescence in individual vials while solutions are on UV illuminator.
4. Measure the emission spectra. For these measurements, we will use 360 nm excitation. But you may set any excitation wavelength close to the absorption maximum and measure emission starting at a wavelength 5–10 nm above the excitation (to make sure you are not scanning through the excitation wavelength). Depending on fluorophore, the typical emission range will be a scan of 200–300 nm.
5. Measure the excitation spectra. Set the observation wavelength close to the emission maximum (preferably on the red side of the emission spectrum) and scan the excitation wavelengths starting at 250 nm (in most experiments, we will not use shorter than 250 nm excitation) and finishing about 10 nm before the observation wavelength.
6. Normalize all spectra and plot them in the described pairs.

RESULTS

In Figures 4-6.1–4-6.3, we present measured spectra with 360 nm excitation for each compound. For each emission spectrum, a corresponding photograph visualizing color is shown on the right. Spectra and the corresponding photographs are arranged in pairs. The colors of fluorescence emission are very similar in each pair while their spectral profiles are distinctly different (bell-shaped vs. structured emissions).

(*Continued*)

EXPERIMENT 4-6 (CONTINUED) STRUCTURED AND UNSTRUCTURED EMISSION SPECTRA

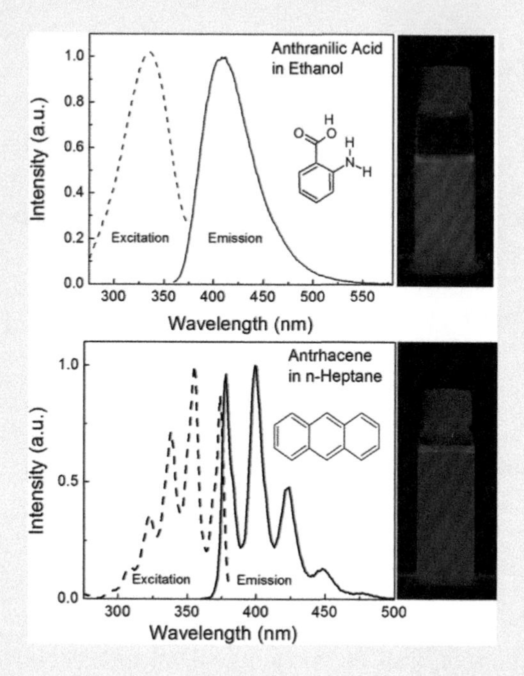

FIGURE 4-6.1 Fluorescence excitation and emission spectra of anthranilic acid in ethanol (top) and anthracene in *n*-heptane (bottom).

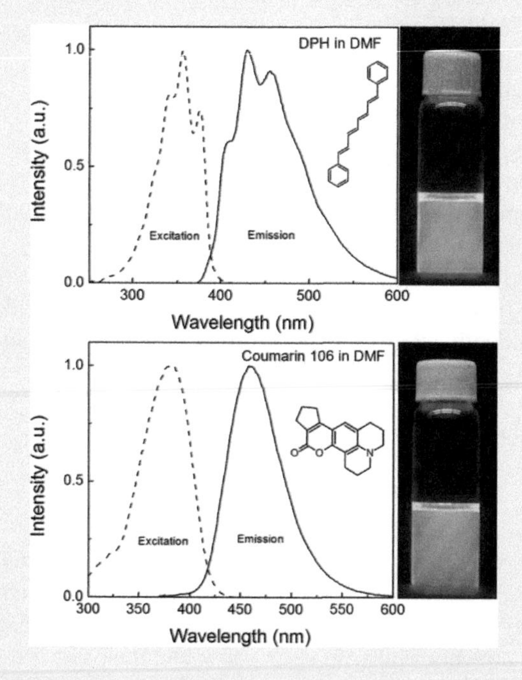

FIGURE 4-6.2 Fluorescence excitation and emission spectra of DPH (top) and C106 in DMF (bottom).

(Continued)

EXPERIMENT 4-6 (CONTINUED) STRUCTURED AND UNSTRUCTURED EMISSION SPECTRA

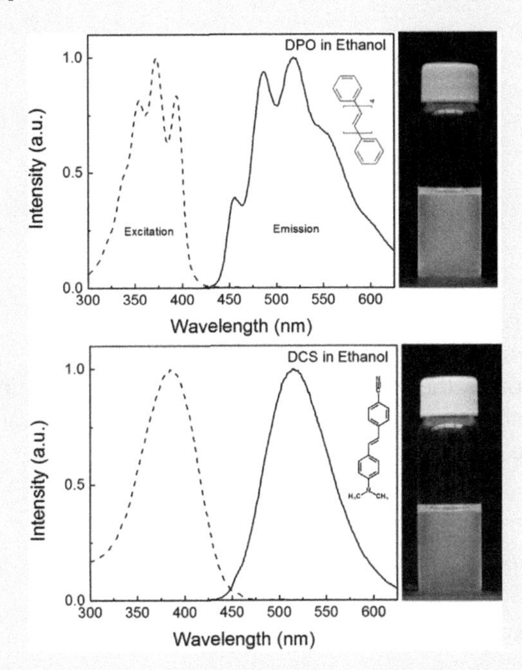

FIGURE 4-6.3 Fluorescence excitation and emission spectra of DPO (top) and DCS in ethanol (bottom).

Note: It is not possible to predict the shape of the fluorescence spectrum from the color of emission. However, the approximate spectral locations can be guessed from the fluorescence colors. It is very convenient for the spectroscopist to properly predict the spectral range of the studied sample just from the color of observed fluorescence.

EXPERIMENT 4-7 DEPENDENCE OF EMISSION ON THE EXCITATION WAVELENGTH

Keywords: emission, fluorescence, organic dyes, spectrofluorometer, spectrophotometer, rhodamine 6G.

One very important property of fluorescence is the fact that emission occurs always from the lowest excited state (azulenes are only very rare exceptions). This means that the emission spectrum should be independent of the excitation wavelength in sense of the spectral profile (normalized spectra should be identical for any excitation wavelength). Of course, the fluorescence intensity will be different because an extinction coefficient is wavelength dependent and the excitation lamp power is different for each excitation wavelength. This will have important consequences later in practical applications, in detection and imaging, and also when looking for mixtures of chromophores where the emission spectrum will depend on the excitation wavelength.

(Continued)

EXPERIMENT 4-7 (CONTINUED) DEPENDENCE OF EMISSION ON THE EXCITATION WAVELENGTH

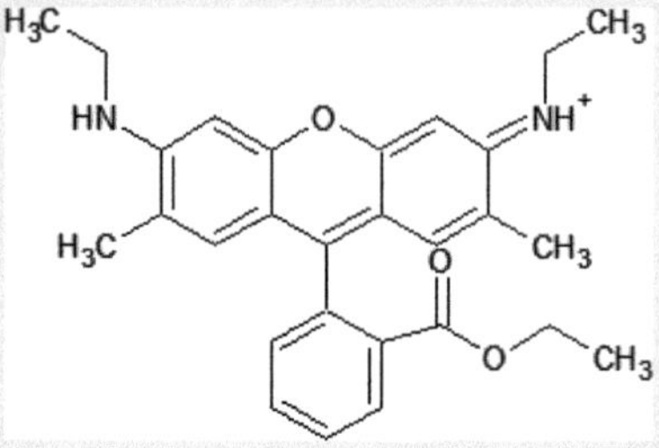

FIGURE 4-7.1 Chemical structure of rhodamine 6G (R6G).

This protocol compares the measurement of excitation spectra in quartz and plastic cuvette for 9-methylanthracene that presents two distinct absorption bands in visible and UV spectral regions (Figure 4-7.1).

MATERIALS

- Rhodamine 6G (R6G)
- Water, ethanol
- 1 cm × 1 cm cuvettes

EQUIPMENT

- Spectrophotometer
- Spectrofluorometer

METHODS

1. Prepare a stock solution of R6G in ethanol.
2. Make a few milliliters of R6G sample in water with an optical density of about 0.1 using a small amount of the stock solution. Place the sample into a cuvette and measure its absorbance.
3. Measure fluorescence of the sample with 500 nm excitation.
4. Set the excitation at 350 mm and measure the emission again. If needed, increase the voltage on the detector to increase the signal. *Do not adjust the slits.*
5. Plot the collected spectra as measured and also normalized.

RESULTS

An absorbance spectrum of rhodamine R6G is shown in Figure 4-7.2.

Figure 4-7.3 shows the emission spectra of R6G in water taken with 500 nm and 350 nm excitations.

The intensity measured with 350 nm excitation is typically lower. But one should remember that overall measured intensity depends on the product of absorption and the lamp intensity at the excitation wavelength. So if we do not correct for relative lamp intensity the measured relative intensities will typically not perfectly reflect the relative absorptions

(Continued)

EXPERIMENT 4-7 (CONTINUED) DEPENDENCE OF EMISSION ON THE EXCITATION WAVELENGTH

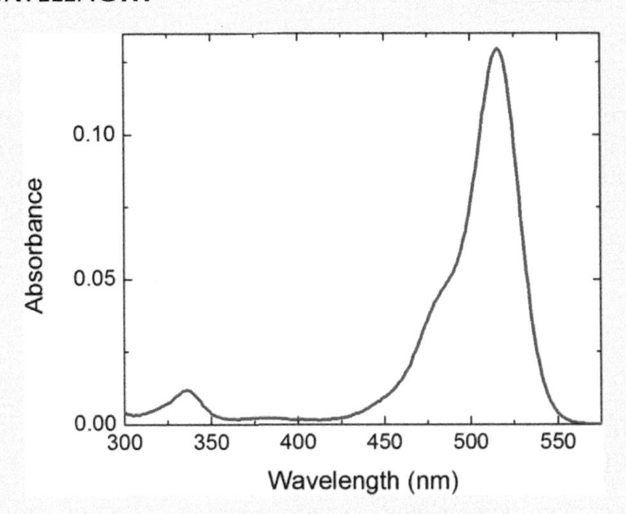

FIGURE 4-7.2 Absorbance spectrum of rhodamine 6G in water.

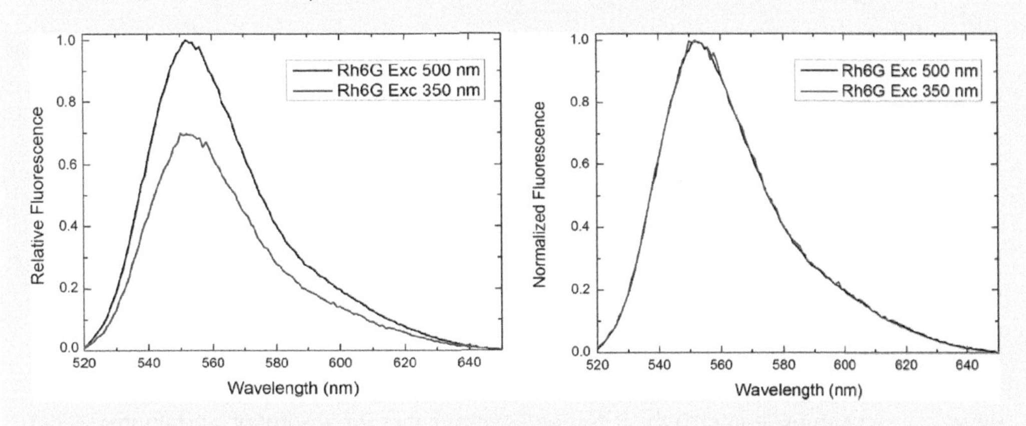

FIGURE 4-7.3 Emission spectra of R6G with different excitation wavelengths. As measured with 500 nm and 350 nm excitations (left) and normalized (right).

measured for two wavelengths. When using different wavelengths for the excitation, we may have to use different slit openings and different voltages on the detector. In order to compare the shapes of the spectra, we kept the same emission and excitation slits for both excitations and used a larger voltage for the 350 nm excitation. Normalized spectra for both excitations are practically identical, see Figure 4-7.3, right. Two emission spectra were normalized by dividing each intensity reading by the maximum intensity of that spectrum. The resulting spectra have a range from 0 to 1. The spectra can be corrected for different intensities of the excitation beam by multiplying each intensity data point by a ratio of respective excitation power.

Note: In a typical spectrofluorometer even when keeping the same slit opening for 350 nm and 500 nm excitation, the lamp energy will be different.

(Continued)

EXPERIMENT 4-7 (CONTINUED) DEPENDENCE OF EMISSION ON THE EXCITATION WAVELENGTH

CONCLUSIONS

The independence of the emission spectrum profile on the excitation wavelength (sometimes called Kasha's rule) is an important characteristic of fluorescence that allows us to use any convenient (or available) excitation wavelength to characterize the emission of a fluorophore and can be used to estimate the purity of the solution (contamination by other emitters). This rule applies only to single fluorophores. The spectral shape of mixtures may strongly depend on the excitation wavelength.

EXPERIMENT 4-8 DETECTION LIMITS FOR ABSORPTION AND FLUORESCENCE MEASUREMENTS

Keywords: absorption, fluorescence, spectrophotometer, spectrofluorometer, rhodamine 101.

Absorption measurements usually precede fluorescence measurements. This is because an absorption measurement can instantly provide the absorption spectrum and/or information on the concentration of the dye if the extinction coefficient for the dye and the thickness of the cuvette are known. In previous experiments, homogeneous illumination of the sample during fluorescence measurements was assured by using a low absorption (below 0.1). On the other hand, absorbance much below 0.1 is a relatively low value and below the optimal spectrophotometer range. We want to test how precision for low absorption rage corresponds to the precision of fluorescence measurement.

In the case of very low concentrations of a fluorophore, a question arises: *what concentrations are still reliable to be measured by absorption and fluorescence?* In Chapter 3 has been shown that absorptions about 0.001 can be reliably measured with proper precautions. Now, we will present an experiment which compares sensitivities for absorption and fluorescence detections.

MATERIALS

- Rhodamine 101 (R101)
- Water
- Glass vials
- 1 cm × 1 cm cuvettes
- Long-wave pass 580 nm glass filter

EQUIPMENT

- Spectrophotometer
- Spectrofluorometer

(Continued)

EXPERIMENT 4-8 (CONTINUED) DETECTION LIMITS FOR ABSORPTION AND FLUORESCENCE MEASUREMENTS

METHODS

1. Make a few milliliters stock solutions of R101 in water. Prepare a 3 mL sample of R101 in water with the absorption of about 0.1 in the maximum. With the extinction coefficient of R101 of about 95,000 L/mol·cm, the concentration of the dye is about 1 μM, see Figure 4-8.1.
2. Prepare progressively diluted samples of R101 in water and measure their absorptions, see Figures 4-8.1 and 4-8.2.
3. Measure the fluorescence spectrum of the sample with the lowest concentration (Figure 4-8.2, 8 nM). For this measurement, you will typically need to reduce the voltage on a spectrofluorometer detector or use narrow excitation/observation slits, see Figure 4-8.3.
4. Make more progressively diluted samples and measure their emission spectra, see Figure 4-8.4. Increase the voltage and open the slits as needed. Also, measure the emission spectrum of the water by itself.
5. Measure excitation fluorescence spectra of R101 at 1 μM and 1 nM concentrations, see Figure 4-8.5.

RESULTS

Figures 4-8.1 and 4-8.2 show the absorptions of diluted solutions of R101 in water. The sample with an R101 concentration of 0.8 μM is easily measurable. However, 100× dilution to the concentration of 8 nM presents significant noise and is rather difficult to distinguish from the water baseline, see Figure 4-8.2. As discussed in Chapter 3, we can improve precision for absorption measurement with a longer-path cuvette but this would be in rather limited in its extent (a specialized absorption cuvette may have 10 cm path that is only 10 times longer than standard 1 cm cuvette).

Next, measure the fluorescence spectrum of the 8 nM sample (lowest concentration measured with the spectrophotometer). This is an easy measurement with the voltage in the low range for the detector and small slits opening on excitation and emission.

Then, dilute further R101 solutions to single-digit picomoles (pM) and measure fluorescence spectra, as shown in Figure 4-8.4. Remember, with progressive dilutions, we will increase the voltage and slit openings.

The lowest possible to measure fluorescence signal is only limited by the background and it is important to relate measured fluorescence to the water (buffer) background. With the spectrofluorometer setting used for the lowest measured concentration (4 pM in our case), we checked the signal just from water. In Figure 4-8.4, the black line represents the signal from the water. The distinguishing feature is a peak at about 700 nm. This is a water Raman signal for used excitation. It is a relatively broad Gaussian-shaped band which is a result of the large slit opening for excitation and emission monochromators.

Sometimes, when absorption is too small and cannot be measured accurately, it could be substituted with the fluorescence excitation spectrum.

Note: In order to measure the excitation spectrum the observation should be set at the wavelength within the emission spectrum, usually at the maximum or slightly longer. A scanning should cover the absorption wavelength range.

(Continued)

EXPERIMENT 4-8 (CONTINUED) DETECTION LIMITS FOR ABSORPTION AND FLUORESCENCE MEASUREMENTS

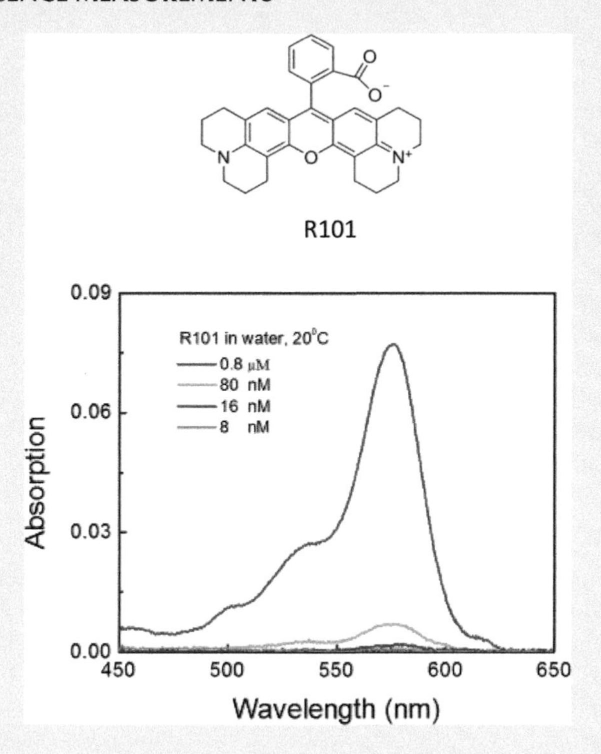

FIGURE 4-8.1 Absorption spectra of progressively diluted samples of R101.

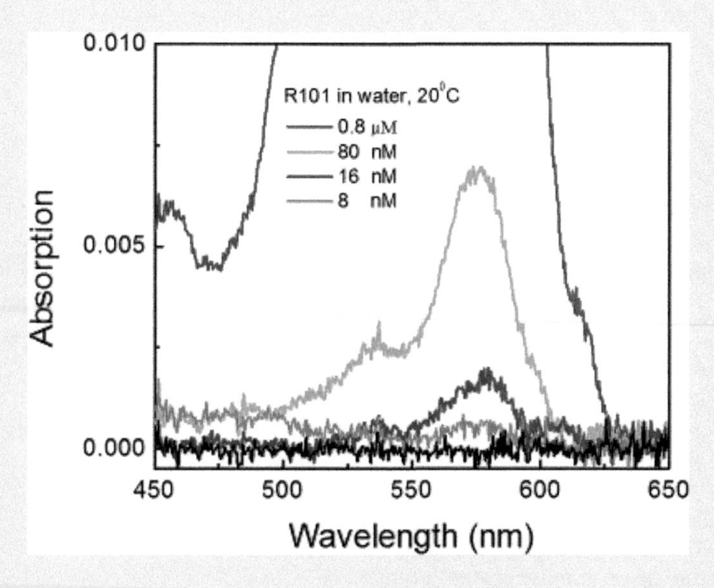

FIGURE 4-8.2 Absorption spectra of progressively diluted samples of R101 in an extended range.

(Continued)

EXPERIMENT 4-8 (CONTINUED) DETECTION LIMITS FOR ABSORPTION AND FLUORESCENCE MEASUREMENTS

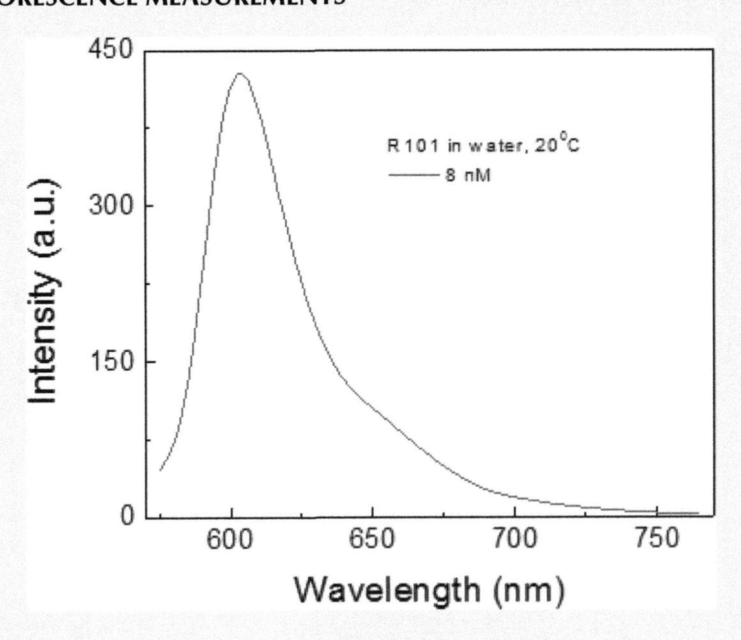

FIGURE 4-8.3 The fluorescence spectrum of 8 nM R101 in water. This is the sample with the lowest concentration in absorption measurements, see Figure 4-6.2. The measurement was done with 10 nm slits on both excitation and emission pathways with a low voltage setting on the photomultiplier detector.

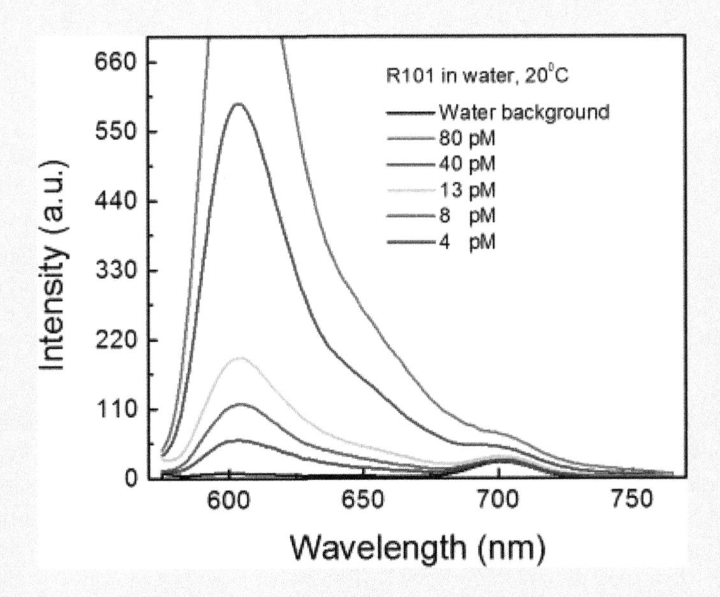

FIGURE 4-8.4 Fluorescence spectra of diluted solutions of R101 in water. These measurements were done with the highest voltage (990 V) allowed and slits 10 nm on both excitation and emission pathways.

(Continued)

EXPERIMENT 4-8 (CONTINUED) DETECTION LIMITS FOR ABSORPTION AND FLUORESCENCE MEASUREMENTS

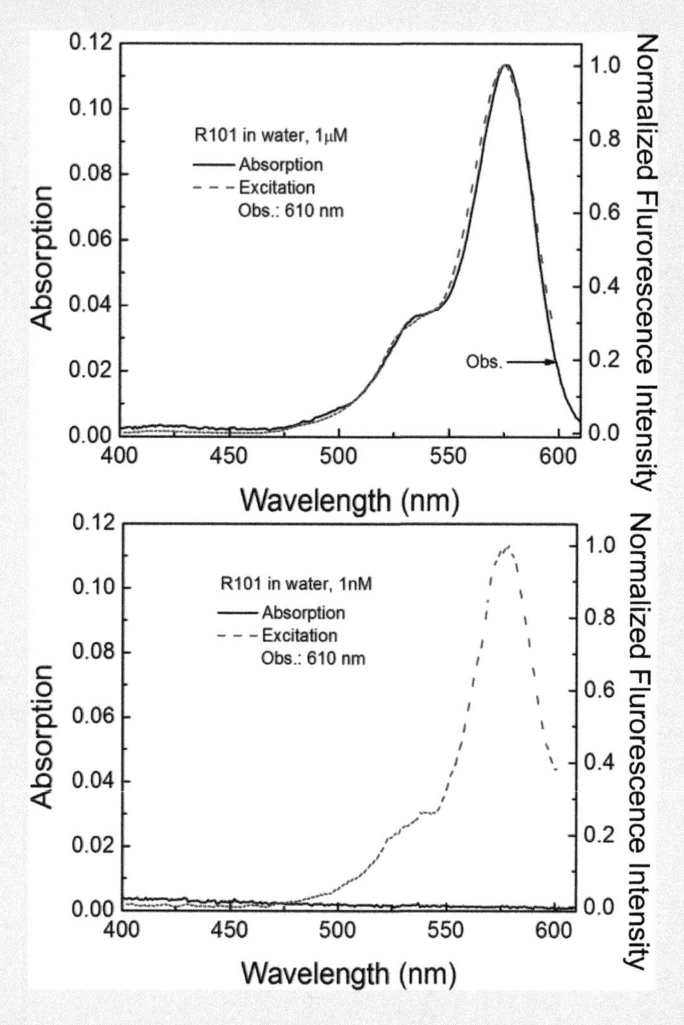

FIGURE 4-8.5 Absorption and fluorescence excitation spectra of R101 in water. Top: Concentration of R101 was 1 μM, both the absorption and excitation spectra can be easily measured. Bottom: The concentration of R101 was 1 nM. The absorption is "invisible" while the fluorescence excitation spectrum is still measurable with reasonable accuracy.

In Figure 4-8.5 (top), we present and compare absorption and fluorescence excitation spectra of 1 μM solution of R101. Within reasonable accuracy the spectra are identical.

Next, we diluted the solution 1,000-fold and measured again, see Figure 4-8.5 (bottom). It is not possible to detect the absorption spectrum but the fluorescence excitation spectrum is easily measurable.

(*Continued*)

EXPERIMENT 4-8 (CONTINUED) DETECTION LIMITS FOR ABSORPTION AND FLUORESCENCE MEASUREMENTS

CONCLUSIONS

The selected fluorophore R101 has a rather high extinction coefficient and almost absolute quantum yield, close to 1. These two molecular parameters are major factors often limiting the detection. From this simple experiment, we can see that fluorescence sensitivity, in this case, is over three orders of magnitude better than absorption. This is the typical difference that we may expect. For fluorophores with lower extinction coefficients and lower quantum yields, the detection limits will be lower (a higher fluorophore concentration will be needed). In the absorption measurements, one can still improve the detection with a longer cuvette, if the chamber of spectrophotometer allows. In fluorescence measurements, one can use a stronger excitation or just multiple passes of the excitation light through the cuvette using a mirror(s). For example, a much stronger excitation (e.g., laser) will again increase detection sensitivity. In general, fluorescence detection is a few thousand times more sensitive than absorption.

EXPERIMENT 4-9 RAMAN SCATTERING OF WATER

Keywords: scattering, Raman scattering, water.

Water is the most common solvent used in biological and biochemical experiments. Water is well transparent in the UV and visible range and does not present fluorescence. However, water molecules (as well as molecules of other solvents) interact with propagating light (electromagnetic field). This is evidenced by the dielectric constant/refractive index of water and out-scattering of the electromagnetic field in the form of Raman scattering. Typically, Raman scattering is a very weak phenomenon, much weaker than fluorescence, and Raman signal is typically negligible compared to the fluorescence signal originating from the same molecules. However, one needs to remember that the concentration of water that is a solvent in this case, as compared to the concentration of typical fluorophores is much higher. The typical chromophore concentrations are in the micromolar range (sometimes even less) while water concentration in a typical buffer is about 55.6 mole. So, in many cases when using very low concentrations of the chromophore, or weakly emitting dyes, the contribution of Raman signal from the solvent (water, ethanol, or other solvents) may need to be considered.

In this experiment, we want to experimentally demonstrate where to expect Raman scattering to be observed. In later experiments, we will present examples of measurements for which Raman is an issue, and needs to be accounted for. This protocol describes the wavelength dependence of Raman scattering from the water. A Raman scattering experiment is performed just like a fluorescence emission measurement where the excitation source, cuvette, and detector form a right angle.

MATERIALS

- Water
- 1 cm × 1 cm cuvette

(Continued)

EXPERIMENT 4-9 (CONTINUED) RAMAN SCATTERING OF WATER

EQUIPMENT

- Spectrofluorometer

METHODS

1. Turn on the fluorometer.
2. Set the sensitivity of the detector high (typically highest).
   a. This is for some instruments setting of the highest voltage.
   b. Also in many cases, we would open the excitation and emission slides to values close to maximum (10 nm or 20 nm).
   c. Set excitation and emission polarizers to vertical. For this polarizer orientation, the Raman signal will be greatest. Later, we will rotate observation polarizer to the magic angle (54.7) and horizontal.
3. Fill a cuvette with water.
4. Place the cuvette in the cuvette holder of the fluorometer.
5. Select an excitation wavelength of 285 nm.
6. Select an observation window between 300 nm and 400 nm.
   a. Always make sure that the excitation light is not observed directly, this can damage the detector. In practice, we typically select the starting wavelength for emission scan 5–20 nm above the excitation wavelength. This will depend on the slit, for larger slides the shift should be larger.
7. Measure emission and record data.
8. Repeat steps 5 through 8, without opening the sample chamber for the following wavelengths.
   a. 300 nm.
   b. 350 nm.
   c. 400 nm.
   d. 450 nm.
   e. 488 nm.
   f. 530 nm.
   g. 550 nm.
   h. 600 nm.

RESULTS

This experiment has been set up as an investigation to measure the dependence of the Raman signal of water on the excitation wavelength. The resulting spectra are presented in Figure 4-9.1. Depending on slit opening for each excitation, we should detect bell-shaped signals that will depend on the excitation wavelength. Measured spectra look like narrow water emission. Furthermore, spectra seem to be dependent on the excitation wavelength and the shapes of the peaks vary. As the excitation wavelength is increased (arrows in Figure 4-9.1), the observed spectral shape redshifts and becomes broader. Also, if we would close slits to smaller values like 2 nm, we will notice that the measured shape will become narrower. This is not the behavior of typical emission. And we know that water does not have an emission. What we have been observing in this experiment is Raman scattering of water. Measured spectral profiles are a result of excitation wavelength distribution (when selecting

(*Continued*)

EXPERIMENT 4-9 (CONTINUED) RAMAN SCATTERING OF WATER

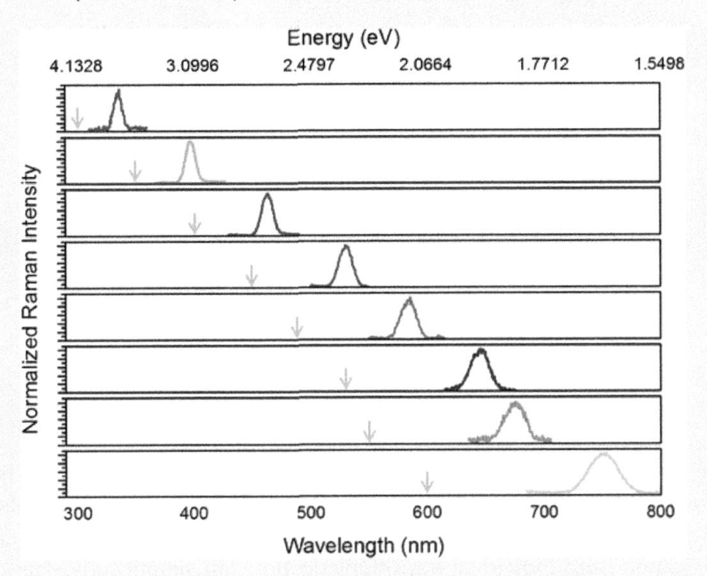

FIGURE 4-9.1 Raman spectrum of water for various excitation wavelengths.

the excitation wavelength the monochromator will not transmit well-defined line but will transmit light of a bandwidth defined by the slits settings) and observation monochromator will transmit scattered light to the detector broadening it further. In effect, we see quite a broadband peak.

Note: There are two important observations. First, the wavelength shift strongly depends on excitation wavelength and it is only about 36 nm for 285 nm excitation, quickly increases with wavelength and for 600 nm excitation, it is already 152 nm. Also because of the wavelength resolution of monochromators, we see band broadening for longer excitation wavelengths. Just for practical applications (as discussed in Chapter 2), Raman scattering of water is shifted of about 3300 cm^{-1} and this energetic shift is constant for each excitation. So, we would expect Raman peak for 280 nm excitation to appear at about 310 nm (280 nm corresponds to ~35,710 cm^{-1} and subtracting 3300 cm^{-1} we have 32,210 that corresponds to about 309 nm and for 350 nm excitation, we would expect Raman to be located at about 397 nm). *Note that 280 nm excitation is typically used for amino acids and protein excitation and Raman peak is very closely located to the expected tyrosine emission.*

The peaks observed in Figure 4-9.1 are correctly fitted with linear combinations of three components corresponding to the three vibrational modes of water. Thus, all three vibrational modes are present. Since 3300 cm^{-1} is an average shift for a linear combination of the three peaks, the observed spectral shape is always shifted by the same amount of energy thus the change in wavelength increases as the excitation wavelength increases. For the data presented in this experiment, the excitation wavelength and observed redshift are presented in Table 4-9.1.

(Continued)

EXPERIMENT 4-9 (CONTINUED) RAMAN SCATTERING OF WATER

TABLE 4-9.1 Measured Scattered Photon Energy Shift

Excitation Wavelength (nm)	Observed Red Shift (nm)	Observed Red Shift (eV)
300	36	0.442
350	47	0.4193
400	63	0.4217
450	80	0.4158
488	98	0.4248
530	115	0.4170
550	125	0.4174
600	152	0.4176

If converted to energy the observed shift is independent of wavelength, however, the relative intensities of the three peaks resulting from different vibrational states may vary which results in a small apparent shift of the one observed spectral shape. The centers of the three peaks resulting from individual transitions do not shift significantly. The $\Delta\varepsilon$ values in Table 4-9.1 are measured for the entire Raman spectral shape.

CONCLUSIONS

The Raman peak usually has much lower intensity than the emission peak; it may be missed as noise in many experiments. However, when working with low fluorophore concentration it is important to identify the Raman peak of water or another solvent (remember—characteristic Raman shift will be different for different solvents) and not be misled by it. An easy way to distinguish Raman from emission is to measure an excitation wavelength dependence of the observed peak if the peak position shifts significantly when we change excitation wavelength it is not emission. Also if excitation is done at a wavelength higher than an emission peak like two-photon excitation, the Raman peak will not be observed.

EXPERIMENT 4-10 RAMAN SCATTERING OF DIFFERENT SOLVENTS

Keywords: Raman scattering, water, methyl alcohol, ethyl alcohol, ethyl acetate, cyclohexane.

We are frequently faced with the question "Is water the only solvent that contributes to Raman scattering?" Raman is a general phenomenon, and therefore different solvents will also present significant Raman scattering. However, the position (shift) and intensity may change. The goal of this experiment is to measure Raman scattering for a few frequently used solvents.

(*Continued*)

EXPERIMENT 4-10 (CONTINUED) RAMAN SCATTERING OF DIFFERENT SOLVENTS

MATERIALS

- 1 cm × 1 cm quartz cuvette
- Water
- Methanol
- Ethanol
- Ethyl acetate
- Cyclohexane

EQUIPMENT

- Spectrofluorometer
- Two linear polarizers (typical fluorometer should be equipped with polarizers).

METHODS

1. Insert cuvette with water into the sample holder
2. Make sure polarizers are in (some instrument will have the possibility to remove polarizers from the light path). Excitation polarizer should be in a vertical position. Depending on the signal level, we can set emission polarizer to vertical (to have maximum signal) or any angle smaller than 90° (an angle greater than 0° will attenuate signal).
3. For excitation, we used 485 nm laser diode but we can select 485 nm from the lamp.
4. Select scanning range for emission monochromator (remember to start few nanometers above the excitation to avoid direct scattering of excitation light into the detector) and set high sensitivity for the system. Raman signal is weak and typically we will have to use the maximum setting.
5. Measure Raman signal. Check the previous experiment for a scanning range.
6. Repeat the measurements for each solvent without changing conditions.

RESULTS

Figure 4-10.1 shows measured spectra (Raman scattering signals) for four different solvents. Water signal presents a typical shift as in the previous experiments. Methanol and ethyl acetate show smaller shifts and comparable intensity. Cyclohexane shows a smaller shift and significantly greater intensity.

We asked the question: "What Raman signal we could expect from the mixture of two different solvents?" In other words, if the Raman signal measured for the mixture will be an additive sum of the respective signals from each individual solvent. For that experiment, we selected water and ethyl alcohol (EtOH). Figure 4-10.2 shows Raman scattering spectra for water, EtOH, and a 50:50 water/EtOH mixture. First, the Raman of EtOH is shifted as compared to water (similar to methyl alcohol shown in Figure 4-9.1) but has about 30% higher intensity. The mixture well approximates contribution of both solvents. But this is not a perfect superposition as we would expect for a 50:50 mixture. When mixing two solvents, the molecular interactions may slightly alter the energy of various states. Also in the case of water/EtOH mixture, the volume gets a little smaller than the cumulative volume of used components.

(*Continued*)

EXPERIMENT 4-10 (CONTINUED) RAMAN SCATTERING OF DIFFERENT SOLVENTS

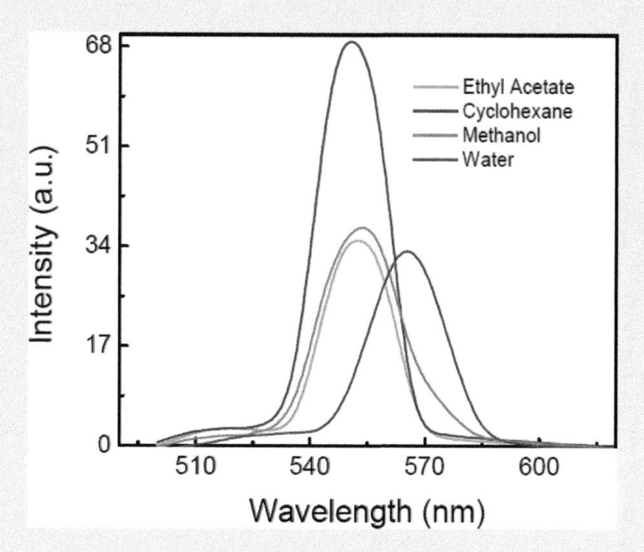

FIGURE 4-10.1 Emission scans for four high purity solvents.

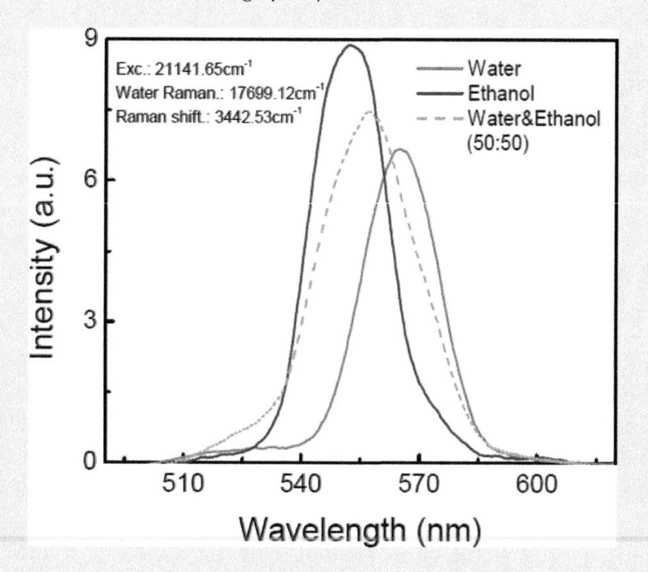

FIGURE 4-10.2 Emission scans for water, EtOH and 50:50 volume:volume mixture.

CONCLUSIONS

Raman transitions are typically very weak much weaker than fluorescence signal from a typical chromophore. However, we need to remember the concentration of the solvent is much higher than the concentration of dyes in solution. For water-based buffers, the water is a dominant component (typically over 55 M) while other contributors like buffer components are in μM to mM concentrations and dye concentrations are even lower than that. In general,

(Continued)

EXPERIMENT 4-10 (CONTINUED) RAMAN SCATTERING OF DIFFERENT SOLVENTS

we will expect a Raman signal from all solvents and in the case of weak fluorescence, we have to account for that. Typically each solvent will present Raman signal at a different position (different shift) and for each solvent relative signal intensity will also be different.

EXPERIMENT 4-11 POLARIZATION DEPENDENCE OF RAMAN SCATTERING FROM WATER

Keywords: Raman scattering, water, polarization.

Raman scattering depends on the excitation wavelength and the shift has a constant value in energy scale (for water about 3,300 cm^{-1}). As discussed in the previous fluorescence emission experiments, a square geometry is used where the excitation source, cuvette, and detector form a right angle. Raman scattering is an instantaneous phenomenon and we expect that it will be highly polarized. To investigate the polarization dependence of Raman emission polarizers are placed between the excitation source and cuvette, and between the cuvette and detector as shown in Figure 4-11.1 (polarization measurements are discussed in more detail in Chapter 6).

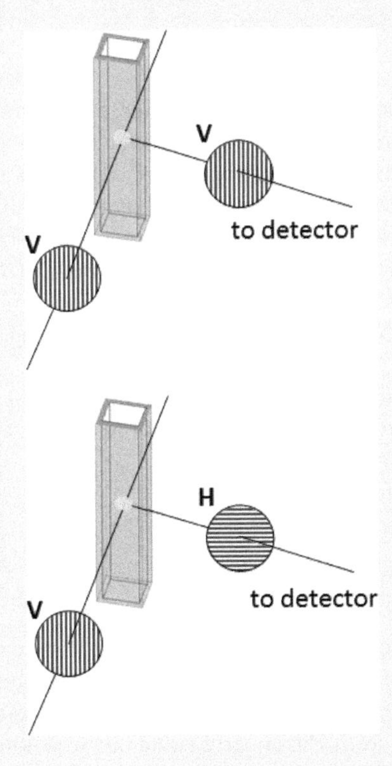

FIGURE 4-11.1 Schematic for the experimental configuration for Raman polarization measurements. Vertically polarized light enters the cuvette with the solvent and out scattered light is observed through polarizers oriented vertically (top), horizontally (bottom).

(Continued)

EXPERIMENT 4-11 (CONTINUED) POLARIZATION DEPENDENCE OF RAMAN SCATTERING FROM WATER

Similar to typical emission and polarization measurements vertically polarized light is used for excitation, "V" and vertical, "V," horizontal, "H," or magic angle "MA" orientation of the polarizer on the observation. The polarizer between the cuvette and detector is shown in a vertical, "V" and horizontal, "H" orientation. The magic angle "MA" orientation is not shown; it is where the angle between vertical and orientation of observation polarizer is 54.7°.

This protocol describes measuring the Raman spectra of water while excitation light is vertically polarized and the excitation polarizer is oriented at 0° (vertical), 90° (horizontal), or 54.7° (MA).

MATERIALS

- 1 cm × 1 cm quartz cuvette
- Water

EQUIPMENT

- Spectrofluorometer
- Two linear polarizers (typical fluorometer should be equipped with polarizers)

METHODS

1. Insert cuvette with water into the sample holder in a square geometry configuration.
2. Make sure polarizers are in (some instrument will have the possibility to remove polarizers from the light path). Excitation polarizer should be in a vertical position.
3. Select the excitation wavelength. Contrary to fluorescence that we should excite in well-defined spectral range (absorption band of chromophore) for Raman measurement, we may select any wavelength. For this experiment, we use UV (280 nm) and visible (485 nm).
4. Select scanning range for emission monochromator (remember to start few nanometers above the excitation to avoid the leak of direct scattering of excitation light into the detector).
5. Measure the Raman signal with a second linear polarizer (polarizer included in the detection line of spectrofluorometer) between the cuvette and detector.
   a. In vertical orientation.
   b. In horizontal orientation.
   c. In magic angle orientation.

RESULTS

First, we run an emission scan with 280 nm excitation. For this scan, we select a broad range 290–500 nm for emission scan. *Note:* This range would typically be too large but 280 nm excitation will excite a lot of potential impurities including any in the cuvette material (some less expensive cuvettes are made from glass/quartz containing unknown compounds that can be excited with UV light). Also, old heavily used cuvettes will contain some partially decompose chemicals that stick or diffuse into the glass and can be excited with UV light. Such cuvette will present a significant fluorescence signal that is not visible to the eye. In addition, water

(*Continued*)

EXPERIMENT 4-11 (CONTINUED) POLARIZATION DEPENDENCE OF RAMAN SCATTERING FROM WATER

frequently used in biology labs is not perfectly pure and can show quite significant fluorescence with 280 nm excitation. So, scanning a broad range using UV excitation will be a good test of purity of our solvent and cuvette and will give us a good feeling on the baseline level.

In Figure 4-11.2, we present emission scans of water solution run with three orientation for observation polarization. The Raman signal is shifted by about 36 nm and is the strongest for vertical orientation of the polarizer on observation. The Raman signal is much lower for MA and even lower for horizontal orientation. Importantly for wavelength larger than 320 nm (of the Raman) measured signal is practically zero indicating that our water and cuvette are clear.

Next, we shift excitation to 485 nm. In this case, we are not worried about purity since in most cases it will be visible in the form of slightly colored cuvette walls or solvent. In Figure 4-11.3, we only present a limited range (560–620 nm). For the used 20 nm slits as expected, we observe rather a broad scattering band. Again scattered light has preferential vertical direction since the VV peak is an order of magnitude higher than the VH peak. The right panel in Figure 4-11.3 shows the three normalized spectra. It is immediately apparent that the VH peak is slightly red-shifted.

It should not be surprising that the Raman signal highly depends on the orientation of observation polarizer. The polarization of scattered light depends on the orientation and symmetry of vibrational modes. In Raman studies, a quantity called the depolarization ratio is sometimes considered. The depolarization ratio, ρ, is defined as the ratio of the emission peak with the polarizers oriented perpendicular and parallel to each other. In this experiment, these were the VV and VH orientation. Thus, the depolarization ratio in this example is.

$$\rho = \frac{I_{\perp}}{I_{\parallel}} = \frac{I_{VH}}{I_{VV}} = \frac{11.15}{1.02} = 10.93$$

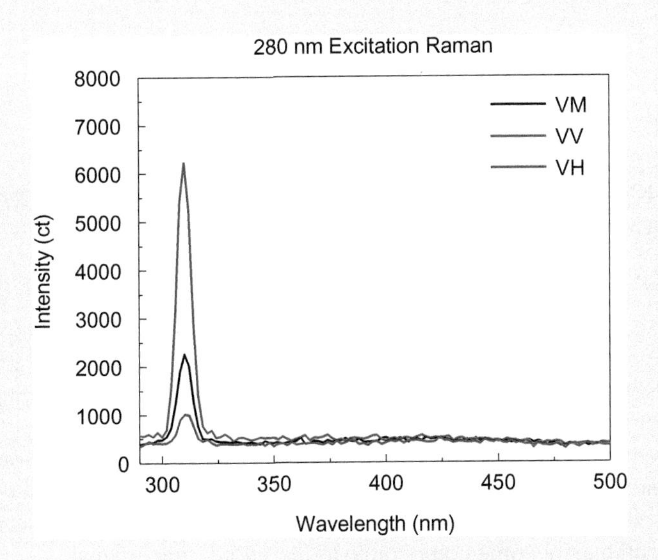

FIGURE 4-11.2 Measured Raman scattering for vertical-vertical (VV), vertical-horizontal (VH), and vertical-magic angle (VM) observation polarizer orientation for water at 280 nm excitation.

(Continued)

EXPERIMENT 4-11 (CONTINUED) POLARIZATION DEPENDENCE OF RAMAN SCATTERING FROM WATER

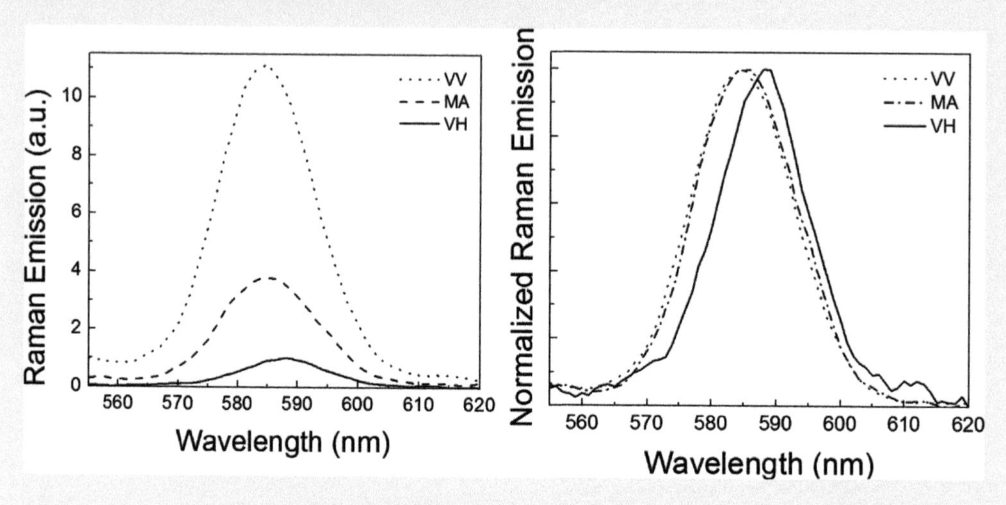

FIGURE 4-11.3 Raman scattering of water at 485 nm excitation as seen through different polarizer orientation. Graph of right presents normalized spectra.

CONCLUSIONS

This experiment clearly demonstrates that Raman scattering is highly polarized. This is the very important property of Raman. This experiment does not show that Raman scattering is also highly directional phenomena with maximum intensity in the direction 90° toward the direction of excitation light propagation and 90° toward the excitation beam polarization. So, the Raman scattering has its maximum value in typical square geometry.

EXPERIMENT 4-12 FLUORESCENCE EMISSION OF LOW CONCENTRATION OF NATA IN WATER

Keywords: Raman scattering, emission, water, polarization, *N*-acetyl tryptophan amide.

Even if Raman scattering is a weak phenomenon, the Raman peaks may appear in fluorescence measurements. Oftentimes the Raman signal is undesired as it may hide spectral features of the fluorescence, present artificial spectral features, or make the fitting of measured emission more difficult.

Raman scattering becomes a problem when using a low concentration of chromophore (frequently required in biological applications) or the chromophore is a poor emitter (as typical UV emitters like DNA bases or amino acids). In such cases, the fluorescence signal can be low and Raman may become comparable to the fluorescence signal.

This protocol describes the measurement of a low concentration of NATA in water and how to quickly identify Raman contribution by changing the polarizer orientation on observation.

(*Continued*)

EXPERIMENT 4-12 (CONTINUED) FLUORESCENCE EMISSION OF LOW CONCENTRATION OF NATA IN WATER

MATERIALS

- 0.4 cm × 1 cm cuvette
- Water
- *N*-acetyl tryptophan amide (NATA)

EQUIPMENT

- Spectrofluorometer equipped with Glen-Thompson polarizers (or polarizers working in UV).

METHODS

1. Prepare a stock solution of NATA in water.
2. Measure the fluorescence emission of NATA with 280 nm excitation and VV polarizer orientation (maximum Raman signal). Dilute the solution with pure water down to the point when Raman peak is clearly visible (we will need to significantly boost instrument sensitivity—typically voltage when diluting the sample). A simple approach is to measure just Raman signal of water adjusting the sensitivity setting at your spectrofluorometer. The measured signal of the NATA solution should be comparable to the Raman signal.
3. For diluted solution, measure emission spectra with three configurations of excitation and observation polarizers:
   a. Vertical-vertical (VV)
   b. Vertical 54.7° (VM)
   c. Vertical-horizontal (VH)

RESULTS

In previous Experiment 4-11, we showed that Raman scattering is strongly polarized and observed signal depends on observation polarizer orientation. Figure 4-12.1 shows the resulting normalized emission spectra of NATA from the above procedure. The narrow peak at about 315 nm clearly visible for vertical observation is the Raman scattering of water (compare to Figure 4-11.2). This peak has significant overlap with the emission of NATA and could be mistaken for the spectral feature of NATA emission. When changing polarizer orientations from vertical to MA, and horizontal the relative intensities of the short-wavelength peaks decreases significantly. This indicates that the signal responsible for a narrow band at ~315 nm and a broadband peak centered at ~360 nm have different origins.

For comparison, in Figure 4-12.1, we are presenting NATA emission that is practically free of Raman contribution (dashed red line). For that, we used a higher concentration of NATA with the absorbance of about 0.1 (this increases the relative contribution of NATA emission). It is clear that for vertical excitation and horizontal observation (blue line) some Raman contribution to the signal is visible. According to Experiment 4-11, the Raman for horizontal observation should decrease about 10-fold.

As shown in Experiment 4-9, the easiest way to identify a Raman peak is to change the excitation wavelength and the Raman peak will shift proportionally to the wavelength

(*Continued*)

EXPERIMENT 4-12 (CONTINUED) FLUORESCENCE EMISSION OF LOW CONCENTRATION OF NATA IN WATER

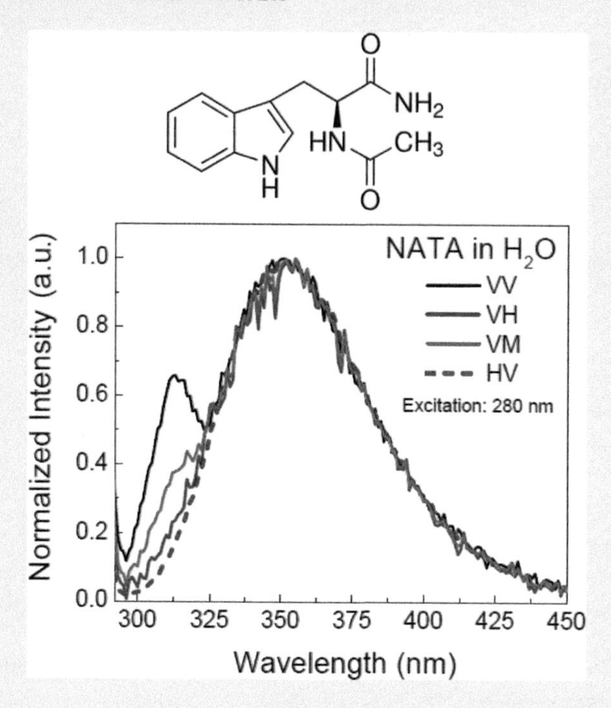

FIGURE 4-12.1 Emission of NATA in water for different polarizers orientation. The reference spectrum (dashed red) shows the emission of higher concentration of NATA with horizontal excitation to further limit Raman contribution.

change. For example, we would recommend shifting excitation to 290 nm. This may lower the overall signal of NATA and will show a significant shift of Raman peak closer to the emission peak. The significant overlap of emission peak and Raman peak is a very big problem in time-resolved measurements. *In the next Experiment 4-13, we will present examples where Raman is located differently in relation to the emission band.* The contribution of Raman scattering can be significantly lowered when using horizontal polarization on excitation (scattering toward the detector will be minimal and fluorescence signal will be dominant). But this is not a proper configuration for measuring emission spectra. Another way to limit Raman contribution in the UV is to use a long-path filter. For UV excitations, Raman peak is typically positioned in front of the main emission band and using long-path filter may eliminate the Raman. As in this case putting a 330 nm long-path filter will sacrifice a short wavelength edge with Raman scattering. This approach is frequently used in experiments where we only want to detect total intensity like sensing or time-resolved measurements.

CONCLUSIONS

Changing the polarizer orientation could be a good way to quickly test for Raman contribution. Also, when using MA conditions for measuring the spectra (a proper way for measuring spectra), we also significantly limit Raman contribution.

(Continued)

EXPERIMENT 4-12 (CONTINUED) FLUORESCENCE EMISSION OF LOW CONCENTRATION OF NATA IN WATER

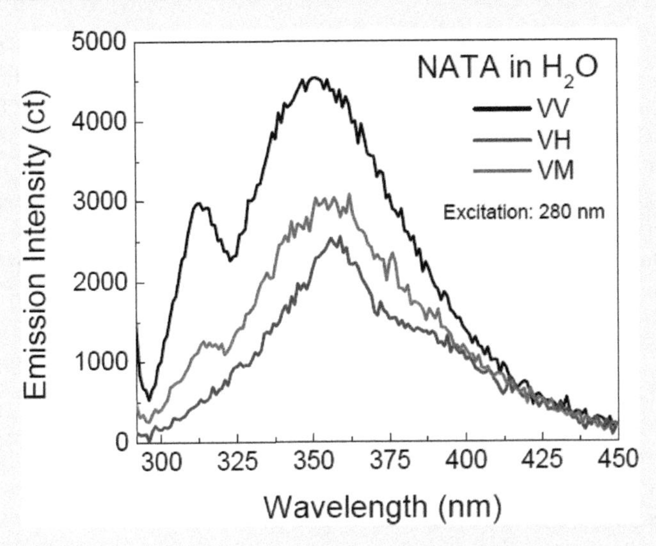

FIGURE 4-12.2 Uncorrected emission spectra measured for NATA solutions from Figure 4-12.1.

Note: It is important to mention that presented emission spectra in Figure 4-12.1 have been corrected for the polarization sensitivity of the detector (we will discuss the correction procedure in the polarization section (Chapter 6). Many systems today may apply such correction automatically without a user knowing that. But if it is not applied the spectra (VV, VH, VM) may look very different. In Figure 4-12.2, we are presenting the uncorrected emission spectra measured for this experiment. This is not unusual for the spectra to be so drastically different. However, it is important to mention that relative contribution of Raman to the emission of NATA will not be affected by the different sensitivity of the detector to different observation polarizer orientation. But can be affected by the different sensitivity of the detector to a different wavelength.

EXPERIMENT 4-13 MEASUREMENTS OF A WEAK FLUORESCENCE COMPARABLE TO RAMAN SCATTERING

Keywords: Raman scattering, emission, fluorescence, organic dyes, spectrofluorometer, anthranilic acid, acridine orange.

Sometimes it is necessary to measure very weak fluorescence. *How to do this correctly?* In order to achieve a measurable signal, the widest slits and the highest voltage on the detector must be used. Under these conditions, the Raman scattering may have an intensity comparable to the fluorescence. Of course, this is assuming the solvent used (water in this experiment) is highly pure and itself does not have any fluorescence. Raman scattering can be attenuated to some extent by using a horizontally oriented polarizer on the emission

(Continued)

EXPERIMENT 4-13 (CONTINUED) MEASUREMENTS OF A WEAK FLUORESCENCE COMPARABLE TO RAMAN SCATTERING

pathway (see previous Experiment 4-11). However, most often it is necessary to use magic angle conditions, and in fluorescence anisotropy measurements it is necessary to use a vertically oriented emission polarizer to measure the vertical emission component. The Raman scattering is shifted from the excitation line by a constant number of reversed centimeters (about 3,300 cm^{-1} in water) and in the short wavelength region (UV-VIS) is usually located at the beginning of the emission spectrum (blue edge). However, at longer wavelengths, it could be located within the emission band or behind the fluorescence peak (red edge). We will demonstrate these cases using anthranilic acid, AA (short wavelengths), and acridine orange, AO (longer wavelengths). It will also become apparent that careful subtraction of Raman scattering sufficiently corrects the emission spectra.

MATERIALS

- 1 cm × 1 cm quartz or glass cuvettes. We can also use disposable fluorescence cuvettes (we will use excitations above 320 nm and such cuvettes are ok).
- Anthranilic acid (AA)
- Acridine orange (AO)
- Water

EQUIPMENT

- Spectrophotometer
- Spectrofluorometer

METHODS

1. Make stock solutions of AA and AO in water. Adjust concentrations for the optical densities to be about 0.1–0.2 and measure absorption spectra using a water baseline.
2. Measure emission spectra for a solution that have absorption of about 0.1 with the emission polarizer oriented at a magic angle.
3. Dilute AA and AO solution about 500–1,000 fold to reach a few nanomoles concentrations.
4. Measure fluorescence emission spectra of diluted solutions and water baseline using 340 nm excitation for AA and 470 nm excitation for AO.

RESULTS

Figures 4-13.1 and 4-13.2 show normalized absorption and emission spectra for AA and AO respectively. For AA, we used 340 nm and for AO 470 nm excitations as indicated by the arrows in Figures 4-13.1 and 4-13.2.

Clearly, both spectra are pure emission and any emission deformations due to secondary effects like scattering or Raman scattering are not visible. To show how Raman influences the emission we will dilute both samples. Figure 4-13.3 shows emission spectra measured for a highly diluted solution of AA and pure water using the same spectrofluorometer setting. For water, a clear Raman band can be seen at about 370 nm. The emission spectrum of AA is clearly distorted as compared to the emission spectrum in Figure 4-13.1. Since water concentration is practically constant in both measurements in Figure 4-13.3, we can subtract from a measured spectrum of AA solution the spectrum of Raman measured just for water (or buffer).

(*Continued*)

EXPERIMENT 4-13 (CONTINUED) MEASUREMENTS OF A WEAK FLUORESCENCE COMPARABLE TO RAMAN SCATTERING

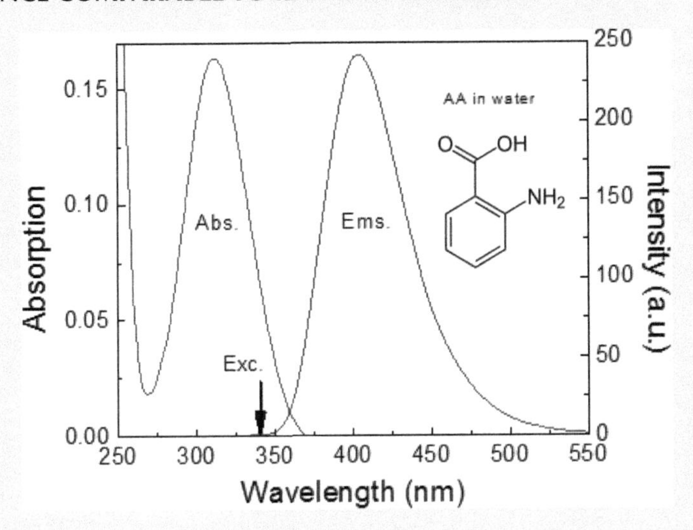

FIGURE 4-13.1 Absorption and fluorescence spectra of anthranilic acid (AA) in water.

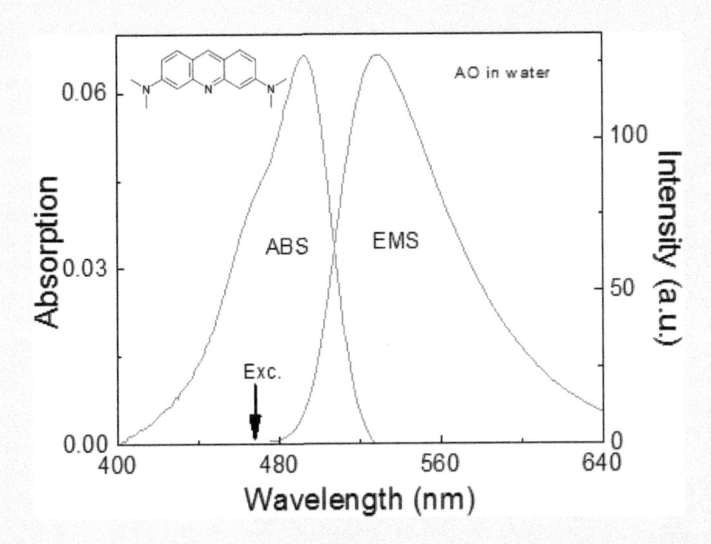

FIGURE 4-13.2 Absorption and fluorescence spectra of acridine orange (AO) in water.

Such correction is presented in Figure 4-13.4. For comparison, we also included the emission spectrum of AA measured for higher concentration at which Raman contribution is completely negligible (red). Both corrected and high concentration spectra very well correspond to each other proving that such correction of emission spectra gives good results.

A similar correction procedure was performed for the AO and is presented in Figures 4-13.5 and 4-13.6. In this case, the Raman is present on the red side of the emission spectrum, impossible to correct by using a long-path filter. Again, the corrected spectrum and measured spectrum for higher concentration are in a good agreement.

(Continued)

EXPERIMENT 4-13 (CONTINUED) MEASUREMENTS OF A WEAK FLUORESCENCE COMPARABLE TO RAMAN SCATTERING

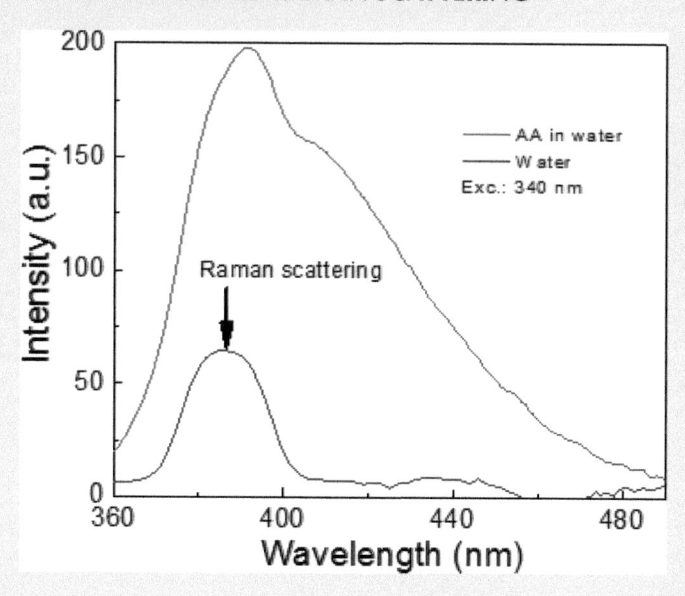

FIGURE 4-13.3 The emission spectrum of diluted AA solution in water and the control of Raman scattering from water.

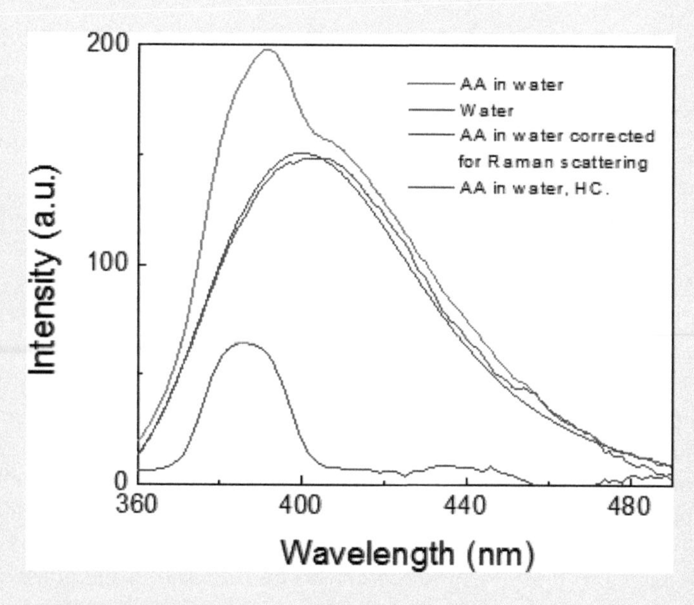

FIGURE 4-13.4 Correction for a Raman scattering. For comparison, the spectrum of higher concentration solution of AA is also shown.

(Continued)

EXPERIMENT 4-13 (CONTINUED) MEASUREMENTS OF A WEAK FLUORESCENCE COMPARABLE TO RAMAN SCATTERING

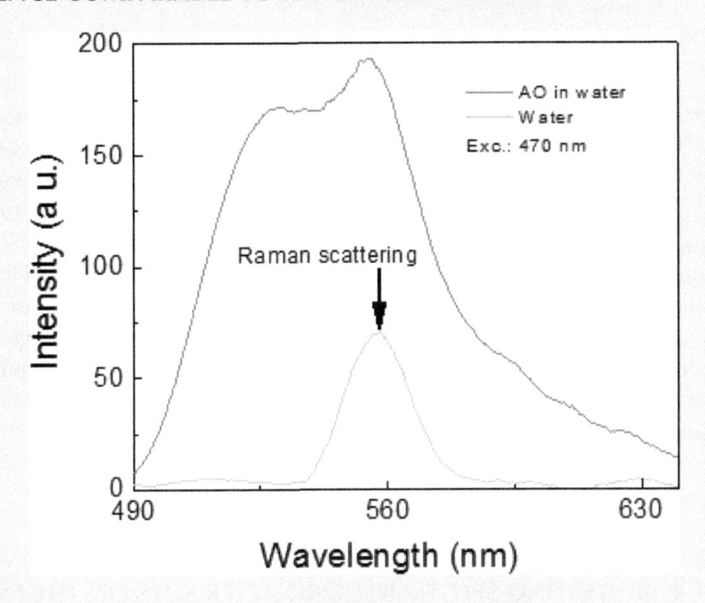

FIGURE 4-13.5 The emission spectrum of a diluted solution of acridine orange (AO) and the control of Raman scattering from water.

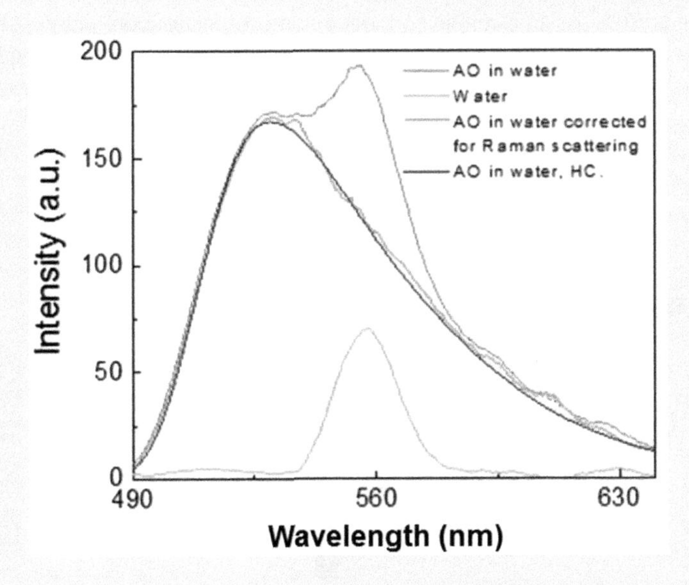

FIGURE 4-13.6 Correction for a Raman scattering. For comparison, the spectrum of higher concentration solution of AO is also shown.

(Continued)

EXPERIMENT 4-13 (CONTINUED) MEASUREMENTS OF A WEAK FLUORESCENCE COMPARABLE TO RAMAN SCATTERING

CONCLUSIONS

These experiments demonstrate that Raman may significantly distort measured emission spectra. However, in many cases, a simple correction procedure (subtraction of measured Raman background) gives good results. In the case when Raman is in the front of dye emission (a typical case for UV excitation), Raman can be also eliminated/reduced by using a proper long-path filter. This will distort the emission spectrum but it is very useful for sensing and fluorescence lifetime measurements.

Another important information coming from this experiment is the fact that for the visible and red spectral range Raman scattering may fall in the middle of emission or into the red edge of the spectrum. Traditionally Raman scattering occurring on the red edge of the emission is less recognized and in many imaging (microscopy) experiments, long-path filters are used that are transmitting Raman signal. This could be a significant problem for time-resolved experiments and FLIM.

EXPERIMENT 4-14 TESTING SPECTROFLUOROMETER SLITS ON THE EXCITATION AND OBSERVATION PATHS

Keywords: emission, fluorescence, spectrofluorometer, Ludox.

Often a spectrofluorometer used in an experiment is thought of as a "black box" which always produces the expected result. In most instances, there is no problem with such a point of view. However, there are situations when a lack of understanding of how an instrument function may lead to corrupted results. As an example consider that most spectrofluorometers use a white light source (lamp) as their excitation source. Out of this white light, the desired wavelength is separated by using a monochromator (grating). Figure 4-14.1 schematically shows an example of the

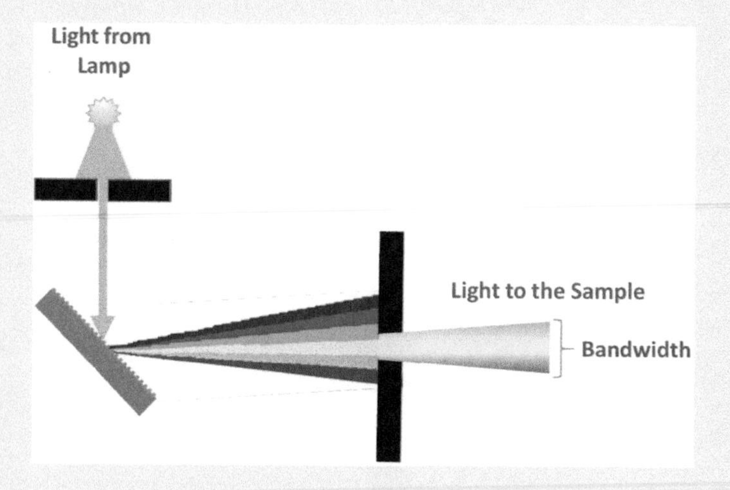

FIGURE 4-14.1 Schematic representation of a grating type monochromator on an excitation side.

(Continued)

EXPERIMENT 4-14 (CONTINUED) TESTING SPECTROFLUOROMETER SLITS ON THE EXCITATION AND OBSERVATION PATHS

wavelength selection. The excitation light from a lamp is directed to a dispersion element (grating) and exit monochromator through an opening (excitation slit). A focusing lens is used to focus light in the middle of a sample holder. The light leaving the monochromator is not monochromatic but consists of a range of wavelengths (band) and the spectral width of the band depending on the slit size. In effect, there is typically a broad range of excitation light frequencies leaving a monochromator and reaching the sample. Exactly the same situation happens with emission light when observed through an analyzing monochromator, and the detector detects a certain bandwidth of wavelengths. Of course, smaller slit means narrower bandwidth. In "2-16 Bandwidth (p. 67)," we briefly presented the concept for the monochromator bandwidth and how it will affect measurements. This experiment and the next one will show how this affects the measurement.

When describing the fluorescence experiments we usually stay: the excitation was at *x* nm and/or the observation was set at *y* nm. In addition, when setting the experimental parameters for fluorescence spectra measurements, one will need to also select the slits for excitation and emission paths. So, for results interpretation, it is important to realize what really *x* nm and *y* nm means. Naturally, the width of the slit (slit opening) is related to the $\Delta\lambda$ of a selected wavelength, but how the number of slits given by the vendor relates to $\Delta\lambda$ (bandwidth) of transmitted light? In the case of the grating as a dispersive element in the monochromator, $\Delta\lambda$ should be independent on the wavelength. Using the Rayleigh scattering, we will check the spectral widths of the light transmitted through the slits on the excitation and emission paths. First, we will use diode lasers for excitation and measure the scattered light using different slit on the emission path. Next, we will use the narrowest slit on the excitation and make the measurements with different slits on the observation at arbitrarily chosen 500 nm excitation light. Then we will set on the observation path, the narrowest slit and record the emission with different slits on the excitation. We will also check the spectral widths at different wavelengths on the excitation.

MATERIALS

- Silica suspension (Ludox, 30 nm)
- Water
- 1 cm × 1 cm cuvettes
- Diode lasers (375 nm, 445 nm, and 660 nm)

EQUIPMENT

- Varian Eclipse, ISS K2, and Picoquant FT300 spectrofluorometers

METHODS

1. Fill a cuvette with a solution of silica suspension in water (add to 3 mL of water a few drops of Ludox).
2. If possible, set the laser excitation and scan emission across the excitation line (not all spectrofluorometers can be used with laser excitation). Make sure the signal is not too high. If necessary, use adequate light attenuation using neutral density filters or lower voltage on the detector. Repeat the scan using different slit openings on emission. Use the smallest scanning step for all measurements.

(*Continued*)

EXPERIMENT 4-14 (CONTINUED) TESTING SPECTROFLUOROMETER SLITS ON THE EXCITATION AND OBSERVATION PATHS

3. Set the smallest slit on the excitation. In the case of the Varian spectrofluorometer, it is 1.5. Set on the observation (emission) path, the largest slit. In the case of the Varian spectrofluorometer, it is 20.
4. Set the excitation wavelength at 500 nm. Measure the emission from 475 nm to 525 nm with the slowest scan speed. Adjust properly the voltage on the detector. Be sure that the signal does not go over the scale. Keep the voltage constant during all measurements.
5. Repeat the scans for all other slits: 10, 5, 2.5, 1.5.
6. Set the observation slit at the smallest number (1.5) and run the scans for all excitation slits (20, 10, 5, 2.5, 1.5).
7. Plot the emission scans for both cases: (a) Constant slit on the excitation (1.5) and various slits on the observation path; (b) Constant slit on the observation path and various slits on the excitation path.
8. Normalize all plots and estimate the HWHM (half with at half maximum) for each combination of excitation/observation slit.
9. Check the spectral widths of excitation light at different excitation wavelengths.

RESULTS

***Using Laser Excitation*:** Typically, diode lasers present a well define emission line typically within 1 nm bandwidth. We used 450 nm and 660 nm laser diodes for excitation in square geometry and regular 1 cm x 1 cm cuvette. As a sample, we used a water suspension of small silica particles (Ludox). We run emission scan starting about 25 nm before and end the scan 25 nm after the excitation wavelength, and detected Rayleigh scattering of the excitation. *Of course, the Rayleigh scattering is exactly at the excitation wavelength, but how it will look through different observation slits?* It is important to carefully monitor the signal intensity and not saturate the detector. We measured both 450 nm and 660 nm (laser lines) using different slit openings on observation. In Figure 4-14.2, we present normalized

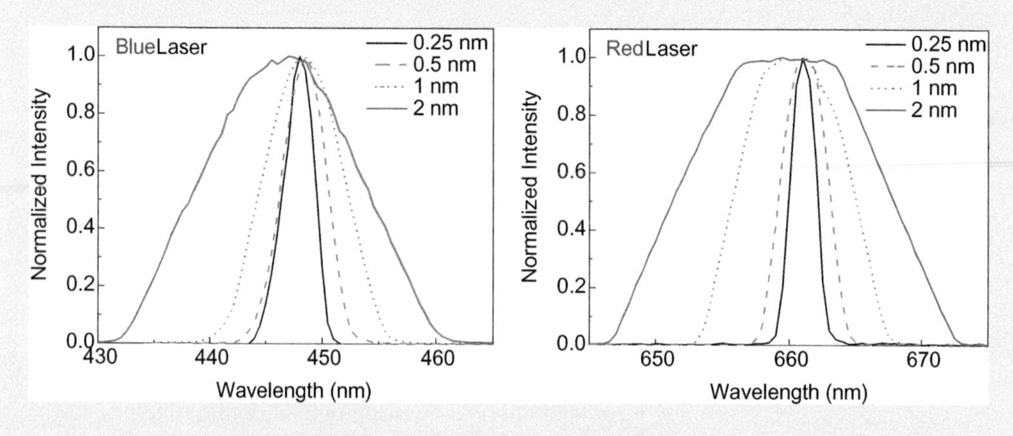

FIGURE 4-14.2 Normalized emission spectra for the scattering of two laser lines measured with different slit openings on emission. Left: 450 nm blue laser diode. Right: 660 nm laser diode.

(Continued)

EXPERIMENT 4-14 (CONTINUED) TESTING SPECTROFLUOROMETER SLITS ON THE EXCITATION AND OBSERVATION PATHS

spectra for both lasers measured in ISS K2 instrument using 0.25, 0.5, 1.0, and 2.0 slits. The bandwidth of laser excitation measured with the smallest slit is about 3 nm and 2 nm for the blue and red lasers, respectively. We assume that the real bandwidth of laser light is about or below 1 nm. The observed 2 nm bandwidth (for 0.25 slit) is the smallest wavelength resolution of the used monochromator. Increasing the slit to 0.5 increases the apparent bandwidth for both lasers to about 5 nm. For slit 1, the apparent bandwidth is about 8 nm for blue and 9 nm for red laser. Finally, a slit 2 results in about 15 nm bandwidth for blue laser and 17 nm for red laser.

Comparison of Laser and Lamp Excitations: In Figure 4-14.3, we are presenting measured scattering of 375 nm laser diode illumination using FT300 spectrofluorometer. For comparison, we also present the scan for 0.5 excitation slit using 480 nm lamp excitation as measured with different emission observation slits. The bandwidth for the 0.5 slit is below 2 nm and is similar for laser diode excitation and lamp excitation with 0.5 nm slit.

Testing Spectrofluorometer with Integrated Slits: Not all spectrofluorometers may utilize laser excitation. Also, some instruments have integrated slits which can be changed only by an operational software. In the case of Varian Eclipse, we first selected the narrowest slit (1.5) on the observation path and scanned the emission within 50 nm of the excitation wavelength using various slits on the excitation, see Figure 4-14.4, left. Next, we set the narrowest (1.5) slit on the excitation and scanned the emission within 50 nm of the excitation wavelength using various slits on the observation, see Figure 4-14.4, right. For each scan, we used the smallest step size (slowest scanning).

Here are a few observations that we would like to point out. The observed signal increases as we increase the slit for both cases on excitation and emission paths. In the bottom, we present normalized detected Rayleigh scattering (scanned intensities). With most narrow slits (1.5), we measured a little over 2 nm bandwidth. With the widest slit on the excitation there is a visible structure in the emission scan, see Figure 4-14.4, left. These peaks are from a xenon lamp used for the excitation (see Section 2-8 "Wavelength Calibration" [p. 56]).

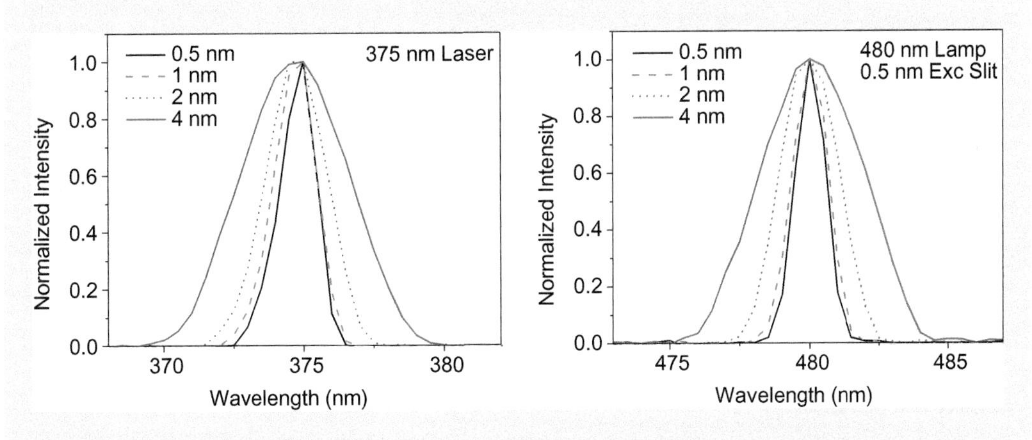

FIGURE 4-14.3 Left: Normalized Rayleigh scattering emission spectra for 375 nm laser lines measured with different slit openings on emission. Right: Emission scan of Rayleigh scattering at 480 nm (lamp excitation) with the excitation slit (0.5) and different slits on the emission channel.

(Continued)

EXPERIMENT 4-14 (CONTINUED) TESTING SPECTROFLUOROMETER SLITS ON THE EXCITATION AND OBSERVATION PATHS

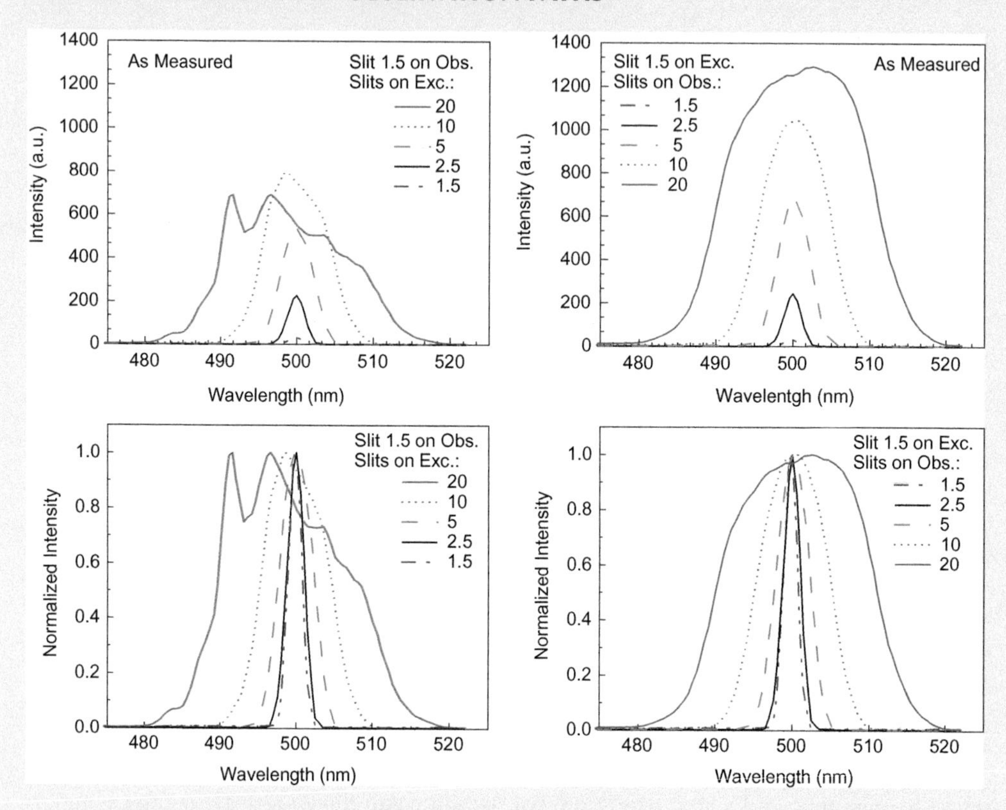

FIGURE 4-14.4 Top: Emission spectra of Rayleigh scattering observed through the narrowest slit (1.5) with different slits on the excitation (left); and observed through different slits with the narrowest slit on the excitation (right). Bottom: Normalized respective emission scans.

Of course, if on the excitation we set a narrower slit, this structure is not seen. Another important observation is that the width of the slit is constant and $\Delta\lambda$ does not depend on the wavelength.

Wavelength Dependence of the Bandwidth: We tested if the measured width dependents on the excitation wavelength. To answer this question, we repeated scans (keeping constant slit on excitation (1.5) and constant slit on emission (5). Measured widths are presented in Figure 4-14.5.

Presented example in Figure 4-14.5 shows that the measured width for different excitation wavelength (between 300 nm and 700 nm) is about 5–6 nm. This is an expected width for the excitation/emission slits 1.5/5 combination.

CONCLUSIONS

Typically the number of the slit describes $\Delta\lambda$ for transmitted light. For most common monochromators, the smallest slit setting yields about 2 nm for transmitted bandwidth. Comparing Figures 4-14.4 (top), it is clear that observed signal strongly increases as we increase the

(Continued)

EXPERIMENT 4-14 (CONTINUED) TESTING SPECTROFLUOROMETER SLITS ON THE EXCITATION AND OBSERVATION PATHS

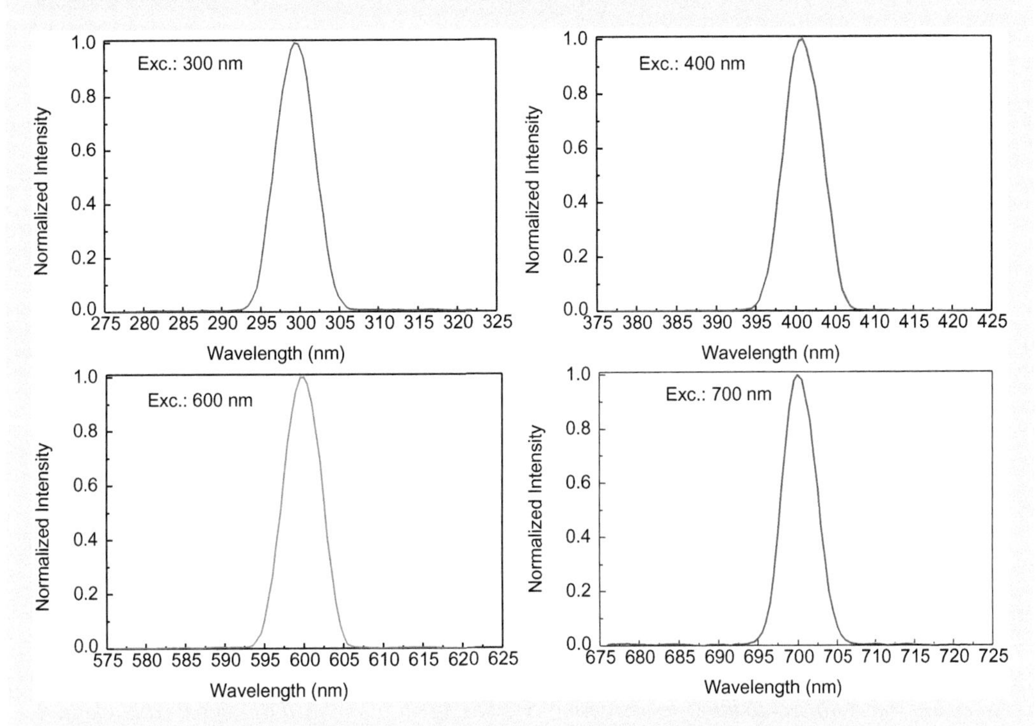

FIGURE 4-14.5 Normalized Rayleigh scattering for 1.5 slit on the excitation and 5 on the observation measured at various excitation wavelengths.

opening of the slit. The increase could be few orders of magnitude when going from smallest to largest slit. However, as we will show in the next experiment, wide slits decrease wavelength resolution that may lead to the loss of important spectral futures.

EXPERIMENT 4-15 EFFECT OF SLIT WIDTH ON MEASURED SPECTRA

Keywords: spectrofluorometer, slit width, fluorescence emission.

In Section 2-16 "Bandwidth" (p. 67), and previous Experiment 4-14, the effect of the monochromator slits opening on the wavelength resolution of the detected signal were discussed. Now we want to test experimentally how setting different slits may affect the shape and intensity of measured emission and/or excitation spectra. Frequently when measuring weak signals we tend to open slits on excitation and emission path to increase the detected signal. This greatly increases the light throughput for the excitation path and emission path but affects the shape of measured emission or excitation spectra. Would opening the slits mask/deform important minute characteristics of the spectra and potentially lead to a loss of information about narrow peaks/bands?

(Continued)

EXPERIMENT 4-15 (CONTINUED) EFFECT OF SLIT WIDTH ON MEASURED SPECTRA

In this protocol, the fluorescence excitation and emission spectra measurements of anthracene, DPH, and rhodamine 6G (R6G) are presented using different slit widths on a spectrofluorometer.

MATERIALS

- Anthracene, diphenylhexatriene (DPH)
- Rhodamine 6G (R6G)
- Ethanol (EtOH), water
- 1 cm × 1 cm cuvettes

EQUIPMENT

- Spectrofluorometer.

METHODS

1. Prepare solutions of anthracene and DHP in EtOH with absorbances of about 0.1. Prepare a solution of R6G in water with absorption below 0.1.
2. Measure excitation spectra for each solution using a fixed observation slit and different slits on excitation. *Note: The solution concentration and observation slit should be sufficient to detect a small but reasonable signal with the smallest excitation slit. Opening the excitation slit will increase signal but we do not want to go over the signal limit when using the largest slits. Each vendor may have a different way of marking slits and we will only use size numbers to indicate increasing/decreasing slit opening. If the signal with a larger slit would go over the limit, we can use a neutral density filter (or lower the voltage).*
3. Measure emission spectra for each solution using a fixed excitation slit and changing the slit on emission. *Note: Again make sure that the concentration and slit opening on excitation are sufficient to detect a reasonable signal with the smallest slit but will not saturate the detector with the largest slits.*
4. Use smallest emission slit and measure emission spectra of anthracene using different slits on excitation.

RESULTS

At first, it may seem like the same result should be duplicated for each solution with each slit and only relative intensity would change. Measuring each solution, the cuvette remains untouched in the instrument and we only change in slit width that in most cases is computer-driven.

In Figure 4-15.1, we present excitation and emission spectra for anthracene solution. When measuring excitation spectra, we kept emission slit fixed to one value and for emission spectra with a fixed excitation slit. We adjusted the solution concentration so we can measure all spectra in an identical setting. In the top of Figure 4-15.1, we present measured spectra and for comparison at the bottom normalized spectra. The overall intensity significantly decreases as we use smaller slits. But most notable change can be seen in normalized spectra. Using the smallest slit opening for excitation and emission scans respectively, the measured spectrum is much weaker but presents well-structured excitation and emission respectively. As we increase the slit, the spectrum loses a structural character and for largest slit opening both excitation and emission scan shows almost single band.

(Continued)

EXPERIMENT 4-15 (CONTINUED) EFFECT OF SLIT WIDTH ON MEASURED SPECTRA

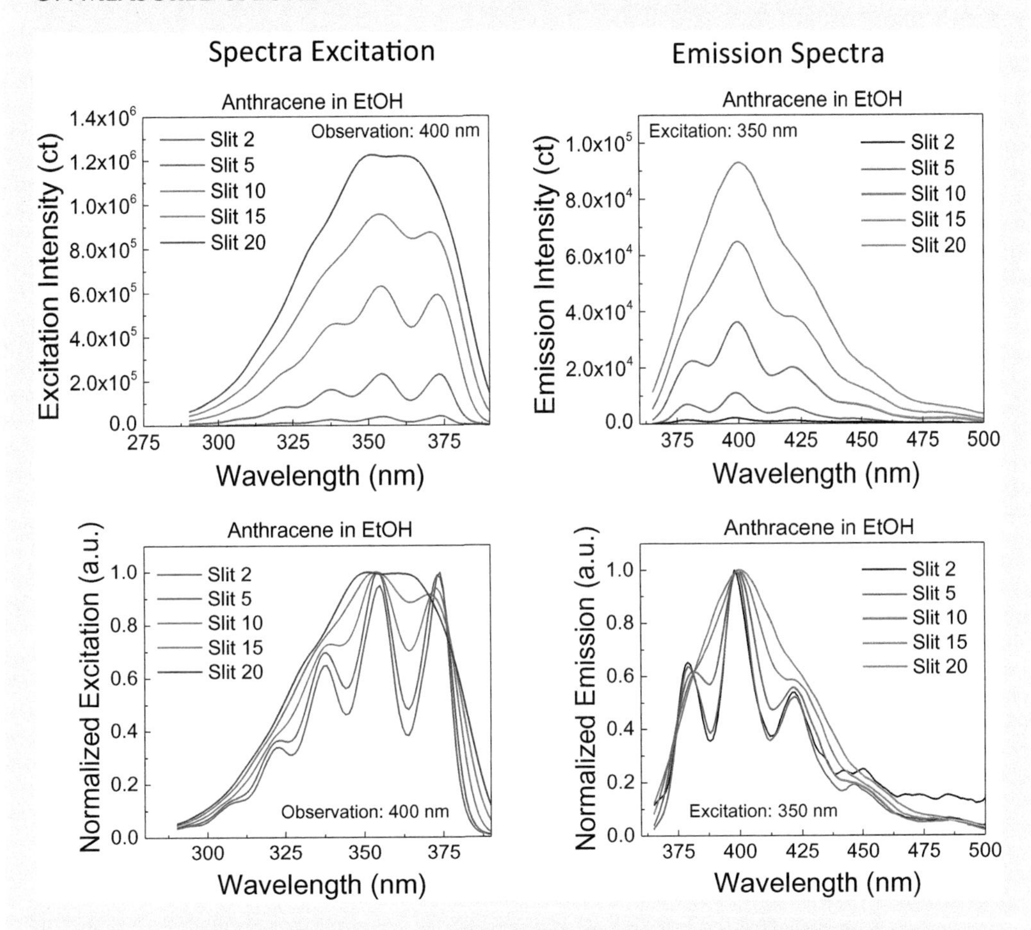

FIGURE 4-15.1 Excitation (left) and emission (right) spectra of anthracene in ethanol measured with different slit openings on excitation and emission, respectively.

Figure 4-15.2 shows similar measurements now made with DPH. DPH still shows structural spectral character but in a much lower extent. Similar to anthracene for larger slit opening, we are losing structural character. It could be a surprise but the spectra for anthracene and DPH measured with the largest slit are very similar in spite of significant difference measured with lowest slits.

In Figure 4-15.3, we present measured excitation and emission spectra for R6G with different slit opening. In contrast to anthracene and DPH, the excitation and emission spectra of R6G are the single bell-shaped band with no apparent structuration. In this case, normalized spectra do not show any significant change as we change the slit width. Only for the largest slit opening, we can see small broadening.

Finally, we want to test if the structure of emission spectra can be affected by the excitation slit (or excitation spectra by the opening of emission slit). For that, we used the anthracene solution and measured emission spectra with the smallest slit on observation using different

(Continued)

EXPERIMENT 4-15 (CONTINUED) EFFECT OF SLIT WIDTH ON MEASURED SPECTRA

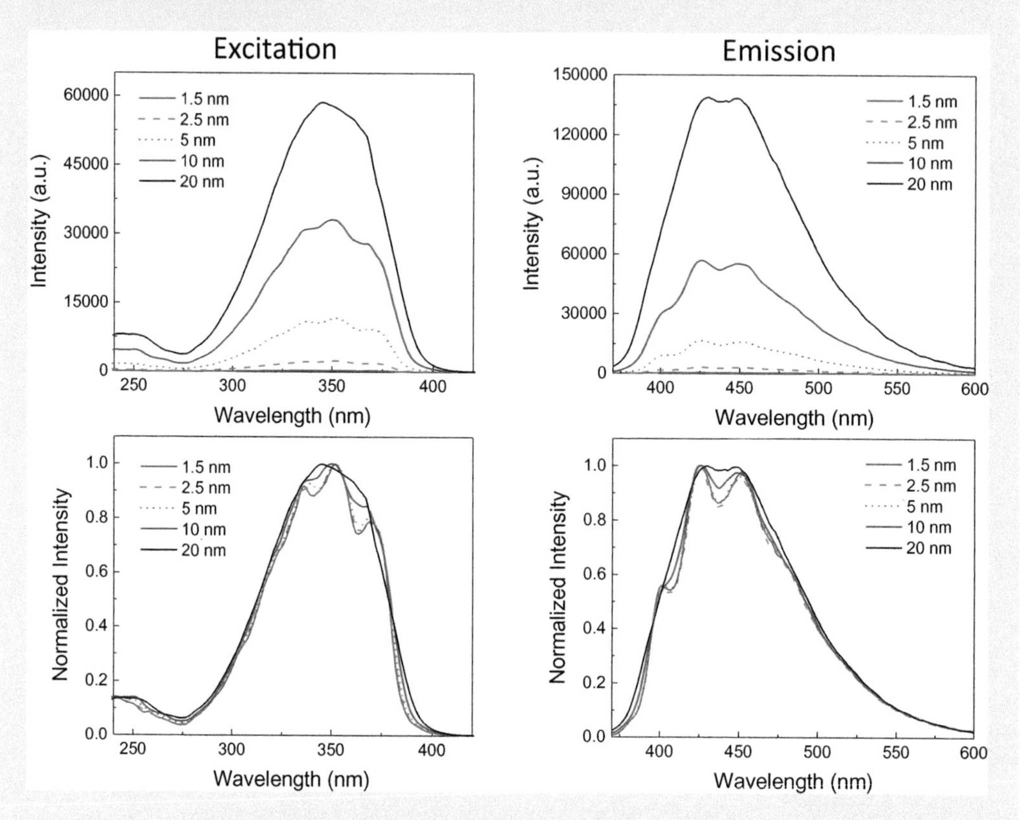

FIGURE 4-15.2 Excitation (left) and emission (right) spectra of DPH in ethanol measured with different slit openings on excitation and emission, respectively.

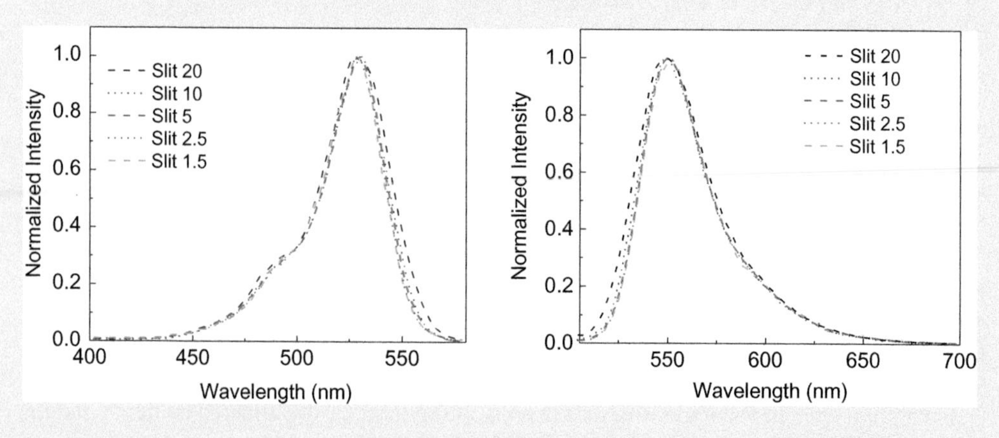

FIGURE 4-15.3 Normalized excitation (left) and emission (right) spectra of R6G in water measured with different slit openings.

(Continued)

EXPERIMENT 4-15 (CONTINUED) EFFECT OF SLIT WIDTH ON MEASURED SPECTRA

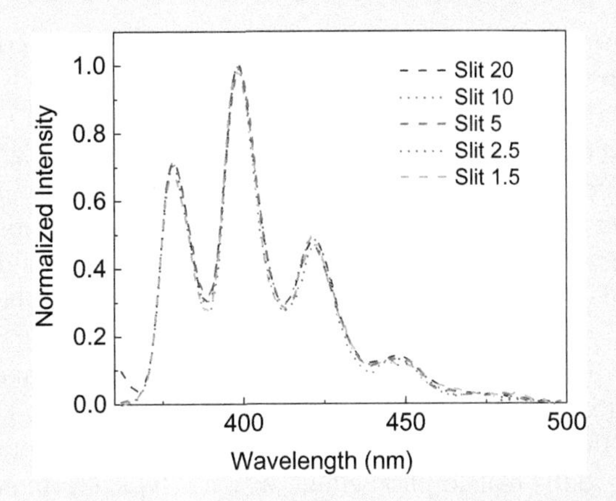

FIGURE 4-15.4 Normalized emission spectra of anthracene in ethanol measured with the emission slit 1.5 and different slit on excitation.

slits on excitation. Figure 4-15.4 left shows actual measured emission spectra of anthracene using a fixed 1.5 slit on emission and different slits on excitation.

From previous measurements, we know that we are exciting with broader and broader bandwidth. Obviously, the signal with the smallest slit on excitation was very low and noisy but even that the normalized shape of all spectra is practically identical. This is not a surprise. As discussed in Chapter 1 according to Kasha's rule, the emission spectrum profile should not depend on the excitation and we can expect the measured emission spectra should be independent of the excitation bandwidth or wavelength.

The important observation is that peaks clearly visible with the smallest slit are not observed using the largest slit width. This happens when the slit opening and the resulting bandwidth that is transmitted becomes larger than the peak bandwidth and peaks separations. This is not a significant problem when measuring broad spectra for which the half-width is larger than bandwidth for largest slit opening. The sampling rate on the spectrofluorometer was always the same (data points were taken at the same intervals—0.5 nm).

EXPERIMENT 4-16 CONCENTRATION DEPENDENCE OF EMISSION OF RHODAMINE 6G—EFFECT OF SAMPLE ABSORPTION

Keywords: emission, fluorescence, organic dyes, spectrophotometer, spectrofluorometer, rhodamine 6G.

Fluorescence is considered as one of the most sensitive detection technologies. Measuring fluorescence spectra is the first step when studying the fluorophore in a solution. These measurements are now done routinely with most fluorescence instruments. However, a great number of details need to be taken into account in order to measure the spectra correctly. This experiment focuses on a very well-known fluorophore (rhodamine 6G—R6G)

(Continued)

EXPERIMENT 4-16 (CONTINUED) CONCENTRATION DEPENDENCE OF EMISSION OF RHODAMINE 6G—EFFECT OF SAMPLE ABSORPTION

and presents the emission spectra measured at different chromophore concentrations. At first glance, one would expect the emission spectrum should increase linearly with concentration. And this happens at a low concentration range, but reaching higher concentrations many anomalous effects begin to emerge. It is important to remember that high dye concentrations for fluorescence are often low from a point of view of a chemist or biologist. A typical mistake made by new students is when making a solution of a well soluble fluorophore, he/she tries to measure the emission spectra. If the chromophore absorbs in UV and is visually colorless (for example indole), it is very easy to get to the concentration range where absorption in the peak is more than 2 in a 1 cm path-length cuvette. In this case, only a fraction of a millimeter can be penetrated by the excitation beam and in typical square geometry, no fluorescence can be seen. Trivial geometrical effects were previously discussed in Experiments 4-4 and 4-5. For fluorophores that absorb in the visible spectral range, a distinct color change can be visually observed due to the reabsorption effect, which is typically an indication of too high a concentration.

MATERIALS

- Rhodamine 6G (R6G)
- Water
- 1 cm × 1 cm cuvettes

EQUIPMENT

- Spectrophotometer
- Spectrofluorometer
- 405 nm and 532 nm laser pointers

METHODS

1. Prepare a stock solution of highly concentrated R6G in water (R6G is well soluble in water and a concentration of 2–3 mM can be easily dissolved in water).
2. Out of the stock solution prepare a few dilutions of R6G in the water down to ~1 μM.
3. Use a concentration of about ~10 μM to measure the absorption spectrum in the 300–600 nm range (see Figure 4-16.1 (top)).
4. Place each solution into a cuvette visually observe fluorescence using 405 nm and 532 nm laser pointers (*be careful not to look on the laser beam directly*). See Figure 4-14.1 (bottom).
5. Measure the fluorescence signal for each dilution.

RESULTS

To visualize the light penetration effect, we selected a common dye rhodamine 6G (R6G) that we prepared at different concentrations ranging from 3 μM to 2 mM. In Figure 4-16.1 (top), we present an absorption spectrum of 12 μM solution. We used two laser pointers for the demonstration; a 532 nm that corresponds almost to the

(*Continued*)

EXPERIMENT 4-16 (CONTINUED) CONCENTRATION DEPENDENCE OF EMISSION OF RHODAMINE 6G—EFFECT OF SAMPLE ABSORPTION

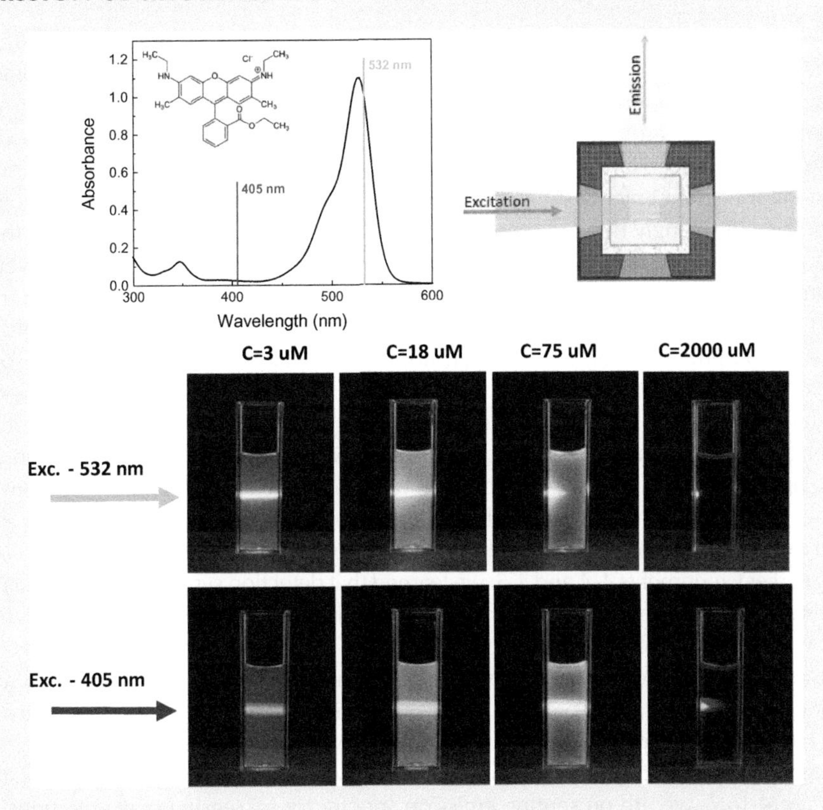

FIGURE 4-16.1 Top left: The absorption spectrum of rhodamine 6G with marked 405 nm and 532 nm excitations used for fluorescence measurements. Top right: Schematic of the cuvette holder with marked excitation and emission path. Bottom: Photographs of cuvettes filled with different concentrations of R6G solutions when excited with the diode laser 532 nm and 405 nm.

maximum absorption and 405 nm where the absorption is much lower (both excitation wavelengths are marked at the absorption spectrum—top). Photographs in Figure 4-16.1 (bottom) present the emission seen from both excitation wavelength. For the lowest concentration (low absorption) for 532 nm laser and 405 nm laser, we can see the uniform line (much brighter for 532 nm excitation). The second concentration already reveals some ununiformed intensity for the 532 nm laser while for the 405 nm the line is uniform and only brighter. For 75 μM concentration for the 532 nm laser, the line ends at the middle while for the 405 nm is still uniform. (*Comment*: Interestingly, the reader may notice that lines ending in the cuvette have a distinct sharp ending. This is due to declining intensity as light is absorbed while traveling through the cuvette. The laser beam has a small diameter, less than 1 mm, and only artificially looks very thick at the beginning since the intensity is very high. But as intensity drops it becomes visually thinner.) Finally, for a very high concentration (2 mM), the 532 nm laser barely enters the cuvette producing a bright spot on the surface. The 405 nm laser penetrates deeper but also ends in the middle.

(*Continued*)

EXPERIMENT 4-16 (CONTINUED) CONCENTRATION DEPENDENCE OF EMISSION OF RHODAMINE 6G—EFFECT OF SAMPLE ABSORPTION

There are a couple of important consequences of these observations. Consider the top-right side of Figure 4-16.1 that presents a schematic of the cuvette holder. The holder will typically have four openings. One for excitation beam (as marked on the figure) and one for the emission (marked in the figure). The openings are typically smaller than the cuvette dimensions (6–8 mm). The detection optics will typically view the center of the cuvette. When monitoring the center of the cuvette (typical setup of a common spectrofluorometer), the emission will strongly depend on absorption. When absorption is much higher than 0.1 (for example, when the absorption at the excitation wavelength is 2 in a 1 cm cuvette the beam intensity in the center is only 10% of that at the entrance to the cuvette and at the end (exit) of the cuvette is only 1%). The absorption at 405 nm excitation is about 15-fold lower than for the 532 nm. So, even when absorption in the peak is much over 2 the absorption for 405 nm wavelength is still within the acceptable range. This is the reason that even for very high concentrations, the fluorescence seen with 405 nm excitation is visually uniform. However, when the concentration increases even when absorption for excitation is low allowing uniform excitation across the cuvette we can notice the color changing from green to orange. This is an inner-filter effect that will depend on the overlap between the absorption and emission spectra for a given dye (see next Experiment 4-18).

In the earlier Experiments 4-4 and 4-5, we learned that detection systems in commercial spectrofluorometers are focused on the center of the sample holder (cuvette) and a couple of millimeters shift will strongly affect the measured signal. From the demonstration in the Figure 4-16.1, it is clear that when the detection system is focused on the center of the cuvette the excitation in the peak of absorption will not produce a detectable signal (fluorescence) in the center of the cuvette. However, the excitation shifted toward the wavelength with a much lower absorption (wavelength for which extinction coefficient is much lower) should give reasonable results.

In Figure 4-16.2, we are presenting emission spectra for increasing concentrations of the R6G solution measured for 405 nm and 532 nm excitations.

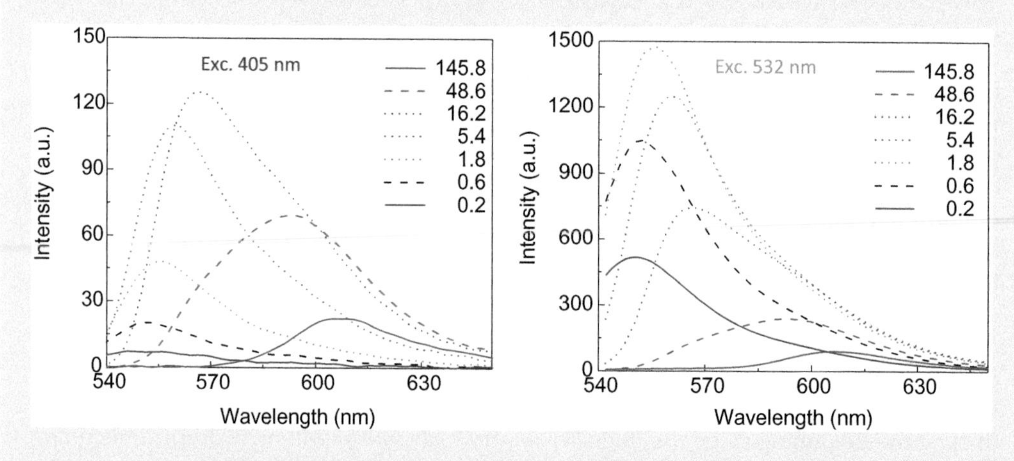

FIGURE 4-16.2 Emission spectra of rhodamine 6G measured with 405 nm excitation (left) and 532 nm excitation (right) for increasing sample absorption at the peak (528 nm) as indicated in each graph from 0.2 to 145.8 as measured in 1 cm path.

(*Continued*)

EXPERIMENT 4-16 (CONTINUED) CONCENTRATION DEPENDENCE OF EMISSION OF RHODAMINE 6G—EFFECT OF SAMPLE ABSORPTION

In the labels of Figure 4-16.2, we marked values for peak absorptions (maximum) as measured for solutions in 1 cm cuvette. *Note: To measure higher absorptions 5.4 and 16.2, we used 1 mm cuvette. Higher absorption were recalculated from the measured absorption at 405 nm (for detail see Experiment 3-8).* For 405 nm excitation at low concentration range (absorption up to 0.6 at the peak), intensity increases proportionally (about 3-fold—blue continuous line and black dashed line). However, when increasing the absorption three times up to 1.8 emission signal in maximum increases a little over two times (*remember absorption at 405 nm is much lower*). Also, emission spectra are clearly shifting toward the red. This is a reabsorption effect (absorption of the emission due to significant absorption-emission spectral overlap—an effect frequently called inner filter effect type II). This is not the case for 532 nm excitation. Already moving from 0.2 absorption to 0.6 absorption, the signal increases less than 2-fold. For higher absorptions, the effect becomes more dramatic and for absorptions over 1.8 the signal at maximum clearly drops down. Also measured spectra are changing dramatically and emission peak shifts from about 540 at low concentration to over 610 nm dramatically losing intensity.

CONCLUSIONS

From this simple experiment, we clearly see that when measuring the emission of samples that have too high absorption (above 0.1) the intensity will not behave linearly and the shape of measured spectra may change. The reason for the first effect is decreasing light intensity that is strongly attenuated before reaching the center of the cuvette (inner filter effect type I). For concentrations below 0.1 absorbance, fluorescence will behave well in a linear fashion. The second effect (spectral change) is due to so call inner filter effect type II (reabsorption—absorption of emitted light by chromophore). This effect is discussed in more detail in Experiment 4-18.

EXPERIMENT 4-17 CONCENTRATION DEPENDENCE OF EMISSION OF INDOLE—EFFECT OF SAMPLE ABSORPTION

Keywords: emission, fluorescence, organic dyes, spectrophotometer, spectrofluorometer, indole.

It is commonly accepted that in order to perform a proper fluorescence experiment, the sample absorption (typically measured for excitation wavelength) should not exceed 0.1. This experiment will demonstrate how the absorption of the excitation beam by the sample may affect the detected signal. As discussed in the previous experiments, it is important to realize that every spectrofluorometer is designed to detect most of the signal from the center of the cuvette (sample holder) (see Experiments 4-4 and 4-5) and that a shift of only a couple of millimeters can make a significant difference. Depending on the system, typically a spot size of 1–2 mm diameter is detected. This was tested previously in Experiment 4-5, and now the same effect can be observed using indole that is well soluble in water and is visually colorless.

(Continued)

EXPERIMENT 4-17 (CONTINUED) CONCENTRATION DEPENDENCE OF EMISSION OF INDOLE—EFFECT OF SAMPLE ABSORPTION

Indole has a relatively small spectral overlap (between absorption and emission spectra, see Figure 4-17.1). This will minimize the change in the emission spectrum due to the inner-filter effect. Also, this molecule is very closely related to a common amino acid—tryptophan and has very similar spectral properties so, the results will directly relate to many experiments with protein emission. In this experiment, we will demonstrate that results may strongly depend on the excitation wavelength, which is a trivial effect of the imperfection in the emission detection. This will be tested using two approaches. First, for excitation, we will use 280 nm that is close to the maximum in the main indole absorption band and 295 nm that is on the red-edge of the first absorption band. The extinction coefficient at 295 nm is almost 10 times lower than in 280 nm, and one can expect that a 295 nm excitation beam will penetrate much deeper into the cuvette producing a uniform excitation across the cuvette for relatively high indole concentrations. Second, a regular 10 mm cuvette will be used as well as with a 2 mm path cuvette that will shift the emission "center of gravity" toward the center of the detection.

MATERIALS

- Indole
- Water
- 1 cm × 1 cm cuvettes
- 0.2 cm × 1 cm cuvettes

EQUIPMENT

- Spectrophotometer
- Spectrofluorometer

METHODS

1. Prepare a stock solution of Indole with a concentration of about 1 mM. Dilute the stock solution to the samples with concentrations ranging from 0.005 mM to 1 mM. In presented results, we used concentrations:
   - 5.6 μM
   - 11 μM
   - 33 μM
   - 100 μM
   - 300 μM
   - 600 μM
   - 1.2 mM
2. Measure the absorbance of one lower concentration (e.g., 100 μM) in the 240–350 nm range.
3. Measure the emission of each solution using 1 cm × 1 cm cuvettes with 280 nm and 295 nm excitations (remember to keep magic angle conditions). Start from the lowest concentration and 280 nm excitation scanning the emission in the 290–500 nm range. For about 5 μM indol solution, most fluorometers should give a good signal. When using 295 nm excitation signal will be about 10 times lower (the exact signal

(*Continued*)

EXPERIMENT 4-17 (CONTINUED) CONCENTRATION DEPENDENCE OF EMISSION OF INDOLE—EFFECT OF SAMPLE ABSORPTION

will depend on absorption and lamp power at a given wavelength) and exactly how much lower will be the reading will depend on the spectrofluorometer (excitation lamp power distribution).

4. Repeat emission measurement using 0.2 cm × 1 cm cuvette when the 2 mm path is used for the excitation.

Note: For lower concentrations, the signal will be lower than that in 1 cm × 1 cm cuvette but higher for higher concentrations.

RESULTS

In Figure 4-17.1, we are presenting normalized absorption and emission spectra of indole in water. For measurement, we can use 1 cm × 1 cm or 0.2 cm × 1 cm cuvette with one of the lower concentrations. The important is that measured absorption at 280 nm for a given cuvette will be about 0.1. Absorption spectrum show some limited structure (^1L_b band can be visible) while emission has a simple bell-shaped spectrum that extends almost to 500 nm. Importantly, the overlap between absorption and emission is minimal. For excitation, we will use 280 nm and 295 nm as marked in the figure. The extinction coefficient at 295 nm is much lower than at 280 and we expect a much lower emission signal.

Figure 4-17.2 shows the emission spectra measured for lowest concentration (~0.0056 mM) with 280 nm and 295 nm excitations. One would notice in the emission spectrum a "bump" at the blue edge especially visible with 295 nm excitation for which the signal is lower. When comparing the emission spectrum to the absorption spectrum it is easy to mistake the blue edge "bump" as a real spectral feature. But this is an artifact resulting from Raman scattering contribution. As discussed in Chapter 2 and Experiments 4-9 through 4-11, Raman scattering of water is shifted of about 3,300 cm^{-1} from the excitation wavelength. For comparison in both graphs (280 nm and 295 nm excitations), we included emission measurements just with water that represents only Raman scattering of water. Simple subtraction of

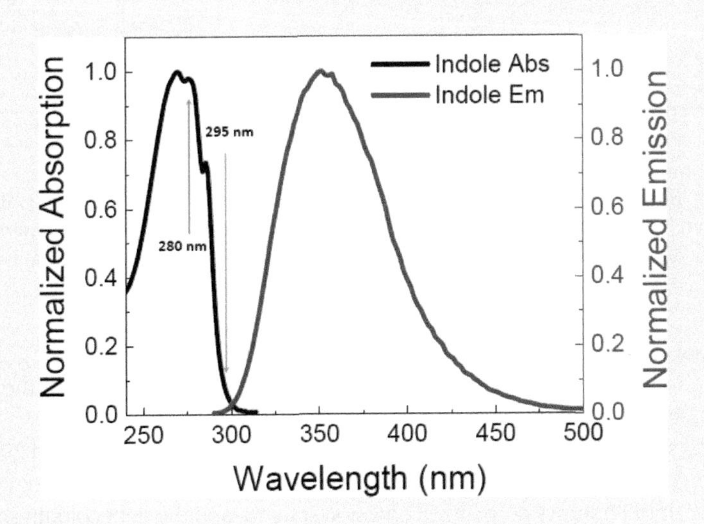

FIGURE 4-17.1 Normalized absorption and emission spectra of indole in water.

(Continued)

EXPERIMENT 4-17 (CONTINUED) CONCENTRATION DEPENDENCE OF EMISSION OF INDOLE—EFFECT OF SAMPLE ABSORPTION

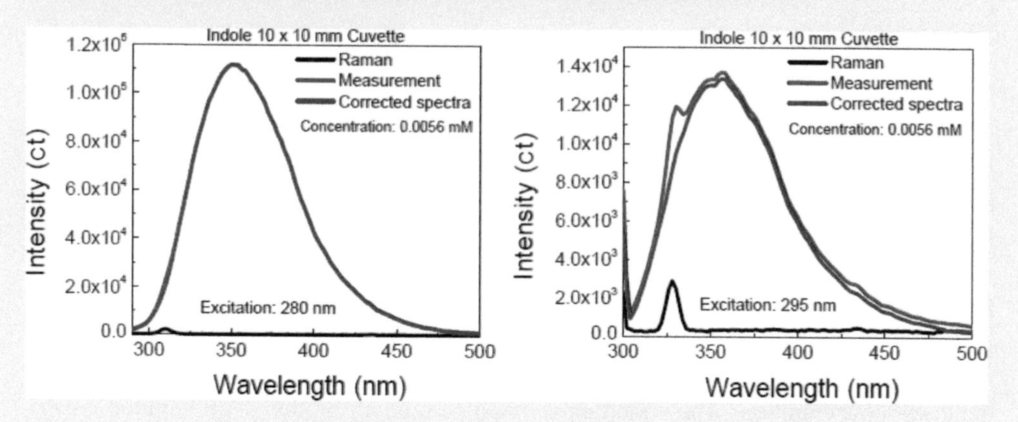

FIGURE 4-17.2 Measured emission of low concentration indole solution with 280 nm (left) and 295 nm (right) excitations. Also included measured water Raman scattering for both excitation and corresponding corrected emission spectra.

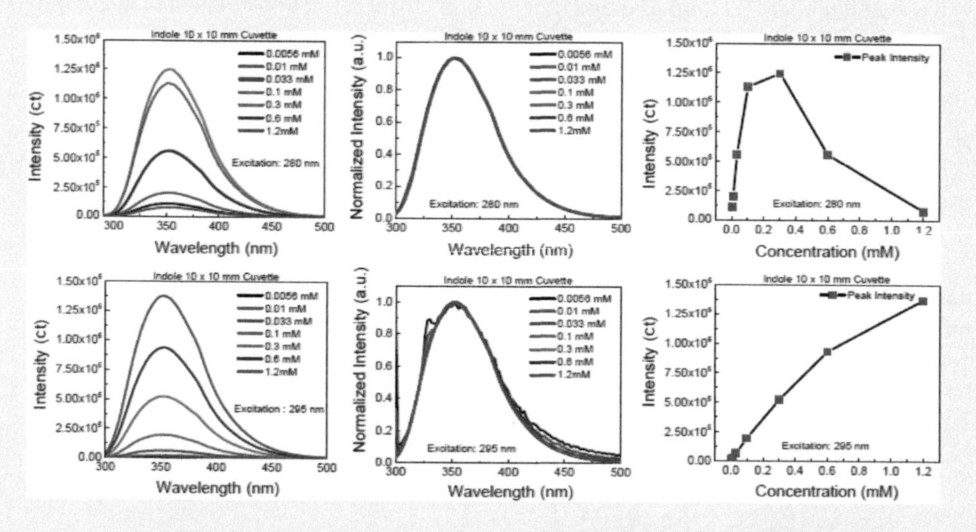

FIGURE 4-17.3 Measured emission spectra for increasing concentrations of indole in water for 280 nm (top) and 295 nm (bottom) excitations in a 1 cm × 1 cm fluorescence cuvette. Middle panels show normalized spectra and right panels show emission intensities at the maximum emission as a function of concentration.

the spectrum of water (Raman) from the emission of indole in water gives a good emission spectrum of indole (shown in blue in the figures). For higher concentrations, the emission of indole dominates and Raman scattering contribution is negligible.

Figure 4-17.3 shows emission spectra measured at 280 nm excitation (top) and 295 nm excitation (bottom) made in a 10 mm × 10 mm cuvette. For both excitations, left column show measured emissions, middle shows normalized emission spectra, and the right column show intensity values measured at the peak (356 nm) for each concentration. There are two important pieces of

(Continued)

EXPERIMENT 4-17 (CONTINUED) CONCENTRATION DEPENDENCE OF EMISSION OF INDOLE—EFFECT OF SAMPLE ABSORPTION

information coming from these measurements. The shape of the emission (normalized spectra) does not depend on indole concentration. This is because the spectral overlap between absorption and emission is very small. Second, the absolute intensity at 280 nm excitation initially (for low concentrations) increases, saturates, and drops quickly for higher concentrations. Interestingly, the highest sample concentration shows the lowest intensity. For 295 nm excitation, this effect is much smaller. Only two of the highest concentrations show a small deviation from linearity.

Note: Compare Figure 4-17.1; at 280 nm, the absorption of indole is much larger and the excitation beam does not penetrate sufficiently the solution for higher concentrations resulting in a much weaker excitation and fluorescence signal from the center of the cuvette. In fact, for the two highest concentrations, we see a clear drop in intensity. At 295 nm, the extinction coefficient of indole is much smaller as compared to 280 nm (about 10 times lower). With this excitation, we see much weaker overall fluorescence but the beam penetrates much deeper even for highest concentrations. In this case, we see an only moderate deviation from linearity for only two highest concentrations. For 295 nm excitation wavelength and emission scan starting at 300 nm, we can see some leaking excitation light for a few initial points. In Figure 4-17.2 and 4-17.3 for 295 nm excitation, the quickly decreasing intensity starting at 300 nm is a contribution of leaking scattered excitation. The excitation light has some bandwidth that is detected (is leaking) through the emission monochromator at a few initial wavelengths and disappears for wavelengths longer than 305 nm. The 280 nm excitation is far from starting 300 nm emission scan and with the same setting for excitation and emission slits is completely not visible.

To prove our point that an observed non-linear concentration behavior is due to the geometrical artifact we repeated measurements of all solutions using a 2 mm × 10 mm cuvette when the excitation beam paths along the shorter path (2 mm). In this way, we are moving emission points toward the center (for highest concentrations the excitation still will only penetrate a few hundred microns but the emission point will be close to the center of the detection. In Figure 4-17.4, we schematically present two cuvettes (1 cm × 1 cm and 0.2 cm × 1 cm) with marked excitation direction and emission observation direction.

Figure 4-17.5 shows measurements made with 280 nm and 295 nm excitations when using 2 mm × 10 mm cuvette. For 280 nm, we still can see a small deviation from linearity for two highest concentrations and for 295 nm excitation, the dependence is clearly linear (the line

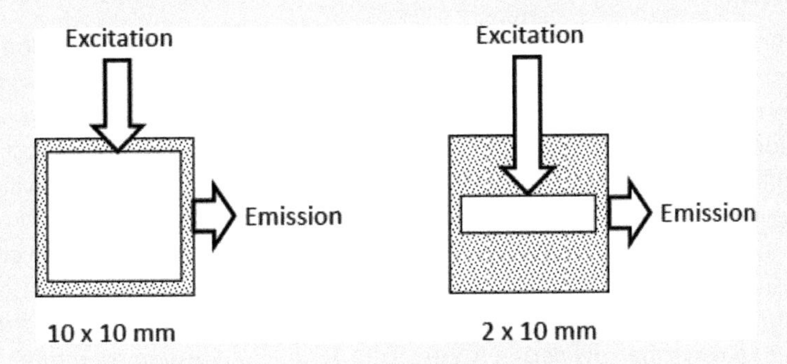

FIGURE 4-17.4 Schematics of two cuvettes used in this experiment (1 cm × 1 cm and 0.2 cm × 1 cm) with marked excitation direction and emission observation.

(Continued)

EXPERIMENT 4-17 (CONTINUED) CONCENTRATION DEPENDENCE OF EMISSION OF INDOLE—EFFECT OF SAMPLE ABSORPTION

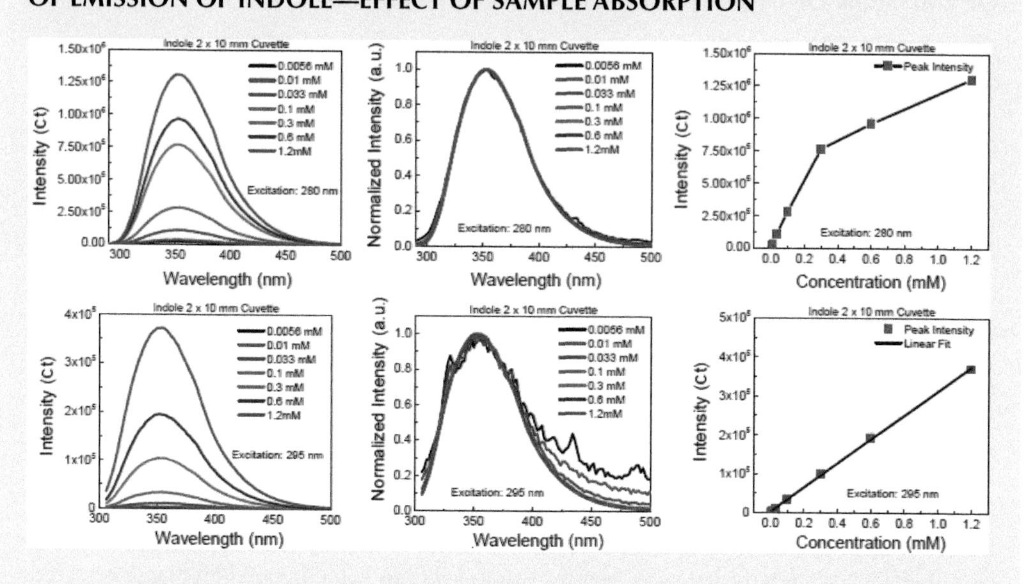

FIGURE 4-17.5 Measured emission spectra for the same solutions as in Figure 4-17.3 of indole in water for 280 nm (top) and 295 nm (bottom) excitations in 0.2 cm × 1 cm fluorescence cuvette. Middle panels show normalized spectra and right panels show emission intensities at the maximum emission as a function of concentration.

represents a linear fit to the data points). Note that emission spectra for lower concentration present significant noise and normalized spectra are not perfect for longer wavelength. This is because the overall emission signal is lower and dark counts/current and experimental noise become comparable to the measured emission signal. To minimize this effect, we could use longer averaging time for each point or multiple repetitive scans and averaging. We should also measure the background with only water (buffer) using the same setting and subtract from the measured indole emission. Such an approach typically gives excellent results correcting not only for the baseline but also for Raman scattering (see Experiment 4-13).

CONCLUSIONS

From these measurements, we can clearly see how too high of sample absorption may affect the measured signal. When using 280 nm excitation (peak of absorption) as indole concentration increases, we observe an increase in signal and then a quick drop. Already for 1 mM solution, the signal is smaller than from 5.6 μM initial solution (almost 200 times lower concentration). We can easily get indole concentration so high that we will practically not see emission. However, shifting the excitation to a wavelength with absorption much lower like in this case at 295 nm, we can see a much more linear behavior. Also changing the cuvette to a smaller path (0.2 cm × 1 cm or even 0.1 cm × 1 cm) significantly helps and for 295 nm excitation solution behaves in a perfectly linear manner. *Note: This is true for chromophores with a very small overlap between absorption and emission spectrum.* This is easy to notice since normalized spectra for all concentrations overlap very well. However, as moving to visible and NIR spectral range it is more common for fluorophores to have significant

(Continued)

EXPERIMENT 4-17 (CONTINUED) CONCENTRATION DEPENDENCE OF EMISSION OF INDOLE—EFFECT OF SAMPLE ABSORPTION

overlap between absorption and emission spectra and increasing concentration will affect measured spectra. In a previous introductory experiment (Experiment 4-16) and some later experiments, we are presenting examples of chromophores with a significant spectral overlap where the emission spectra are strongly affected. The concentration effect is dramatic and it is easy to prepare indole concentration of a few mM concentration. When asking the student to measure indole fluorescence she/he may not intentionally make a very high concentration. The solution visually will be clear but at 280 nm excitation, the signal will not be measured or will be very weak. Further increase in concentration will only decrease the signal. The conclusion can be that indole is not a good fluorophore. Just checking the solution with 300 nm (or longer) excitation or using thinner (1 mm) cuvette will show good emission.

EXPERIMENT 4-18 CONCENTRATION DEPENDENCE OF EMISSION: RHODAMINE 6G AND ANTHRACENE—EFFECT OF EMISSION REABSORPTION

Keywords: emission, fluorescence, organic dyes, spectrophotometer, spectrofluorometer, rhodamine 6G, anthracene.

In a couple of previous experiments, the effects of sample concentration (absorption) on the fluorescence signal were discussed. Specifically, indole that has only minimal overlap between absorption and emission spectra shows only minute emission spectrum distortion. Chromophores that have a small overlap between absorption and emission are not that common (the majority of them are UV chromophores) and most fluorophores in visible and red spectral range will have significant overlap between absorption and emission spectra. This leads to a significant change in the emission spectrum as chromophore concentration increases. This effect is known as the inner filter effect type II. For this experiment, we will use two chromophores. Rhodamine 6G (R6G) in water, which has a bell-shaped absorption and emission spectra, and anthracene in EtOH, which shows a clear structured absorption and emission spectra.

In Experiment 4-16, we presented R6G which exhibited both effects, sample absorption and sample reabsorption affecting the measured signal. The goal of this experiment is to show how measured emission (total intensity and spectral profile) is dependent on sample absorption. As mentioned above, the emission spectrum will depend on spectral overlap between absorption and emission. So, for this protocol, R6G will be used as it has significant spectral overlap as shown in Figure 4-18.1 (see also Figure 4-16.1). This time we will correlate the emission spectral shift with sample absorption as measured in 1 cm path cuvette. To minimize the effect of excitation attenuation a 2 mm × 10 mm cuvette will be used for emission measurements.

MATERIALS

- Rhodamine 6G (R6G)
- Anthracene
- Water and EtOH
- 1 cm × 1 cm, 0.4 cm × 1 cm, and 0.2 cm × 1 cm cuvettes.

(Continued)

EXPERIMENT 4-18 (CONTINUED) CONCENTRATION DEPENDENCE OF EMISSION: RHODAMINE 6G AND ANTHRACENE—EFFECT OF EMISSION REABSORPTION

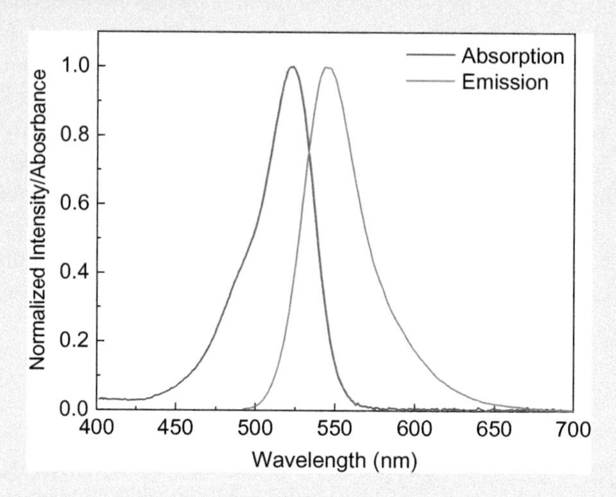

FIGURE 4-18.1 Normalized absorption and emission spectra of rhodamine 6G.

EQUIPMENT

- Spectrophotometer
- Spectrofluorometer

METHODS

1. Prepare four solutions of R6G with increasing concentration ranging from 2 μM to 200 μM.
   - 2.5 μM
   - 25 μM
   - 100 μM
   - 200 μM
2. Measure the absorbance of each solution using 1 cm × 1 cm cuvette.
   - We will use an excitation wavelength range from 380 to 450 nm where absorbance is significantly lower than 1. This ensures the sufficient beam intensity reaches the center of the cuvette.
3. Measure the emission of each solution in a 0.2 cm × 1 cm cuvette using excitation wavelength range from 380 to 450 nm. For this particular measurement, we used 405 nm excitation.
4. Prepare few solutions of anthracene in EtOH that have different absorptions ranging from less than 0.1 to about 10 in the long-wavelength absorption band. It is important to realize that the reabsorption effect is due to physical sample absorption, not sample concentration. This is the reason why it is better to refer to sample absorption, not concentration.
5. Similar to R6G (point 2) measure absorption in the 280–400 nm range. Measure emission spectra for each solution using 310 nm excitation.

(*Continued*)

EXPERIMENT 4-18 (CONTINUED) CONCENTRATION DEPENDENCE OF EMISSION: RHODAMINE 6G AND ANTHRACENE—EFFECT OF EMISSION REABSORPTION

RESULTS

Rhodamine 6G: In Figure 4-18.2, we are presenting results for R6G measurements. Presented absorption spectra measured in 1 cm path show that for concentrations above 25 μM, the absorption at the maximum is too high to recover proper absorption at the long-wavelength peak.

A short-wavelength absorption peak at about 350 nm shows a progressive increase in measured absorption. We can safely use for any excitation wavelength from 380 nm to 450 nm range. Since for the highest concentration absorption at 405 nm is about 0.6 in 1 cm path, the absorption in 0.2 cm cuvette at this wavelength should be significantly below 0.2 for all concentrations. In this case, we may neglect excitation light absorption as it travels to the center of the 2 mm cuvette. Importantly, using short-wavelength excitation we can measure full emission range starting at 500 nm. Obviously measured intensities for different concentrations will be different but at this point, we are only interested in the spectral profile. In Figure 4-18.2 (right upper panel), we present normalized emission spectra to

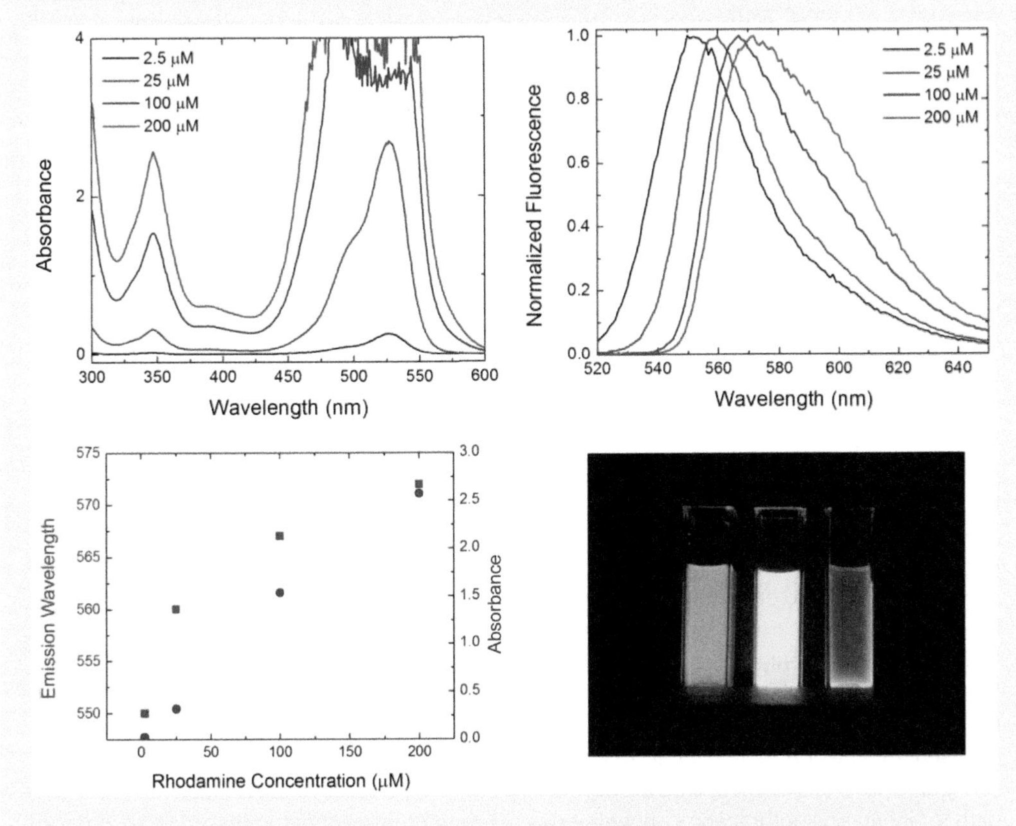

FIGURE 4-18.2 Absorbance and emission spectra of different concentration of R6G (top). Bottom-left shows the absorption as measured at 350 nm (blue circles) and the position of emission maximum (red square) for increasing R6G concentration. Photograph of used solutions in cuvettes illuminated with UV illumination stage (bottom-right).

(Continued)

EXPERIMENT 4-18 (CONTINUED) CONCENTRATION DEPENDENCE OF EMISSION: RHODAMINE 6G AND ANTHRACENE—EFFECT OF EMISSION REABSORPTION

facilitate emission peak center shift and changes in the spectral shape. The photograph in the bottom right shows three solutions (2.5 μM, 25 μM, and 200 μM) in 1 cm × 1 cm cuvettes illuminated with UV light (positioned on UV illuminator). It is clear that these solutions have significantly different colors that are changing from green (lowest concentration) to practically red (highest concentration). However, the change in the maximum emission is rather small; just about 20 nm (from 552 nm to about 572 nm). However, the apparent shape of a red edge of measured spectra changes with increasing concentration. It could be seen that the concavity of the emission curve to the right of the peak changes with increasing concentrations (we can see appearing a bump on the red spectral edge). The camera (or human eye) detects a total number of emitted photons that corresponds to the area under the emission curve and it may be thought of as a histogram of how many photons of a particular color are emitted by the sample. Red color starts at a wavelength of about 580–600 nm, which is not close to the peak center, however comparing the relative intensities (number of photons) at 600 nm of the 2.5 and 200 μM the proportion of red photons emitted is a lot larger in the 200 μM sample. This is the reason for the apparent reddish color in the photograph.

The graph on the bottom left of Figure 4-18.2 presents measured absorption values at the 350 nm peak (we cannot measure absorption in the long-wavelength peak) and measured emission peak maximum as a function of concentration. Measured at 350 nm absorption and wavelength for maximum emission are plotted on the same scale of Rh6G concentration, same independent axis but different dependent axis. Absorbance appears that it is a linear function of concentration, as expected ($Ab(C) = \varepsilon \times l \times C$) for measured absorptions below 2. As discussed in the absorption experiments (Chapter 3; Experiment 3-8) in the absorption range below 3 the dependence is expected to be linear. The peak intensities of the emission maximum definitely do not have a linear relationship with concentration. This is sufficient to show that the correspondence between peak maximum and concentration is not a direct function. Note that in 0.2 cm cuvette and excitation wavelength 405 nm, we do not have any significant attenuation due to excitation beam extinction as it travels through the cuvette. Contrary to the indole result in Experiment 4-16, this emission signal attenuation is due to reabsorption effect (absorption of emitted photons by chromophore as they travel through the solution).

Anthracene: Figure 4-18.3 shows normalized absorption spectrum (for intermediate sample concentration) and emission spectrum as measured for lowest sample concentration. Anthracene is a good emitter and measuring fluorescence for a sample that has an absorption below 0.1 at the maximum (about 357 nm) with 310 nm excitation should not be a problem. Doing this experiment, we would recommend trying a few different excitation wavelengths (for example, each absorption peak or valley). The total intensities will vary but normalized spectra will overlap perfectly. It is clear that the long-wavelength absorption peak and short-wavelength emission peak overlap significantly.

Figure 4-18.4 shows measured absorption spectra in a 4 mm cuvette recalculated to a 1 cm path for increasing anthracene concentrations. *Note: It is impossible to measuring absorption over 3 using a standard spectrophotometer. So, frequently when using high concentration we can use shorter path cuvette where absorption can be measured and recalculate it into*

(*Continued*)

EXPERIMENT 4-18 (CONTINUED) CONCENTRATION DEPENDENCE OF EMISSION: RHODAMINE 6G AND ANTHRACENE—EFFECT OF EMISSION REABSORPTION

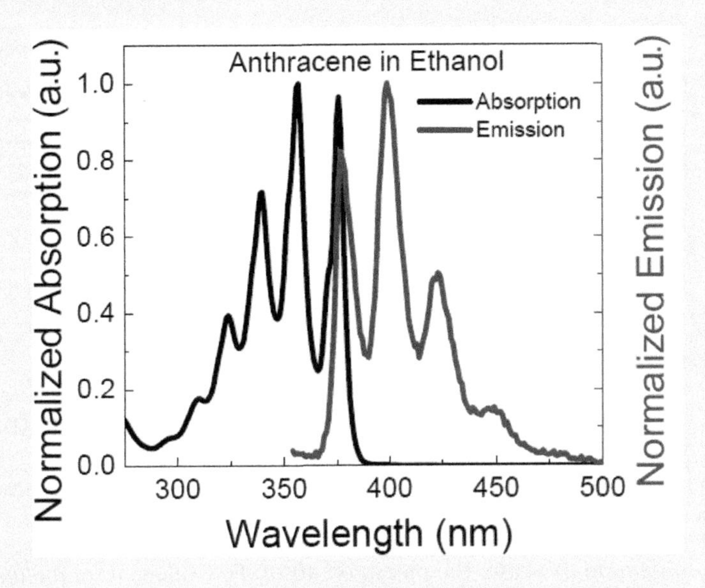

FIGURE 4-18.3 Normalized absorption and emission spectra of anthracene in EtOH.

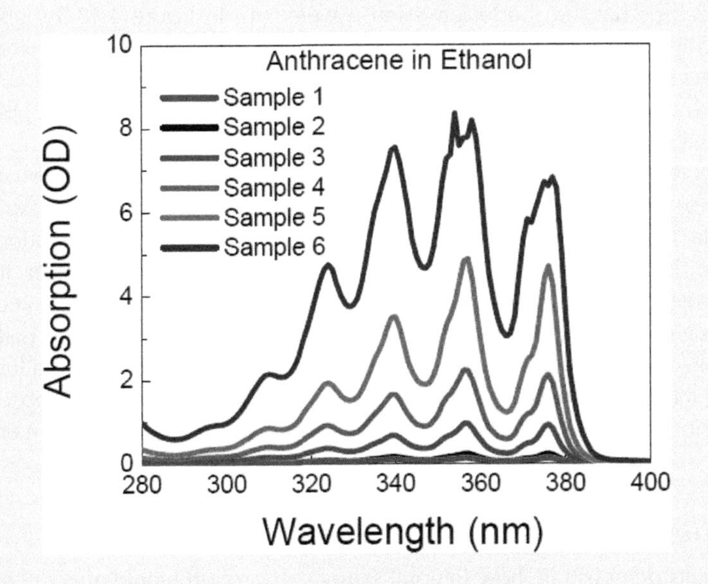

FIGURE 4-18.4 Measured absorption spectra of anthracene in a 4 mm path cuvette recalculated to a 1 cm path.

(Continued)

EXPERIMENT 4-18 (CONTINUED) CONCENTRATION DEPENDENCE OF EMISSION: RHODAMINE 6G AND ANTHRACENE—EFFECT OF EMISSION REABSORPTION

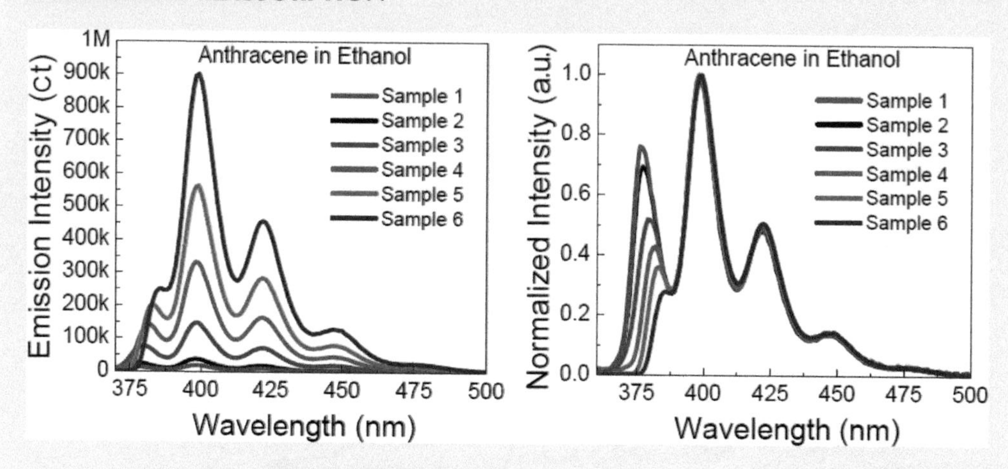

FIGURE 4-18.5 Emission spectra of anthracene in EtOH. Left: measured emission spectra; Right: normalized spectra.

a standard 1 cm path. In this case, the measured absorption values were multiplied by 2.5. The anthracene concentration was adjusted to double in each step that can be clearly seen from absorption values for wavelength below 350 nm (linear range).

Figure 4-18.5 left shows measured emission spectra in a 2 mm × 10 mm cuvette (2 mm path for excitation) excited at 310 nm. Measured emission spectra for higher concentrations are clearly deformed at the shorter emission wavelength. In Figure 4-18.5 (right), we show normalized emission spectra already presented on the left. It is clear that the first emission peak quickly disappears as the concentration increases. But the second emission peak is not effaced and the emission spectra over 390 nm are perfectly overlapping. For the first 4 concentrations (samples 1–4), the emission signals at a wavelength above 390 nm increase proportionally to the concentration—measured absorptions (doubles at each step). Interestingly, for the last two steps, the increase is lower than expected. To explain this, we refer the reader to earlier Experiments 4-4 and 4-5. For the highest concentration at 310 nm, the absorption in the 2 mm path is about 0.4 what leads to a significant excitation beam attenuation as it paths through the cuvette. This results in a shift of the emission center toward the wall through which the excitation beam enters the cuvette. Even if the path length for excitation is only 2 mm and the emission spot position will not be displaced a lot we already see observed intensity increase is less than expected. However, we do not observe any significant emission spectrum distortion above 390 nm (only the total relative intensity can be affected).

CONCLUSIONS

Presented results demonstrate how internal sample absorption would affect observed emission spectra. It is clear, that even for good absorption at the excitation wavelength (below 0.1), we can observe significant distortion of emission spectra. One interesting observation is that normalized emission spectra of R6G are shifting the maximum toward the red and they

(Continued)

EXPERIMENT 4-18 (CONTINUED) CONCENTRATION DEPENDENCE OF EMISSION: RHODAMINE 6G AND ANTHRACENE—EFFECT OF EMISSION REABSORPTION

evidently change the shape. In contrast, anthracene emission spectrum changes at the blue edge but normalized emission spectra perfectly overlap at the red emission edge. This is due to the limited extent of spectral overlap and the emission shape. For R6G, the absorption extends beyond the emission maximum. If so, as we increase the concentration reabsorption will shift the maximum (by absorbing longer wavelength photons) and normalization is to the different spectral point. For anthracene, the absorption does not reach a distinct emission maximum that is a sharp peak at about 410 nm. Then the normalization is done to the same undistorted peak. The red side of the spectrum does not change with respect to the maximum. However, we want to remind that the relative intensity increase in the undistorted emission spectral range can be affected by the geometrical factor leading to not linear intensity increase.

EXPERIMENT 4-19 SECOND ORDER TRANSMITTANCE LEAKING THROUGH THE MONOCHROMATOR

Keywords: emission, fluorescence, organic dyes, spectrophotometer, fluorometer, *N*-acetyl tryptophan amide.

Highly sensitive fluorescence systems (spectrofluorometers) are widely available. A very high sensitivity can, in some instances, be a problem and may lead to unexpected mistakes. In this experiment (as well as several of the following), a rather trivial effect will be presented, that may arise from the fact that most current fluorometers use grating monochromators. It is well known that this type of monochromator would transmit a higher-order harmonics. A lesser-known fact is that this can be quite a significant effect. Simply put, a 300 nm photon will freely pass through the monochromator set at 300 nm wavelength, but will also have a significant probability of passing through the monochromator set at 600 nm. This experiment shows how significant this effect can be and how to eliminate such an artifact.

In this protocol, measurements will be taken of the emission spectra of water solutions of NATA and bovine serum albumin (BSA). We will deliberately use small concentrations.

MATERIALS

- *N*-acetyl tryptophan amide (NATA) and bovine serum albumin (BSA)
- Water
- 0.4 cm × 1 cm quartz cuvettes
- 305 nm long-pass filter (305 nm LP)
- Piece of optical glass (e.g., microscope slide)

EQUIPMENT

- Spectrophotometer
- Spectrofluorometer

(Continued)

EXPERIMENT 4-19 (CONTINUED) SECOND ORDER TRANSMITTANCE LEAKING THROUGH THE MONOCHROMATOR

METHODS

1. Prepare a low concentration water solution of NATA. Make sure the water is highly purified. Since we will use UV excitations any impurity would contribute to the emission. Typically, we would prepare the concentration that would result in an absorption below 0.04 (for some systems, we will need to use lower concentrations).
2. Measure absorbance of the NATA solution.
3. Put a cuvette with a solution of NATA into the fluorometer. Make sure to set magic angle conditions, proper voltage, and slits, etc. to ensure good emission signal. We will use three excitation wavelengths (270 nm, 280 nm, 290 nm) using the scanning emission range 300–800 nm.

RESULTS

Figure 4-19.1 shows an absorption spectrum measured for NATA in a 0.4 cm path cuvette. Maximum for measured absorption is below 0.04.

Figure 4-19.2 shows three emission spectra of NATA excited at 270 nm, 280 nm, 290 nm. We are using a good concentration for NATA (absorption above 0.01) and we expect to measure good emission. A more experienced experimentalist would end the scan at about 500 nm (see previous Experiment 4-12). However, for a student, it would be educational to make a measurement in a broader range. Figure 4-19.2 shows the emission measurements in 300–800 nm range.

The emission bands at 300–500 nm are what we would expect. Since the fluorescence signal is high, the Raman scattering contribution is not visible. However, spectral features observed above 500 nm are unexpected. Sometimes students report they found red emission out of NATA, Tryptophan, protein, or another UV emitter. Sharp peaks at 540 nm, 560 nm, and 580 nm are easy to identify since these are doubles of used excitation wavelengths.

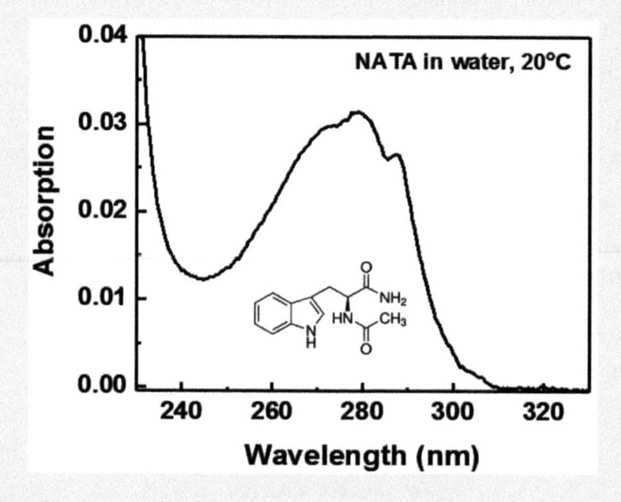

FIGURE 4-19.1 Absorption of NATA solution in water. Chemical structure of NATA is shown under the curve.

(Continued)

EXPERIMENT 4-19 (CONTINUED) SECOND ORDER TRANSMITTANCE LEAKING THROUGH THE MONOCHROMATOR

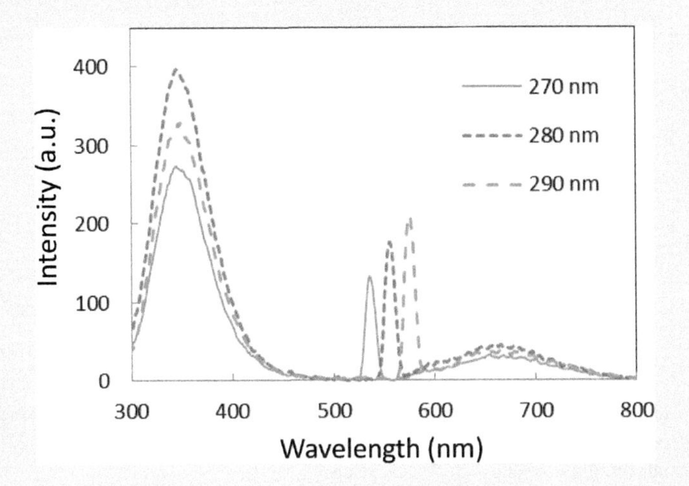

FIGURE 4-19.2 Emission spectra of NATA in water measured with 270 nm, 280 nm, and 290 nm excitations.

Any solution will scatter excitation light and scattered excitation light will be transmitted through the emission monochromator. What may not be anticipated is that the intensity can be significant as compared to a weak emission. Another more troubling finding is a broadband of 600–800 nm range that has all characteristics of a typical emission. It changes intensity with a different excitation wavelength but does not shift peak position or changes the shape. Again, its intensity is significant as compared to the main emission peak at 350 nm. To eliminate scattered light, we typically use filters. In Figure 4-19.3, we show transmittance

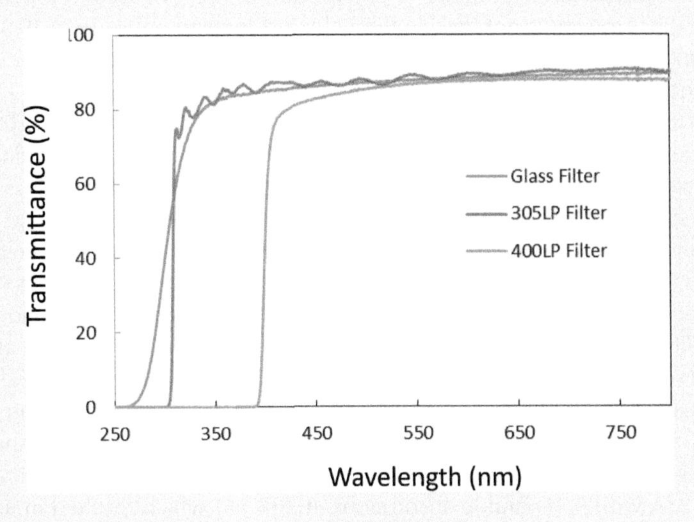

FIGURE 4-19.3 The transmittance of 305 nm and 400 nm long-pass filters and transmittance of optical quality glass.

(*Continued*)

EXPERIMENT 4-19 (CONTINUED) SECOND ORDER TRANSMITTANCE LEAKING THROUGH THE MONOCHROMATOR

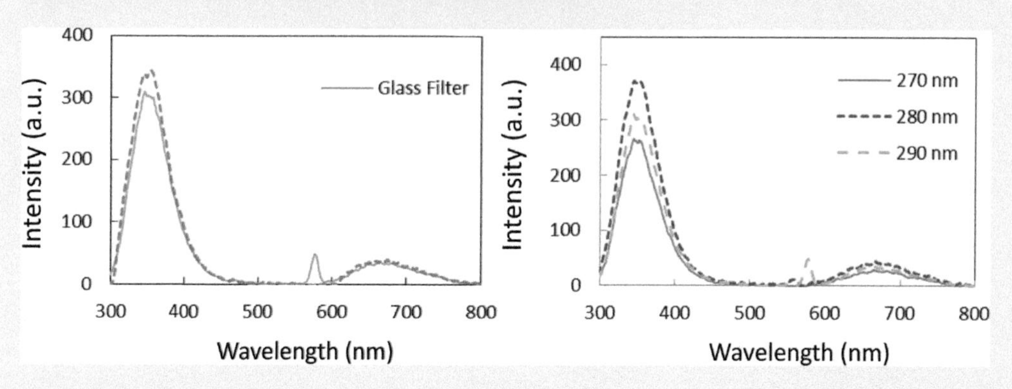

FIGURE 4-19.4 Left: Measured emission of NATA with 290 nm excitation and observation through the 305 nm long-pass filter and a glass filter. Right: Measured emission of NATA with 270 nm, 280 nm, and 290 nm through the glass filter.

measured for two long-pass filters (305 nm and 400 nm). Also in this figure, we included transmission of optical glass that is widely available. It is well known that glass does not transmit UV light but it is the less common knowledge that we can use it as a good long-pass filter.

Placing a long-pass filter on the observation should eliminate scattered light. Figure 4-19.4 (left) shows measured emission with 290 nm excitation using a 305 nm long-pass filter or just a glass plate as a filter on observation. The 305 nm LP filter completely eliminated the second-order transmission of the excitation light from the emission spectrum. The glass filter significantly lowered the second-order but the peak is still visible (some 290 nm leaks through the glass). In Figure 4-19.4 (right), we present emission spectra measured through the glass but using three excitations (290 nm, 280 nm, and 270 nm). It is clear that second-order transmission for 280 nm excitation almost disappeared and for 270 nm we do not observe second-order transmission (the excitation 270 nm light is not leaking through). Just looking on Figure 4-19.3, we can see that transmittance of glass does not decline very quickly like the 305 nm LP filter (305 nm filter is an interference filter that has a sharp rising edge) but extends over the 270–330 nm range. Transmittance is about 20% for 290 nm, less than 10% for 280 nm, and close to 0% at 270 nm.

An interesting observation can be made when using a long-pass filter starting at a longer wavelength, like 350 nm LP or shown in Figure 4-19.3 400 nm LP filter. We will observe the emission spectrum starting at about 400 nm (we will only observe the red-edge of NATA emission). At longer wavelength (second-order) the observed spectral feature will start at the doubled wavelength also (700 nm or 800 nm) just transforming the observed emission.

We checked an effect of the second order of grating dispersion with another spectrofluorometer (different brand), see Figure 4-19.5. For this experiment, we only used 290 nm excitation and solution of NATA and bovine serum albumin. To eliminate second-order transmission for the emission monochromator, we used a 305 cut-off filter placed on observation. Figure 4-19.5 shows normalized measured emission spectra for NATA (left) and BSA (right) without the filter and with the 305 LP filter on observation. We prepared comparable samples of NATA and BSA with comparable absorptions at 280 nm which resulted in a significantly better signal from NATA with 290 nm excitation. In effect for NATA, we used a smaller slit on observation. In this case, the second-order leak is smaller (compare normalized signals for

(Continued)

EXPERIMENT 4-19 (CONTINUED) SECOND ORDER TRANSMITTANCE LEAKING THROUGH THE MONOCHROMATOR

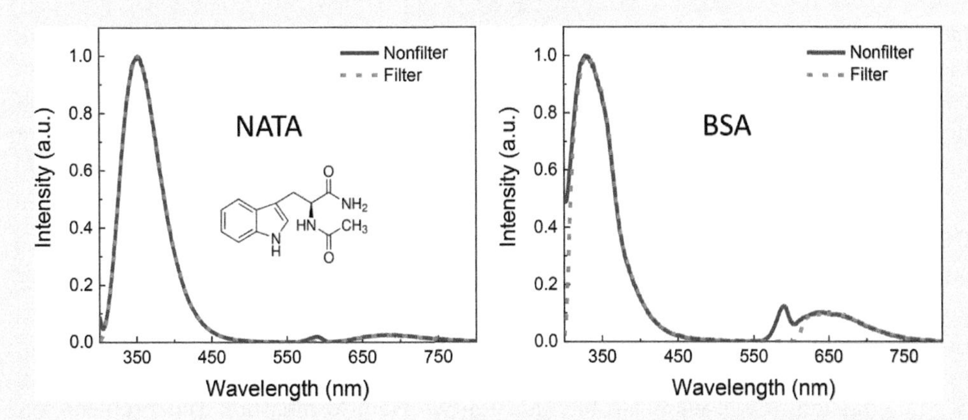

FIGURE 4-19.5 Emission spectra of NATA and BSA in water measured different spectrofluorometer with 290 nm excitations, with and without long-pass filter (305 LP). This instrument is also equipped with the grating monochromator.

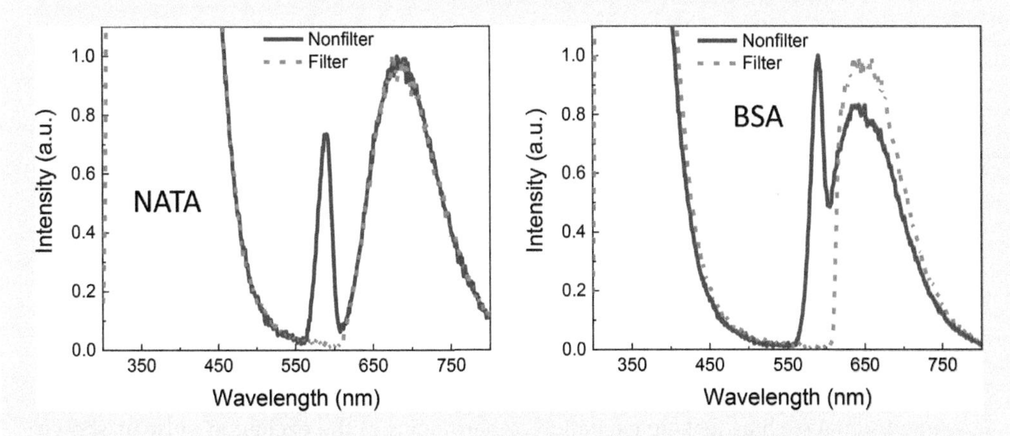

FIGURE 4-19.6 Emission spectra of NATA and BSA in water from Figure 4-17.5 normalized to the observed maximum at the red range with and without 305 LP filter.

NATA and BSA at 600–750 nm range. Also, the emission of BSA has an emission maximum shifted toward the blue as compared to NATA. This is mostly due to the fact that tryptophan is carried in protein structure and shielded from the water environment. In Figure 4-19.6, we present the same spectra but now normalized to the peaks at wavelengths above 550 nm. In both cases, the second transmission of excitation is clearly visible at 580 nm. The spectra measured with the cut-off filter does not present second-order transmission. For BSA, we can clearly see short-wavelengths emission spectrum is also cut-off.

Note: One would observe that emission spectra measured with filter and glass start from the much lower value as compare to Figure 4-19.2 at wavelengths close to 300 nm while the maximum value at about 350 nm (the peak) is only slightly affected.

(Continued)

EXPERIMENT 4-19 (CONTINUED) SECOND ORDER TRANSMITTANCE LEAKING THROUGH THE MONOCHROMATOR

Interestingly, the long-wavelength broadband has not been affected by the filter or glass. We could consider it real but in such case, the solution when excited with UV light would present a red color (the eye does not see light below 400 nm but is very sensitive to red color). We do not see a red color from NATA, tryptophan, or indole solutions. The measured broadband is a transmittance of a real UV emission band as a second-order through the grating monochromator.

CONCLUSIONS

It is very important to be alert that most monochromators will transmit in the second-order and when measuring in a broad range we have to be able to account for that. Second-order transmittance of excitation is easy to eliminate since the long-pass cut-off filter will not transmit it. But we cannot use such a filter to eliminate second-order transmittance of real emission. It will eliminate the emission we want to measure. This problem can be very relevant when measuring the emission of donor-acceptor systems in FRET experiments. In this case, we are using shorter wavelength donor excitation and when observing emission in the acceptor emission range we can see the leak of excitation and acceptor emission.

EXPERIMENT 4-20 MEASURING EMISSION USING FILTERS: CORRECTING FOR FILTER DISTORTION

Keywords: filter, spectrophotometer, fluorescein, transmission.

For highly fluorescent samples, monochromators in most systems are sufficient to collect good emission spectra. But quite often we are dealing with a rather low fluorescence (weakly fluorescent molecules) and a very strong laser or lamp excitation must be used. In such a case, it is necessary to use a filter(s) placed in the emission path to remove unwanted signal such as leaking excitation or harmonics of the excitation pulse as shown in the previous experiment. Filters are not perfect on/off switches for pathing light, often they have complicated absorbance/transmittance characteristics. When using filters one must remember that most filters will change/influence transmitted light, and when analyzing spectra it is necessary to consider how filters affect measured spectra and how this can be corrected.

In this protocol, the emission of fluorescein is determined with and without a filter, and the two results are shown to be the same.

MATERIALS

- 1 cm × 1 cm cuvette
- Water
- Fluorescein
- Filter which transmits between 480 nm and 700 nm

(*Continued*)

EXPERIMENT 4-20 (CONTINUED) MEASURING EMISSION USING FILTERS: CORRECTING FOR FILTER DISTORTION

EQUIPMENT

- Spectrophotometer
- Spectrofluorometer

METHODS

1. Prepare a solution of fluorescein in a cuvette.
2. Measure the absorbance of fluorescein. Adjust the concentration to have absorption at the peak of about 0.05 (below 0.1).
3. Measure the absorbance/transmittance of the selected filter.
4. Use an excitation wavelength below 470 nm and measure the emission of fluorescein without a filter.
5. Place the filter in the emission path and measure the emission of fluorescein with a filter in the beam path.

RESULTS

Fluorescein is a very good emitter and for the absorption of 0.05, it should present an excellent emission signal.

Figure 4-20.1 shows the transmission of the used filter. We purposely selected a filter that has no uniform transmittance in the region of interest (range of fluorescein emission). The filter does not transmit below 470 nm (it will properly eliminate leaking/scattered excitation light) but it does transmit light in the entire region above 480 nm.

In the experiment, we are using good fluorescein concentration resulting in an excellent emission so, measured spectra will not require additional corrections for leaking excitation or Raman scattering. Figure 4-20.2 shows measured normalized emission spectra without (blue) and with the filter (green). The two spectra look different, the center is shifted and the concavity on the right side of the plot is changed. These spectra are different enough to suggest that they come from different dyes or for example fluorescein in buffers of different pH. This difference may be enough to change the interpretation of the experimental results. However, this is the same solution (the solution was not changed in-between the two measurements)—the cuvette was not even touched.

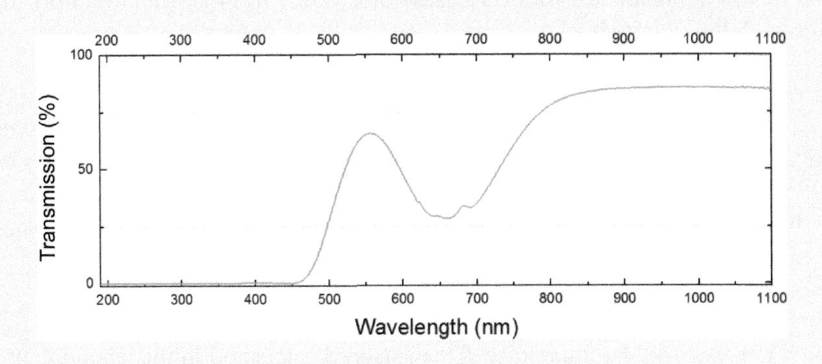

FIGURE 4-20.1 Transmission of the used filter.

(Continued)

EXPERIMENT 4-20 (CONTINUED) MEASURING EMISSION USING FILTERS: CORRECTING FOR FILTER DISTORTION

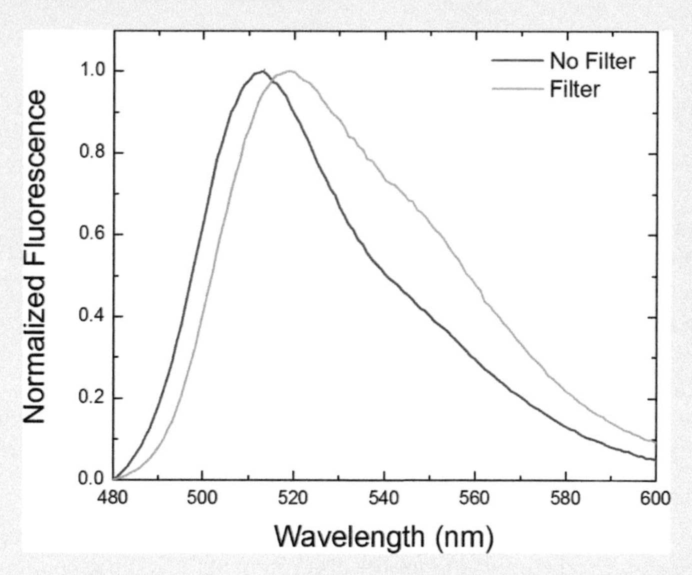

FIGURE 4-20.2 Fluorescence emission measured with (green) and without (blue) filter.

The change in measured spectra is due to the presence of the filter. Because the transmission of the filter is known it is possible to correct for the light attenuation due to the filter. In the presence of the filter measured light intensity at each wavelength (λ) is $I(\lambda) = I_0(\lambda) * T(\lambda)$ and the true emission spectrum ($I_0(\lambda)$) can be calculated as:

$$I_0(\lambda) = I(\lambda)/T(\lambda)$$

This simple procedure will be shifting every individual point on the vertical axis to account for emission light attenuation due to the filter absorption (light extinction as it is pathing through the filter) such that the difference in transmission due to the filter is equal at every point. There are multiple approaches to accomplish this, one such procedure is presented below. Find a function (or series of points), $f(\lambda)$, such that when the transmission, $T(\lambda)$ is multiplied by $f(\lambda)$ the result is 1 for every value of λ, ($T(\lambda) * f(\lambda) = 1$). Then multiply the spectrum taken with the filter by $f(\lambda)$.

1. Let $f(\lambda) = 1/_{T(\lambda)}$ for each λ in the emission spectrum (It's ok to find more).
   a. This results in non-physical values for transmission, but this is just a math work step.
2. Multiply each point in the emission spectrum taken with the filter by $f(\lambda)$. This will shift the value up or down.
   a. If the data points in the spectrum and $f(\lambda)$ have different λ values interpolate $f(\lambda)$ to find these values.
3. Normalize (or scale) the resulting spectrum.

Figure 4-20.3 shows $T(\lambda)$, $f(\lambda)$, and $T(\lambda) * f(\lambda)$ measured/calculated in this experiment. In this procedure, $f(\lambda)$ is just the inverse of the transmission (taken point by point).

(Continued)

EXPERIMENT 4-20 (CONTINUED) MEASURING EMISSION USING FILTERS: CORRECTING FOR FILTER DISTORTION

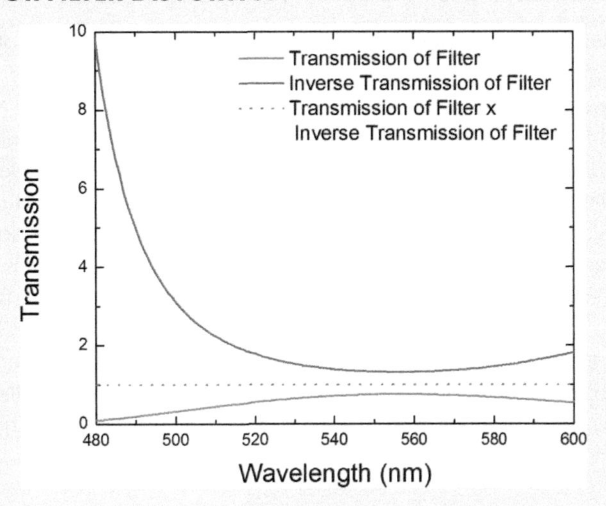

FIGURE 4-20.3 The transmission of the used filter, its inverse and line 1 plotted on the same wavelength scale as the spectrum from Figure 4-20.2.

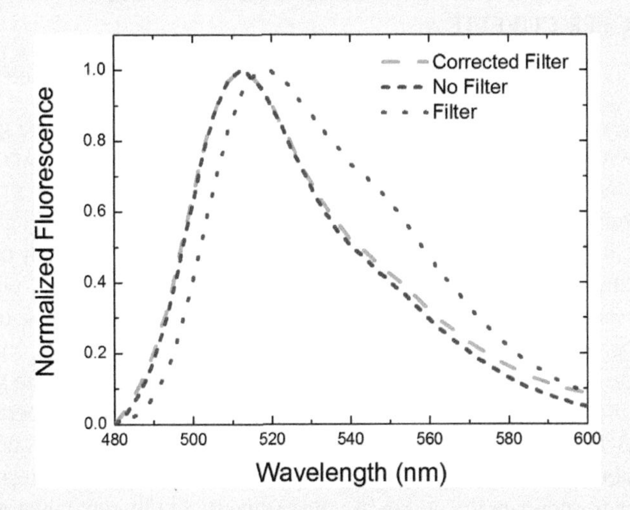

FIGURE 4-20.4 Spectra fluorescence without filter, with filter, and corrected emission with filter.

In this procedure, the resulting spectrum is normalized at the end to be easily compared with spectra wit and without filter. The resulting spectrum, along with the original spectra with and without the filter, are plotted in Figure 4-20.4.

The corrected emission taken with the filter is almost identical to the spectrum taken without the filter, the centers are the same and the line shapes are the same at most wavelengths. There is some variation at higher wavelengths resulting from interpolation and rounding errors. In this example, the data of the measured spectrum was also interpolated to allow for easier graphing. The correction was applied to actual data (not a fit) so noise is present in all spectra.

(Continued)

EXPERIMENT 4-20 (CONTINUED) MEASURING EMISSION USING FILTERS: CORRECTING FOR FILTER DISTORTION

It is conceptually wrong to fit a spectrum taken through a filter as it does not relate to properties of the chromophore.

This example was possible because the use of a filter was not necessary. There are no undesired spectral features in the wavelength of interest (480–600 nm). Such undesired features usually arise due to the interaction of the excitation light with optical elements in its path.

Most times a filter is used out of necessity and there is no spectrum without a filter known to compare. This is often the case for microscopy. If emission and transmission data are taken at the same wavelengths interpolation is not necessary and transmission values $T(\lambda)$ can be used (or alternatively function ($f(\lambda)$ may be found point by point). Since, as shown in this experiment, a filter can alter the spectral shape and the center of a peak it is necessary to correct for a filter if the spectral shape is to be studied. It is often sufficient to observe the presence of a spectrum and analysis will not be required. Interference filters present many small oscillations in the transmittance spectrum (see Experiment 3-15 and Figure 4-19.3) which will result in oscillations in the measured spectrum. Any filter artifacts are corrected in the same way.

EXPERIMENT 4-21 MEASURING IN UV SPECTRAL RANGE: USING THE PROPER CUVETTE

Keywords: emission, fluorescence, NATA, spectrophotometer, fluorometer.

Many biological experiments require measurements in the UV spectral range. Amino acids (phenylalanine, tyrosine, and tryptophan) or nucleic acids require UV excitation (typically below 300 nm) and emit fluorescence in the range below 400 nm. For an inexperienced researcher, such measurements in the UV region can be problematic. UV light is invisible to the human eye, so it is difficult to visually evaluate the sample, cuvette, or any optical elements. They will be visually clear and transparent. As an example in Figure 4-21.1, we are presenting a photograph of various cuvettes (quartz, glass, and plastic). A standard spectrofluorometer would typically be equipped with UV optics (optical elements like lenses and windows that are transparent for wavelength above 230 nm. However, judging a cuvette while unable to measure its transmission can be difficult even for the most experienced experimentalist.

In Experiment 3-16, we tested these cuvettes in absorption measurements and evaluated their use in spectrophotometry. In this experiment, some common examples are presented of measuring the emission of tryptophan analog NATA when using three different cuvettes. In most spectroscopy/biochemistry, there are quite a few cuvettes that are made for different purposes, but over time get mixed and sometimes are used improperly. In this experiment, we are presenting and comparing measurements done in quartz, optical glass, and fluorescence plastic cuvettes.

MATERIALS

- NATA (Trp analog)
- Water
- 1 cm × 1 cm quartz cuvettes
- 1 cm × 1 cm glass cuvettes
- 1 cm × 1 cm plastic cuvettes

(*Continued*)

EXPERIMENT 4-21 (CONTINUED) MEASURING IN UV SPECTRAL RANGE: USING THE PROPER CUVETTE

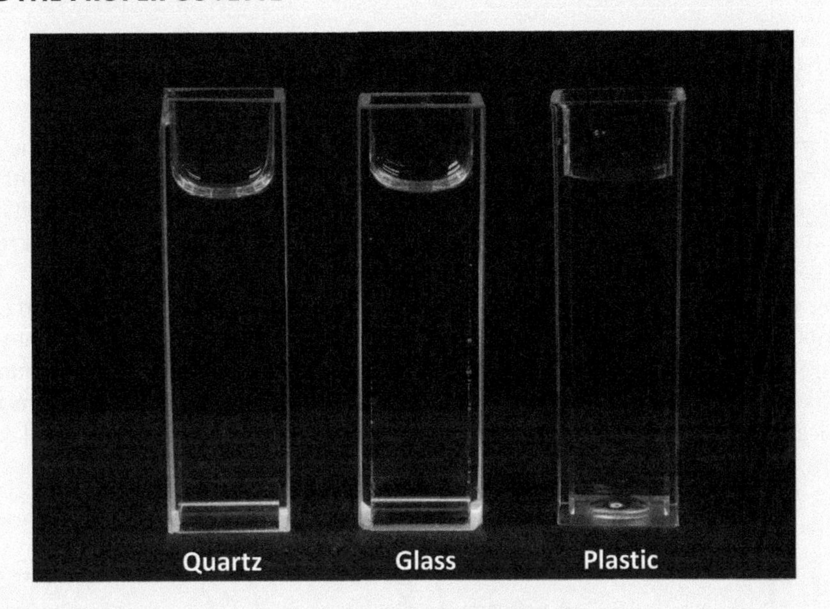

FIGURE 4-21.1 Photograph of three cuvettes used in this experiment. For an inexperienced user, it is difficult to judge which cuvette will work in the UV spectral range.

EQUIPMENT

- Spectrophotometer
- Spectrofluorometer

METHODS

1. Prepare a water solution of NATA. Make sure the water is highly purified. Since we will use UV excitations any impurity would contribute to the emission.
2. Start spectrophotometer and measure blank with air. For measurements of transmission and absorption refer to Chapter 3.
3. Fill each cuvette (quartz, glass, and plastic) with purified water and measure its transmittances in the range 200–800 nm. For details see Experiment 3-16.
4. First, use a quartz cuvette. Fill it with water and measure blank. Measure the absorbance of NATA solution in a quartz cuvette and adjust the concentration to have absorption close to 0.1 at the maximum (~278 nm). Make sure you have 10 mL or more of NATA solution that will be enough for all three cuvettes (quartz, glass, and plastic).
5. Put the selected cuvette with a solution of NATA into the spectrofluorometer. Make sure to set magic angle conditions, proper voltage, slits, etc. to ensure good emission signal. We will use three excitation wavelengths (270 nm, 280 nm, and 290 nm) using the emission range 300–800 nm. Measure fluorescence spectra of NATA in each cuvette.

(Continued)

EXPERIMENT 4-21 (CONTINUED) MEASURING IN UV SPECTRAL RANGE: USING THE PROPER CUVETTE

RESULTS

Figure 4-21.2 show transmittances measured for each cuvette filled only with water. On the left, we present a full range of 200–800 nm and on the right only selected 230–430 nm range. On the right, we marked three excitations (270 nm, 280 nm, and 290 nm). The quartz cuvette has an excellent transmittance in the full range. Below 330 nm, the transmittance of glass cuvette decreases faster than a plastic one. However, below 260 nm transmittance of plastic cuvette abruptly drops to zero making it not transparent while glass still has residual transmittance down to 240 nm.

In Figure 4-21.3 (left), we present measured absorption for NATA solution in quartz, glass, and plastic cuvettes in 230–330 nm range. On the right, we only present the 250–330 nm range. As discussed in Experiment 3-16, measured absorptions when viewed in the 250–330 nm range is practically identical. But we need to remember that the absorption measurement is relative to the background (reference). So, even a barely transparent cuvette transmits some small amount

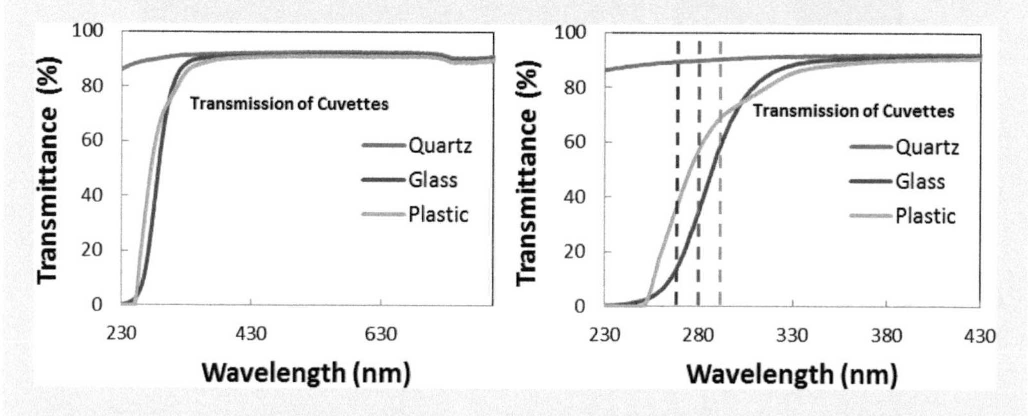

FIGURE 4-21.2 Transmittances of quartz, glass, and plastic cuvettes. Left full spectral range and right expanded UV range. On the right panel, vertical lines mark three excitations (270 nm, 280 nm, and 290 nm).

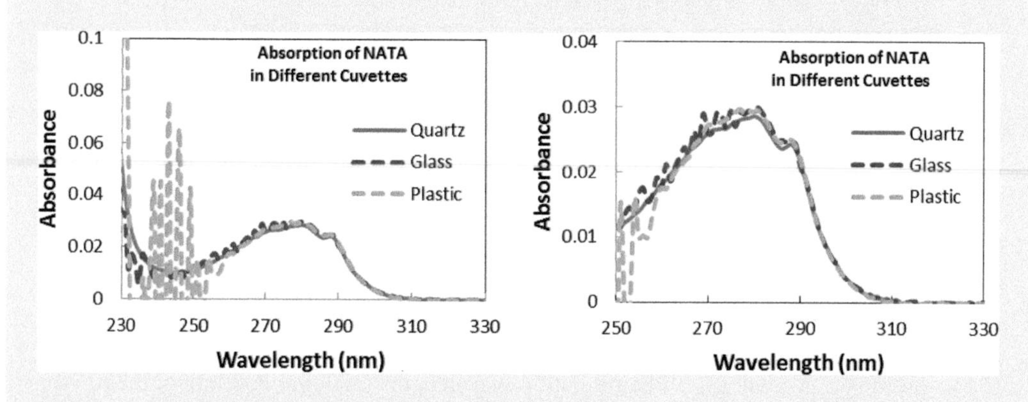

FIGURE 4-21.3 Absorption spectra for NATA in water measured in quartz, glass, and plastic cuvettes. Left full range 230–330 nm and on the right expanded range 250–330 nm.

(Continued)

EXPERIMENT 4-21 (CONTINUED) MEASURING IN UV SPECTRAL RANGE: USING THE PROPER CUVETTE

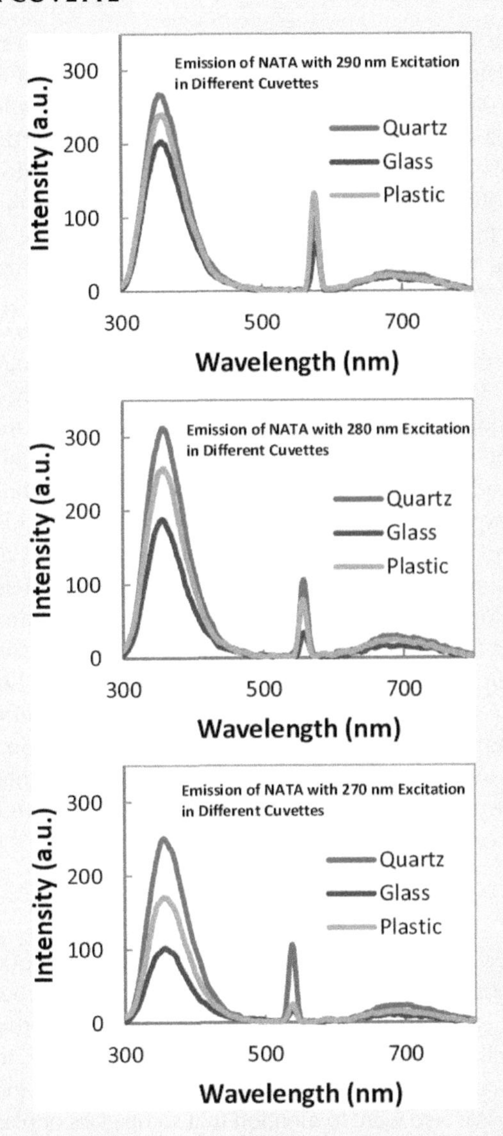

FIGURE 4-21.4 Measured emission spectra in quartz, glass, and plastic cuvettes with 270 nm (bottom), 280 nm (middle), and 290 nm (top) excitations. Note that spectra are presented up to 800 nm and measured signal above 500 nm is an artifact due to second-order transmittance of emission monochromator.

of light and absorption measurement should be acceptable. As the transmittance of glass and plastic decreases in the UV range, we still see reasonable absorbance but with highly increasing noise. This is because less and less light reaches the detector. Also in the range below 250 nm plastic is practically not transparent resulting in very high noise.

Figure 4-21.4 shows emission spectra measured for NATA solutions in water for three excitations (270 nm, 280 nm, and 290 nm) using quartz, glass, and plastic cuvettes.

(Continued)

EXPERIMENT 4-21 (CONTINUED) MEASURING IN UV SPECTRAL RANGE: USING THE PROPER CUVETTE

For 290 nm excitation (top), the signal from quartz is the highest while from glass is the lowest. The signal from the plastic cuvette is about only 10% lower and from glass cuvette a little over 20% lower as compared to the quartz cuvette. Transmittances for three cuvettes in Figure 4-21.2 for 290 nm are about 88% for quartz, 68% for plastic, and only 55% for glass. This may look like a significant discrepancy. However, when exciting fluorescence the excitation beam travels only through the one wall of the cuvette (transmittance is measured for two walls, the beam enters on one side and exits on opposite side of the cuvette) and the beam attenuation on a single wall is only 94% for quartz, 82% for plastic, and about 74% for glass.

So, we should see relative signals for plastic to be about 87% of that for quartz and for a glass to be about 78% of that for quartz. If we take into account that at the maximum emission (~350 nm) transmittance for glass is slightly lower than for plastic, we have a really good agreement between measured and expected values. This is even more evident for shorter wavelength excitation. For 270 nm, the transmittance for a single quartz wall is about 92%, for plastic about 61%, and for glass just only 37% and we would expect a signal from the plastic cuvette to be about 66% of the signal from quartz and for glass 40% of that for quartz. These numbers are in good agreement with measured emissions shown in the bottom panel (270 nm excitation) in Figure 4-21.4. It is interesting also to look at second-order transmittance. The numbers for second-order transmittance of scattered excitation are in a good agreement with measured transmittances of each cuvette (transmittance for two walls). This should not be a surprise since the excitation light has to enter the cuvette and scattered light (of the same wavelength) has to exit through the other wall. So, the excitation light travels through two cuvette walls similar to the spectrophotometer beam. Note that the second-order transmittance of emission light reflects relative signals intensities as measured for the first order (direct emission).

CONCLUSIONS

From this experiment, we can conclude that we can measure tryptophan absorption in glass or plastic cuvette and obtained values will be in a reasonable agreement with the true value measured in the quartz cuvette with only much larger noise. Measurements of emission will be very different and will depend on the excitation wavelength. For long-wavelength excitations, 290 nm and above the error will be small but it will quickly increase toward shorter wavelength excitations. We also want to mention that some glass or plastic cuvettes will have very low transmittance below 300 nm and we will not measure any signal. See the example in the next experiment where we used an ordinary glass cuvette. We also want to mention that many glass cuvettes when exciting with UV light may present autofluorescence that would corrupt the emission measurement. In summary, the measurements in the UV region should be done in quartz cuvettes. *Warning*: It is important to remember that signal above 500 nm is an artifact and apparent emission band centered about 700 nm is a result of the second-order transmission of the emission monochromator. This will not be observed with a prism-based monochromator.

EXPERIMENT 4-22 EMISSION SPECTRA AND SECOND ORDER MONOCHROMATOR TRANSMISSION FOR NATA AND BSA

Keywords: emission, fluorescence, NATA, BSA, spectrophotometer, fluorometer.

In previous experiments, we presented the emission spectra of NATA in the broad wavelength range. In such a case second-order transmittance of excitation can be a significant problem. This problem becomes more evident when the sample emission is weak and/or sample scattering is larger. There are many reasons for the increased scattering from liquid solutions. First, different molecules/particles will scatter differently. For example, tryptophan and NATA that are small molecules will scatter much less than proteins that are typically much larger (a few nanometers particles). Also, quantum dots or any silica particle will scatter even more. The solution may also frequently contain larger particles like protein aggregates or even air bubbles.

In this experiment, we will use NATA and bovine serum albumin (BSA) and monitor tryptophan emission. Later to show that this effect depends on solution scattering, a low concentration of Ludox (silica particles ~30 nm diameter) commonly used as a scattering reference in time-resolved measurements, will be introduced.

MATERIALS

- NATA (Trp analog)
- Bovine serum albumin (BSA)
- Water
- Ludox
- 1 cm × 1 cm quartz cuvettes

EQUIPMENT

- Spectrophotometer
- Spectrofluorometer

METHODS

1. Prepare water solutions of NATA and BSA. Make sure the water is highly purified. Since we will use UV excitations any impurity would contribute to the emission.
2. Start the spectrophotometer and measure a blank with water in a 10 mm × 10 mm quartz cuvette. For measurements of transmission and absorption, refer to experiments in Chapter 3.
3. Measure absorbance for each solution and adjust the concentration to have an absorption close to 0.1. Make sure you have 10 mL or more of each solution (NATA and BSA).
4. Put a cuvette with a solution of NATA into the spectrofluorometer. Make sure to set magic angle conditions, proper voltage, slits, etc. to ensure good emission signal. We will use 278 nm excitation wavelength and measure emission in the range of 300–800 nm.
5. Repeat the same for BSA.
6. Prepare a low concentration of Ludox solution. When using a commercially available stock solution of Ludox typically a small drop added to 10 mL of solution gives significant scattering. Prepare additional solutions of Ludox by diluting (~3 to 4-fold) with water already prepared Ludox.

(Continued)

EXPERIMENT 4-22 (CONTINUED) EMISSION SPECTRA AND SECOND ORDER MONOCHROMATOR TRANSMISSION FOR NATA AND BSA

7. Fill the cuvette with water first and measure emission using the same setup as for NATA and BSA. Scan only the second-order transmittance order, from 520 nm to 600 nm.
8. Repeat measurements with low Ludox concentration and then next with a higher one.

Note: If you use just one cuvette, it is important to start with water and go with increasing Ludox concentrations. This is because washing Ludox may take time and the beginning experiment with a higher Ludox concentration would require extra cleaning of the cuvette.

RESULTS

Figure 4-22.1 shows the absorption spectra measured for NATA and BSA solutions. Measured absorptions are slightly different but at 278 nm they are well matched.

Figure 4-22.2 presents the emission spectra of NATA, BSA, and pure water measured in the 300–800 nm range. We can clearly see the emission spectra of NATA and BSA in the 300–450 nm range. Note that emissions of NATA and BSA are different. This is expected since tryptophan inside a protein would have a different environment than one completely exposed to water as in NATA. The second-order scattering is clearly visible and for BSA is clearly much higher. For comparison, we are presenting an emission scan for a water solution. Scattering out of water depends mostly on a number of air bubbles (more precisely nano-bubbles) that are difficult to eliminate. So, the signal would vary and occasionally for water can be even higher than for NATA or other organic molecules dissolved in water. But we want to show that the water solution will typically significantly scatter excitation light. *Note:* Water second-order scattering is much stronger than the Raman signal. It is easy to see while taking the cuvette out and vigorously shaking and measuring again. The signal will go up and after some time drop back to normal. We could wait much longer and water would contribute to scattering much less. Note that second-order emission transmittance is

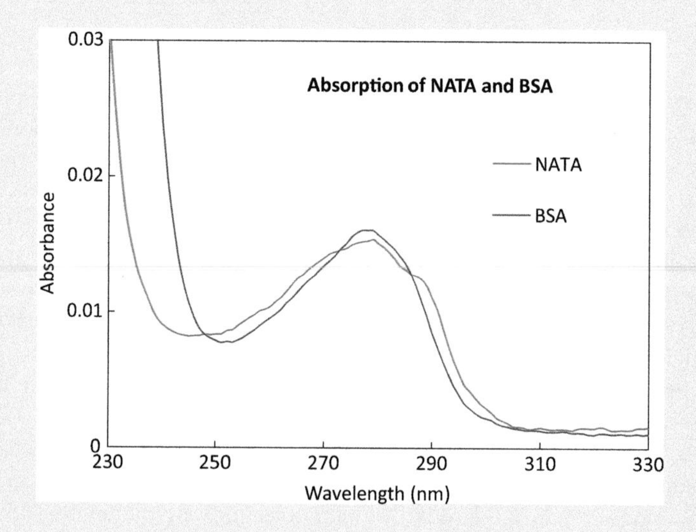

FIGURE 4-22.1 Absorption spectra for NATA and BSA solutions used in this experiment.

(Continued)

EXPERIMENT 4-22 (CONTINUED) EMISSION SPECTRA AND SECOND ORDER MONOCHROMATOR TRANSMISSION FOR NATA AND BSA

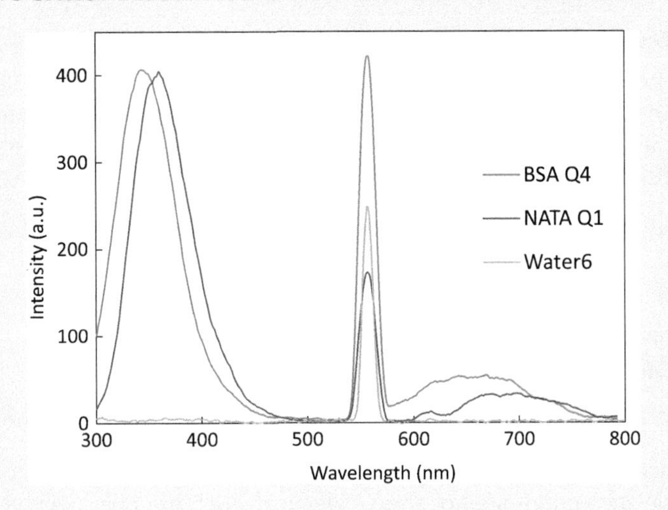

FIGURE 4-22.2 Measured emission spectra for NATA, BSA, and water with 278 nm excitation.

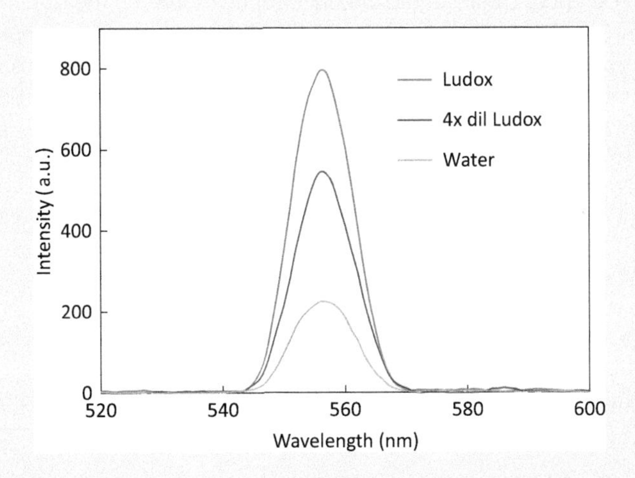

FIGURE 4-22.3 Measured second-order transmittance of scattered excitation for Ludox solutions and water.

significant and proportionally shifted (the wavelength shift between a maximum of emission of BSA and maximum of emission of NATA are doubled in the second-order).

For 278 nm excitation, we expect Raman scattering to occur at about 306 nm. This can be seen as a small bump about 300 nm for water spectrum and is completely masked by emission. Interestingly in a second-order, a small bump in front of emission can be identified.

In Figure 4-22.3, we present measurements for two different solutions of Ludox in water and just pure water for excitation wavelength 278 nm. The scattering from an even lower concentration of Ludox results in clearly higher scattering.

It is clear that a minimal increase of scattered light by very low Ludox concentration affects a lot of measured signal.

(Continued)

EXPERIMENT 4-22 (CONTINUED) EMISSION SPECTRA AND SECOND ORDER MONOCHROMATOR TRANSMISSION FOR NATA AND BSA

CONCLUSIONS

It is important to remember that second-order transmission of scattered excitation light can corrupt emission spectra, especially for large particles. When measuring the fluorescence of proteins, cells, and other biological specimens we should remember about that. Also when measuring particles this could be a factor.

EXPERIMENT 4-23 CORRECTION OF FLUORESCENCE FOR THE SAMPLE ABSORPTION

Keywords: emission, fluorescence, quinine sulfate, spectrophotometer, fluorometer.

In many fluorescence experiments, there is a need to compare fluorescence of various samples, and sample absorptions are not always identical. Such examples will be found in later experiments on quantum yield determination. Higher absorption of the sample usually results in a stronger fluorescence. Details of absorbance (optical density) measurements have been discussed previously in Chapter 3. *How to properly correct fluorescence emission when using samples with different absorptions?* A frequent answer is to correct for absorption, but this only works in a very narrow range. In this experiment, the fluorescence of quinine sulfate will be measured using samples with different optical densities and how to apply proper corrections.

MATERIALS

- Quinine sulfate (QS)
- Water
- 0.1 N H_2SO_4
- Glass vials
- 1 cm × 1 cm cuvettes

EQUIPMENT

- Spectrophotometer
- Spectrofluorometer

METHODS

1. Prepare a stock solution of quinine sulfate in 0.1 N H_2SO_4.
2. Prepare a reference (R) solution with an optical density OD (absorbance) of about 0.1.
3. Make sample solutions with various ODs, about the same as the reference, half of the reference and about twice higher (we want to have broader absorbance distribution).
4. Measure absorptions of the reference and all samples; adjust concentrations of QS as needed.
5. Measure fluorescence spectra of the reference and all samples.

(Continued)

EXPERIMENT 4-23 (CONTINUED) CORRECTION OF FLUORESCENCE FOR THE SAMPLE ABSORPTION

RESULTS

From years of experience, we know that many students and researchers use a "shortcut" while correcting fluorescence spectra for the sample absorption, e.g., simply dividing the fluorescence spectrum by absorption. This is incorrect even when absorptions are low. The correction should consider the amount of absorbed light. According to the absorption law, the absorbed by the sample intensity is proportional to the factor $(1–10^{-OD})$, not to OD itself.

In Figure 4-23.1, we are presenting the results of measurements made for three solutions. Top: absorption of both reference and sample are the same (remember we are using QS as a reference and sample so the profile of absorption and emission are identical; the observed

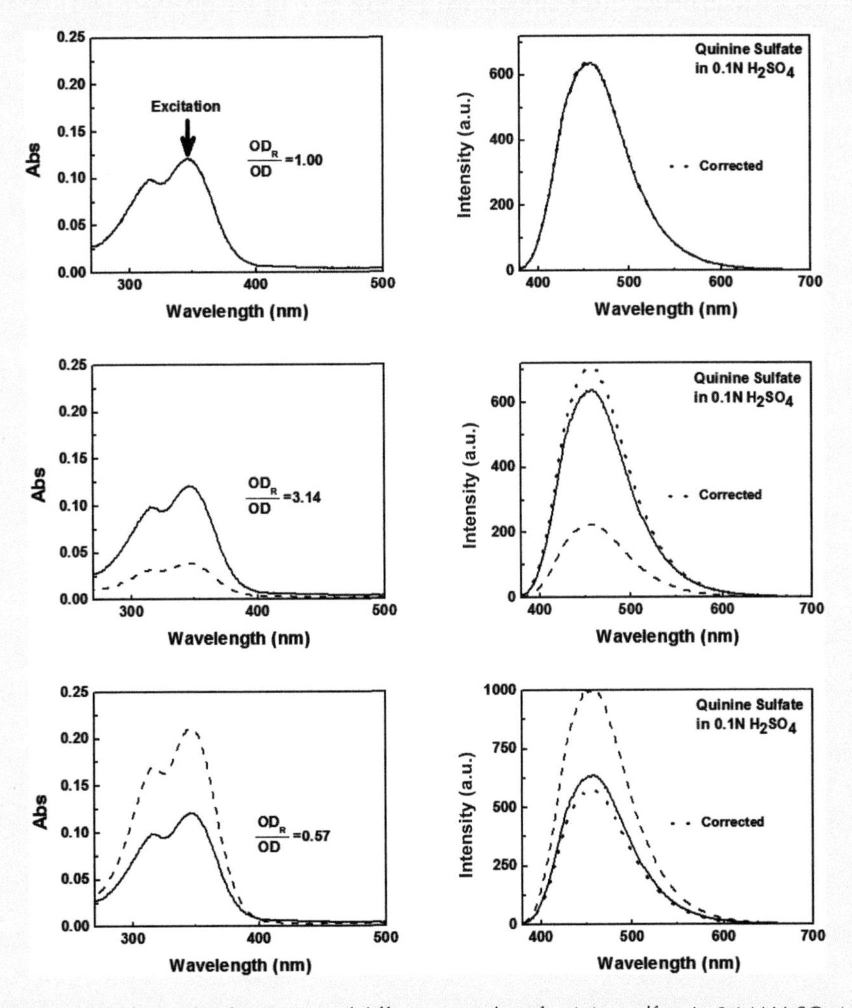

FIGURE 4-23.1 Left panels: absorptions of different samples of quinine sulfate in 0.1 N H_2SO_4 (dashed lines). The absorption of the reference solution is shown by solid lines. Right panels: corresponding fluorescence spectra; as measured (dashed lines) and corrected by optical densities (dotted lines). The fluorescence spectrum of the reference is shown by solid lines.

(Continued)

EXPERIMENT 4-23 (CONTINUED) CORRECTION OF FLUORESCENCE FOR THE SAMPLE ABSORPTION

difference is only due to the absorption difference). Middle: the reference signal is 3.14 times higher. Bottom: the reference has a lower absorbance.

On the right, in Figure 4-23.1, we present measured emission spectra for each case (solid and dashed lines correspond to the sample and reference respectively). A dotted line is calculated/corrected emission spectrum using just an absorption. The correction is just a ratio of measured absorptions as indicated in the figure. After the correction, the emission of the sample should be identical to the emission reference. It is evident that the corrected emission for higher reference concentration is too high and for lower sample absorption is too low.

On Figure 4-23.2, we are presenting the same results but now as the correction factor, we will use the ratio between the amount of light absorbed by the reference and the sample.

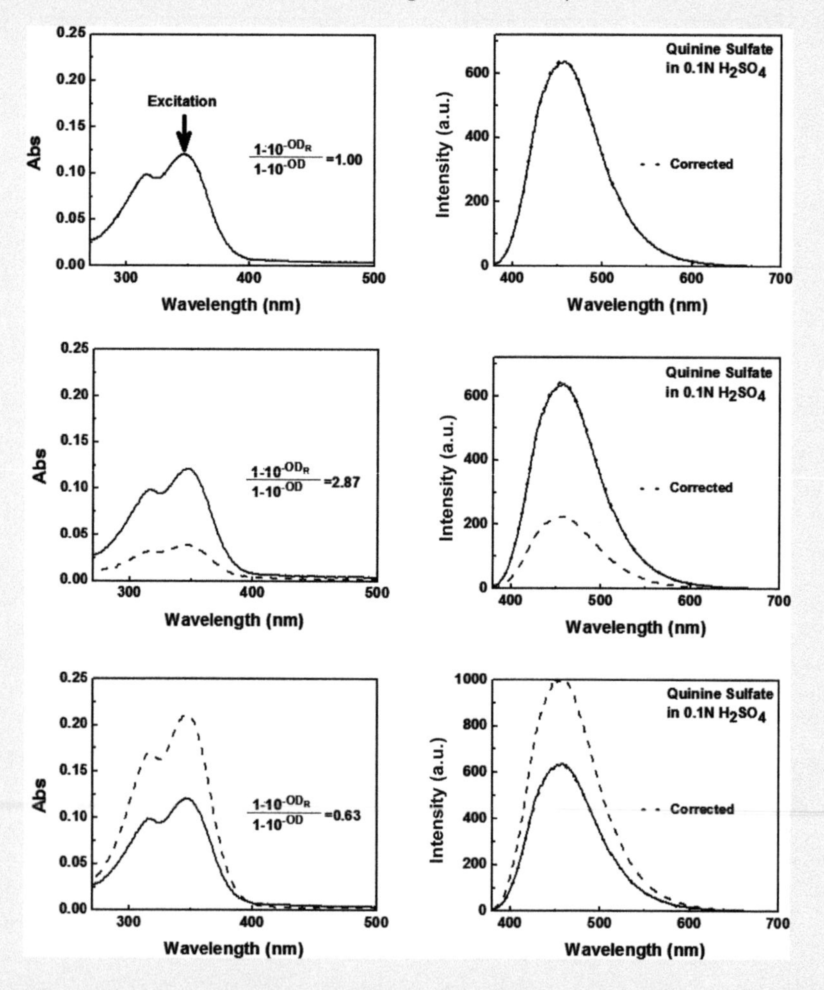

FIGURE 4-23.2 Left panels: absorptions of different samples of quinine sulfate in 0.1N H_2SO_4 (dashed lines). The absorption of the reference solution is shown by solid lines. Right panels: corresponding fluorescence spectra; as measured (dashed lines) and corrected by the amounts of the absorbed light (dotted lines). The fluorescence spectrum of the reference is shown by solid lines.

(*Continued*)

EXPERIMENT 4-23 (CONTINUED) CORRECTION OF FLUORESCENCE FOR THE SAMPLE ABSORPTION

TABLE 4-23.1 Assumed Optical Densities, Correction Factors, and Potential Errors

OD_R	OD	$\frac{OD_R}{OD}$	$\frac{1-10^{-OD_R}}{1-10^{-OD}}$	Pot. Error (%)
0.5	0.1	5	3.32	33
0.2	0.1	2	1.80	10
0.05	0.01	5	4.79	4
0.05	0.04	1.250	1.235	0.3

In Figure 4-23.2, we indicated appropriate factors within the figure. In all cases, the corrected spectrum well overlaps with the reference.

In Table 4-23.1, we are presenting the measured optical density, corrected values and made an error in the sample signal evaluation.

CONCLUSIONS

The simplest way to avoid errors is to adjust absorptions at the excitation wavelength of studied samples to the absorption of the reference. A small absorption, below 0.05, suggested by many authors still can allow a few percent errors if the optical density of the sample is not matched with an optical density of the reference. For example, absorption of 0.02 and 0.08 are both in the good range below 0.1 but the amounts of absorbed light are very different when calculating using measured absorbance or the amount of absorbed light ($1-10^{-OD}$). *Warning*: It is necessary to check the absorption of used solvents and buffers. For example, 1N H_2SO_4, if not fresh can often show not a negligible absorption in UV.

EXPERIMENT 4-24 DISTINGUISHING BETWEEN EMISSIONS FROM MULTIPLE CHROMOPHORES IN A SOLUTION

Keywords: rhodamine 6G, rhodamine B, cuvette, emission spectrum.

In some experiments, it is important to distinguish whether the observed signal originates from a single chromophore or multiple chromophores. This may be necessary to separate different emissive species, to distinguish potential resonance energy transfer, or just to test if there is a significant background contribution. According to Kasha's rule, the fluorescence signal originates from the lowest excited state independently to what level the fluorophore was originally excited to (independently of the wavelength used for excitation). Different chromophores would also have different absorption and emission spectra. What these facts mean in practice is that the emission spectrum for a single chromophore should not change the spectral shape when we use different excitation wavelengths. The only change would be the relative intensities of the measured spectra. The same should hold true for excitation

(Continued)

EXPERIMENT 4-24 (CONTINUED) DISTINGUISHING BETWEEN EMISSIONS FROM MULTIPLE CHROMOPHORES IN A SOLUTION

spectra when observed at different wavelengths. This offers a powerful way for distinguishing if given emission or excitation spectra originating from a single chromophore or are a mixture of two or more fluorophores. This should likewise hold true when using special conditions like additional filters, as long as for all other conditions remain unchanged when making measurements (e.g., the same filters, slits openings, etc.).

MATERIALS

- 10 mm × 10 mm absorption cuvette
- 2 mm × 10 mm fluorescence cuvette
- Ethanol
- Rhodamine 6G (R6G)
- Rhodamine B (RhB)

EQUIPMENT

- Spectrophotometer
- Spectrofluorometer

METHODS

1. Prepare solutions (about 10 mL each) with the same/comparable absorbance (optical densities) below 0.05 (these are very good fluorophores and we would like to avoid artifacts from too high absorption or interactions. *Note: In Experiment 3-21, we show how to define interacting species in the mixture from absorption measurements.*
   a. Rhodamine 6G
   b. Rhodamine B
2. Prepare 1:1 mixture of rhodamine 6G and rhodamine B solution by mixing the same amounts of R6G and RhB solutions.
3. Measure the absorption for each solution.
4. Measure the emission spectra of each solution using two different excitation wavelengths (e.g., 490 nm and 310 nm) (remember to keep magic angle conditions).
5. Measure the excitation spectra of each solution using two different observation wavelengths (e.g., 540 nm and 565 nm) and magic angle conditions.
6. Construct ratios for emission and excitation spectra measured for each solution.

RESULTS

In Figure 4-24.1, we are presenting absorption spectra for each solution. It is not surprising that absorptions for the mixture are a little lower (this is a dilution effect). The absorption spectra are slightly shifted. We purposely selected similar dyes to demonstrate the effect can be significant even if the difference is small.

In Figure 4-24.2, we present emission spectra measured for 490 nm excitation for RhB, R6G, and 1:1 RhB:R6G solution mixture. When looking on individual emission spectra each of them could be considered as an emission of a single species. If the spectra are not separated by much the mixture of the two bell-shaped spectra will appear as a single bell-shaped emission that is

(*Continued*)

EXPERIMENT 4-24 (CONTINUED) DISTINGUISHING BETWEEN EMISSIONS FROM MULTIPLE CHROMOPHORES IN A SOLUTION

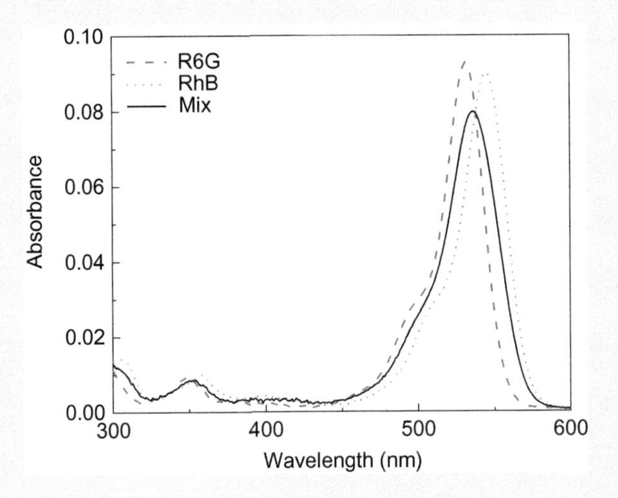

FIGURE 4-24.1 Absorption spectra for rhodamine 6G, rhodamine B, and their mixture in ethanol.

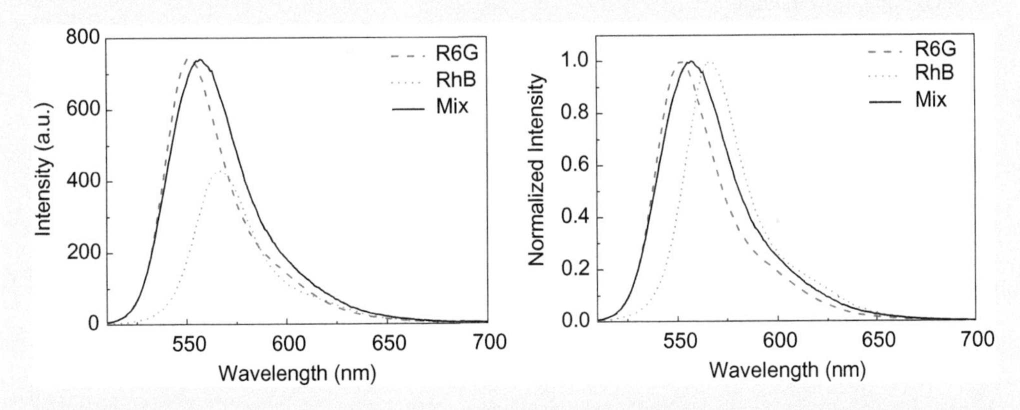

FIGURE 4-24.2 Emission spectra for rhodamine 6G, rhodamine B, and mixture in ethanol excited at 490 nm (left). On the right, we present normalized emission spectra.

only a little broader than the individual components. Since in general, the emission spectra have not a defined width it is difficult to judge without the prior knowledge that a broader spectrum is a mix and not just another molecule that just has a broader emission spectrum.

If two mixed components have different absorptions, exciting the mixture solution with two wavelengths will yield two emission spectra with different contributions of two components in the mixture. From the absorption spectra (Figure 4-24.1) of R6G and RB, we can predict that the excitation at 490 nm and for example 310 nm will result in relatively different populations of R6G and RhB excited molecules. We cannot tell it from absorption of the mixture only. The constructed intensity ratios of measured emissions for each solution excited with 490 nm and 310 nm are presented in Figure 4-24.3. For two individual components (R6G and RhB), the lines are relatively flat indicating that the shape of emission does not depend on excitation. In contrast, the ratio for the mixture shows a clearly changing ratio across the emission spectrum.

(Continued)

EXPERIMENT 4-24 (CONTINUED) DISTINGUISHING BETWEEN EMISSIONS FROM MULTIPLE CHROMOPHORES IN A SOLUTION

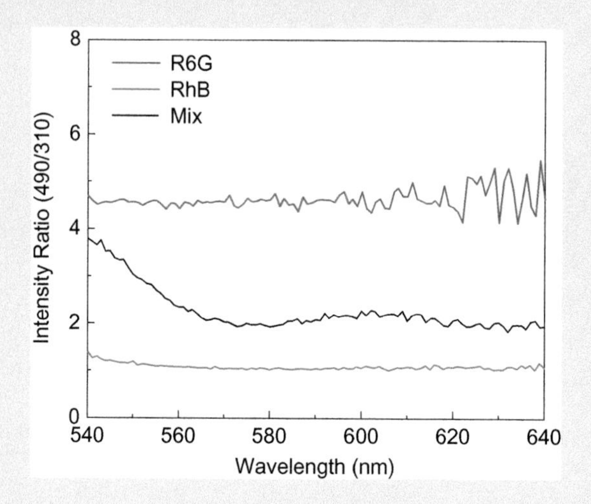

FIGURE 4-24.3 Emission intensity ratios for spectra measured with 490 nm and 310 nm excitations for R6G, RhB, and a mixture.

Note: Knowing the absorption of both components, we were able to select excitation wavelengths that yield a significant difference. For unknown mixtures, we would recommend constructing ratios for a few different pairs of excitation wavelengths. This is to avoid excitation at two wavelengths for which the ratio of absorptions could be similar.

Next, we have chosen two emission wavelengths 540 nm and 565 nm close to emission maxima of R6G and RhB respectively and measured two excitation spectra for each solution. In Figure 4-24.4, we are presenting calculated intensity ratios of excitation spectra for RhB, R6G, and a mixture. For two solutions with a single dye, we observe two flat lines representing

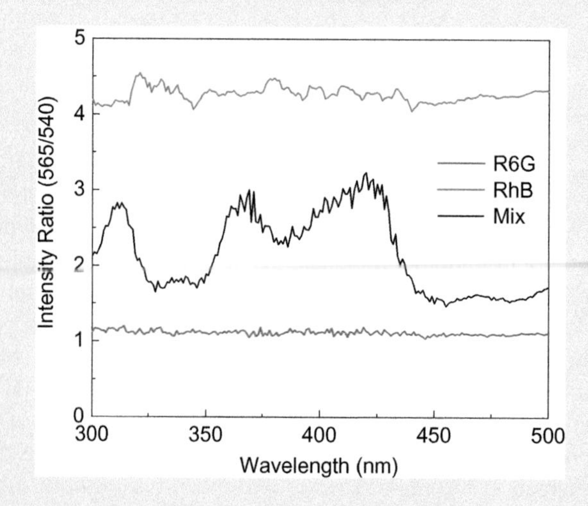

FIGURE 4-24.4 Excitation intensities ratios for spectra measured with 540 nm and 565 nm observations for R6G, RhB, and a mixture.

(Continued)

EXPERIMENT 4-24 (CONTINUED) DISTINGUISHING BETWEEN EMISSIONS FROM MULTIPLE CHROMOPHORES IN A SOLUTION

the difference in emission for each observation. In contrast, the solution of mixture clearly indicates drastically changing behavior across excitation wavelengths. The observed changes will obviously depend on selected observations. When working with an unknown mixture we would also recommend to use a few different pairs for observation wavelengths.

CONCLUSIONS

Two or three randomly selected excitations for a single emissive species will result in the same emission spectra that should only differ in relative intensities (the relative observed intensity will only depend on relative extinction coefficient and lamp power at given excitation wavelength). The ratio of emission spectra or excitation spectra will yield a flat line across the wavelength scale. However, even a small addition of a different species with different spectra will produce distinct variations in observed ratios. One practical application could be a simple check for compound purity. A chemical decomposition or photodegradation typically result in different products with different absorption and emission. This simple check can be used to determine sample stability. But we want to note that if the products are not emissive or significantly separated spectrally we may not detect the difference and it could be necessary to use similar detailed procedure also for absorption spectra.

EXPERIMENT 4-25 HOW EMISSION OF CHROMOPHORE DEPENDS ON SOLVENT POLARITY—EMISSION SPECTRA OF DCS AND DPH

Keywords: buffer, polarity, fluorescence emission, 4-dimethylamino-4′-cyano stilbene, diphenylhexatriene.

The solvent is a fundamental component of the majority of fluorescence experiments and it is important to consider its properties and how it will affect fluorophore response. Different solvents may alter the emissive properties of some chromophores, while not altering those of others. There are no general rules to which solution properties all chromophores are sensitive to. However, determining solvent effects on chromophores experimentally is possible.

In this protocol, two chromophores, DCS and DPH, with a reasonably similar chemical structure, are suspended in solutions with different polarity indices, and their fluorescence emissions are measured.

MATERIALS

- 4-Dimethylamino-4′-cyano stilbene (DCS)
- Diphenylhexatriene (DPH)
- Heptane
- Cyclohexane
- Toluene
- Ethyl acetate
- EtOH

(Continued)

EXPERIMENT 4-25 (CONTINUED) HOW EMISSION OF CHROMOPHORE DEPENDS ON SOLVENT POLARITY—EMISSION SPECTRA OF DCS AND DPH

- Acetonitrile
- Dimethylformamide
- 1 cm × 1 cm cuvettes

EQUIPMENT

- Spectrofluorometer

METHODS

1. Using your preferred method prepare solutions of DCS and DPH in heptane, cyclohexane, toluene, ethyl acetate, EtOH, acetonitrile, and dimethylformamide. Typically, we would recommend preparing stock solutions of each dye in a different solvent and out of the stock solution to prepare adequate concentrations so the fluorescence signals will be comparable.
2. Measure the fluorescence of each solution.

RESULTS

Figure 4-25.1 shows the normalized spectra for DCS measured in all selected solvents. An immediate observation is that respective emission spectra are spread through a large range from 420 nm to over 550 nm. Another interesting feature is that distinct peaks are visible in

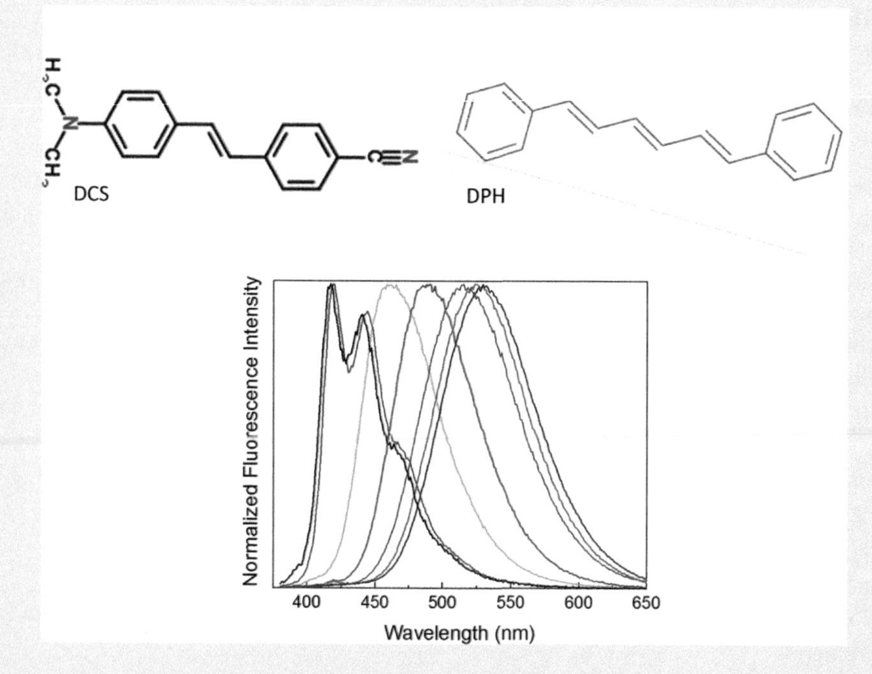

FIGURE 4-25.1 Normalized fluorescence spectra of DCS in solutions with varying polarity indices. See Table 4-25.1 for details. Top: Structures of DCS and DPH.

(Continued)

EXPERIMENT 4-25 (CONTINUED) HOW EMISSION OF CHROMOPHORE DEPENDS ON SOLVENT POLARITY—EMISSION SPECTRA OF DCS AND DPH

TABLE 4-25.1 Solvents Used in This Experiment, Their Polarity Parameters, and Observed Fluorescence Emission Centers of DCS in Each Solution

Solvent	Polarity Index	Dipole Moment (D)	Peak DCS Emission (nm)
Heptane	0.0	0.0	416, 439
Cyclohexane	0.2	0.0	419, 444
Toluene	2.4	0.31	460
Ethyl acetate	4.4	1.87	488
EtOH	5.2	1.66	515
Acetonitrile	5.8	3.44	525
Dimethylformamide	6.4	3.72	530

the short wavelengths spectral range while for spectra at longer wavelength only one peak is visible (single bell-shaped emission band).

The emission wavelength of DCS increases as the polarity of the solvent it's suspended in increases. The results are summarized in Table 4-25.1. Thus, from left to right, the peaks in Figure 4-25.1 are the emission of DCS in heptane, cyclohexane, toluene, ethyl acetate, EtOH, acetonitrile, and dimethylformamide.

Figure 4-25.1 and Table 4-25.1 summarize the behavior of a single chromophore as a function of the polarity index of the solvent it is dissolved in. In Figure 4-25.2, we are presenting a photograph of DCS dissolved in all seven solvents while illuminated with the UV illuminator lamp. Progressive change in color is clearly visible. The wavelength changes presented by DCS are significant enough to be observed with the naked eye. *Note:* All used solvents themselves do not fluoresce and are colorless (not shown in the photograph).

In Figure 4-25.3, we present emissions of DCS and DHP in just two solvents with the largest change in polarity index: heptane (0) and dimethylformamide (6.4). The spectra of DCS have very little overlap and not only a wavelength shift of about 100 nm but the shape of the peak changed drastically. In contrast, spectra of DHP in the same solvents show no noticeable shift in wavelength and only minor changes in shape.

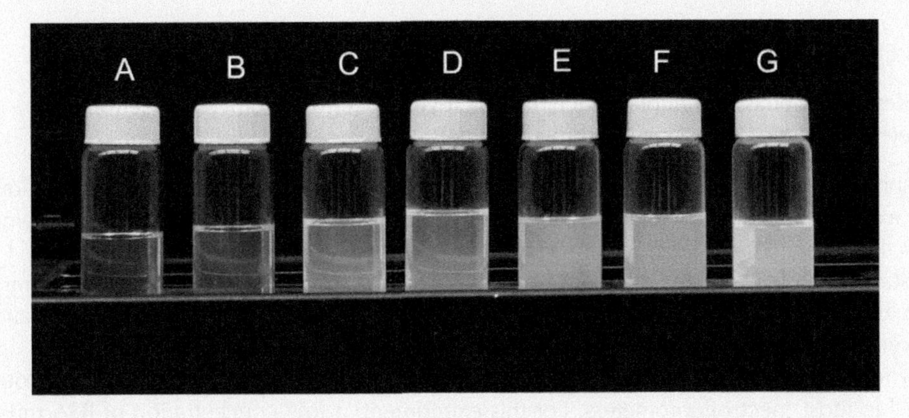

FIGURE 4-25.2 DCS in solutions with varying polarity indices. See Table 4-25.1 for details.

(Continued)

EXPERIMENT 4-25 (CONTINUED) HOW EMISSION OF CHROMOPHORE DEPENDS ON SOLVENT POLARITY—EMISSION SPECTRA OF DCS AND DPH

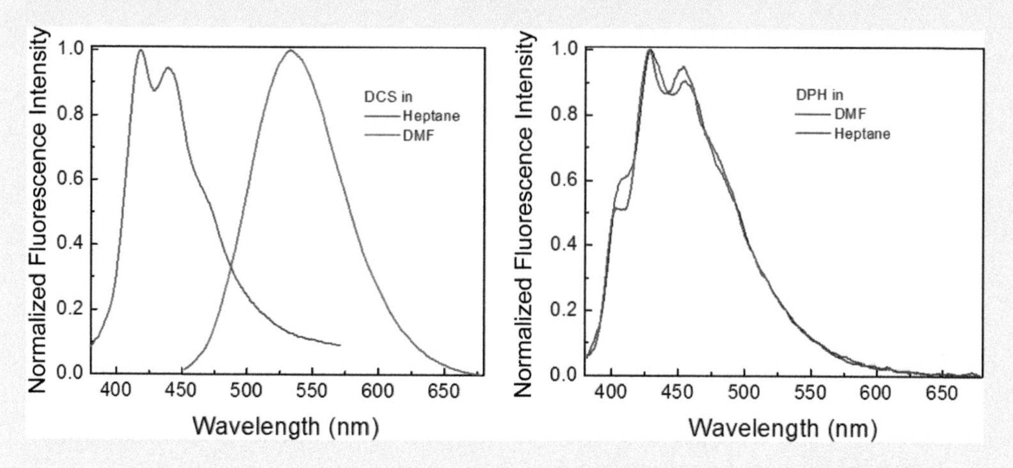

FIGURE 4-25.3 Left: DCS in heptane and dimethylformamide; Right: DPH in heptane and dimethylformamide.

CONCLUSIONS

It is obvious that different fluorophores would/can present distinct spectral changes as a function of different solvent polarities. But a large number of available chromophores with no standard behaviors in different types of solvents should not be seen as a confusing problem to be solved with memorization. Instead, the experimenter should be aware of these differences and be able to exploit them when designing experiments. DCS allows for the measurement of certain property (polarity index) quite easily, but if isolating the response of a system ignoring polarity is desired other chromophores such as DHP may be used. It is equally important to remember that if the changes of emission due to solvent polarity or being not considered while working with DCS experimental results may be contaminated and wrong conclusions may be reached.

EXPERIMENT 4-26 MEASURING EMISSION SPECTRA WITH UV EXCITATION

Keywords: emission, fluorescence, organic dyes, spectrophotometer, spectrofluorometer, rhodamine 6G, bovine serum albumin (BSA).

In some experiments, emission spectra of visible dyes are measured with UV excitations. For example, in biological experiments, UV excitation is used to follow the tryptophan emission. If a sample is also labeled with a visible fluorophore we may want to utilize a single UV excitation to follow amino acid emission and dye emission simultaneously. Another example would be measuring FRET when measuring acceptor emission with the donor excitation wavelength.

The goal of this experiment is to alert the reader to potential errors that may be encountered in most spectrofluorometers. For this experiment, a low concentration of BSA mixed with R6G is used.

(*Continued*)

EXPERIMENT 4-26 (CONTINUED) MEASURING EMISSION SPECTRA WITH UV EXCITATION

MATERIALS

- Rhodamine 6G (R6G)
- Bovine serum albumin (BSA)
- Water
- 1 cm × 1 cm quartz cuvettes
- Long-pass cut-off filters 305 nm and 400 nm, and a glass filter.

EQUIPMENT

- Spectrophotometer
- Spectrofluorometer

METHODS

1. Prepare solutions of R6G and BSA in water. Make sure the water is highly purified. Since we will use low concentrations and UV excitations any impurity would contribute to the emission.
2. Start the spectrophotometer and measure a blank with air. For measurements of transmission and absorption refer to Chapter 3.
3. Measure the transmittance of selected filters and a glass slide.
4. Fill the quartz cuvette with water and measure blank.
5. Fill the cuvette with the BSA solution and measure absorbance. Make sure that it is below 0.1 at 290 nm excitation (dilute sample as needed).
6. Measure the emission spectrum of BSA in the 300–800 nm range (remember to keep magic angle conditions).
7. Measure emission of the R6G solution with 290 nm excitation using the same spectral range. Dilute R6G solution with water to make sure the emission peak of R6G (~560 nm) is comparable to the emission peak of BSA (~330 nm). *Note:* It is difficult to use exact concentrations since various instruments may have relatively different sensitivities in the UV and visible range. The easiest is to measure emission signals.
8. Mix both solutions about 1:1 and measure emission with 290 nm excitation without the filter on observation and with various filters.
9. Measure emission of a mixed solution with 480 nm excitation.

RESULTS

Figure 4-26.1 show transmittances for 305 nm and 400 nm filters. In the figure, we also included transmission of the glass that can be used as a UV filter. In Figure 4-26.2, we are presenting emission spectra measured with excitation 290 nm and collected without a filter, with a 305 nm long-pass filter, and a 400 nm long-pass filter.

The emission spectrum measured without the filter shows an evident sharp peak at 580 nm. From previous experiments (Experiments 4-19 and 4-22), we know this is the second harmonics of the excitation leaking through the monochromator. We want to mention that relative heights of the leaking scattering will depend on BSA and R6G concentrations. If the peak is weak we recommend diluting solution (also we may add some Ludox

(*Continued*)

EXPERIMENT 4-26 (CONTINUED) MEASURING EMISSION SPECTRA WITH UV EXCITATION

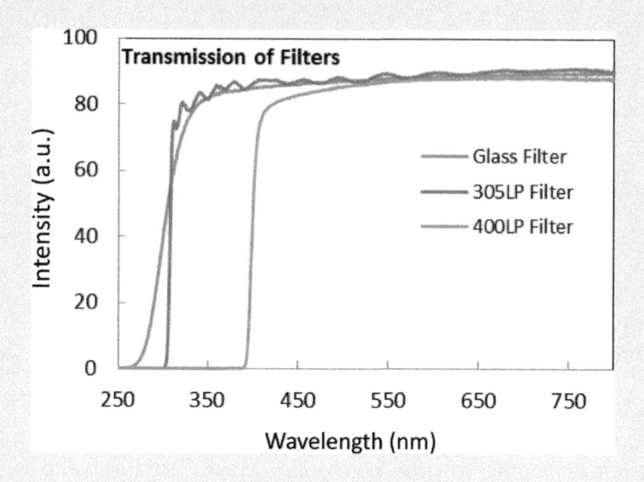

FIGURE 4-26.1 Transmission of filters used in the experiment.

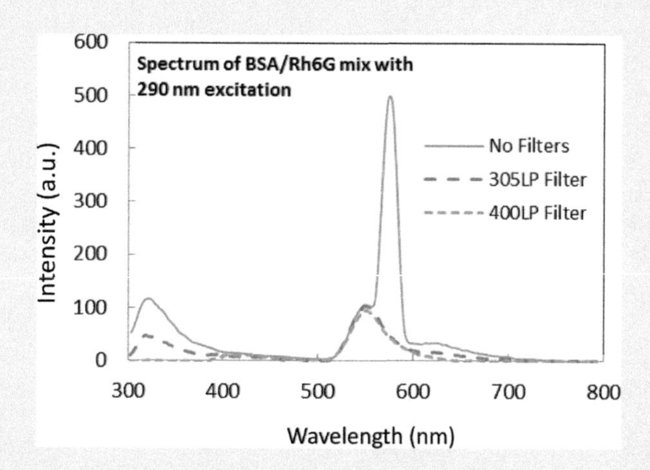

FIGURE 4-26.2 Measured emission spectra of BSA and R6G mixture.

to increase sample scattering). Putting a long-pass 305 nm filter eliminates the second harmonic from the emission spectrum. However, a small bump is present on the red side of R6G emission. This is the second harmonics of BSA emission leaking through the monochromator that cannot be eliminated by the 305 nm cut-off filter. The dashed line in the figure is an emission spectrum measured with 400 nm long-pass filter on emission observation. In this case, the BSA emission is eliminated and the bump on the red side of R6G spectrum disappeared. To better visualize the effect in Figure 4-26.3, we are presenting normalized spectra for measurements made with different excitation wavelengths and different filters on observation.

(*Continued*)

EXPERIMENT 4-26 (CONTINUED) MEASURING EMISSION SPECTRA WITH UV EXCITATION

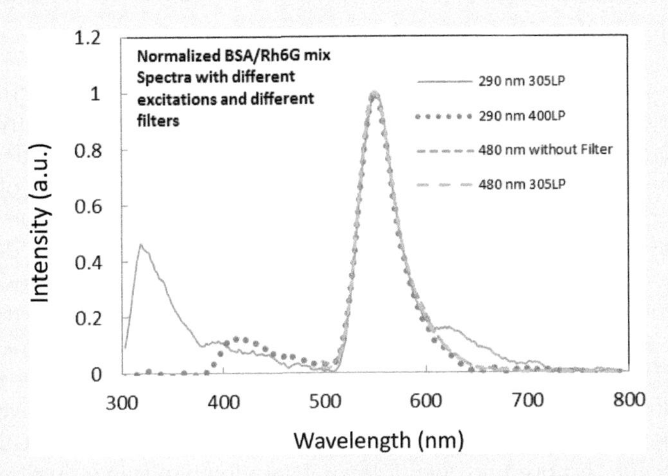

FIGURE 4-26.3 Emission spectra measured for a mixture of BSA and R6G with 290 nm excitation with 305 nm long-pass filter and 400 nm long-pass filter. For comparison, we also presented R6G emission spectrum with 485 nm excitation.

When using the 305 nm long-pass filter we see partial emission of BSA and R6G emission shows the clear bump on the red-side of emission band. When using the 400 nm LP filter that severely cuts BSA emission we clearly eliminate the long-wavelength bump on R6G emission. In the figure, we also plotted emission of R6G excited with 480 nm with and without the 305 nm LP filter. These normalized spectra are identical confirming that the 305 nm LP filter does not alter the measured R6G emission. Also with 480 nm excitation where R6G extinction coefficient is much higher, the signal is much stronger and there is no need to use any extra filters to limit the excitation or Raman scattering. Comparing the emission spectrum measured with 290 nm excitation and 400 nm LP filter with emission spectrum measured with 480 nm excitation we can conclude that the 400 nm LP filter eliminates spectral artifacts resulting in correct R6G emission. However no filter or using only a 305 nm LP filter results in incorrect/corrupted R6G emission spectra.

CONCLUSIONS

When measuring emission spectra of a two dye system with UV excitation it is important to use proper filters to prevent second-order transmission of excitation and emission of the short-wavelength emitting components. When looking for tryptophan emission, the transmission of the second-order tryptophan emission will appear in the 600–750 nm range that overlaps with the emission of most red and NIR dyes. The leak of excitation is easy to recognize and can be easily eliminated by a short wavelength cut-off filter. However, the emission of a short wavelengths emitting component requires the filter that will eliminate the original short wavelengths emission spectrum. The measurement for a short and long wavelengths emitting species should then be done separately using two different filters.

EXPERIMENT 4-27 MEASURING EXCITATION SPECTRA OF 9-METHYLANTHRACENE IN QUARTZ AND PLASTIC CUVETTES

Keywords: 9-Methylanthracene, cuvette, emission spectrum.

The excitation spectrum is an intrinsic property of a chromophore. After correction for the excitation lamp efficiency (typical lamp presents not a uniform energy distribution across the spectrum but many present spectrofluorometers will use a separate detector to correct for lamp efficiency) excitation spectrum should reflect the absorption of the chromophore. However, the measured excitation spectrum can be affected by the experimental setup, especially the cuvette. The transmissions of different cuvettes will depend on the type of cuvette, and in the UV range may even depend on the vendors (two fused silica cuvettes from different vendors may show significantly different transmittance at deep UV—below 280 nm). In the previous Experiment 4-21, we presented different cuvettes and how they affect absorption and emission spectra. In this experiment, we will show how the measurement of the excitation spectrum can be affected by the cuvette. This protocol compares the measurement of excitation spectra in quartz and plastic cuvette for 9-methylanthracene that presents two distinct absorption bands in visible and UV.

MATERIALS

- 1 cm × 1 cm UV plastic cuvette
- 1 cm × 1 cm UV quartz cuvette
- 9-Methylanthracene (9-MA)
- Methanol

EQUIPMENT

- Spectrophotometer
- Spectrofluorometer

METHODS

1. Prepare a solution of 9-methylanthracene in methanol that has an absorbance of about 0.02 at maximum (~365 nm).
2. Measure the emission spectrum using magic angle conditions in either cuvette. Note the location of emission peaks.
3. Measure the excitation spectra (remember about MA) for one of the emission peaks observed in step 2.
   a. 388 nm.
   b. 413 nm.
   c. 435 nm.

RESULTS

First, we measure absorption and fluorescence spectra of 9-MA in methanol, see Figures 4-27.1 and 4-27.2.

Figure 4-27.2 shows the normalized emission spectrum of 9-methylanthracene excited at 365 nm. The measured spectra should be the same in both cuvettes.

(*Continued*)

EXPERIMENT 4-27 (CONTINUED) MEASURING EXCITATION SPECTRA OF 9-METHYLANTHRACENE IN QUARTZ AND PLASTIC CUVETTES

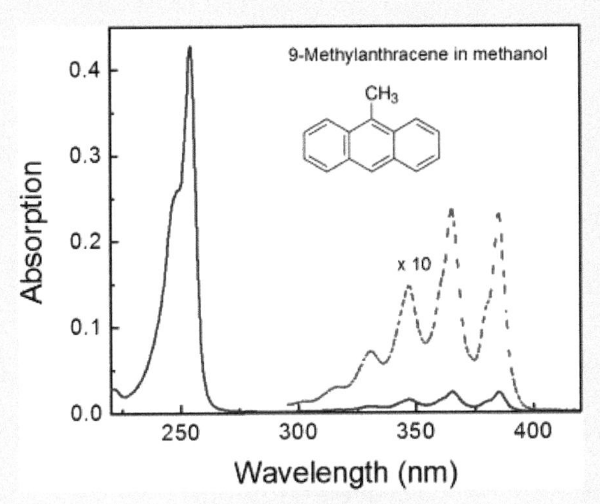

FIGURE 4-27.1 Absorption spectrum of 9-MA in methanol. The short wavelength band (S_0–S_2 transition) is much stronger than the long-wavelength absorption band.

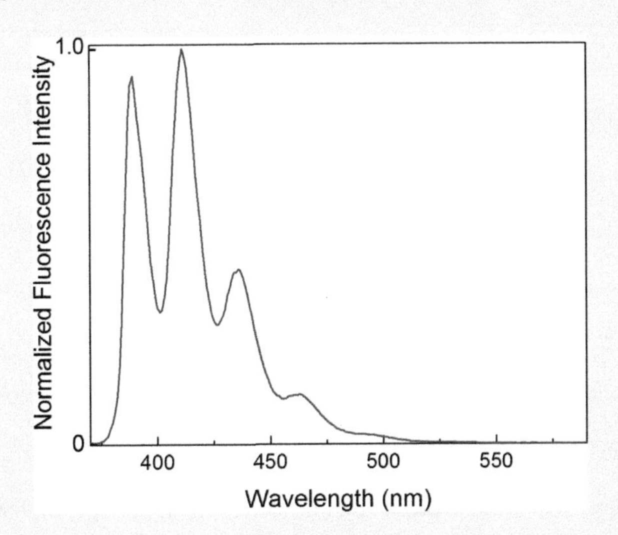

FIGURE 4-27.2 Emission spectrum of 9-MA in methanol.

It is interesting to evaluate a mirror symmetry between absorption and fluorescence spectra, see Figure 4-27.3.

The mirror symmetry is clearly visible. The excitation spectrum of 9-MA is almost identical to the absorption spectrum, see Figure 4-27.4 and mirror spectral symmetry is more pronounced.

Figure 4-27.5 presents normalized excitation spectra measured in quartz (top) and UV grade plastic (bottom) cuvettes respectively. Measured three excitation spectra with the observation wavelengths positioned at the three tallest peaks from the emission spectrum

(Continued)

EXPERIMENT 4-27 (CONTINUED) MEASURING EXCITATION SPECTRA OF 9-METHYLANTHRACENE IN QUARTZ AND PLASTIC CUVETTES

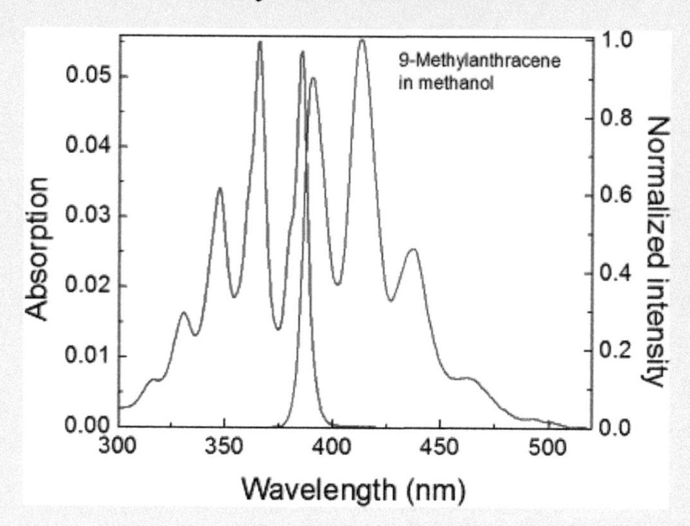

FIGURE 4-27.3 Normalized absorption and fluorescence spectra of 9-MA in methanol.

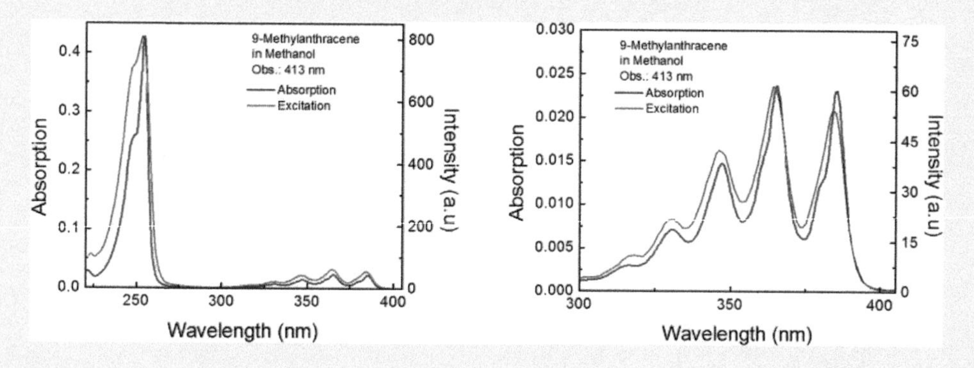

FIGURE 4-27.4 Excitation fluorescence spectrum of 9-MA in methanol and its comparison with the absorption; normalized at the short-wavelength peak. The right figure contains only a normalized long-wave part of the spectra.

(~388 nm, ~412 nm, ~435 nm) are presented from left to right. The spectra have been normalized to allow for direct comparison of the relative line intensity shapes. In a previous experiment Figure 4-21.2, we presented transmittance of different cuvettes. The plastic cuvette shows a much lower transmittance than the quartz cuvette below 300 nm. Using the plastic cuvette (with low transmittance in the UV) results in the peak at 260 nm being lower than the group of peaks centered at 420 nm. This happens because the plastic cuvette attenuates the excitation light by absorbing most of UV excitation. Just comparing transmittances for quartz and plastic cuvettes in Figure 4-21.2, we can conclude that the observed intensity for excitation in the UV peak (~260 nm) should be about 10 times weaker. *Note:* Remember presented transmittance in Figure 4-21.2 corresponds to two walls and attenuation of excitation is about half—transmittance twice larger). In Figure 4-21.2, the quartz cuvette transmittance is about

(Continued)

EXPERIMENT 4-27 (CONTINUED) MEASURING EXCITATION SPECTRA OF 9-METHYLANTHRACENE IN QUARTZ AND PLASTIC CUVETTES

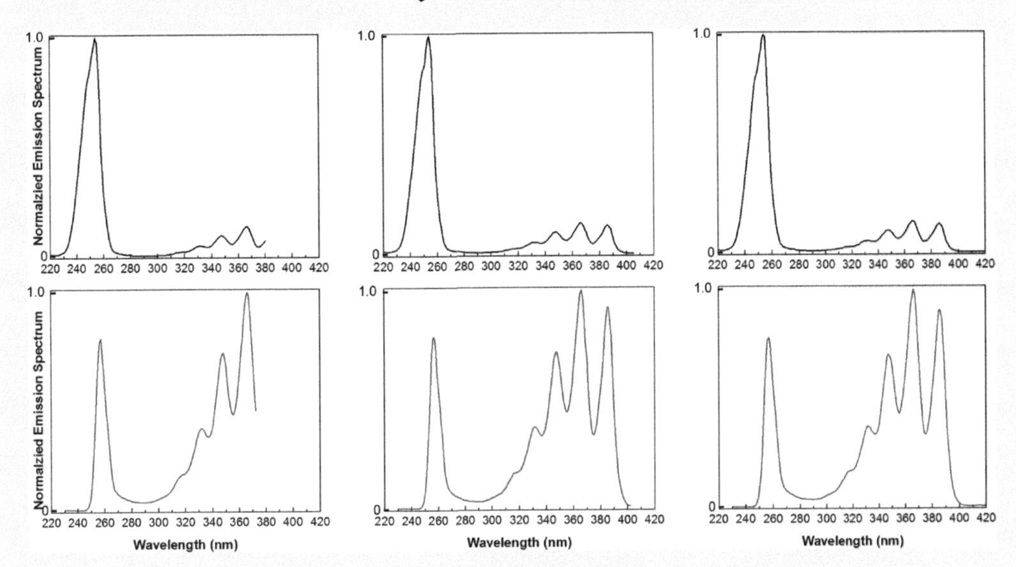

FIGURE 4-27.5 Normalized excitation spectra in quartz (top) and plastic (bottom) cuvette measured at 388 nm, 412 nm, and 435 nm observations, respectively.

87% and the plastic cuvette only about 8% for light traveling through two walls. So, excitation light attenuation by a single quartz wall would be about 6% (94% transmittance) and for single plastic cuvette wall about 84% (16% transmittance). In effect relative change in blue and UV excitation bands should be about 6 to 7-folds that is in good agreement with presented values in Figure 4-27.2. The emission attenuation is minimal for both cuvettes (transmittances above 350 nm are high and similar).

CHAPTER 5

Steady-State Fluorescence

Applications

INTRODUCTION

Practical applications are the most important drivers for fluorescence technology development. Many applications of fluorescence are based on fluorescence signal change resulting from fluorophores interacting with the surrounding environment. Various molecular interactions can result in a change in the fluorescence. The signal change is typically an increase or decrease in fluorescence intensity, or sometimes also a shift in the emission peak position. A process that decreases observed fluorescence intensity (quantum yield of the fluorophore) is called fluorescence quenching. There are many processes that lead to a decrease in observed fluorescence intensity. Most common are: collisional quenching (process when the excited fluorophore collides with another molecule and dissipates access of excitation energy in a non-radiative way), excited-state reaction (a process where excitation changes physicochemical properties of the fluorophore enabling chemical reaction like proton transfer), molecular rearrangements (a process where excited molecules undergo conformational change), ground state complex formation (where the fluorophore forms the complex with other molecule and changes its properties), and energy transfer (FRET—a long-range radiationless excitation energy transfer to a suitable acceptor). The opposite process, an increase of intensity, is observed when an interaction with the environment leads to a decrease in quenching. Typical examples are the binding of 8-anilino-1-naphthalenesulfonic acid (ANS) to the protein or ethidium bromide (EtBr)-binding (intercalating) to DNA. Both ANS and EtBr are heavily quenched in a water environment and weakly fluorescent. However, upon binding the fluorophore becomes shielded from the water environment increasing its fluorescence brightness.

Many practical biomedical applications are measurements of biological specimens that are optically dense, such as blood or tissue presenting very high absorption and scattering. Working with such samples leads to many perturbations of the fluorescence signal. These include fluorophore quenching by physiological components, background absorption like that of hemoglobin or tissue, scattering, or just emission from physiological components that add to the fluorophore signal. High absorption and scattering of hemoglobin and tissue can lead to an apparent signal decrease of the sample that is not due to the fluorophore interacting with the so-called target. The apparent decrease in fluorescence due to the absorption of emitted light by absorbing background or attenuation of excitation by highly absorbing/scattering background are undesirable perturbations.

In this chapter, we will present a set of steady-state experiments in which the fluorescence intensity of a fluorophore increases/decreases as a result of interactions with other molecules. These experiments are structured into three general groups. First, we will present experimental examples where the observed fluorescence intensity changes due to "trivial" effects like reabsorption of emitted light or absorption of excitation light. It is important to realize that the ignoring and not correcting for these effects may lead to an over-interpretation of experimental results. Next, we will present experiments for studying fluorophores binding to protein or deoxyribose nucleic acid (DNA). The general experimental concept will be explained, but it is important to remember that these processes are at the base of many DNA sequencing and forensic applications (detecting proteins). Finally, this chapter will conclude with a variety of quenching experiments, mostly resulting from collisional encounters between the fluorophore and quenchers. We will also present some discussion on how to interpret quenching measurements. Additionally, experimental examples of static quenching will be presented, which can be a valuable source of information about binding between the fluorescent probe and the quencher.

Experimental Configurations

Most of the experiments are done in a square configuration (square geometry). In this case, the excitation line and emission line form a precise 90° angle. Because the excitation light and fluorescence travel substantial distances in the cuvette, the low-optical density of samples is required, below 0.1. These are very severe limitations considering the practical applications of fluorescence-based detection. For example, blood or just hemoglobin has an optical density in the order of hundreds per 1 cm thickness, and using square geometry will not be possible. One of the first assignments for new students is to measure the fluorescence of indole in water. A typical response is that indole has no fluorescence. Indole is a highly fluorescent compound, well soluble in water. It is easy to make too high concentration, which will have very high absorption and will present fluorescence only on the surface that is blocked by the sample holder. In Experiments 4-16 and 4-17, we have shown that already absorption above 1 is a huge problem even when using a 1 mm path cuvette. For such highly absorbing and scattering samples (we call optically dense), a front-face configuration should be used. Also, many solid samples have shape and size making a square geometry unusable, and the only way to see fluorescence is to detect the signal from the surface. Another different group of examples are thin layer samples. For example, when studying membranes, a sample will typically be deposited on a surface (substrate). Most of the detection assays are surface-confined (antibodies or DNAs immobilized on the surface). Another important group is fluorophores embedded in polymer films. Such thin polymer samples are frequently used as sensors. But detection should again be in a front-face configuration.

In Figure 5.1 (top), we schematically present a concept for front-face configuration. This subject was already discussed in Chapter 1. Shortly, the excitation comes from the left under an angle α and is either p-polarized (electric field in the plane of incident) or s-polarized (electric field perpendicular to the plane of incident). It is important to realize that as light enters the sample that has a higher refractive index it will refract. Consequently, the p-polarization direction will change. The s-polarized light will not change the direction of its polarization. So, it will be convenient to use s-polarized light for excitation since its polarization will not be affected by the light refraction, and excitation light will preserve the direction of polarization. We observe fluorescence under the angle β and again the light escaping the sample will also refract. But for the emission both s- and p-polarizations (for anisotropy measurements) need to be observed. The main problem arises from the reflectivity on the interface. The reflectivity on the dielectric interface depends on the angle and light polarization. So, s- and p-polarized light beams are reflected differently. Reflectivity curves for water–air interface for s- and p-polarized light are presented in the middle of Figure 5.1. For angles close to 0° (orthogonal to the dielectric surface), both polarizations are equally reflected. As the angle increases, the reflectivity of s- and p-polarized light changes. For angles greater than 20°, the difference becomes significant, especially for angles close to the Brewster angle where p-polarized light is not reflected. The ratio of reflectivity is presented at the bottom of Figure 5.1. This will very strongly affect the polarization of observed fluorescence. Only for an angle 0°, both s- and p-polarized components remain unperturbed (both components will only be slightly but equally attenuated by reflection). Only this (0° to the normal) will constitute proper observation for fluorescence.

Front-Face Adapters

In Figure 5.2, we are presenting a schematic concept for sample compartment adapted to front-face experiments. The excitation light enters from the top, is reflected by two mirrors and excites the sample surface under the angle. The surface of the sample is orthogonal to the observation direction. The excitation light is vertically polarized (orthogonal to the figure surface). For such observation, the magic angle is satisfied by an observation polarizer oriented under 54.7° (as in square geometry). The sample stage allows the sample to move along the observation direction. The sample holder typically will allow holding a flat sample. An example of the demountable cuvette is shown in the left corner. Such demountable cuvettes may have an optical path from 100 μm up to 1 mm. They require a minimal amount of sample volume and are very easy to clean. Another example can be a microscope slide with a sample deposited on the surface or a polymer film.

In Figure 5.3, we are presenting a photograph of the front-face adapter used in Varian Eclipse spectrofluorometer. On the left shown is just a stage (platform) and on the right stage mounted in the spectrofluorometer.

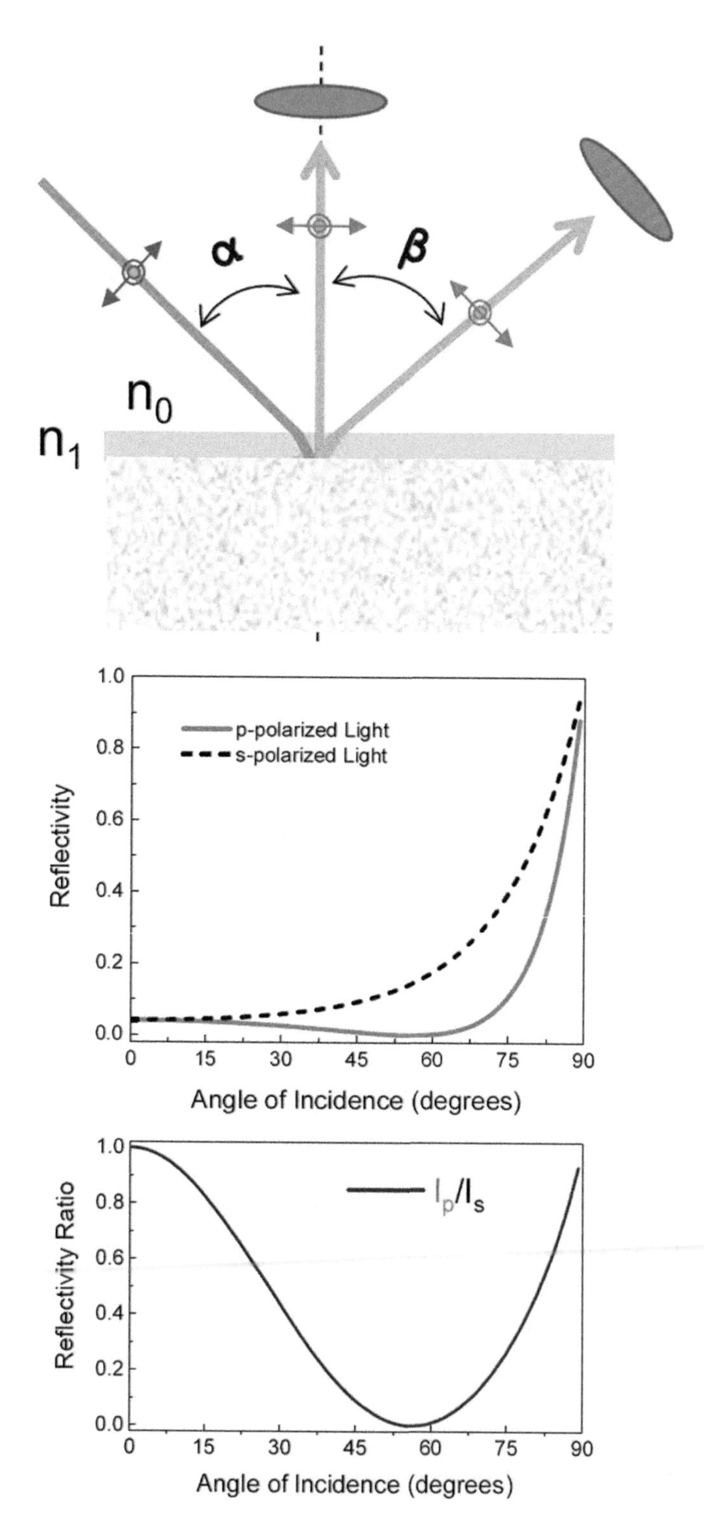

FIGURE 5.1 Concept for front-face configuration with different observation angles. Bottom part shows reflectivity curves for orthogonal and parallel component to the excitation plane polarization of emitted light as a function of observation angle, β.

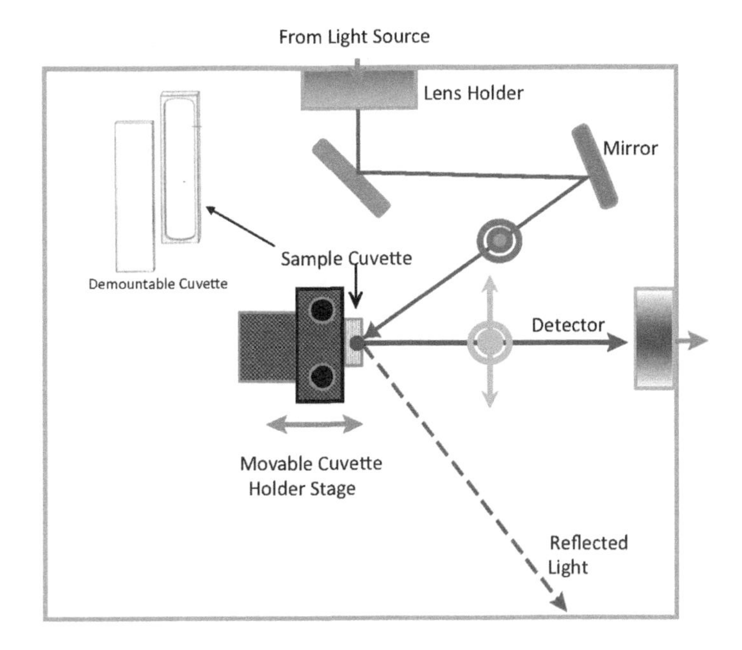

FIGURE 5.2 Schematic representation of front-face configuration in the spectrofluorometer chamber.

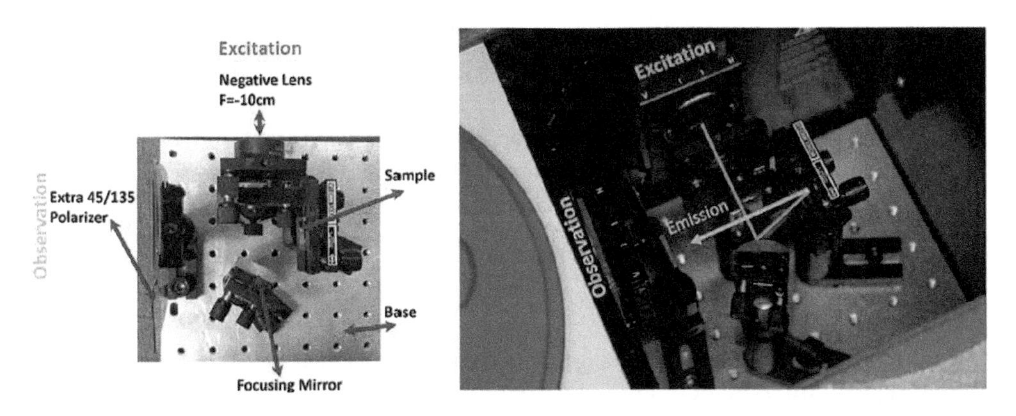

FIGURE 5.3 Photograph of front-face setup for Varian Eclipse spectrofluorometer. On the left, a stage platform; and on the right, a stage mounted in the spectrofluorometer.

In Figure 5.4, we present a photograph for front-face configuration for Picoquant FT300 and ISS K2 systems designed and built by us. Typically, there are multiple ways of adopting a spectrofluorometer chamber for front-face measurements. The important issue is that the surface of the sample must be orthogonal to the direction of observation and position of the emitting spot that has to be very precisely adjusted in the focal point of the observation lens.

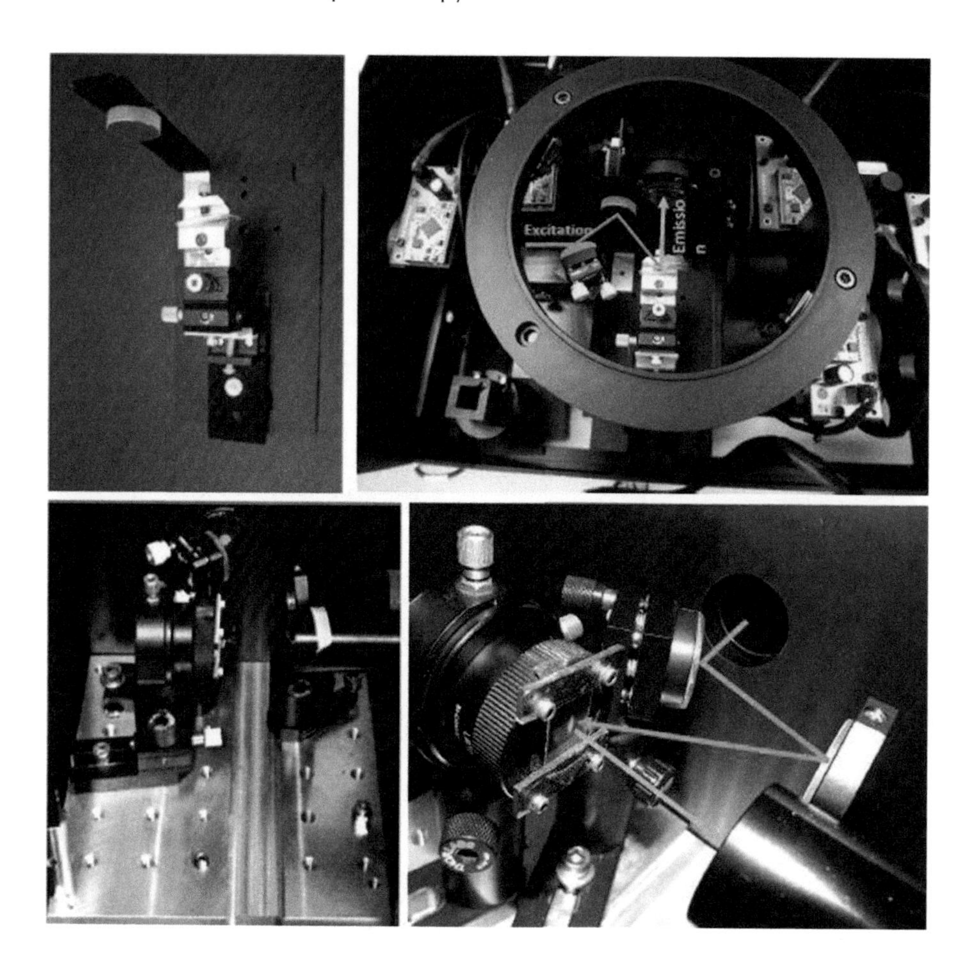

FIGURE 5.4 Photograph of front-face setups for Picoquant FT300 (top) and ISS K2 (bottom) spectrofluorometer. On the left are presented stage platforms and on the right are photographs of mounted stages in both spectrofluorometer chambers.

EXPERIMENT 5-1 CORRECTION FOR ABSORBING SPECIES IN THE FLUOROPHORE EMISSION REGION

Keywords: fluorescence, inner filter, spectrofluorometer.

Often fluorescence measurements are performed with samples that present highly absorptive background at the fluorophore emission range. Simply speaking an intrinsic sample absorption overlaps with the emission spectrum of a fluorophore (probe). These include measurements in cells and tissue and those done in the presence of blood or cellular components. Also, often we have to deal with solutions of chromophores in solvents or buffers that absorb in the fluorophore mission wavelength range. The fluorescence emitted by a given fluorophore must pass through a layer of the sample (buffer) that will significantly absorb emission

(Continued)

EXPERIMENT 5-1 (CONTINUED) CORRECTION FOR ABSORBING SPECIES IN THE FLUOROPHORE EMISSION REGION

light. For example, fluorescence light originating in the center of a 1 cm × 1 cm cuvette travels an average of 0.5 cm through the sample layer. This, of course, will attenuate and typically deform the fluorescence spectrum. *How to correct for this effect?* The following experiment demonstrates a simple procedure for correcting measurements completed under conditions in which a medium absorbs light in the emission range.

MATERIALS

- 7-Amino-4-methylcoumarin (C120)
- 3,3′-Diethyloxacarbocyanine iodide (DOCI)
- Ethanol, glass vials
- 1 cm × 1 cm cuvettes, 0.5 cm × 1 cm cuvette

EQUIPMENT

- Spectrophotometer (Cary 50, Varian, Inc.)
- Spectrofluorometer (Eclipse, Varian, Inc.)

METHODS

1. Make stock solutions of C120 and DOCI in ethanol, sonicate.
2. Prepare about 10 mL of C120 with an optical density (OD) of about 0.2 at 350 nm and DOCI with OD of about 2 at 490 nm.
3. Prepare samples: 2 mL C120 + 2 mL water, 2 mL C120 + 2 mL DOCI, and 2 mL water + 2 mL DOCI. These samples are: fluorophore only, a fluorophore with absorbing background in the emission, and absorbing background control, respectively. Measure the absorption spectra of the samples (see Figure 5-1.1).
4. Measure the fluorescence of the samples at 350 nm excitation, see Figure 5-1.2.
5. Measure a transmission spectrum of the fluorophore with absorbing background in the 0.5 cm × 1 cm cuvette (0.5 cm pathway), see Figure 5-1.3. Alternatively, recalculate the transmission from the absorption in Figure 5-1.1.
6. Correct fluorescence for absorbing background (divide fluorescence emission spectrum by transmission in the 0.5 cm pathway, see Figure 5-1.2).

RESULTS

Absorption spectra of C120, DOCI, and mixture C120 + DOCI are presented in Figure 5-1.1. In the figure are also presented the chemical structures of C120 and DOCI. Note that in the maximum of fluorophore (C120) excitation (~350 nm) absorption of DOCI is minimal and in this example, we will not have to worry about absorption of the absorber (this we will discuss in next experiments). Because of a minimal excitation, the emission of DOCI is very weak and will not contribute to the emission of C120. This was a very careful selection to demonstrate only the correction for the fluorophore (C120) emission absorption.

(*Continued*)

EXPERIMENT 5-1 (CONTINUED) CORRECTION FOR ABSORBING SPECIES IN THE FLUOROPHORE EMISSION REGION

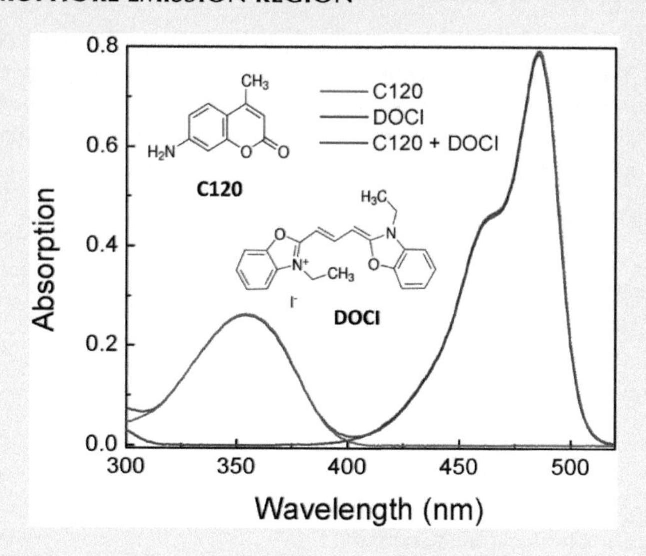

FIGURE 5-1.1 Absorption spectra of C120, DOCI, and a mixture of C120 + DOCI in ethanol.

In Figure 5-1.2, we are presenting emission spectra of C120, DOCI, and a mixture of C120 + DOCI measured with 350 nm excitation. In addition, we also present a corrected emission of the mixture. The fluorescence emission of C120 (orange) is represented by a clear, bell-shaped spectrum. Emission of DOCI only is negligible, just a baseline (blue). Measured emission of a mixture shows a very visible "valley" near 480 nm and in overall is significantly deformed from the emission of C120 only.

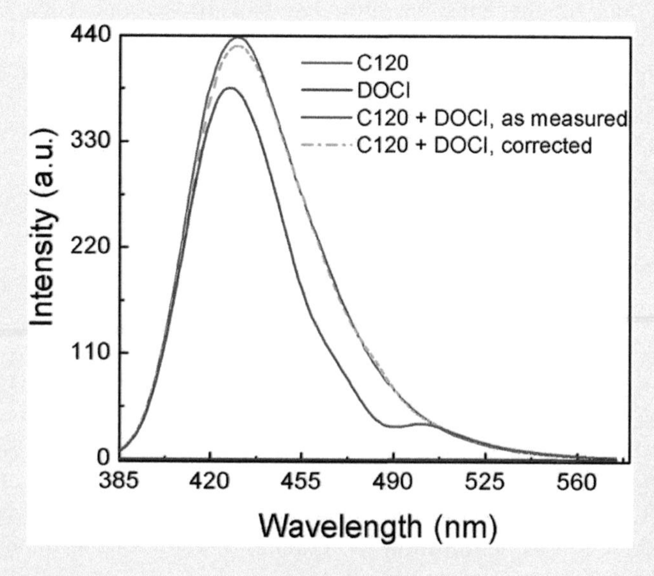

FIGURE 5-1.2 Fluorescence spectra of C120, DOCI, and a mixture of C120 + DOCI in ethanol at 350 nm excitation.

(Continued)

EXPERIMENT 5-1 (CONTINUED) CORRECTION FOR ABSORBING SPECIES IN THE FLUOROPHORE EMISSION REGION

Measured absorption spectra in Figure 5-1.1 show that absorption of DOCI has a clear maximum at about 480 nm. So, the absorption by this compound is quite significant in the C120 emission range. The absorption maximum about 0.8 corresponds to slightly more than 15% transmittance and even 0.4 absorbance means only 40% transmittance. Such light attenuation is significant and results in a substantial emission spectra distortion. We can do the correction in two ways. First, knowing absorption (Figure 5-1.1) we can calculate expected transmittance. However, we need to remember that fluorescence originates in the center of the cuvette and in average, emission light needs to travel only through half of the cuvette width (5 mm in our case—see explanation *Note* at the end). Another approach can be to measure the transmittance of the mix solution directly using 5 mm path cuvette. In Figure 5-1.3, we are presenting transmittance measured for C120 + DOCI solution in a 0.5 cm cuvette (solid red) and as recalculated from measured absorption spectrum using the Beer–Lambert law ($T = 10^{-A/2}$), Figure 5-1.1 (dashed blue). Both corrections are practically identical. The corrected C120 spectrum can be obtained by dividing the measured C120 + DOCI spectrum by the correction profile (transmittance).

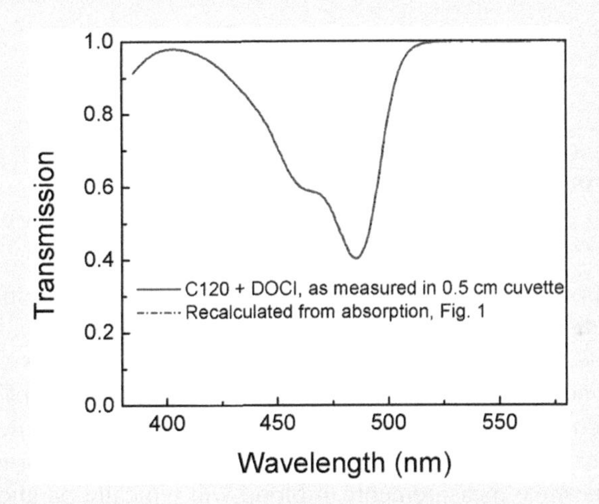

FIGURE 5-1.3 Transmission of the mixture C120 + DOCI measured in 0.5 cm pathway cuvette and recalculated transmission from the absorption measured in Figure 5-1.1.

Note: The calculated correction assumes that the emission is due to a point excitation in the center of the cuvette (so the emission light will path exactly 0.5 cm distance in the solution before exiting the cuvette. Figure 5-1.4a shows schematically the side and top views of the cuvette with the marked excitation beam and observation direction. Such a case is possible with laser excitation but many spectrofluorometers use the lamp for excitation and will not have point excitation. In fact, Varian Eclipse has quite a broad elongated beam as schematically shown in Figure 5-1.4b. In this case, parts of fluorescence light travel shorter distance (fluorophores are closer to the right cuvette wall) and parts longer distance before exiting the cuvette. To calculate the measured fluorescence light intensity, we will need to average through all distances (for all emitters within the excitation beam). For uniform light intensity distribution within the beam (uniform excitation spot), the measured emission intensity attenuation due to absorption will be equivalent to that as whole light would be emitted from the central point as in the case A.

(Continued)

EXPERIMENT 5-1 (CONTINUED) CORRECTION FOR ABSORBING SPECIES IN THE FLUOROPHORE EMISSION REGION

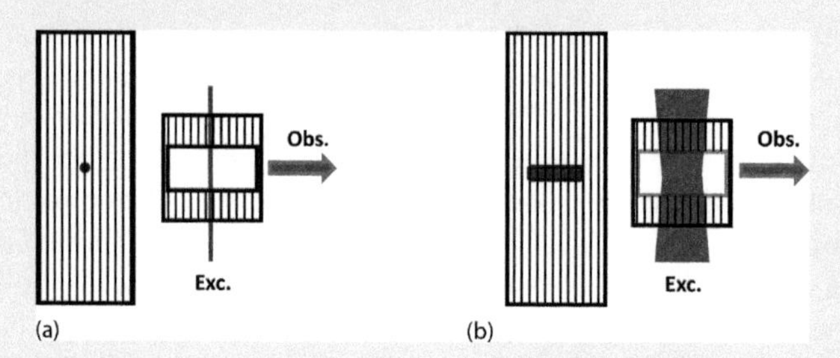

FIGURE 5-1.4 Schematics of side and top views of 0.5 cm cuvette with marked excitation spots for point excitation (a); and large spot excitation (b).

We can avoid the need for the correction procedure by using a small cuvette (e.g., 1 mm × 1 mm) that is typically difficult, or by using a front-face geometry and 1 mm thin flat cuvette.

EXPERIMENT 5-2 CORRECTION FOR ABSORBING SPECIES IN THE EXCITATION REGION

Keywords: fluorescence, inner filter, spectrofluorometer.

In the previous Experiment 5-1, the effect of an absorbing background in the emission spectral range was discussed. Even more common (but less recognized) is the sample (buffer) absorption at the fluorophore excitation range. In many fluorescence experiments, an absorbing background at the excitation range may significantly corrupt results. This effect is much more difficult to recognize and is often ignored. Organic solvents or biological samples may have their natural absorption within the excitation range used for fluorophores. For example, fluorescence measurements in blood will typically be affected by hemoglobin absorption. In solutions with multiple chromophores, it is often difficult to selectively excite a single fluorophore. For example, in FRET measurements, acceptor molecules always absorb excitation light (at the donor excitation wavelength) and are directly excited. Another example would be quenching of UV fluorophores (e.g., tryptophan) by acrylamide, which presents a significant absorption below 290 nm. Because part of the excitation light is absorbed by the background, fewer photons are being absorbed by fluorophores. If the optical density of the background is high, the excitation of fluorophores is weaker than expected and the fluorescence signal of interest decreases. This can be very easily misinterpreted as a fluorophore quenching. Also, the highly absorptive background will significantly change excitation beam intensity as it propagates through the sample, making the fluorescence signal much higher at the beginning (entry side of the cuvette) than at the exit side. This will shift the center of gravity for emission, which may have a significant additional effect on observed signal (see Experiments 4-4 and 4-5). *How to correct for this effect?* The following experiment demonstrates the correction procedure required when measurements are made in the presence of a background which absorbs light in the excitation range.

(*Continued*)

EXPERIMENT 5-2 (CONTINUED) CORRECTION FOR ABSORBING SPECIES IN THE EXCITATION REGION

MATERIALS

- 4-Dicyanmethylene-2-methyl-6-(p-dimethylaminostyryl)-4H-pyran, (DCM); trisodium (4*E*)-5-oxo-1-(4-sulfonatophenyl)-4-[(4-sulfonatophenyl)hydrazono]-3-pyrazolecarboxylate; aka tartrazine (T)
- Ethanol
- Glass vials
- 1 cm × 1 cm cuvettes, 0.4 cm × 1 cm cuvettes, 0.2 cm × 1 cm cuvettes

EQUIPMENT

- Spectrophotometer
- Spectrofluorometer

METHODS

1. Make stock solutions of DCM and T in ethanol, sonicate.
2. Prepare about 10 mL of DCM with optical density (OD) of no more than 0.2 at 470 nm and T with OD of about 2 at 425 nm.
3. Prepare samples: 2 mL DCM + 2 mL ethanol, 2 mL DCM + 2 mL T, 2 mL ethanol + 2 mL T. These samples are: fluorophore only, the fluorophore with absorbing background in the excitation region, and absorbing background control, respectively. Measure the absorptions of the samples (see Figure 5-2.4).
4. Measure the fluorescence of the samples at 425 nm excitation, see Figure 5-2.5.
5. Correct emission spectra for the absorption of T. We will present a simplified approach to account for media absorption.

RESULTS

In the introductory Chapters 1 and 2, we presented the arguments proving that absorption is an additive property of the sample (examples of absorption additivity are presented in Experiment 3-21). So, if we have our fluorophore of interest mixed with another absorber that contributes to the beam attenuation we should account for light intensity attenuation due to the absorber (we assume absorber and fluorophore do not interact).

First, we will theoretically consider absorption and emission in the system—fluorophore + absorber. Let consider a sample in the cuvette that contains two chromophores (fluorophore and absorber). The overall/total absorption of the sample in the cuvette is A that is a contribution of the fluorophore (A_1) and absorber (A_2). To calculate total signal from the fluorophore embedded in the highly absorbing sample let's consider the system in Figure 5-2.1. The light of intensity I_0 enters the cuvette and is absorbed as it propagates through the cuvette. The fluorescence signal of the fluorophore is proportional to the intensity of incoming light (excitation light) and the amount of light absorbed by the fluorophore at any given depth. For a given very thin layer of the sample of thickness d_x, the amount of absorbed light dI_x is:

$$dI_x = I_x \cdot \left(1 - 10^{-A\,dx/l}\right) = I_x \cdot \left(1 - 10^{-(A_1+A_2)\,dx/l}\right) \tag{5-2.1}$$

(Continued)

EXPERIMENT 5-2 (CONTINUED) CORRECTION FOR ABSORBING SPECIES IN THE EXCITATION REGION

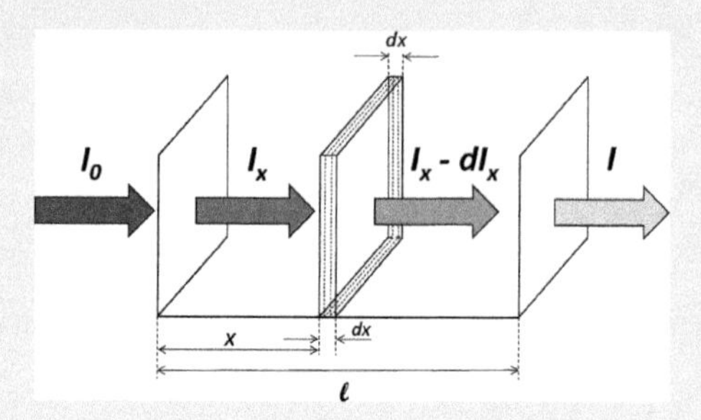

FIGURE 5-2.1 Absorption of light by a thin layer at depth x. The total thickness of the cuvette is *l*.

where $I_x = I_0 \cdot 10^{A x/l}$ is the intensity of light coming to the given layer, *dx*. The observed fluorescence signal dF_x from the layer *dx* is proportional to the amount of light absorbed by fluorophore:

$$dF_x = I_x \cdot \left(1 - 10^{-A_1 dx/l}\right) = I_0 \cdot 10^{-A x/l} \cdot \left(1 - 10^{-A_1 dx/l}\right) \quad (5\text{-}2.2)$$

The fluorescence signal from the whole cuvette (thickness *l*) can be calculated by simple integration or just for the purpose of numerical analysis, we can divide the cuvette path to *n* layers where absorption of a single layer, $\Delta A = A/n$ and respective absorptions for fluorophore and for the absorber are $\Delta A_1 = A_1/n$ and $\Delta A_2 = A_2/n$. The fluorescence signal from the *i*-th layer will be proportional to fluorophore absorption in this layer:

$$F_i = I_0 \cdot 10^{-(i-1)(\Delta A_1 + \Delta A_2)} \cdot \left(1 - 10^{-\Delta A_1}\right) \quad (5\text{-}2.3)$$

Note that the signal from each layer will depend on two factors: absorption of the fluorophore in the layer (ΔA_1) and cumulative excitation (light attenuation) up to the *i*-th layer (remember both fluorophore and absorber contribute to the cumulative attenuation). And the total observed fluorescence signal from the cuvette will be:

$$F = \sum_{i=1}^{n} F_i \quad (5\text{-}2.4)$$

The fluorescence signal will then depend on fluorophore and absorber absorptions. For a given absorption of the fluorophore (A_1), we can calculate a correction factor as the function of absorber absorption, A_2. We define the correction factor as:

$$K = \frac{F(A_2)}{F(0)} \quad (5\text{-}2.5)$$

where $F(A_2)$ is a fluorescence signal measured for a given absorber absorption, A_2 at the excitation wavelength and $F(0)$ is a fluorescence signal that should be observed with no absorber (zero absorber absorption). Figure 5-2.2 shows how the total fluorescence signal depends on the absorption of the absorber for different values of absorption of the fluorophore (emitter). To correct the measured fluorescence signal in the presence of the absorber to the signal,

(Continued)

EXPERIMENT 5-2 (CONTINUED) CORRECTION FOR ABSORBING SPECIES IN THE EXCITATION REGION

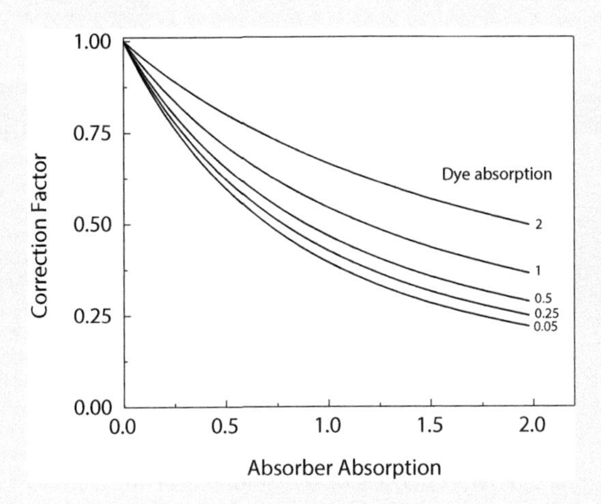

FIGURE 5-2.2 The correction factor for fluorescence signal attenuation as a function of absorber absorption. The relative fluorescence signal attenuation due to excitation light absorption by the absorber is shown for five different fluorophore absorption.

we would expect with no absorber absorption F, we would divide the emission signal at each wavelength by the correction factor K:

$$F = \frac{F(A_2)}{K} \tag{5-2.6}$$

For no absorber, the correction factor is 1 (no signal modification due to absorber absorption). As absorption of the absorber increases the factor quickly decreases and observed fluorescence signal should be corrected proportionally. Interestingly, for low fluorophore absorptions (up to 0.2), the correction factor changes very little (the signal attenuation is solely due to absorber absorption). Only when fluorophore absorption increases to the values comparable to the absorption of absorber, we observe more significant changes. Intuitively, for high absorptions of the absorber and high absorption of the fluorophore, the absorption of light should split equally between absorber and fluorophore. Indeed, for absorber absorption 2 and fluorophore absorption 2, the correction factor is close to 0.5 and the expected fluorescence intensity would drop to 50%.

We want to stress that the correction factor K is calculated only for the absorber absorption at the excitation wavelength. The correction due to the emission reabsorption by the absorber should be applied independently (see previous Experiment 5-1). The correction for emission reabsorption we will discuss in separate experiments.

To visualize the effect in Figure 5-2.3, we are presenting a photograph of a cuvette containing DCM and T excited with the blue laser (405 nm) and the green laser (532 nm). On the left is the cuvette with DCM/T mixture and on the right cuvette with only DCM of identical concentration to that in the mixture. The concentration of T was adjusted to give the absorption of about 1 (at 405 nm), what yields 10-fold excitation beam attenuation when it travels through the cuvette (intensity at the point of entry for excitation beam (right side of

(*Continued*)

EXPERIMENT 5-2 (CONTINUED) CORRECTION FOR ABSORBING SPECIES IN THE EXCITATION REGION

FIGURE 5-2.3 Photographs of DCM and DCM + T solutions illuminated with blue and green laser pointers. Green 532 nm light is out of T absorption range. Blue light is strongly attenuated by absorption of T.

the left cuvette) and at the point of excitation beam exit (left side of the left cuvette)). For the mixture excited with the blue laser, we can clearly see beam attenuation as it propagates into the solution (cuvette on the left) while for just DCM, the fluorescence is uniform along the cuvette path. However, for the green laser that is off the absorption of T both mixture and corresponding solution of the DCM only show identical signals.

Figure 5-2.4 shows the absorption spectra measured in 1 cm path cuvette for DCM, T, and mixture of DCM/T. The absorption of DCM is about 0.1 in maximum and 0.062 at 425 nm what is appropriate for fluorescence measurements. However, absorption of T is about 1 at 425 nm excitation and it is strongly overlapping with the absorption spectrum of DCM. We selected tartrazine as an absorbing medium since it does not show any significant fluorescence (we consider T as non-fluorescent compare to DCM in this experiment). In this case, we can only consider an effect of T exclusively as an absorption (attenuation) of the excitation beam.

Figures 5-2.4 through 5-2.7 show emission spectra measured for DCM, Tm and DCM/T mixture in three different cuvettes excited at 425 nm. In each figure, we also introduced the corrected curve (points) according to the presented model.

For the 2 mm cuvette, the corrected emission very well corresponds to the experimentally measured DCM only. The results are a little worse for the 4 mm cuvette and quite off for the 10 mm cuvette.

Why could we not correct a measurement that was done in a 10 mm cuvette? To answer this question, we refer the reader to Experiment 4-4 where we showed the dependence of the fluorescence signal on the position of the fluorescence spot. When looking at Figure 5-2.3 (left), we can see that the "center of gravity" for fluorescence is shifted to the right side of the cuvette (excitation beam entrance). We repeated the slit measurements with DCM solution excited at 532 nm, so the signal across the cuvette should be uniform, and measure the fluorescence signal for central slit position and position shifted about 3 mm toward the left (see Figure 5-2.8). It is clear that shifting the position results in

(Continued)

EXPERIMENT 5-2 (CONTINUED) CORRECTION FOR ABSORBING SPECIES IN THE EXCITATION REGION

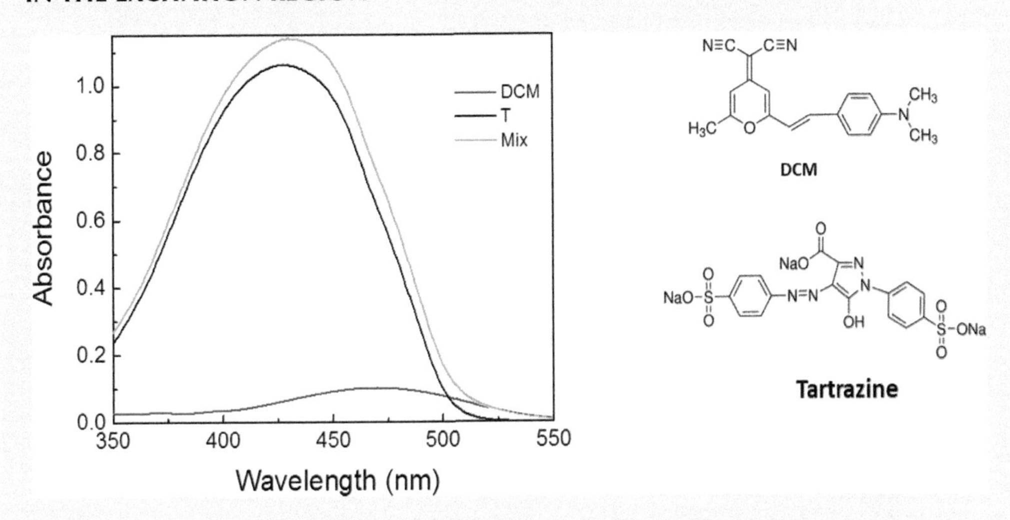

FIGURE 5-2.4 Absorption spectra of DCM, T, and mixture DCM + T in ethanol.

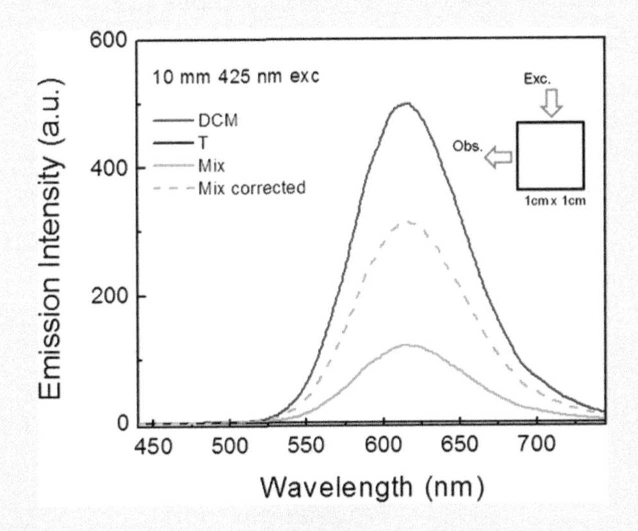

FIGURE 5-2.5 Fluorescence emission spectra of DCM, T, and mixture DCM + T in ethanol measured in a 1 cm × 1 cm cuvette. The corrected spectrum of DCM in the presence of absorber (green dashed line) was obtained using the correction factor (0.383) from Figure 5-2.2.

significant signal attenuation due to the geometry of detection. In the case of the 10 mm cuvette where most of the signal is generated at the cuvette side where the excitation beam is entering, the measured signal is artificially attenuated due to the geometry. So, a significant part of the fluorescence signal due to the geometrical factor (shift of emission center of gravity) that it is impossible to account for in a theoretical way. The maximum

(*Continued*)

EXPERIMENT 5-2 (CONTINUED) CORRECTION FOR ABSORBING SPECIES IN THE EXCITATION REGION

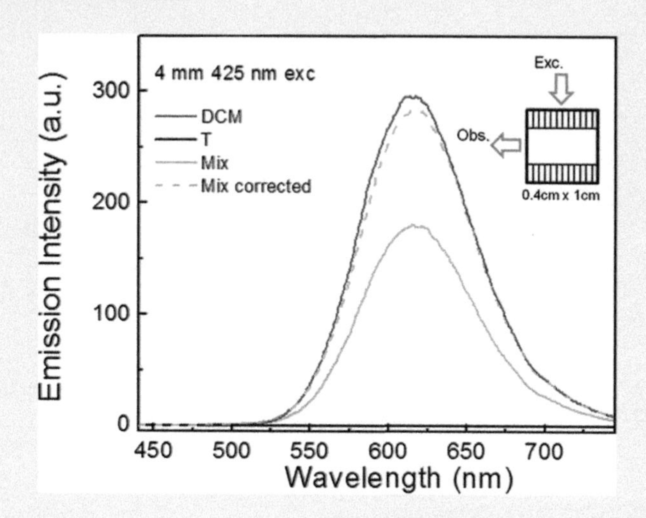

FIGURE 5-2.6 Fluorescence emission spectra of DCM, T, and mixture DCM + T in ethanol measured in a 0.4 cm × 1 cm cuvette. The corrected spectrum of DCM in the presence of absorber (green dashed line) was obtained by using the correction factor (0.636) from Figure 5-2.2.

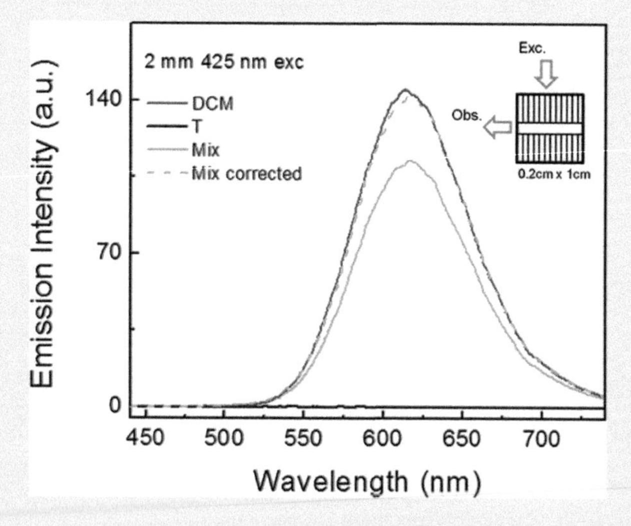

FIGURE 5-2.7 Fluorescence emission spectra of DCM, T, and mixture DCM + T in ethanol measured in a 0.2 cm × 1 cm cuvette. The corrected spectrum of DCM in the presence of absorber (green dashed line) was obtained by using the correction factor (0.792) from Figure 5-2.2.

shift could be up to 5 mm as compared to the cuvette center. For 4 mm cuvette, the maximum shift is only 2 mm and the geometrical factor is much smaller leading to the quite good corrected result. For 2 mm cuvette, the maximum shift is only 1 mm and practically all fluorescence is in the geometrical detection center. For this cuvette and any thinner one, the correction will work perfectly.

(Continued)

EXPERIMENT 5-2 (CONTINUED) CORRECTION FOR ABSORBING SPECIES IN THE EXCITATION REGION

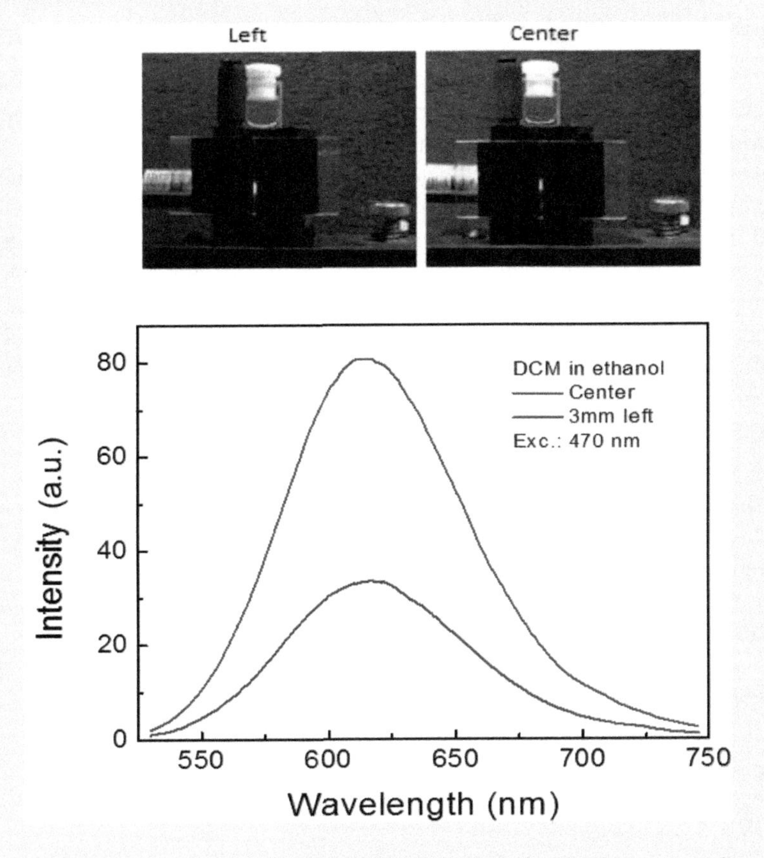

FIGURE 5-2.8 Fluorescence spectra of DCM solution observed for the slit positioned on the center and shifted by 3 mm, as shown in the photograph (top).

CONCLUSIONS

In the case of absorbing background, the measurements should be done in narrow (2 mm or smaller path) cuvettes. This is true also for just high concentrations of fluorophore that will shift fluorescence "center of gravity" to a side of the cuvette. This will be very important when studying fluorescence quenching where the absorption of the quencher can be very high, FRET in solution when the concentration of acceptor is high, and many other cases.

Just for practice, we would recommend preparing solutions of different concentrations (absorptions) of T and DCM. So, the absorption of T will be about 1 and DCM below 0.1 at 1, 0.4, and 0.2 cm cuvettes, respectively. This will result in much larger signal attenuations in 0.4 and 0.2 cm cuvettes than that presented in Figures 5-2.5 through 5-2.7. The correction factor will still work properly yielding corrected spectra almost identical to that measured for a DCM only.

EXPERIMENT 5-3 CORRECTION FOR ABSORBING SPECIES IN THE EXCITATION REGION: EXPERIMENTAL APPROACH

Keywords: fluorescence, inner filter, spectrofluorometer.

In the previous Experiment 5-2, a theoretical model was described for correcting fluorescence emission affected by background absorption at the excitation wavelength. Although the corrections worked well for thin cuvettes, it failed in the case of a regular 1 cm × 1 cm cuvette. This discrepancy can be explained by the geometrical change in the detection in the presence of an absorber, because the fluorescence spot appears at the edge of the cuvette, while in the absence of the absorber entire width (1 cm) of the cuvette is illuminated (see Figure 5-2.3). The theory does not include this difference (does not account for geometrical factors). However, one can empirically test and correct observed fluorescence in the presence of the absorbing background. The experimentally evaluated correction factor will include the signal distortion due to both geometrical factor and competitive photons absorption by the background.

MATERIALS

- 4-Dicyanmethylene-2-methyl-6-(p-dimethylaminostyryl)-4H-pyran, (DCM); trisodium (4*E*)-5-oxo-1-(4-sulfonatophenyl)-4-[(4-sulfonatophenyl)hydrazono]-3-pyrazolecarboxylate; aka tartrazine (T)
- Ethanol, glass vials
- 1 cm × 1 cm cuvettes
- Pipettes

EQUIPMENT

- Spectrophotometer
- Spectrofluorometer

METHODS

1. Make a few milliliters of stock solutions of DCM (with absorption about 1) and T in ethanol (absorption of ~2), sonicate.
2. Prepare a series of absorber solutions (3 mL) with different absorptions, from 0 (just ethanol) up to about 1.5 at 425 nm, by mixing the stock solution of T with ethanol. Measure the absorptions of all prepared samples, see Figure 5-3.1.
3. Add 30 μL of the stock solution of DCM to all samples. The sample without T will be the reference with the absorption of about 0.1 at the maximum and 0.06 at 425 nm, see Figure 5-3.1. The same amount of DCM should be in all samples.
4. Measure fluorescence of the samples at 425 nm excitation, see Figures 5-3.2.
5. Divide each spectrum by the reference spectrum (without absorber), see Figure 5-3.3. Read the values at 425 nm (excitation wavelength). These are correction factors. In order to correct the spectrum, one needs to divide the measured spectrum by the correction factor. Alternatively, one can divide the entire spectrum by the proper ratio from Figure 5-3.3.
6. Plot correction factors (use intensities at the maximum) as a function of absorber absorption, see Figure 5-3.4. This dependence is only for the 0.06 absorption of the dye at the 425 nm excitation.
7. Use a measured correction factor to correct the spectrum from the Experiment 5-2, Figure 5.2-5 (1 cm × 1 cm cuvette, absorption of T 1.06 at 425 nm), see Figure 5-3.5.

(*Continued*)

EXPERIMENT 5-3 (CONTINUED) CORRECTION FOR ABSORBING SPECIES IN THE EXCITATION REGION: EXPERIMENTAL APPROACH

RESULTS

Figures 5-3.1 and 5-3.2 show absorption and fluorescence spectra of all samples prepared. There was no detectable fluorescence from T samples (when using spectrofluorometer setting as for DCM).

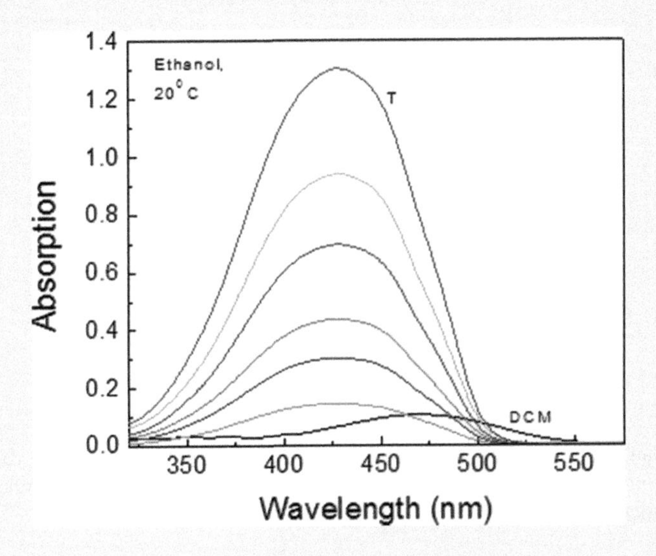

FIGURE 5-3.1 Absorption spectra of DCM and T used in the experiment. The absorption of DCM at 425 nm is 0.062.

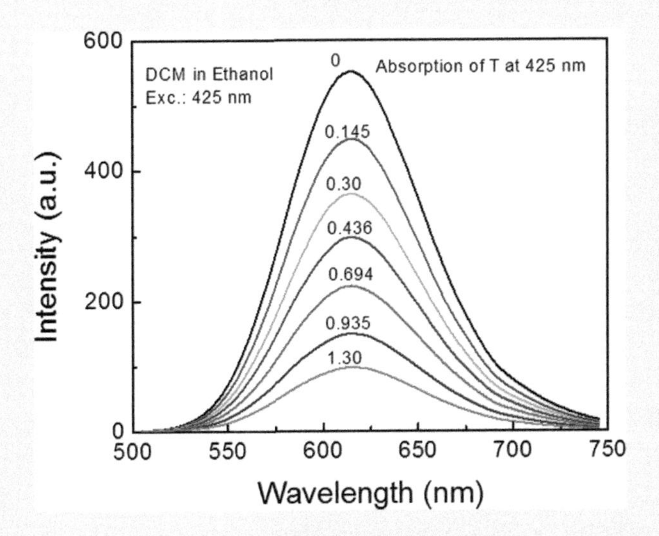

FIGURE 5-3.2 Fluorescence spectra of DCM in the absence and presence of T. The same concentration of DCM was in all samples.

(Continued)

EXPERIMENT 5-3 (CONTINUED) CORRECTION FOR ABSORBING SPECIES IN THE EXCITATION REGION: EXPERIMENTAL APPROACH

Figure 5-3.3 show the ratios of spectra. Each spectrum has been divided by the reference spectrum of DCM without T. These are horizontal lines which indicate no spectral shifts in the presence of T.

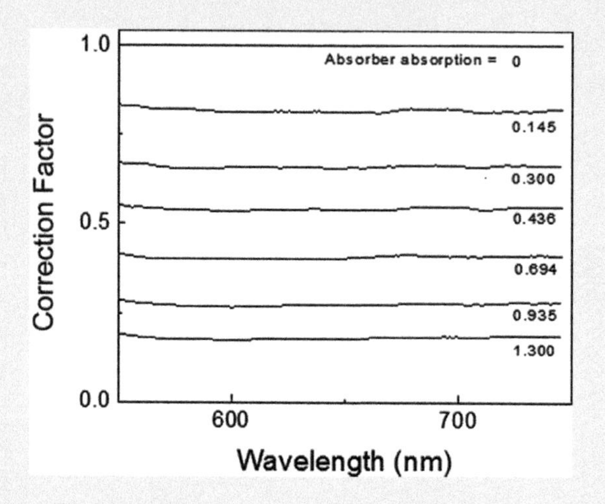

FIGURE 5-3.3 Ratios of spectra for different concentrations (absorption) of the absorber. Spectra were divided by the reference spectrum without T. The lines are fairly horizontal which indicates no spectral shifts in the presence of absorber.

Next, we read the ratio values at the 615 nm (maximum) and plot them as a function of T absorption, see Figure 5-3.4. The empirical correction factors are lower than theoretically predicted (dashed line) in Experiment 5-2.

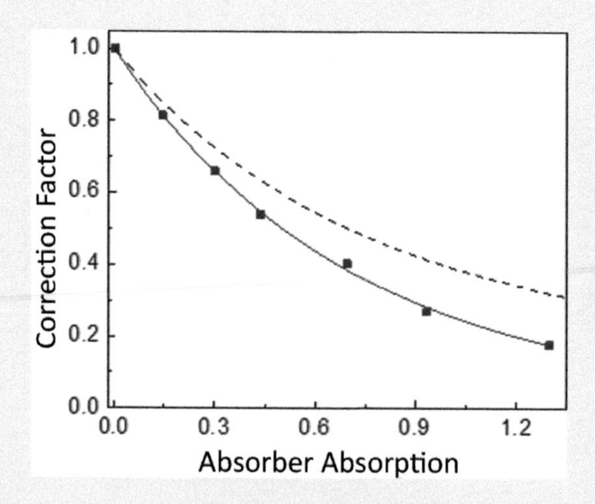

FIGURE 5-3.4 Correction factors as a function of absorber absorption. The absorption of dye was 0.062. For the absorber absorption of 1.06, the correction factor is 0.238. The dashed line is for the theoretical correction described in the Experiment 5-2.

(Continued)

EXPERIMENT 5-3 (CONTINUED) CORRECTION FOR ABSORBING SPECIES IN THE EXCITATION REGION: EXPERIMENTAL APPROACH

Figure 5-3.5 presents the spectra of DCM and mixture measured in 10 mm cuvette as shown in the Experiment 5-2. The empirically evaluated correction factor well corrects the measured spectrum, see dashed green line.

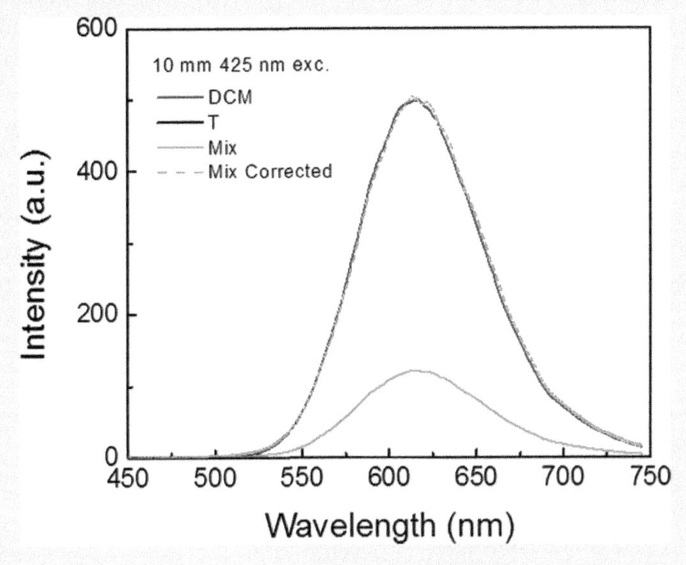

FIGURE 5-3.5 Fluorescence spectra of DCM reproduced from Experiment 5-2. The dashed line represents the DCM spectrum with T (green solid line) corrected with the experimental correction factor of 0.238 from Figure 5-3.4.

CONCLUSIONS

The empirical correction, similar to G-factor or spectral corrections, works only for a specific spectrofluorometer, cuvette, and precisely defined values of fluorophore absorption. Of course, it will work for other cuvettes from Experiment 5-2 but the correction factors have to be measured for each cuvette. In the case of 1 cm × 1 cm cuvette, it is important to check the cuvette holder if the entire 1 cm length is being observed. Some holders block a small fraction (~millimeter) from each side of the cuvette. In the case of presence of absorber, this will block a significant part of fluorescence spot.

Note: It could be valuable to measure correction factor empirically for shorter path cuvettes. The experimental correction factor perfectly agrees with the theoretical for 2 mm and smaller cuvettes (data not shown).

EXPERIMENT 5-4 CORRECTION OF FLUORESCENCE SPECTRA FOR WAVELENGTH SENSITIVITY

Keywords: fluorescence, spectra, fluorescence standards.

Fluorescence in solutions occurs from the lowest vibrational level (0) of the lowest excited state S_1 of the fluorophore. Electronic transitions to various vibrational levels of the ground state S_0 are all allowed, which results in a broad emission spectrum. Unfortunately, photodetectors and hardware optics are wavelength dependent and observed spectra are distorted from the real spectra. To overcome this problem, it is necessary to correct the measured spectra to obtain the true emission spectrum of the fluorophore. A number of dyes have been investigated and proposed as standards. Comparison of a measured spectrum with standard spectrum at each wavelength allows the determination of the wavelength-dependent correction factors. This experiment demonstrates how to create a correction curve for the spectrofluorometer using commonly accepted fluorescence standards. Four reference compounds have been selected covering visible range of the spectrum. For more details refer to: *Phys. Chem. Chem. Phys.*, 2009, 11, 9850–9860 by Kengo Suzuki et al.

MATERIALS

- 2-Aminopyridine (2-APy), quinine sulfate (QS), *N,N*-dimethylaminonitrobenzene (DMANB), 4-dimethylamino-4′-nitrostilbene (DMANS)
- 0.1N H_2SO_4, benzene, hexane, o-dichlorobenzene
- Glass vials
- 1 cm × 1 cm quartz cuvettes

EQUIPMENT

- Spectrophotometer
- Spectrofluorometer
- Cell phone or camera
- UV illuminator

METHODS

1. Make solutions of all four fluorophores in adequate solvents: 2-AP and QS in 0.1N H_2SO_4, DMANB in a mixture of benzene/hexane 3:7 (v:v), and DMANS in o-dichlorobenzene. Make photographs of observed fluorescence in vials while solutions are on UV illuminator (Figure 5-4.1). *Note:* We do not need to worry about differences in solvents refractive indexes like in case of measuring quantum yield. The profile of the emission spectrum is defined for used solvent.
2. Prepare (diluted) samples of all four fluorophores into optical densities below 0.1 and measure absorption spectra (Figure 5-4.2).
3. Prepare samples for fluorescence: 2-AP and QS in 0.1N H_2SO_4 with concentrations of 0.01 mM; DMANB in the mixture of benzene/hexane 3:7 with concentration 0.1 mM and DMANS in o-dichlorobenzene with the concentration of 1 mM.
4. Measure emission spectra (use excitation wavelengths near maxima of absorptions) and compare with the spectra of standards.
5. Calculate correction factors for each fluorophore using appropriate standard spectrum. Create a correction curve for the entire range of the spectrum.

(*Continued*)

EXPERIMENT 5-4 (CONTINUED) CORRECTION OF FLUORESCENCE SPECTRA FOR WAVELENGTH SENSITIVITY

RESULTS

Figure 5-4.1 shows all prepared stock solutions placed on a UV illuminator. Absorption spectra of diluted solutions are shown in Figure 5-4.2. Note that it is not necessary to have exactly equal absorptions. The emission spectra profiles at low concentration range should not be dependent on absorption (concentration).

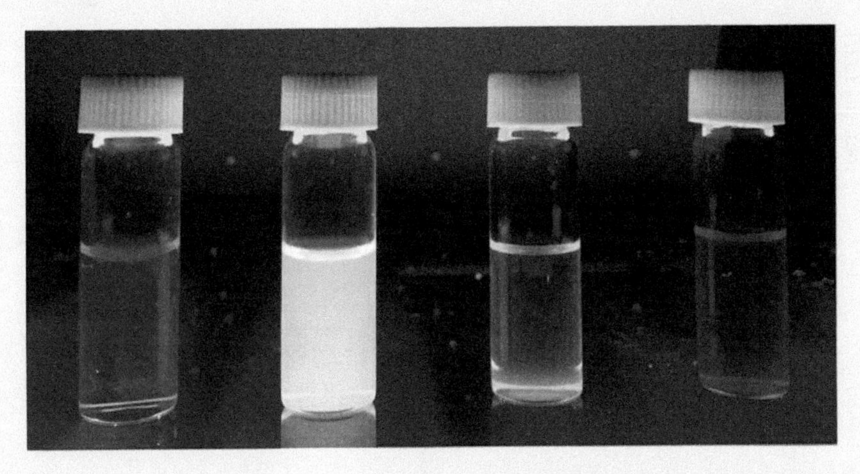

FIGURE 5-4.1 Photograph of fluorescence standards stock solutions. From left to right: 2-AP, QS, DMANB, and DMANS.

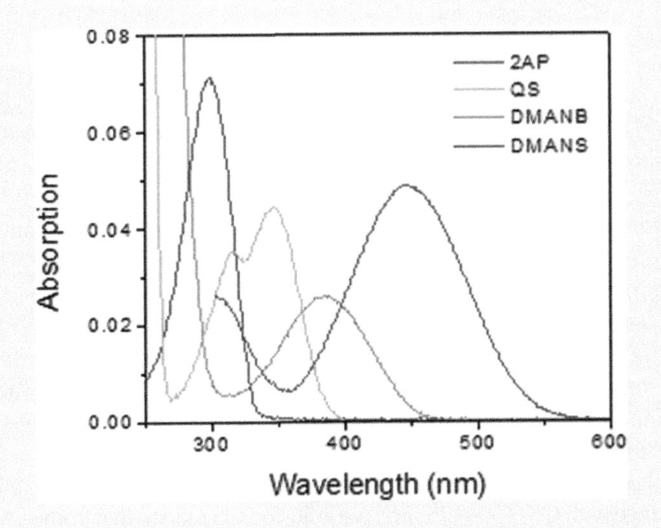

FIGURE 5-4.2 Absorption spectra of standard fluorophores: 2-AP in 0.1 N H_2SO_4, QS in 0.1 N H_2SO_4, DMANB in mixture benzene/hexane 3:7 (v: v), and DMANS in o-dichlorobenzene.

(Continued)

EXPERIMENT 5-4 (CONTINUED) CORRECTION OF FLUORESCENCE SPECTRA FOR WAVELENGTH SENSITIVITY

Next, we measure the fluorescence emission of prepared solutions and compare each with a normalized spectrum from the literature (S). In Figure 5-4.3, we are presenting measured spectra (dashed line) and corresponding standard spectra (see *Phys. Chem. Chem. Phys.*, 2009, 11, 9850–9860 by Kengo Suzuki et al. and references there).

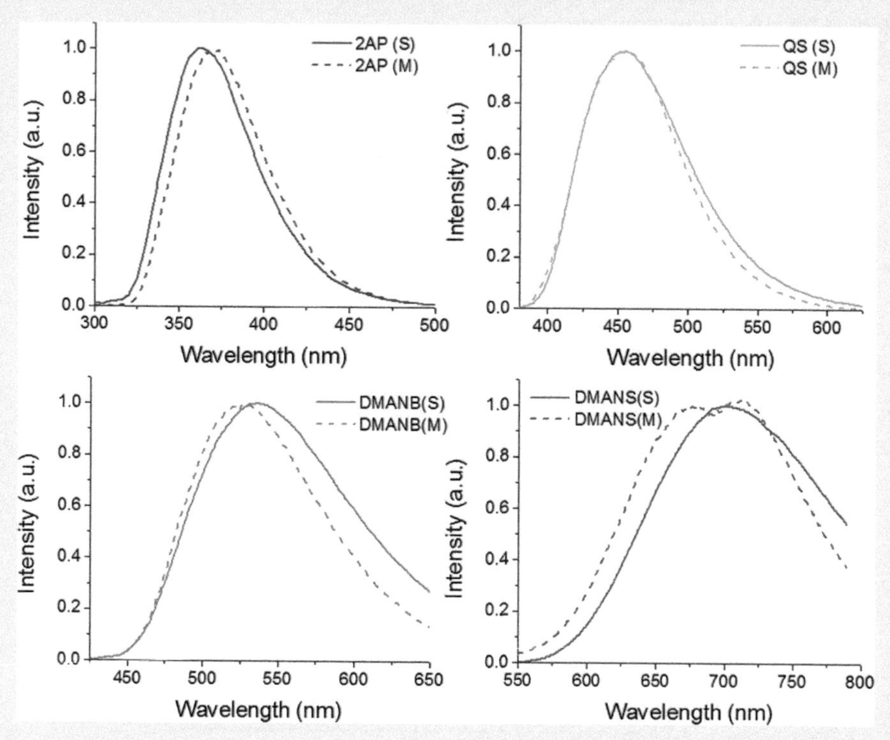

FIGURE 5-4.3 Fluorescence spectra of standard fluorophores emitting in various spectral regions. Upper left: 2-AP (0.01 mM in 0.1 N H_2SO_4), upper right: QS (0.01 mM in 0.1 N H_2SO_4, lower left: DMANB (0.1 mM in mixture benzene/hexane 3:7 v:v), lower right: DMANS (1 mM in o-dichlorobenzene). Solid lines (S) are for fluorescence standards (taken from literature) and dashed lines (M) are for measured solutions.

Clearly, measured spectra differ from standard spectra. In order to obtain the corrected intensities, the measured intensity I_M should be multiplied by a factor I_S/I_M. These ratios are presented in Figure 5-4.4.

Correction curves of individual fluorophores cover only a limited range of wavelengths. It is not a surprise that individual correction curves present different ratios. In each spectral range, these are relative values. Each individual ratio can be rescaled by any adjusting factor. When connecting ("sawing") different intensity ratios across wavelengths, we need to adjust all values to the one range.

Figure 5-4.5 shows one correction curve for a full range. We started with the ratio for 2-AP and in overlapping range with QS (400–475 nm), we adjusted QS ratio to that from 2-AP. We repeated the procedure for the overlapping range between QS and DMANB (500–530 nm) and then for DMANB and DMANS (590–620 nm). The overlapping ranges can be seen in Figure 5-4.5.

(Continued)

EXPERIMENT 5-4 (CONTINUED) CORRECTION OF FLUORESCENCE SPECTRA FOR WAVELENGTH SENSITIVITY

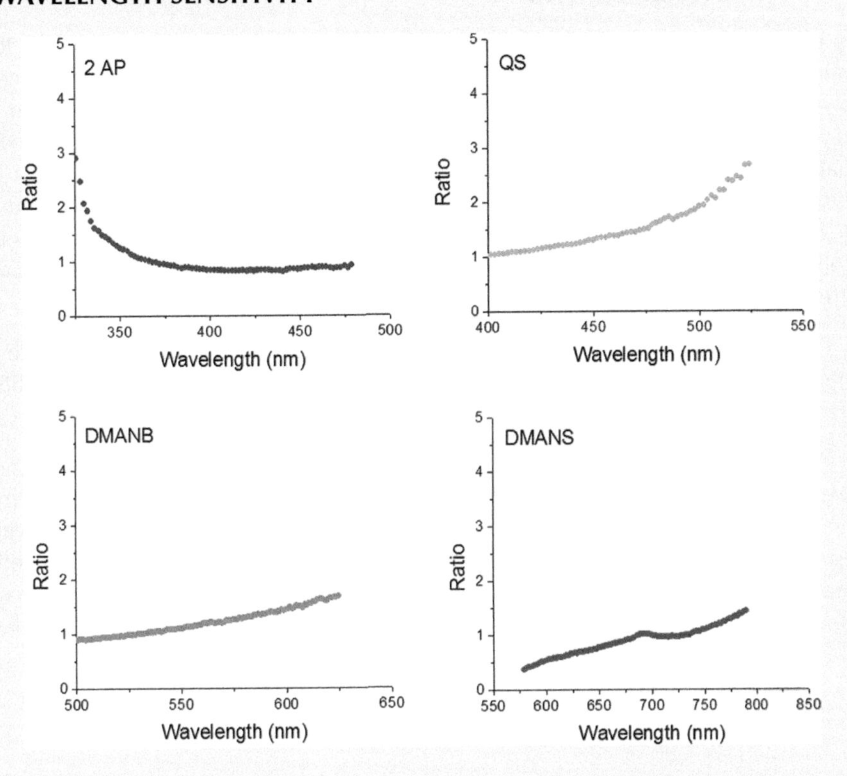

FIGURE 5-4.4 The wavelength-dependent ratios IS/IM of standards intensities to measured intensities of individual fluorophores.

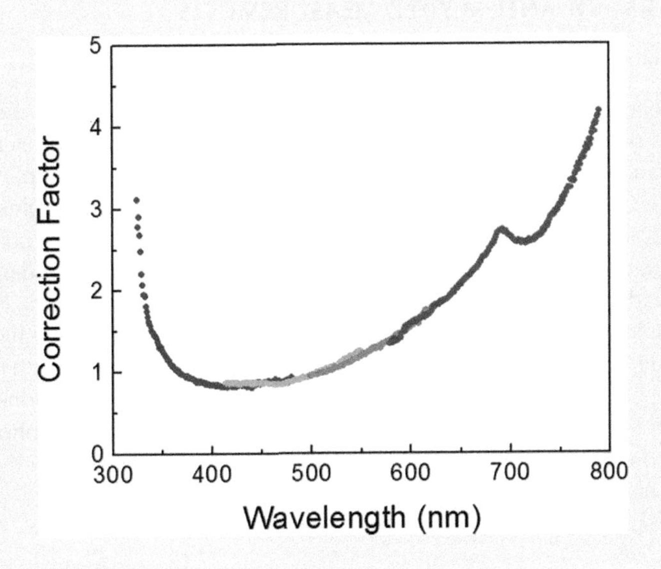

FIGURE 5-4.5 Correction curve for fluorescence spectra.

(Continued)

EXPERIMENT 5-4 (CONTINUED) CORRECTION OF FLUORESCENCE SPECTRA FOR WAVELENGTH SENSITIVITY

Any measured spectrum with this instrument (Varian, Inc.) should be corrected for the spectral sensitivity with the determined correction factors in Figure 5-4.5. In practice, you multiply the measured fluorescence intensity at a given wavelength by the correction factor for this wavelength. Between 350 and 550 nm, the correction factors are close to 1. However, in UV and red and near-infrared regions, corrections are significant.

We selected these four fluorophores for a historical reason. These dyes have been proposed as spectral standards in the 1950s and are still commonly used. There are recently many other standards proposed for various spectral ranges. However, the procedure for spectra corrections will be always the same.

Note: The corrections curve (Figure 5-4.5) applies only to the instrument used. Each spectrofluorometer has a specific photodetector and optical hardware. The "dip" around 680 nm is characteristic for the metallic dispersion grating (plasmonic effect). In old literature, it has been known as "Wood's effect."

For many recent spectrofluorometers, the correction curve could already be predetermined and integrated into the software. In this case, the measured and standard spectra will well overlap. But we still recommend checking the correction curve occasionally. With the instrument being used over time many factors changes and the correction factors may also change.

EXPERIMENT 5-5 QUANTUM YIELD MEASUREMENTS

Keywords: fluorescence, quantum yield, Rhodamine 6G, Rhodamine 101.

Quantum efficiency or quantum yield (QY) has a straightforward definition, the ratio of emitted to absorbed photons. However, this is one of the most difficult parameters to measure. Absolute QYs can be measured with integrated spheres which require complex instrumentation. To estimate QY of unknown/new fluorophore quantum yield, most commonly, we will use standards. Comparing the fluorescence signal from the standard and unknown fluorophore, the unknown QY can be determined. The most established QY standards are quinine sulfate (QS) in acidic water (0.1N H_2SO_4) and Rhodamine 6G (R6G) in ethanol. In several independent studies, it was established that R6G in ethanol at room temperature has a QY = 0.95 and QS in 0.1N H_2SO_4 has a QY = 0.54. When measuring QY it is important to remember that the emission spectrum represents energy (light intensity) distribution as a function of wavelength (or wavenumber). So, to recalculate this to the number of emitted photons, one will need to take into consideration the exact spectral range for the emission. In this experiment, QY measurements will be made using R6G in ethanol as a reference. It will also be shown how the lack of spectral correction alters the recovered QY value.

(Continued)

EXPERIMENT 5-5 (CONTINUED) QUANTUM YIELD MEASUREMENTS

MATERIALS

- Rhodamine 6G (R6G), Rhodamine 101 (R101)
- Ethanol
- Glass vials
- 1 cm × 1 cm cuvettes

EQUIPMENT

- Spectrophotometer
- Spectrofluorometer

METHODS

1. Prepare stock solutions of R6G and R101 in ethanol.
2. Prepare a reference (R6G) and sample (R101) solutions with ODs of below 0.1.
3. Measure absorptions of the reference and sample, see Figure 5-5.1.
4. Measure fluorescence spectra of the reference and sample using 544 nm excitation wavelength (intersection of both spectra where ODs are the same) and next using 531 nm excitation where absorptions are different, see Figures 5-5.2 and 5-5.3.
5. Correct fluorescence spectra for the spectral sensitivity of the instrument (see Experiment 5-4); see Figures 5-5.4 and 5-5.5) and calculate the quantum yield of R101 with R6G reference (*R*). Use the following equation:

$$Q = Q_R \frac{I}{I_R} \frac{\left(1-10^{-OD_R}\right)}{\left(1-10^{-OD}\right)} \frac{n^2}{n_R^2}$$

   where Q is the quantum yield, Q_R is the quantum yield of reference, I is the total integrated intensity of chromophore fluorescence, I_R is the total integrated intensity of the reference fluorescence, n is the refractive index of the solvent in which fluorophore is dissolved, n_R is the refractive index of the solvent in which the reference is dissolved, and OD and OD_R are the optical densities of sample and reference, respectively.

We refer readers to Section "2-6 Measuring Emission (p. 45) for a detailed derivation of the quantum yield equation. There are a few important factors that the experimentalist will need to be careful when evaluating quantum yield of unknown/new compound. (1) It is important to realize that the amount of absorbed light (absorbed photons) is proportional to the factor $(1–10^{-OD})$ not just absorption (OD). The factor $(1–10^{-OD})$ describes the fraction of absorbed excitation light, see Experiment 5-2. (2) A small difference in refractive indexes will play a significant role. For example, when using the reference in ethanol that has refractive index of 1.36 and sample in propylene glycol that has refractive index of 1.44, the error is quite large about 12%. And for very minor difference between water (1.33) and ethanol, it is almost 5%. So, using the same solvent helps significantly. (3) To calculate the total number of photons emitted in a given spectral range, we have to be careful how to integrate emission spectra

(Continued)

EXPERIMENT 5-5 (CONTINUED) QUANTUM YIELD MEASUREMENTS

(integrated intensities). For a system with grating monochromator (majority of commercial systems), the wavelength scale is linear and the set bandpass $\Delta\lambda$ is constant across all wavelengths. So, after correcting for detection system wavelength sensitivity (Experiment 5-4), the integral should be calculated with the wavelength scale. For a detailed explanation, we refer the reader to earlier Chapters 1 and 2, and the excellent explanation/derivation can be found in "Molecular Fluorescence Principles and Applications" B. Valeur (2001, Wiley-VCH Verlag, GmbH).

RESULTS

From years of experience, we know that many students and researchers use a "shortcut" while correcting fluorescence spectra for absorption by simply dividing the fluorescence spectrum by measured absorption (OD). This is incorrect, even when absorptions are low. The correct calculation should consider the amount of absorbed light by the sample. According to the absorption law, the absorbed by the sample light is proportional to the incoming intensity and factor $1-10^{-OD}$, not the *OD* itself.

We are going to measure R101 quantum yield in ethanol using R6G also in ethanol as a reference with QY = 0.95.

In Figure 5-5.1 we are presenting absorption spectra of R6G and R101 in Ethanol. For convenience the chemical structures are also presented in the figure.

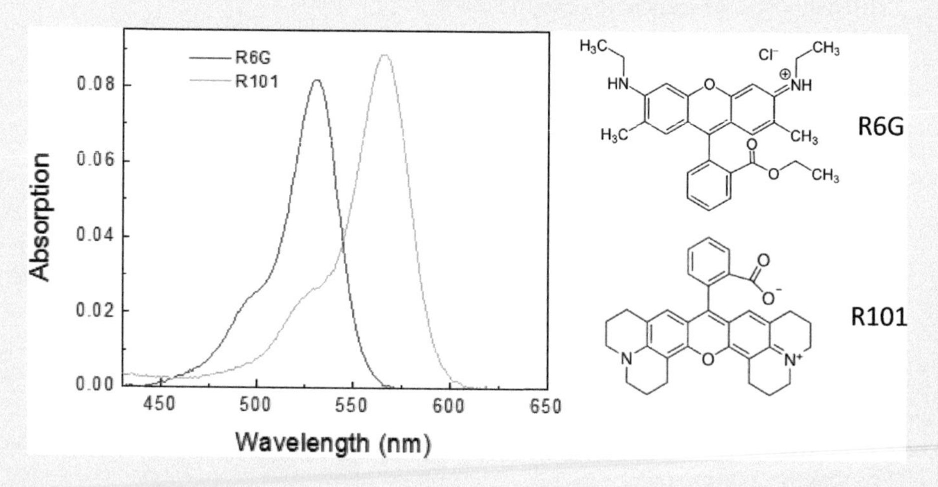

FIGURE 5-5.1 Absorption spectra of R6G reference and R101 sample in ethanol.

At 544 nm absorptions of both the reference and sample are equal. We will use this wavelength for the excitation. Since both the sample and reference are in the same solvent (same refractive index), the equation for the quantum yield simplifies to:

$$Q = Q_R \frac{I}{I_R}$$

(Continued)

EXPERIMENT 5-5 (CONTINUED) QUANTUM YIELD MEASUREMENTS

The measured fluorescence spectra at 544 nm excitation are presented in Figure 5-5.2. These spectra are not corrected for the wavelength sensitivity of spectrofluorometer.

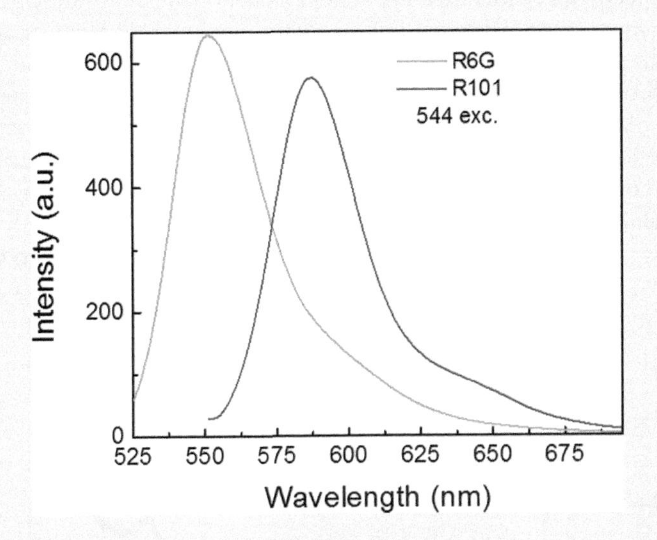

FIGURE 5-5.2 Fluorescence spectra of R6G reference and R101 sample in ethanol, as measured at 544 nm excitation. The spectra are not corrected for the wavelength sensitivity.

For comparison, we show emission spectra measured with excitation at 531 nm (Figure 5-5.3). At this excitation wavelength, both absorptions are below 0.1 but they are significantly different. Again, these spectra are not corrected for the wavelength sensitivity of the detection system.

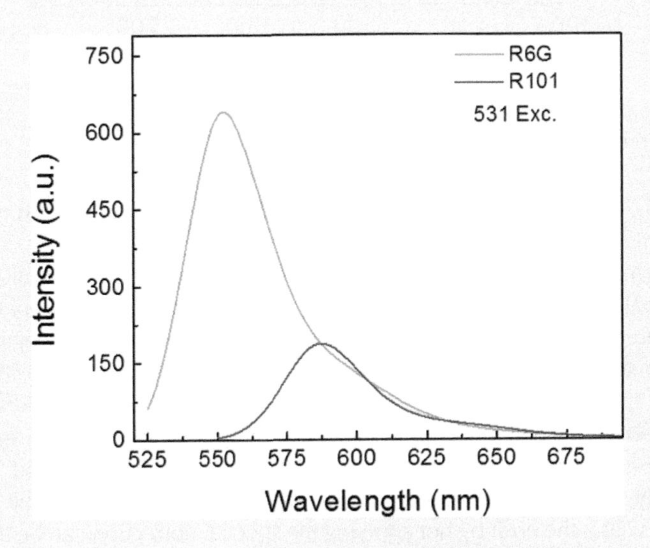

FIGURE 5-5.3 Fluorescence spectra of R6G reference and R101 sample in ethanol, as measured at 531 nm excitation. The spectra are not corrected for the wavelength sensitivity.

(Continued)

EXPERIMENT 5-5 (CONTINUED) QUANTUM YIELD MEASUREMENTS

The emission spectrum of R101 is shifted toward the red (~40 nm) as compared to the emission spectra of the R6G standard. In order to compare both signals, we need to correct both emission spectra for wavelength-dependent detection system sensitivity.

We should remind the reader that this correction includes all factors/components that are between the sample and detector output. This will include optical elements (like lenses, monochromator, and filters) and the detector itself. For some measurements, especially in deep UV, we should also include the cuvette wall that may have wavelength-dependent transmittance (see Experiments 3-3 and 3-4). For the presented examples, we will only use a correction curve as determined in the previous Experiment 5-4. To correct spectra, we multiply measured fluorescence intensities at each wavelength by the correction factor from Figure 5-5.4. Both spectra (reference and sample) need to be corrected.

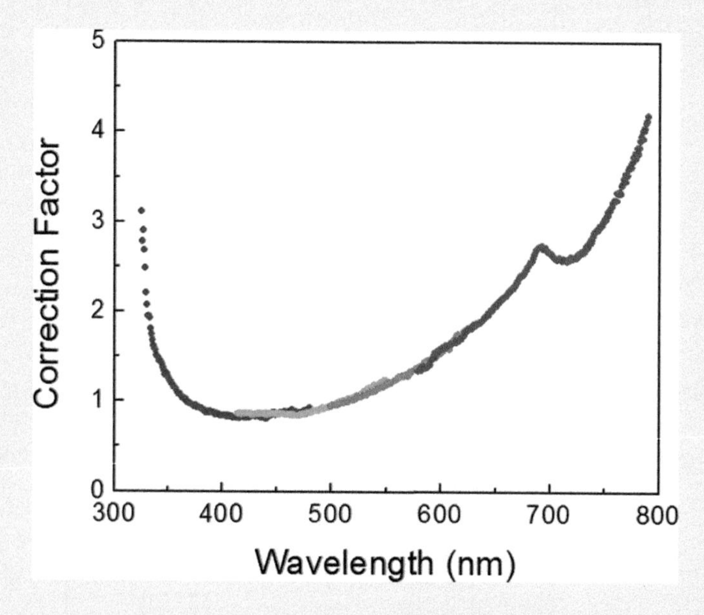

FIGURE 5-5.4 Correction curve for fluorescence spectra (from Experiment 5-4).

In Figures 5-5.5 and 5-5.6, we are presenting corrected emission spectra from Figures 5-5.2 and 5-5.3.

There are two major sources of potential errors in QY measurements/calculation. First, correction for the spectral sensitivity of the instrument (see Experiment 5-4) is necessary if reference and sample spectra are at different wavelength ranges. Depending on the wavelength range, even a small few nanometers shifts can have a meaningful impact. The integrated intensities (areas under the fluorescence spectrum) are different before and after spectral correction. For the calculation of QY, it is needed to calculate the ratio of integrated intensities I/I_R. In the case of uncorrected spectra (Figures 5-5.2 and 5-5.3), these ratios are 0.888 and 0.295 for 544 and 531 nm excitations, respectively whereas corrected spectra (Figures 5-5.5 and 5-5.6) give the ratios of 1.045 and 0.355. In this case, the error by not applying the spectral correction can be as high as 15%.

The second source of the potential error is the use of OD instead of $(1-10^{-OD})$. In the case of 544 nm excitation, both ODs are identical and the problem is avoided. However,

(Continued)

EXPERIMENT 5-5 (CONTINUED) QUANTUM YIELD MEASUREMENTS

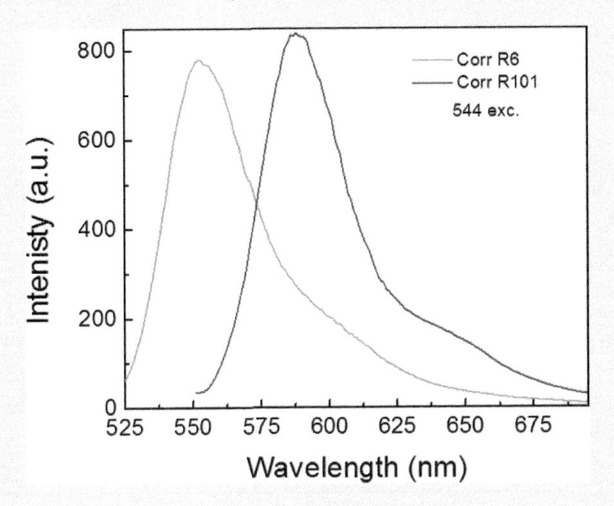

FIGURE 5-5.5 Corrected fluorescence spectra of R6G reference and R101 sample in ethanol at 544 nm excitation. Compare spectra on Figures 5-4.2.

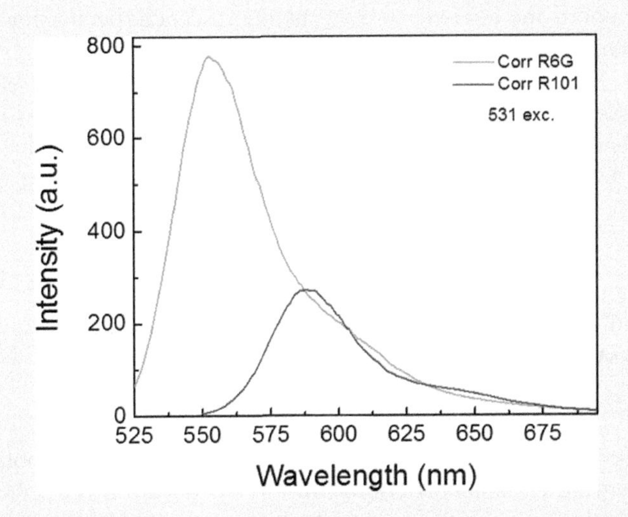

FIGURE 5-5.6 Corrected fluorescence spectra of R6G reference and R101 sample in ethanol at 531 nm excitation. Compare spectra on Figures 5-5.3.

at 531 nm excitation, the ratio of ODs is 3.05 and the ratio of $(1–10^{-OD})$ is only 2.90 what gives about 5% error. *Note, that for both excitations absorptions are significantly below 0.1.*

The QY of R101 in ethanol at 544 nm excitation (calculated from the equation above) is equal to $0.95 \times 1.045 = 0.99$. If one uses uncorrected spectra will obtain $0.95 \times 0.888 = 0.844$.

At the excitation 531 nm, the QY of R101 is $0.95 \times 0.355 \times 2.90 = 0.98$. Without both corrections, the calculation results in $0.95 \times 0.295 \times 3.05 = 0.85$. So, when not properly correcting spectra and absorptions we can get significantly different results.

(Continued)

EXPERIMENT 5-5 (CONTINUED) QUANTUM YIELD MEASUREMENTS

The summary of QY measurements is presented in the table below.

QY of R101			
$\lambda_{exc.} = 544\,\text{nm}$		$\lambda_{exc.} = 531\,\text{nm}$	
the same solvent, identical ODs		the same solvent, different ODs	
$Q = Q_R \frac{I}{I_R}$		$Q = Q_R \frac{I}{I_R} \frac{\left(1-10^{-OD_R}\right)}{\left(1-10^{-OD}\right)}$	
I/I_R from **uncorrected** spectra	I/I_R from **corrected** spectra	I/I_R from **uncorrected** spectra	I/I_R from **corrected** spectra
0.888	1.045	0.295	0.355
		the ratio of ODs	the ratio of $(1-10^{-OD})$
		3.05	2.9
QY = 0.84	**QY = 0.99**	**QY = 0.85**	**QY = 0.98**

CONCLUSIONS

QY measurements require careful measurements of absorption and fluorescence spectra of the studied fluorophore and a reference dye. The fluorescence spectra should be corrected for detection system sensitivity of spectrofluorometer and for a number of absorbed photons, proportional to $(1-10^{-OD})$. If different solvents are used for the sample and reference, the correction for refractive indexes should be applied $\left(\frac{n^2}{n_R^2}\right)$. Since refractive index ratio comes as a square, the difference can be significant. For example, when using water that has a refractive index of 1.33 for the sample and glycerol for reference (1.47), we can make over 22% error.

EXPERIMENT 5-6 MEASURING FLUORESCENCE OF HIGHLY ABSORBING SAMPLES

Keywords: fluorescence, reabsorption, Rhodamine 123.

While making fluorophore solutions, it is important to keep the fluorophore concentration low. However, there are a number of reasons why this will not always be possible. Sometimes physiological conditions require the use of a highly concentrated solution, such as protein polymeric systems that will dissociate upon dilution. Or when using a high concentration of the acceptor in FRET experiments. Using highly concentrated solutions will present problems such as insufficient penetration for the excitation beam, internal reabsorption of emission resulting in apparent spectral shifts; as both the excitation and emission beams have to travel/across significant path length of the regular cuvette, which is usually in millimeters. One way to deal with these problems is to use a thin layer of the sample in specialized cuvettes to keep the optical density low. This can be achieved with demountable cuvettes with spacers and front-face configuration for fluorescence measurement. The demountable precise cuvettes are expensive. Often it is more convenient to just use glass slides with the solution "sandwiched" between them. Such slides have a calculated thickness of a few

(Continued)

EXPERIMENT 5-6 (CONTINUED) MEASURING FLUORESCENCE OF HIGHLY ABSORBING SAMPLES

microns and using a front face geometry set-up, one can avoid the problems mentioned above even with highly concentrated solutions.

In this experiment, we will discuss different experimental approaches to measure the fluorescence of highly absorbing (high OD) samples. The different configurations are schematically presented in Figure 5-6.1.

MATERIALS

- Rhodamine 123
- Propylene glycol (PG)
- Scintillation vials
- 0.2 cm × 1 cm, and 1 cm × 1 cm glass cuvettes (two each for high concentration and low concentration solutions)
- Glass microscope slides, 25 mm × 75 mm × 1 mm

EQUIPMENT

- Spectrophotometer
- Spectrofluorometer with a custom-made front-face attachment

METHODS

1. Make stock solutions of Rhodamine 123 in propylene glycol, sonicate. Using the stock solution, prepare a few milliliters of samples with low and high absorptions, such that the absorbance of the low concentrated solution is around 0.1 and that of high concentration solution is around 1.0 when measured using a 2 mm path cuvette (about 5 in 10 mm path cuvette).
2. Measure absorption (Figures 5-6.2 and 5-6.3) and emission (Figures 5-6.4 and 5-6.5) spectra for 0.2 cm × 1 cm and 1 cm × 1 cm cuvettes with square geometry set-up.
3. Prepare samples for front-face configuration. Place 10 μL of each solution between microscope slides; gently squeeze a drop of solution between two slides and wait about 10 minutes until solutions are distributed homogeneously. Measure emission spectra. *Note:* We are using propylene glycol solutions that do not evaporate quickly and is homogeneously distributed between slides. When using ethanol or water be sure that solutions will not dry too quickly. One solution for that could be using a vacuum grease to seal the sample between slides.

RESULTS

In Figure 5-6.1, we are presenting three basic configurations used for the experiment. The first two are square geometry with two different cuvettes. Standard 1 cm × 1 cm cuvette and thin layer 0.2 cm × 1 cm cuvette with excitation along the short path. The third is the conceptual schematic of the front-face excitation configuration with a sandwiched sample.

In Figures 5-6.2 and 5-6.3, we present measured absorption spectra for high (HC) and low (LC) concentrations of R123 in PG measured in standard 1 cm × 1 cm cuvette and 0.2 cm × 1 cm cuvette respectively. The low concentration presents very low absorption at

(Continued)

EXPERIMENT 5-6 (CONTINUED) MEASURING FLUORESCENCE OF HIGHLY ABSORBING SAMPLES

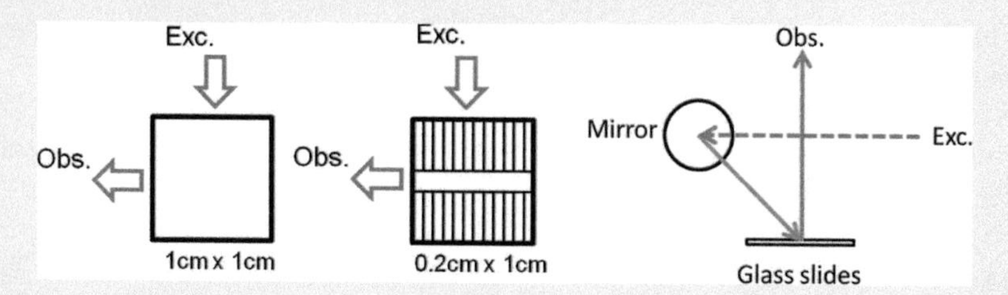

FIGURE 5-6.1 Schematic representation of square and front face geometry set-up to measure emission spectra.

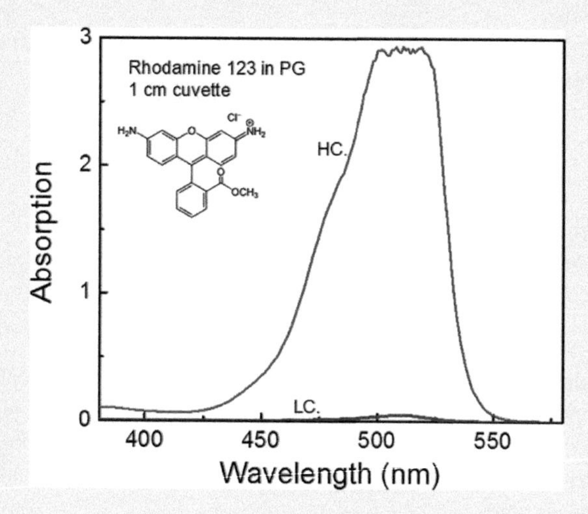

FIGURE 5-6.2 Absorption spectra of Rhodamine 123 in propylene glycol—high and low concentration solutions in 1 × 1 cm cuvette.

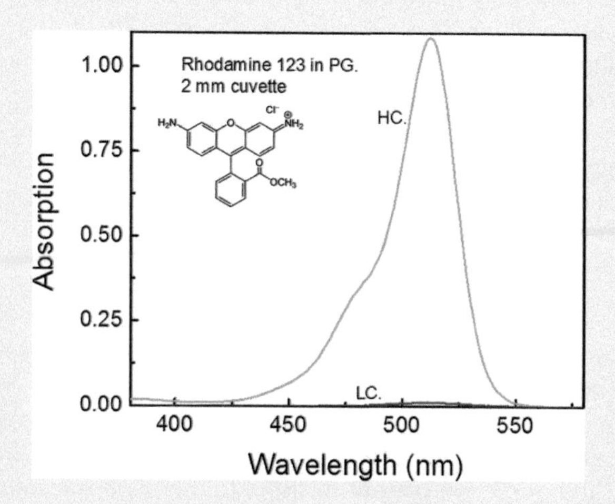

FIGURE 5-6.3 Absorption spectra of Rhodamine 123 in propylene glycol—high and low concentration solutions in a 0.2 cm × 1 cm cuvette. The low concentration solution has an absorbance of 0.013.

(Continued)

EXPERIMENT 5-6 (CONTINUED) MEASURING FLUORESCENCE OF HIGHLY ABSORBING SAMPLES

maximum about 0.1. The high concentration shows very high absorption and in the absorption, maximum measured absorption is over the range of the spectrophotometer (OD > 3).

Next, we used the 0.2 cm × 1 cm cuvette and measured absorption spectra through a shorter, 2 mm path. Absorption of the HC sample is still very high but easily measurable and absorption of LC is proportionally five times lower than in 1 cm path but still easily measurable.

We then measured fluorescence spectra of HC and LC using 1 cm × 1 cm cuvettes (Figure 5-6.4). Spectra are clearly distinct, which is a result of reabsorption occurring in HC sample. Also, the intensity difference in the maximum is only about 5-fold while the concentration difference is about 60-fold (when comparing absorptions). This is basically due to two effects. One, as discussed in previous Experiments 4-4 and 5-3, most of the excitation light in a 1 cm cuvette is absorbed in the first couple of millimeters and excitation intensity in the middle of the cuvette that is seen by the detector is much lower for HC solution. Second is reabsorption of the emitted light. Most of the emission light originates in the center of the cuvette and has to pass through 5 mm of the solution. Just comparing the absorption spectrum with the emission for LC solution, one would notice that emission already starts below 500 nm while absorption extends up to 550 nm. In this so-called spectra, overlapping region emission light will be reabsorbed. In Figure 5-6.4, we also present the emission spectrum of the LC solution scaled/normalized to the maximum emission of HC solution (dashed line). The emission of HC solution is significantly shifted toward the longer wavelength (we call it red-shifted). For HC, the maximum emission is about 550 nm about 20 nm shifted as compared to 530 nm emission maximum for LC. The emission spectrum of HC is significantly lowered/reduced just where R123 absorption extends.

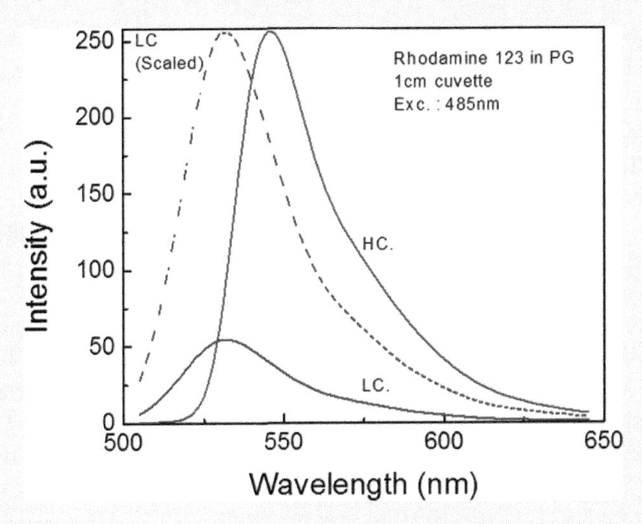

FIGURE 5-6.4 Emission spectra of Rhodamine 123 in 1 cm × 1 cm cuvette. The dotted spectrum is the scaled spectrum of low concentration to better represent the red-shift in spectra between the two solutions.

Next, we repeated the measurements for LC and HC solutions in 2 mm × 10 mm cuvette. As presented in Experiments 4-4, 4-5, and 5-3, the emission attenuation due to excitation light absorption is much lower (practically negligible). Indeed measured emission signal (Figure 5-6.5) in 2 mm path is now almost 30-fold lower than that in 10 mm path. Still not as

(Continued)

EXPERIMENT 5-6 (CONTINUED) MEASURING FLUORESCENCE OF HIGHLY ABSORBING SAMPLES

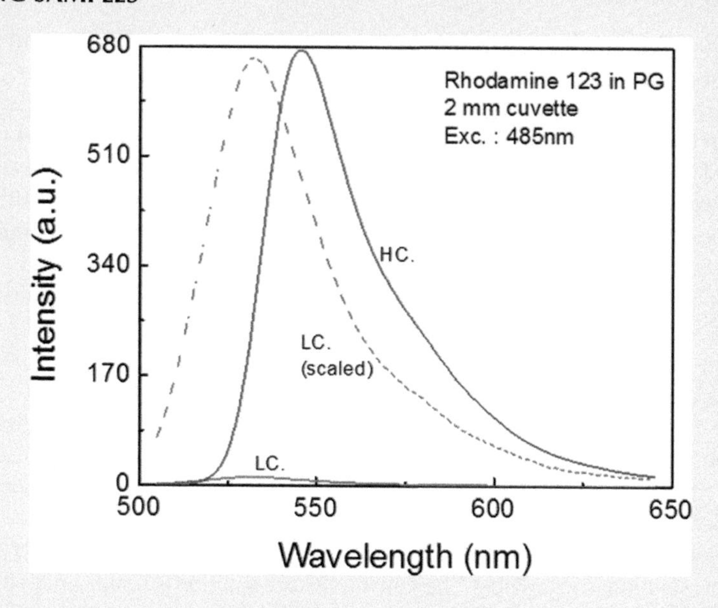

FIGURE 5-6.5 Emission spectra of Rhodamine 123 in 0.2 cm × 1 cm cuvette. The dotted spectrum is the scaled spectrum of low concentration to better represent the red-shift in spectra between the two solutions.

expected from absorption difference of about 60-fold. But looking on the rescaled/normalized emission of LC, it is clear that the reabsorption effect is the same (as expected since the path for emission light did not change (~5 mm).

The fluorescence spectra of high concentration sample (HC) in both cuvettes, 1 cm × 1 cm and 0.2 cm × 1 cm, are distorted at shorter wavelengths and artificially shifted to longer wavelengths. In a square geometry set-up, these measurements are highly unreliable. Highly concentrated solutions require very thin sample layers in both excitation and emission paths.

These can be achieved with a front-face set-up (see Figure 5-6.1, right).

In Figure 5-6.6, we are presenting emission spectra measured in a thin layer (~10 μm) in the front-face configuration. For such a thin layer sample, absorption is much lower. Measured emission spectra are now very different in intensity (close to 60-fold as expected from relative concentrations and measured absorptions) and spectra are shifted very little for LC and HC solutions. Clearly, with front-face set-up and a thin sample layer, the reabsorption in the HC sample is negligible.

(*Continued*)

EXPERIMENT 5-6 (CONTINUED) MEASURING FLUORESCENCE OF HIGHLY ABSORBING SAMPLES

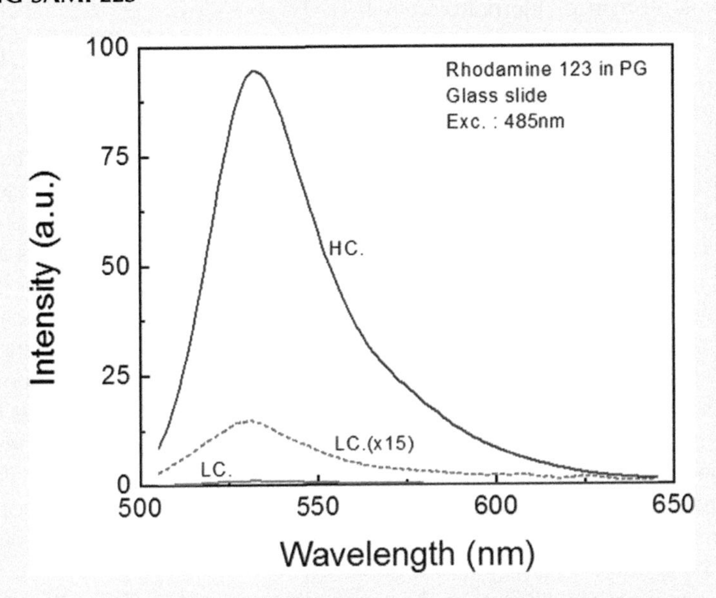

FIGURE 5-6.6 Emission spectra of Rhodamine 123 in glass slide sandwich measured with front face geometry set-up. The dotted spectrum is the scaled spectrum of low concentration to better represent no spectral shift between the low and high concentrated solutions.

CONCLUSIONS

Samples with high optical densities should not be measured in the square-geometry configuration because spectra are distorted by inner filter effects and can be artificially shifted toward longer wavelengths. Front-face configuration allows to keep low absorptions and provides correct spectral measurements.

It is not easy to measure the thickness of the layer between the two slides. This can be done in two ways. First, when knowing concentration and absorption in 1 mm path, one can measure the absorption of the sample between the slides (make sure the reference is from two slides with only solvent in-between). The ratio of measured absorption should be proportional to the ratio of thicknesses. In the second approach, the apparent thickness of the solution layer between slides can be estimated by depositing a calibrated volume (about 10 μL) on the slide and then measuring the area of the spot between slides (highly concentrated solution will present clear color and LC still can be visible).

EXPERIMENT 5-7 BINDING OF FLUOROPHORES TO PROTEINS

Keywords: fluorescence, fluorophore binding, ANS, proteins.

Many fluorescence applications are aiming at macromolecular detections. Typically, it will be protein and/or DNA detection with many practical biomedical applications. In this experiment, one of the most common examples of protein detection is discussed. The protein surface exposes various amino acid groups on the surface that are formed to hydrophobic and hydrophilic patches/pockets on the protein surface. Naturally, the patches attract or repel water molecules that may affect protein conformation and dictate protein function. The characteristic properties/shapes of the pocket may work as binding sites for different molecules or molecular assemblies/proteins called ligands. Chromophores that have proper size and charge (usually not easily soluble in water) will locate themselves in such hydrophobic pockets and stay bound in place. A change from a polar water environment to a highly hydrophobic surrounding upon binding to the protein often greatly increases fluorescence of the chromophore. The equilibrium between bound and unbound fluorophores in the solution depends on the protein and fluorophore concentrations and the fluorophore affinity for binding to the protein. If unbound and bound fluorophores present different fluorescence signals, the measure of the fluorescence signal (increase/decrease) as a function of protein concentration can be utilized for detecting protein presence and/or evaluating binding affinity at various conditions.

This particular experiment will study the binding of 1, 8-amino-anilino-sulfonic acid (ANS) to protein. ANS is very weakly fluorescent in water and upon binding to protein (e.g., human serum albumin (HSA)) becomes highly fluorescent.

MATERIALS

- Human serum albumin
- 8-anilino-1-naphthalenesulfonic Acid (ANS)
- PBS buffer, pH = 7.4
- Glass vials
- 1 cm × 1 cm disposable cuvettes

EQUIPMENT

- Spectrophotometer
- Spectrofluorometer
- Cell phone or camera
- UV illuminator

METHODS

1. Make stock solutions of ANS and HSA in PBS buffer (use glass vials).
2. Prepare 10–20 plastic cuvettes containing fixed volume V_0 (2–3 mL) of HSA solution (equal volume in each cuvette) with a concentration of HSA changing from 0 (just a buffer) to 70 μM.
3. Knowing the concentration of ANS in the stock solution calculate the needed volume of ANS solution to be added to each cuvette containing HSA to have a final concentration of ANS (C_A) of about 25–30 μM. Typically, you would like to have a

(Continued)

EXPERIMENT 5-7 (CONTINUED) BINDING OF FLUOROPHORES TO PROTEINS

highly concentrated stock solution of ANS to minimize the dilution effect. If initial solution volume in each cuvette is V_0 and the corresponding concentrations of HSA in each cuvette is C_i^0 when adding a constant volume, V of concentrated ANS that has an initial concentration C_A^0 we can calculate concentrations of ANS and HSA in each cuvette, respectively:

$$C_A = \frac{C_A^0 \cdot V}{V + V_0}$$

$$C_i = \frac{C_i^0 \cdot V_0}{V + V_0}$$

From this, we can see that if $V_0 >> V$ when adding ANS dilution of protein is negligible and dilution of ANS is high. It is important to remember that the final concentration of ANS, C_A should result in ANS absorption about or below 0.1.

4. Measure emission of each cuvette using excitation wavelength from the 330–370 nm range.
5. Put a set of prepared cuvettes on the UV illumination stage and make photographs of observed fluorescence while solutions are on UV illuminator.

RESULTS

Figure 5-7.1 shows measured emission spectra for 18 solutions for which the concentration of HSA increases from 0 to 70 μM and concentration of ANS is kept constant at 26 μM. The measured intensity of fluorescence quickly increases with HSA concentration.

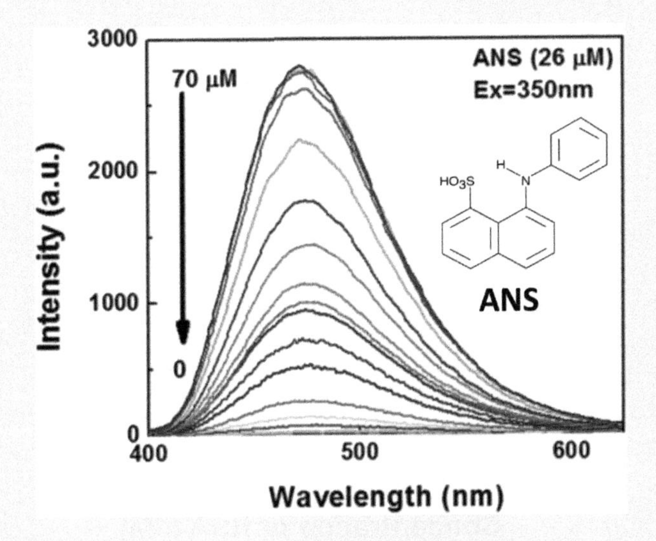

FIGURE 5-7.1 Fluorescence spectra of ANS in PBS buffer in absence and presence of HSA. Concentrations of HSA were: 0, 0.02, 0.1, 0.2, 0.4, 0.8, 1.3, 1.7, 1.9, 2.1, 3, 4, 8, 16, 25, 40, 50, and 70 μM.

(*Continued*)

EXPERIMENT 5-7 (CONTINUED) BINDING OF FLUOROPHORES TO PROTEINS

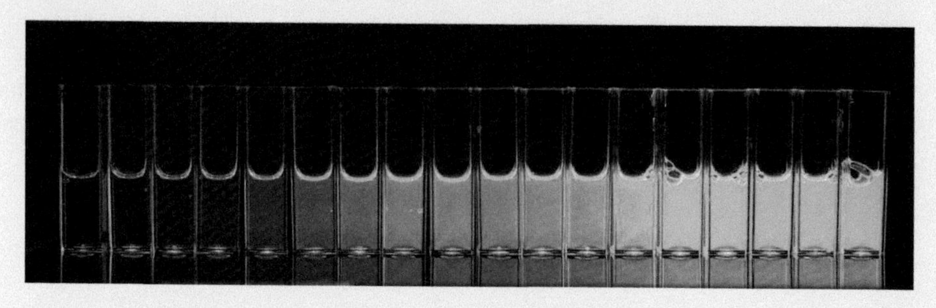

FIGURE 5-7.2 Photograph of all 18 measured samples (HSA concentrations from 0 μM to 70 μM) placed on the UV illuminator.

To visually present the effect in Figure 5-7.2, we are showing a photograph of 18 cuvettes with solutions under the uniform UV illumination.

Figure 5-7.3 shows the dependence of measured fluorescence of ANS as a function of HSA concentration using fixed 26 μM concentration of ANS. There are multiple ways of drawing measured points. We could just use single point intensity at a given wavelength (typically maximum) or use integrated intensity over part of the emission spectrum or just the entire spectrum. In the presented Figure 5-7.3, we used just average value as measured at maximum. The other approach, to use integrated emission spectrum, can be more precise especially when measured signals are very low and experimental noise is significant. But in this case, the experimentalist should take proper correction for background (dark counts and residual fluorescence).

Presented in Figure 5-7.3, experimental points can be fitted with a model of choice for ANS binding. In the presented example, the theoretically fitted curve was satisfactory with just a single binding constant (solid line).

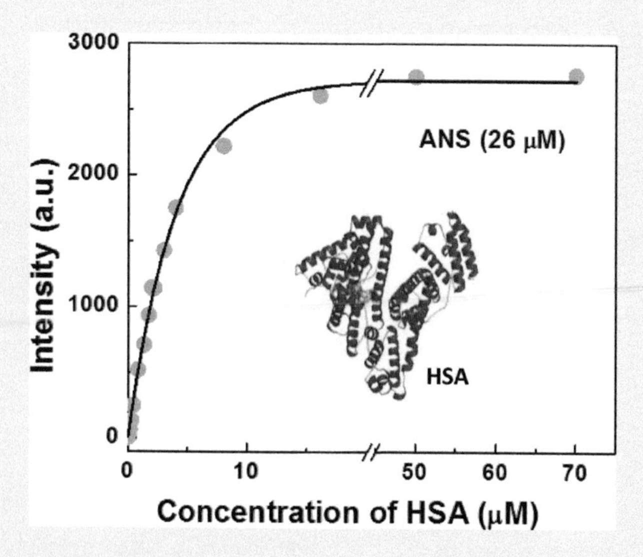

FIGURE 5-7.3 Dependence of ANS fluorescence intensity on the HSA concentration.

(Continued)

EXPERIMENT 5-7 (CONTINUED) BINDING OF FLUOROPHORES TO PROTEINS

Note: As seen in Figures 5-7.1 through 5-7.3, ANS fluorescence dramatically depends on the HSA concentration. ANS fluorescence in water in the absence of HSA is very weak. This very pronounced signal dependence for ANS (and some other dyes) has been used for protein detection including detection of blood traces in the case of forensic applications. Alternatively, you can use bovine serum albumin (BSA).

EXPERIMENT 5-8 BINDING OF FLUOROPHORES TO DNA

Keywords: fluorescence, DNA, Hoechst.

In the previous experiment, an example of ANS binding to the protein was presented. In the case of DNA, there are multiple places (groves) where some dyes can bind to DNA, usually intercalating between nucleic bases. This binding shields fluorophores from a water phase, stabilize the excited state and immobilize the fluorophore. Often this results in a significant increase in the brightness of the fluorophores. Such increase of fluorescence has been broadly utilized for DNA detection (in so-called DNA arrays) and various forensics applications are based on this phenomenon. This experiment will study the binding of Hoechst33258 to DNA.

MATERIALS

- Calf thyme DNA
- Hoechst33258
- PBS buffer, pH = 7.4
- Glass vials
- 1 cm × 1 cm disposable cuvettes

EQUIPMENT

- Spectrophotometer
- Spectrofluorometer

METHODS

1. Make stock solutions of Hoechst33258 and DNA in PBS buffer (use a glass vial).
2. Prepare samples containing a constant concentration of Hoechst33258 (optical density of about 0.1) and various concentrations of DNA (from 0 to 80 μM). The practical method to prepare such solution will be analogical to previous experiment (Experiment 5-7, point 3).
3. Measure fluorescence spectra as a function of DNA concentration.
4. Create the dependence of observed Hoechst33258 fluorescence intensity on DNA concentration.

(*Continued*)

EXPERIMENT 5-8 (CONTINUED) BINDING OF FLUOROPHORES TO DNA

RESULTS

In Figure 5-8.1, we are showing emission spectra measured for increasing concentration of DNA in the solution.

Fluorescence of Hoechst fluorophore strongly increases with DNA concentration. The binding of Hoechst to DNA is illustrated in Figure 5-8.2.

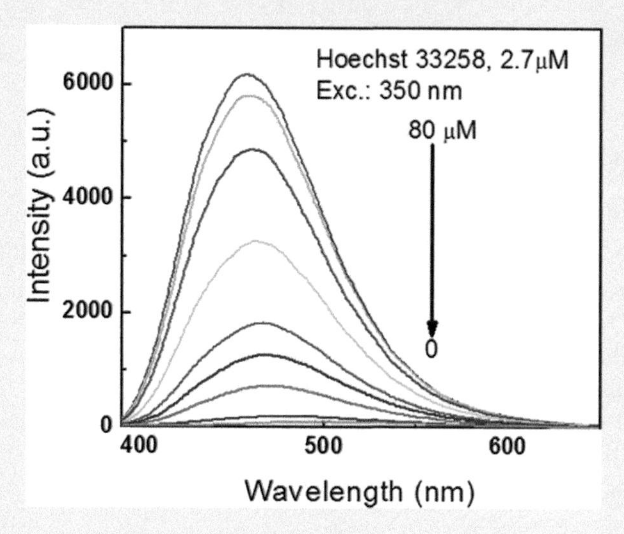

FIGURE 5-8.1 Hoechst 33258 fluorescence spectra in the function of DNA concentration. The DNA concentrations were: 0, 0.4, 0.8, 8, 10, 20, 40, 60, and 80 μM.

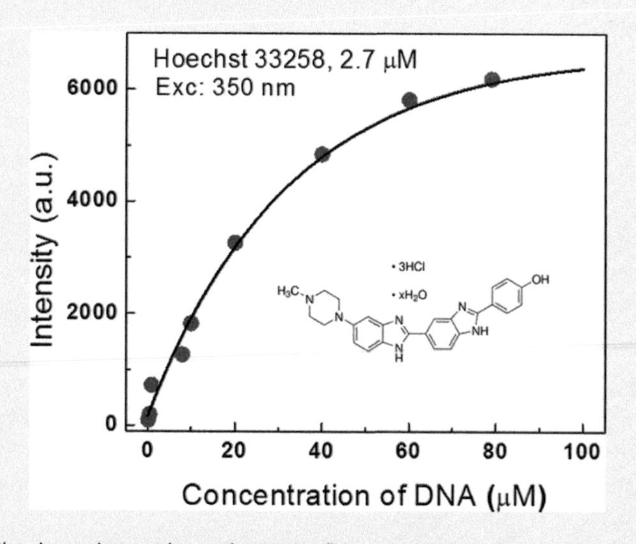

FIGURE 5-8.2 The dependence of Hoechst33258 fluorescence intensity on DNA concentration.

Note: Alternatively, other fluorophores like DAPI or ethidium bromide (EtBr) can be used in this experiment. In fact, the large palette of available different color dyes has been developed for genomic/forensic applications.

EXPERIMENT 5-9 EFFECT OF INTRINSIC HEAVY ATOMS ON FLUORESCENCE

Keywords: fluorescence, heavy atoms, Rhodamine 123, Eosin Y, Erythrosine B.

There are many factors that may affect fluorescence, and it is not easy to predict the fluorescence signal from the structure of the chromophore alone. Fluorophores that are very similar in chemical structures may have different efficiencies of fluorescence emissions (different quantum yields). There are multiple reasons for that but one common and an easily predictable cause is the presence of heavy metal atoms in the molecule structure. For this experiment, three well-known fluorophores of very similar structures are investigated, as shown in Figure 5-9.1. The main difference is the presence of the heavy atom, bromide (Br) or iodide (I).

MATERIALS

- Rhodamine 123 (R123)
- Eosin Y
- Erythrosine B (ErB)
- Water
- 1 cm × 1 cm cuvettes

EQUIPMENT

- Spectrofluorometer
- Spectrophotometer

METHODS

1. Make stock solutions of studied three compounds: R123, Eosin Y, and ErB, see Figure 5-9.1. Use a few milliliters of water, glass, or plastic vials. If the compound did not dissolve completely, add more water and sonicate the solution.
2. Prepare samples with absorptions about 0.1 using stock solutions.
3. Measure absorption spectra using a water baseline (Figure 5-9.2). Having the absorption of ErB about 0.1, adjust the concentrations of R123 and Eosin Y solutions that all three absorptions will cross at a single point at about 500 nm. *Note:* This is not necessary but will eliminate the need for absorption correction and greatly simplify the analysis.
4. Measure fluorescence spectra with the excitation wavelength corresponding to the same absorption (~500 nm, Figure 5-9.3). Remember to use magic angle conditions.

RESULTS

We purposely adjusted the concentrations to have the same absorption at a given wavelength (500 nm in this case). The same absorption of all studied compounds assures that in each measurement, the same number of fluorophore molecules is being excited (the same number of photons is absorbed). If the absorptions are not the same at the excitation wavelength, the measured emission spectra should be adequately corrected for the amount of absorbed light (photons). In practice, each emission spectrum should be divided by the factor $(1-10^{-A})$ at the excitation wavelength (see quantum yield Experiment 5-5).

(Continued)

EXPERIMENT 5-9 (CONTINUED) EFFECT OF INTRINSIC HEAVY ATOMS ON FLUORESCENCE

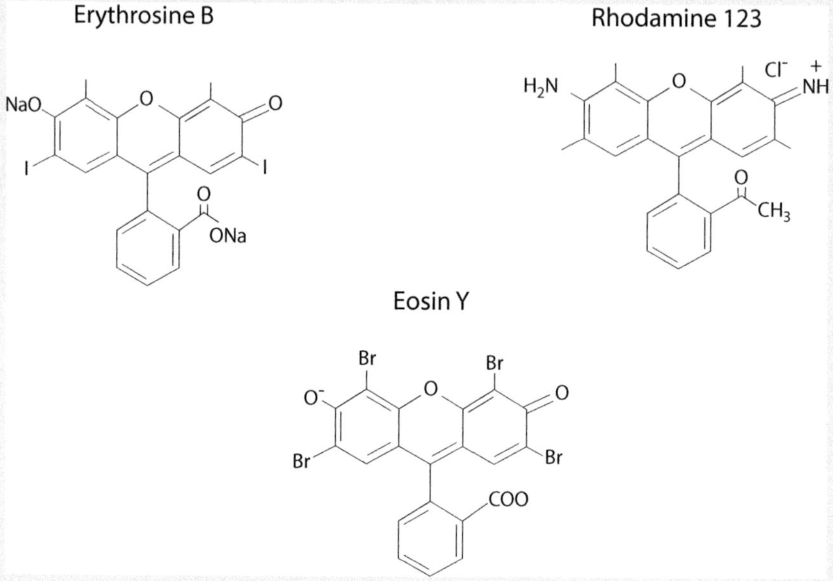

Erythrosine B

Rhodamine 123

Eosin Y

FIGURE 5-9.1 Chemical structures of ErB, Eosin Y, and Rh123.

Figure 5-9.2 shows measured absorption spectra for all three compounds. The concentrations have been adjusted to have an identical optical density (OD) at about 500 nm. Note, that at the excitation wavelength absorption is about 0.05 ensuring a uniform sample excitation. Alternatively, one can use a 480 nm excitation where absorptions are also the same.

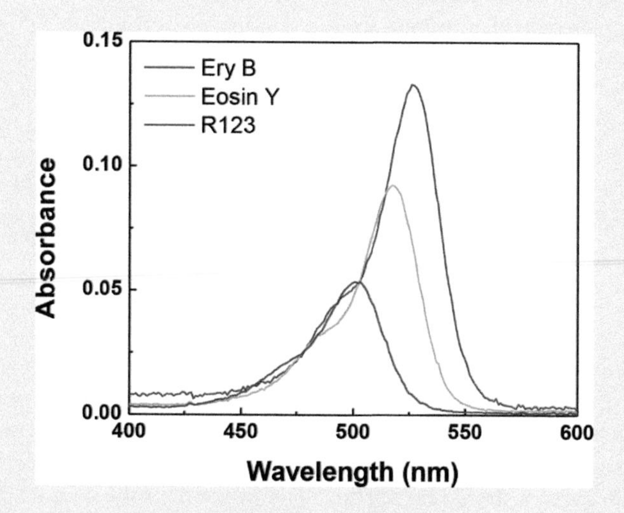

FIGURE 5-9.2 Absorption spectra of ErB, Eosin Y, and R123. Concentrations were about 1.3, 0.8, and 0.6 μM, respectively. At 500 nm, all compounds have the same absorption of about 0.05.

(Continued)

EXPERIMENT 5-9 (CONTINUED) EFFECT OF INTRINSIC HEAVY ATOMS ON FLUORESCENCE

In Figure 5-9.3, we are showing measured emission spectra. R123 has an excellent quantum yield and shows strong emission. The Eosin Y that contains multiple Br atoms shows much weaker fluorescence and ErB that contain in the same positions multiple I atoms shows the weakest fluorescence. Using the excitation wavelength of 500 nm (same OD for each sample) in a simple way ensures that we are exciting the same number of molecules and the observed difference is only due to significant differences in quantum yields.

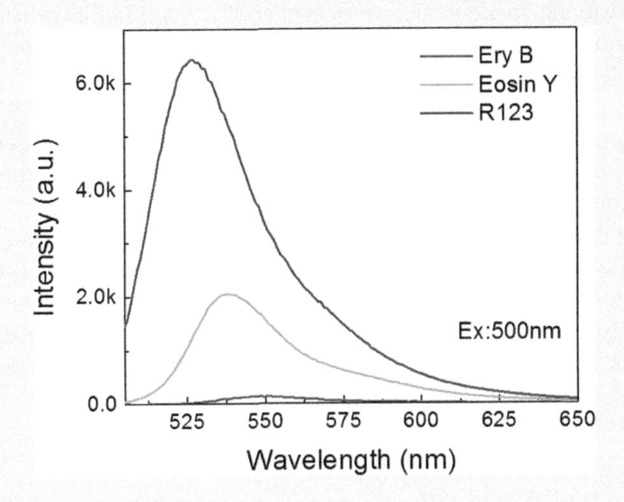

FIGURE 5-9.3 Fluorescence spectra of ErB, Eosin Y, and Rh123 at 500 nm excitation.

Note: In some cases, the fluorescence efficiency can be predicted from the chemical structure of the fluorophore. Presence of heavy atoms such as Cl, Br, and I within the chromophore structure usually internally quench the fluorescence and such chromophores will present much lower quantum efficiency.

EXPERIMENT 5-10 FLUORESCENCE QUENCHING OF BRIGHT FLUOROPHORES

Keywords: fluorescence, quenching, acrylamide, 2-aminopurine.

The diffusion process in liquid samples results in multiple collisions between molecules and fluorophores that will interact (a brief contact) with molecules present in the solution. When in-contact with fluorophores, some molecules may lead to fast radiationless deactivation of the fluorophore's excited state. In effect, the fluorophore will lose its excitation energy without emitting a photon and the apparent fluorescence signal will decrease. The process that leads to the decrease of the fluorescence signal is called quenching, and molecules that produce the quenching are called quenchers. The probability for quenching depends on the quencher concentration, and to study quenching it is typically necessary to vary the quencher concentration (*Note:* Quenching also depends on other factors like temperature or solvent viscosity that needs to be controlled during the experiment). So, in quenching experiments, a series

(*Continued*)

EXPERIMENT 5-10 (CONTINUED) FLUORESCENCE QUENCHING OF BRIGHT FLUOROPHORES

of samples need to be prepared with a constant concentration of the fluorophore and different concentrations of the quencher. Having identical concentrations of the fluorophore in all samples will allow a simple comparison of the fluorescence signals that will only depend on quencher concentration. *How to prepare such samples?* Several methods of preparing desirable samples were discussed previously. But let's assume that we want to prepare few samples for the experiment, 5 mL of each. Samples should have constant (identical) concentration of fluorophore with varying concentration of quencher. It is important to precisely monitor the concentration of quencher in each sample since the fluorescence signal will be plotted as a function of quencher concentration). To do so, you may:

1. Combine proper volumes of the stock solutions of sample and quencher and the solvent. By mixing these three ingredients, one can obtain samples with desired concentrations.
2. Use a stock solution of the fluorophore and pre-prepared series of a given volume, V_0 (e.g., 5 mL) of n samples with adequate quencher concentrations. Then, add to each sample the same small volume (V) of the fluorophore stock solution. Similar to the procedure in the Experiment 5-7, we should be aware of the volume increase upon fluorophore addition. The final concentrations of the fluorophore (C_F) and a quencher (C_q) will be:

$$C_F = \frac{C_F^0 \cdot V}{V + V_0} \text{ and } C_q = \frac{C_q^0 \cdot V_0}{V + V_0}$$

   where C_F^0 and C_q^0 are initial stock concentrations of fluorophore and quencher, respectively. When the volume of prepared quencher solutions is much larger than the volume of an added fluorophore ($V_0 >> V$), the dilution of the quencher will be minimal but the dilution of the fluorophore will be high.

   For example, if we want to prepare 5 mL of the sample with 1 M quencher concentration, we are going to add 0.5 mL of the fluorophore solution, the volume will increase to 5.5 mL and the quencher concentration will decrease by a factor of 5/5.5. Therefore, the initial concentration of the quencher should be higher by a factor of 5.5/5, e.g., instead 1 M it should be 1.1 M.
3. Alternatively, we can prepare larger volumes of the fluorophore solution and the highest concentration of quencher with a fluorophore. Let's assume that we want the highest quencher concentration of 1 M, and we have 2 M quencher stock solution. In order to obtain the 20 mL sample of 1 M quencher with a fluorophore, we need to mix 10 mL of the stock solution of the quencher with 10 mL of fluorophore solution. The reference sample without the quencher will be obtained by mixing 10 mL of the fluorophore solution with 10 mL of the solvent (*Note:* Mixing two solutions of equal volumes result in concentrations dropping by half. So, the initial concentration of the quencher will be 1 M (half of 2 M stock concentration) and fluorophore concentration will be half of fluorophore stock solution concentration). Then, any quencher concentration can be achieved by mixing these two solutions (the highest quencher concentration and reference sample without quencher). In this preparation, the fluorophore concentration in each sample will be constant and quencher concentrations will be different.

(*Continued*)

EXPERIMENT 5-10 (CONTINUED) FLUORESCENCE QUENCHING OF BRIGHT FLUOROPHORES

Acrylamide is a known fluorescence quencher often used in UV and visible spectral regions. In this experiment, we will demonstrate an efficient quenching of 2-aminopurine (2AP), a nucleotide analog used in DNA/RNA studies.

MATERIALS

- 2-Aminopurine
- Acrylamide
- PBS buffer, pH 7.4
- 1 cm × 1 cm cuvettes
- Pipettes

EQUIPMENT

- Spectrophotometer
- Spectrofluorometer

METHODS

1. Prepare 5 mL stock solutions of 2AP (with an optical density of about 1 at the maximum of absorption) and 1 M acrylamide in PBS buffer.
2. Prepare samples (2 mL) of 2AP (optical density of about 0.1) with various concentrations of acrylamide up to 0.15 M in PBS buffer. For example, in order to get the sample (2 mL) with acrylamide concentration 0.1 M, you need to mix 0.2 mL 2AP stock solution, 0.2 mL acrylamide stock solution, and 1.6 mL PBS buffer.
3. Measure fluorescence emission of all samples, starting with the sample without acrylamide (*remember to use magic angle conditions*). Use the excitation wavelength at 310 nm, outside the absorption of acrylamide.
4. Create a Stern–Volmer plot and estimate the Stern–Volmer quenching constant.

RESULTS

In quenching experiments, many quenchers will have high absorption in the UV spectral range and it is important to check (control) absorption of the quencher. If the quencher absorbs at the fluorophore's excitation wavelength, the correction is needed (see Experiment 5-2 and later Experiment 5-11). Figure 5-10.1 shows the absorption of 0.2 M acrylamide solution in 1 cm cuvette. The absorption of acrylamide below 300 nm becomes significant. In the figure, we also marked the used excitation wavelength (310 nm) that is clearly off the acrylamide absorption range. In the presented experiment, we will change acrylamide concentration from 0 M up to 0.15 M. In Figure 5-10.2, we present the absorption spectrum of 2AP in PBS buffer. Again, we used 310 nm excitation wavelength that is off acrylamide absorption (Figure 5-10.3).

The fluorescence quenching of 2AP by acrylamide is very effective, 25 mM concentration of acrylamide reduces the fluorescence intensity to half.

(Continued)

EXPERIMENT 5-10 (CONTINUED) FLUORESCENCE QUENCHING OF BRIGHT FLUOROPHORES

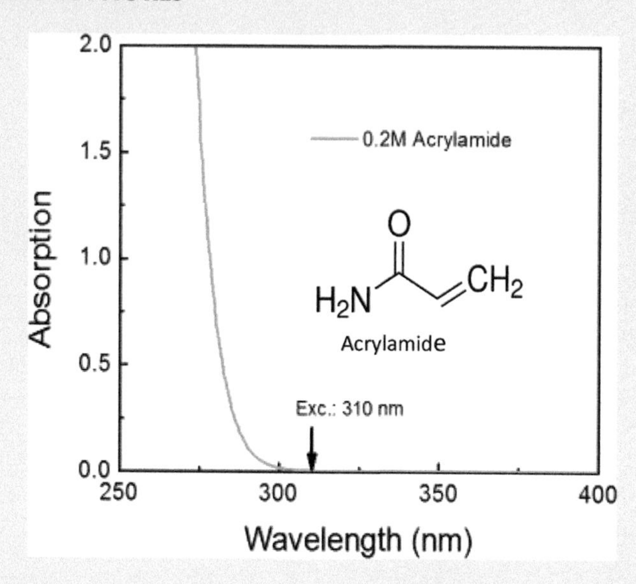

FIGURE 5-10.1 The absorption spectrum of 0.2 M acrylamide in PBS buffer.

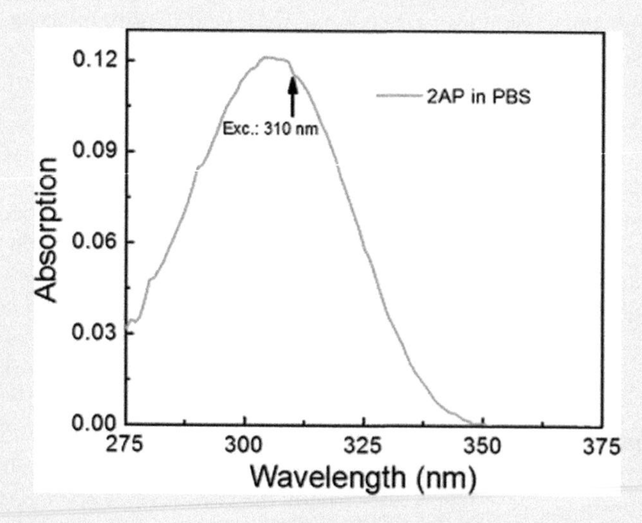

FIGURE 5-10.2 The absorption spectrum of 2-aminopurine in PBS buffer. At the 310 nm excitation, the absorption of acrylamide is negligible.

A simplest quenching theory of Stern–Volmer predicts the dependence:

$$\frac{F_0}{F} = 1 + K_{sv}[Q]$$

where Q is the quencher concentration and K_{sv} is a Stern–Volmer constant.

(*Continued*)

EXPERIMENT 5-10 (CONTINUED) FLUORESCENCE QUENCHING OF BRIGHT FLUOROPHORES

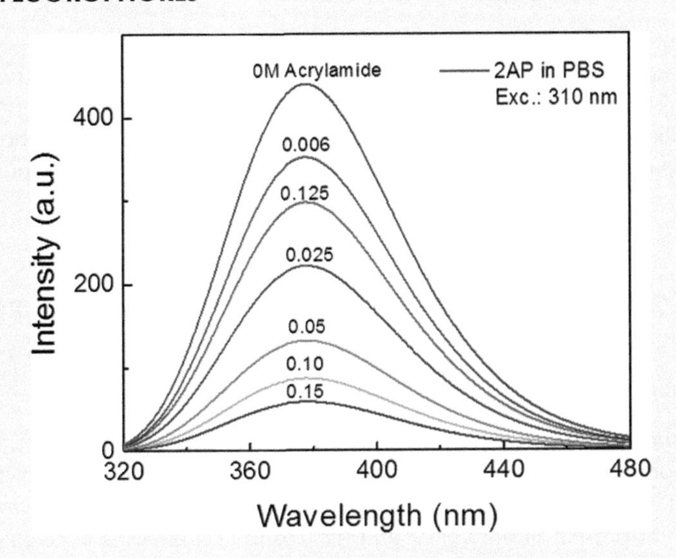

FIGURE 5-10.3 Fluorescence spectra of 2-aminopurine in absence and presence of acrylamide.

The quenching process is described by the Stern–Volmer constant K_{sv}. In order to determine K_{sv}, one should plot the intensity ratio as a function of the quencher concentration and find the slope (Figure 5-10.4).

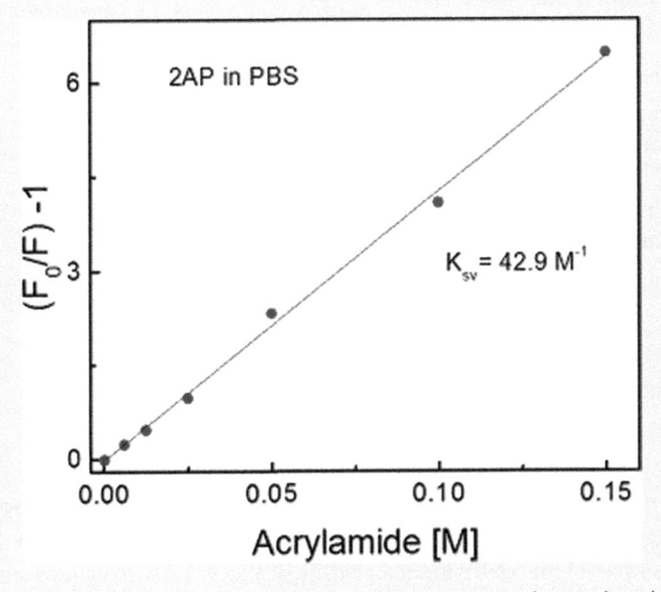

FIGURE 5-10.4 Stern–Volmer plot for the quenching of 2-aminopurine by acrylamide.

(Continued)

EXPERIMENT 5-10 (CONTINUED) FLUORESCENCE QUENCHING OF BRIGHT FLUOROPHORES

Note: In any quenching experiments, it is important to excite the sample outside of the quencher absorption. Also, always measure the absorption spectrum in the presence of the quencher. The appearance of a new absorption band will indicate a new absorbing species, most likely as a result of a chemical reaction. In such a case, one should be aware of the presence of static quenching. This experiment can also be done in pure water.

EXPERIMENT 5-11 EFFECT OF ABSORBING QUENCHER ON MEASURED FLUORESCENCE QUENCHING

Keywords: fluorescence, quenching, acrylamide, tryptophan.

In some experiments, especially working in UV, quenchers or even buffers may present some residual absorption (background). When using high concentrations of the quencher the overall quencher (background) absorption can be significant. In many cases, the quencher will not present fluorescence and the absorption effect is neglected which may lead to a significant problem. A popular dynamic quencher, acrylamide, is often used for the quenching of tryptophan fluorescence in proteins. However, it has a substantial absorption below 300 nm. The maximum of tryptophan absorption and excitation is near 280 nm. So, shorter wavelength excitation gives higher fluorescence signal that is easier to separate from leaking excitation and students are often tempted to use a shorter excitation wavelength. This experiment will compare a long-wavelength excitation at 298 nm with a short-wavelength at 278 nm while studying the quenching of tryptophan fluorescence. The goal of this experiment is to demonstrate how fluorescence quenching is affected by the intrinsic absorption of the quencher.

MATERIALS

- L-tryptophan (Trp)
- Acrylamide (ACR), potassium iodide (KI)
- Water
- 1 cm × 1 cm and 0.2 cm × 1 cm quartz cuvettes
- Pipettes and vials

EQUIPMENT

- Spectrophotometer
- Spectrofluorometer

METHODS

1. Prepare 10 mL stock solutions of Trp (with an optical density of 1 at the maximum of absorption) and 1 M acrylamide in PBS buffer.
2. Prepare samples (2 mL) of Trp (optical density of about 0.1) with various concentrations of acrylamide up to 0.4 M. Measure absorption spectra in the 1 cm × 1 cm quartz cuvette.

(Continued)

EXPERIMENT 5-11 (CONTINUED) EFFECT OF ABSORBING QUENCHER ON MEASURED FLUORESCENCE QUENCHING

3. Measure fluorescence emission of all samples, starting with the sample without acrylamide. Make fluorescence measurements in the 0.2 cm × 1 cm cuvette. Use two excitation wavelengths, 298 and 278 nm. Keep the same conditions for both excitations except the voltage on the detector.
4. Create Stern–Volmer plots for both excitations.
5. Calculate the correction factors for absorptions of Trp with acrylamide at both excitation wavelengths. Create corrected Stern–Volmer plots.

RESULTS

Measured absorptions in the 1 cm path cuvette for Trp buffer, 0.4 M acrylamide solution and Trp in 0.4 M acrylamide solution are presented in Figure 5-11.1. On the left, we are presenting an expanded scale with marked absorptions for 0.4 M acrylamide at 278 and 298 nm. At 298 nm, the absorption of acrylamide only is small and at 278 nm is quite significant ~1.93. The Trp absorptions at 278 and 298 nm are marked on the right-side graph.

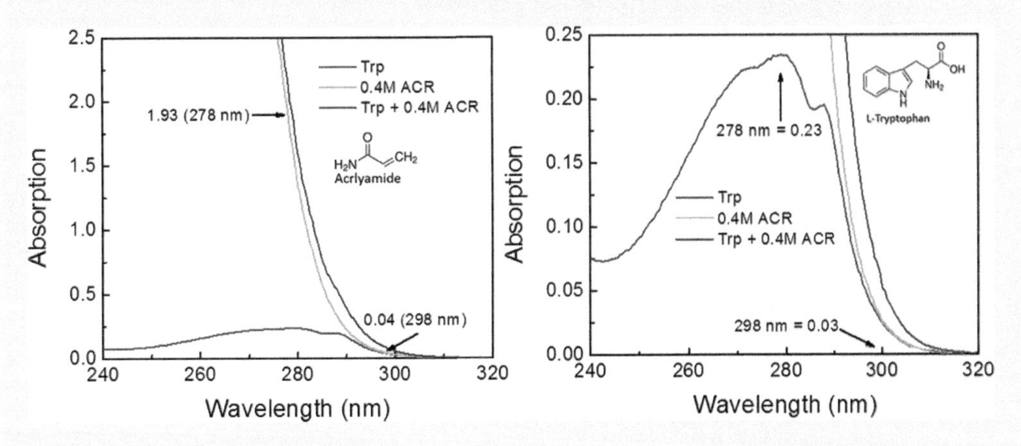

FIGURE 5-11.1 Absorption spectra of Trp, ACR, and Trp + ACR. On the right, we present an expanded scale to visualize Trp absorption. The chemical structures for acrylamide (left) and Trp (right) are shown as inserts.

In Figure 5-11.2, we are presenting measured emission spectra of Trp as a function of acrylamide concentration with 298 nm excitation (left) and 278 nm excitation (right). Since the absorption of acrylamide at 278 nm is significant, we will use 2 mm path cuvette. As discussed in Experiments 5-2 and 5-3, this is necessary to avoid errors due to the relative shift of the emission spot. The spectrofluorometer used in this experiment has a good tolerance for a 1 mm shift so, a 2 mm cuvette will not be a problem (in some instruments, we would recommend only 1 mm cuvette; how to test your instrument we discussed in earlier experiments). To measure the initial emission (with no quencher), we adjusted the spectrofluorometer voltage to have similar initial intensities so, intensity decreases can be directly compared. It is clear that for a few initial concentrations of acrylamide (up to 0.2 M), the intensity changes are very similar. However, the observed change increases significantly for 278 nm excitations for higher acrylamide concentrations and for 0.4 M it is almost twice larger quenching

(Continued)

EXPERIMENT 5-11 (CONTINUED) EFFECT OF ABSORBING QUENCHER ON MEASURED FLUORESCENCE QUENCHING

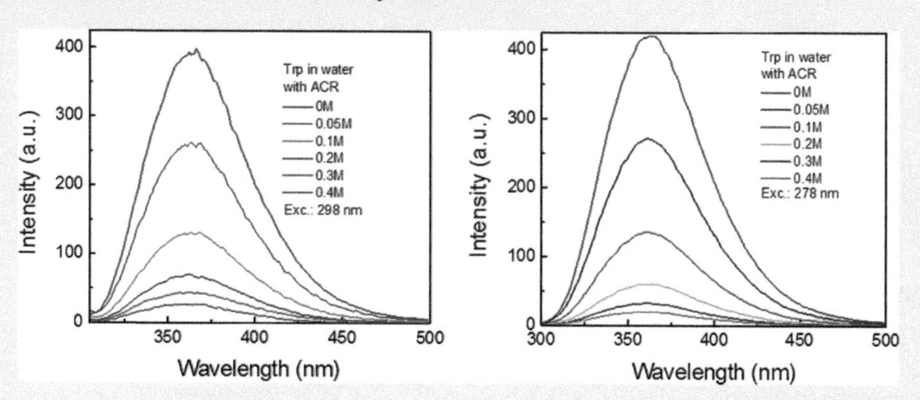

FIGURE 5-11.2 The intensity of Trp fluorescence in the absence and presence of ACR for 298 and 278 nm excitations.

TABLE 5-11.1 Quenching of Trp Fluorescence by Acrylamide at 278 and 298 nm Excitations

Exc. - 278 nm	Int. 360 nm	$(F_0/F)-1$	Exc. - 298 nm	Int. 360 nm	$(F_0/F)-1$
0	418.67	0	0	386.90	0
0.1	135.94	2.07	0.1	129.20	1.99
0.2	59.79	6.00	0.2	68.74	4.62
0.3	32.85	11.74	0.3	43.21	7.95
0.4	19.75	20.19	0.4	26.74	13.46

Note: Absorption of acrylamide (0.4 M) in 1 cm × 1 cm cuvette:
298 nm - 0.043
278 nm - 1.930.

(the emission is almost 50% of that with 298 nm excitation). Just for direct comparison in Table 5-11.1, we present tabularized values of measured intensities and calculated quenching: $(F_0/F)-1$.

In Figure 5-11.3, we are presenting Stern–Volmer plots for 298 nm (left) and 278 nm (right) excitations, respectively. Clearly, the dependence measured at 278 nm is different from that measured at 298 nm excitation. But the quenching should be independent of excitation wavelength (since spectra and fluorescence lifetimes are independent of excitation, the observed quenching should be identical).

In Experiment 5-2, we discussed the procedure for correcting for the absorber at the excitation wavelength. In Table 5-11.2, we are presenting: absorbance, correction factors (see Experiment 5-2), measured and corrected fluorescence intensities at 278 and 298 nm excitations for different acrylamide concentrations.

In Figure 5-11.4 (left), we are presenting Stern–Volmer plots as measured and after applying the correction. We are presenting corrected Stern–Volmer plots for 278 nm (left) and on

(*Continued*)

EXPERIMENT 5-11 (CONTINUED) EFFECT OF ABSORBING QUENCHER ON MEASURED FLUORESCENCE QUENCHING

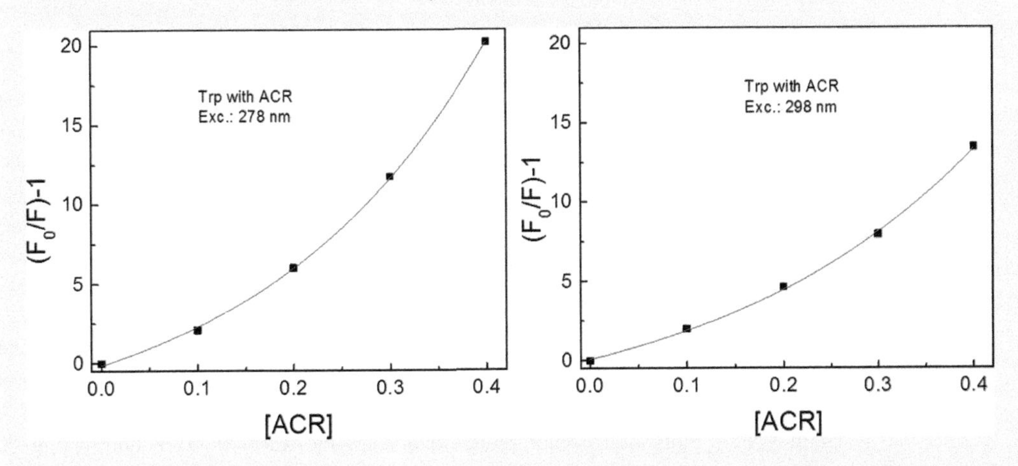

FIGURE 5-11.3 Stern–Volmer plots for Trp fluorescence quenching by ACR with 278 and 298 nm excitations.

TABLE 5-11.2 Correction Factors for Absorbing Quencher (see Experiment 5-2) and Corrected Trp Intensities for 278 and 298 nm Excitations

[ACR]	Abs.	Correction	I_M	I_C	I_{0C}/I_C-1
278 nm					
0	0	1	418.67	418.67	0
0.1	0.096	0.904	135.94	150.3	1.79
0.2	0.193	0.821	59.79	72.8	4.75
0.3	0.29	0.749	32.85	43.87	8.54
0.4	0.386	0.685	19.75	28.82	13.52
298 nm					
0	0	1	386.9	386.9	0
0.1	0.0022	0.998	129.2	129.46	1.98
0.2	0.0045	0.995	68.74	69.085	4.6
0.3	0.006	0.993	43.21	43.515	7.89
0.4	0.009	0.99	26.74	27.01	13.3

the right for 298 nm excitations. It is clear that both represent the same quenching. On the right graph, we also indicated corrected 278 nm excitation as dotted blue line that perfectly overlaps with 298 nm excitation.

Next, we measured Trp quenching by potassium iodide (KI) which is not absorbing in the excitation region, see Figure 5-11.5. In both cases (278 and 298 nm excitations), the measured quenching is equal. There is no need for the correction (we recommend for the new experimentalist to do measurements).

(Continued)

EXPERIMENT 5-11 (CONTINUED) EFFECT OF ABSORBING QUENCHER ON MEASURED FLUORESCENCE QUENCHING

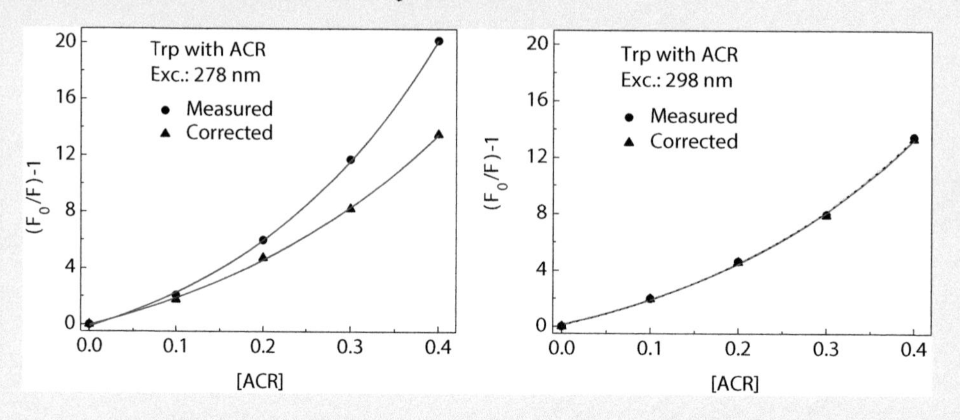

FIGURE 5-11.4 Stern–Volmer plots for Trp fluorescence quenching by ACR with correction for the quencher absorption.

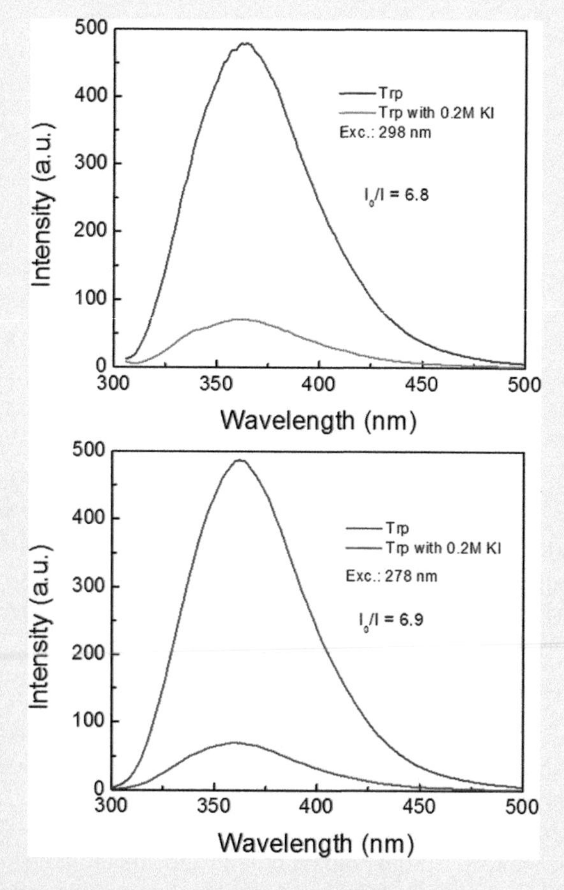

FIGURE 5-11.5 Trp fluorescence quenching by not absorbing quencher KI. There is no difference between 298 and 278 nm excitations.

(*Continued*)

EXPERIMENT 5-11 (CONTINUED) EFFECT OF ABSORBING QUENCHER ON MEASURED FLUORESCENCE QUENCHING

CONCLUSIONS

It is very easy to mix fluorescence quenching and excitation light absorption by the quencher. This, of course, will lead to an artificial overestimation of the quenching constant. The measured values should either be corrected as shown above or the sample should be excited at the wavelength where quencher does not absorb. A similar situation also occurs in FRET measurements when the acceptor can absorb the excitation light.

EXPERIMENT 5-12 EFFICIENCY OF FLUORESCENCE QUENCHERS

Keywords: fluorescence, quenching, heavy atoms, Rhodamine 110, potassium chloride, potassium iodide.

Fluorescence quenching is a result of interactions between fluorophores and molecules or atoms called quenchers. This interaction (quenching) leads to a decrease in the fluorescence signal. The process when the quencher forms a stable complex with the chromophore in a ground state is called static quenching. The process when the quencher only briefly contacts the fluorophore during a molecular collision, we call collisional or dynamic quenching. So, it is no surprise that quenching efficiency could be different for various molecules.

Collisional quenching depends on quencher concentration, diffusion constant, temperature, and quencher effectiveness. Because of that, quenching efficiency will be different for different quenchers. Some quencher molecules are more effective than others. In this experiment, quencher molecules that are similar at a first glance will be shown to quench the same fluorophore differently.

MATERIALS

- Rhodamine 110 (R110)
- Potassium chloride (KCl)
- Potassium bromide (KBr)
- Potassium iodide (KI)
- Water
- 1 cm × 1 cm cuvettes
- Pipettes

EQUIPMENT

- Spectrophotometer
- Spectrofluorometer

METHODS

1. Prepare a 5 mL stock solution of R110 in water with the absorption of about 1 at the maximum.
2. Prepare 5 mL stock solutions (1 M) of KCl, KBr, and KI in water.

(Continued)

EXPERIMENT 5-12 (CONTINUED) EFFICIENCY OF FLUORESCENCE QUENCHERS

3. Make 2 mL samples of R110 fluorophore with various concentrations of quencher, up to 0.2 M in water. Desired concentrations of the quencher (KCl, KBr, or KI) can be achieved by mixing appropriate amounts of stock solutions of R110, quencher, and water, keeping fluorophore concentration, the same in all samples (OD of about 0.1). For sample preparation details, see Experiments 5-9 through 5-11.
4. Measure the fluorescence emission of each sample at magic angle conditions. Start with the sample without the quencher.
5. Construct Stern–Volmer plots for each quencher. We will use Stern–Volmer equation in the form:

$$\frac{F_0}{F} - 1 = K_{SV} \cdot [Q]$$

where F_0 is the fluorescence intensity with no quencher present and F fluorescence intensity at a given quencher concentration. When plotting $F_0/F - 1$ in a function of concentration, $[Q]$ the Stern–Volmer quenching constant, K_{SV} will be represented by the slope of the line. For the brief introduction to quenching processes, we refer the reader to Chapter 1 and more conceptual details to "*Molecular Fluorescence Principles and Applications*" by B. Valeur (2001, Wiley-VCH Verlag, GmbH) and "*Principles of Fluorescence Spectroscopy*" by *J.R. Lakowicz* (2006, Springer Science + Business Media, LLC).

RESULTS

In Figure 5-12.1, we are presenting absorption spectra of the three selected quenchers at 0.2 M concentration in a 1 cm × 1 cm cuvette dissolved in water. Above 300 nm, none of the quenchers absorb the light. In Figure 5-12.2, we show the absorption spectrum of the

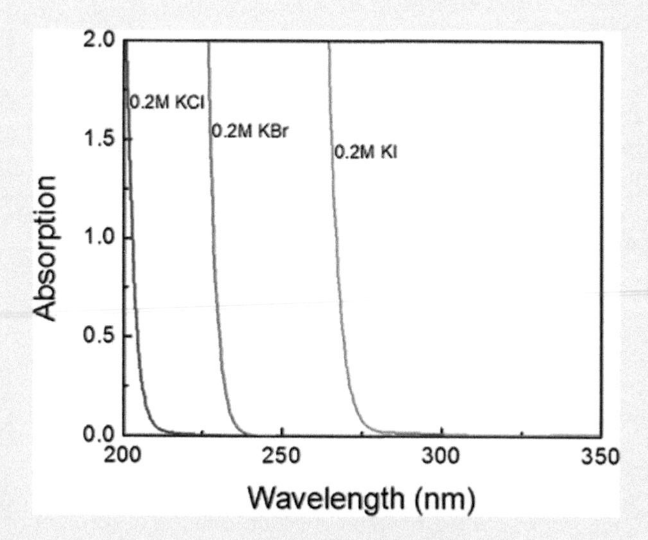

FIGURE 5-12.1 Absorption spectra of quenchers. These quenchers are transparent in the UV-visible spectral region.

(*Continued*)

EXPERIMENT 5-12 (CONTINUED) EFFICIENCY OF FLUORESCENCE QUENCHERS

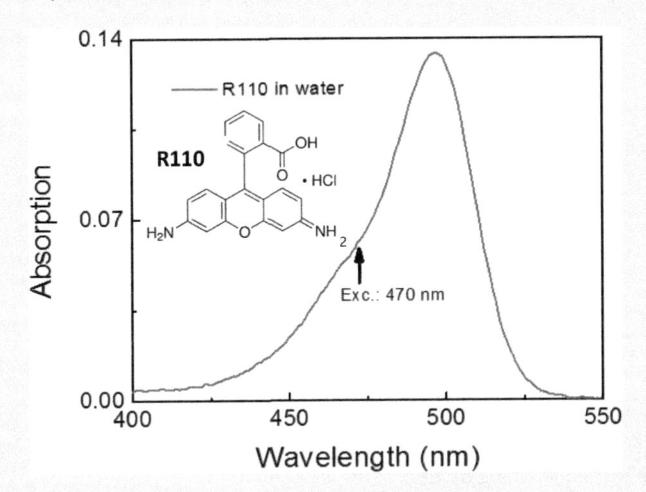

FIGURE 5-12.2 The absorption spectrum of R110 fluorophore. The arrow shows the excitation used in fluorescence measurements.

fluorophore R110. For the excitation, we used wavelength 470 nm that is entirely out of quenchers absorptions.

Figure 5-12.3 left shows the measured emission spectra of R110 for various potassium chloride concentrations. The emission spectrum changes very little what it means KCl very weakly quenches R110. In Figure 5-12.3 (right), we show measured fluorescence intensity change as a function of KCl concentration with line fitted according to a Stern–Volmer equation.

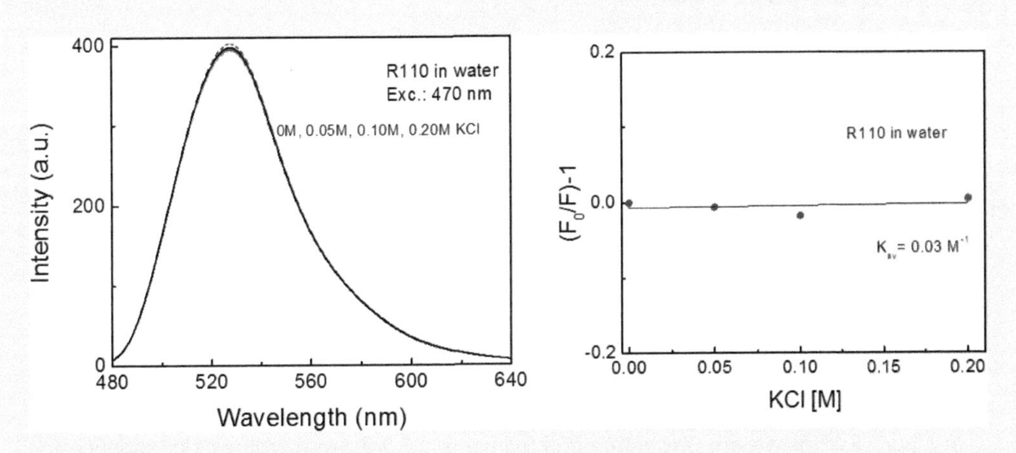

FIGURE 5-12.3 Left: The emission spectrum of R110 in the absence and presence of KCl. Right: Stern–Volmer plot for the R110 quenching by KCl.

The quenching effect of potassium bromide is shown in Figure 5-12.4 (left) and 5-12.4 (right). As the concentration of KBr increases, we observe a modest decrease of fluorescence, indicating reasonable quenching of R110 by KBr.

(Continued)

EXPERIMENT 5-12 (CONTINUED) EFFICIENCY OF FLUORESCENCE QUENCHERS

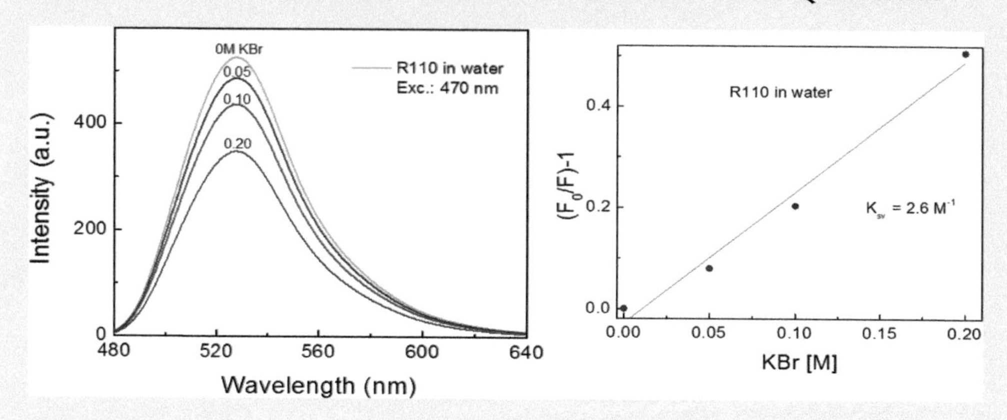

FIGURE 5-12.4 Left: The emission spectrum of R110 in the absence and presence of KBr. Right: Stern–Volmer plot for the R110 quenching by KBr.

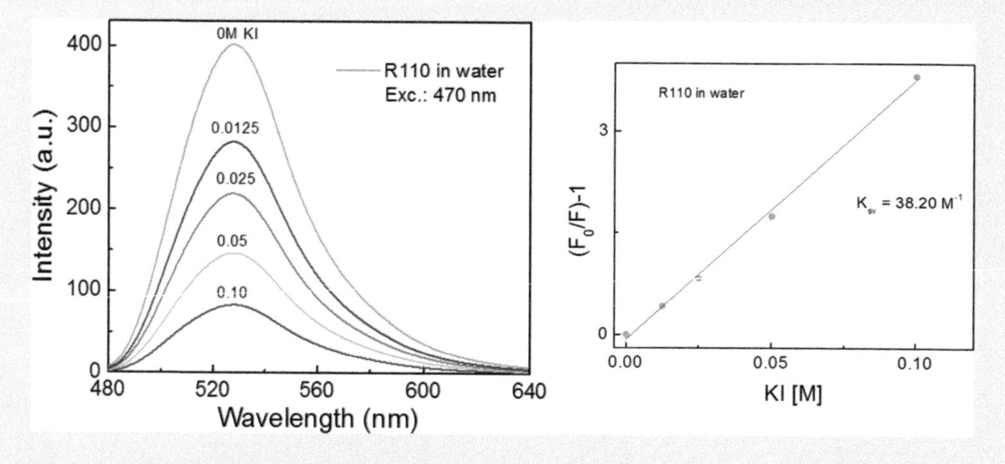

FIGURE 5-12.5 Left: The emission spectrum of R110 in the absence and presence of KI. Right: Stern–Volmer plot for the R110 quenching by KI.

Finally, in Figure 5-12.5, we are presenting quenching by potassium iodide. As indicated in Figure 5-12.5, the fluorescence signal decreases significantly with increasing concentration of KI. Clearly, KI is a very efficient quencher of R110.

In Figure 5-12.6, we are presenting cumulative graph for Stern–Volmer dependence for all three quenchers. In this cumulative plot, it easy to compare slopes for various quenchers. The conclusion is that R110 is quenched very effectively by KI, much less by KBr, and practically not quenched by KCl.

In Figures 5-12.3 through 5-12.5, we indicated the fitted values of Stern–Volmer constant K_{SV}. These values change dramatically from 0.03 M^{-1} to about over 38 M^{-1} when going from KCl to KI.

(Continued)

EXPERIMENT 5-12 (CONTINUED) EFFICIENCY OF FLUORESCENCE QUENCHERS

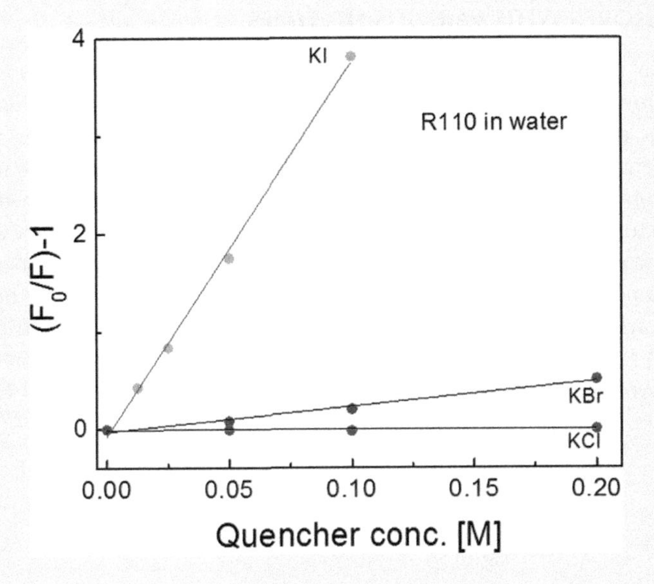

FIGURE 5-12.6 Stern–Volmer plot for the R110 quenching by various quenchers.

CONCLUSIONS

Both static and dynamic quenching requires molecular contact between the fluorophore and quencher. The requirement of molecular contact for quenching results in the numerous biochemical applications of quenching phenomenon. For example, quenching measurements reveal the accessibility of fluorophores to quenchers. In general, KI quenches fluorophores stronger than KBr and KCl. However, first of all, the quenching depends also on the fluorophore. For example, KCl strongly quenches quinine sulfate. This experiment can be also done with other dyes like Rhodamine 6G, Rhodamine 123, Rhodamine B, etc.

EXPERIMENT 5-13 FLUORESCENCE QUENCHING OF FLUOROPHORES WITH VARIOUS LIFETIMES

Keywords: fluorescence, quenching, heavy atoms, Rhodamine 123, Eosin Y, Erythrosine B, potassium iodide.

In most cases, we will be dealing with collisional (dynamic) quenching. As discussed previously, quenching efficiency depends on factors like the type of quencher and solvent viscosity but also depends on the fluorescence lifetime of the chromophore. In the previous Experiment 5-12, we compared the quenching efficiencies of different quenchers on the selected fluorophore (Rhodamine 110). It is intuitively obvious that for a fixed quencher concentration and solvent viscosity, the lifetime of the fluorophore will directly affect the number of fluorophore-quencher encounters. Simply put, the longer the fluorescence lifetime of the chromophore, the more time the quencher will have to diffuse to the

(Continued)

EXPERIMENT 5-13 (CONTINUED) FLUORESCENCE QUENCHING OF FLUOROPHORES WITH VARIOUS LIFETIMES

chromophore in the excited state. It is also expected that various fluorophores will be quenched differently by the same quencher due to, for example, an electron donor-acceptor properties of the dye, and a charge repulsing-attraction for the fluorophore-quencher pair. In this experiment, we will compare quenching efficiencies for similar size molecules with different quantum yields and lifetimes. The difference is a heavy atom (Br or I) in the molecular structure. The addition of a heavy atom in a molecular framework results in quenching of the excited state and a decrease of the fluorophore quantum yield. This also results in concomitant shortening of fluorescence lifetimes. The same fluorophores used in Experiment 5-9 will be applied here, with and without heavy atoms. An intrinsic quenching by heavy atoms reduces lifetimes of fluorophores, as will be shown later in Chapter 7. We selected a very efficient quencher KI but the experiment will work well also with other quenchers.

MATERIALS

- Rhodamine 123 (Rh123)
- Eosin Y (Eos)
- Erythrosine B (ErB)
- Potassium iodide (KI)
- Water
- 1 cm × 1 cm cuvettes

EQUIPMENT

- Spectrophotometer
- Spectrofluorometer

METHODS

1. Prepare stock solutions of Rh123, Eosin Y, and ErB in water, sonicate to make sure whole compounds have been dissolved.
2. Prepare a stock solution of KI in water.
3. Prepare samples of fluorophores with various concentrations of KI up to 0.3 M in water. Desired concentrations of KI can be achieved by mixing appropriate amounts of stock solutions of KI, sample, and water, keeping fluorophore concentration (OD < 0.1), the same in all samples (for details, see previous Experiments 5-9 through 5-11).
4. Measure fluorescence emission of each fluorophore at magic angle conditions. Make Stern–Volmer plots.

RESULTS

All three fluorophores are structurally similar but the brightness of each of them is different. As it has been discussed in Experiment 5-9, this is the effect of the heavy atom.

(Continued)

EXPERIMENT 5-13 (CONTINUED) FLUORESCENCE QUENCHING OF FLUOROPHORES WITH VARIOUS LIFETIMES

It is important to keep the same experimental conditions for each individual fluorophore. However, different set-up (usually voltage) will be used for Rh123, Eos, or ErB.

In Figures 5-13.1 through 5-13.3, we are presenting quenching by KI of R123, Eosin Y, and ErB. The brightest R123 is strongly quenched while the weakest ErB is barely quenched by KI. In Figure 5-13.4, we are presenting quenching of all three dyes in the form of Stern–Volmer dependence.

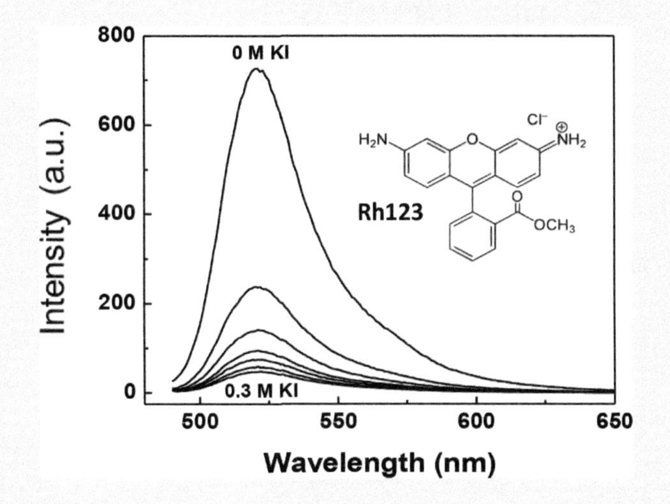

FIGURE 5-13.1 Quenching of Rh123 in water by KI.

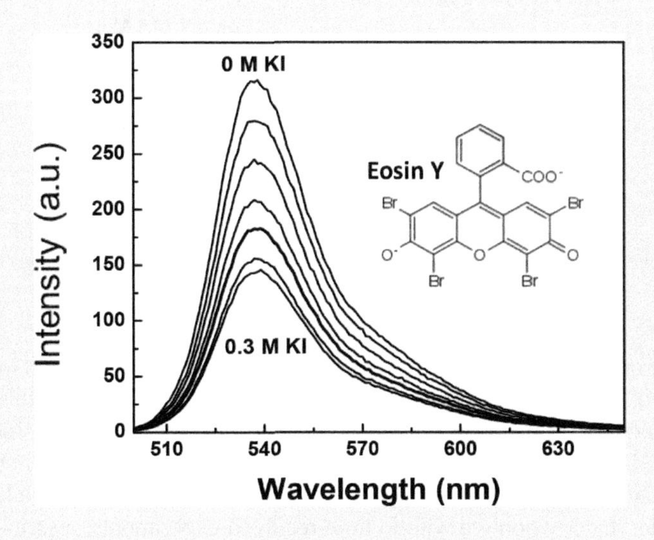

FIGURE 5-13.2 Quenching of Eos in water by KI.

(Continued)

EXPERIMENT 5-13 (CONTINUED) FLUORESCENCE QUENCHING OF FLUOROPHORES WITH VARIOUS LIFETIMES

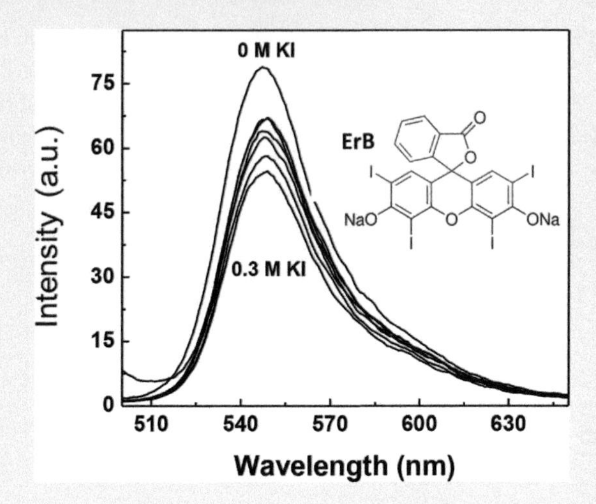

FIGURE 5-13.3 Quenching of ErB in water by KI.

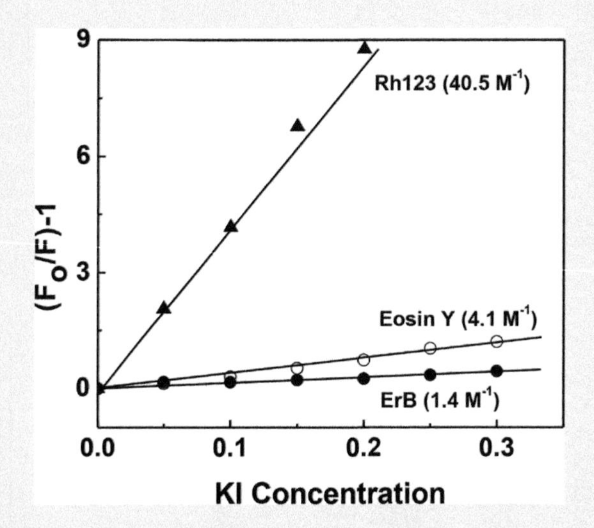

FIGURE 5-13.4 Stern–Volmer plots for KI quenching of Rh123, Eos, and ErB in water at 20°C.

CONCLUSIONS

It is evident that quenching efficiency strongly depends on the lifetime, and in this case also quantum yield, of a fluorophore. We will get to a detailed explanation of this effect later in the advanced section. In this experiment, we are adding additional collisional quenching mechanism that is an additional competitive process to the internal quenching by heavy atom in Eosin Y and ErB. In effect, the apparent efficiency of collisional quenching for Eosin Y and ErB is smaller. Later when we will do time-resolved experiments, a simpler explanation of this phenomenon will be the shorter fluorescence lifetime of the heavy atom substituted molecules that lead to a lower chance for the quencher and fluorophore to collide during the time fluorophore is in the excited state.

EXPERIMENT 5-14 DEPENDENCE OF FLUORESCENCE QUENCHING ON DIFFUSION

Keywords: fluorescence, quenching, Coumarin 152, potassium iodide.

Few previous experiments clearly indicated that collisional quenching directly depends on fluorophore lifetime and quencher effectiveness. A fluorophore with a longer fluorescence lifetime simply has more time in the excited state to encounter the quencher. Conversely, lower viscosity increases diffusion which also increases the probability of encountering the quencher during the time the fluorophore is in the excited state. In this experiment, we want to demonstrate how a solvent viscosity affects the quenching of the fluorophore. Coumarin 152 (C152) has similar fluorescence lifetimes in ethanol and propylene glycol, about 1.6 ns (in Chapter 7, we will present time-resolved experiments but for a moment we need you to trust us). But these two solvents have very different viscosities. Ethanol is a low viscosity solvent and diffusion should be very fast, while propylene glycol has a high viscosity and diffusion will be strongly limited.

MATERIALS

- Coumarin 152 (C152)
- Potassium iodide (KI)
- Ethanol (EtOH), water, propylene glycol (PG)
- 1 cm × 1 cm cuvettes
- Pipettes

EQUIPMENT

- Spectrophotometer
- Spectrofluorometer

METHODS

1. Prepare stock solutions of C152 in EtOH and PG with absorptions of about 1.0.
2. Prepare a 2 M stock solution of KI in water. Do not use ethanol or PG for KI stock solution.
3. Make 2 mL samples of C152 dye with various concentrations of KI up to 0.2 M in both solvents, EtOH and PG. Desired concentrations of KI can be achieved by mixing appropriate amounts of stock solutions of KI, a stock solution of the sample, and the solvent/buffer. For details of sample preparation see previous experiments. For example, in order to get the sample with 0.2 M KI you need to mix 0.2 mL of C152 stock solution, 0.2 mL KI stock solution, and 1.6 mL of the solvent. Keep fluorophore concentration (OD of about 0.1), the same in all samples and references (no KI). Alternatively, you can make larger volumes (10 mL) of the sample with highest KI concentration (0.2 M) and the sample without KI, and mix these two solutions in order to get a proper KI concentration.
4. Measure fluorescence emission of the C152 fluorophore in each solvent at magic angle conditions.
5. Construct Stern–Volmer plots of C152 quenching by KI in both solvents.

(Continued)

EXPERIMENT 5-14 (CONTINUED) DEPENDENCE OF FLUORESCENCE QUENCHING ON DIFFUSION

RESULTS

Absorption spectra of C152 in PG and EtOH are shown in Figure 5-14.1. The spectra are very similar to close absorptions.

Fluorescence spectra of C152 in EtOH in absence and presence of KI are shown in Figure 5-14.2 (left). On the right, we are showing the corresponding Stern–Volmer plot. Fitted with Stern–Volmer quenching constant $K_{SV} = 14.3\ M^{-1}$.

Similar Figure 5-14.3 shows the fluorescence of C152 as a function of KI concentration (left) and corresponding Stern–Volmer plot. The calculated value of Stern–Volmer

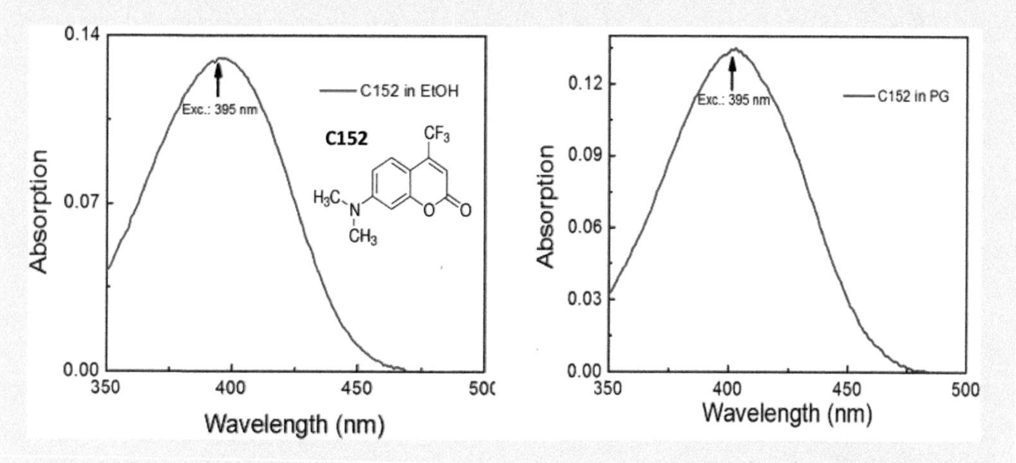

FIGURE 5-14.1 The absorption spectra of C152 in EtOH (left, red) and PG (right, blue). The arrows indicate the excitation wavelength used in fluorescence measurements.

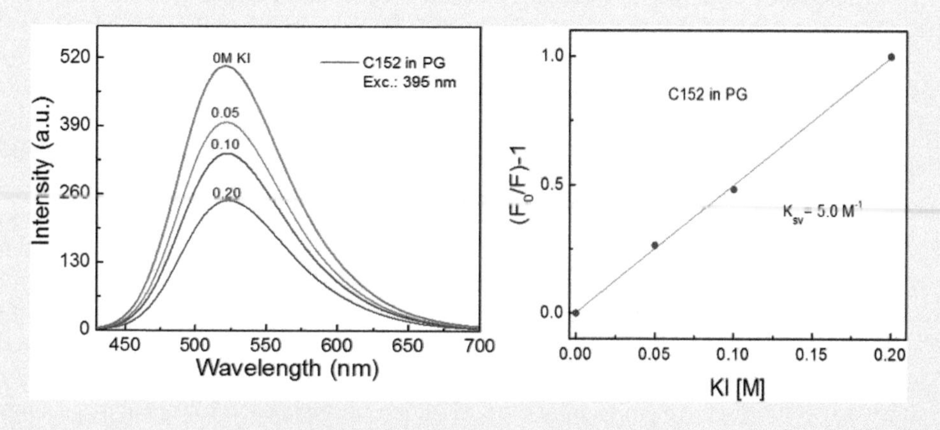

FIGURE 5-14.2 Left: Fluorescence spectra of C152 in PG in the absence and presence of KI. Right: Stern–Volmer plot for C152 quenching by KI in PG.

(Continued)

EXPERIMENT 5-14 (CONTINUED) DEPENDENCE OF FLUORESCENCE QUENCHING ON DIFFUSION

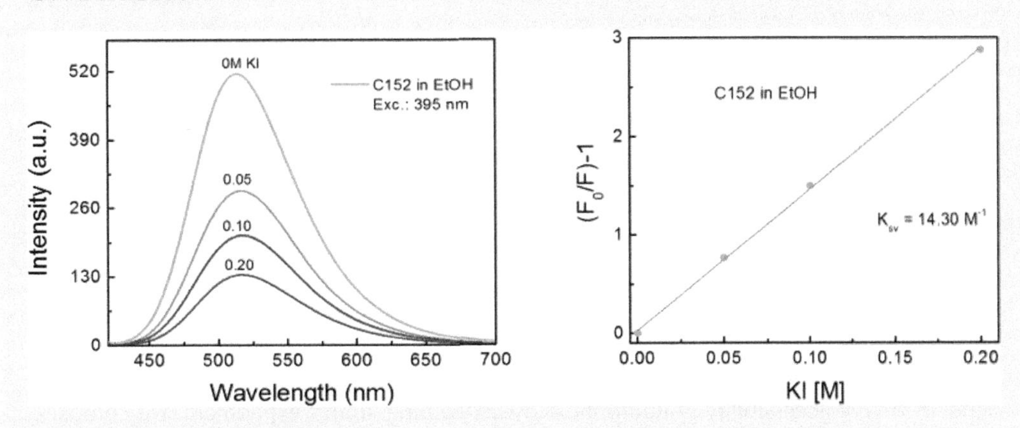

FIGURE 5-14.3 Left: Fluorescence spectra of C152 in EtOH in the absence and presence of KI. Right: The Stern–Volmer plot for C152 quenching by KI in EtOH.

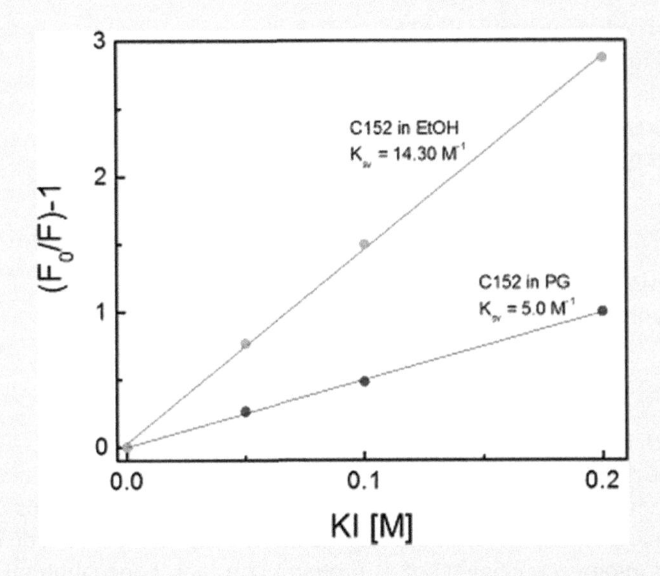

FIGURE 5-14.4 The Stern–Volmer plots for C152 quenching by KI in various solvents.

quenching constant in PG is $K_{SV} = 5\ M^{-1}$ that is almost three times smaller than for EtOH. Just for comparison, we are showing both Stern–Volmer plots in Figure 5-14.4. Visually, the slopes are very different.

Note: With similar lifetimes in both solvents, the difference of the quenching efficiency is due to different solvent viscosities. PG is more viscous than EtOH, therefore, the quenching of C152 in PG is less effective than in the EtOH.

EXPERIMENT 5-15 ACCESSIBILITY OF INTRINSIC FLUOROPHORES BY THE QUENCHER

Keywords: fluorescence, quenching, acrylamide, human serum albumin (HSA), *N*-acetyl-tryptophan amide (NATA).

Fluorescence quenching requires molecular contact between the chromophore molecule and the quencher molecule. Because of this, quenching may have numerous applications by revealing important information about biochemical systems. For example, quenching measurements can reveal the accessibility of fluorophores to quenchers. Consider a fluorophore that is a part of a protein (e.g., tryptophan) or is bound to a protein. If the protein is impermeable to the quencher, and the fluorophore is located in the interior of the macromolecule, then neither collisional nor static quenching can occur. For this reason, quenching studies can be used to reveal the localization of fluorophores in proteins. Acrylamide is a known dynamic quencher. This protocol describes how quenching efficacy depends on the accessibility of fluorophore by acrylamide. In this experiment, we compare quenching of a free tryptophan analog (*N*-acetyl-tryptophan amide) to an intrinsic tryptophan moiety in protein human serum albumin (HSA).

MATERIALS

- *N*-Acetyl-tryptophan amide (NATA)
- Human serum albumin (HSA)
- Acrylamide
- PBS Buffer (pH 7.4)
- 1 cm × 1 cm quartz cuvettes

EQUIPMENT

- Spectrophotometer
- Spectrofluorometer

METHODS

1. Make stock solutions of NATA and HSA protein in PBS buffer.
2. Make a stock solution of 2 M acrylamide in PBS buffer.
3. Prepare samples containing NATA (constant concentration) and various concentrations of acrylamide: 0, 0.05, 0.10, 0.15, 0.20, 0.25, and 0.30 M of acrylamide in PBS.
4. Prepare samples containing HSA protein (constant concentration) and various concentrations of acrylamide: 0, 0.05, 0.10, 0.15, 0.20, 0.25, and 0.30 M of acrylamide in PBS.
5. Measure fluorescence spectra of NATA series using excitation wavelength 298 nm at magic angle conditions. We recommend long-wavelength tryptophan excitation due to increased absorption of acrylamide toward the UV. In general, shorter wavelength excitation should work but we will need to make sure to correct for quencher absorption (see Experiment 5-2).
6. Measure fluorescence spectra of HSA protein series using excitation wavelength 298 nm at magic angle conditions.

(*Continued*)

EXPERIMENT 5-15 (CONTINUED) ACCESSIBILITY OF INTRINSIC FLUOROPHORES BY THE QUENCHER

RESULTS

In Figure 5-15.1, we present fluorescence spectra of NATA measured at different concentrations of acrylamide. The decrease in fluorescence is high indicating easy access of quencher to tryptophan moiety.

Figure 5-15.2 shows the effect of increasing acrylamide concentration on tryptophan fluorescence buried in the protein structure. It is known that tryptophan residue in HSA is

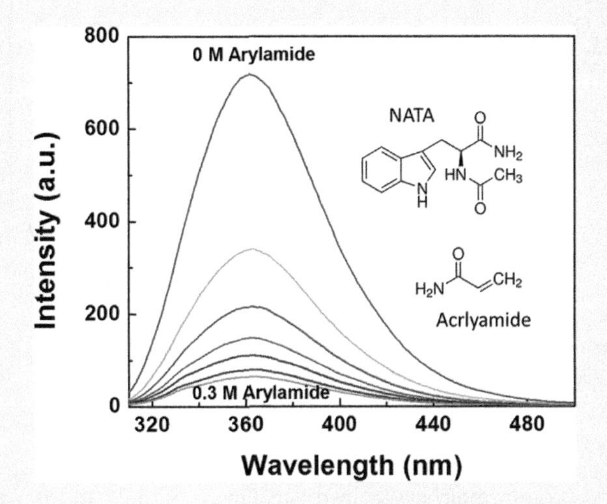

FIGURE 5-15.1 Acrylamide quenching of NATA. The concentrations of acrylamide were: 0, 0.05, 0.10, 0.15, 0.20, 0.25, and 0.30 M.

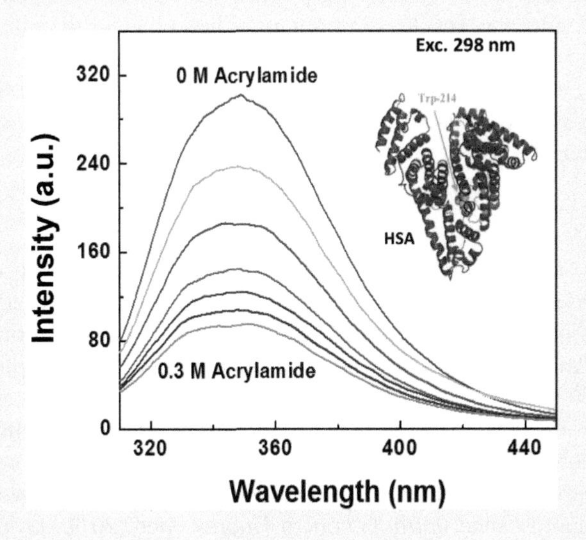

FIGURE 5-15.2 Acrylamide quenching of HSA tryptophan emission. The concentrations of acrylamide were: 0.05, 0.10, 0.15, 0.20, 0.25, and 0.30 M.

(Continued)

EXPERIMENT 5-15 (CONTINUED) ACCESSIBILITY OF INTRINSIC FLUOROPHORES BY THE QUENCHER

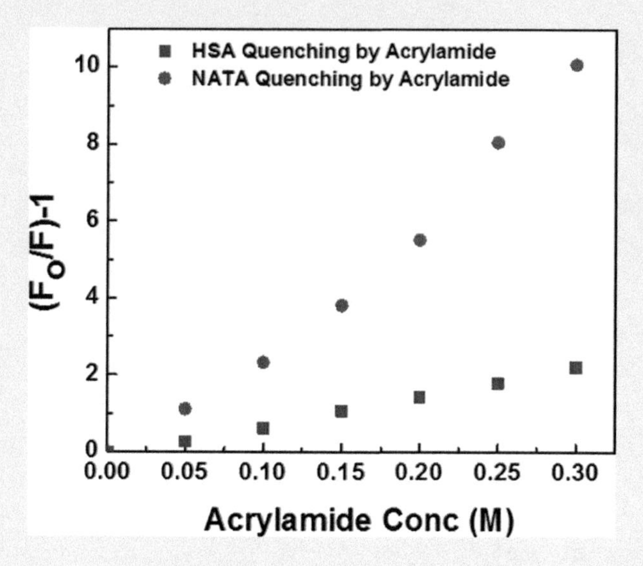

FIGURE 5-15.3 Stern–Volmer plots of NATA and HSA fluorescence quenching by acrylamide. The Stern–Volmer constant for HSA acrylamide quenching is several folds smaller than for NATA.

located deep in the protein matrix (see insert in Figure 5-15.2) and is well shielded from water molecules. In both cases, NATA and HSA we used the same concentrations of acrylamide. In the case of NATA, the first addition (0.05 M) produced significant quenching while in case of HSA the effect is much less pronounced. It is a reasonable expectation that the quenching of tryptophan residues in proteins is less efficient than in the case of a free fluorophore.

In Figure 5-15.3, we are presenting Stern–Volmer dependencies for both NATA and HSA. It is clear that slopes are very different clearly revealing a very different value for the Stern–Volmer quenching constants.

CONCLUSIONS

In this experiment, it is important to excite at a long-wavelength side of the tryptophan absorption, 298 nm or longer. This is because of a residual quencher (acrylamide) absorption which can attenuate the excitation at higher quencher concentrations. For details, see Experiments 5-11 and 5-12. As a precaution, always check the absorption of the highest concentration of the quencher.

The acrylamide quenching of NATA displays an upper-curvature, an effect often observed in quenching studies. The more advanced model of an active sphere can satisfactorily fit such data. For more advanced models of quenching phenomena, we refer readers to recent books on fluorescence (*Introduction to Fluorescence, 2014, Taylor & Francis CRC, D. Jameson; Molecular Fluorescence. Principles and Applications, 2001, Wiley-VCH Verlag, GmbH, B. Valeur; Principles of Fluorescence Spectroscopy, 2006, 1999, 1983, Springer Science+Business Media, LLC, J.R. Lakowicz*).

EXPERIMENT 5-16 FLUORESCENCE QUENCHING: ACCESSIBILITY OF EXTRINSIC FLUOROPHORE BY THE QUENCHER

Keywords: fluorescence, quenching, fluorescein, fluorescein NHS ester, dimethylformamide, human serum albumin (HSA), potassium iodide (KI).

In the previous experiment, the dynamic quenching of intrinsic tryptophan fluorescence by acrylamide was introduced. A similar quenching effect will occur for an extrinsic fluorophore bound to the protein. Depending on where the binding site for the fluorophore is located, the quenching of the free and bound fluorophore should be different. At the same time, a probe located on the surface and exposed to the solution will still be partially shielded by the protein body. Similar to acrylamide, potassium iodide (KI) is a known dynamic quencher. In this experiment, we will label the protein with fluorescein NHS ester and compare relative fluorescence of a free dye in solution to that of the dye bound to the protein at different concentrations of quencher (KI). This protocol describes how the quenching efficacy depends on the accessibility of fluorescein by iodide when the dye is free or labeled/attached to the protein human serum albumin (HSA).

MATERIALS

- Fluorescein (Fl)
- Fluorescein NHS ester
- Dimethylformamide (DMF)
- 5 mL pre-packed Sephadex G-25 column
- Human serum albumin (HSA)
- Potassium iodide (KI)
- Buffer (pH9, TRIS)
- 1 cm × 1 cm cuvettes

EQUIPMENT

- Spectrophotometer
- Spectrofluorometer

METHODS

1. Label HSA with fluorescein NHS ester. Stock solutions of fluorescein and fluorescein NHS ester were prepared in dry DMF. The fluorescein sample for KI quenching was prepared by adding a stock solution to the 5 mL of TRIS buffer pH-9 until an optical density of the sample was 0.05. The Fl-labeled HSA sample was prepared as follow: To 0.5 mL of 10 μM HSA solution was added fluorescein NHS ester, in less than 10 μL of DMF. The sample was protected from light and gently shaken for 3 hours. The labeled protein was separated on a 5 mL pre-packed Sephadex G-25 column using TRIS buffer pH 9. The extent of labeling was around 2–3 fluorescein molecules per HAS (see Experiment 3-22 for detailed labeling procedure). The labeled protein concentration was diluted with TRIS buffer to 0.25 μM to keep sample optical density around 0.05. Desired concentrations of KI were achieved by adding appropriate amounts of a stock solution of KI (1 M) and buffer to the sample, keeping the same fluorophore concentration in all samples.

(Continued)

EXPERIMENT 5-16 (CONTINUED) FLUORESCENCE QUENCHING: ACCESSIBILITY OF EXTRINSIC FLUOROPHORE BY THE QUENCHER

2. Make stock solutions of fluorescein NHS ester and labeled HSA protein in TRIS buffer (pH-9).
3. Make a stock solution of 2 M KI in TRIS buffer.
4. Using procedures analogical to previous experiments prepare samples containing fluorescein NHS ester (constant concentration) and various concentrations of KI (in our case 0, 0.025, 0.04, 0.07, 0.11, 0.180, and 0.30 M) of KI in TRIS pH-9. Do the same for labeled HSA protein.
5. Measure fluorescence spectra for series of both solutions using excitation wavelength in the 470–490 nm range (remember to use magic angle conditions).

RESULTS

In order to deactivate an excited state of the fluorophore, quencher molecules must approach closely (practically get in contact with the fluorophore). In the solution of low viscosity, many encounters of fluorophore–quencher molecules occur during the time scale of the fluorophore lifetime. The accessibility of fluorophore by the quencher plays an important role in the observed quenching process.

In Figure 5-16.1, we are presenting fluorescence intensities measured for fluorescein free in the solution and bound to HSA for various KI concentrations. The quenching of free fluorescein is significant while quenching of fluorescein bound to protein is much weaker.

In Table 5-16.1, we present calculated values of Stern–Volmer parameters and in Figure 5-16.2, we are presenting Stern–Volmer plots for both systems.

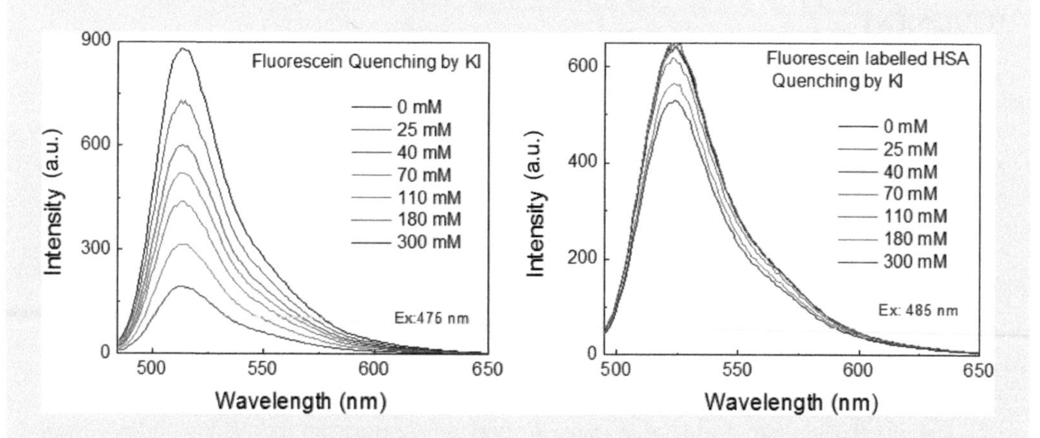

FIGURE 5-16.1 Quenching of the fluorescein fluorescence by potassium iodide (left) and fluorescein-labeled HSA (right).

(Continued)

EXPERIMENT 5-16 (CONTINUED) FLUORESCENCE QUENCHING: ACCESSIBILITY OF EXTRINSIC FLUOROPHORE BY THE QUENCHER

TABLE 5-16.1 Stern–Volmer Parameters of Fluorescein and Fluorescein-HSA Quenching

KI Concentration	Fluorescein	Fluorescein-HAS
0	0.00	0.00
25	0.22	0.00
40	0.40	0.01
70	0.86	0.09
110	1.27	0.13
180	1.79	0.22
300	2.98	0.35

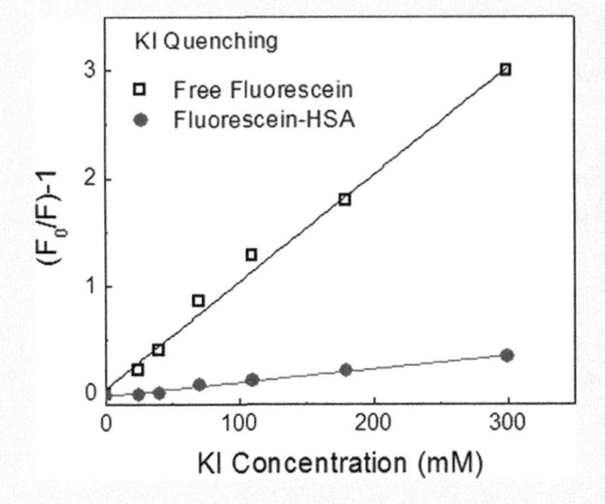

FIGURE 5-16.2 Stern–Volmer plots for the KI quenching of free fluorescein and labeled on HSA protein.

CONCLUSIONS

As predicted, free fluorescein is very effectively quenched by the KI. At the room temperature, water is a relatively low viscosity solvent and for fluorescence lifetime of free fluorescein close to 4 ns, an efficient quenching is expected. In the same time, the fluorescein molecules that are attached to the surface of the protein are effectively shielded from the quencher. It could be a surprise that the decrease in quenching is far more than twice. There are a few reasons for that. The protein surface is typically well folded and chromophores frequently have a tendency to bind into the well-formed pockets where the quencher accessibility is limited. Also, bound chromophores can be quenched by proximal amino acids shortening fluorescence lifetime. We would typically observe fluorescence lifetime to be 20–50% shorter for fluorescein bound to the protein.

Note: We purposely kept labeling efficiency low (~2 fluoresceins per protein) to limit fluorescein self-quenching.

EXPERIMENT 5-17 FLUORESCENCE QUENCHING OF QUININE SULFATE BY SALT

Keywords: fluorescence, quenching, quinine sulfate, salt (NaCl).

In Experiment 5-12 (*Efficiency of Fluorescence Quenchers*), it was demonstrated that potassium chloride practically did not quench rhodamine fluorescence. The quenching was sensitive to the heavy metal and followed the direction Cl < Br < I, being strongest for potassium iodide. In general, this is always true, KI is an excellent quencher for most fluorophores and KCl is not. However, there are fluorophores sensitive to halogens, including chloride. The primary example is the oldest known fluorophore—quinine sulfate. It is probably the most studied fluorescence standard for spectra and its quantum yield is very bright in an acidic solution of sulfuric acid. The first aim of this experiment will be to show that in hydrochloric acid, the fluorescence of quinine sulfate is weak compared to sulfuric acid. The second, that a salt (just kitchen salt) is an efficient quencher of quinine sulfate.

MATERIALS

- Quinine sulfate (QS)
- Sulfuric acid, hydrochloric acid
- Salt (NaCl)
- Water
- 1 cm × 1 cm cuvettes, vials
- Camera or cell phone, 405 nm laser pointer

EQUIPMENT

- Spectrophotometer
- Spectrofluorometer

METHODS

1. Make a stock solution of QS in water or ethanol.
2. Prepare 10 mL solutions of 0.1 M, H_2SO_4, and HCl. Add the same amount of QS stock solution to both. Measure absorption of both samples, see Figure 5-17.1.
3. Measure fluorescence spectra of both samples using 350 nm excitation wavelength, see Figure 5-17.2. Using the cell phone and 405 nm laser pointer, make a photograph of both samples illuminated simultaneously, see Figure 5-17.2, right.
4. Prepare 1 M stock solution of NaCl. Prepare samples of QS in 0.1 N H_2SO_4 containing various amounts of NaCl, from 0 to 0.1 M NaCl. Measure absorption spectra of prepared samples, see Figure 5-17.3.
5. Measure fluorescence spectra of the series of samples using a 350 nm excitation wavelength, see Figure 5-17.4. Make the Stern–Volmer plot, see Figure 5-17.5.

RESULTS

Absorption spectra of QS in 0.1 M sulfuric and hydrochloric acids are very similar, see Figure 5-17.1. Both samples contain the same amount of acid molecules (0.1 M) and the same concentration of QS.

(*Continued*)

EXPERIMENT 5-17 (CONTINUED) FLUORESCENCE QUENCHING OF QUININE SULFATE BY SALT

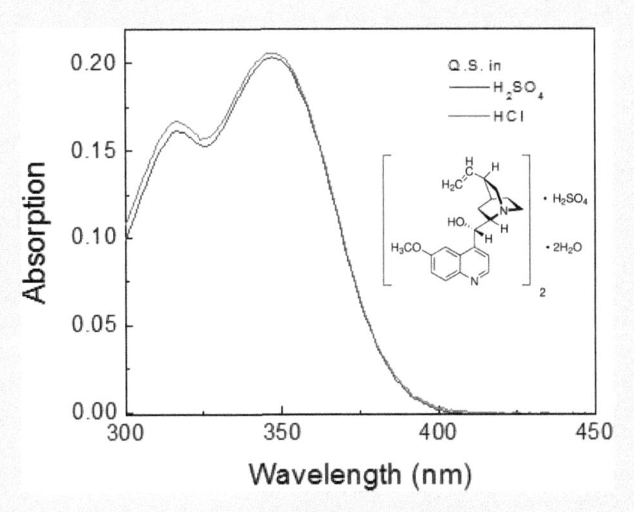

FIGURE 5-17.1 Absorption spectra of QS in 0.1 M H_2SO_4 and in 0.1 M HCl.

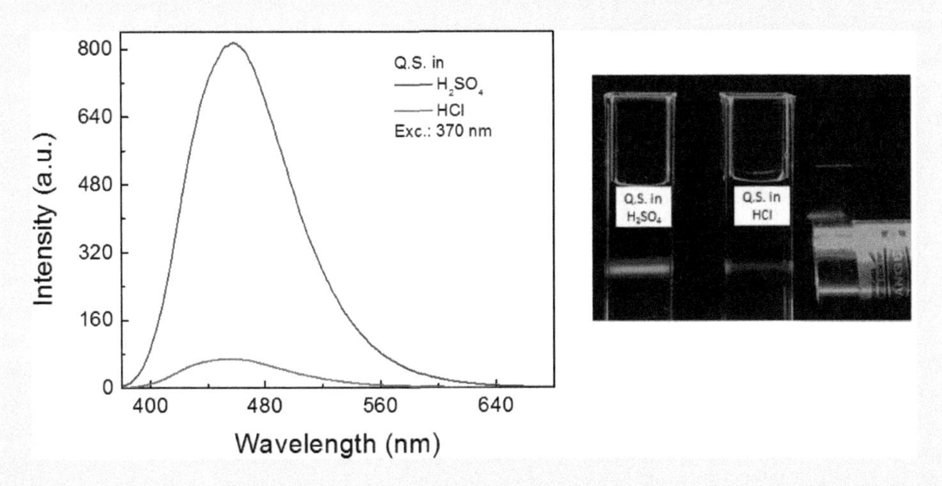

FIGURE 5-17.2 Left: Fluorescence spectra of QS in 0.1 M H_2SO_4 and in 0.1 M HCl. Right: Photograph of the samples used upon illumination from 405 nm blue laser pointer.

With adjusted QS concentrations in both samples, we measured fluorescence spectra, see Figure 5-17.2. The emission of QS solution in HCl is much weaker even if absorptions are practically identical. To visualize this effect on Figure 5-17.2 (right), we are showing a photograph of QS solutions in H_2SO_4 and HCl when simultaneously illuminated with 405 laser pointer.

Next, we measure absorption and fluorescence spectra of QS in 0.1 N H_2SO_4 in the absence and presence of NaCl, see Figures 5-17.3 and 5-17.4. The absorption spectra do

(Continued)

EXPERIMENT 5-17 (CONTINUED) FLUORESCENCE QUENCHING OF QUININE SULFATE BY SALT

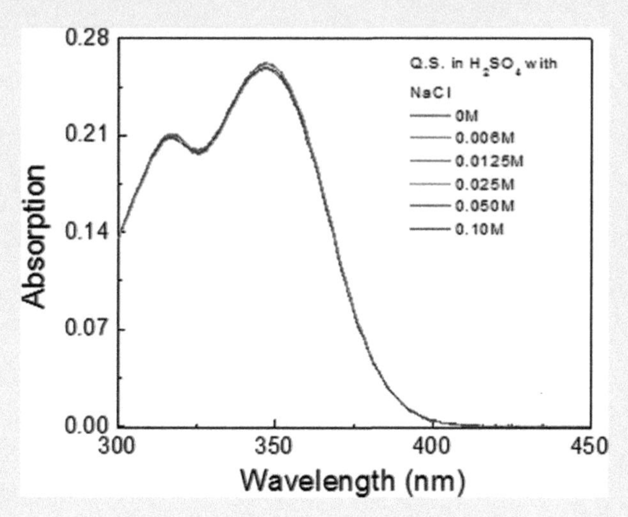

FIGURE 5-17.3 Absorption spectra of QS in 0.1 N H_2SO_4 in the absence and presence of NaCl.

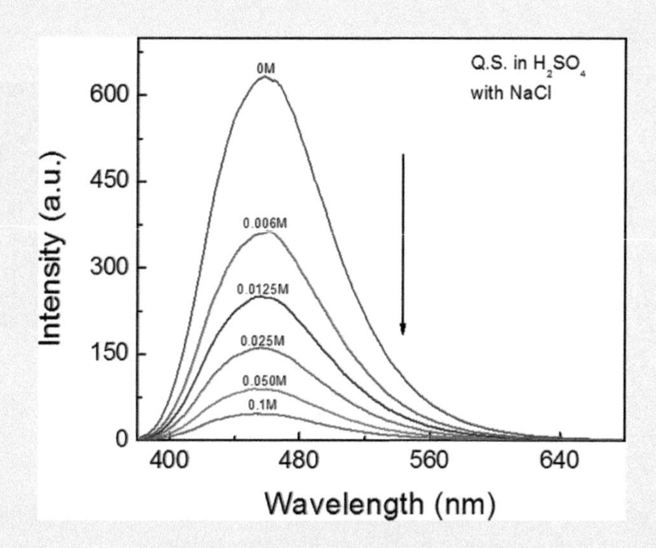

FIGURE 5-17.4 Fluorescence spectra QS in 0.1 N H_2SO_4 in the absence and presence of NaCl.

not change upon addition of NaCl up to 0.1 M. In contrast, fluorescence spectra as shown in Figure 5-17.4 strongly depend on the presence of NaCl.

The Stern–Volmer quenching constant is high and the dependence is linear, see Figure 5-17.5. The value of K_{SV} is significantly higher than constants recovered for rhodamins quenched by potassium iodide (see Experiments 5-12 and 5-15). The reason for this is a longer lifetime of QS.

(Continued)

EXPERIMENT 5-17 (CONTINUED) FLUORESCENCE QUENCHING OF QUININE SULFATE BY SALT

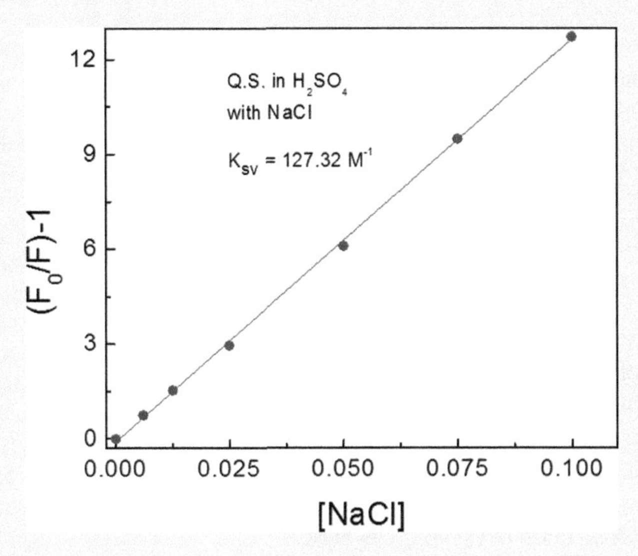

FIGURE 5-17.5 The Stern–Volmer plot of QS fluorescence quenching by NaCl.

CONCLUSIONS

The QS absorption spectra in the presence of NaCl do not show any new bands indicating that there are not ground state complexes with different absorption. Comparison of QS emission in H_2SO_4 and HCl shows much weaker emission in HCl. Also, addition of NaCl to the solution of QS in H_2SO_4 results in a significant decrease of fluorescence (quenching) with a significant Stern–Volmer constant. Potassium chloride acts very similar to NaCl which indicates that chloride ions are responsible for the very effective quenching.

EXPERIMENT 5-18 FLUORESCENCE QUENCHING OF QUININE SULFATE BY HALOGENS

Keywords: fluorescence, quenching, quinine sulfate, potassium iodide.

In Experiment 5-17 (*Fluorescence quenching of Quinine Sulfate by Salt*), fluorescence quenching by kitchen salt was observed, proving to be efficient. However, kitchen salt is not usually considered to be a good quencher. In this experiment, we will evaluate the fluorescence quenching properties of quinine sulfate by a series of halogens using potassium fluoride, chloride, bromide, and iodide.

(*Continued*)

EXPERIMENT 5-18 (CONTINUED) FLUORESCENCE QUENCHING OF QUININE SULFATE BY HALOGENS

MATERIALS

- Quinine sulfate (QS)
- Sulfuric acid
- Potassium fluoride (KF), potassium chloride (KCl), potassium bromide (KBr), potassium iodide (KI)
- Water
- 1 cm × 1 cm cuvettes, vials

EQUIPMENT

- Spectrophotometer
- Spectrofluorometer

METHODS

1. Make a stock solution of QS in 0.1 N H_2SO_4.
2. Prepare 100 mL solutions of QS in 0.1 N H_2SO_4 with the absorption of about 0.1 at maximum, see Figure 5-18.1.
3. Prepare 10 mL of solutions (0.1 M) of each quencher (KF, KCl, KBr, and KI) using QS solution prepared in #2. By mixing the QS solution (#2) with quencher solution, one can obtain any desired concentration of the quencher keeping the same concentration of QS. For example, in order to get 10 mM concentration of the quencher you need to mix 1.8 mL of QS solution with 0.2 mL of the quencher solution for total 2 mL of the sample volume. Of course, you can use any other method for samples preparation described in Experiment 5-10.
4. Make samples with 10 mM concentration of each quencher and measure fluorescence spectra of all samples. Start with QS solution without any quencher, see Figure 5-18.2.
5. Prepare samples with various concentrations of the chosen quencher and measure fluorescence spectra at 350 nm excitation. Always start with the unquenched sample, see Figures 5-18.3 through 5-18.6.
6. Construct Stern–Volmer plots for each quencher, see Figure 5-18.7.

RESULTS

Absorption spectra of QS in the presence of 0.1 M of any KF, KCl, KBr, and KI do not differ from each other, see Figure 5-18.1. However, the fluorescence spectra are significantly different, see Figure 18-5.2.

Whereas KF practically does not quench QS, other halogens are efficient quenchers.

(Continued)

EXPERIMENT 5-18 (CONTINUED) FLUORESCENCE QUENCHING OF QUININE SULFATE BY HALOGENS

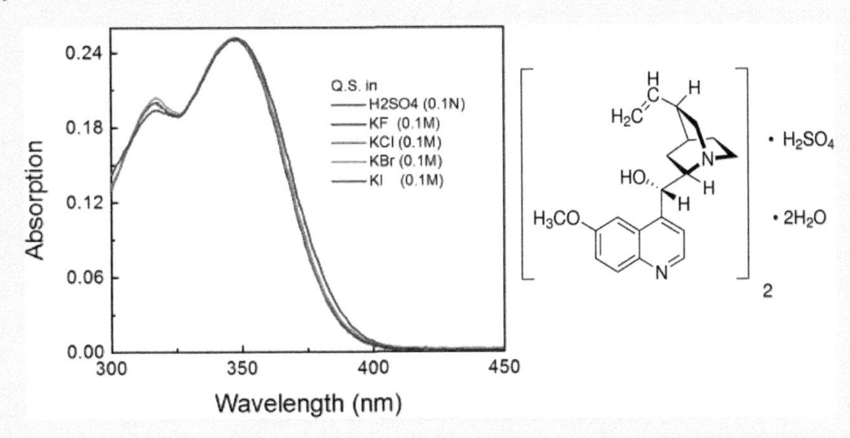

FIGURE 5-18.1 Absorption spectra of QS in 0.1 N H_2SO_4 and in 0.1 M of halogens.

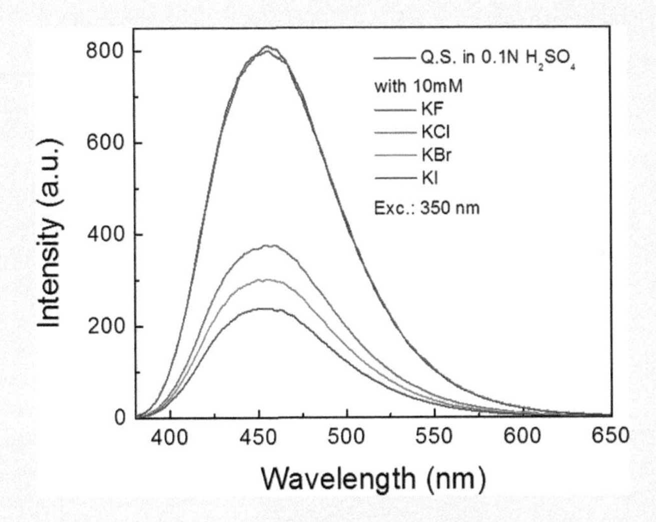

FIGURE 5-18.2 Fluorescence spectra of QS in 0.1 M H_2SO_4 without (red) and with 10 mM of different quenchers.

Next, we measure quenching of QS in the presence of various concentrations of halogens, see Figures 5-18.3 through 5-18.6.

Note, that concentrations of the quenchers are different for each halogen.

Next, we construct Stern–Volmer plots for QS quenching by halogens, see Figure 5-18.7.

(Continued)

EXPERIMENT 5-18 (CONTINUED) FLUORESCENCE QUENCHING OF QUININE SULFATE BY HALOGENS

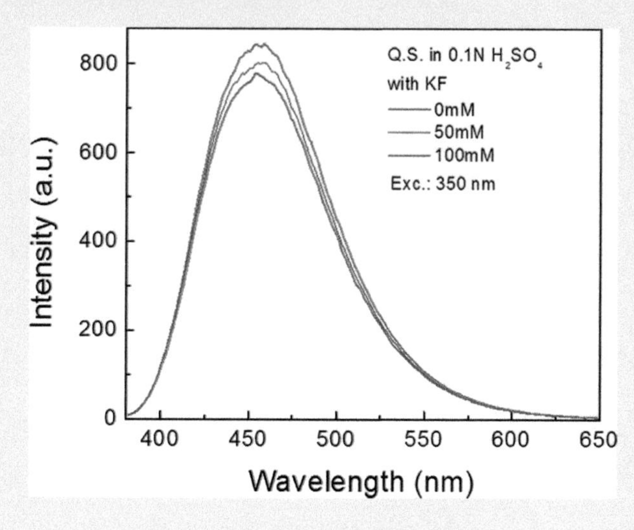

FIGURE 5-18.3 Fluorescence spectra of QS in 0.1 M H_2SO_4 without and with KF.

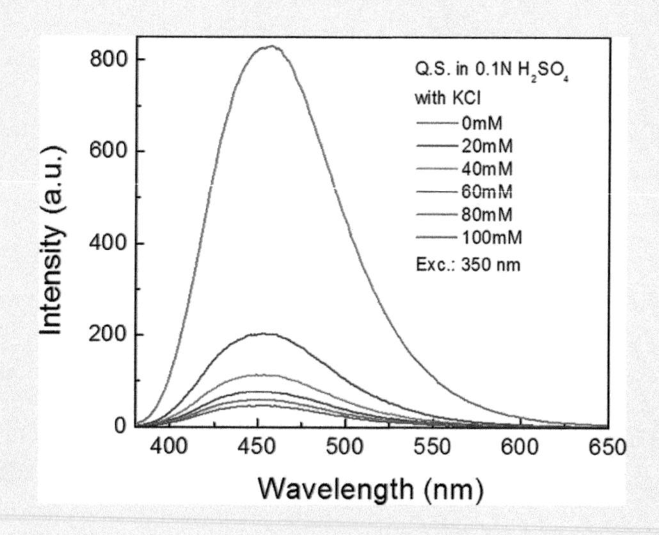

FIGURE 5-18.4 Fluorescence spectra of QS in 0.1 M H_2SO_4 without and with KCl.

(Continued)

EXPERIMENT 5-18 (CONTINUED) FLUORESCENCE QUENCHING OF QUININE SULFATE BY HALOGENS

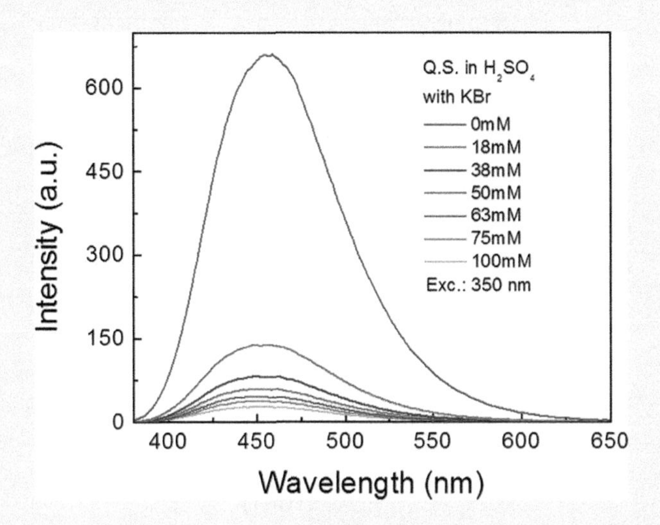

FIGURE 5-18.5 Fluorescence spectra of QS in 0.1 M H_2SO_4 without and with KBr.

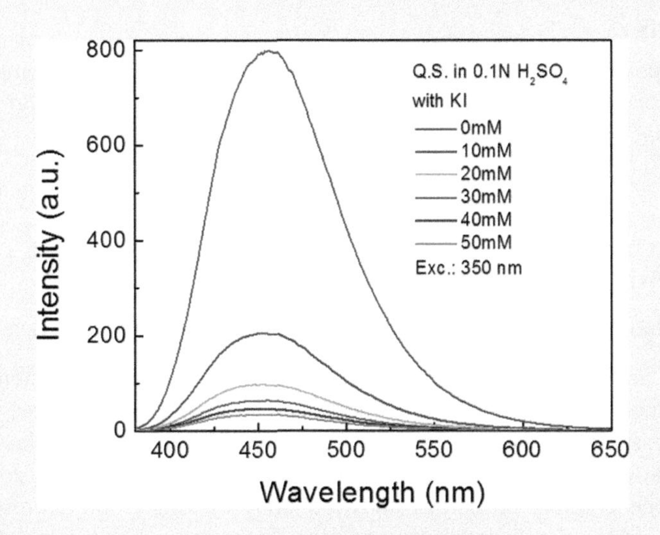

FIGURE 5-18.6 Fluorescence spectra of QS in 0.1 M H_2SO_4 without and with KI.

(Continued)

EXPERIMENT 5-18 (CONTINUED) FLUORESCENCE QUENCHING OF QUININE SULFATE BY HALOGENS

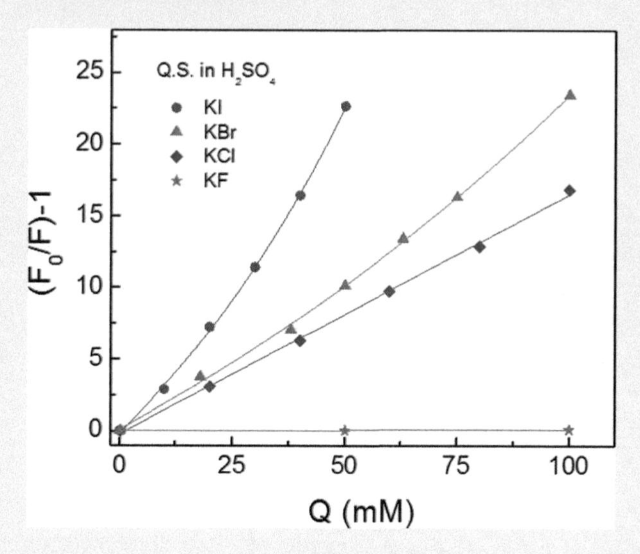

FIGURE 5-18.7 Stern–Volmer plots for QS fluorescence quenching by various quenchers.

CONCLUSIONS

Halogens, except KF are efficient quenchers of QS. The efficiency of quenching goes in order from KF to KI, for iodide being highest. *Warning: One should be careful with using QS with buffers. Many buffers contain salts that result in significant quenching.*

EXPERIMENT 5-19 STATIC FLUORESCENCE QUENCHING

Keywords: fluorescence, quenching, Erythrosine B, mercury (1) nitrate dehydrate.

The addition of another chemical compound to the fluorescent solution often results in the interaction of fluorophores with the extra added molecules. This, of course, will affect the observed fluorescence. In contrast to collisional (dynamic) quenching, the interaction of added molecules with fluorophores already occurs in the ground state. The *complex* formed may change spectroscopic properties and frequently becomes non-fluorescent. What is observed is in effect a significant decrease in the fluorescence. Such interaction of fluorophores and quenchers in the ground state is called *static quenching.* The quencher molecules do not need to diffuse to the excited fluorophores, they already are near and forming complexes. Fluorophores involved in the complexes with quencher molecules do not contribute to the observed fluorescence (they are quenched). Therefore, fluorescence lifetimes are not affected by static quenching. Since the complex is formed in a ground state one can also expect changes in the absorption spectra.

This experiment demonstrates the static quenching using mercury (1) nitrite dihydrate as a static quencher.

(*Continued*)

EXPERIMENT 5-19 (CONTINUED) STATIC FLUORESCENCE QUENCHING

MATERIALS

- Erythrosine B (ErB)
- Mercury (1) nitrate dihydrate ($Hg_2(NO_3)_2 \cdot 2H_2O$)
- Water
- Ethanol
- 1 cm × 1 cm cuvettes, plastic or glass vials

EQUIPMENT

- Spectrophotometer
- Spectrofluorometer
- Camera or cell phone, blue laser pointer

METHODS

1. Prepare a stock solution of ErB in water, sonicate to make sure whole compounds have been dissolved.
2. Prepare a stock solution (0.2 mM) of $Hg_2\ (NO_3)_2 \cdot 2H_2O$ in mixture water–ethanol (1:1). This compound has limited solubility in water. Be sure that everything has been dissolved.
3. Prepare samples of ErB in water (optical density of about 0.1) with various concentrations of $Hg_2\ (NO_3)_2 \cdot 2H_2O$. Keep the ErB concentration constant in all solutions.
4. Measure fluorescence emission of all samples at magic angle conditions. Make a Stern–Volmer plot.
5. Using a blue laser (405 nm) pointer make photographs of unquenched and quenched solutions.
6. Measure absorptions of all samples.

RESULTS

Small amounts of mercury nitrate quench effectively ErB fluorescence, see Figure 5-19.1.

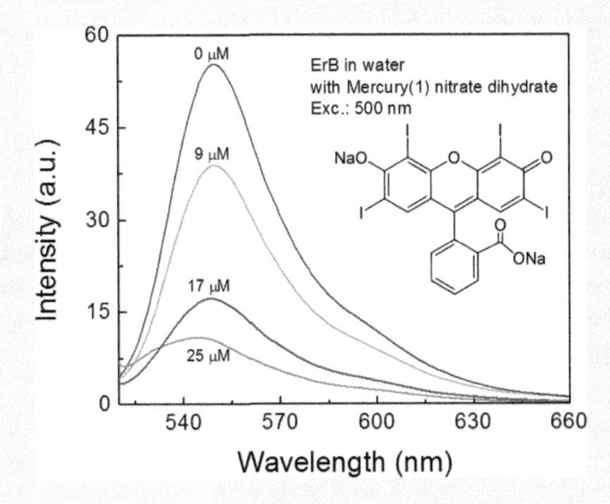

FIGURE 5-19.1 Fluorescence spectra of ErB in water with various concentrations of $Hg_2\ (NO_3)_2 \cdot 2H_2O$.

(Continued)

EXPERIMENT 5-19 (CONTINUED) STATIC FLUORESCENCE QUENCHING

Both dynamic and static quenching can be described by the Stern–Volmer equations. In the case of the static quenching, the relative change in the fluorescence intensity is equal:

$$F_0/F = 1 + K_S \times C$$

where F_0 is the fluorescence intensity in the absence of the quencher, C is the quencher concentration and K_S is the association constant for complex formation.

A Stern–Volmer constant for the ErB quenching is over 100,000 M^{-1} (see Figure 5-19.2) which is many orders of magnitude more than constants observed in a dynamic quenching.

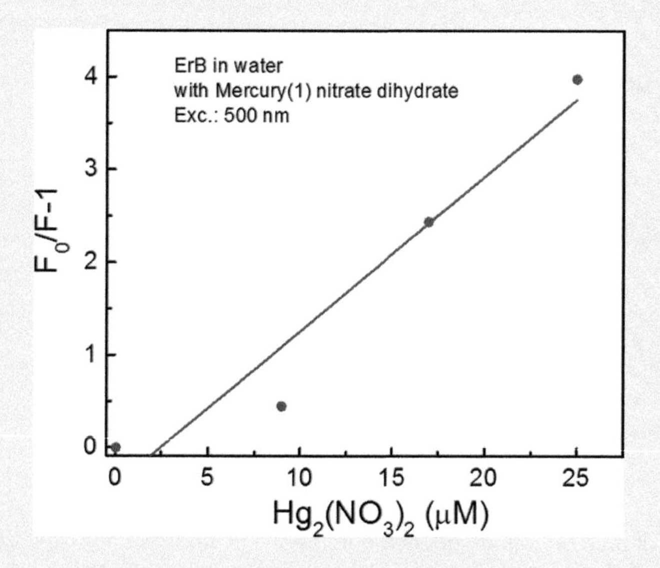

FIGURE 5-19.2 Stern–Volmer plot for ErB quenching by $Hg_2(NO_3)_2 \cdot 2H_2O$.

The strong quenching of ErB is visualized in Figure 5-19.3.

What really happens after addition of $Hg_2(NO_3)_2 \cdot 2H_2O$ to the ErB solution?

The absorption spectra in the presence of mercury nitrate are changed, see Figure 5-19.4. There is evident presence of a new fluorophore–quencher complex. The concentration of free ErB molecules decreases and, of course, the observed fluorescence is weaker, which can be interpreted as a quenching.

(Continued)

EXPERIMENT 5-19 (CONTINUED) STATIC FLUORESCENCE QUENCHING

FIGURE 5-19.3 Photograph of ErB solutions without (left) and with $Hg_2(NO_3)_2 \cdot 2H_2O$ (right).

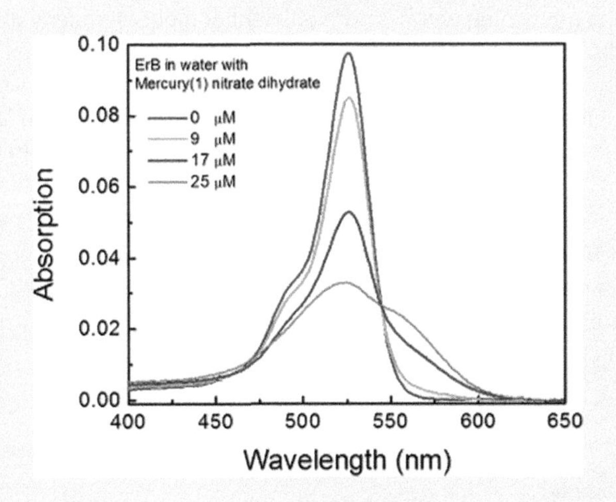

FIGURE 5-19.4 Absorption spectra of ErB in water with and without $Hg_2(NO_3)_2 \cdot 2H_2O$.

(Continued)

EXPERIMENT 5-19 (CONTINUED) STATIC FLUORESCENCE QUENCHING

Note: The static quenching can be recognized from changes in absorption spectrum, not changed lifetime and from the temperature dependence. The association constant will decrease with the temperature and the static quenching, in contrast to dynamic, will decrease. Often, both quenching mechanisms are present and observed changes do not fulfill the simple Stern–Volmer dependence, usually upper-curvature deviation from Stern–Volmer linear dependence is observed.

What if someone will try to treat presented above quenching as a collisional? It will be instructive to calculate the bimolecular quenching constant. For this, estimate the Stern–Volmer constant from Figure 5-19.2 and assume the ErB lifetime as 0.1 ns. Such calculated "bimolecular quenching constant" significantly exceeds the diffusion limit.

EXPERIMENT 5-20 EFFECT OF TEMPERATURE ON FLUORESCENCE QUENCHING

Keywords: fluorescence, quenching, Rhodamine 110, potassium iodide, erythrosine B, mercury nitrate.

The dependence of fluorescence quenching on temperature is rather complex. First, the fluorescence of the fluorophore of interest may more or less depend on the temperature. Second, the solvent viscosity also depends on temperature. Finally, the process of the quenching itself may depend on the temperature. At higher temperatures, the diffusion is faster, and the contact of fluorophore with the quencher molecule is more probable. It is reasonable to expect that collisional (dynamic) quenching will be more efficient at higher temperatures. In contrast, the static quenching (discussed in the previous experiment) involves a complex formation of dyes with quencher molecules in the ground state, which is more probable in lower temperatures. Both will be investigated, the collisional quenching of R110 by potassium iodide (similar system as in Experiment 5-12), and erythrosine B quenched by mercury (like in Experiment 5-19).

MATERIALS

- Rhodamine 110 (R110), erythrosine B (ErB)
- Potassium iodide (KI), mercury (1) nitrate dihydrate $[Hg_2(NO_3]_2 \cdot 2H_2O$ (Hg)
- Water, ethanol
- 1 cm × 1 cm cuvettes, plastic or glass vials
- LWP495 glass filter

EQUIPMENT

- Spectrophotometer
- Spectrofluorometer
- Temperature attachment

METHODS

1. Prepare a 5 mL stock solutions of R110 and ErB in water with absorptions of about 1 at the maxima.

(Continued)

EXPERIMENT 5-20 (CONTINUED) EFFECT OF TEMPERATURE ON FLUORESCENCE QUENCHING

2. Prepare 5 mL stock solutions (1 M) KI in water, and 0.2 mM of $Hg_2 (NO_3)_2 \cdot 2H_2O$ in mixture water-ethanol (1:1).
3. Make 2 mL samples of R110 fluorophore without KI and with 7 mM KI in water. Keep the concentration of R110 constant and absorption below 0.1. For details of the sample preparation, see Experiments 5-9 through 5-11. These two samples will be used to study the collisional quenching.
4. Make 2 mL samples of ErB fluorophore without Mercury and with 10 μM of mercury (1) nitrate dehydrate. Keep the concentration of ErB constant and absorption below 0.1. We will use these two samples to study the static quenching.
5. Measure the fluorescence spectra of each sample at various temperatures between 10°C and 60°C. Use the excitation of 480 nm.
6. Plot the dependence of fluorescence intensities in the function of temperature.

RESULTS

There are two ways to pursue temperature-dependent experiments. One is to measure each sample for all temperatures, and then compare the sample with and without the quencher. Another way is to measure both samples at a given temperature then change the temperature and measure again. Both ways are equivalent. The important point is to stabilize the sample at a given temperature. With electronically controlled temperature chamber it is easy because the controller will show when the sample temperature is stabilized. Measurements without the electronic controller should be done carefully and with patience. It takes about 5–10 minutes to stabilize the sample temperature with the change by 10°C. Of course, it is preferable to measure directly the sample temperature.

Collisional Quenching. As a fluorophore-quencher pair, we selected R110 and KI, the system used in Experiment 5-12.

Fluorescence emission spectra of R110 without KI are shown in Figure 5-20.1 (left). The change of the fluorescence intensity is minimal within the 50°C change. In the presence of KI (Figure 5-20.1 (right)), the change is clearly visible.

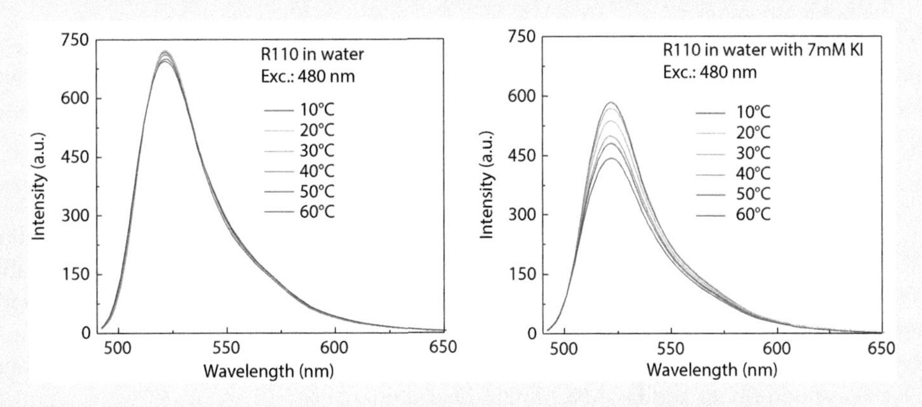

FIGURE 5-20.1 Fluorescence spectra of R110 in the water at various temperatures. Left: Without KI. Right: With 7 mM KI.

(Continued)

EXPERIMENT 5-20 (CONTINUED) EFFECT OF TEMPERATURE ON FLUORESCENCE QUENCHING

The temperature-dependent fluorescence intensities are shown in Figure 5-20.2. This presentation visualizes the temperature-induced changes in the fluorescence intensity.

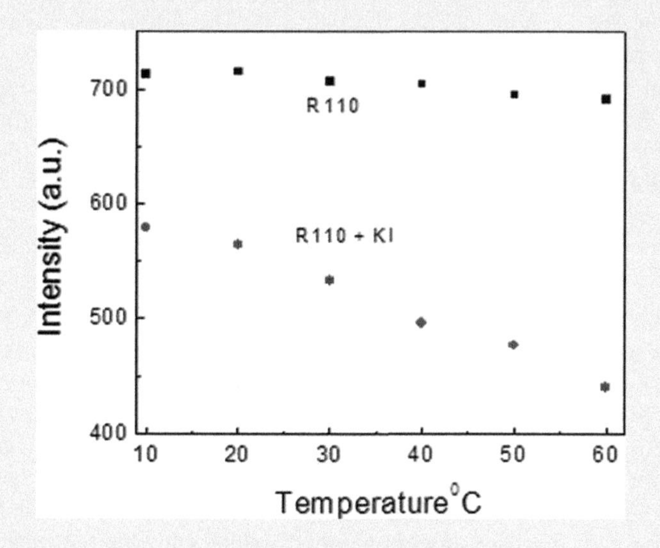

FIGURE 5-20.2 Dependence of fluorescence intensities of R110 with and without KI on temperature.

The collisional quenching can be described with Stern–Volmer dependence: $\frac{I_0}{I} = 1 + K_{SV} \cdot [Q]$, where $[Q]$ is the quencher concentration and K_{SV} is the Stern–Volmer constant, measured in M^{-1}.

It is clear that the Stern–Volmer constant, K_{SV}, is higher in higher temperatures (compare the ratios of R110/R110+KI intensities). The collisional quenching is more efficient in higher temperatures.

Static Quenching. The mechanism of static quenching is very different than in the case of collisional quenching. In the case of static quenching, complexes of fluorophore and quencher are formed already in the ground state. This can be seen in the changes of the absorption spectrum, see Experiment 5-19.

The complexation reaction can be described by similar equation to Stern–Volmer:

$F_0/F = 1 + K_S \times [Q]$, where $[Q]$ is the quencher concentration and K_S the association constant. In contrast to collisional quenching, the association reactions are more probable in lower temperatures. Therefore, we expect different temperature-dependence for static quenching than for collisional.

For temperature-dependent static quenching, we selected the system studied in Experiment 5-19, ErB as fluorophore and $Hg_2(NO_3)_2 \cdot 2H_2O$ as a quencher.

In fact, the temperature-fluorescence spectra show this, see Figure 5-20.3. In the presence of mercury, the ErB fluorescence significantly increases with temperature.

(*Continued*)

EXPERIMENT 5-20 (CONTINUED) EFFECT OF TEMPERATURE ON FLUORESCENCE QUENCHING

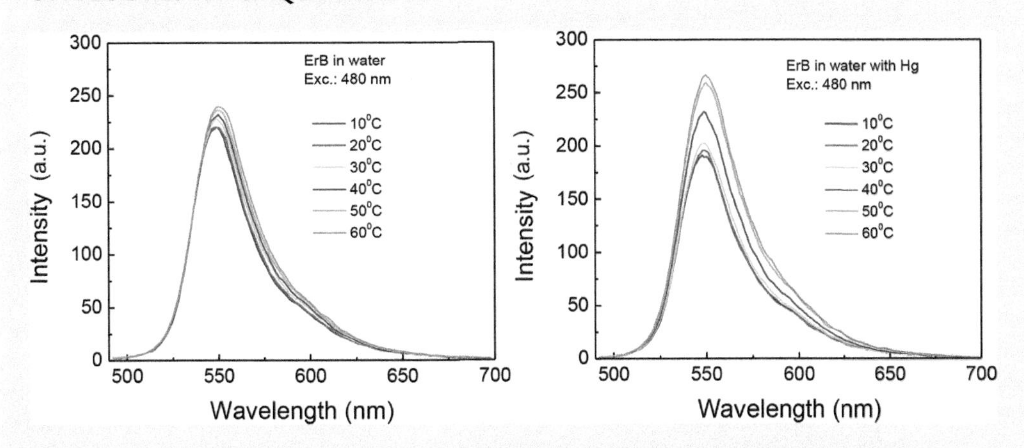

FIGURE 5-20.3 Fluorescence spectra of ErB in water in the function of temperature. Left: Without Hg. Right: In the presence of Hg.

The fluorescence intensities in the function of temperature are presented in Figure 5-20.4. This dependence clearly shows that association constant K_S decreases with temperature. Please, note that intensity of ErB increases slightly with temperature. It is an interesting unusual observation because in most cases, the fluorescence intensity decreases with temperature.

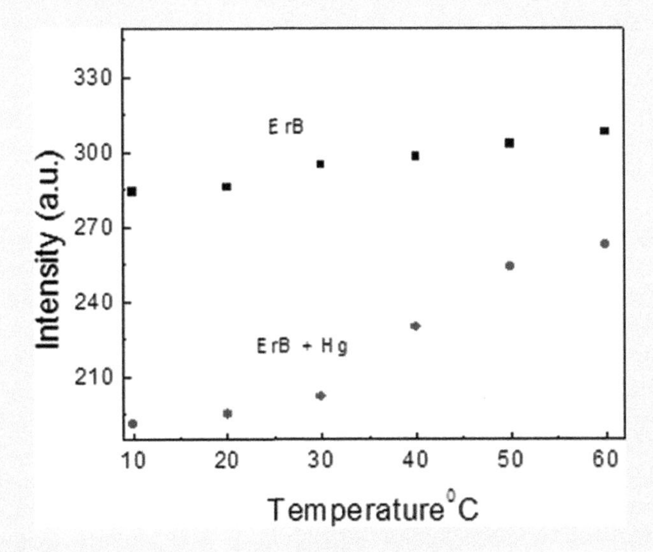

FIGURE 5-20.4 Dependence of fluorescence intensities of ErB with and without Hg on temperature.

Note: Dynamic and static quenching show different dependence on temperature. While dynamic quenching increases with temperature, the static quenching decreases. It is an easy test on the quenching mechanism.

CHAPTER 6

Steady-State Fluorescence Polarization

Anisotropy

INTRODUCTION

Polarization is a very important property of electromagnetic radiation which led to many practical applications, from studying molecular orientation and mobility to 3D glasses. As a simple consequence of Maxwell's equations, oscillations of electric and magnetic fields are orthogonal to each other, and both are orthogonal to the direction of wave (light) propagation. We discussed this in more detail in Chapters 1 and 2. A single-photon will have a very well-defined direction of electric field oscillation. But in most cases, what we are dealing with is a large collection of photons. For example, a typical laser pointer with 1 mW of power will emit in the order of 10^{15} photons per second. Even the moonlight is about 10^{12} photons per second per m^2. Figure 6.1 (left) presents a simplified view of so-called isotropic light where the electric vectors are randomly distributed around the direction of light propagation, Z. Linearly polarized light is an ensemble of photons for which the electric field is oriented in one plane (Figure 6.1, right). Interestingly, isotropic light will have the planes of the electric field oscillations randomly distributed only when viewed from the direction of propagation. However, when viewed from the direction orthogonal to the direction of propagation (X or Y) the projection of all vectors will present as a single direction (like linearly polarized light).

Looking from the direction of light propagation, we can define light polarization by defining a ratio of intensities polarized along X (I_X) and Y (I_Y) directions, respectively. We can define the ratio of intensities $d = I_X/I_Y$ or frequently used a definition of light polarization, P as:

$$P = \frac{I_{\parallel} - I_{\perp}}{I_{\parallel} + I_{\perp}} \tag{6.1}$$

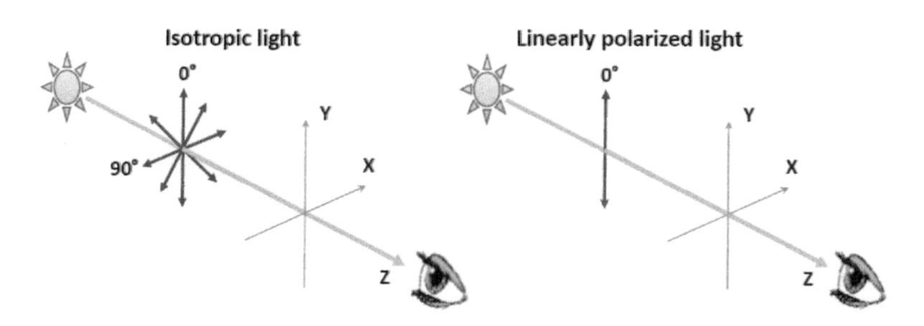

FIGURE 6.1 Schematics of isotropic and polarized light traveling in the Z direction.

Note that in the case of light propagating in one direction the sum $I_{\parallel} + I_{\perp}$ represents total light intensity. However, this is not true when defining the polarization of fluorescence (emission field polarization).

The concept of fluorescence polarization is almost 100 years old. For linear transition moments (practically all organic molecules), absorption of light by a molecule depends on the relative orientation of its transition moment in relation to the direction of light polarization. Probability of light absorption is the highest for a parallel orientation of transition moment and zero for a perpendicular orientation in respect to the direction of light polarization. In effect, the absorption process of polarized light results in a non-isotropic distribution of excited molecules, a process called *photoselection*. Consequently, emitted light (fluorescence) will not be isotropic. Uneven fluorescence intensities along the coordinate axes were first described by F. Weigert in 1920. This phenomenon has been called *fluorescence polarization*, a definition adopted from the description of incoming light polarization. Theories of fluorescence polarization were developed by Vavilov and Lewshin (who also collected extensive experimental data), and later Jablonski and Perrins. In the late 1950s, Jablonski noticed that in some more complicated cases, the description of fluorescence polarization becomes limited and requires additional definitions for excitation and observation directions. Consequently, in 1957, he introduced a more general concept of *fluorescence anisotropy*. In the following years, others started using the term fluorescence anisotropy. Although fluorescence polarization and fluorescence anisotropy can be used alternatively, the latter is more general because it describes the radiation field rather than just state of polarization of incoming emission light toward the direction of observation. Equations involved in the theory of fluorescence polarization become incomparably simpler when anisotropy notation is being used. Therefore, nowadays most researchers use the term fluorescence anisotropy.

When linearly polarized excitation light is used, determination of fluorescence anisotropy is reduced to the measurement of two fluorescence intensities through a polarizer oriented parallel and perpendicular to the excitation light polarization. For such a system (that has cylindrical symmetry of excited molecules transition moments), a ratio out of two measured orthogonal light intensities called $I_{\parallel}$ (parallel) and $I_{\perp}$ (perpendicular) will define anisotropy, r as:

$$r = \frac{I_{\parallel} - I_{\perp}}{I_{\parallel} + 2I_{\perp}} \tag{6.2}$$

The important aspect of fluorescence anisotropy (and polarization) is that this is a ratiometric parameter. The direct ratio of two light intensities $d = I_{\parallel} / I_{\perp}$ yields:

$$r = \frac{d-1}{d+2} \quad \text{or polarization} \quad P = \frac{d-1}{d+1} \tag{6.3}$$

Because of that, anisotropy (polarization) is a parameter independent of excitation light intensity (assuming the intensity is below the saturation level), from the total signal intensity, and other factors affecting light intensity. This enables many practical applications of fluorescence polarization, from detecting molecular rotation, macromolecular binding, high throughput screening, and polarization-based sensing.

HOW TO SELECT AND CHECK POLARIZERS

The first step for measuring polarized absorption, and much more frequently polarized emission, is to obtain polarized light. Measuring emission polarization is more routine and a typical spectrofluorometer should be equipped with polarizers (one on excitation and one on emission). A standard spectrofluorometer also needs polarizers to ensure the magic angle (MA) conditions necessary for most solution measurements. In contrast, standard spectrophotometers will not be equipped with a polarizer. So, before measuring polarized absorption and/or emission polarization, we should obtain polarizers, or in the case of a spectrofluorometer check if adequate polarizers are available in the instrument/system. Occasionally to save on cost, some spectrofluorometers can be equipped with sheet polarizers so the results can be misleading when measuring in the UV or NIR range.

Today, we have many polarizers available in the market. A good fluorescence system will typically be equipped with Glan-Thompson polarizers. The Glan-Thompson polarizer is the best type of polarizer but requires a very well collimated beam with minimal divergence. Besides this, it is quite expensive and in many cases, polarizer sheets are used. Also, polarizer sheets are readily available, inexpensive, and most importantly do not require for the beam to be well collimated.

It can be a little surprising for many experimentalists, but the beam in a typical spectrophotometer is partially polarized, and in some cases, the beam polarization can be significant. This is due to multiple reflections and the necessary monochromator in the light path. The beam polarization is not important for measuring isotropic samples (almost all measurements) and vendors do not even specify the polarized properties of the beam in spectrophotometers. To adopt the spectrophotometer to perform polarized absorption measurements, we will need to introduce the polarizer into the beam path. If the beam is partially polarized, the polarizer should be oriented in such a way so as to minimize losses of available beam intensity.

How to know what would be a good polarizer for a particular application? The most common polarizers are sheet polarizers, Glan-Thompson polarizers, or recently grid polarizers became more popular. Some commercial polarizers will typically be marked to indicate the direction of polarization (the direction that transmits light polarized parallel to that direction). Having one polarizer of known polarization it is easy to align any other polarizer by crossing them. When two polarizers are crossed, the transmitted intensity should be at the lowest value, with two good polarizers close to 0. But typically, it would be at least 50 to much over 100-fold attenuation when going from vertical-vertical (VV) to vertical-horizontal (VH) polarizer orientations. Just looking at a polarizer cannot show what the plane of light polarization is since the human eye is not sensitive to light polarization. If we do not know a polarizer orientation, the simplest way to find an approximate direction for the polarization plane is to look at a flat dielectric surface reflecting light. For example, a well-polished floor, a shiny wall, or a piece of flat glass on the table that reflects light. A parallel unpolarized beam of light (or light from a very distant source) when reflecting from a dielectric surface will become partially polarized. For the incident angle close to the Brewster angle (for typical dielectrics like glass about 50°–65° of incident angle), the reflected light will be well (almost completely) polarized in the plane of the reflecting surface. In fact, the component of light polarized perpendicular to the plane will penetrate the dielectric, while part of light polarized in the plane of reflecting surface will be significantly reflected. Figure 6.2 shows the calculated reflectivity for s (in a surface plane) and p (orthogonal to the surface plane) polarized light components (electric vector parallel and perpendicular to the dielectric surface). In a broad range (30°–70°) the parallel to the surface plane component will dominate reflected light intensity. Looking at the light reflected from the dielectric flat surface through the polarizer (under an angle close to 60° of the normal to the surface) and rotating the polarizer, one would notice that intensity of reflected light changes significantly. It is easy to adjust the polarizer to see the minimum intensity of reflected light (it is much better to look for a minimum than maximum since the eye will distinguish this better). When observing maximum attenuation (minimum intensity) the polarization plane of a polarizer is orthogonal to the dielectric plane. For example, looking on the reflection of a distant ceiling light on the floor of a long corridor for one polarizer orientation, the light reflection will almost

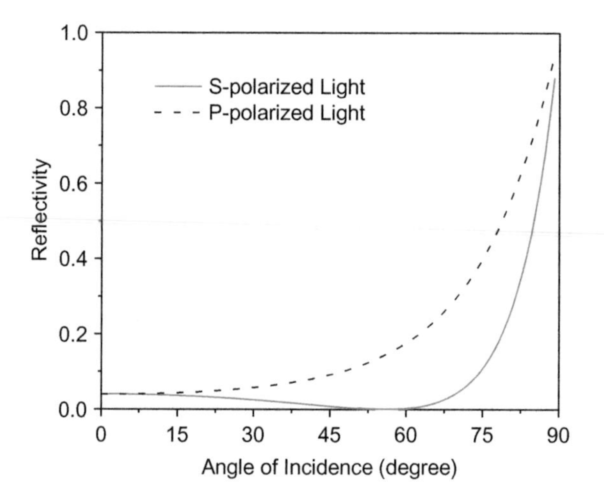

FIGURE 6.2 Relative intensity of reflected light from the glass dielectric surface for s- and p-polarized light.

disappear completely. This is when the plane of polarization is oriented orthogonally to the floor. In this position, the polarizer orientation is vertical. Another simple way will be to look through a polarizer on an old LCD display. This display is typically highly polarized at 45° to the vertical. Looking through the polarizer on the LCD display and rotating the polarizer will reveal a distinct intensity decrease (almost complete) for a 45° polarizer orientation (remember polarizer orientation is ±90°).

Obviously, visually we can only achieve limited precision in assigning the plane of polarization for a given polarizer. When doing this very carefully and repeating the routine a few times, we were able to get the plane of polarization for a sheet polarizer within a 2°–3° precision. This is a very good level of precision (remember one step of the second arm in a wristwatch or clock is 6°). Regardless of quite good precision, we always recommend using a commercially calibrated polarizer to check such an estimate.

The next important step is to check the spectral range in which the polarizer transmits and sufficiently polarizes the light. In many cases, efficiency for light polarization will strongly depend on wavelength. Most of the polarizer sheet will lose its ability to polarize light below 400 nm and above 700 nm. In the first experiments in this chapter, we will show how to check the working range of polarizers.

TRANSITION MOMENTS

An important observable parameter responsible for light-matter interaction is the transition moment, M. The transition moment is the product between two states of a molecule that are described by real wave functions and a real part of an electric dipole moment transition operator. The electric dipole transition moment is a vector operator (M) which is linked to the molecular framework coordinates. Practically, all transitions (for absorption and emission) in UV, VIS, and NIR range are dipole transitions with well-defined orientations. Depending on the molecular symmetry, that dictates properties of the two wave functions, each electronic transition will be associated with an electric dipole operator and will have a well-defined orientation of a transition moment in the molecular framework. For planar (most of the organic molecules has atoms arranged in one plane) highly symmetrical molecules like C_{2V} and D_{2h}, the transition moments will always be oriented along the main axis of symmetry. So for molecules like anthracene, pyrene or perylene, the transition moment will always be in the molecular plane oriented along the main symmetry axis (which are orthogonal at these cases). For non-symmetrical molecules, like fluorescein, Rhodamine 6G, etc. the transition moment will be in the molecular plane, but its orientation will depend on other factors.

Absorption and emission transitions are linear with well-defined directions always fixed in the molecular plane. Typically, a molecule has only one emission transition (emits from the lowest excited state) and its transition moment orientation is always fixed in the molecular framework (has one well-defined direction). In contrast, absorption can also be to higher energy levels and each transition has its own transition moment that may have different orientations. *As discussed later, this is the main reason why emission polarization (anisotropy) depends on the excitation wavelength. In typical polarization experiments, we always try to use the longest possible excitation wavelength to make sure that the limiting anisotropy will be the highest.*

POLARIZED ABSORPTION

While describing light and chromophore interactions, polarization must be considered. Electromagnetic radiation for which the oscillation of the electric field (light polarization) is along the molecule's transition moment will be preferentially absorbed while the light for which light polarization is orthogonal to the transition moment will not interact with the molecule at all. Since emission is always from the same excited state, the polarization of emitted light will always be along the emission transition moment independently of the excitation.

Absorption Transition Moments. Consider a very simple case of an ensemble of molecules, like anthracene molecules perfectly oriented the same way (Figure 6.3). Such a condition is possible in a crystal. Anthracene has two distinguished absorption transitions, a long-wavelength absorption band (320–380 nm) and a short wavelength band (250–260 nm) range. The long-wavelength transition is oriented (polarized) along the short symmetry axis, y; while the short wavelength along the long symmetry axis, z. If the polarization of incoming light is along the z-axis, it will be efficiently absorbed by the short-wavelength transition and not absorbed by the long-wavelength transition. The measured absorption spectrum will be $A_Z(\lambda)$. If the incoming light will be polarized along the Y-axis it will be efficiently absorbed by the long-wavelength transition and not absorbed by the short-wavelength transition. The measured absorption for this light polarization will be $A_Y(\lambda)$. These two absorptions will be very different. Associating the laboratory system with the light polarization the Z-axis is called parallel (∥) and the

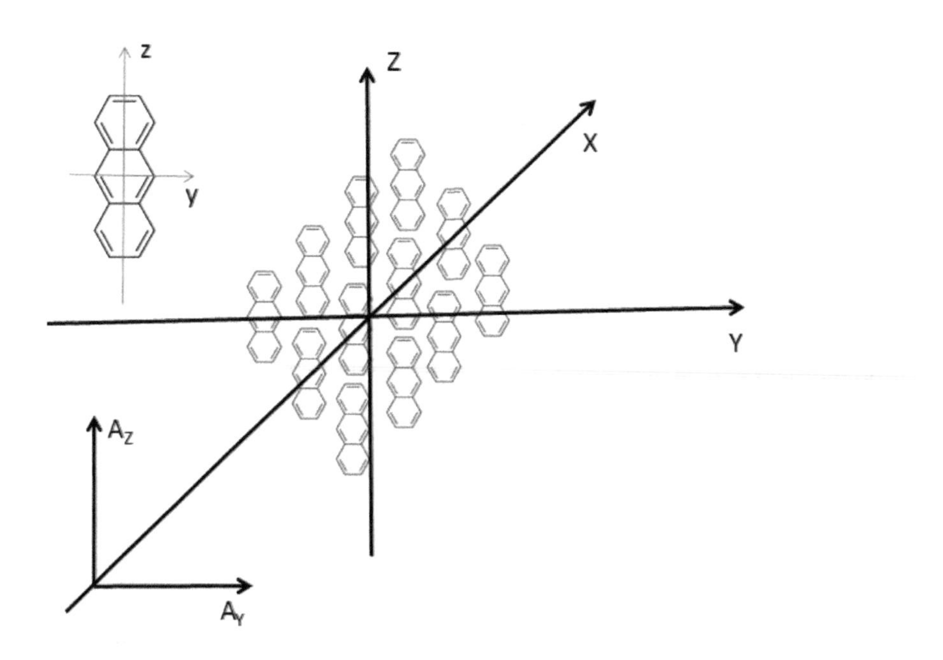

FIGURE 6.3 Schematic of anthracene molecules oriented in the space.

Y-axis perpendicular ($\perp$). The difference between the two orthogonal absorption components is called dichroism.

$$D(\lambda)=\frac{A_{\parallel}(\lambda)-A_{\perp}(\lambda)}{A_{\parallel}(\lambda)+A_{\perp}(\lambda)}=\frac{d(\lambda)-1}{d(\lambda)+1} \tag{6.4}$$

where the quantity $d(\lambda)= A_{\parallel}(\lambda)/A_{\perp}(\lambda)$ is the dichroic ratio. Additionaly, more general, concepts have been developed to generalize the description of absorption dichroism for non-isotropic systems. These include reduced linear dichroism LD^r:

$$LD^r(\lambda)=\frac{A_{\parallel}(\lambda)-A_{\perp}(\lambda)}{\frac{1}{3}\left(A_{\parallel}(\lambda)+2A_{\perp}(\lambda)\right)}=\frac{A_{\parallel}(\lambda)-A_{\perp}(\lambda)}{A_{iso}(\lambda)} \tag{6.5}$$

And absorption anisotropy, K:

$$K(\lambda)=\frac{A_{\parallel}(\lambda)-A_{\perp}(\lambda)}{A_{\parallel}(\lambda)+2A_{\perp}(\lambda)} \tag{6.6}$$

For isotropic systems where the absorbers are randomly distributed $\left(A_{\parallel}(\lambda)=A_{\perp}(\lambda)\right)$, the dichroism, reduced linear dichroism, and absorption anisotropy are all equal to 0, independent of the observation wavelength.

It is important to realize that if one has an oriented system of molecules (partially oriented can be sufficient) by measuring polarized absorption components, it is possible to recover the orientation of various transition moments associated with different absorption transitions. These kinds of measurements are called linear dichroism and are made on oriented molecular systems. To orient molecules stretched polymers, liquid crystals or crystals are frequently used. We will present a few experiments on how to measure linear dichroism and interpret the results.

EMISSION POLARIZATION (ANISOTROPY)

There are two important consequences of transition moments being linear and having well-defined orientations in the molecular frame. Excitation with linearly polarized light will preferentially select molecules that have an absorption transition moment parallel to the direction of light polarization. This process is called photoselection. In effect, even for an isotropic ensemble of molecules (solution), the population of excited molecules will have a well-defined preferential orientation. Second, the light emitted by each molecule will be polarized in the plane defined by the direction of the emission transition moment. And if the excited molecules are oriented or partially oriented, the emitted light will be polarized (partially polarized). The polarization of emission light can be measured and may provide very important information about molecular orientation and changes in molecular orientation between the act of absorption and act of emission. In this case, we can study what is happening with the emitter during its time in the excited state (lifetime). As mentioned earlier by analogy to light polarization (Equation 6.1), the fluorescence polarization, P has been originally used to describe emission polarization:

$$P = \frac{I_{\parallel} - I_{\perp}}{I_{\parallel} + I_{\perp}} \tag{6.7}$$

This describes emission polarization as observed from one given direction and, for an ensemble of molecules does not give a full emission field description. In 1957, Jablonski introduced emission anisotropy (r) to describe the emission field of an ensemble of emitting fluorophores in a more general and additive term. For excitation with linearly polarized light (most cases) for which excited molecules are arranged (photoselected) in a cylindrical symmetry, the emission anisotropy, r is defined:

$$r = \frac{I_{\parallel} - I_{\perp}}{I_{\parallel} + I_{\perp} + I_{\perp}} = \frac{I_{\parallel} - I_{\perp}}{I_{\parallel} + 2I_{\perp}} = \frac{I_{\parallel} - I_{\perp}}{I} \tag{6.8}$$

where I is the total emission intensity. Having the total fluorescence intensity in the denominator is an important conceptual difference between the definitions of emission polarization (Equation 6.7) and emission anisotropy (Equation 6.8). Both contain the difference in emission intensities in the numerator, but the normalization factors (denominators) are different. In the case of anisotropy, the denominator represents total emission intensity while polarization contains only partial intensity. As we will show later, anisotropy is a simple additive function while polarization always needs to be properly defined.

To describe the emission polarization detected from an ensemble of molecules each molecule is assumed to have an absorption transition moment, $\boldsymbol{A}$, and an emission transition moment, $\boldsymbol{F}$, fixed in the molecular frame. The geometry to describe light interacting with an arbitrary chromophore is presented in Figure 6.4. The excitation light is represented as an electric field vector (purple) denoted as the $\boldsymbol{E}$ vector traveling in the Y direction polarized along the Z-axis. The absorption transition moment, $\boldsymbol{A}$ is oriented under the angle ω_1 to the vertical axis of the laboratory system. The interaction between a Z plane-polarized wave with absorption dipole $\boldsymbol{A}$ depends on the angle ω_1 between the light amplitude vector $\boldsymbol{E}$ and the direction of the absorption transition moment $\boldsymbol{A}$. This is proportional to the projection of vector $\boldsymbol{A}$ on the direction of vector $\boldsymbol{E}$, $\sim\cos\omega_1$. Since the intensity of light is proportional to the square of the amplitude of electric vector E^2, the probability for the absorption of light by absorption vector $\boldsymbol{A}$ is $p \sim \cos^2\omega_1$. Probability for light absorption by any given absorption dipole will depend on dipole orientation with regard to the incoming light polarization. Effectively, the distribution function of excited molecules, $f_e(\omega_1)$ is a product of a distribution function of molecules in the ground state $f_g(\omega_1)$ and the probability of absorption, p that depends on the angle between electric vector, $\boldsymbol{E}$ and dipole moment, $\boldsymbol{A}$:

$$f_e(\omega_1)\,d\omega_1 = f_g(\omega_1)\cos^2\omega_1\,d\omega_1 \tag{6.9}$$

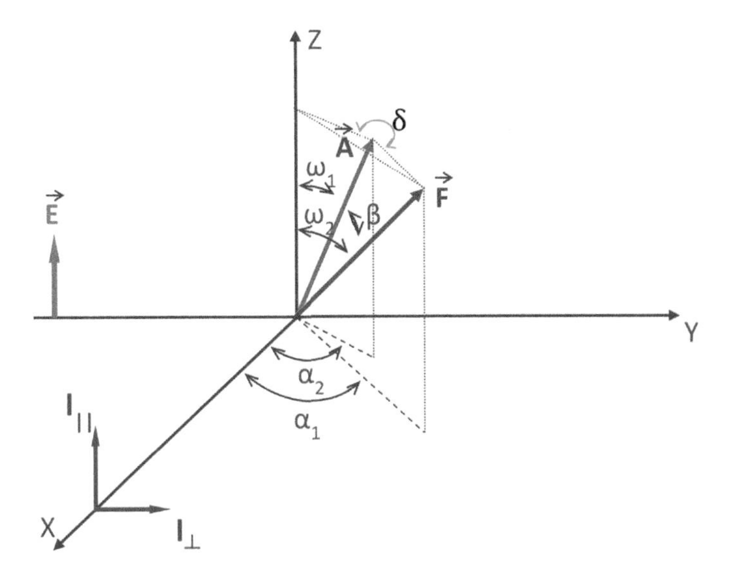

FIGURE 6.4 Coordinate system and angle definitions for calculating emission anisotropy. Excitation light comes from the left. Emission observation is along the *X*-axis.

For isotropic (random) distribution of molecules in the ground state ($f_g(\omega_1)d\omega_1 = \sin\omega_1 d\omega_1$), the distribution of excited vectors is:

$$f_e(\omega_1)d\omega_1 = \cos^2\omega_1 \sin\omega_1 d\omega_1 \tag{6.10}$$

Equation (6.10) describes the probability for photon absorption as a function of the angle between the direction of light polarization and molecule transition moment orientation. Different probability for absorption as a function of transition moment orientation results in an anisotropic distribution of excited molecules (photoselection). The distribution of absorption transition moments excited with linearly polarized light (Equation 6.10) is shown in Figure 6.5. Figure 6.5, shows two projections for the excited transition moments distribution function, along the *X*-axis (A) and along *Z*-axis (B). The distribution projection along the *X*-axis has maximum probability along the light polarization direction and zero probability for directions orthogonal to the light polarization. It is important to realize that the distribution function is symmetrical around the *Z*-axis (direction of the excitation light polarization) and the distribution projection along the *Z*-axis is fully symmetrical (B). This is a consequence of the fact that light absorption depends only on angle ω_1 and does not depend on angle α around the *Z*-axis. Such systems (distributions) are described as having cylindrical symmetry. Now, we move on to fluorescence observed along *X*-axis (in fact any observation in the XY plane will give the same results because of the cylindrical symmetry of the system). We observe the intensity of light parallel ($I_{\parallel}$) and perpendicular ($I_{\perp}$) to the excitation light polarization. In general, the orientation of emission transition moment $\boldsymbol{F}$ in the moment of emission can be different than the orientation of the absorption transition moment $\boldsymbol{A}$ in the moment of absorption. This difference might

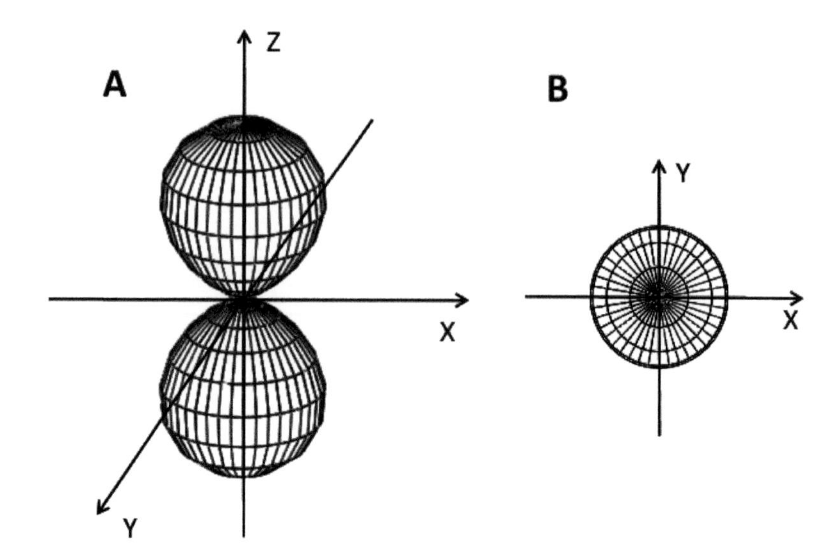

FIGURE 6.5 Calculated transition moments distributions after excitation with Z axis polarized light viewed along *Y*-axis (A) and *Z*-axis (B).

result from a number of different factors. Both transitions can inherently be different when absorption is to a different state that the emissive state (for example excitation to higher absorption bands) or the molecule reorients before the emission due to Brownian motion or other processes/movements. If the angle between the emission transition moment, ***F*** and *Z*-axis is ω_2, the intensities parallel, $I_{\|}$, and perpendicular, $I_{\perp}$, will be:

$$I_{\|} \sim \cos^2\omega_2 \quad \text{and} \quad I_{\perp} \sim \sin^2\omega_2 \tag{6.11}$$

We have to remember that in the XY plane there are two equivalent directions along the *X* and *Y* axes and two perpendicular components $I_{\perp}^{x}$ and $I_{\perp}^{y}$. The emission anisotropy (Equation 6.8) in the simplified case of cylindrical symmetry (excitation with linearly polarized light ($I_{\perp}^{x} = I_{\perp}^{y}$)) becomes:

$$r = \frac{I_{\|} - I_{\perp}}{I_{\|} + I_{\perp} + I_{\perp}} = \frac{I_{\|} - I_{\perp}}{I_{\|} + 2I_{\perp}} = \frac{I_{\|} - I_{\perp}}{I} \tag{6.12}$$

where *I* is total emission intensity.

As we will show later, having the total fluorescence intensity in the denominator makes anisotropy a simple additive function while polarization always needs to be properly defined.

Taking into account that $I = I_{\|} + 2I_{\perp}$, the Equation (6.8) can be rewritten for a single emission transition moment oriented at angle ω_2:

$$r(\omega_2) = \frac{I_{\|} - I_{\perp}}{I_{\|} + 2I_{\perp}} = \frac{3}{2}\frac{I_{\|}}{I} - \frac{1}{2} = \frac{3}{2}\cos^2\omega_2 - \frac{1}{2} \tag{6.13}$$

And for an ensemble of many randomly oriented molecules, we have to average over all possible orientations (directions) of ω_2

$$r(\omega_2) = \frac{3}{2} \ll \cos^2\omega_2 \gg - \frac{1}{2} \tag{6.14}$$

where $<<>>$represents the average value. The ω_2 angle depends on angles ω_1, the angle between absorption and emission transition moments, β, and azimuthal angle, δ that represents molecular plane rotation.

To make averaging possible, one needs to represent the ω_2 as a function of ω_1, β, and δ:

$$\cos\omega_2 = \cos\omega_1\cos\beta + \cos\omega_1\sin\beta\cos\delta \tag{6.15}$$

Squaring and averaging (Equation 6.15) over the azimuthal angle δ ($<\cos\delta> = 0$ and, $<\cos^2\delta> = ½$) we have:

$$\left\langle\cos^2\omega_2\right\rangle = \cos^2\omega_1\cos^2\beta + \frac{1}{2}\sin^2\omega_1\sin^2\beta = \left(\frac{3}{2}\cos^2\omega_1 - \frac{1}{2}\right)\left(\cos^2\beta - \frac{1}{3}\right) + \frac{2}{6} \tag{6.16}$$

where β is constant (fixed) for given transitions in a molecular framework and to obtain the mean value $<<\cos^2\omega_2>>$, the averaging needs to be done only over ω_1. Substituting the (6.16) into (6.14), we get:

$$r(\omega_1,\beta) = \left(\frac{3}{2} < \cos^2\omega_1 > - \frac{1}{2}\right)\left(\frac{3}{2}\cos^2\beta - \frac{1}{2}\right) \tag{6.17}$$

Using Equation 6.16, we can calculate the average as:

$$< \cos^2\omega_1 > = \frac{\int_0^{\pi/2} \cos^2\omega_1 f_g(\omega_1)\, d\omega_1}{\int_0^{\pi/2} f_g(\omega_1)\, d\omega_1} = \frac{\int_0^{\pi/2} \cos^2\omega_1\cos^2\omega_1\sin\omega_1\, d\omega_1}{\int_0^{\pi/2} \cos^2\omega_1\sin\omega_1\, d\omega_1} \tag{6.18}$$

where $f_g(\omega_1)d\omega_1$ is the distribution function of absorption transition moments in the medium. For isotropic systems (typical solutions), this is just random distribution. We need to remember that in general, the distribution function of molecules in the ground state does not have to be random, leading to a more complicated description. For a random distribution of molecules in the ground state, the average $<\cos^2\omega_1>$ becomes:

$$< \cos^2\omega_1 > = \frac{\int_0^{\pi/2} \cos^2\omega_1\cos^2\omega_1\sin\omega_1\, d\omega_1}{\int_0^{\pi/2} \cos^2\omega_1\sin\omega_1\, d\omega_1} = \frac{3}{5} \tag{6.19}$$

And the expected emission anisotropy only depends on the angle between absorption and emission transition moments in the time of absorption and time of emission:

$$r(\beta)=\frac{2}{5}\left(\frac{3}{2}\cos^2\beta-\frac{1}{2}\right)=0.4\left(\frac{3}{2}\cos^2\beta-\frac{1}{2}\right)=0.4\frac{3\cos^2\beta-1}{2} \tag{6.20}$$

For co-linear transition moments $\beta = 0°$, the expected anisotropy is 0.4 (polarization 0.5). In the absence of any depolarizing factors, the ratio of observed intensities polarized along Z ($I_{\parallel}$) and X ($I_{\perp}$) axes is 3, which provides the highest fluorescence anisotropy in an isotropic system of 0.4 (polarization of 0.5). For orthogonal transition moments, orientation $\beta = 90°$, the expected anisotropy is −0.2. For any other angle between transition moments, the emission anisotropy must be in the range of −0.2 to +0.4 (this is for a random distribution of molecules in the ground state). The maximum value of emission anisotropy, 0.4 is termed fundamental anisotropy. Depending on the angle β, the maximum emission anisotropy will be different. For example, for 10°, the $r = 0.382$, for 20°, the $r = 0.33$, for 30°, the $r = 0.25$, for 54.7°, the $r = 0$, and for 80°, the $r = -0.182$. Such maximum values are called limiting anisotropies. It is important to stress that perfect co-linearity of transition moments ($\beta = 0°$) is very difficult to observe in an experimental setting. The excitation typically leads to molecular perturbation and some molecule orientational displacements and measured values of emission anisotropy are below 0.4. In Equation 6.14, we considered that angle β is the angle between the absorption transition moment at the moment of absorption and emission transition moment at the time of emission. Interestingly, there is no reason to assume that the angle β is just an intrinsic difference in the transition moments orientations in the molecular framework or molecule reoriented during the fluorescence lifetime. In fact, multiple processes like the intrinsic orientation of the transition moments in the molecular framework, molecule reorientation (rotation), or even energy transfer lead to a change in apparent orientation. If there are multiple processes contributing to transition moment reorientation, each with its own angle β_i using the analogical approach to this we just used, we will obtain:

$$r=0.4\left(\frac{3}{2}\cos^2\beta_1-\frac{1}{2}\right)\left(\frac{3}{2}\cos^2\beta_2-\frac{1}{2}\right)\ldots\left(\frac{3}{2}\cos^2\beta_i-\frac{1}{2}\right) \tag{6.21}$$

Each angle independently contributes to the emission anisotropy decrease (depolarization). For example, a fluorescent probe attached to the macromolecule will be depolarized by the intrinsic difference in the orientation of transition moments for absorption and emission, β, probe rotation, φ as well as by overall macromolecular tumbling, θ. The observed anisotropy can then be represented:

$$r=0.4\left(\frac{3}{2}\cos^2\beta-\frac{1}{2}\right)\left(\frac{3}{2}\cos^2\varphi-\frac{1}{2}\right)\left(\frac{3}{2}\cos^2\theta-\frac{1}{2}\right) \tag{6.22}$$

FUNDAMENTAL AND LIMITING ANISOTROPIES

The fundamental anisotropy of 0.4 (0.5 for polarization) comes from averaging in Equation 6.19. It is the maximum anisotropy value that can be observed for an isotropic system of chromophores when the absorption and emission transition moments are co-linear. But can the fundamental anisotropy be higher or lower than the fixed value of 0.4? To have fundamental anisotropy higher than 0.4, one should use an oriented system where the average in Equation 6.19 will have a different value. Good examples of such systems are crystals or chromophores oriented in an anisotropic medium like oriented/stretched polymers or liquid crystals. For a system perfectly oriented along one axis, the limiting anisotropies are 1.0 for co-linear transition moments, and −0.5 for transition moments orthogonal to the direction of orientation. These values are reached when one component vanishes completely ($I_{\perp}$ for 1.0 and $I_{\parallel}$ for −0.5). Indeed, anisotropies close to such limiting values have been measured. When is the fundamental anisotropy lower than 0.4 for an isotropic ensemble of molecules? For this to happen, the emission state should be degenerated in a way that more than one emitting transition moments are contributing to emission. This is the case with highly symmetrical molecules that have two or more equivalent symmetry axis, such as a benzene molecule. Its hexagonal ring has three equivalent symmetry axes and emission can occur along any of them. Excitation of the benzene molecule with linearly polarized light will be through one of the absorption transition moment (three absorption transition moments are equivalent), but emission probability will be equally distributed among three of the possible emission transitions. In this case, averaging over angle ω_2 must be first done in the molecular plane leading to the fundamental anisotropy of 0.1. There are other high symmetry molecules like coronene that also presents lower fundamental anisotropy. As the fundamental anisotropy reflects polarization for co-linear transition moments, the limiting anisotropy is the anisotropy when the absorption and emission transition moments are not co-linear and are oriented under the angle β relative to each other. In effect, it is a fundamental anisotropy multiplied by a depolarization factor $\left(\frac{3}{2}\cos^2\beta_1 - \frac{1}{2}\right)$ as in Equations 6.21 and 6.22. In a solution, typically we can measure only limiting anisotropies since the vibrational states of the fluorophore leads to small depolarization and the highest measured anisotropy values are typically lower than 0.4.

One very important property of anisotropy is that this quantity is simply additive. For example, two species in the solution that contribute with fractional intensities f_1 and f_2 have anisotropies r_1 and r_2, respectively, will show an average anisotropy r:

$$r = f_1 r_1 + f_2 r_2 \tag{6.23}$$

This becomes very important if one wants to study the fluorophore (ligand) binding to proteins or DNA. Free fluorophore in solution rotates very quickly and between acts of absorption and emission (fluorescence lifetime) is able to lose a big part of anisotropy, so its anisotropy is typically close to zero. However, after binding, the fluorophores moves with the macromolecule much slower and the loss of anisotropy is lower during the fluorophore lifetime. This fraction contributes much larger anisotropy.

The anisotropy lost due to the molecular rotation was first described by Perrin (Lakowicz, Velure & Barbaros, Jameson):

$$r = \frac{r_0}{1 + \tau/\tau_c} \tag{6.24}$$

where r_0 is the limited anisotropy (anisotropy that would be measured if the molecule did not move), τ the fluorescence lifetime, and τ_c is the so-called correlation time that describes how fast the molecule rotates (often the symbol Θ is used instead τ_c).

Taking into account, the definition of emission polarization, P and emission anisotropy, r we can show the relationship between both quantities for systems with cylindrical symmetry (ensemble of randomly oriented molecules excited with linearly polarized light):

$$r = \frac{2P}{3-P} \tag{6.25}$$

CONSEQUENCES OF LINEAR TRANSITION MOMENTS

There are important consequences of linear transition moments for chromophores and consequent polarized absorption and emission. We will discuss a few of them that are relevant to practical experiments. A question often asked by students is: *What will be fluorescence anisotropy if the excitation light is not polarized (excitation by isotropic light)?*

First of all, one needs to realize that there is no unpolarized light that could be used for sample excitation. Even so-called isotropic light (Figure 6.1) that has electric vectors equally distributed around the Z-axis (direction of propagation) is polarized in a plane orthogonal to the propagation direction. Because of that, molecules oriented along the direction of propagation will not be excited. It is possible to calculate that the distribution of excited transition moments will be the "eight-shaped distribution" rotated around the Z-axis. In effect, the fundamental anisotropy will be half of that expected for excitation with linearly polarized light ($r = 0.2$). *Note: Alternatively, we can consider isotropic excitation as a combination of two polarized excitations (vertical and horizontal) with equal intensities. Horizontal excitation gives anisotropy zero while vertical 0.4. Observed anisotropy is an average of two equal fractions of anisotropy, 0 and 0.4 resulting in 0.2 average value.* So, in most cases, the observed fluorescence would be significantly polarized.

Second, we need to remember that polarization of 0.4 in practice means that the parallel component of emission is three times larger than the orthogonal component ($I_{\parallel} = 3I_{\perp}$). To properly measure both components is a fundamental requirement for measuring emission anisotropy. A typical experimental system is equipped with many optical elements (lenses, monochromator, detector, etc.) for which the response may depend on light polarization. The monochromator is a good example, wherein many cases the polarization of incoming light is important and the transmittance may distinctly depend on light polarization. So, such systems will respond differently to different light polarization. Therefore, to obtain a proper polarization value (anisotropy) the instrument response needs to be corrected. To correct for different sensitivities to parallel and perpendicular light polarizations, it is necessary to know the so-called "G-factor." The G-factor is a number that adjusts the response of parallel (or orthogonal) components to correct for different sensitivities to the two light polarizations (0° and 90°). The G-factor typically depends on the observation wavelength and in practice, it is always wavelength dependent. This correction is only for the detection path. Since the G-factor depends on the experimental configuration and may change if some optics have been modified, most companies do not supply the G-factor values.

We, therefore, recommend that for each set of polarization measurements, the *G*-factor should be determined independently. It will typically not change dramatically if exactly the same configuration is used repeatedly, but we recommend a minimum weekly check of the *G*-factor.

DETERMINING THE *G*-FACTOR FOR SPECTROFLUOROMETER

Fluorescence anisotropy involves the measurement of two orthogonally polarized light intensities. The important experimental consequence of this is that the detection system (monochromator, PMT detector, or multiple glass elements) is biased to light polarization. It responds differently to incoming vertical and horizontal light polarization, requiring an extra measurement to calibrate the system. In practice, we only need to calibrate the sensitivity for one component relative to the another. In practice, there are two main approaches for obtaining the *G*-factor, for both of which the observed polarization (anisotropy) is equal to zero. Zero polarization means the emission for which parallel and perpendicular intensities are equal. This can be due to the geometry of the system where both observed intensity components are equal or due to very long fluorescence lifetime during which all excited fluorophores rotate many times and are able to completely randomize before emission.

Square Geometry

The first, simplest approach (for square geometry only) is to rotate the excitation polarization to the *X* direction (horizontal excitation polarization, H). The distribution of excited molecules seen from the observed direction will be symmetrical, as shown in Figure 6.6. When viewed along the *X* direction, the distribution is symmetrical, and both *Z* and *Y*-polarized components must be equal. So, as shown in Figure 6.6, $I_{\parallel} = I_{\perp}$ and the ratio

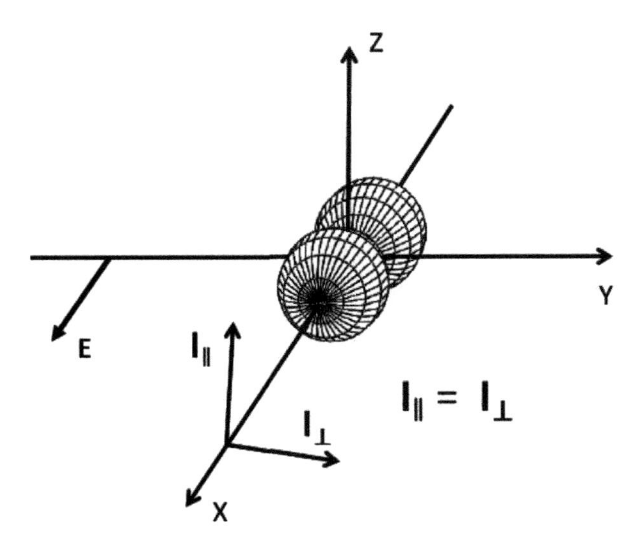

FIGURE 6.6 Distribution of excited transition moments with horizontal excitation as seen from the *X* direction.

$I_{\parallel}/I_{\perp}$ should be equal to 1. The G-factor can be defined as $G = I_{\parallel}/I_{\perp}$ and the measured anisotropy will be calculated as:

$$r = \frac{I_{\parallel} - GI_{\perp}}{I_{\parallel} + 2GI_{\perp}} \tag{6.26}$$

Rotation of the light polarization has important advantages. It does not require the intrinsic chromophore polarization to be zero, and the same fluorophore that is studied can be used to evaluate the G-factor. In this case, a spectral range for the G-factor will be exactly as needed. However, it is important to stress that this works exclusively for square geometry where the observation is orthogonal to the direction of excitation. This method will not work for any other direction of observation, even in the XY plane with the observation different than observations orthogonal to the Y direction and in the XY plane.

Rotation of excitation polarization will also not work for front-face, HTS plate, microscopy, or non-square geometry in general. To find the G-factor in such cases, we need a fluorophore with a known polarization (standard) emitting in the same range as the chromophore under investigation. Since the polarization of any fluorophore may depend on conditions (e.g., temperature, viscosity, etc.), the most practical is the fluorophore that presents long fluorescence lifetime as compared to the fluorophore correlation time (speed of rotation).

One possible approach to get 0 polarization is to use a fluorophore with an intrinsically present polarization equal to zero. A long-lived fluorophore in a low viscosity solvent will show very low (close to 0) polarization. For example, a metal-ligand complex $[Ru(pby)_3]^{+2}$ that has about 400 ns fluorescence lifetime in ethanol or water will have steady-state anisotropy equal to 0.00. Even shorter-lived fluorophores like fluorescein in the water at pH 9 (about 4 ns lifetime) will have steady-state anisotropy close to 0. A more advanced experimentalist may also use a specific excitation where anisotropy due to the orientation of absorption transition moment or mixing of two differently orientated transition moments exhibits a value of emission anisotropy close to zero. *It is important to remember that the G-factor refers to the detection channel that is completely independent of excitation wavelength (the detection system has no information what excitation wavelength has been used).* So the measured G-factor does not depend on the excitation wavelength as long as the emission exhibits zero polarization. But we also want to alert the reader that even a slightly higher anisotropy than 0 for the standard (or not fully symmetrical distribution of excited transition moments) would significantly corrupt the calculated anisotropy.

Determining G-Factor with Fluorophore of Known Anisotropy

If the known chromophore with known anisotropy, r_s is used as a reference, the G-factor will be:

$$G = \frac{I_{\parallel}(1 - r_s)}{I_{\perp}(2r_s + 1)} \tag{6.27}$$

And for $r_s = 0$ the $G = I_{\parallel}/I_{\perp}$ as previously defined. Equation 6.27 makes it possible to evaluate the G-factor when using a standard of known polarization. However, it is unlikely to have such perfect circumstances. Emission polarization strongly depends on factors like temperature and solvent viscosity and many other difficult to control factors and this approach is only recommended for advanced researchers.

Other Methods for Determining the G-Factor

Horizontal excitation and zero polarization standards are the most convenient ways of determining the G-factor of an instrument. However, more and more researchers from different fields have started using polarization assays in HTS or microscopy, and they typically do not have access to zero polarization standards. Also, frequently we need to evaluate the G-factor in the full spectrofluorometer range, and we may not have access to a wide range of fluorophores of zero or known polarization emitting between 300 nm and 900 nm. Thus, we discuss a couple of less known but very useful approaches for determining/estimating the G-factor in a full spectrofluorometer range.

Using synchronous scanning. All standard spectrofluorometers have an option for a synchronous scan. This is a simultaneous scan of excitation and emission with a fixed wavelength shift between both channels. The wavelength shift can be set to zero. In this case, we will look at scattered light (Caution! The excitation and emission wavelengths are the same and the signal can be very high. Make sure to set low voltage so as not to burn your detector). We can use Ludox solution which will produce sufficient scattering. Such scattering off dielectric particles is highly polarized (close to 100%). We can also use Raman scattering if we want/can set the shift in kK. Vertical excitation is used to measure the signal with vertical (parallel) observation and another different observation angle (e.g., 54.7° observation). As a precaution, check the quality of the polarizer. Also, measure the horizontal (perpendicular) component. The perpendicular component should be close to zero. Typically, good polarizers will give a ratio of $I_{\|}/I_{\perp} > 50$. If this ratio is lower than 10, there is a problem with the polarizer or scattering set up. Scattering can be contaminated with some emission, the concentration of the scattering particles being too high, or there may be multiple scattering which changes the plane of light polarization (frequently observed in tissue scattering). If polarizer sheets are used at wavelengths below 300 nm and above 700 nm, where they purely polarize light, the $I_{\|}/I_{\perp}$ will be lower than 50. When using a Glan-Thompson polarizer, the beam may not be well collimated. It should be well collimated (as close as possible to parallel) when entering the polarizer.

Let's consider vertically polarized light (as scattering is in this case). We measure the parallel intensity $I_{\|}$ and adjust it to be in the optimal spectrofluorometer detection intensity range. Obviously, $I_{\perp} << I_{\|}$, and the horizontal component cannot be measured. The vertical polarizer sees the maximum intensity. We measure the entire range of interest with vertical observation. Then we set our polarizer to a fixed angle, α (e.g., 54.7°—magic angle). Now the intensity seen by a perfect detector at an angle α should be:

$$I_{\alpha} = I_{V} \cos^2 \alpha \tag{6.28}$$

This is the intensity transmitted through the polarizer oriented at an angle α where I_v is the intensity measured with V (parallel) polarizer orientation. For real detectors (not perfect) this intensity should be decomposed into two components as seen by the detector; parallel (vertical) to the excitation light and perpendicular (horizontal) to the excitation light. These contributions will be:

$$I_{\parallel} = I_{\alpha} \cos^2 \alpha \quad \text{and} \quad I_{\perp} = I_{\alpha} \sin^2 \alpha \tag{6.29}$$

where $I_{\parallel}$ and $I_{\perp}$ are the intensities components of the intensity transmitted by a polarizer projected in the parallel and perpendicular directions, respectively. The intensity seen by the detector (measured intensity at the angle α) will be I_{α}^{m}:

$$I_{\alpha}^{m} = I_{\parallel} + I_{\perp}/G \tag{6.30}$$

Our system detects the vertical component as modified by the G-factor which we defined previously as $G = I_{\parallel}/I_{\perp}$. We can now calculate the G-factor as:

$$G = \frac{I_{\perp}}{I_{\alpha}^{m} - I_{\parallel}} = \frac{I_{V} \cos^2 \alpha \sin^2 \alpha}{I_{\alpha}^{m} - I_{V} \cos^2 \alpha \cos^2 \alpha} \tag{6.31}$$

It seems easy to measure G-factor using this approach, but there are a few obstacles that make it rather difficult to apply in practice. For example, this approach is very sensitive to the precision of the observation angle. Since the numerator and denominator contain squares of a trigonometric function, a small 1–2 degree offset may produce a significant error. Even if most recent fluorometers have automatic polarizer controls, such precisions may be a concern. Also, from our experience, we would only recommend using observation angles larger than 20° and smaller than 70°. For smaller or larger angles, an experimental error becomes too large.

Using an additional polarizer for observation. As mentioned previously, a detection channel has no information on how zero polarization has been generated. It just expects that the parallel and perpendicular components are equal. One way to evaluate the G-factor is to use an additional polarizer oriented at 45° in the emission channel immediately after the sample. Independently of intrinsic sample polarization, the light reaching the emission polarizer will be polarized at 45°. This allows us to use any chromophore that we want to measure polarization for (similar to horizontal excitation in square geometry). Also, we are not limited to square geometry and can use any geometry for excitation. With an emission polarizer, we measure parallel and perpendicular intensities:

$$I_{\parallel} = I_0 \cos^2 45 \quad \text{and} \quad I_{\perp} = I_0 \sin^2 45 \tag{6.32}$$

And the G-factor can be evaluated as previously defined $G = I_{\parallel}/I_{\perp}$. As in the previous method where we used a synchronous scan precision for determining 45° angle, the angle is crucial and any possible error will have a huge impact. In Figure 6.7, we present a ratio of $\cos^2 \alpha/\sin^2 \alpha$ in the angle range 38–52° (solid line). For 45°, the value is one, but a 5° error produces over 40% error, and even just 2° will give 15% deviation. This is very large and setting a polarizer at an exact 45° angle could be a challenge. This is one of the reasons that this approach has not been frequently used, even if it appears very convenient.

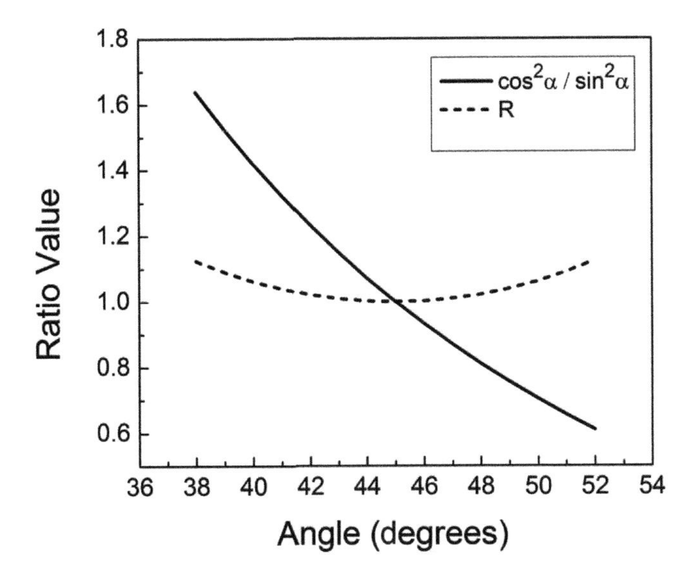

FIGURE 6.7 Calculated relative change in the G-factor as a function of the precision for 45° angle orientation (solid line). Dashed line represents an average value calculated from two complementary angles.

A simple trick can greatly improve precision. A look at the solid curve in Figure 6.7 shows that the ratio crosses the value 1 at 45° and almost linearly increases for angles below 45° and decreases for angles above 45°. If we install our 45° polarizer in a square block that we can rotate this polarizer exactly 90° we can do two measurements for 45° and 45° + 90°. The second angle (45°+ 90°) must be complementary and will result in an oppose error. Now if we take ratios as defined:

$$R_1 = \frac{\cos^2 45}{\sin^2 45} \text{ and } R_2 = \frac{\cos^2 45+90}{\sin^2 45+90} \text{ and define average value } R = \frac{R_1 + R_2}{2} \tag{6.33}$$

The error for estimating the G-factor becomes much smaller. In Figure 6.7, the dashed line represents the values of R. Even a 5° error in setting an angle 45° results in an error of only 6%. In Experiment 6-3, we present a measurement of G-factor using this approach.

Note: It is important to remember that the angles must be complementary, 45° + 90° and a simple flip of the polarizer. The variation in precision should be in the opposite direction.

Selecting Fluorophores for G-Factor Determination

Depending on the experimental configuration, we may need to use different approaches to evaluate the G-factor. In most cases using a fluorophore is the most suitable approach. For that, we would need proper fluorophores to cover the needed spectral range. Also, it is convenient to evaluate the G-factor in the full spectrofluorometer range. So, the first step is to select dyes that will cover the spectral range we will be working in. This will depend

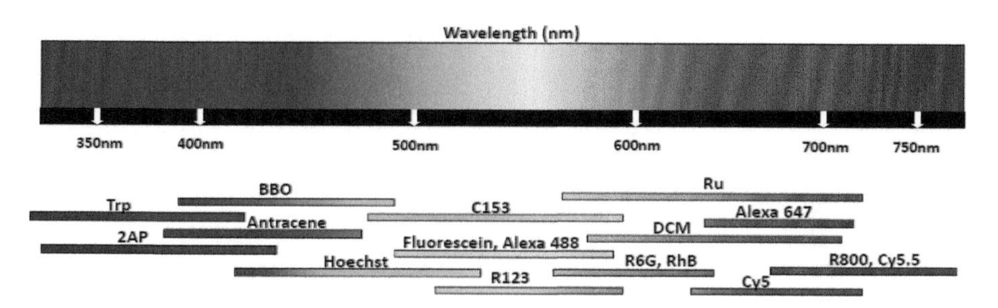

FIGURE 6.8 Colored spectral range 300–800 nm and some common fluorophores covering this spectral range.

on the fluorophore (probe) we want to study. The standard for the reference should be in the same or larger spectral range. For square geometry when we can rotate the excitation light polarization to horizontal (90°), we can conveniently use the same fluorophore as is being studied. However, on the edges of the emission spectrum (short wavelength and long wavelength), the signal will be low and using different chromophores could help improve the precision for G-factor determination. Occasionally, we will do G-factor determination for the entire spectral range of the spectrofluorometer. In this case, we will look for the group of fluorophores that will cover the entire spectral range. Figure 6.8 presents some fluorophores selected in such a way that their spectral ranges significantly overlap, cumulatively covering the full spectral range. It is important for there to be significant spectral overlap, since the signals on the edges are much lower, yielding a much higher error. It is important to stress that this is only a few dyes and many more can be identified/used.

Magic Angle

We already briefly discussed magic angle in Chapters 4 and 5. This is a very important constraint for many fluorescence measurements and it is always important to keep magic angle conditions for emission spectra measurements. This is a direct consequence of photoselection, emission polarization, and fluorophore mobility. So, when explaining the basic concept of fluorescence polarization, this is a good place to better expound on the concept of MA.

Imagine the system which has anisotropy $r = 0.4$ (no depolarization due to transition moments reorientation or molecular rotation). If one wants to measure emission spectra with the polarizer oriented vertically ($I_{\|}$) or horizontally ($I_{\perp}$), the emission spectrum measured with vertical polarization will turn out to be three times greater than with horizontal polarization. In fact, the emission spectrum will change (will be dropping down) as the observation polarizer rotates from 0° to 90°. Contrary, when the anisotropy for the same system is negative (e.g., −0.2 when the excitation wavelength has been selected to excite the orthogonal transition as compares to emission) the spectrum measured with a vertically oriented polarizer ($I_{\|}$) will be smaller than with a horizontally oriented one ($I_{\perp}$). And when the initial anisotropy is zero (due to excitation wavelength or/and fast molecular rotation) measured intensity will not be dependent on the observation polarizer orientation (assuming the G-factor is 1 or spectrum was corrected for the G-factor). If one would like to compare true intensities of two samples that present different anisotropy, *what would be the proper way for measuring the emission spectra?* To avoid such problems one should measure total emission intensity. Typically, it is not possible to collect all photons emitted by the sample, but for samples excited with polarized light the emitting molecules will have symmetrical distribution (cylindrical symmetry) and light will be emitted according to such distribution of transition dipoles. Knowing the symmetry of emitted light is enough to collect fluorescence intensities along the three major axes of symmetry to calculate total emission. In the case of cylindrical symmetry directions X, Y, Z (as in Figure 6.4) are the main symmetry axes. Then the total intensity, I_T is directly proportional to the sum $I_X + I_Y + I_Z$. For a coordinate system as in Figure 6.4, it will be $I_T = I_\parallel + I_\perp + I_\perp = I_\parallel + 2I_\perp$. Obviously making measurements along two or three axes would not be convenient, and we would prefer to find a polarizer position (orientation) for which observed intensity will only be proportional to a total intensity. The simplest approach is to find the angle (called the magic angle) for which we will measure the intensity that is equal to the average contribution of intensities measured along three symmetry axes. In this case:

$$I_{MA} = (I_\parallel + 2I_\perp)/3 \tag{6.34}$$

This will happen for a polarizer orientation in which the contribution from the parallel and perpendicular intensity components ($I_\parallel$ and as $I_\perp$) are equal ($I_\parallel = I_\perp$) (remember there are two $I_\perp$ components contributing to the total intensity and only one parallel component $I_\parallel$). This happens only for the polarizer orientation is equal to 54.7° (for a system with cylindrical symmetry), the above-mentioned magic angle (MA). Physically, the magic angle corresponds to the condition where averaging over the angle $\cos^2\theta$ leads to the value of 1/3 (each component contributes 1/3 intensity). Observation emission intensity under the MA will be independent of sample polarization and independent from processes leading to sample depolarization (e.g., molecule rotation and reorientation).

It is important to stress that the observed fluorescence signal at the magic angle is independent of sample polarization and will not be affected by depolarization processes like rotational diffusion. Any displacement in transition moment orientation will equally (proportionally) contribute to changes in the parallel ($I_\parallel$) and perpendicular ($I_\perp$) intensity components. This becomes very important when measuring fluorescence lifetimes. When the time for rotation of chromophore (correlation time) is comparable to its fluorescence lifetime, the fluorophores reorientation will affect the observed intensities decays when viewed by parallel or orthogonally oriented polarizers. Only when the observation polarizer is oriented under the magic angle the observed intensity decay is unaffected by molecular rotation. As for fluorescence spectra, an improper polarizer orientation will only affect absolute intensity, but will not change the spectral profile. However, as we will discuss in the next chapter, when measuring fluorescence lifetime the effect can be very dramatic, resulting in significant changes of observed intensity decays. So, it becomes very important for fluorescence intensity decay (fluorescence lifetime) measurements to always use magic angle observation. For a more detailed experimental explanation for magic angle conditions, see Experiments 4-3 and 7-10.

In the following section, experiments will be presented demonstrating how to check polarizers, how to measure polarized absorption of chromophores embedded into an anisotropic media, how to determine *G*-factor for various instruments (spectrofluorometers), and how to measure emission/excitation polarization spectra. Different examples of measuring emission (excitation) anisotropy in various isotropic systems will be presented, as well as some examples of such measurements in oriented systems.

EXPERIMENT 6-1 HOW TO SELECT AND CHECK POLARIZERS USING A SPECTROPHOTOMETER

Keywords: light polarization, polarizers, sheet polarizers, Glan-Thompson polarizer, absorption, water, polarization.

For any polarization experiment/measurement, it is fundamental to have polarized light and be able to measure the state of light polarization (emitted or transmitted). There are various ways for measuring light polarization but the simplest one (most convenient) is to use a light polarizer. In Chapter 1 and the "*Introduction*" to Chapter 6, we discussed polarized light, how polarizers work, and how to estimate the plane of polarization for an unknown polarizer. One very important less known aspect is finding the spectral range in which the polarizer transmits and polarized light. There are two factors that we need to consider. First, the polarizer needs to be transparent in the spectral range where we want to use it. Second, in this range, the polarizer must transmit and polarize light (transmit only one light polarization). *Note: We stress transmit and polarize since in many instances the polarizer would transmit electromagnetic radiation but not polarize it completely.* Already looking through two crossed polarizers on a bright light source (sunlight or a lamp for example), we will notice that some light goes through (the field is not completely black) and transmitted light has some color (frequently bluish and/or reddish).

We will use a spectrophotometer to precisely find in what spectral range polarizers can be used and if they polarize light in this range adequately. Obviously, if the polarizer is not transparent it cannot be used. And if it is transparent (semi-transparent) but not polarize light sufficiently (does not transmit light of a single polarization plane), it will not fulfill its role either.

Just looking at a polarizer sheet everybody will notice that it is dark like a pair of sunglasses. This means a significant part of the light is not transmitted by the sheet. In addition, the losses due to beam polarization could add another level of attenuation. Polarizers should always be used in spectrofluorometers and they will typically be a part of standard equipment. The excitation light would typically be vertically polarized (the direction, we frequently call parallel) while the emission polarizer will have at least three positions; vertical (parallel), horizontal (perpendicular), and magic angle (54.7°). Also, many spectrofluorometers will have a full capability to orient both polarizers (on excitation and emission) at any angle between 0° and 90°. In contrast, a typical spectrophotometer will not be equipped with a polarizer and the polarizer will not even be available as an option. So, it will have to be added "in-house" by the user.

When equipping the spectrophotometer with the polarizer as a first step (after defining the polarization plane for the polarizer), we would recommend to estimate the beam polarization in the spectrophotometer. The direction of beam polarization would be the optimal polarizer orientation (position for which the polarizer transmits maximum light intensity).

(*Continued*)

EXPERIMENT 6-1 (CONTINUED) HOW TO SELECT AND CHECK POLARIZERS USING A SPECTROPHOTOMETER

Most experimentalists working with spectrophotometers do not think about the spectrophotometer beam polarization. It does not matter for measuring isotropic samples and the general assumption is that light is not polarized. But this is typically not true and the spectrophotometer beam will present a significant polarization. So, the light intensity transmitted by a single polarizer may strongly depend on polarizer orientation. To optimize our measurement, we need to ensure that by inserting a polarizer into the beam we will not attenuate the beam intensity too much. For perfectly isotropic (non-polarized beam), the attenuation should be over 60% loss (50% attenuation due to light polarization plus residual losses due to absorption and two reflective surfaces). So, in such a case, the polarizer sheet would transmit about or less than 40% of light (a perfect Glan-Thompson polarizer could transmit more, close to 45% of isotropic light). But if our beam has residual polarization, depending on the degree of polarization and polarizer orientation this loss could be less or much more leaving only a few percent of the original beam intensity. To evaluate the beam polarization, we will perform a simple set of measurements. First, we will use a sheet polarizer since such a polarizer does not require good beam collimation and we can use it in any spectrophotometer without the need to collimate the beam. Take two polarizers and look at a bright light source like a lamp, ceiling light, etc. and try to cross them while rotating one of the polarizer sheets. You will notice that when polarizers have crossed the field of view would become black but you may see some deep blue and/or red color leaking through. This polarizer does not work in the blue and/or red spectral range which is typical in the case for simple polarizer sheets. But in the visible spectral range, this polarizer will typically work well.

MATERIALS

- Various polarizers (sheet, Glan-Thompson, grid)
- Stands, holders, and a rotary mount.
- Various holders for mounting custom stand to the spectrophotometer.

EQUIPMENT

- Spectrophotometer

METHODS

1. Start the spectrophotometer, wait at least 5 min to warm-up and measure the baseline without the polarizer (typically an empty spectrophotometer). In the Varian Cary 50 spectrophotometer that has a lock-in type detection, the measurement can be done with the chamber open, likewise for the diode array Agilent 8453 spectrophotometer. The low-level external light will not perturb the measurement. In this case, we can measure transmittance without the need to close the chamber.
2. Install the polarizer holder in the spectrophotometer (see Experiments 3-24 and 3-25 for how to install the holder). Check the baseline by measuring absorption/transmission (for example of an installed platform in Agilent Cary 60 see Figure 6-1.1a). The baseline should not change and measured absorbance should be 0 (100% transmission since the polarizer is not yet installed and light should freely go through).

(*Continued*)

EXPERIMENT 6-1 (CONTINUED) HOW TO SELECT AND CHECK POLARIZERS USING A SPECTROPHOTOMETER

3. Insert one polarizer into the holder in the beam path (Figure 6-1.1a). Measure the transmittance for a few orientations of the polarizer. You will notice that measured transmittance depends on the polarizer orientation. To find the optimal polarizer orientation, measure the transmittance starting from the vertical polarizer orientation and rotate the polarizer every 10–30° up to 180° (six or more positions). Figure 6-1.2 shows measured transmittances for different polarizer orientation in the Agilent Cary 60 spectrophotometer when rotating the polarizer every 10°. This measurement can be also done for a single wavelength for which the polarizer is working well (typically any wavelength in the visible range 450–600 nm is good for the majority of optical polarizers) or a narrow spectral range like 500–600 nm. Figure 6-1.3 shows single-point measurements. As seen from Figures 6-1.2 and 6-1.3, the transmittance may change over three-fold. The Agilent Cary 60 spectrophotometer has a relatively strong intrinsic beam polarization and most other spectrophotometers would have a less polarized beam. The maximum transmittance is for a polarizer orientated at about 90°.
4. Now we want to estimate more precisely what is the beam polarization in the spectrophotometer (to find an optimal orientation of the polarizer we should choose which would transmit maximum light intensity). In this case, we can measure transmittance at a single wavelength and move the polarizer with small steps starting from a position close to the maximum transmittance found in the previous point. We recommend starting from a position close to maximum transmittance (in this case, we start from the position at about 84° (close to 90°) and rotate the polarizer every 2–5° in one direction (left or right). Figure 6-1.4 shows measured transmittances at 550 nm observation wavelength when the polarizer is rotated from 84° in the clockwise direction every 2°. Steps are marked with arrows. For the first four steps, we observe an increase and for the next two, a decrease of transmittance. We come back in the seventh step to 92° (angle after which the first decrease was observed) and continue with a 1° step going back. Final orientation is marked in red in Figure 6-1.4. In practice, the precision of 1° in a simple test experiment is very good and we would not decrease the step further.
5. Use the first polarizer in the spectrophotometer oriented at an angle allowing maximum transmittance. *See previous points of the experiment.* Measure transmittance in the full range. Figure 6-1.5 (black line) shows the transmittance for the polarizer used in the previous experiment in 200–1,100 nm range. We would conclude that above 300 nm range, the polarizer sheet has acceptable transmittance. We arbitrarily assume that more than 10% transmittance is acceptable (this would depend on a spectrophotometer but for most 10% transmittance that corresponds to an absorption of 1 is in the acceptable range).
6. Insert the second polarizer with an orientation parallel to the first one (see photograph in Figure 6-1.1b). For the second polarizer oriented parallel to the first one, the transmittance drops slightly due to extra reflections and polarizer sheet absorption—Figure 6-1.5 (blue line).
7. Rotate the polarizer to 90° and measure transmittance, Figure 6-1.5 (red line). Good polarizers will now be crossed and the transmitted intensity should be zero. This is true for the range below 750 nm. However, above 750 nm transmittance through crossed polarizers quickly increases. This is the indication that in the range above 750 nm polarizers quickly loses its ability to polarize the light.

(Continued)

EXPERIMENT 6-1 (CONTINUED) HOW TO SELECT AND CHECK POLARIZERS USING A SPECTROPHOTOMETER

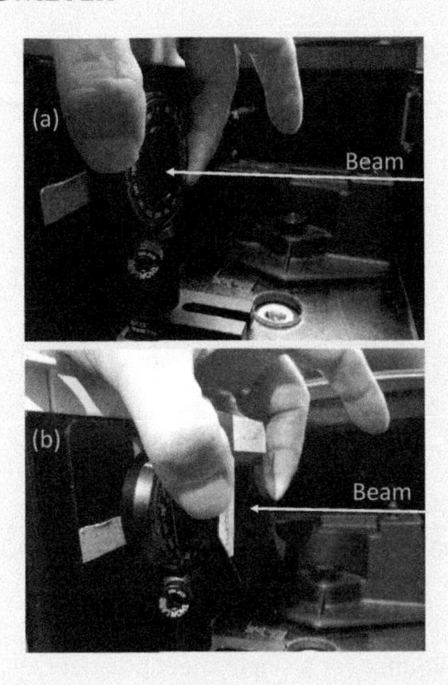

FIGURE 6-1.1 (a) Chamber of Agilent 60 spectrophotometer with a mounted polarizer on a rotator. (b) Spectrophotometer chamber with the second polarizer in the beam path.

RESULTS

Figure 6-1.2 shows measured transmittances for polarizer orientation changing from 0° (vertical) to 180° at 10° intervals (180° overlay with 0°). It appears that polarizer transmits light for wavelength longer than 300 nm. But transmitted intensity strongly depends on the wavelength and polarizer orientation. For wavelengths over 850 nm transmittance is high. But for the visible range (400–700 nm) vertical polarizer orientation transmits only ~20% making the absorption measurement difficult. For the horizontal (91°) polarizer orientation, the transmittance in the visible range is about 60% allowing for three times more light to reach the detector as compared to the vertical polarizer orientation. In general, the extent of intensity lost depends on intrinsic beam polarization and orientation of inserted polarizer but it will always be significant. So it is important to optimize polarizer orientation to minimize the light intensity losses. For a good orientation of the polarizer, we can expect about 50% of light transmitted for the isotropic intrinsic beam light or more (up to 90%) when the beam is fully polarized. We always want to use the first polarizer orientation for which we have maximum transmittance (the intensity loss is minimal).

Looking in Figure 6-1.2, we see that the polarizer transmits light with wavelengths longer than 350 nm and we can expect that for wavelengths longer than 350 nm this will be a good polarizer. *But this is not the case.* To test the range for which the polarizer polarizes the light, we will use a second identical polarizer that we will orient at 0° and 90° as compared to the orientation of the first polarizer. In Figure 6-1.1b, we present a picture of how this can be done. In Figure 6-1.5, we presented the results. For the parallel orientation, the intensity drops down about 20%. This is due to reflections on two interfaces and some absorption of the second polarizer (the absorption contributes to the shorter wavelength range below 400 nm).

(Continued)

EXPERIMENT 6-1 (CONTINUED) HOW TO SELECT AND CHECK POLARIZERS USING A SPECTROPHOTOMETER

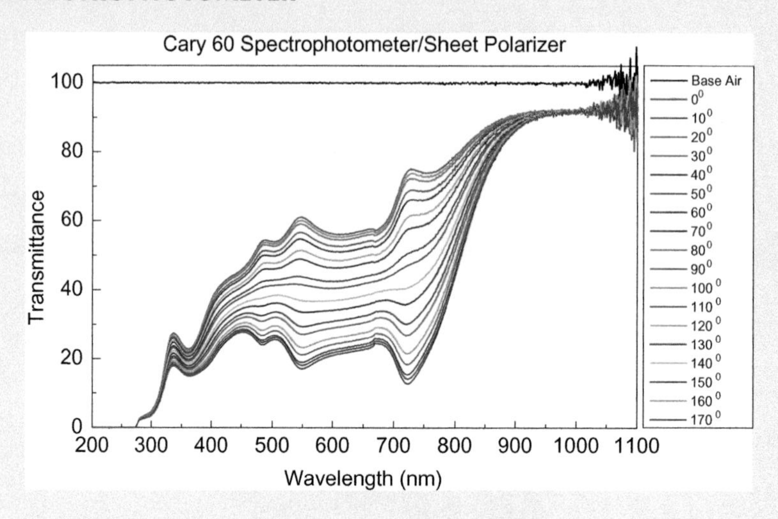

FIGURE 6-1.2 Measured transmittance for a single sheet polarizer as a function of angle.

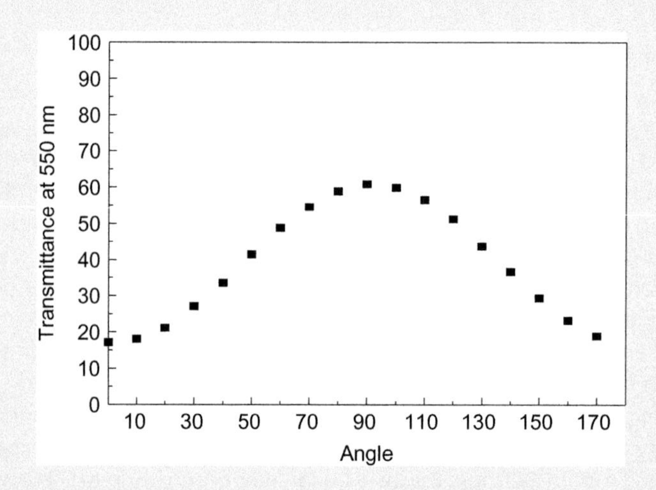

FIGURE 6-1.3 Measured transmittance of a sheet polarizer at 550 nm as a function of angle.

Rotating the polarizer to 90° results in practically zero transmittance up to 720 nm confirming that polarizers are crossed and polarize the light well. But above 720 nm, the transmittance quickly increases that is a clear indication that in this range transmitted light is not perfectly polarized. For wavelengths longer than 950 nm, the transmittance is the same as for a parallel polarizer orientation indicating that for this range the polarizer does not polarize light at all. This polarizer sheet can be used in the 320–740 nm range. Below 320 nm, the polarizer does not transmit a sufficient amount of light and above 750 nm, the polarizer transmits light well but does not polarize it.

(Continued)

EXPERIMENT 6-1 (CONTINUED) HOW TO SELECT AND CHECK POLARIZERS USING A SPECTROPHOTOMETER

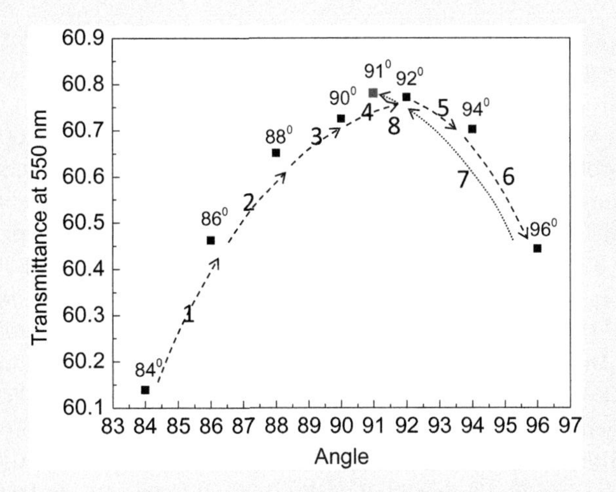

FIGURE 6-1.4 The measured transmittance at 550 nm from 84° adjusted at 2° step clockwise (six steps). Then back at 1° step counterclockwise.

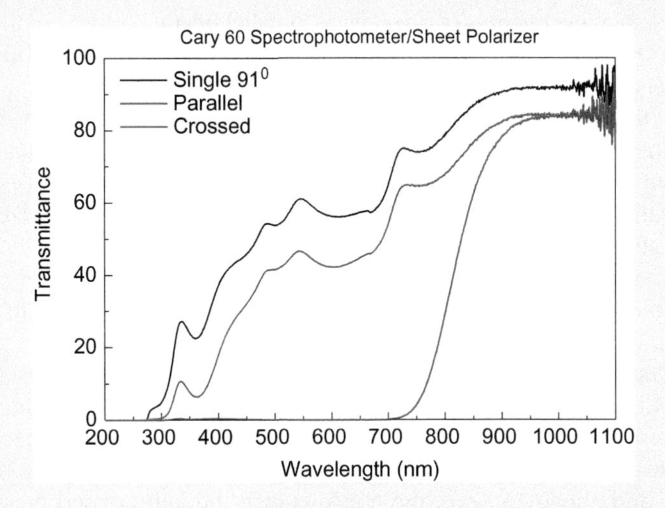

FIGURE 6-1.5 Measured transmittance for a single polarizer (black) and second polarizer inserted parallel (blue) and perpendicular—crossed (red).

In the next step, we will measure the transmittance range for different polarizers and find the spectral range for which polarizers efficiently polarize the light. *Remember that a polarizer may transmit light but does not polarize it.* To do this, we set the spectrophotometer wavelength range to maximum (full wavelength range of a spectrophotometer). As previously for this experiment, we should use two identical polarizers for each of the exercises. We will test Glan-Thompson, grid polarizer, and three different sheet polarizers.

(*Continued*)

EXPERIMENT 6-1 (CONTINUED) HOW TO SELECT AND CHECK POLARIZERS USING A SPECTROPHOTOMETER

In the following measurements, we consider different polarizers. First, we present an experiment where we test the Glan-Thompson polarizers. *Note: Make sure that the beam in the spectrophotometer is properly collimated (it is important for the Glan-Thompson polarizers for the beam to be parallel—with minimal divergence).* The Agilent Cary 60 does not require extensive/special collimation for the beam but some spectrophotometers may require an additional set of lenses to collimate the beam. In Figure 6-1.6a, we present measured transmittances for Glan-Thompson polarizers. A single Glan-Thompson polarizer reduces transmittance to below 80% in the visible and more in the UV range but still for 300 nm and above the transmittance is significantly over 20%. *Note: It is important to remember that polarizer is oriented parallel to the spectrophotometer beam polarization that in this case is close to 90° but different.* This polarizer can be easily used in the 250–1,100 nm range. An addition of another Glan-Thompson polarizer that is parallel to the previous reduces transmittance another 10–15% across the whole range (this is mostly due to reflection from the surface like in the case of the glass plate and cuvette wall—Experiments 3-1 and 3-2. For a polarizer rotated to 90°, measured transmittance should drop below 1%. In fact, we should vary a little of the angle around assumed 90° to find the minimum transmittance. A barely visible red line (close to zero) shows the excellent crossing of these two polarizers. The dynamic range for the Glan-Thompson polarizer is very good when the beam is properly collimated with not more than a few degrees of divergence (parallel beam). Just for comparison, we are showing the 45° relative polarizer's orientation for which transmittance is about 50% of two parallel polarizers (cyan line).

In Figure 6-1.6b, we present a grid polarizer. The grid polarizers also transmit in the full 250–1,100 nm range. Crossing the polarizers (90°) show a significant attenuation across the whole range. But in the UV part (below 350 nm), we notice a higher than 2% transmittance. When transmittance through crossed polarizers is comparable to parallel and 45° orientation, this is the indication that polarizers do not polarize sufficiently. We would not recommend these polarizers to be used for the range below 320 nm.

Finally, in Figure 6-1.6c, we present a simple inexpensive sheet polarizer. Its transmittance does not extend below 420 nm. But when crossing we can see that this is a good polarizer for the spectral range 420–750 nm.

In Figure 6-1.6d, we present another set of sheet polarizers. Transmittance for a single polarizer is acceptable for a spectral range above 240 nm. This could be an excellent polarizer. Crossing two polarizers yield very good attenuation in the 300–750 nm range. But below 300 nm and above 750 nm, the transmittance through crossed polarizers quickly increases. Below 300 nm and above 750 nm polarizer losses its ability to polarize light. This polarizer can be safely used in the 300–750 nm range. This is a very thin polarizer and for longer wavelength (above 750 nm), we can observe small characteristic oscillations. This is a common interference effect characteristic of very thin films. *Note that in the* 250 *nm range the polarizer has good transmittance but does not polarize the beam effectively.*

After these experiments, we should know how to estimate beam polarization of our spectrophotometer and we should know what range we can use our polarizers. In general, most sheet polarizers are good in the visible range. It could be expected that in UV their performance will not be good. But it is less expected that above 750 nm range where the transmittance is very good the polarizers quickly lose its ability to polarize light.

(Continued)

EXPERIMENT 6-1 (CONTINUED) HOW TO SELECT AND CHECK POLARIZERS USING A SPECTROPHOTOMETER

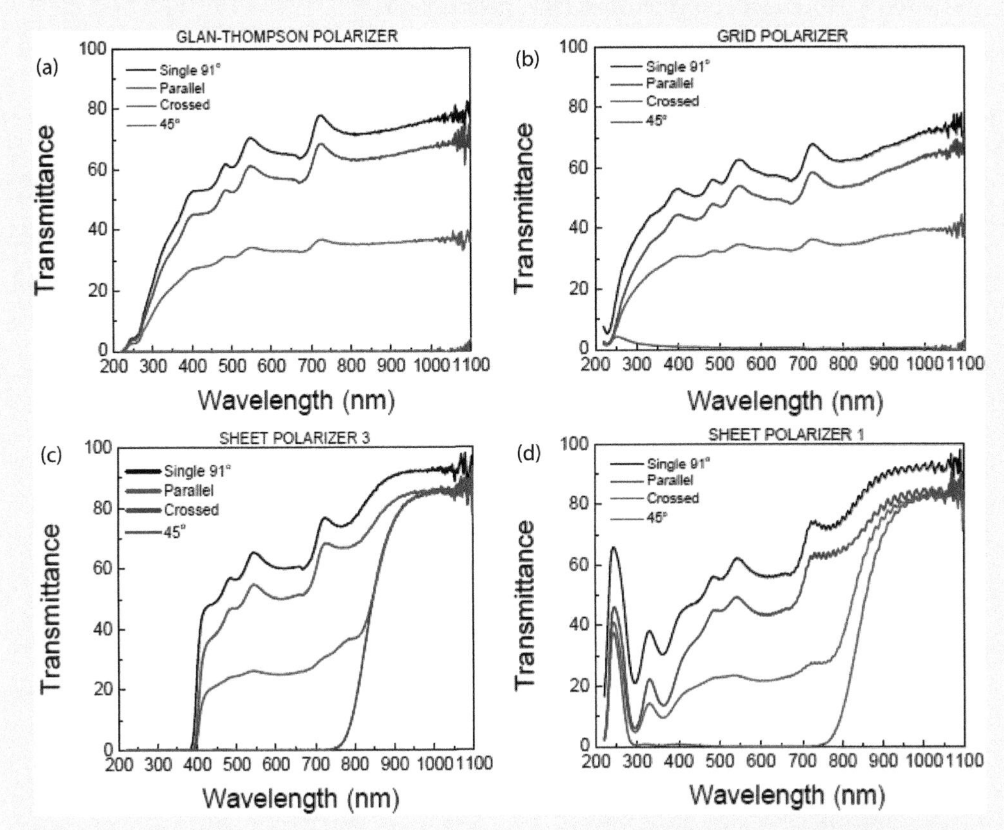

FIGURE 6-1.6 Measured transmittances for four different polarizers. (a) Glan-Thompson, (b) Grid, (c) Flexible sheet, (d) UV sheet. As in Figure 6-1.5, the black line presents transmittance of a single polarizer, the blue line is transmittance when the second identical polarizer is inserted parallel, the red line is for second polarizer perpendicular—crossed, and the pink line is for the second polarizer oriented at 45° to the first one. Note that parallel is 91° in this case and 45° and crossed (90°) is to the first polarizer orientation.

CONCLUSIONS

Presented in this exercise experiments allow defining the transmission range for the polarizer and the range in which polarizers can be used. Also, we learned how to check the beam polarization of our spectrophotometer. Typically sheet polarizers perform well in the visible range and they are easy to use. As it is expected that the sheet polarizer loses transparency in the UV and will not work in the UV range. It is rather surprising that they have excellent transmittance in the NIR (700–1,000 nm) but they completely lose the ability to polarize light. The experimentalist should remember about that since we have quite a lot of dyes in this spectral range and many polarization assays have been developed for such dyes.

EXPERIMENT 6-2 ORIENTING POLYMER FILMS: STRETCHED PVA FILMS

Keywords: absorption, polymer stretching, polarization.

In the previous Experiment 6-1, we learned how to polarize the beam in the spectrophotometer and how to check the working range of any given polarizer. The ability for polarizing and testing the beam polarization allows for studying anisotropic properties of absorbing and/or emitting systems. In fact, it allows studying individual molecules and molecular ensembles that are oriented in any anisotropic matrix. In Experiments 3-24 and 3-26, we also discussed how to embed organic molecules into PVA films. In isotropic (unstretched) PVA films, chromophores are randomly oriented and absorption measurements when using polarized light give the same result independently of the direction of light polarization or film orientation (assuming film surface is perpendicular to the spectrophotometer beam). This situation changes when the film has been oriented by for example a unidirectional stretching. Most effective for orienting the ghost molecules is unidirectional stretching that results in a significant alignment of elongated and planar molecules.

PVA is a very flexible polymer and when the stretching force (in one direction) is applied the polymer film will increase its length. Like most flexible materials (e.g., polyethylene or just gum), the total volume of the film cannot change and as the polymer is stretched its width and thickness decrease proportionally. In effect, polymer chains are being oriented parallel to the stretching direction and aligned closely together. To stretch PVA film we will warm up the film to about 80–90°C what greatly increases polymer flexibility.

The goal of this experiment is to show a simple but effective way for stretching/orienting the PVA film.

MATERIALS

- Polyvinyl alcohol (e.g., Aldrich Mowiol—Mw 130,000 (other sizes that are over 50,000 are also OK))
- Water, flask, small bottles, Petri dish, and paper tape

EQUIPMENT

- Mechanical stretching device
- Heater

METHODS

1. For details on how to prepare PVA films, see Experiment 3-24.
2. Having polymer films cut a small square piece of PVA film (about 1 in × 0.5 in).
3. Wrap with paper tape (or good scotch tape) two ends of each film (see Figure 6-2.1) and mark two points separated by 5–10 mm on the film edge in the film center (the separation marks could be closely spaced when using smaller film).
4. Mount the film in the stretching machine (an example is shown in Figure 6-2.2, left).
5. Warm up the film. This can be done in a few ways depending on the capability you have in the laboratory. A typical approach would be embedding the stretching device in a closed container with heating and warming it up. A simple way would be placing the stretching device on a heated plate, or by gently blowing hot air on the film.

(Continued)

EXPERIMENT 6-2 (CONTINUED) ORIENTING POLYMER FILMS: STRETCHED PVA FILMS

The important thing is to uniformly increase the temperature over the entire film. Increased humidity helps stretching.

6. Gently turn the stretching screw and stretch the film. *Note:* It is relatively easy to stretch the film up to four-fold when it is warm without breaking it. However, stretching the film up to five-fold requires more experience. The maximum stretching that we were able to achieve using the highly specialized stretching mechanism was 6.5-fold. In Figure 6-2.2 (right), we are showing the stretched film.
7. Slowly cool down the film and gently dismount the film. Be careful as the film is much thinner and more flexible. *Note:* Dismounting the film before it fully cooled down to room temperature will result in partial shrinking of the film.

To stretch the polymer film, we will cut a square piece of PVA film (about 1 in × 0.5 in) and the two ends of each film will wrap with paper tape. Example of PVA films prepared for stretching is presented in Figure 6-2.1 (left). We will typically mark two points on a film side.

RESULTS

When stretching the film the distance between two marked points will increase and by measuring the distance between these two points before and after stretching we can measure how many folds the film has been stretched. On Figure 6-2.1 we present examples of PVA films before stretching—top and after stretching—bottom. Before stretching, we mark sides of the film at fixed distance (e.g., 5 mm or 10 mm). After the stretching, we measure the distance between two marks and divide the distance after stretching by the distance before the

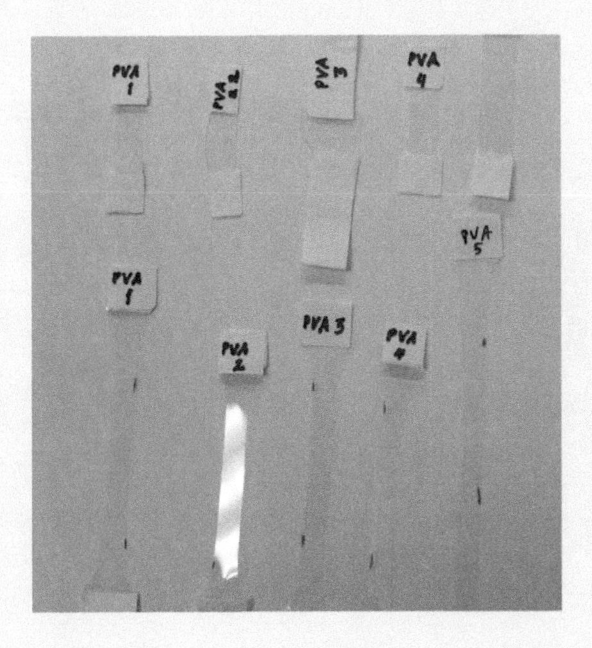

FIGURE 6-2.1 Picture of PVA films prepared for stretching on the top and films after stretching on the bottom.

(Continued)

EXPERIMENT 6-2 (CONTINUED) ORIENTING POLYMER FILMS: STRETCHED PVA FILMS

stretching to determine how many folds we stretched the film, N. Presented in Figure 6-2.1 films were stretched about four-fold. During stretching, the film width and thickness decreases and to describe orientation in a stretched film, we define a parameter called a stretching ratio, R_s which is different from stretching fold (4). To understand the definition of stretching ratio, imagine a small sphere of radius r in the polymer volume (in practice, draw the circle of radius r on the isotropic film before stretching). As we stretch the film, the sphere/circle will elongate forming an ellipsoid and the stretching ratio, R_s is defined as $R_s = a/b$. Where a is a semi-major axis and b is a semi-minor axis of the ellipsoid formed from an imaginary sphere in polymer during uniform stretching. To understand the meaning of stretching ratio, we have to realize that during stretching the total volume of the film cannot change. So, an increase in length is at the expense of film width and film thickness. Comparing an imaginary volume of the sphere $4\pi r^3/3$ to the volume of a formed ellipsoid $4\pi ab^2/3$, we can calculate a stretching ratio as a function of how many folds the film was stretched ($R_s = N^{3/2}$). For example, if we stretch the polymer film four-fold, the sphere of radius 1 will deform to an ellipsoid with a semi-major axis of 4 and semi-minor axis of 1/2 and $R_s = 8$ and if we stretch film five-fold, the stretching ratio will be $R_s = 11.2$.

The simple stretching device is shown in Figure 6-2.2. The PVA film has been mounted in jaws (left) and by rotating the screw, the stretching force is applied. *Note: We present in the picture PVA films containing chromophore (colored) for better visibility.* Typically to obtain best results, the film should be stretched at an elevated temperature (about 80–90°C). At higher temperature, the PVA polymer becomes more elastic and is easier to stretch. A paper tape also protects the film from breaking in the contact places/points with the holding jaws during

FIGURE 6-2.2 Manual stretching device with mounted PVA film (left) and four-fold stretched film on the right.

(Continued)

EXPERIMENT 6-2 (CONTINUED) ORIENTING POLYMER FILMS: STRETCHED PVA FILMS

stretching (see Figure 6-2.2). On the right in Figure 6-2.2, we show a blank PVA film stretched about four-fold. Examples of already stretched PVA films are shown also in Figure 6-2.1 (right).

CONCLUSIONS

In this experiment, we presented a simple way for stretching polymer films. PVA films can be stretched up to five-fold (ratio of the length of the film after stretching to the length before stretching—the distance between two points marked on the film edge). Already 4–5 fold stretching gives an almost maximal possible film alignment. Stretching the film over five-fold is very difficult and films frequently break. The largest stretching, we were able to achieve was about 6.5-fold after hundreds of trials. But even this did not give us much better results than film stretched 4.5-fold. Stretched polymer films can be used to measure absorption with polarized light for two film orientations; parallel and perpendicular to light polarization. *Note:* We always mark two dots on the film to evaluate the stretching and we do not use the distance between two moving jaws holding the film before and after stretching. Distance between the dots must be smaller than the distance between jaws. As seen in the Figure 6-2.1, the stretched film is not uniform (the width of the film increases on the two ends) and the stretching is maximal and uniform only in the central part of the film. *Note:* For many quick experiments, we can just stretch the polymer film by holding it between two plyers over the heated hot-plate. The needed force when the film is heated to about 80–90°C is not big and can be done just by hand.

EXPERIMENT 6-3 POLARIZED ABSORPTION OF STRETCHED PVA FILMS

Keywords: absorption, water, polarization.

In the previous Experiment 6-2, we presented the method for stretching/orienting polymer films. Such films can be used to orient embedded dye molecules and measure the polarized absorption (linear dichroism) or polarized emission. But as discussed in Experiment 3-25, the PVA film itself will attenuate transmitted light intensity by reflecting a small part of the light and absorbing the light in the blue/UV spectral range. Absorption of PVA films is one of the smallest among the different common polymers and practically only PVA and to some extent polyethylene can be comfortably used in the range below 350 nm. Similar to isotropic films, we need to characterize the reflectivity and absorption of oriented films. Since the polymer is oriented (anisotropic), we need to control its properties for two light polarizations; parallel and perpendicular to the film stretching direction. It is not a surprise that these properties should be different for both polarizations since the polymer chains are now oriented and will reflect and absorb differently each light polarizations. Also, these properties will depend on stretching and when using films with a different stretching ratio we should make independent standards for each of them.

(Continued)

EXPERIMENT 6-3 (CONTINUED) POLARIZED ABSORPTION OF STRETCHED PVA FILMS

In this experiment, we will show how to evaluate spectroscopic properties of the oriented film (reflectivity and absorption) for parallel and orthogonal light polarization using four-fold stretched film (stretching ratio 8) as an example. We will use an approach similar to that for the isotropic film (Experiment 3-25) but now we will use a polarized beam and measure these properties for parallel and perpendicular orientation of the light polarization to the film stretching direction. Since the film is oriented, we expect absorption parallel and perpendicular to the film orientation will be different and a different correction will have to be considered when embedded dyes are measured. In Experiments 6-1 and 6-2, we presented a simple way how to install into the spectrophotometer chamber polarizer and a rotatory stage in which we can mount a stretched film.

MATERIALS

- Glan-Thompson or grid polarizer
- Stands, holders, and rotatory mount for the film
- Blank PVA film stretched four-fold

EQUIPMENT

- Spectrophotometer with an installed polarizer

METHODS

1. Start the spectrophotometer, wait at least 5 minutes to warm-up.
2. Install the polarizer and make sure it is in the optimal position (see previous Experiment 6-1).
3. Install the rotatory stage for the film (see Experiment 6-1) and make sure the beam passes through the center of the hole, Figure 6-3.1.
4. Measure the baseline with the polarizer in and no film installed in the rotatory mount. Repeat it for four positions (0°, 90°, 180°, 270°) of the rotatory stage as compared to the direction of light polarization. For a properly installed stage, each reading should be identical (very close) within the experimental error.
5. Mount your film and measure absorption for the film with the stretching direction oriented parallel and perpendicular to the polarizer polarization plane. This will include four measurements:
   - Stretching direction parallel to the polarizer polarization plane (0° and 180°)
   - Stretching direction perpendicular polarizer polarization plane (90° and 270°).

RESULTS

Figure 6-3.1 shows the spectrophotometer sample chamber (left), stretched PVA film mounted in the film holder (top right), and the film holder mounted on the rotatory stage. The mounted polarizer can be seen on the left side of the chamber. We recommend rotation of the film while keeping the fixed position of the polarizer. There are a few reasons for that. First, the beam in a spectrophotometer is typically polarized and we will encounter

(Continued)

EXPERIMENT 6-3 (CONTINUED) POLARIZED ABSORPTION OF STRETCHED PVA FILMS

FIGURE 6-3.1 Left: Spectrophotometer chamber with a mounted polarizer and rotatory stage. The beam entry is on the right side and detector on the left side. Right: Stretched PVA film mounted in the film holder (top). Film holder mounted on the rotatory stage (bottom).

a significant loss in intensity lowering the precision of the measurement. Second, when rotating the polarizer, we will have to measure two baselines since the detection is typically sensitive to light polarization. Instead, when rotating the film, we do not change light polarization or any detection system configuration. *Note:* PVA is an amorphous polymer that does not change/rotates the plane of light polarization when the film is rotated. In practice, it does not matter if the film is exposed to an already polarized light (light goes through the polarizer first and then the sample film) or light passes first through the film and then through a polarizer to select polarization. It is sometimes convenient to install the polarizer on the end closer to the detector.

Figure 6-3.2 shows measured absorptions for light polarized parallel and perpendicular the stretching direction of the film ($R_s \approx 8$). It is clear that absorption depends on light polarization. The film-oriented perpendicular to the plane of beam polarization absorbs much less. Also as presented in Figure 6-3.2 (bottom) (expanded view), the film reflectivity depends on film orientation. This is not a surprise since we would expect that for oriented (anisotropic) system/film properties would depend on light polarization. Similar to isotropic film (Experiment 3-25), the light extinction is due to film reflectivity and in the range below 450 nm due to PVA absorption. We will use the same procedure described in Experiment 3-25 to find the reflectivity part and the part due to PVA absorption. For simplicity, we assume that the reflectivity function for PVA used in Experiment 3-25 is the same for parallel and perpendicular light polarization (this is not really true and the function is slightly different but we will only consider high absorption and the potential error is negligible). We used the spectral range of 500–700 nm and fitted the respective reflectivity components.

(Continued)

EXPERIMENT 6-3 (CONTINUED) POLARIZED ABSORPTION OF STRETCHED PVA FILMS

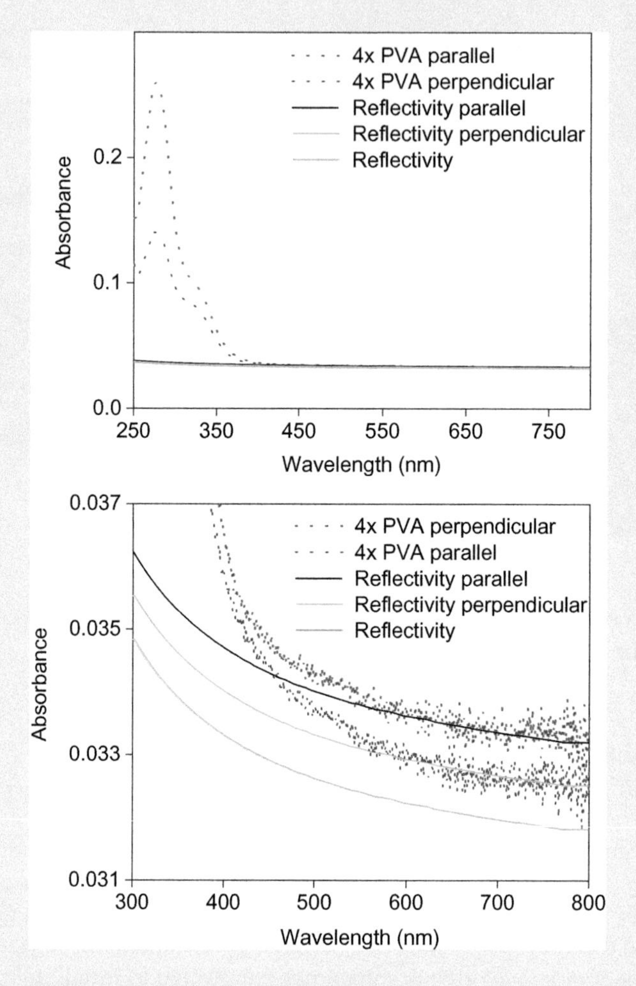

FIGURE 6-3.2 Measured parallel (blue dots) and perpendicular (red dots) absorption components for the stretched film. Green line represents film reflectivity from Experiment 3-25. Black and cyan solid lines are fitted reflectivity for parallel and perpendicular light polarization. Bottom: expanded scale.

Blue and red points in Figure 6-3.2 represent experimentally measured absorptions for parallel and perpendicular light polarization, respectively. The green solid line represents PVA reflectivity function as in Experiment 3-25. Black and cyan solid lines represent fitted film reflectivity for parallel and perpendicular light polarization respectively. When fitting the function, a constant number needed to be added to account for the shift (the constant shift is due to light attenuation by film imperfections or some invisible impurities). The shift and the amount are marked in the figure by vertical arrows. This shift is film dependent

(Continued)

EXPERIMENT 6-3 (CONTINUED) POLARIZED ABSORPTION OF STRETCHED PVA FILMS

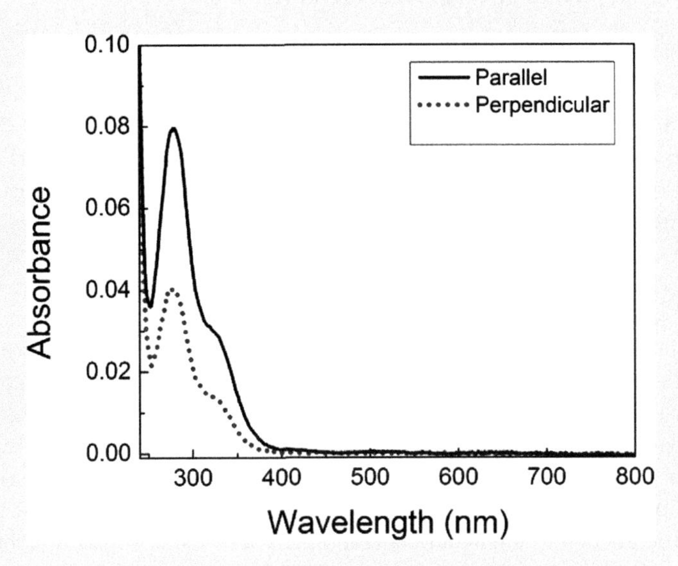

FIGURE 6-3.3 The absorbance of four-time stretched PVA film for the light polarized parallel (solid) and perpendicular (dotted) to the stretching direction of the film.

and will vary from film to film or will change if we use a different spot in the same film. As discussed in Experiment 3-25, such a shift should be very small below 0.005. Using the procedure described in Experiment 3-25, we extracted parallel and perpendicular absorption components for PVA film (Figure 6-3.3). The absorption of four-fold stretched PVA film-oriented parallel to the direction of light polarization is over two times higher than absorption for the perpendicular orientation. We will have to consider such a difference when measuring the absorption of dyes embedded into PVA.

CONCLUSIONS

The absorption of oriented PVA film depends on the light polarization and for parallel orientation of stretching direction to the light polarization is over two times larger than for perpendicular orientation. In our experiment, we only considered the film for which $R_s \approx 9$. In fact, the dichroism of absorption for PVA film changes very little for stretching ratios $R_s > 8$. The biggest change is in the initial stage of stretching and after stretching the film fourfold ($R_s = 8$), we only observe minor improvements in dichroism. Because of some uncertainty in film reflectivity, the embedded dye concentration should be sufficient to ensure the absorption significantly higher than light attenuation due to reflectivity on two film interfaces (surfaces). In practice, in the visible and red spectral range absorptions over 0.05 are already sufficient. In the UV range, we need to consider film absorption and for the film thickness of 0.1 mm we would like to have an absorbance over 0.1.

EXPERIMENT 6-4 POLARIZED ABSORPTION OF CHROMOPHORES ORIENTED IN PVA FILMS: LINEAR DICHROISM

Keywords: absorption, PVA, oriented system, water, linear dichroism.

In the previous Experiments 6-2 and 6-3, we presented the method for stretching/orienting polymer films and the method to evaluate polarized absorption of stretched PVA films. We will use the same concept and experimental setup to measure the polarized absorption of chromophores oriented in the PVA films. The method to produce PVA films with various dyes was discussed in Experiments 3-24 and 3-26, we presented examples of measuring the absorption of chromophores in isotropic PVA film. To measure polarized absorptions of a chromophore in an oriented film, we will use a similar concept. Now, we will measure two orthogonally polarized absorption components (parallel and perpendicular) of the chromophore in the PVA film and subtracting the respective PVA contributions (reflections on film surfaces and PVA absorption). For measurement, we will use a spectrophotometer equipped with a polarizer (see Experiment 6-2) and will measure polarized absorption of the film-oriented parallel and perpendicular to the plane of light polarization. Two absorption components for the chromophore will be recovered. One, when the stretching direction is parallel to the polarizer orientation (polarization of light) and second when the stretching direction is perpendicular to the light polarization. To describe absorption dichroism we can use a dichroic ratio $R_d = A_{\parallel}/A_{\perp}$ or frequently used reduced linear dichroism (LD^r) or absorption anisotropy (K):

$$LD^r = \frac{A_{\parallel} - A_{\perp}}{A_{\parallel} + 2A_{\perp}} = \frac{R_d - 1}{R_d + 2} \text{ and } K = \frac{A_{\parallel} - A_{\perp}}{3(A_{\parallel} + 2A_{\perp})} = \frac{R_d - 1}{3(R_d + 2)}$$

All three descriptions (R_d, LD^r, and K) are interchangeable and equivalent.

In this experiment, we will show how to measure parallel and perpendicular absorption of the chromophore in an oriented PVA film and how to subtract the respective PVA film contributions (reflection and absorption). We will use an approach similar to that for the isotropic film (Experiment 3-26) but now we will use a polarized beam and consider film properties for parallel and perpendicular light polarization as compared to the film stretching direction. Since the stretched film enforces the chromophore's orientation, we expect the absorption parallel and perpendicular to the film orientation will be different reflecting the chromophore orientation and the orientation of chromophore absorption transition moment's orientation within the molecular framework. We will present a simple case of an elongated molecule and planar molecules and demonstrate significant chromophore orientation.

MATERIALS

- Glan-Thompson or grid polarizer
- Chromophores. Pyridine 1 (PY-1), LDS, ethidium bromide (EtBr)
- Polyvinyl alcohol (e.g., Aldrich Mowiol—Mw 130,000 (other sizes that are over 50,000 are OK)
- Water, flask, small bottles, Petri dish, paper tape.
- Stands, holders, and rotatory mount for the film.
- Blank PVA film stretched four-fold.

(Continued)

EXPERIMENT 6-4 (CONTINUED) POLARIZED ABSORPTION OF CHROMOPHORES ORIENTED IN PVA FILMS: LINEAR DICHROISM

EQUIPMENT

- Spectrophotometer with an installed polarizer
- Mechanical stretching device
- Heater

METHODS

1. Start the spectrophotometer, wait at least 5 minutes to warm-up.
2. Install the polarizer and rotatory stage for the film as described in Experiment 6-1.
3. Make sure the beam passes through the center of the hole. Figure 6-3.1 in the previous experiment.
4. Stretch the films (PVA with chromophore and reference blank PVA if needed). See Figure 6-4.1
5. Mount the film with the chromophore and measure absorption for the film with the stretching direction oriented parallel and perpendicular to the polarizer polarization plane. This will include four measurements:
   - Stretching direction parallel to the polarizer polarization plane (0° and 180°)
   - Stretching direction perpendicular polarizer polarization plane (90° and 270°)
6. Correct the measured absorption for blank film absorption and reflectivity. Use measured in the previous experiment reflectivity components and absorptions components for parallel and perpendicular light polarization respectively. The reflectivity does not depend on the film thickness and absorption is proportional to the film thickness. We can also measure reference film independently and use the previously described procedure (Experiment 6-3) to evaluate reflectivity and absorbance. Measuring the film properties for each PVA preparation is a good practice since PVA properties may slightly vary from batch-to-batch and for different preparations (polymerization conditions).

RESULTS

Figure 6-4.1 shows three films with chromophores before stretching (left) and after four-fold stretching (right). The selected chromophores represent different shapes and will orient differently in the stretched PVA. In Figure 6-4.2, we are presenting molecular structures for all

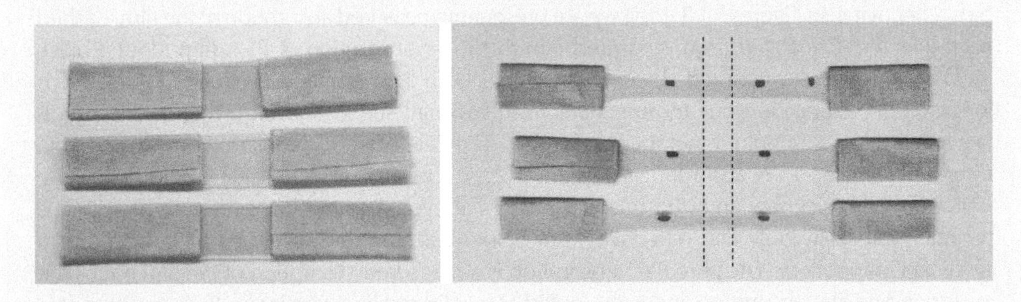

FIGURE 6-4.1 PVA films containing different chromophores. Left: Isotropic films before stretching. Right: four-fold stretched films.

(Continued)

EXPERIMENT 6-4 (CONTINUED) POLARIZED ABSORPTION OF CHROMOPHORES ORIENTED IN PVA FILMS: LINEAR DICHROISM

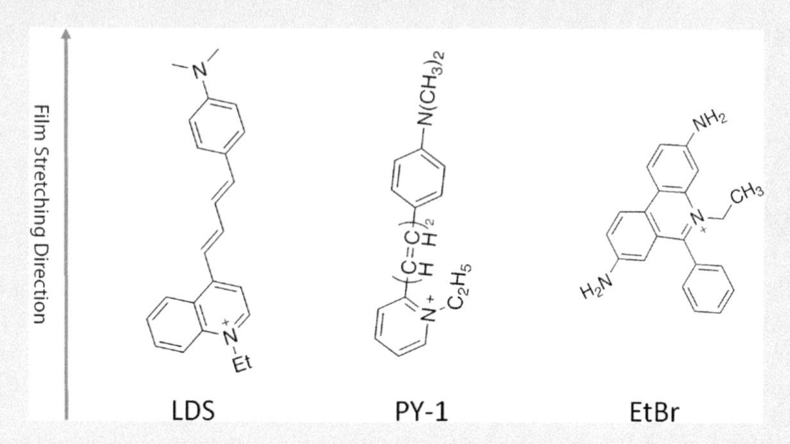

FIGURE 6-4.2 Schematic chemical structures of chromophores used for orientation. Arrow marks stretching direction for the PVA films and chromophores are positioned to reflect potential orientation axis.

three chromophores with a marked film stretching direction. Each chromophore is oriented so the long molecular axis (the axis that is expected to be preferential chromophore orientation) is oriented along the stretched direction of the film. The very elongated LDS molecule will orient very efficiently, PY-1 will be less oriented, and EtBr does not have clear long molecular axis and will be the least oriented. But we want to stress that even EtBr will significantly orient its molecular plane during stretching (it is a planar molecule).

Measured absorption for the isotropic film is independent of the spot at which we will measure absorption but for the stretched film we will use the center part between two marked vertical dashed lines. This is typically the best oriented and most uniform place. *Note: Remember the absorption beam cross-section (diameter) should be significantly smaller than film width to avoid the film edges.*

Stretched films were installed in the holder as in Figure 6-3.1 (previous experiment) and measured absorption was measured in the 250–800 nm range. Figure 6-4.3 shows an example of measured absorptions for Pyridine 1 (PY-1) in isotropic and stretched PVA films (two- and four-fold). The absorption in the isotropic PVA film is a sum of PY-1 absorption and PVA absorption (plus reflectivity). In Figure 6-4.3 (top), we are presenting measured absorption of the film (dashed black line) and calculated from measured film thickness absorption of PVA film (dotted black line). *Note:* The total measured absorption includes both light attenuations due to PVA absorption and light attenuation due to reflections from two film surfaces (even if reflectivity is typically negligible, we should remember about that). The calculated PVA absorption was obtained using an estimated PVA absorption in Figure 3-25.5 (Experiment 3-25) and multiplied by the measured PVA thickness (in 100 μm). In the range above 400 nm, the contribution of PVA is minimal (practically only film reflectivity) but below 350 nm, the PVA absorption becomes a significant component. The pure PY-1 absorption is a measured absorption of the film minus the corresponding film absorption and is presented as a solid red line. Similar for two- and four-fold stretched films, we measured polarized absorptions of each film (parallel—dashed black and

(Continued)

EXPERIMENT 6-4 (CONTINUED) POLARIZED ABSORPTION OF CHROMOPHORES ORIENTED IN PVA FILMS: LINEAR DICHROISM

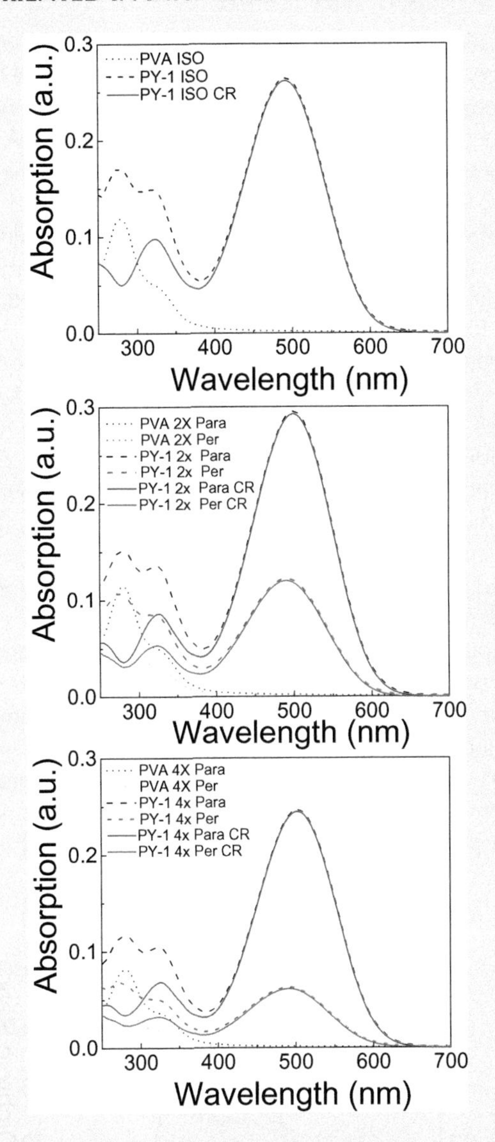

FIGURE 6-4.3 Top: Measured absorption for isotropic PVA film containing pyridine 1 chromophore (dashed black line), corresponding PVA absorption (dotted black), and recovered absorption of PY-1 (solid red). Middle: Measured polarized absorption components for two-fold stretched PVA film containing PY-1 (parallel: dashed black and perpendicular: dashed blue). Corresponding PVA absorption components are shown in dotted black (parallel) and dotted blue (perpendicular). Recovered PY-1 absorption components are shown in solid red (parallel) and solid pink (perpendicular). Bottom: Measured polarized absorption components for four-fold stretched PVA film containing PY-1 (parallel: dashed black and perpendicular: dashed blue). Corresponding PVA absorption components are shown in dotted black (parallel) and dotted blue (perpendicular). Recovered PY-1 absorption components are shown in solid red (parallel) and solid pink (perpendicular).

(*Continued*)

EXPERIMENT 6-4 (CONTINUED) POLARIZED ABSORPTION OF CHROMOPHORES ORIENTED IN PVA FILMS: LINEAR DICHROISM

perpendicular—dashed blue) and subtracted corresponding (calculated) polarized absorptions of the films. The polarized PVA absorptions are presented in a dotted black line (parallel) and dotted blue line (perpendicular). In Figure 6-4.3 (middle and bottom), we are presenting polarized absorption components of PY-1 in two- and four-fold stretched PVA films. The solid red line represents the parallel component and the solid pink line represents the perpendicular component of PY-1 absorption, respectively. The difference between parallel absorption component and perpendicular absorption component quickly increases with the film stretching ratio.

An interesting observation can be made in Figure 6-4.3. Since all measurements are made using the same PVA film (isotropic, two-fold, and four-fold stretched), we can compare the measured absorbance for each component for different stretching ratios. The initial PY-1 absorbance in the isotropic film in the maximum is about 0.26. For two-fold stretched film parallel and perpendicular components are 0.29 and 0.12, respectively giving an isotropic absorption ($(A_{\|} + 2A_{\perp})/3)$) of 0.177. For four-fold stretched film, the respective components are 0.25 and 0.07 yielding isotropic PY-1 absorption of 0.13. To compare absorptions of the isotropic and stretched film, it is important to remember that during the stretching film thickness decreases as the square root of R_d. So, we need to multiply the calculated isotropic absorption of the two- and four-fold stretched films by 1.44 and 2, respectively. The agreement between measured absorbance for an isotropic film (unstretched) and calculated isotropic absorbance for each stretched film is very good.

In Figure 6-4.4 are presented, the recovered absorption components (parallel and perpendicular) for PY-1 in two- and four-fold stretched film. The circles represent the dichroic ratio $R_d = A_{\|}/A_{\perp}$ for two- and four-fold stretched films. The maximum value for R_d is over 4 for four-fold stretched film.

In a similar way, films containing LDS have been stretched and parallel and perpendicular absorption components have been recovered. Recovered absorption components (parallel and perpendicular) are presented in Figure 6-4.5. Again, the solid line represents parallel

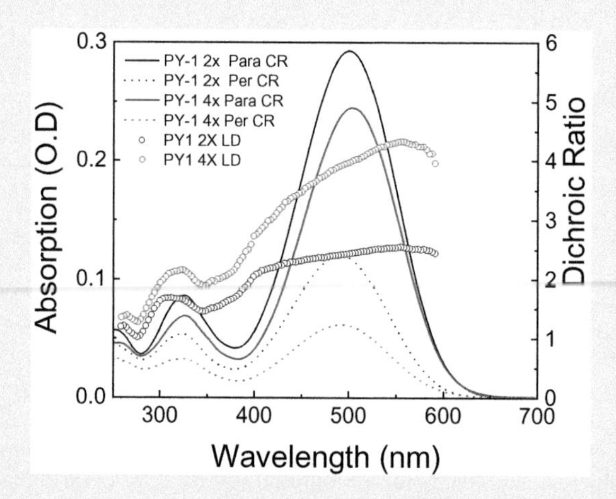

FIGURE 6-4.4 Measured polarized absorption components for PY-1 is two-fold (black) and four-fold (blue) stretched films. Solid line represents parallel and dotted perpendicular components, respectively. Circles represent calculated the dichroic ratio (black for two-fold and blue for four-fold stretched films).

(*Continued*)

EXPERIMENT 6-4 (CONTINUED) POLARIZED ABSORPTION OF CHROMOPHORES ORIENTED IN PVA FILMS: LINEAR DICHROISM

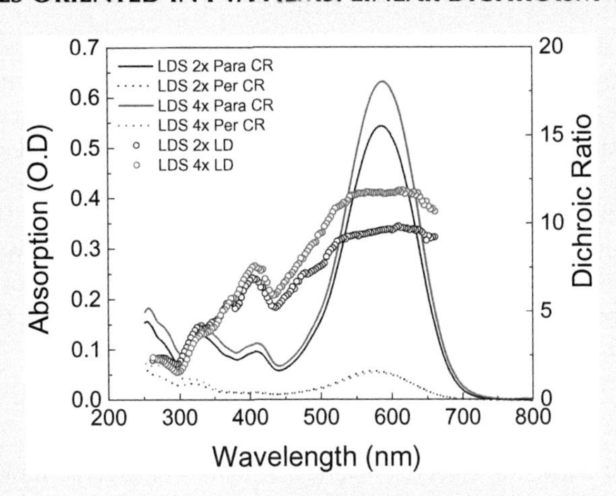

FIGURE 6-4.5 Measured polarized absorption components for LDS in two-fold (black) and four-fold (blue) stretched films. Solid line represents parallel and dotted perpendicular components, respectively. Circles represent calculated the dichroic ratio (black for two-fold and blue for four-fold stretched films).

absorption components and dotted lines perpendicular components (black two-fold and blue four-fold stretched films, respectively). LDS is an elongated molecule that very effectively orients in stretched PVA film. The measured dichroic ratio for the two-fold stretched film is already close to 10 and for a four-fold stretched film it is over 12. This indicates that the transition moment for long-wavelength absorption transition is oriented along the orientation molecular axis.

As a different example, Figure 6-4.6 presents the measured absorption components for EtBr molecule oriented in the four-fold stretched PVA film. This is a planar molecule that has

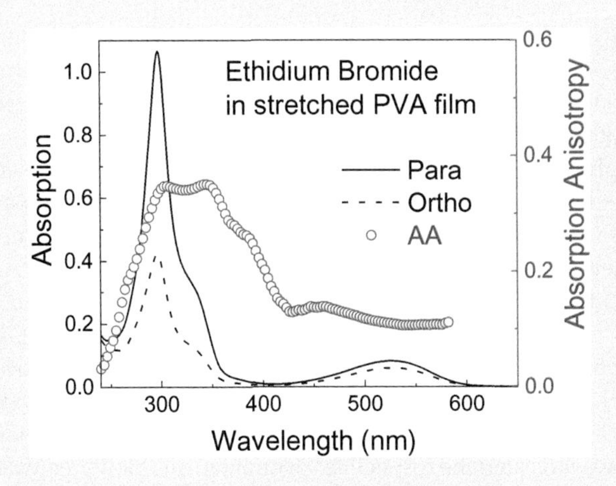

FIGURE 6-4.6 Measured polarized absorption components for EtBr in four-fold stretched films. Solid line represents parallel and dotted perpendicular components, respectively. Circles represent calculated reduced dichroism.

(Continued)

EXPERIMENT 6-4 (CONTINUED) POLARIZED ABSORPTION OF CHROMOPHORES ORIENTED IN PVA FILMS: LINEAR DICHROISM

no well-defined orientation axis. However, the molecular plane (this is a planar molecule) orients well. In Figure 6-4.6, the calculated absorption anisotropy is presented. A long-wavelength absorption transition shows low LD^r value of about 0.1 that corresponds to a dichroic ratio of 1.33. This value could suggest that this is planar and symmetrical (circular) molecule similar to benzene or coronene. However, for a short wavelength absorption transition, the LD^r value approaches 0.4 that correspond to a dichroic ratio close to 3. This could indicate that these molecules have not only well-oriented molecular planes (as would be expected for planar molecules) but also significantly orients along the longest molecular axis in the molecular plane. The long-wavelength transition forms an angle close to 45° with respect to the long orientation axis.

CONCLUSIONS

Stretched PVA films allow for a very efficient orientation of organic molecules. Measured polarized absorption components allow estimating the orientation of absorption transition moments. For elongated molecules that are very effectively oriented in the stretched film, the orientation of transition moments can be well estimated with respect to the long molecular orientation axis. More planar molecules orient less effectively and transition moments directions can only be estimated with limited precision.

EXPERIMENT 6-5 MEASURING *G*-FACTOR: SQUARE GEOMETRY CASE

Keywords: emission, polarization, spectra.

Practically, every instrument is sensitive to light polarization (responds differently to vertically and horizontally polarized light). So, the first step before measuring emission anisotropy is to evaluate the instrument sensitivity to vertical and horizontal light polarization. Since the emission anisotropy is proportional to the ratio of measured light intensity polarized vertically to the light intensity polarized horizontally, it is enough to evaluate the relative sensitivity for detecting horizontally polarized light in relation to the vertically polarized one. The correction factor is called *G*-factor that is typically wavelength dependent (different for each wavelength).

In this experiment, we will present steps for *G*-factor measurements for three instruments that well represent most of the existing systems nowadays. To measure *G*-factor in a broad range (full instrument range), we need to select a set of dyes that will cover the full UV-Vis spectral range. In Figure 6.8, in the introduction to this chapter, we show a few examples of different fluorophores covering a broad spectral range that can be used for *G*-factor determination. For each dye, we indicated the respective spectral ranges. However, we want to stress that many more dyes can also be used and in the case of square geometry and horizontal excitation we do not need to worry about dye polarization. Generally, to cover the full UV-VIS spectral range, we will need to select 4–6 different dyes. In the presented experiment, we will use (2AP,

(*Continued*)

EXPERIMENT 6-5 (CONTINUED) MEASURING *G*-FACTOR: SQUARE GEOMETRY CASE

BBO, C153, and Ru), but any other combination from Figure 6.8 or different dyes emitting in the appropriate spectral range will work fine as long as they cover the required spectral range. We also want to stress that in many instances, we will only need to check the *G*-factor for the range where the dye of interest emits. And in the case of square geometry (most cases), *G*-factor evaluation can be conveniently done with the same dye you intend to study.

In this experiment, we are presenting *G*-factor measurement for the most common set up, square geometry, schematically shown in Figure 6-5.1 (top view). The horizontally polarized excitation beam comes from the bottom and we observe emission in the left detection channel. In such a case, excited (photoselected) dyes will present a symmetrical distribution (circle) as viewed from the direction of observation (shown in the top left corner). *It is very important to maintain a 90° angle between the excitation and emission lines.* In such a configuration, measured vertical and horizontal emission light intensities should be equal. Measured differences are due only to the different sensitivity of the detection system to vertically and horizontally polarized light.

In this protocol, the spectra of selected dyes are measured by using horizontally polarized excitation light and measuring emission through vertically and horizontally oriented polarizers. We will use different instruments (ISS K2, Varian Eclipse, and Picoquant FT 300). For an FT300 system, we also show measurements made with two different detectors: Microchannel Plate (MCP-PMT) and a Photomultiplier tube (PMT).

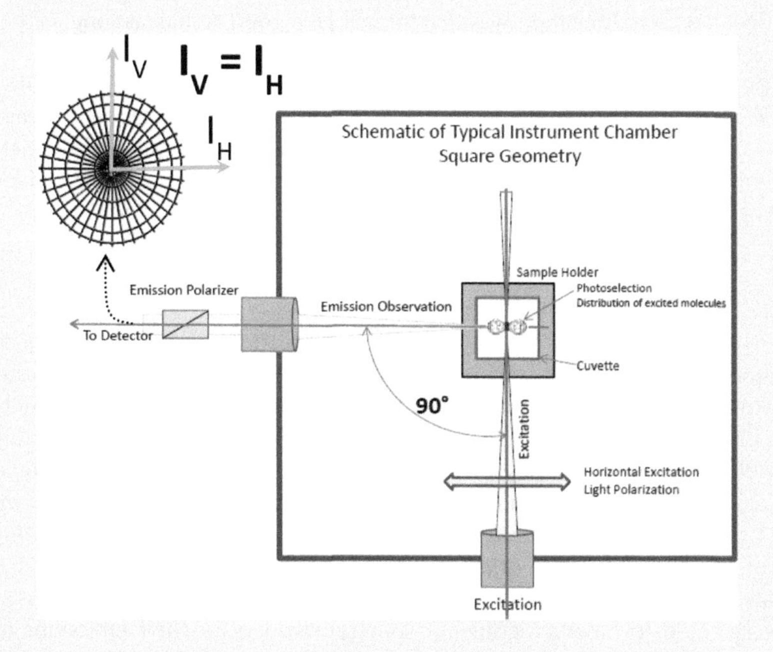

FIGURE 6-5.1 Schematic of a top view for the square geometry detection setup. Horizontal polarization of excitation light and 90° emission observation. Top left corner: Intensity distribution as seen by the left detection channel.

(*Continued*)

EXPERIMENT 6-5 (CONTINUED) MEASURING *G*-FACTOR: SQUARE GEOMETRY CASE

MATERIALS

- Cuvette
- Water and EtOH
- 2-Aminopurine (2AP), anthranilic acid (AA), 2,5-bis-(4-biphenylyl)-oxazol (BBO), Coumarin (C153), and ruthenium bipyridyl ($[Ru(pby)_3]^{+2}$)
- 1 cm x 1 cm cuvette

EQUIPMENT

- Fluorometers with PMT detector (ISS K2, Varian Eclipse, and Picoquant FT 300)
- Fluorometer with MCP-PMT detector (Picoquant FT 300)

METHODS

1. Prepare the solution of selected dyes. Remember the absorption should be below 0.1 as measured in the cuvette, we are doing the experiment in. It is not required but we use EtOH as a solvent. Typically, we prefer to use low viscosity solvents in which anisotropy will be low (close to zero). In this case, any mistake in the orientation of excitation polarizer (not perfectly horizontal) will not have a significant impact.
2. Orient your excitation polarizer to horizontal (90°) and select the optimal excitation wavelength. This is typically a long absorption band. We want to stress that in this configuration any excitation wavelength will give good results as long as the signal is strong enough.
3. Measure the emission spectra for two emission polarizer orientations vertical (parallel)—0° and horizontal—90°. It is recommended to repeat these measurements a few times and average to get less variation due to instrumental noise. Remember that the error in the evaluated *G*-factor will propagate to the error for calculated anisotropy. So, we want to determine this parameter with the highest precision.

RESULTS

Note that for an anisotropy measurement we do not need to correct spectra for the quantum efficiency of the detector and other instrumental factors. Calculated *G*-factor and later anisotropy/polarization reflects the signal ratio and is independent of such factors. Figures 6-5.2 presents the measured emission spectra for selected dyes (a mixture of 2AP and AA, BBO, C153, and $[Ru(pby)_3]^{+2}$) for parallel and orthogonal orientations of observation (emission) polarizer measured in ISS K2 fluorometer. *Note: We used a mixture of 2AP and AA (anthranilic acid) to extend the emission range. The shape of the emission spectrum may look surprising but this does not affect G-factor determination.*

Calculated *G*-factors $G = \frac{I_{\parallel}}{I_{\perp}}$ for different dyes are presented in Figure 6-5.3. Used dyes were selected to have a significant spectral overlap for their emissions to clearly see overlying *G*-factors as we move from one dye to another. *Note: The G-factor itself is not a relative parameter as it was for spectral correction for wavelength sensitivity (see Experiment 5-4). Values of wavelength-dependent G-factors determined with different*

(Continued)

EXPERIMENT 6-5 (CONTINUED) MEASURING *G*-FACTOR: SQUARE GEOMETRY CASE

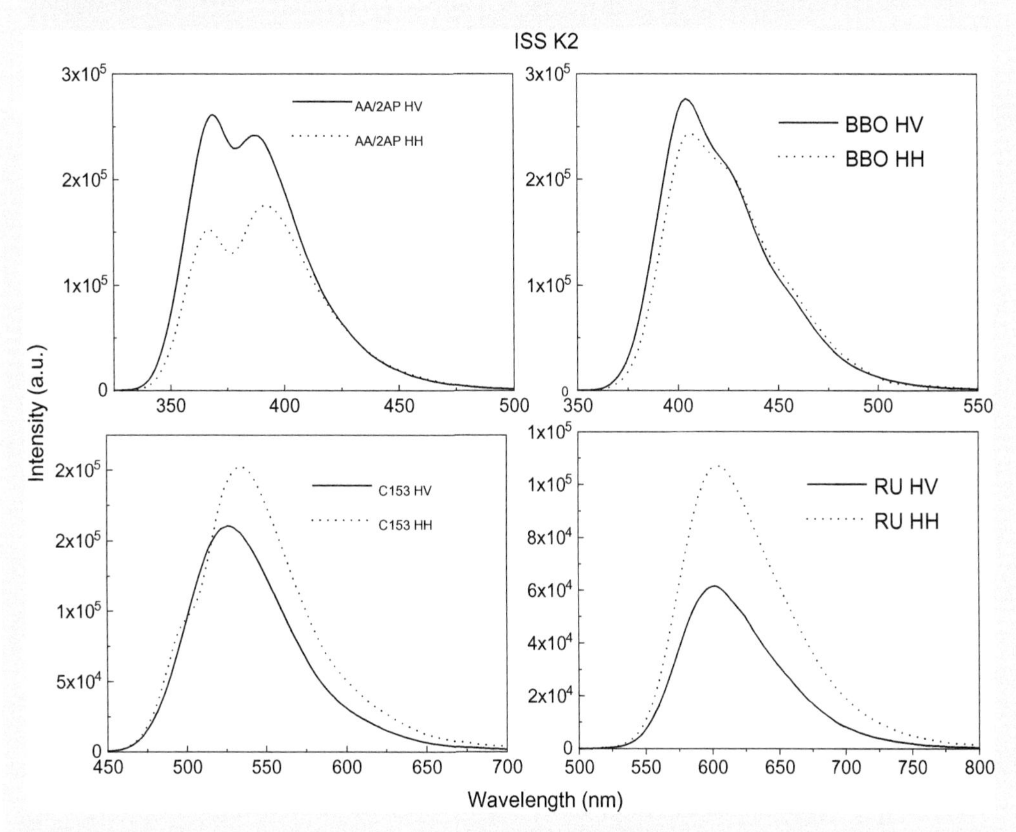

FIGURE 6-5.2 Parallel and perpendicular emission components measured with horizontal polarization of excitation for four selected dyes on ISS K2 instrument.

dyes should match exactly. Figure 6-5.4 shows all four measured *G*-factors in one graph. The *G*-factors for different spectrally overlapping dyes must overlay well since *G*-factor is the instrument characteristic and cannot depend on used dyes. It is normal to observe a larger statistical variation on the edges of emission spectra where the measured signal is lower. Figure 6-5.5 presents measured *G*-factors for the Varian Eclipse and FT300 (with the MCP and PMT detector). *G*-factors measured for ISS K2 and FT300 follow a similar pattern while the *G*-factor for Varian Eclipse is completely different. This is due to the fact that the Varian Eclipse uses a different type of grading monochromator and the K2 and FT300 use different monochromator brands but equipped with a similar configuration for diffraction grading. It is interesting to note that *G*-factors for MCP and PMT measured with the FT300 are very similar but only slightly shifted. MCP and PMT use the same monochromator and the difference is due to a small difference in detectors and mostly due to the different light path when going to PMT or MCP (the difference being one more reflecting mirror).

(Continued)

EXPERIMENT 6-5 (CONTINUED) MEASURING G-FACTOR: SQUARE GEOMETRY CASE

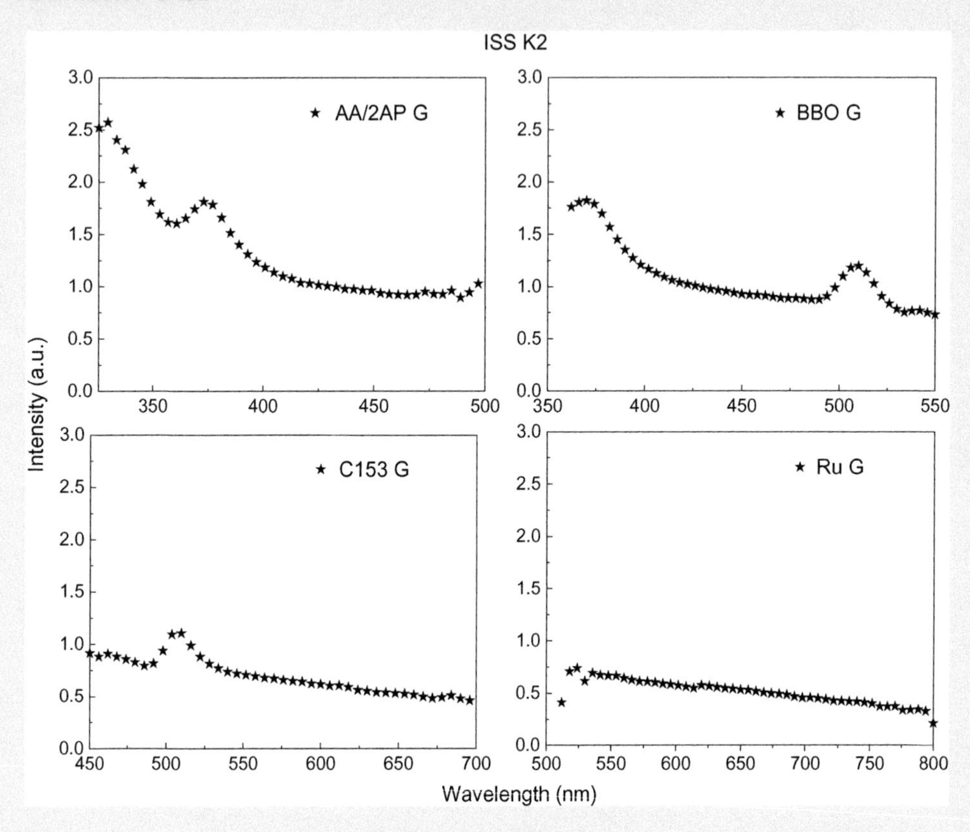

FIGURE 6-5.3 *G*-factor measured for different dyes on ISS K2 spectrofluorometer.

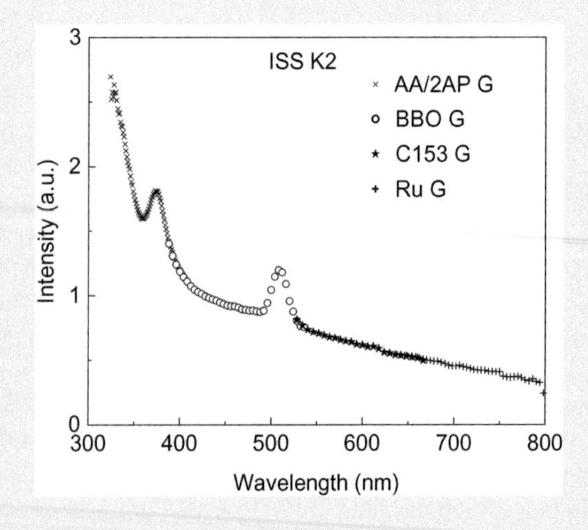

FIGURE 6-5.4 Combined *G*-factors from all four dyes. It represents *G*-factor for the 330–800 nm range.

(Continued)

EXPERIMENT 6-5 (CONTINUED) MEASURING *G*-FACTOR: SQUARE GEOMETRY CASE

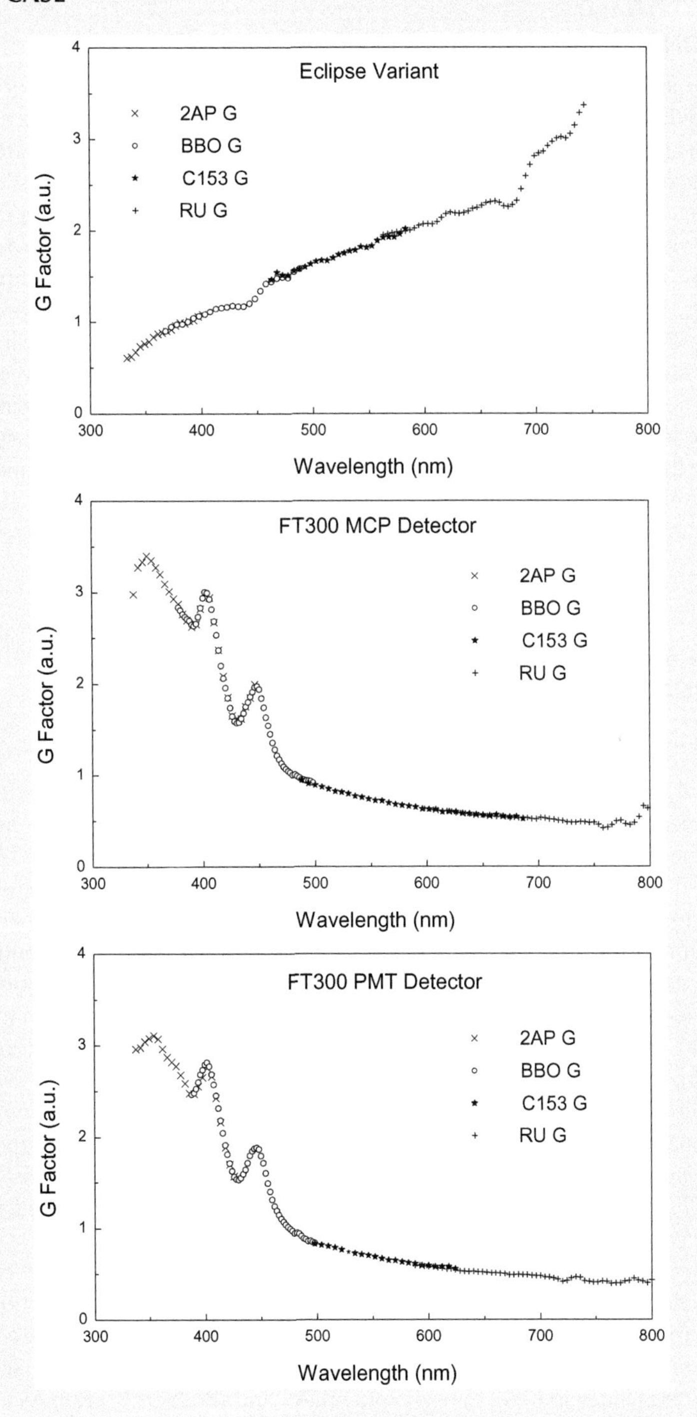

FIGURE 6-5.5 Respective *G*-factors measured for Varian Eclipse, FT300 (PMT), and FT300 (MCP).

(*Continued*)

EXPERIMENT 6-5 (CONTINUED) MEASURING *G*-FACTOR: SQUARE GEOMETRY CASE

CONCLUSIONS

It is clear that for all instruments, the *G*-factor strongly depends on the observation wavelength. The difference between UV and red spectral range can be as large as four times. In addition, in all systems due to the diffraction grating, we will observe characteristic bumps (most diffraction grading are based on thin-film technology and these drastic changes in reflectivity for some wavelengths are plasmonic effects). When working in the range close to (overlapping with) these bumps, it is important to make sure that the monochromator resolutions (typically set by the slit width) for *G*-factor evaluation and anisotropy measurements are the same. The plasmonic effect is at a precisely defined wavelength (to a fraction of a nanometer) and increasing the resolution (decreasing monochromator slit) will increase the effect while lowering the resolution (larger slit) will smooth out the bump. We want to stress that it is very important to properly evaluate *G*-factor with minimum error since this error will propagate to measured anisotropies. We would also suggest, since it is easy, to evaluate (or just check) the *G*-factor for every sample since a small change in beam position or cuvette displacement may slightly affect the *G*-factor of the system.

EXPERIMENT 6-6 EVALUATING INSTRUMENTAL *G*-FACTOR USING LONG-LIVED REFERENCE FLUOROPHORES

Keywords: fluorescence, anisotropy, *G*-factor, long-lived fluorophores.

The theoretical bases for anisotropy measurements are described in the *Introduction* part. To properly measure emission anisotropy, it is crucial to properly evaluate an instrument *G*-factor. In the previous Experiment 6-5, we presented a simple way for *G*-factor determination for square geometry systems. However, in many instances, we may not be able to rotate the excitation polarization or frequently measurements are not made in square geometry. Examples of such configurations were discussed in Chapter 2 and in Figure 2-6.3, we presented examples of different configurations. In the case of front-face configuration or configuration in-line (microscope), we cannot rotate the excitation polarization to obtain zero polarization. In such cases, we need to use polarization standards (fluorophores of known emission polarization) or use a different approach.

In this experiment, we are presenting an example for an alternative way for the *G*-factor estimation using a zero anisotropy standard. *Note: We want to stress that emission anisotropy of any chromophore depends on many factors (e.g., temperature, viscosity, etc.) typically contributing to emission depolarization (lowering the anisotropy). So, the best standards would be molecules that present zero anisotropy and further potential depolarization will not have any effect.* According to the Perrin equation ($r_o/r = 1 + t/\theta$), the measured anisotropy r approaches zero when the ratio of t/θ increases. For example, if this ratio is about 50, the maximum measured anisotropy is less than 0.01. A high ratio of t/θ can be achieved with small fast-rotating fluorophores which have long fluorescence lifetimes. We selected three such dyes—anthranilic acid (AA), dansyl amide (DA), and $[Ru(bpy)_3]^{2+}$ (Ru). As a solvent, we chose low viscous ethyl acetate. For each of these fluorophores, the ratio t/θ is above 100,

(*Continued*)

EXPERIMENT 6-6 (CONTINUED) EVALUATING INSTRUMENTAL *G*-FACTOR USING LONG-LIVED REFERENCE FLUOROPHORES

resulting in fluorescence anisotropy smaller than 0.005. Fluorescence spectra of these fluorophores are not structured and cover a wide wavelength range from 380 nm to 720 nm. We will use each dye individually and evaluate *G*-factors for each of them, and connect *G*-factors between different overlapping spectral ranges as it was done in the previous Experiment 6-5. We will also use a mixture of these three fluorophores, which covers the entire visible wavelength range and a single excitation at 360 nm. It is not easy to select such a mixture and in many instances, fluorophores may interact leading to a significant change in emission properties. Finally, we will use horizontal excitation to evaluate the *G*-factor and compare both *G*-factors as evaluated from vertical and horizontal excitation. *Note: We want to stress that the goal of this experiment is to show that completely depolarized emission (zero) can be used for G-factor evaluation. In this case, the result is independent of light polarization used for excitation. But, if we are not sure that the fluorophore will completely depolarize, we warn that vertical excitation may lead to a significant error in G-factor evaluation.*

MATERIALS

- Dansyl amide (DA), anthranilic acid (AA), ruthenium bipyridyl, $[Ru(bpy)_3]^{2+}$ (Ru)
- Ethyl acetate (EA)
- Glass or quartz cuvettes, 1 cm × 1 cm

EQUIPMENT

- Spectrophotometer (Cary 50, Varian Inc.)
- Spectrofluorometer (Eclipse, Varian Inc.)

METHODS

1. Prepare stock-solutions of individual fluorophores in water, sonicate and make sure, the dyes dissolved completely.
2. Dilute the solutions appropriately to achieve optical densities of about 0.15 at maxima.
3. Measure absorption and fluorescence spectra, see Figures 6-6.1 and 6-6.2. *Note: Absorption at 360 nm (excitation wavelength) is much below 0.1.*
4. First, use 360 nm horizontal excitation and measure vertical and horizontal emission components for each dye. Evaluate *G*-factor for each range (day emission range) as it was done in the previous experiment.
5. Rotate polarization to vertical and measure vertical and horizontal emission components (I_{VV} and I_{VH}) for each dye. Using *G*-factor from the previous step evaluate and plot emission anisotropies for each individual dye. *Note: Since the lifetime of each dye is much longer than its correlation time the calculated anisotropy should be close to zero (see Figure 6-6.3).*
6. Use the prepared solution and make a mixture of all three fluorophores (1:1:1) and measure its absorption spectrum, see Figure 6-6.4.
7. Use horizontally polarized excitation at 360 nm and measure emission spectra with an observation polarizer oriented vertically (I_{HV}) and horizontally (I_{HH}). As in the previous Experiment 6-5, calculate *G*-factor for the entire spectral range.

(*Continued*)

EXPERIMENT 6-6 (CONTINUED) EVALUATING INSTRUMENTAL *G*-FACTOR USING LONG-LIVED REFERENCE FLUOROPHORES

8. Use vertically polarized excitation at 360 nm and measure emission spectra for vertical (I_{VV}) and horizontal (I_{VH}) observations. Evaluate *G*-factor defined as $\boldsymbol{I}(_{VV})/\boldsymbol{I}(_{VH})$ in a full range.
9. Plot both *G*-factors as evaluated from horizontal and vertical excitation.

RESULTS

In Figure 6-6.1, we present measured absorption spectra for all three dyes. Normalized emission spectra of dyes are shown in Figure 6-6.2. Emission spectra cover a broad range (370–750 nm) and present a significant spectral overlap.

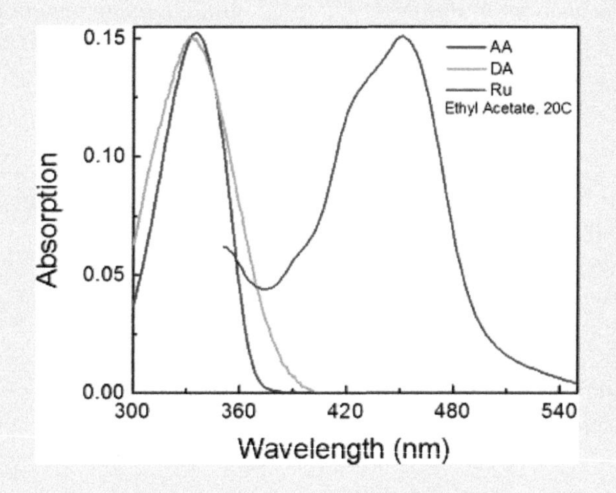

FIGURE 6-6.1 Absorption spectra of three selected fluorophores in EA.

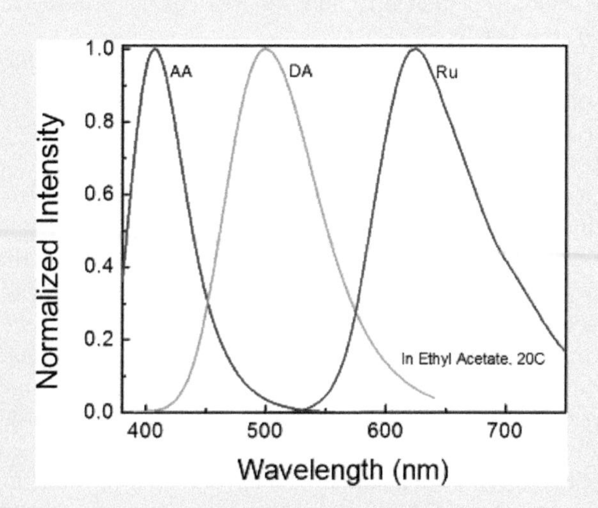

FIGURE 6-6.2 Emission spectra of three selected fluorophores.

(Continued)

EXPERIMENT 6-6 (CONTINUED) EVALUATING INSTRUMENTAL *G*-FACTOR USING LONG-LIVED REFERENCE FLUOROPHORES

Using horizontal excitation, we measured the *G*-factor with each dye separately for each respective spectral range (see Experiment 6-5). Next, using vertical excitation we measured I_{VV} and I_{VH} emission components, corrected for each dye I_{VH} component with the respective *G*-factor, and calculated emission anisotropy. Figure 6-6.3 presents emission anisotropies as evaluated for AA (top), DA (middle), and Ru (bottom). As expected all anisotropies are very close to zero. However, each of these anisotropies covers only limited ranges of wavelengths.

In order to extend the wavelength range, we mixed all three solutions (1:1:1) and measure absorption and fluorescence spectra. Figure 6-6.4 presents, the cumulative absorption of the mixture and Figure 6-6.5 presents the measured emission spectrum with 360 nm excitation observed under the magic angle (54.7°). The emission spectrum of the mixture shows three distinct peaks and in the entire range, the measured signal is strong in a good spectrofluorometer range.

Next, we used horizontal excitation and measured I_{HV} and I_{HH} emission components. Figure 6-6.6 presents measured intensities I_{HV} and I_{HH}. Figure 6-6.7 presents the *G*-factor as evaluated from the mixture of three fluorophores (I_{HV}/I_{HH}).

We repeated the *G*-factor measurement using vertical excitation. Figure 6-6.8 shows measured intensities I_{VV} and I_{VH}. Figure 6-6.9 shows the *G*-factor calculated as I_{VV}/I_{VH}. Because all fluorophores have long fluorescence lifetimes the emission polarization for each of them should be close to zero. For fluorophore anisotropy intrinsically equal to zero both ratios, I_{VV}/I_{VH} and I_{HV}/I_{HH} should be identical leading to the same values of *G*-factors. Note that absolute intensities detected for vertical and horizontal excitations would typically be different (compared to Figures 6-6.6 and 6-6.8). So, using vertical excitation for *G*-factor estimation should yield similar values as when using horizontally polarized excitation. In fact, in Figure 6-6.10, we show both *G*-factors in one graph. Indeed both values are very close.

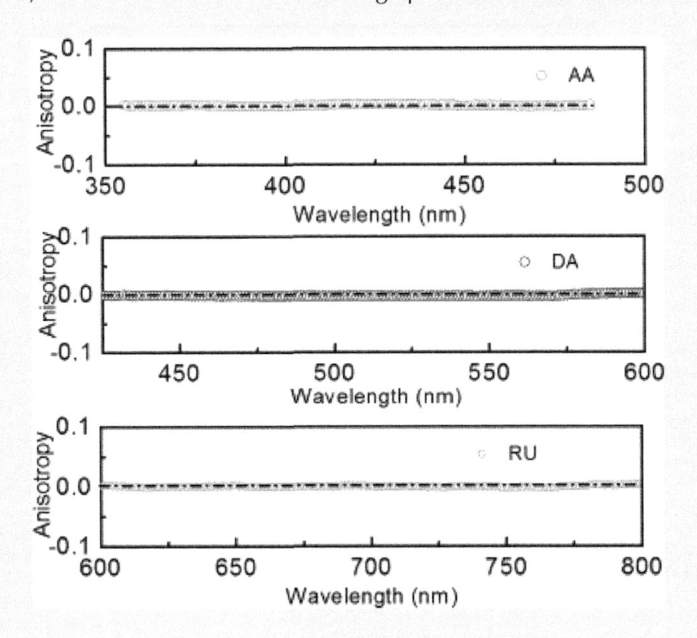

FIGURE 6-6.3 Fluorescence emission anisotropies of AA, DA, and Ru in EA measured individually.

(Continued)

EXPERIMENT 6-6 (CONTINUED) EVALUATING INSTRUMENTAL *G*-FACTOR USING LONG-LIVED REFERENCE FLUOROPHORES

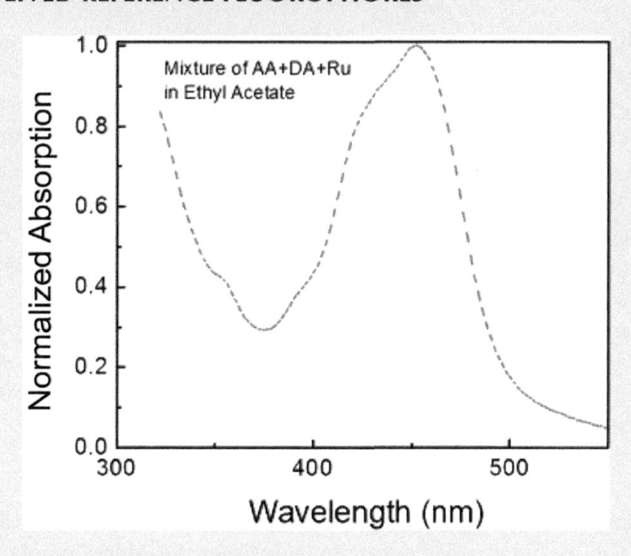

FIGURE 6-6.4 Absorption spectrum of the mixture AA + DA + Ru in EA. Ru fraction is the largest because of its low quantum yield.

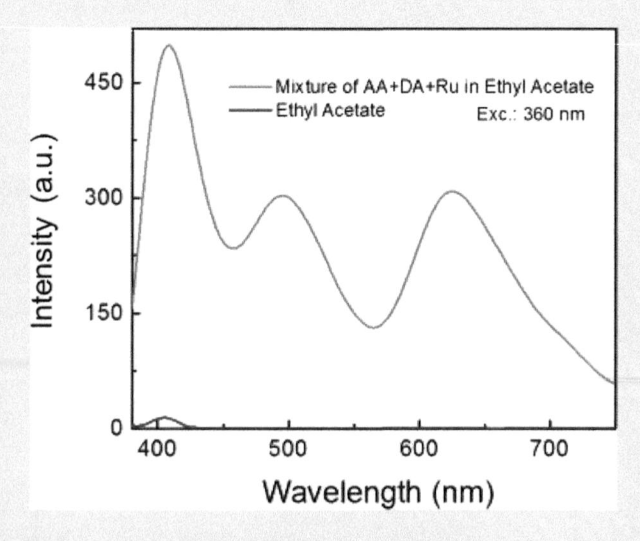

FIGURE 6-6.5 Emission spectrum of the mixture AA + DA + Ru in EA excited at 360 nm at magic angle conditions. The background signal from EA only is negligible.

(*Continued*)

EXPERIMENT 6-6 (CONTINUED) EVALUATING INSTRUMENTAL *G*-FACTOR USING LONG-LIVED REFERENCE FLUOROPHORES

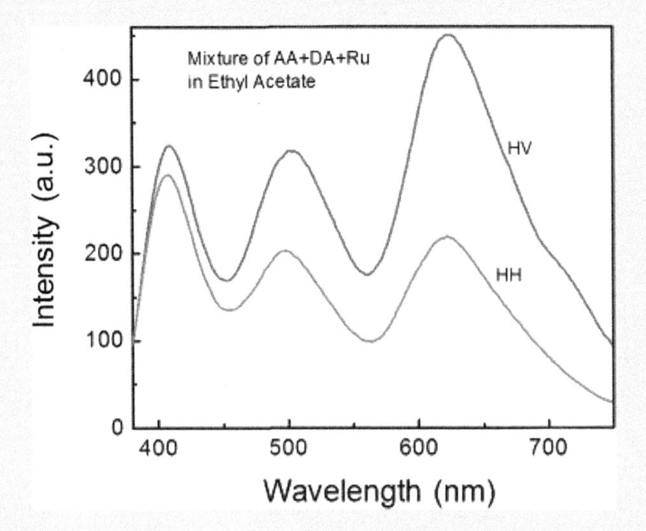

FIGURE 6-6.6 Emission spectra of the mixture with HV and HH polarizers orientation. In the square geometry, the H polarization on the excitation results in isotropic intensities of the observed fluorescence, meaning the V and H components of the emission are equal.

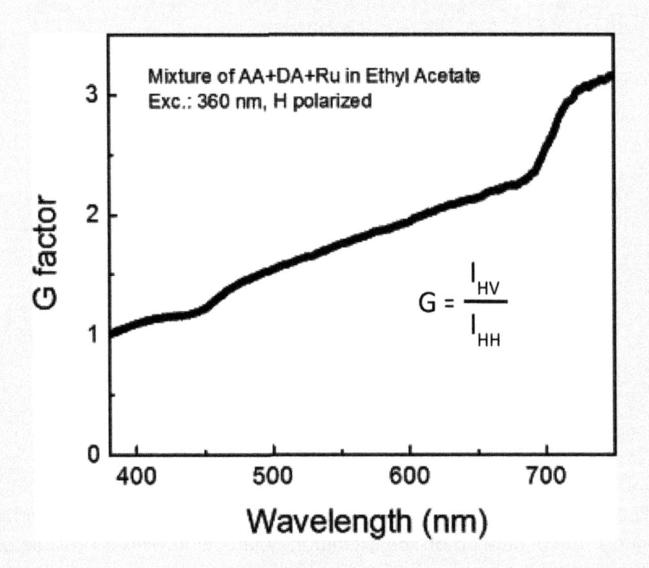

FIGURE 6-6.7 The *G*-factor calculated for horizontally (H) oriented polarization of the excitation path.

(Continued)

EXPERIMENT 6-6 (CONTINUED) EVALUATING INSTRUMENTAL *G*-FACTOR USING LONG-LIVED REFERENCE FLUOROPHORES

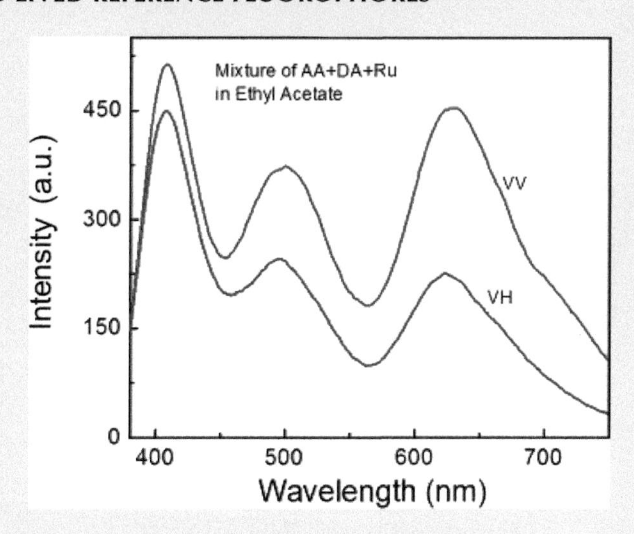

FIGURE 6-6.8 Fluorescence anisotropy measurements of the mixture. A zero anisotropy for the observed fluorescence is expected for long-lived and quickly rotating of fluorophores.

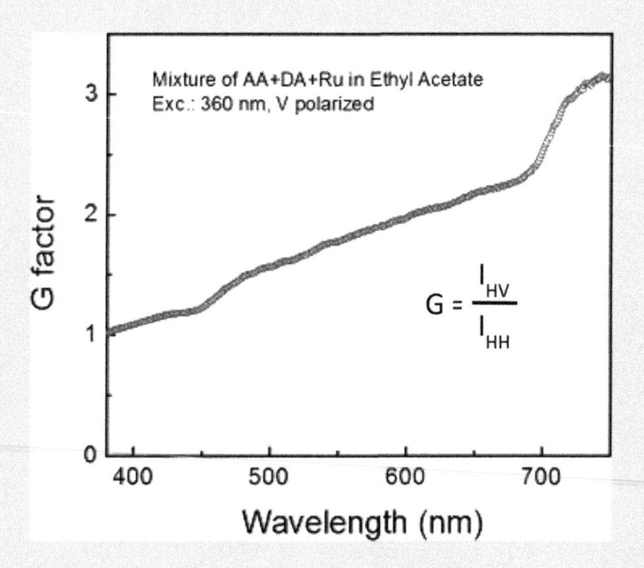

FIGURE 6-6.9 The *G*-factor calculated from I_{VV} and I_{VH} components. It is expected that V and H polarized components do not depend on the excitation polarization. This is possible only for solutions with a zero anisotropy.

(*Continued*)

EXPERIMENT 6-6 (CONTINUED) EVALUATING INSTRUMENTAL *G*-FACTOR USING LONG-LIVED REFERENCE FLUOROPHORES

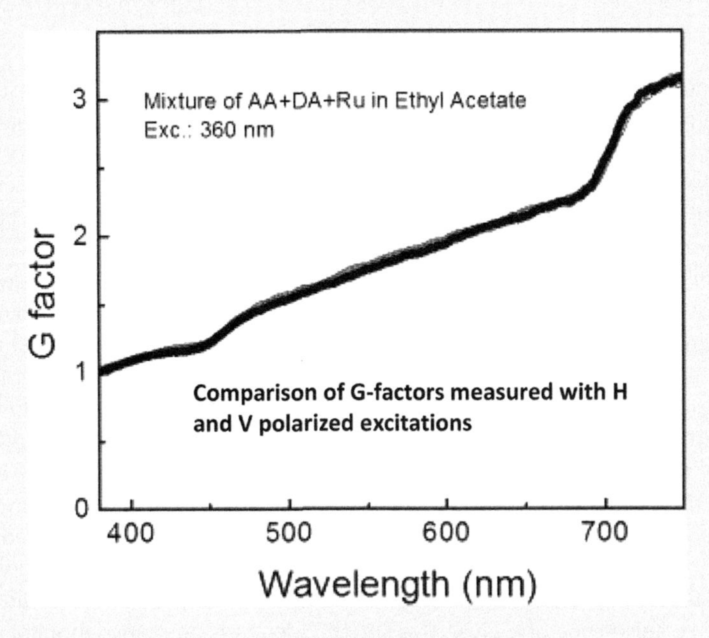

FIGURE 6-6.10 Comparison of *G*-factor obtained with H polarized excitation light with correction factor obtained for quickly rotated long-lived fluorophores excited with V polarized excitation light.

CONCLUSIONS

Quickly rotating fluorophores with long-lifetimes will have intrinsic anisotropy close to zero and can be used as references for zero fluorescence anisotropy. As shown in Figure 6-6.10, the evaluated *G*-factor will be independent of excitation light polarization. However, as we had mentioned at the beginning, the anisotropy of a fluorophore depends on many factors. It may change due to temperature, viscosity, or interactions between fluorophores (e.g., quenching) so, in square geometry, we recommend to use horizontal excitation. Only if using horizontal excitation is not possible, we would use vertical (or any other) excitation polarization but make sure that the fluorophore is really presenting zero anisotropy. The need for using zero anisotropy standards is clearly justified in the front-face configuration as well as in microscopy.

Note: We also caution the reader that the presented example where we were able to mix fluorophores and the measured absorption and emission is a simple sum of all components is rather rare. Many fluorophores will interact and form complexes (aggregates) leading to the change in fluorescence spectra and fluorescence lifetimes. This typically also may affect anisotropy.

EXPERIMENT 6-7 MEASURING *G*-FACTOR IN THE CASE WHEN WE CANNOT USE HORIZONTAL EXCITATION

Keywords: emission, *G*-factor, anisotropy.

Using horizontal excitation is a very convenient and common approach to measure *G*-factor. But there are multiple examples where this cannot be done. Even classic square geometry in some cases may use laser excitation for which light polarization is fixed to vertical and polarization rotation is possible but requires special devices. Other common examples are front-face configuration and microscopy frequently used for polymer films, surfaces, or very high concentrated solutions. In such a case, we could use standards of known anisotropy or zero anisotropy standards but appropriate standards covering the needed emission range may not be available. Also, in front-face configuration (especially microscopy) using zero anisotropy standards can present an additional problem. In both front-face and microscopy, a significant portion of the signal can be generated from the chromophore layer that sticks to the surface. Such surface-bound fluorophores would present high anisotropy. In these cases, we simply cannot achieve conditions in which horizontal excitation will yield an isotropic distribution of excited transition moments (vertical and horizontal intensities are equal) when viewed from the direction of emission observation. In such a case, a possible option is to use standards for which we know anisotropies. However, for an inexperienced researcher, this could be difficult. First, it is difficult to find such standards and second, polarization of most standards strongly depends on temperature that in many instances could be difficult to control. Since anisotropy of a given dye strongly depends on many laboratory conditions (e.g., temperature), typically we will look for a fluorophore that has long fluorescence lifetime and use it in a low viscosity solvent. Fast rotation of fluorophores before emitting a photon randomizes their distribution and emitted light has no preferential orientation (is randomized) yielding equal intensities for parallel and perpendicular observation. A very good standard in the red spectral range is ruthenium metal-ligand complex $[Ru(bpy)_3]^{2+}$ that will have about 400 ns fluorescence lifetime in water or ethanol. Also in shorter spectral ranges, we may use for example pyrene (especially pyrene excimer emission) or dansyl that in some low viscosity solvent (e.g., ethanol) have fluorescence lifetimes over 10 ns much longer than its correlation time of few hundred picoseconds. Also, in many cases, an experienced experimenter may use the excitation wavelength for which initial anisotropy is close to zero. Many fluorophores will have multiple (typically weaker) bands where initial anisotropy is low. Zero anisotropy will be observed when the absorption transition moment and emission transition moment form an angle close to 54.7°. For that, typically we will have to have prior knowledge about the fluorophore excitation anisotropy spectrum.

Having a zero-anisotropy standard, we can measure emission with any excitation light polarization and use exactly the same calculations of *G*-factor as for horizontal excitation in Experiment 6-1.

When zero anisotropy standards are not available or the experimentalist is not sure if the anisotropy of a fluorophore is zero, we recently realized we could use a 45° polarizer on the observation/detection line. A polarizer oriented at 45° to vertical (horizontal) that is inserted in the observation line immediately after the sample (before polarizer on emission line) should transmit light polarized at 45° to vertical. This light as projected on the detector has vertical and horizontal light intensity equal ($\cos^2 45° = 1/2$). This is a rather novel approach not used in anisotropy studies before because a small error (in the order of 1°) results in

(*Continued*)

EXPERIMENT 6-7 (CONTINUED) MEASURING *G*-FACTOR IN THE CASE WHEN WE CANNOT USE HORIZONTAL EXCITATION

FIGURE 6-7.1 Examples of polarizers cut at 45°. Two left are mounted on a square block. Right is an example of how to cut 45° polarizer from a polarizer shit that will create 45° and 135° complementary angles.

a significant error for estimating *G*-factor. But as we discussed in the introduction to this chapter measuring emission spectra for two complementary orientations (45° and 135°) and averaging gives very good results for *G*-factor estimation. A simple device produced in the laboratory or a precisely cut polarizer that can be used is shown in Figure 6-7.1.

Using such devices makes routine checks of *G*-factor very convenient and can be done with very high precision. This is especially convenient for front-face configuration and solid samples (e.g., polymer films or surface deposited dyes) where measuring the *G*-factor with standard dyes is very inconvenient and sometimes even impossible. This also includes in-line configuration as used in microscopy and this method can be easily adapted to microscopy. To understand the concept, we refer readers to introduction to this chapter.

The 45°/135° (45°/(45°+90°)) polarizer should be installed on the emission/detection line between the sample and the instrument emission polarizer (typically immediately after sample as shown in Figure 6-7.2). In principle, when using a sheet polarizer, it can be placed before the collimating lens or just after the lens and before the instrument polarizer analyzer (when an additional 45°/135° polarizer requires collimated parallel beam (like Glan-Thompson), it should be installed right after a collimating lens and before the system polarizer). However, we do not recommend using Glan-Thompson since it is very difficult to properly align.

The goal of this experiment is to test how well the 45°/135° polarizer works and how precisely we can recover *G*-factor. For that, we will use square geometry where we already evaluated the *G*-factor (see Experiment 6-1) and we will repeat previous measurements but using vertical excitation and the additional 45°/135° polarizer on emission (in fact, we do not have to use a polarizer on excitation at all). The recovered *G*-factor, in this case, should be independent of excitation polarization (or emission polarization). Later, we will utilize this approach to a front-face configuration.

MATERIALS

- Polarizer with precisely marked 45° orientation
- Water and EtOH
- 2AP, BBO, C153, and Ru
- 1 cm × 1 cm cuvette

(Continued)

EXPERIMENT 6-7 (CONTINUED) MEASURING *G*-FACTOR IN THE CASE WHEN WE CANNOT USE HORIZONTAL EXCITATION

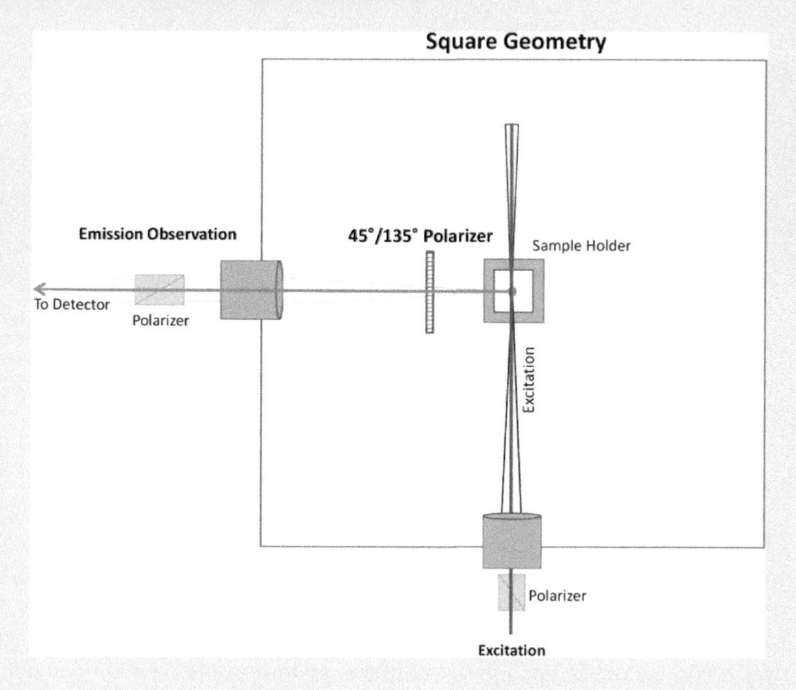

FIGURE 6-7.2 Schematic of square geometry configuration with a marked position for 45°/135° polarizer.

EQUIPMENT

- Spectrofluorometer with MCP-PMT detector
- Smartphone with flashlight option

METHODS

1. As in the previous experiment, prepare a solution of selected dyes.
2. Orient your excitation polarizer to vertical (0°) and select an optimal excitation wavelength for each fluorophore respectively. *Note: In practical experiments, we would recommend using horizontal orientation for the excitation polarizer. This will minimize potential polarization.* However, in this demonstration, we will use vertical to prove that the measured *G*-factor with this method does not depend on sample polarization.
3. Insert the 45°/135° polarizer as indicated in Figures 6-7.2 and 6-7.5. It is important to have the possibility to easily rotate the polarizer by 90° (*Important: Do not flip polarizer front back since this is not a compatible configuration*).
4. Measure the emission spectra for parallel—0° and horizontal—90° orientation of observation polarizer.
5. Rotate the 45° polarizer by 90° and repeat the measurements for parallel—0° and horizontal—90° orientation of observation polarizer.
6. It is recommended to repeat these procedures (#4 and #5) a few times and average to get less variation. Remember the error in the evaluated *G*-factor will propagate to the

(Continued)

EXPERIMENT 6-7 (CONTINUED) MEASURING *G*-FACTOR IN THE CASE WHEN WE CANNOT USE HORIZONTAL EXCITATION

error in calculating anisotropy. So, we want to determine this parameter with the highest possible precision.

7. Graph calculated *G*-factors for 45° and 135° of the additional polarizer orientation.
8. Average both measurements. The average should give proper *G*-factor.

RESULTS

In our experiment, we used a single polarizer cut with the highest precision to a 45° angle that we installed in a square aluminum block (see Figure 6-2.1), so a 90° rotation can easily be achieved. Typically when done precisely, we can get about 0.5° precision. This precision does not sound great but to achieve it in normal laboratory situation you need to practice, it took us quite a few trials. Next, we tested 4° and about 1° off polarizers (we used polarizers from trial cuts we made to get first correct polarizer). *Note: 4° is less than a single step of a second arm in an analog watch.* For a precise method for estimating polarization plane orientation, see Experiment 3-26.

In Figure 6-7.3, we are presenting two repeats of measurements done with a polarizer that is off by 4° (left), 1° (center), and then 0.5° (right). For the polarizer that is 4° off, we can clearly see two traces (top and bottom) for 45° and 135° polarizer orientation that are significantly different. Averages of 45° and 135° measurements are in the middle. Measurements with the polarizer that is about 1° off are much better. Measurements with a correct 45° polarizer (less than 0.5° off) gave us a set of six lines that are within the experimental noise. The average value from the two measurements (45° and 135°) are very close for all three measurements. It is clear that average value gives excellent results in a relatively broad range of angle estimation (up to 4°). Just for comparison, in Figure 6-7.4, we are showing *G*-factor measured with a correctly oriented polarizer, whereas on the figure we indicated *G*-factor measured using Ru (circles) as determined in Experiment 6-1. We would like to stress that the average gives a very good result (perfectly overlapping) while a single measurement with 4° error in the orientation produces a dramatic difference in the *G*-factor. In fact, a measurement with a polarizer that has a precision of 1–2° (this is less than one-third of a single step of a second arm in a watch) disqualifies *G*-factor determination.

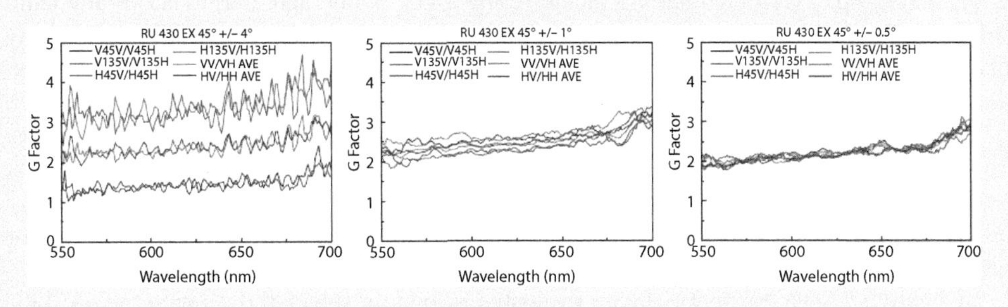

FIGURE 6-7.3 *G*-factors measured with 45°/135° additional polarizer with offset 4°, 1°, and 0.5° and the average value.

(Continued)

EXPERIMENT 6-7 (CONTINUED) MEASURING G-FACTOR IN THE CASE WHEN WE CANNOT USE HORIZONTAL EXCITATION

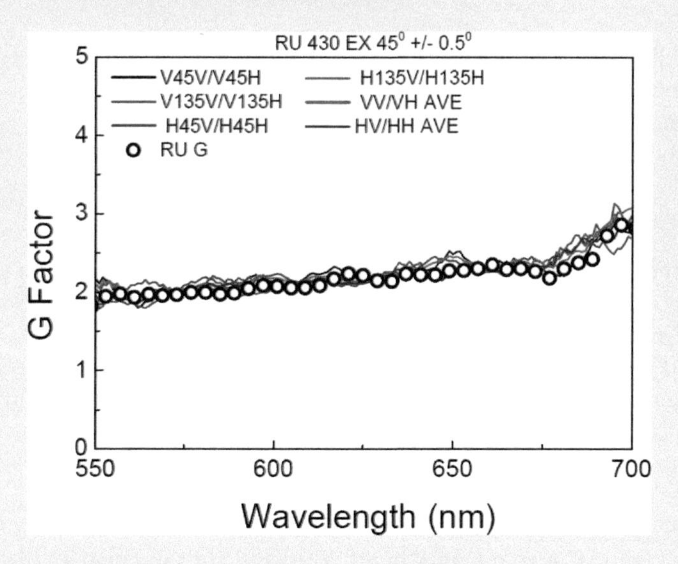

FIGURE 6-7.4 Comparison of G-factors measured with correct 45°/135° (continuous six lines) and using conventional approach and long-lived ruthenium dye (circles).

It is important to realize that when using an additional 45°/135° polarizer, the result is completely independent from the state of light polarization for emission. The transmitted light will be always 45°/135° polarized and may only differ in the total intensity depending on incoming light polarization (for completely vertically polarized light we will detect only half of the intensity). In this case, an interesting possibility emerges. We can use any white light source and use a scattering solution instead of the sample. In this case, the scattered light will obviously be highly polarized but this will not be a problem. We can use an excitation lamp and synchronous scan with the observation and excitation wavelength set to the same value. We would typically recommend a small few nanometers shift between emission and excitation monochromator to reduce the light intensity reaching detector or we can use any external source for systems that do not have lock-in type of detection. *Note: It is important to remember to use neutral density filters to appropriately attenuate light intensity to avoid saturating the detector.*

As an example, we determine a G-factor using a cell phone light (flash LED or any lamp source). In Figure 6-7.5, we present a picture of the actual configuration. On the left is a side view of an open FT300 compartment and cell phone installed at the left entrance to the fluorometer. On the right is the top view of the compartment with an indicated cuvette and 45°/135° polarizer.

The LED light is collimated on the cuvette containing the scattering solution (Ludox). The scattered light crosses the 45°/135° polarizer and we record intensity through vertical and horizontal polarizer analyzer on the emission channel. Depending on the light intensity and scatterer concentration, we should use appropriate attenuation filters on excitation or emission channel.

In Figure 6-7.6, we present G-factor evaluated using four selected dyes with the 45°/135° polarizer that covers a full range of the fluorometer. On this graph, we also included a G-factor determined when using cell phone light illumination and scatterer.

(Continued)

EXPERIMENT 6-7 (CONTINUED) MEASURING *G*-FACTOR IN THE CASE WHEN WE CANNOT USE HORIZONTAL EXCITATION

Cell Phone
Cuvette with Scatterer
45°/135° Polarizer

Side View
Top View

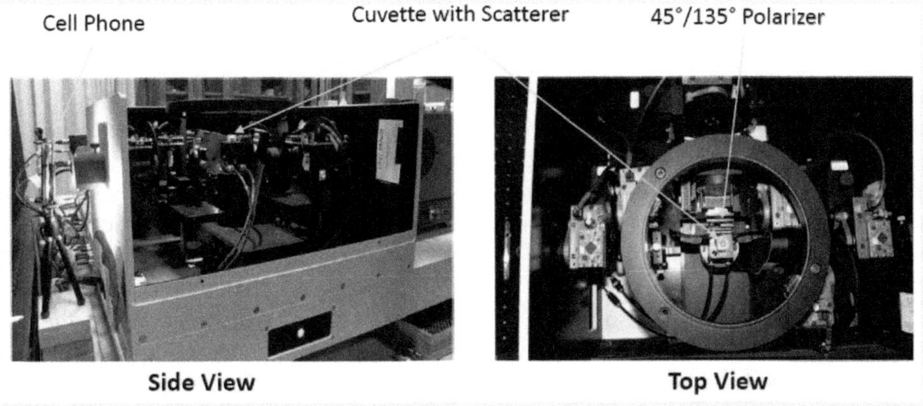

FIGURE 6-7.5 Left: Side view of fluorometer with the cell phone installed on the left entrance into the fluorometer. Right: Top view of open sample compartment with an indicated position of the cuvette and 45°/135° polarizer.

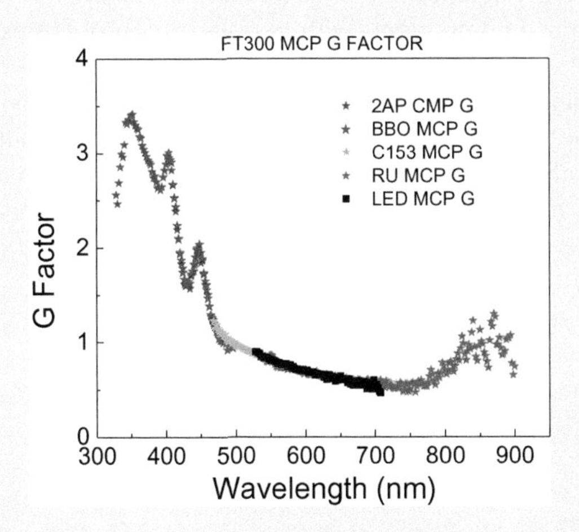

FIGURE 6-7.6 Cumulative *G*-factor determined using 45°/135° polarizer. At the 500–700 nm range, we overlay *G*-factor determined using cell phone LED light.

CONCLUSIONS

When we cannot use horizontal excitation in square geometry or the geometry of our system is not a square one, using an additional 45°/135° polarizer could be the most convenient approach. As demonstrated in Figure 6-7.6, using a white LED light could be a simple way for a routine *G*-factor check.

Note: It is important to use a cuvette with a scattering solution since this will preserve the geometry of fluorescence. Mounting LED (or cell phone) inside the chamber would typically lead to a significant error since the geometry of the emitting spot is different/changing as compared to a typical emission spot.

EXPERIMENT 6-8 MEASUREMENTS OF STEADY-STATE EMISSION ANISOTROPIES: A SIMPLE CASE OF SQUARE GEOMETRY

Keywords: emission, emission anisotropy, *G*-factor.

In most cases, the anisotropy measurements will be done in a square-geometry format. Measurements of fluorescence anisotropy require the excitation with polarized light and the detection of fluorescence intensity observed through a parallel and perpendicular oriented polarizer as compared to the excitation light polarization. Usually, the excitation is vertically (V) oriented and the observation is observed through either vertically (V) or horizontally (H) oriented polarizer. Notation VV means vertical excitation and vertical (parallel) observation, VH says vertical excitation and horizontal (perpendicular) observation. In the emission anisotropy measurements, the excitation is fixed (usually) at the maximum of the long-wavelength excitation/absorption peak and the emission is scanned across the fluorescence spectrum. In the case of highly overlapped absorption and emission spectra, the excitation can be set at shorter wavelengths. All spectrofluorometers require a correction for uneven transmission/sensitivity to V and H polarized light because optics, especially its dispersive components, is sensitive (response differently) to the polarization of the detected light. The correction factor (*G*-factor) should be evaluated independently. We discussed in detail *G*-factor measurements in three previous experiments (Experiments 6-5 through 6-7). In the square-geometry configuration, the easiest way to determine the *G*-factor is to measure V and H fluorescence components with H polarization of the excitation. In this case, the observed excited state population is symmetrical (seen as a circle), as shown in previous experiments and by definition, V and H measurements should provide the same fluorescence intensities. If not, the H component must be adjusted to match the V component, and the adjusting factor is the *G*-factor. Equivalently, the V component can be adjusted to the H component. It does not matter as long as we are consistent. In this experiment, we will guide the reader on how to measure correctly the emission anisotropy spectra in the simplest square geometry case.

MATERIALS

- Ethanol (EtOH) and glycerol
- Rhodamine 6G (R6G)
- Glass or plastic vials
- 1 × 1 cm cuvettes

EQUIPMENT

- Spectrophotometer (Cary 50, Varian Inc.)
- Spectrofluorometer (Eclipse, Varian Inc.)

METHODS

1. Make a high concentration stock solution of R6G in EtOH. Be sure, all the dye is diluted and the solution is clear.
2. Prepare a 5 mL sample of R6G in glycerol by adding a small drop of the R6G stock solution. Shake well until the sample solution is homogeneous. Measure absorption of the sample using glycerol as a reference baseline. Be sure that the absorption of R6G is

(*Continued*)

EXPERIMENT 6-8 (CONTINUED) MEASUREMENTS OF STEADY-STATE EMISSION ANISOTROPIES: A SIMPLE CASE OF SQUARE GEOMETRY

lower than 0.1. *Note: It is important to minimize the amount of added EtOH to maintain the highest possible viscosity of R6G solution.*

3. Measure the fluorescence emission spectra of the R6G sample at the magic angle (MA) conditions using 510 nm excitation wavelength.
4. Measure fluorescence emission spectra of the R6G sample at HH and HV polarizers' orientations (excitation polarizer set to horizontal (90°) orientation). Calculate the wavelength-dependent *G*-factor across fluorescence spectrum. We recommend to repeat the measurement two to three times and calculate the average *G*-factor to minimize the error. The error in *G*-factor will propagate to calculated anisotropy.
5. Measure fluorescence emission spectra of the R6G sample at VV and VH polarization conditions (vertical orientation of excitation polarizer). Apply the *G*-factor correction.
6. Plot the emission anisotropy of R6G in glycerol.

RESULTS

In Figure 6-8.1, we present absorption and emission (measured under magic-angle conditions) spectra for R6G in glycerol at 20°C. R6G presents a significant spectral overlap so, the excitation wavelength for the emission scan was set at 510 nm to allow scanning of the full emission spectrum.

Next, we measured *G*-factor. For that, we use 510 nm excitation wavelength and set the excitation polarizer to horizontal (90°) and measure emission spectra with vertical (V) and horizontal (H) orientation of the emission polarizer in the 520–680 nm range. Two measured intensity profiles are presented in Figure 6-8.2. It is clear that the measured intensities are very different. Since we expect them to be identical (for horizontal excitation, we expect to see an isotropic distribution of excited dipoles as presented on the right in Figure 6-8.2).

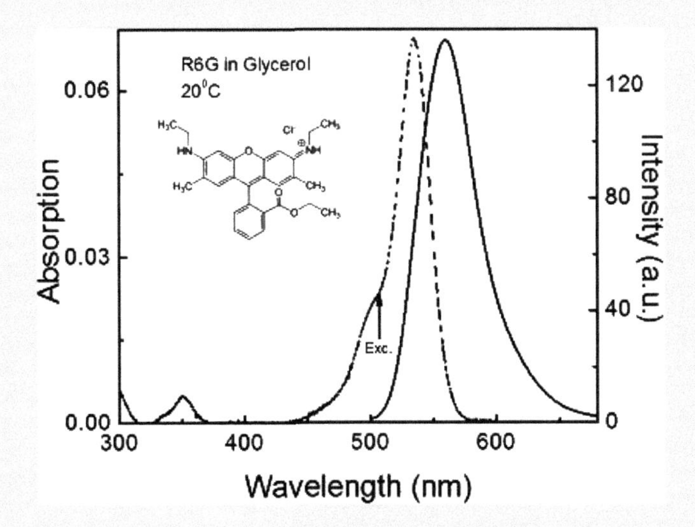

FIGURE 6-8.1 Absorption and fluorescence spectra of R6G in glycerol. In order to measure anisotropies across the entire fluorescence emission, the excitation wavelength was set at 510 nm.

(Continued)

EXPERIMENT 6-8 (CONTINUED) MEASUREMENTS OF STEADY-STATE EMISSION ANISOTROPIES: A SIMPLE CASE OF SQUARE GEOMETRY

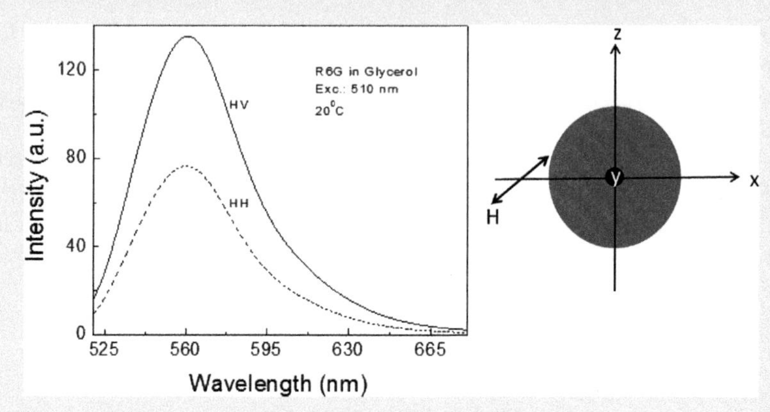

FIGURE 6-8.2 Polarized emission intensity components of R6G fluorescence with horizontally (H) polarized excitation. In an ideal detection system, HV and HH components should be equal (see intensity distribution on the right). The correction (*G*-factor) is the ratio of I_{HV}/I_{HH} at each wavelength and should be applied to the I_{HH} component (I_{HH} × *G*-factor) in order to equalize I_{HV} and I_{HH} measured intensities.

There is no written role which component should be corrected but traditionally, we define the wavelength-dependent *G*-factor as $G(\lambda) = I_{HV}(\lambda)/I_{HH}(\lambda)$. Using this definition to calculate anisotropy, we will multiply the I_{VH} emission intensity component by the *G*-factor or divide I_{VV} emission intensity component by *G*-factor. When defining *G*-factor as $G = I_{HH}/I_{HV}$, we would multiply the I_{VV} emission intensity component by *G*-factor.

In Figure 6-8.3, we are presenting calculated wavelength-dependent *G*-factor as $G(\lambda) = I_{HV}(\lambda)/I_{HH}(\lambda)$ from measured intensity components in Figure 6-8.2.

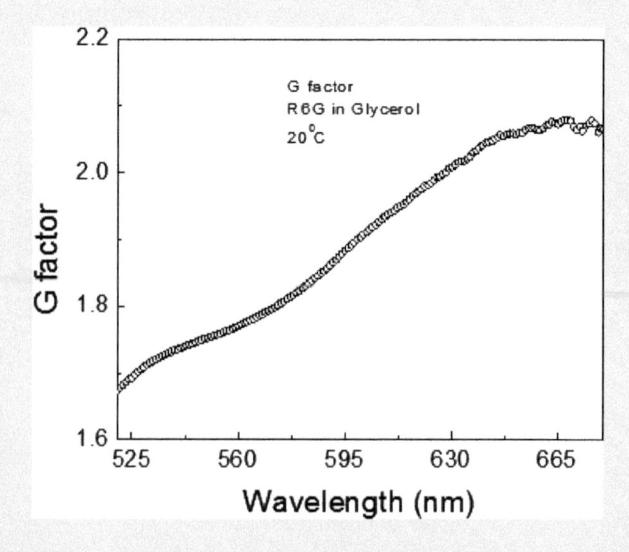

FIGURE 6-8.3 Calculated wavelength-dependent *G*-factor from I_{HV} and I_{HH} intensity components as measured in Figure 6-8.2.

(Continued)

EXPERIMENT 6-8 (CONTINUED) MEASUREMENTS OF STEADY-STATE EMISSION ANISOTROPIES: A SIMPLE CASE OF SQUARE GEOMETRY

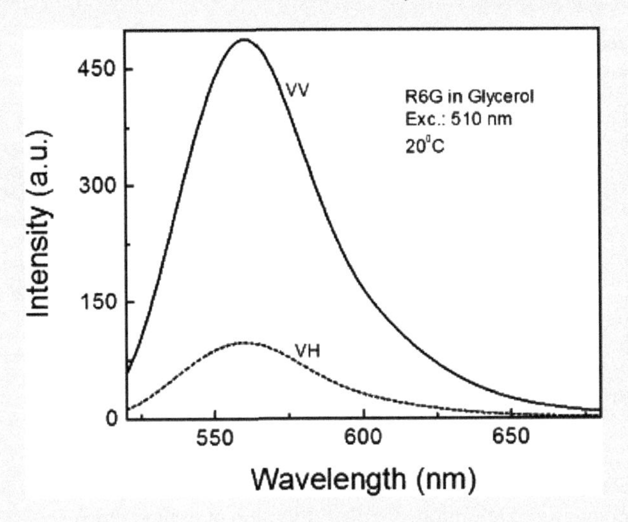

FIGURE 6-8.4 I_{VV} and I_{VH} intensity components of R6G fluorescence before *G*-factor correction.

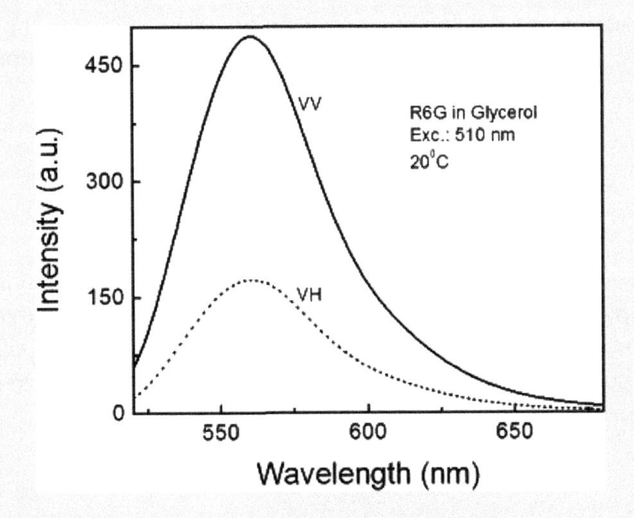

FIGURE 6-8.5 I_{VV} and I_{VH} intensity components of R6G fluorescence after *G*-factor (from Figure 6-8.3) correction.

Just for comparison in Figure 6-8.4, we are presenting measured fluorescence intensity components with vertical excitation without the *G*-factor correction (as measured in the instrument) and in Figure 6-8.5 after applying the *G*-factor correction. *Note: The G-factor is a wavelength-dependent parameter and the emission intensity I_{VH} needs to be multiplied at which wavelength by a corresponding G-factor value.*

(Continued)

EXPERIMENT 6-8 (CONTINUED) MEASUREMENTS OF STEADY-STATE EMISSION ANISOTROPIES: A SIMPLE CASE OF SQUARE GEOMETRY

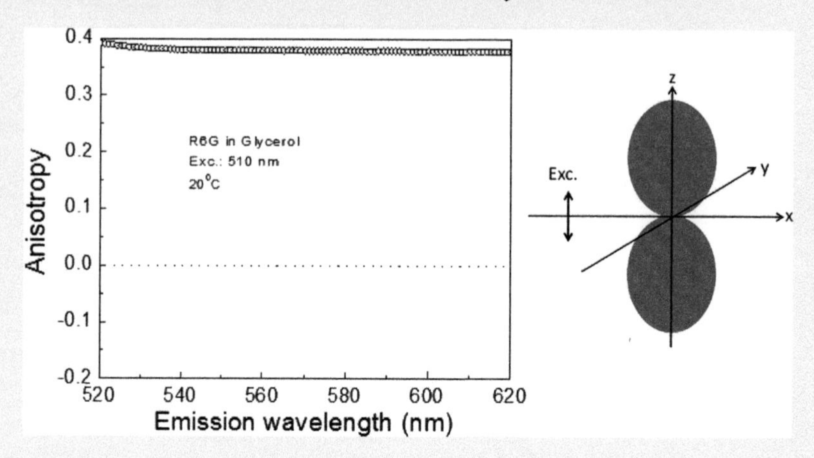

FIGURE 6-8.6 Fluorescence emission anisotropy of R6G measured in glycerol for 510 nm excitation. Right: Calculated for measured anisotropy (0.36) distribution of excited dipoles.

Using measured uncorrected fluorescence intensities (Figure 6-8.4) or the corrected horizontal component, I^c_{VH} from Figure 6-8.5 we can calculate emission anisotropy as:

$$r(\lambda) = \frac{I_{VV}(\lambda) - G(\lambda) \cdot I_{VH}(\lambda)}{I_{VV}(\lambda) - G(\lambda) \cdot 2I_{VH}(\lambda)} = \frac{I_{VV}(\lambda) - I^c_{VH}(\lambda)}{I_{VV}(\lambda) + 2I^c_{VH}(\lambda)}$$

Calculated fluorescence anisotropy is presented in Figure 6-8.6. We presented it in the full scale for anisotropy (−0.2 to 0.4) for an isotropic system. The emission anisotropy for R6G in glycerol is high, over 0.36. On the right side of Figure 6-8.6, we show a population of excited molecules that results in an anisotropy of 0.36. Compare this distribution to the distribution observed for horizontal excitation (Figure 6-8.2).

CONCLUSIONS

In this example, we presented details for measuring emission anisotropy. The selected dye (R6G) in glycerol at 20°C presents a very high anisotropy (>0.36) and in most cases, we will measure smaller anisotropy values. Also, measured anisotropy is higher at shorter wavelengths slowly decreasing toward longer wavelengths. This is a common behavior that is mostly due to the process called excited-state relaxation. Relaxed molecules would emit at longer wavelengths. Since the relaxation process takes time, the emitting molecules have more time to reorient. In addition, the transition moment orientation in a relaxed state may have a slightly different orientation than the transition moment for absorption (unrelaxed molecule). The shorter wavelength emission range is dominated by unrelaxed molecules quickly emitting before any reorientation occurs.

EXPERIMENT 6-9 MEASUREMENTS OF FLUORESCENCE EXCITATION ANISOTROPIES: A SIMPLE CASE OF SQUARE GEOMETRY

Keywords: emission, excitation anisotropy, *G*-factor.

In the previous experiment, we presented an example of measuring emission anisotropy. We also mentioned that emission anisotropy will typically depend on the excitation wavelength. Contrary to emission spectra that are typically a single transition from the lowest energy excited state (S_1) to a ground state (S_0), the absorption spectra represent multiple transitions to higher excitation levels (S_n). These transitions will typically have different orientations of transition moments in the molecular frame. Consequently, measured excitation anisotropy will typically depend on the excitation wavelength.

For the excitation anisotropy measurements, the observation wavelength is fixed at one wavelength, usually close to the maximum intensity of the fluorescence spectrum and the excitation is scanned across the fluorophore absorption range (spectrum). Similar to fluorescence emission anisotropy, measurements of fluorescence excitation anisotropy require the excitation with polarized light and detection of fluorescence intensity observed through a parallel and perpendicular oriented polarizer as compared to the excitation light polarization. Usually, the excitation is vertically (V) oriented and the observation is through either a vertically (V) or horizontally (H) oriented polarizer. VV means vertical excitation and vertical (parallel) observation, VH stays for vertical excitation and horizontal (perpendicular) observation. In the case of the fluorescence excitation spectra, the *G*-factor is only calculated for the observation wavelength (single value at a given observation wavelength) that in square geometry is equal to I_{HV}/I_{HH} at the given observation wavelength. The choice of excitation wavelength for the *G*-factor measurement is irrelevant and typically we will select an excitation wavelength that gives us an optimal signal to minimize the error. The measured excitation wavelength-dependent horizontal intensity component I_{VH} should be multiplied by a single value of *G*-factor in order to correct for the difference in detection sensitivity.

For this experiment, we will use the same solution as in the previous Experiment 6-8.

MATERIALS

- Ethanol (EtOH) and glycerol
- Rhodamine 6G (R6G)
- Glass or plastic vials
- 1 × 1 cm cuvettes

EQUIPMENT

- Spectrophotometer
- Spectrofluorometer

METHODS

1. Make a high concentration of stock-solution R6G in EtOH. Be sure that all the dye is diluted and the solution is clear.
2. Prepare a 5 mL sample of R6G in glycerol by adding a small drop of R6G stock solution. Shake well until the sample solution is homogeneous. Measure absorption of the sample using glycerol as a reference baseline. Be sure that the absorption at each

(*Continued*)

EXPERIMENT 6-9 (CONTINUED) MEASUREMENTS OF FLUORESCENCE EXCITATION ANISOTROPIES: A SIMPLE CASE OF SQUARE GEOMETRY

wavelength is lower than 0.1. *Note:* It is important to minimize the amount of added EtOH to maintain the highest possible viscosity of R6G solution.

3. As in Experiment 6-8, measure the fluorescence emission spectra of the R6G sample at magic angle (MA) conditions using 510 nm excitation wavelength.
4. Measure the fluorescence emission intensities of the R6G sample using I_{HH} and I_{HV} polarizers' orientations at 580 nm. Calculate the *G*-factor at 580 nm observation. *Note: We need the G-factor value only for the observation wavelength. So, it is possible to measure emission intensities I_{HH} and I_{HV} only at this single wavelength multiple times and average them to obtain a G-factor value as precisely as possible.* The excitation wavelength is not relevant and we typically will use the most optimal (giving best signal). But make sure the scattered excitation light is not leaking to the observation channel (e.g., 510 nm excitation wavelength and 580 nm observation wavelength are significantly separated so we do not expect any scattering of 510 nm leaking to 580 nm observation).
5. Measure the fluorescence excitation spectra (emission intensities) of R6G sample at MA, VV, and VH polarization conditions. Apply the *G*-factor correction. *Note:* Remember, we will need only one value of *G*-factor for the observation wavelength (580 nm in our case) and multiply the excitation wavelength-dependent I_{VH} intensity component at each wavelength by this single number.
6. Plot the excitation anisotropy of R6G in glycerol.

RESULTS

The absorption and emission spectra of R6G are presented in Figure 6-9.1. In the figure, we marked the observation wavelength.

We used the 510 nm excitation and evaluated the G-factor to be equal to 1.82. This value should be in agreement with the previously estimated G-factor. In Figure 6-9.2, we are

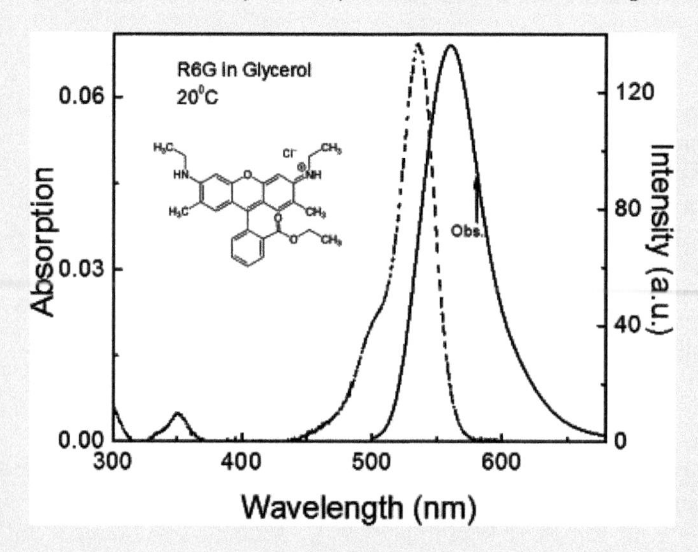

FIGURE 6-9.1 Absorption and fluorescence spectra of R6G in glycerol. In order to observe entire excitation spectra, the observation was set at 580 nm.

(Continued)

EXPERIMENT 6-9 (CONTINUED) MEASUREMENTS OF FLUORESCENCE EXCITATION ANISOTROPIES: A SIMPLE CASE OF SQUARE GEOMETRY

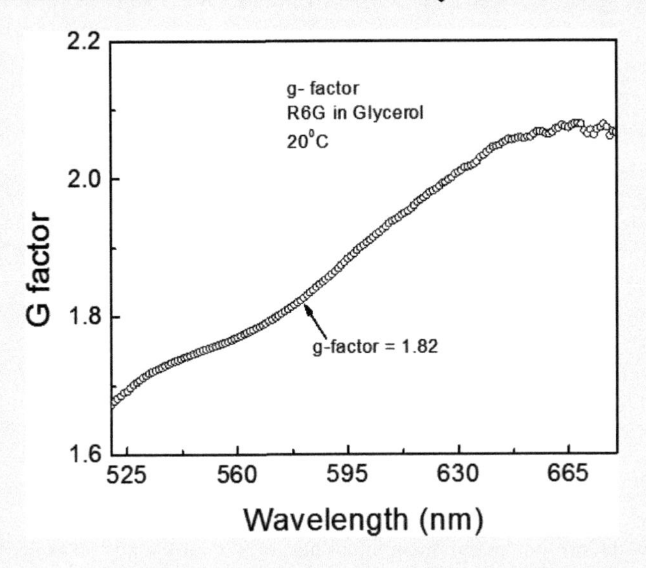

FIGURE 6-9.2 The G-factor calculated from I_{HV} and I_{HH} measurements as in Experiment 6-3.

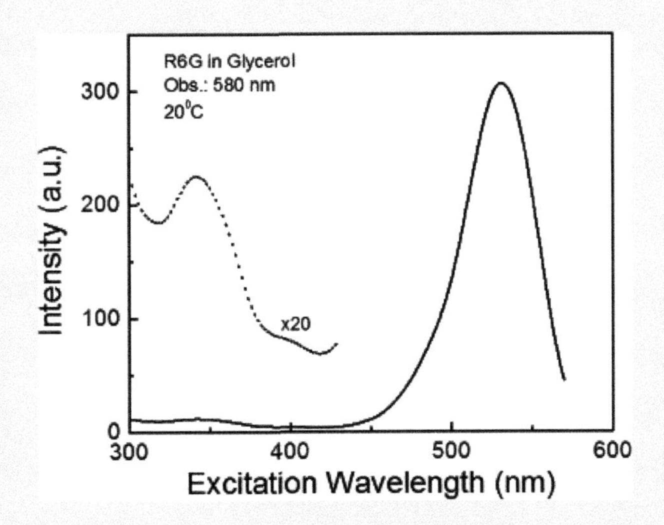

FIGURE 6-9.3 Excitation spectrum of R6G fluorescence with 580 nm observation and MA condition.

presenting the G-factor as it has been evaluated in the previous experiment where we marked the G-factor value as measured for 580 nm observation.

Figure 6-9.3 presents the excitation spectrum measured under MA conditions. A long-wavelength absorption band is clearly visible. This is a single transition to the first excited state. Much weaker short-wavelength absorption bands involve multiple transitions as seen in an expanded view for wavelengths below 450 nm.

(Continued)

EXPERIMENT 6-9 (CONTINUED) MEASUREMENTS OF FLUORESCENCE EXCITATION ANISOTROPIES: A SIMPLE CASE OF SQUARE GEOMETRY

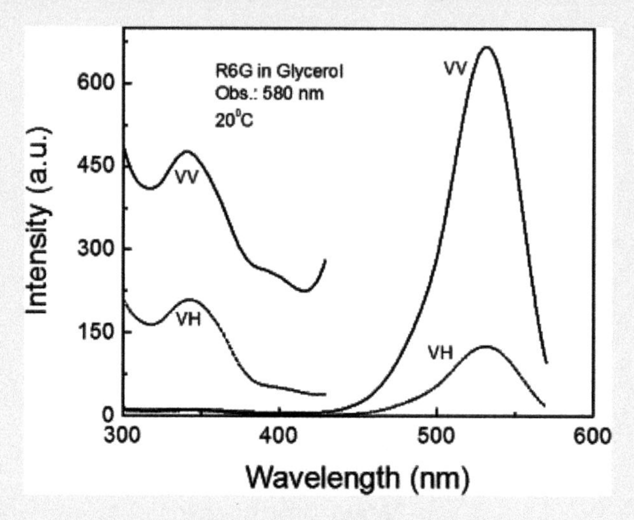

FIGURE 6-9.4 Excitation spectra (emission intensities) of R6G at VV and VH oriented polarizers, as measured (not corrected for *G*-factor).

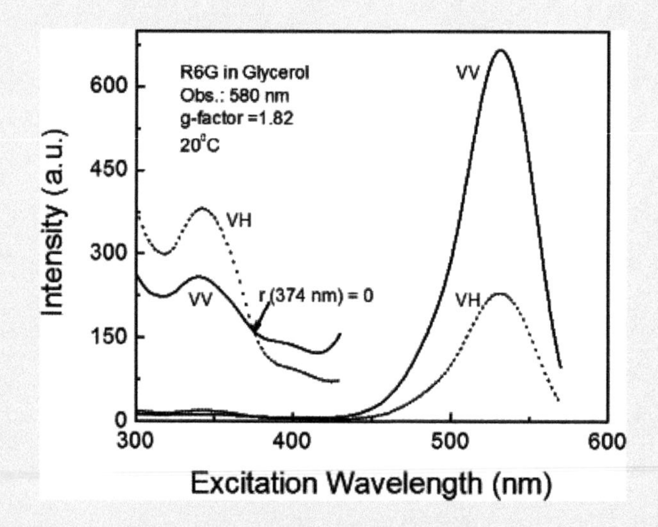

FIGURE 6-9.5 Excitation spectra of R6G at VV and VH oriented polarizers, corrected for *G*-factor.

In Figure 6-9.4, we present I_{HH} and I_{HV} intensity components as measured across the absorption/excitation spectrum with 580 nm observation. In Figure 6-9.5, we applied the *G*-factor correction (multiplied I_{VH} component by a factor of 1.82). This significantly shifts up the horizontal intensity component. Two intensity components I_{HH} and I_{HV} cross (have an identical value) at 374 nm. This is the point of equal intensities that corresponds to zero anisotropy.

(*Continued*)

EXPERIMENT 6-9 (CONTINUED) MEASUREMENTS OF FLUORESCENCE EXCITATION ANISOTROPIES: A SIMPLE CASE OF SQUARE GEOMETRY

Out of two intensity components (I_{HH} and I_{HV}) in Figure 6-9.4, we can calculate the excitation anisotropy r as:

$$r(\lambda)=\frac{I_{VV}(\lambda)-G\cdot I_{VH}(\lambda)}{I_{VV}(\lambda)+2G\cdot I_{VH}(\lambda)}=\frac{I_{VV}(\lambda)-I_{VH}^{c}(\lambda)}{I_{VV}(\lambda)+2I_{VH}^{c}(\lambda)}$$

As compared to the previous experiment, we are now using a single value of G-factor for 580 nm observation wavelength. We also can use corrected values of the horizontal component, I_{VH}^{c} from Figure 6-9.5 and calculate anisotropy directly. In Figure 6-9.6, we present excitation anisotropy measured with 580 nm observation wavelength. The anisotropy strongly depends on excitation wavelength changing values from 0.37 down to −0.14. This is a pretty large range demonstrating multiple different absorption transitions with different orientations of transition moments. The marked zero anisotropy at 374 nm corresponds to the apparent transition moment orientation of 54.7°.

The lowest measured anisotropy (−0.14) corresponds to the 350 nm excitation wavelength. To realize how this affects measured intensity components in Figure 6-9.7, we present measured uncorrected emission intensities (I_{VV} and I_{VH}) with 350 nm excitation. It is didactic to compare Figure 6-9.7 with Figure 6-8.4 in the previous experiment.

G-factor corrected emission intensities are presented in Figure 6-9.8 (we used wavelength-dependent G-factor as shown in Figure 6-9.3 in the previous experiment). These intensities are completely opposite to that presented in Figure 6-8.5 in the previous experiment. It is interesting to calculate emission anisotropy for 350 nm excitation wavelength. The presented emission anisotropy in Figure 6-9.9 is practically constant across the emission spectrum but very different from that presented in Figure 6-8.6 in the previous experiment.

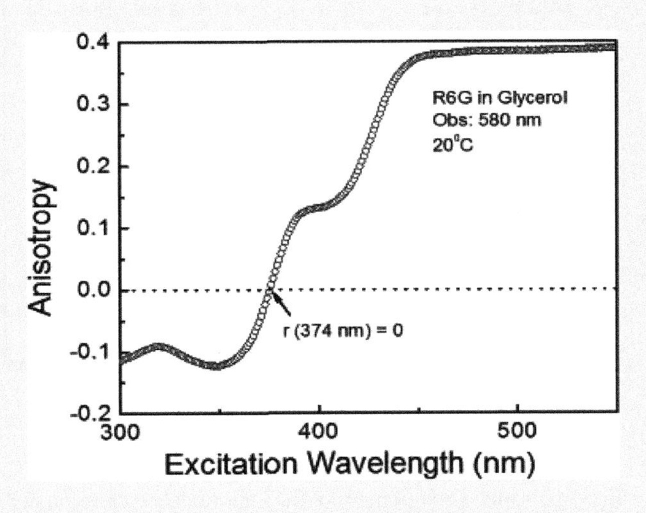

FIGURE 6-9.6 Fluorescence excitation anisotropy spectrum of R6G in glycerol.

(Continued)

EXPERIMENT 6-9 (CONTINUED) MEASUREMENTS OF FLUORESCENCE EXCITATION ANISOTROPIES: A SIMPLE CASE OF SQUARE GEOMETRY

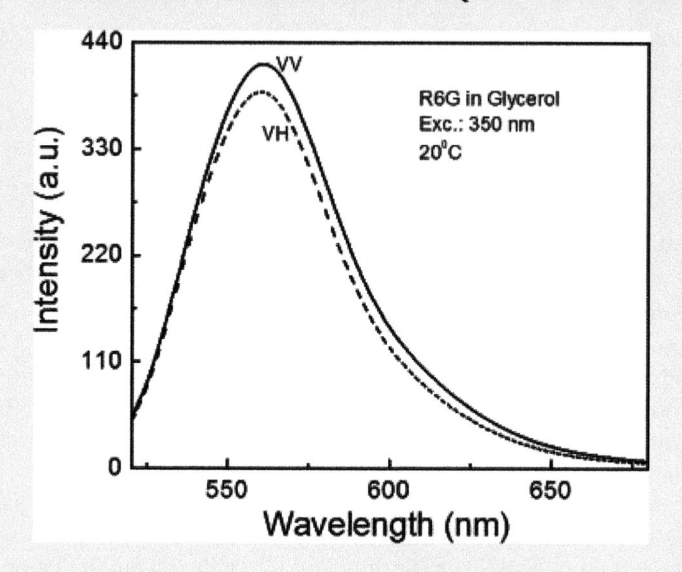

FIGURE 6-9.7 I_{VV} and I_{VH} polarized intensity components of R6G fluorescence emission at 350 nm excitation, not corrected for G-factor.

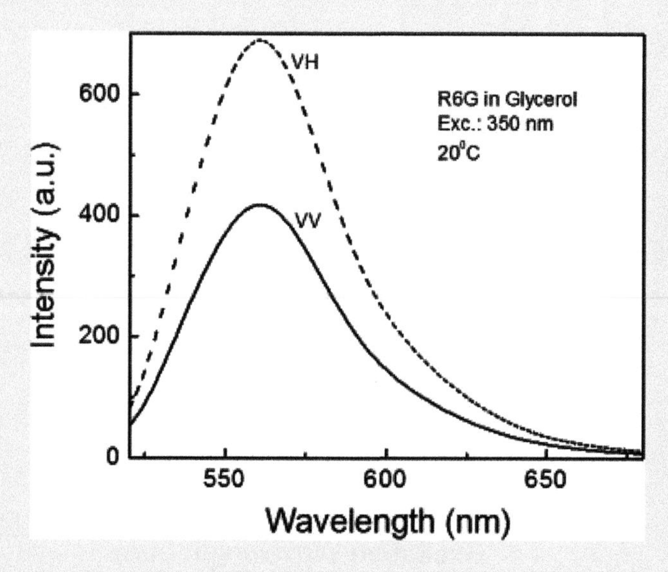

FIGURE 6-9.8 I_{VV} and I_{VH} polarized components of R6G fluorescence emission at 350 nm excitation corrected for wavelength-dependent G-factor as in Figure 6-9.3.

(Continued)

EXPERIMENT 6-9 (CONTINUED) MEASUREMENTS OF FLUORESCENCE EXCITATION ANISOTROPIES: A SIMPLE CASE OF SQUARE GEOMETRY

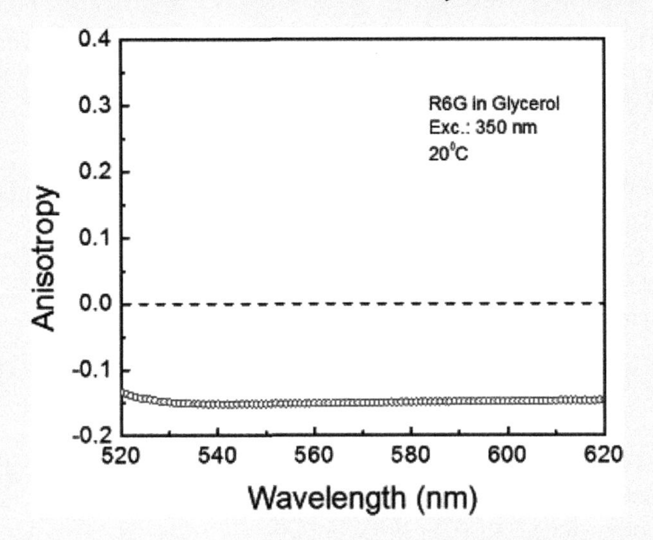

FIGURE 6-9.9 Fluorescence emission anisotropy spectrum of R6G in glycerol at 350 nm excitation.

CONCLUSIONS

It is important to realize that excitation anisotropy is wavelength dependent and in some cases can be used to identify different absorption transitions (similar to linear dichroism discussed in Chapter 3). To calculate excitation anisotropy, we will use the *G*-factor determined for a single observation wavelength. And when changing the observation, we need to remember to use an adequate *G*-factor.

EXPERIMENT 6-10 FLUORESCENCE EMISSION ANISOTROPY—DEPENDENCE ON THE SOLVENT VISCOSITY

Keywords: fluorescence, anisotropy, solvent viscosity.

Rotational diffusion of the excited fluorophore is a major factor contributing to fluorescence depolarization. Rotational correlation time is a time in which the molecule rotates one radian (57.3°). For low viscosity solvents in which rotational diffusion is high (speed for molecular rotation is high) and correlation time is much shorter than fluorescence lifetime, molecules will lose information about orientation and measured anisotropy will be low, close to zero. In high viscosity solvents, the dragging force is much higher (frictional drag coefficient is high) and molecules rotate much slower leading to a much higher anisotropy.

In this experiment, we will measure anisotropy of R6G in two solvents. One of low viscosity (water) and second of much higher viscosity (glycerol). In the following

(*Continued*)

EXPERIMENT 6-10 (CONTINUED) FLUORESCENCE EMISSION ANISOTROPY—DEPENDENCE ON THE SOLVENT VISCOSITY

Experiment 6-11, we will show how to modify the viscosity of the solvent and present results for a few different viscosity systems. The goal of this experiment is to show details for polarization measurements.

MATERIALS

- Water and glycerol
- Rhodamine 6G (R6G)
- Glass or plastic vials
- 1 cm × 1 cm cuvettes
- Polarizer
- Cell phone or camera

EQUIPMENT

- Spectrophotometer
- Spectrofluorometer

METHODS

1. Prepare stock-solutions of R6G in EtOH.
2. Prepare samples of R6G in water and glycerol with optical densities not exceeding 0.1. Measure fluorescence spectra of individual fluorophores at magic angle (MA) conditions.
3. Measure fluorescence anisotropies (VV and VH emission components) of individual fluorophores. If you need to evaluate the *G*-factor using horizontal excitation).
4. Make photographs of polarized fluorescence components.

RESULTS

High emission anisotropy close to 0.4 means that the I_{VV} intensity component is almost threefold greater then I_{VH} component (it is a 3:1 ratio for $r = 0.4$). When fluorescence anisotropy approaches zero, the I_{VV} and I_{VH} components do not differ (they are the same). By measuring anisotropy, we are essentially measuring the ratio of both intensity components. A typical instrument (spectrofluorometer) can do it with very high precision. However, when the fluorescence anisotropy is high, even the human eye can easily distinguish different brightness of I_{VV} and I_{VH} components when viewed through a polarizer sheet. Since the human eye is not sensitive to light polarization, we view the solution through a differently oriented polarizer to visually distinguish the intensity differences. Also, the same solution excited with horizontal light polarization and viewed from the side will present zero anisotropy and both intensities I_{VV} and I_{VH} should be identical.

First, we measured intensity components for R6G in water. The measured and corrected for *G*-factor intensities I_{VV} and I_{VH} are presented in the left of Figure 6-10.1. Water is a low viscosity solvent and molecules are quickly rotating randomizing the orientation of excited molecules. Simplified distribution of excited molecules leading to measured low anisotropy is presented in the inset. The measured intensities I_{VV} and I_{VH} are only

(*Continued*)

EXPERIMENT 6-10 (CONTINUED) FLUORESCENCE EMISSION ANISOTROPY—DEPENDENCE ON THE SOLVENT VISCOSITY

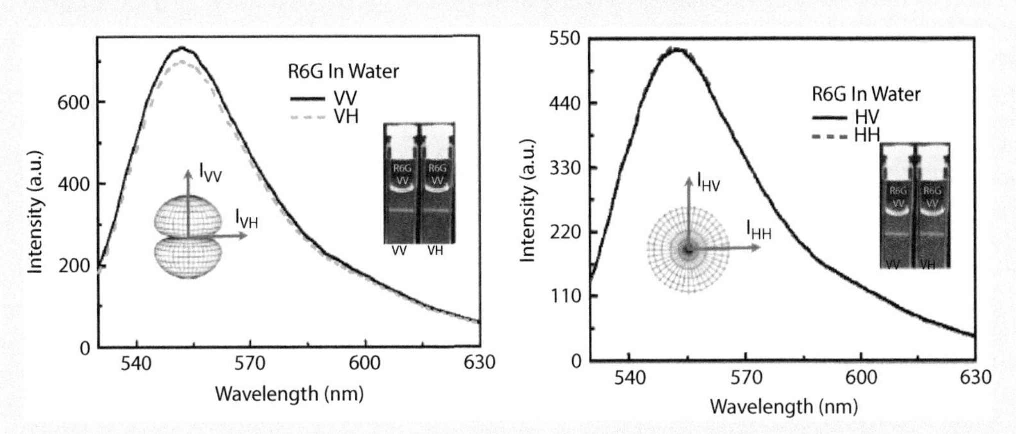

FIGURE 6-10.1 Left: Corrected fluorescence spectra of R6G emission in water measured with vertical excitation and vertical and horizontal observations respectively. The photographs in the inset were taken through a vertical and horizontal polarizer when the sample was excited with vertically polarized light and viewed from a side. Right: Corrected fluorescence spectra of R6G emission in water measured with horizontal excitation and vertical and horizontal observation respectively. The I_{HV} and I_{HH} intensity components are equally bright (as shown in the schematic inset for horizontal excitation, the distribution leads to equal I_{HV} and I_{HH} intensities).

slightly different resulting in low anisotropy. The photography inset shows the cuvette viewed through vertical and horizontal polarizers respectively. For comparison on the right of Figure 6-10.1, we are presenting measured intensities for vertical and horizontal observation when excited with horizontal polarization for excitation. I_{HV} and I_{HH} intensity components are equal. The inset shows a predicted distribution of excited molecules. The photograph inset shows a cuvette viewed through vertical (left) and horizontal (right) polarizer. In both photographs, the emission brightness seen through the vertical and horizontal polarizer is practically the same.

Next, we repeated previous measurements but now with R6G in glycerol. The viscosity of glycerol is very high and expected anisotropy should be close to 0.4. In Figure 6-10.2, we present measured and corrected I_{VV} and I_{VH} intensities for R6G in glycerol. As expected, they are significantly different. The schematic inset represents the distribution of excited molecules. A photograph's inset shows two photographs of emission viewed through a vertical and horizontal polarizer. The fluorescence brightness in two cuvettes is significantly different. For comparison in Figure 6-10.2 (right), we are presenting measured intensities for R6G in glycerol but now with horizontal excitation I_{HV} and I_{HH}. Measured intensities as expected from molecular distribution in the excited state (schematic inset) are equal. The photographs inset show two cuvettes viewed through the vertical and horizontal polarizer. The observed fluorescence brightness is equal in both cuvettes.

In Figure 6-10.3, we are presenting normalized emission spectra (MA conditions) and measured anisotropies of R6G in water (left) and in glycerol (right).

(Continued)

EXPERIMENT 6-10 (CONTINUED) FLUORESCENCE EMISSION ANISOTROPY—DEPENDENCE ON THE SOLVENT VISCOSITY

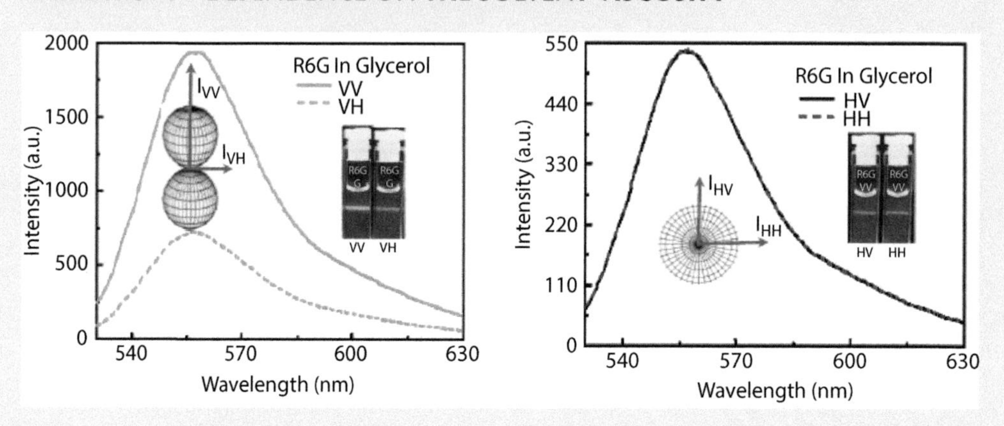

FIGURE 6-10.2 Left: Corrected fluorescence spectra of R6G emission in glycerol measured with vertical excitation and vertical and horizontal observation, respectively. The photographs in the inset were taken through a vertical and horizontal polarizer when the sample was excited with vertically polarized light. Right: Corrected fluorescence spectra of R6G emission in glycerol measured with horizontal excitation and vertical and horizontal observation, respectively. The I_{HV} and I_{HH} intensity components are equal (as shown in schematic inset for horizontal excitation the distribution leads to equal I_{HV} and I_{HH} intensities). The photographs inset show equally bright fluorescence.

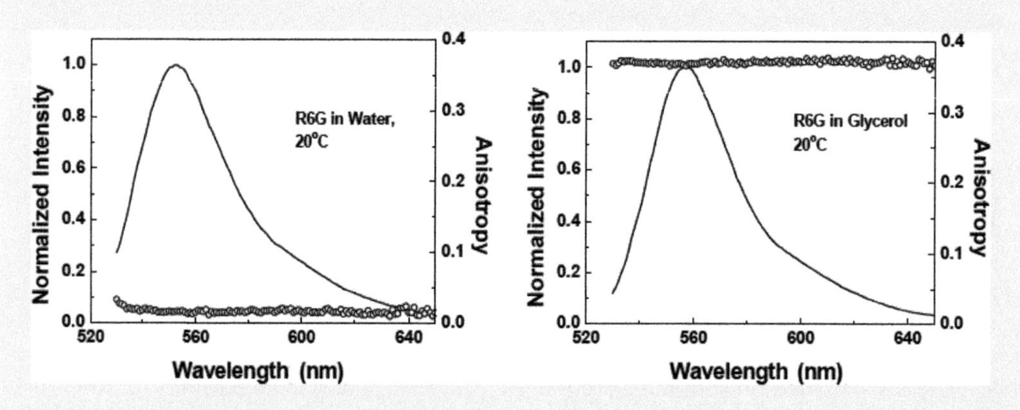

FIGURE 6-10.3 Fluorescence emission anisotropy of R6G in water (left) and glycerol (right). Fluorescence spectra were measured at magic angle conditions.

CONCLUSIONS

A solvent viscosity is an important factor controlling molecular diffusion. The speed of rotational diffusion dictates how high fluorescence anisotropy we will observe. Most fluorophores have lifetimes in the range of a few nanoseconds and in low viscosity solvents can freely move/rotate (correlation times for typical fluorophores in low viscosity solvents are in picoseconds). Fast rotational motion of fluorophore molecules in a low viscosity solvent will result in randomization

(Continued)

EXPERIMENT 6-10 (CONTINUED) FLUORESCENCE EMISSION ANISOTROPY—DEPENDENCE ON THE SOLVENT VISCOSITY

of emission transition moment orientations induced by excitation photoselection. Therefore, the observed fluorescence is depolarized and close to 0. In contrast, in a high viscosity solvent like glycerol, fluorophore molecules during the lifetime will move very little and observed fluorescence will be highly polarized reflecting transition moments distribution after excitation (photoselection). Some fluorophore molecules are used for viscosity (or microviscosity) measurements because measured anisotropies carry the information about an environment/solvent viscosity.

EXPERIMENT 6-11 DEPENDENCE OF FLUORESCENCE EMISSION AND EXCITATION ANISOTROPY ON THE SOLVENT VISCOSITY

Keywords: fluorescence, anisotropy, solvent viscosity.

In the previous experiment, we showed that measured steady-state anisotropy depends on solvent viscosity. There are many factors that may affect the rotational diffusion of the excited fluorophore. Solution temperature is one important factor that will change rotational diffusion and solvent viscosity so, it is important to control the temperature when comparing different solvents. Various solvents will affect the fluorophore differently and typically have different viscosity. We can create a desirable viscosity by mixing the high viscosity solvent and low viscosity solvent (like glycerol and water or ethanol) in a proper ratio. This can be a very convenient approach if we have two solvents in which the fluorescence properties of the fluorophore are similar especially fluorescence lifetimes are comparable (the same). In this experiment, we will measure R6G in different viscosity systems and present a set of experiments demonstrating how these factors change the observed polarization. We will consider both fluorescence emission anisotropy and fluorescence excitation anisotropy.

MATERIALS

- Water, ethyl alcohol (EtOH), propylene glycol (PG), glycerol
- Rhodamine 6G (R6G)
- Glass or plastic vials
- 1 cm × 1 cm cuvettes

EQUIPMENT

- Spectrophotometer
- Spectrofluorometer

METHODS

1. Prepare stock-solutions of R6G in EtOH.
2. Prepare samples of R6G in water, EtOH, propylene glycol, glycerol, and EtOH/glycerol mixture (in our case 7:3 but it could be a little different). Measure the absorption spectra. Remember, optical densities not exceeding 0.1.

(*Continued*)

EXPERIMENT 6-11 (CONTINUED) DEPENDENCE OF FLUORESCENCE EMISSION AND EXCITATION ANISOTROPY ON THE SOLVENT VISCOSITY

3. Use a long-wavelength excitation close to the point where the emission starts (~500 nm in the case of R6G) and measure the fluorescence spectra of each solution at magic angle (MA) conditions.
4. As described in earlier experiments, use horizontal excitation and measure the *G*-factor of your system (see Experiments 6-5 through 6-7).
5. Set the excitation polarization to vertical and excitation wavelength to about 500 nm (or the wavelength you want to use for your fluorophore). Measure I_{VV} and I_{VH} intensity components.
6. Correct I_{VH} for the wavelength-dependent *G*-factor and calculate emission anisotropy.
7. Set observation wavelength above the point where the absorption ends (~580 nm for R6G), set excitation polarization to vertical and measure the excitation spectra (I_{VV} and I_{VH}) while scanning the excitation wavelength for vertical and horizontal observation polarization. Make sure you do not scan through the observation.
8. Correct the I_{VH} component for *G*-factor (remember that excitation anisotropy correction is a single value for observation wavelength—580 nm in our case—multiply I_{VH} component by a single value of *G*-factor measured at 580 nm observation). Calculate the excitation anisotropy.

RESULTS

As in previous experiments, we will start by measuring a *G*-factor. *Note: It is a good custom before any anisotropy/polarization measurement to check the G-factor even if we know it.* We should measure the *G*-factor in the entire emission range and for excitation anisotropy will select a value for an appropriate wavelength (580 nm in our case). For the *G*-factor measurement, we can use any of our R6G solutions, rotate excitation to horizontal and measure I_{HV} and I_{HH} and calculate the *G*-factor (for details see Experiment 6-5). To measure anisotropy for each solvent, we set the vertical polarizer orientation and excitation wavelength to 510 nm and measured I_{VV} and I_{VH} in the emission range (520–650 nm). To correct for *G*-factor, we multiplied I_{VH} at each wavelength by the corresponding *G*-factor values (at each wavelength) and then calculated sample anisotropies. To measure excitation anisotropies, we set the observation at 580 nm and scanned I_{VV} and I_{VH} components within the absorption range (300–550 nm in our case). The I_{VH} components have been corrected for the *G*-factor at 580 nm and excitation anisotropies were calculated.

In Figure 6-11.1, we are presenting normalized absorptions and emissions spectra of R6G, and the corresponding emission and excitation anisotropies for different solvents (water, EtOH, PG, and glycerol) at room temperature ~20°C. The viscosity of water and EtOH are low and the measured anisotropies (excitation and emission) are flat and close to zero. It is not a surprise to see the small variation (noise) of excitation anisotropy measured in the 380–430 nm range where the absorption of R6G is very low. Going to PG, the viscosity increases and in PG we are measuring a higher and

(*Continued*)

EXPERIMENT 6-11 (CONTINUED) DEPENDENCE OF FLUORESCENCE EMISSION AND EXCITATION ANISOTROPY ON THE SOLVENT VISCOSITY

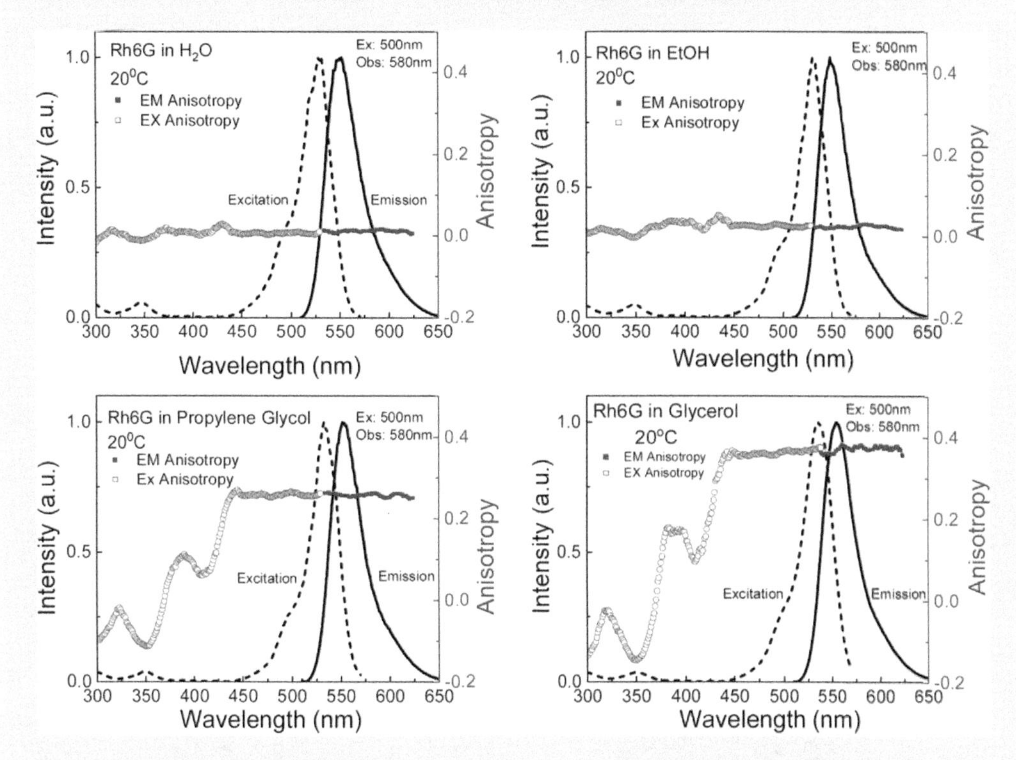

FIGURE 6-11.1 Normalized absorptions and emission spectra with emission and excitation anisotropies of R6G in water, EtOH, PG, and glycerol.

flat emission anisotropy. The excitation anisotropy changes dramatically. It is not flat anymore, at a longer wavelength, it increases and at a shorter wavelength, it decreases too much lower and even negative values. For glycerol that has the highest viscosity, the effect is even larger. We can clearly see that while emission anisotropy is always flat, the excitation anisotropy shows a distinct structured character. It is flat at long wavelength, quickly drops below 450 nm to slightly recover (increases) at 400 nm, and drops to about −0.15 for 350 nm.

One simple way to change viscosity is to mix two solvents (that mix well) like glycerol and EtOH. In Figure 6-11.2, we are presenting emission and excitation anisotropy as measured for glycerol, EtOH, and mixtures of 5:5 and 7:3 EtOH to glycerol.

Another way to change solvent viscosity is to change the temperature. The Brownian rotation strongly depends on temperature quickly increasing at a higher temperature. In Figure 6-11.3, we are presenting anisotropy measurements for R6G in glycerol at 0°C, 20°C, 40°C, and 60°C. The drop of anisotropy in the 60° range is significant.

(Continued)

EXPERIMENT 6-11 (CONTINUED) DEPENDENCE OF FLUORESCENCE EMISSION AND EXCITATION ANISOTROPY ON THE SOLVENT VISCOSITY

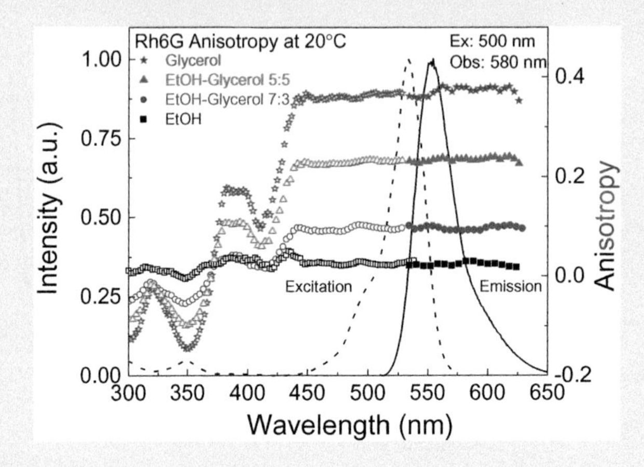

FIGURE 6-11.2 Normalized absorptions and emission spectra with emission and excitation anisotropies of R6G in EtOH, glycerol, 5:5 EtOH:glycerol, and 7:3 EtOH:glycerol.

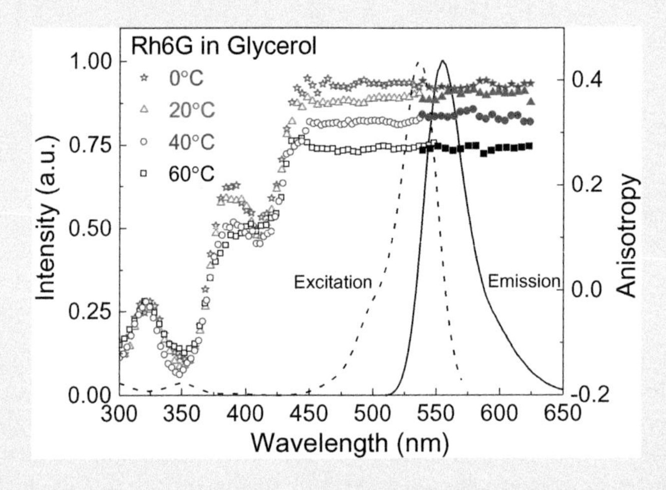

FIGURE 6-11.3 Normalized absorptions and emission spectra with emission and excitation anisotropies of R6G in glycerol measured at different temperatures.

CONCLUSIONS

The measured anisotropy strongly depends on solvent viscosity. In the selected example (R6G), the fluorescence lifetime does not change significantly between different solvents and the observed change is mainly due to a significant change in viscosity. For low viscosity solvents like water or EtOH, the measured anisotropy is very low—close to zero. But as the solvent viscosity increases, the emission anisotropy increases (or decreases in the case of negative anisotropy). We can change the viscosity of a solution by mixing solvents or just by changing the temperature. In many cases, the anisotropy dependence on temperature is significant and it is very important to precisely control sample temperature during the experiment.

EXPERIMENT 6-12 FLUORESCENCE ANISOTROPY OF MULTIPLE EMITTING SPECIES

Keywords: fluorescence, anisotropy, a mixture of fluorophores, anisotropy additivity.

Although emission polarization and emission anisotropy describe equivalently, the measured state of polarization of fluorescence (more precisely distribution of emitting transition moments) and often are used alternatively, the concept of emission anisotropy is more general. Polarization describes light (an electromagnetic wave) incoming to the detector viewed from the direction of the detector whereas the anisotropy describes an emission field of fluorescence in all directions. In the case of a mixture of fluorophores with different individual polarizations/anisotropies, the effective/resulting polarization of the emission of the mixture is rather complex when expressed in emission polarization. However while expressed in emission anisotropy, it becomes much simpler. In the latter case, as we discussed in Chapter 1, the total fluorescence anisotropy of the mixture is the simple sum of individual anisotropies (r_i) weighted with their fractional intensities (f_i):

$$r = f_1 r_1 + f_2 r_2 + \ldots$$

In this experiment, we will show the fluorescence anisotropy measurements for the mixture of Rhodamine B and Erythrosine B. We will compare the total measured anisotropy with the calculated from individual fraction and anisotropies.

MATERIALS

- Rhodamine B (RhB)
- Erythrosine B (ErB)
- Water and ethanol (EtOH)
- Glass or plastic vials
- 1 cm × 1 cm cuvettes

EQUIPMENT

- Spectrophotometer
- Spectrofluorometer

METHODS

1. Prepare stock-solutions of RhB and ErB in EtOH.
2. Make individual water solutions of RhB and ErB with optical densities not exceeding 0.1 for ErB and 0.01 for RhB (RhB is a brighter fluorophore then ErB).
3. Measure fluorescence spectra of individual fluorophores at magic angle (MA) conditions.
4. Measure the fluorescence anisotropies of individual fluorophores.
5. Make a mixture of both fluorophores and measure the fluorescence spectrum at MA conditions.
6. Use a simple deconvolution procedure to resolve the fluorescence spectrum of the mixture into two individual spectra of RhB and ErB as their linear combination.
7. Measure the fluorescence anisotropy of the mixture.

(*Continued*)

EXPERIMENT 6-12 (CONTINUED) FLUORESCENCE ANISOTROPY OF MULTIPLE EMITTING SPECIES

RESULTS

We intended to make a mixture of RhB and ErB for which the fluorescence signals from both components will be comparable for a given excitation wavelength (this will obviously correspond to different molar concentrations since the quantum yield of these dyes and absorptions are different). In practice, we would select the excitation wavelength to be favorable (larger absorption) for the lower quantum yield ErB. We may adjust concentration and excitation wavelength to measure very close the emission intensities (measured at MA) for both dyes. When making a mixture of two fluorophores, we can easily make a perfect 50:50 ratio. Later, we will show how to solve an exact mixture composition. First, we will measure the absorption spectra of all 3 (RhB, ErB, and RhB/ErB mix) solutions to make sure that the absorption is below 0.1 (spectra not showed). In the case of RhB and ErB, both dyes can be effectively excited with about 490 nm excitation giving a comparable signal (you may adjust the excitation wavelength to equalize signals). Fluorescence anisotropies of RhB and Erb are distinctively different. In Figure 6-12.1, we present the corrected I_{VV} and I_{VH} emission intensity components for RhB (left) and ErB (right) with 490 nm excitations. As the emission signals are comparable, the ratio I_{VV}/I_{VH} are different clearly indicating different polarizations.

In Figure 6-12.2, we present the corrected emission spectra measured under MA conditions with marked anisotropies calculated from curves in Figure 6-12.1. The emission spectra are shifted and measured anisotropies are distinct.

Now we measured emission intensity components (I_{VV} and I_{VH}) for the mixture. As presented in Figure 6-12.3. The spectral profiles (I_{VV} and I_{VH}) clearly indicate that both spectra are a composition of RhB and ErB but the ratio I_{VV}/I_{VH} clearly changes across the spectral profile being high at the short wavelength and decreasing toward the longer wavelength.

Next, we measured the emission spectrum of the mixture using MA conditions. *Note: It is also possible to recalculate the MA intensity profile from the measured polarized intensity components* ($I_{MA} = (I_{VV} + 2I_{VH})/3$). The measured MA condition emission spectrum of the mixture is presented in Figure 6-12.4 as a solid line. The solid line well overlays with the open

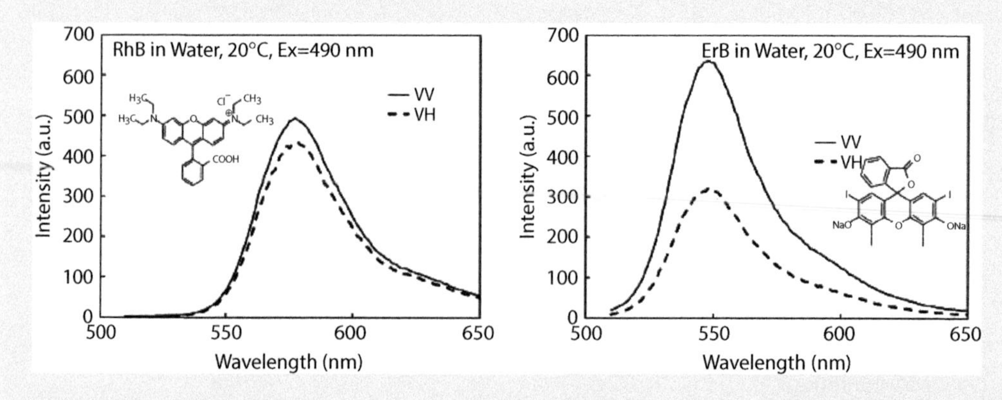

FIGURE 6-12.1 Measurements of fluorescence intensity components for individual fluorophores. The I_{VH} components were corrected for the *G*-factor of the instrument (see description of *G*-factor measurements).

(Continued)

EXPERIMENT 6-12 (CONTINUED) FLUORESCENCE ANISOTROPY OF MULTIPLE EMITTING SPECIES

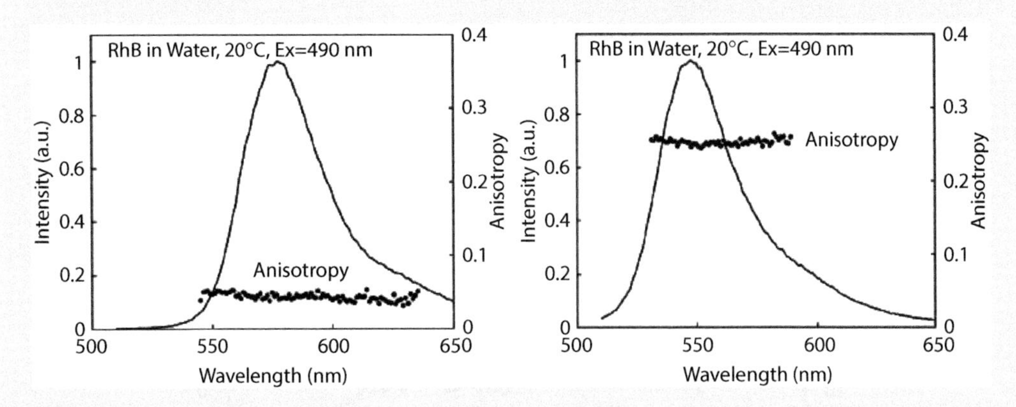

FIGURE 6-12.2 Fluorescence spectra of RhB (left) and ErB (right) measured at MA conditions and fluorescence anisotropies of individual fluorophores as calculated from the corrected I_{VV} and I_{VH} intensities in Figure 6-12.1.

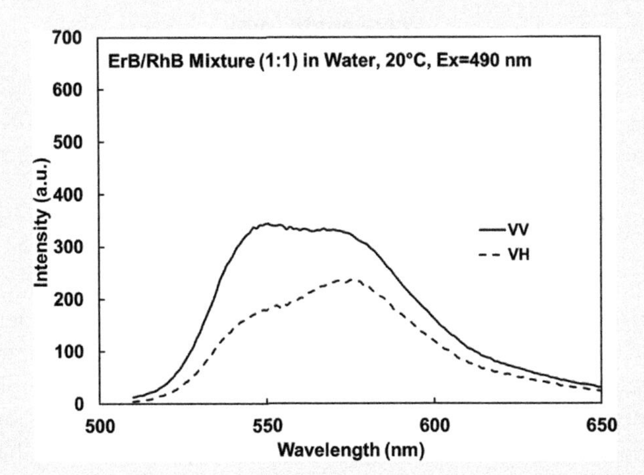

FIGURE 6-12.3 Emission spectra of RhB and ErB mixture measured with vertical (solid) and horizontal (dashed) observation (I_{VV} and I_{VH}). I_{VH} component was corrected for G-factor.

circles that have been fitted using a linear combination of emission spectra of RhB and ErB presented in Figure 6-12.2. Recovered emission spectra of RhB and ErB are presented as two solid bell-shaped peaks.

To calculate the expected anisotropy from the mixture at any given wavelength, we can use values for recovered emission intensity of RhB and ErB at any given wavelength and calculate the expected anisotropy as:

$$r(\lambda) = \frac{I_{RhB}(\lambda) \cdot r_{RhB}(\lambda) + I_{ErB}(\lambda) \cdot r_{ErB}(\lambda)}{I_{RhB}(\lambda) + I_{ErB}(\lambda)}$$

(Continued)

EXPERIMENT 6-12 (CONTINUED) FLUORESCENCE ANISOTROPY OF MULTIPLE EMITTING SPECIES

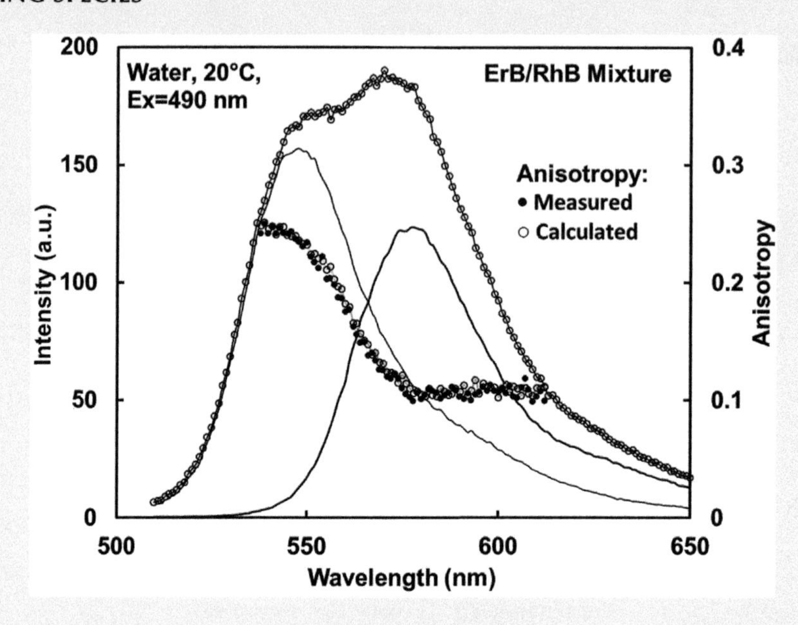

FIGURE 6-12.4 Fluorescence spectrum of the mixture of RhB and ErB and deconvoluted emission spectra of individual fluorophores (using spectra from Figure 6-12.2). The measured anisotropies are marked as closed (solid) circles and calculated from individual fractions as open circles.

where $I_{RhB}(\lambda)$ and $I_{ErB}(\lambda)$ are recovered in Figure 6-12.4 intensity components at any given wavelength λ for RhB and ErB, respectively and $r_{RhB}(\lambda)$ and $r_{ErB}(\lambda)$ are corresponding anisotropies at the wavelength λ for RhB and ErB, respectively as shown in Figure 6-12.2. Calculated and measured anisotropies are presented in Figure 6-12.4 as open and closed circles, respectively. Both anisotropies are very well overlapping indicating that the anisotropy is an additive parameter.

CONCLUSIONS

Emission anisotropy is an additive parameter. For a typical mixture of non-interacting molecules with different anisotropies, the observed anisotropy will be a simple sum of each individual anisotropies weighted by their emission intensity contributions.

Note: Not many fluorophores mix without any interactions. Some fluorophores can quench each other. For example, tryptophan is a known static quencher. Therefore, it is important to deconvolute the spectrum of the mixture into individual spectra of mixed components. A good agreement between theoretically predicted emission and measured spectrum of the mixture proves the absence of any unexpected fluorescence (from an eventual complex), also pre-assumed fractions do not have to be exact.

EXPERIMENT 6-13 DEPENDENCE OF FLUORESCENCE ANISOTROPY ON TEMPERATURE IN VISCOUS SOLUTIONS

Keywords: fluorescence, anisotropy, solvent viscosity, propylene glycol.

Viscosities of some solvents strongly depend on temperature. It is difficult to pour glycerol from a bottle in room temperature but when it is heated it flows smoothly. Propylene glycol with the viscosity of about 60 cP at room temperature increases its viscosity almost three-fold at 0°C. In such a modest change of temperature, many fluorophores do not change lifetimes significantly. Such fluorophores (according to the Perrin equation in introduction) should substantially change their fluorescence anisotropy with temperature. We will demonstrate this for both emission and excitation anisotropies of Rhodamine 110.

MATERIALS

- Propylene glycol (PG)
- Rhodamine 110 (R110)
- Glass or plastic vials
- 1 cm × 1 cm cuvettes
- Glass filters (long pass 495 and 520 nm)

EQUIPMENT

- Spectrophotometer
- Spectrofluorometer equipped with a temperature-controlled cell

METHODS

1. Prepare stock-solutions of R110 in EtOH.
2. Prepare the sample of R110 in PG with an optical density not exceeding 0.1, see Figure 6-13.1.
3. Measure fluorescence emission and excitation spectra at magic angle (MA) conditions, see Figure 6-13.2.
4. Measure the fluorescence emission anisotropy of R110 in PG with the excitation wavelength of 490 nm at various temperatures, from 0°C to 80°C. For this, you need to measure the VV and VH components. Apply the *G*-factor correction (from measured HV and HH components). *Note: G-factor should not depend on solution temperature so a one-time evaluation of the G-factor should be sufficient.*
5. Measure fluorescence excitation anisotropy with the observation at 530 nm. Use the *G*-factor for 530 nm.

RESULTS

Make sure that the prepared solutions are homogeneous. Figure 6-13.1 shows the absorption spectrum of R110 in PG at 20°C. At the shorter wavelength range, below 350 nm, appears another absorption band representing a transition to a higher electronic state. Often these transitions are orthogonal (or forming a significant angle) to the $S_0 \rightarrow S_1$ transition direction, in this case around 500 nm. The fluorescence anisotropy resulted from the excitation to a higher electronic state can be very different, frequently negative.

(*Continued*)

EXPERIMENT 6-13 (CONTINUED) DEPENDENCE OF FLUORESCENCE ANISOTROPY ON TEMPERATURE IN VISCOUS SOLUTIONS

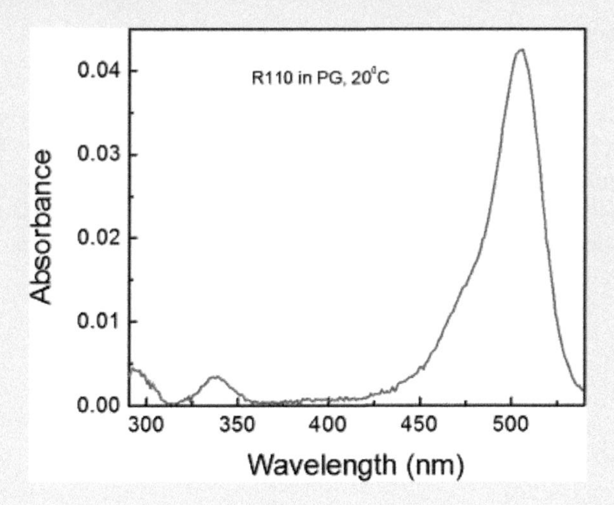

FIGURE 6-13.1 Absorption spectrum of R110 in PG.

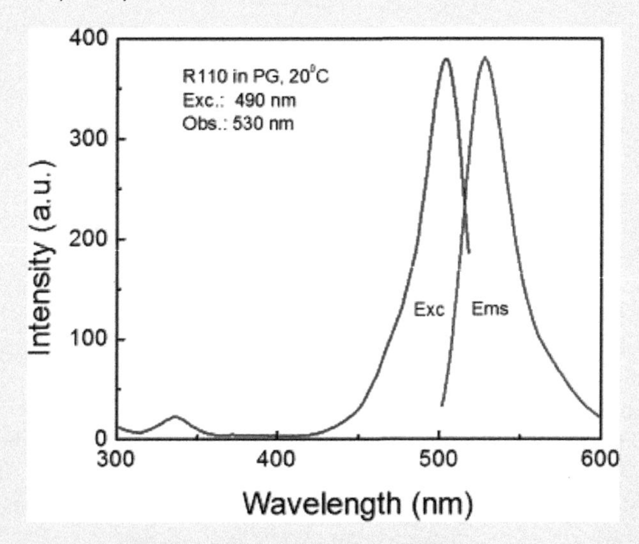

FIGURE 6-13.2 Fluorescence emission and excitation spectra of R110 in PG.

The excitation spectrum (Figure 6-13.2) confirms the presence of the excitation to the higher electronic state.

Next, we measured the emission anisotropies at different temperatures with the excitation wavelength of 490 nm and an observation wavelength of 530 nm. In addition, on the emission line, we used an LP 495 filter on the observation to attenuate (limit) a potential leak of the excitation to the observation. *Note: When measuring polarization we need to take special effort to make sure there is no excitation leaking to the detector. For vertical observation, the scattering is much stronger than for a horizontal or even magic angle.* Measured intensity components VV and VH for 0°C and 80°C are shown in Figure 6-13.3. The VH component has been

(Continued)

EXPERIMENT 6-13 (CONTINUED) DEPENDENCE OF FLUORESCENCE ANISOTROPY ON TEMPERATURE IN VISCOUS SOLUTIONS

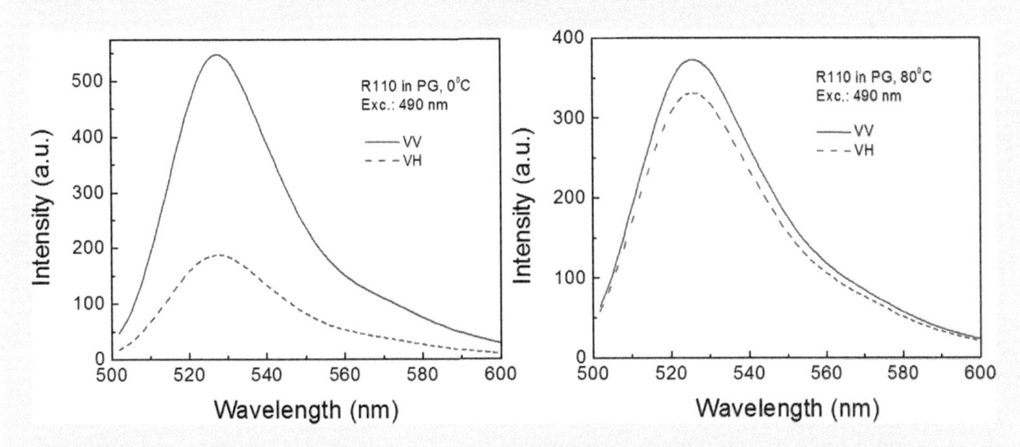

FIGURE 6-13.3 Polarized components of R110 fluorescence in PG at 0° and 80°C. VH components have been corrected for G-factor.

corrected for G-factor. To evaluate the G-factor, we can measure HV and HH components (recommended with all anisotropy measurements). Clearly, the ratio of VV/VH components strongly depends on the temperature. If fluorescence anisotropy would be 0, the VV and VH components will be identical. For the anisotropy of 0.4, the ratio VV/VH will be 3.

Using the G-factor corrected polarized components VV and VH calculate the emission anisotropies of R110 in PG at various temperatures (see right portion of Figure 6-13.4). The emission anisotropy changes from about 0.35 to 0.05, which is a lot.

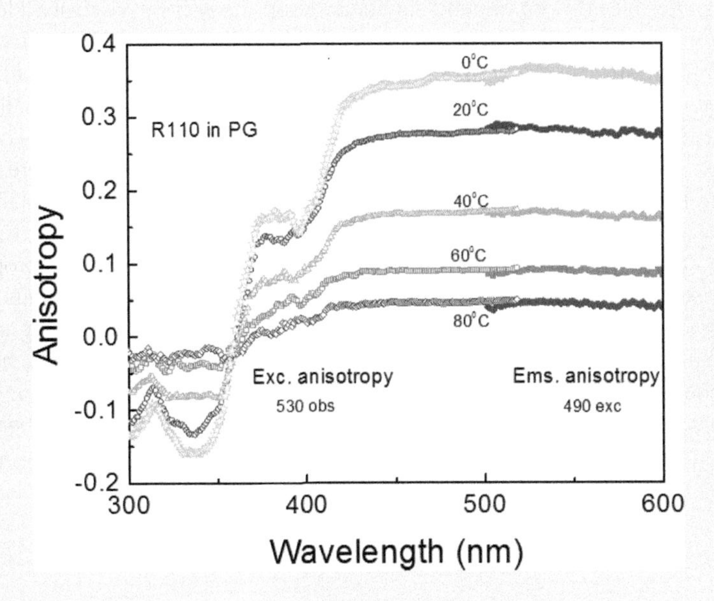

FIGURE 6-13.4 Fluorescence excitation and emission anisotropies of R110 in PG at various temperatures.

(Continued)

EXPERIMENT 6-13 (CONTINUED) DEPENDENCE OF FLUORESCENCE ANISOTROPY ON TEMPERATURE IN VISCOUS SOLUTIONS

Figure 6-13.4 shows measured emission and excitation anisotropies at different temperatures. For the excitation anisotropy, we used an observation at 530 nm and a long-pass filter LP 520 on the observation. To calculate excitation anisotropies, we used the *G*-factor value for 530 nm to correct the VH component.

The emission anisotropy quickly decreases with the temperature dropping from 0.35 at 0°C to 0.05 at 80°C. The excitation anisotropies at the long-wavelength range decrease with temperature (in the same way as emission anisotropy), while the short wavelengths (negative) anisotropies increase.

CONCLUSIONS

Strong changes of the fluorescence anisotropies with temperature can be expected only for fluorophores which lifetimes are not sensitive to temperature. Often fluorophores lifetimes decrease with temperature which compensates the solvent viscosity change and observed anisotropy changes only modestly. In the case of R110, the change of fluorescence anisotropy as a function of temperature is significant. This molecule in glycerol can be used as a temperature sensor. You are encouraged to plot the dependence of r versus T and estimate the range and accuracy of anisotropy-based temperature measurements.

EXPERIMENT 6-14 ANISOTROPY-BASED VISCOSITY SENSING

Keywords: fluorescence, anisotropy, solvent viscosity.

Lifetimes of some dyes do not depend significantly on the solvent viscosity. However, free rotation of a molecule in the solution depends on the solvent viscosity (high viscosity solvents induce much stronger the dragging force slowing down molecular rotation). These dyes can be conveniently used to detect/monitor solvent viscosity. This is especially useful for detecting viscosity in a microenvironment, e.g., cells or cellular compartments/components. The bases for such viscosity sensing emerge from the Perrin equation where the relative change of the fluorescence anisotropy is proportional to the ratio of molecular correlation time (speed of rotation) to the fluorescence lifetime. With the constant lifetime, the measured fluorescence anisotropy depends on the molecule correlation time that is proportional to the solvent viscosity. As the sizes of fluorophores are small, the measurements refer rather to the local microviscosity, not a typical viscosity that can be measured using a classic viscometer. The microviscosity in homogeneous solvents is equal to the viscosity, however in a non-homogeneous environment, it could be different in different locations. For example, in the cells, the viscosity will be different in membranes, in mitochondria, or other organelles.

(*Continued*)

EXPERIMENT 6-14 (CONTINUED) ANISOTROPY-BASED VISCOSITY SENSING

The fluorescence-based viscosity sensing can apply to both micro and macro objects. In this experiment, we will demonstrate how to calibrate and apply an anisotropy-based viscosity sensing using Rhodamine 110 as a fluorophore.

MATERIALS

- Ethanol (EtOH) and glycerol
- Rhodamine 110 (R110)
- Glass or plastic vials
- 1 cm × 1 cm cuvettes

EQUIPMENT

- Spectrophotometer
- Spectrofluorometer

METHODS

1. Prepare stock-solutions of R110 in EtOH.
2. Prepare mixtures of glycerol/ethanol (v/v) with various fractions of glycerol.
3. Estimate viscosities of prepared mixtures using viscosity data from the literature (Figure 6-14.1). The viscosity of a solvent can be precisely measured with the viscometer. A glycerol/EtOH mixture is a uniform solution and microviscosity would directly correspond to macroviscosity.
4. Prepare samples of R110 in various glycerol/ethanol mixtures with the absorption of about 0.05 in all samples. Measure absorption and fluorescence spectra, see Figure 6-14.2.
5. Measure VV and VH components for all samples. Correct the VH component for *G*-factor (for this you need to measure HV and HH components), see Figure 6-14.3.
6. Plot the emission anisotropies and estimate anisotropy values at 530 nm, see Table 6-14.1.
7. Construct the dependence of anisotropy on the viscosity, see Figure 6-14.5.
8. Look at Experiment 6-13 and estimate the viscosity of propylene glycol using the value of a measured anisotropy. Compare with viscosities of propylene glycol from the literature.

RESULTS

First, we will prepare solutions with known viscosities. For this, we will use mixtures of glycerol/ethanol with different fractions of glycerol. Viscosities of such mixtures are well described and reported in the literature, see the inset in Figure 6-14.1. The dependence presented in Figure 6-14.1 will allow us to estimate the viscosity of any glycerol/ethanol mixture if we know the glycerol fraction.

(*Continued*)

EXPERIMENT 6-14 (CONTINUED) ANISOTROPY-BASED VISCOSITY SENSING

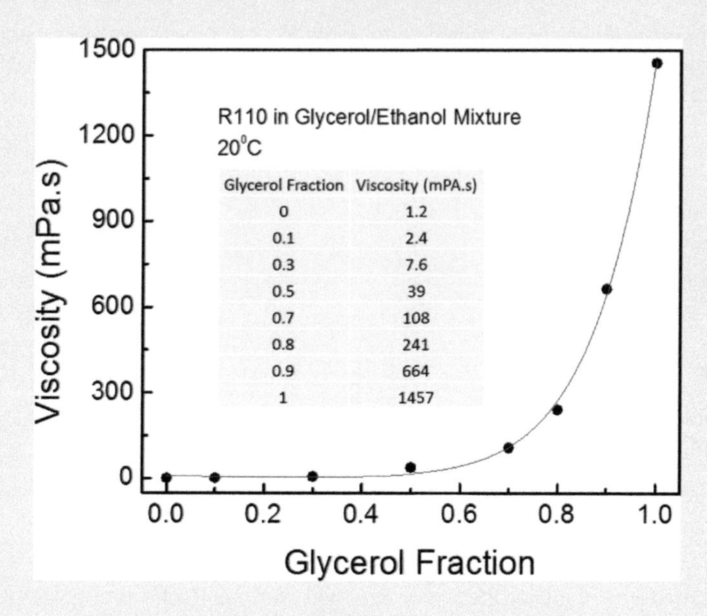

FIGURE 6-14.1 Dependence of viscosity on the glycerol fraction in mixtures of glycerol/ethanol (v/v) from the literature.

We prepared mixtures as described in Table 6-14.1, but you are free to prepare different glycerol fractions. Figure 6-14.2 shows the absorption and fluorescence spectra of solutions with extreme viscosities (ethanol and glycerol) measured with magic angle conditions.

TABLE 6-14.1 Viscosities of Solutions Used in this Experiment

Glycerol Fraction	**Viscosity (mPA·s)**
0	1.2
0.0625	1.85
0.125	2.854
0.25	5.986
0.375	15.48
0.5	39.12
0.6	70.08
0.7	107.87
0.8	241
1	1457

Next, measure the anisotropies of all samples. VV and VH components for extreme viscosities are shown in Figure 6-14.3. Emission anisotropies of all samples are shown in Figure 6-14.4.

(*Continued*)

EXPERIMENT 6-14 (CONTINUED) ANISOTROPY-BASED VISCOSITY SENSING

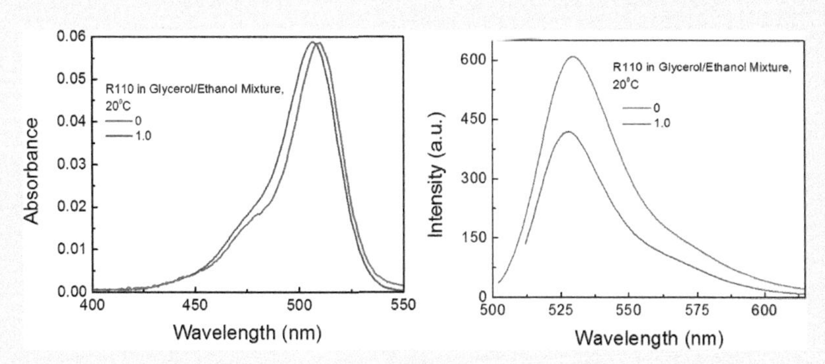

FIGURE 6-14.2 Absorption (left) and fluorescence (right) spectra of R110 in ethanol (green) and glycerol (violet and red) at 20°C.

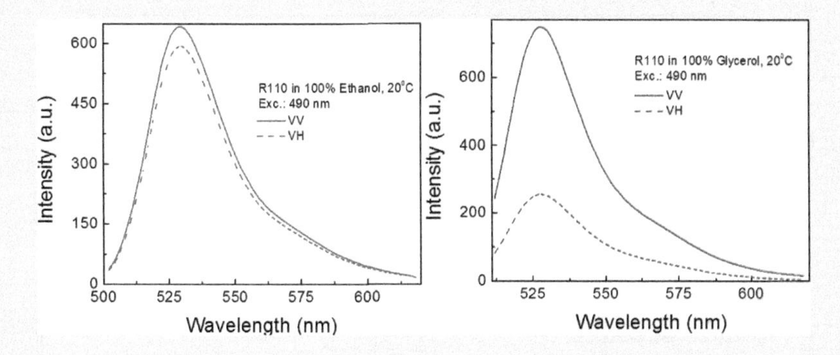

FIGURE 6-14.3 VV and VH components of R110 fluorescence emission in ethanol (left) and glycerol (right). The VH components were corrected for G-factor.

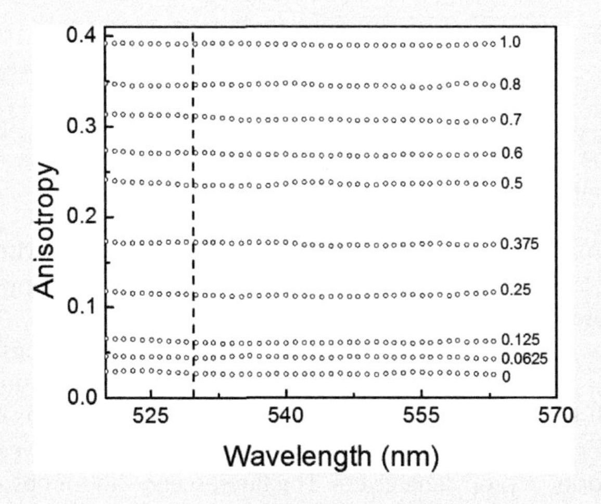

FIGURE 6-14.4 Fluorescence emission anisotropies of R110 in glycerol/ethanol mixtures at 20°C. The excitation was 490 nm. The vertical dashed line marks 530 nm.

(Continued)

EXPERIMENT 6-14 (CONTINUED) ANISOTROPY-BASED VISCOSITY SENSING

TABLE 6-14.2 Values of R110 Fluorescence Anisotropies at 530 nm in Various Glycerol/Ethanol Mixtures (from Figure 6-14.4)

Glycerol Fraction	Viscosity mPA·s	Anisotropy at 530 nm
0	1.2	0.027
0.0625	1.85	0.044
0.125	2.854	0.061
0.25	5.986	0.114
0.375	15.48	0.172
0.5	39.12	0.235
0.6	70.08	0.275
0.7	107.87	0.312
0.8	241	0.346
1	1457	0.390

The measured values of anisotropies at different mixtures are presented in Table 6-14.2 together with the viscosity values estimated from Figure 6-14.1 (literature data). The dependence of R110 fluorescence anisotropies on solvent viscosities is shown in Figure 6-14.5.

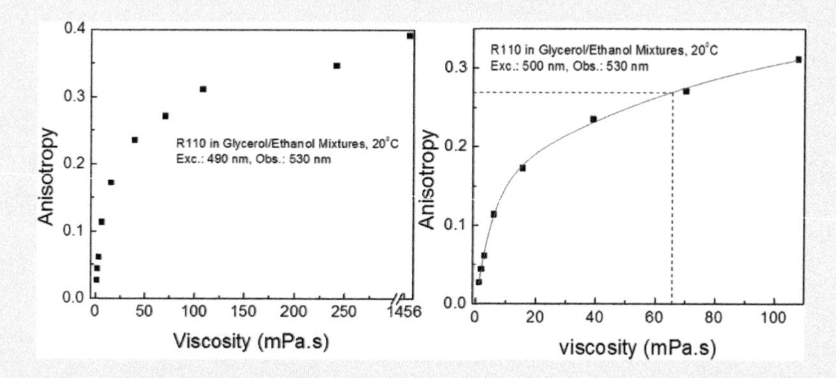

FIGURE 6-14.5 Dependence of R110 fluorescence anisotropy on the viscosity. The values of anisotropy were taken at 530 nm observation. Left: Expanded range of viscosity (scale 0–300 mPa·s), right: limited range (0–100 mPa·s).

Using the experimental dependence of measured anisotropy versus viscosity presented in the Figure 6-14.5 (right), a single point measurement of R110 anisotropy at 490 nm excitation and 530 nm observation will give a viscosity of any given solution.

Our strategy now is to recover the solvent viscosity from the previously measured fluorescence anisotropy of R110 in propylene glycol. Absorption and emission spectra of R110 in propylene glycol are similar to that of R110 in glycerol and ethanol. Also, fluorescence lifetimes do not differ much and we could use a calibration curve for glycerol/EtOH mixture to estimate the viscosity of propylene glycol. The fluorescence anisotropy of R110 in propylene glycol at 20°C is 0.27 (see the Experiment 6-13, *Dependence of Fluorescence Anisotropy on Temperature in Viscous Solutions*). Using the graph on Figure 6-14.5 (right), one can find

(Continued)

EXPERIMENT 6-14 (CONTINUED) ANISOTROPY-BASED VISCOSITY SENSING

that this anisotropy value corresponds to the viscosity range of about 60 mPa·s, whereas the literature value of propylene glycol viscosity at 20°C is about 60 mPa·s.

CONCLUSIONS

In some specific cases when fluorescence lifetime does not depend on the solvent and/or solvent viscosity, we can use a simple measurement of anisotropy to quickly determine viscosity. This could be useful for a local viscosity distribution and especially useful for a micro-environment characterization where viscosity is not homogeneous. For example, anisotropy-based cell images can reveal local viscosities in organelles.

Note: As we will demonstrate later a much more precise piece of information can be recovered when one is able to monitor probe fluorescence lifetime and time-resolved anisotropy.

EXPERIMENT 6-15 ANISOTROPY-BASED ASSAYS: ACRIDINE ORANGE BINDING TO DNA

Keywords: fluorescence, anisotropy, solvent viscosity.

Binding of dyes to macromolecules (proteins, DNA) have been demonstrated already in Chapter 5. In many cases, the fluorescence efficiency of bound fluorophores increases many folds. In such cases, the fluorescence intensity accurately represents binding kinetics. However, in the case when fluorescence intensity changes only modestly upon the binding, the intensity itself is not a good measure for the binding kinetics. Also, in many practical assays, samples are turbulent and not perfectly transparent leading to significant perturbations in measured intensities. A more accurate way will be to measure the parameter/response that would be independent of overall fluorescence intensity. Example of such a response could be fluorescence anisotropy or fluorescence lifetime for which the change is independent of total signal intensity. In this experiment, we will present an example whereupon binding the fluorescence intensity does not change significantly. The bound fluorophores are immobilized in the macromolecule matrix of DNA and cannot rotate freely showing a significant change in the measured fluorescence anisotropy. We will present anisotropy-based measurements of acridine orange (AO) when binding to DNA.

MATERIALS

- Acridine orange (AO), HBSS buffer (buffer)
- Calf thymus DNA (DNA)
- Glass vials
- Quartz cuvettes and pipettes

(Continued)

EXPERIMENT 6-15 (CONTINUED) ANISOTROPY-BASED ASSAYS: ACRIDINE ORANGE BINDING TO DNA

EQUIPMENT

- Spectrophotometer
- Spectrofluorometer

METHODS

1. Prepare a stock-solution of AO in HBSS buffer. Estimate the concentration of the AO using extinction coefficient 27,000 cm^{-1} M^{-1} at 490 nm. Measure the absorption of AO solution used in the experiment, see Figure 6-15.1.
2. Prepare a stock solution (about 0.5 mM) of DNA in HBSS buffer. Measure the absorption of the DNA stock solution in UV. For this, you may either use a 1 mm micro-cuvette or dilute the stock solution several folds, see Figure 6-15.2. Use an extinction coefficient of 13,500 cm^{-1} M^{-1}/base pair at 259 nm.
3. Prepare samples (about 7–8) of AO in HBSS buffer with various concentrations of DNA. Keep the concentration of AO constant in all samples. A convenient way for the preparation is to mix adequate amounts of the AO stock solution, DNA stock solution, and an adequate amount of buffer to keep the same total volume. In this case, the amount of AO stock solution will always be the same and amounts of DNA and buffer will vary, see the inset in Figure 6-15.7.
4. Measure the absorption and fluorescence spectra of extreme concentrations of DNA, see Figures 6-15.3 and 6-15.4.
5. Measure emission anisotropies of all samples. For this, you need to measure VV and VH components of fluorescence and correct the VH component for *G*-factor (the *G*-factor can be measured using horizontal sample excitation). Plot emission anisotropies, see Figure 6-15.6.
6. Read anisotropy values at 525 nm (see inset in Figure 6-15.7) and plot the dependence of fluorescence anisotropy as a function of DNA concentration.

RESULTS

We start with the preparation of the stock solution and its characterization. Figures 6-15.1 and 6-15.2 show measured absorptions for prepared solutions of AO and DNA.

Next, we will use the prepared series of samples with different concentrations of DNA, from 0 to about 0.3 mM and the fixed concentration of AO. Figure 6-15.3 shows the measured sample absorption for no DNA and with 0.309 mM DNA present. We can see a small shift in the absorption spectrum toward longer wavelengths and shows a modest hypochromic effect.

Keeping a constant AO concentration at about 2 μM, we measured the emission spectra for extreme DNA concentrations; see Figures 6-15.3 to 6-15.4.

The efficiency of the fluorescence emission, when bound to DNA, is about two-fold higher than in the buffer.

Next, we measured the emission anisotropies of all samples using 480 nm excitation. The VV and VH components for extreme DNA concentrations are shown in Figure 6-15.5. After corrections for *G*-factor, it is clear that fluorescence anisotropy is significantly higher at the presence of DNA. If the VV and VH component would be equal, the fluorescence anisotropy will be zero.

(*Continued*)

EXPERIMENT 6-15 (CONTINUED) ANISOTROPY-BASED ASSAYS: ACRIDINE ORANGE BINDING TO DNA

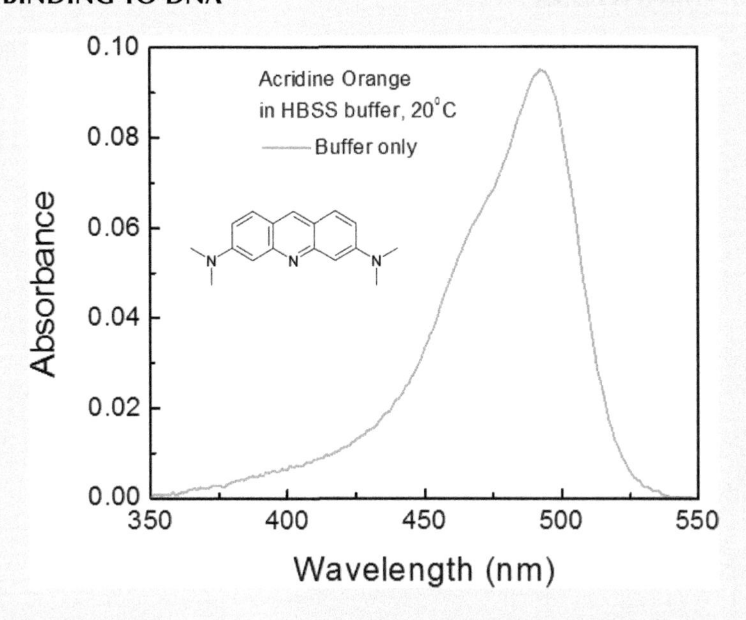

FIGURE 6-15.1 Absorption spectrum of AO in HBSS buffer. The stock solution of AO was diluted 50-fold. The concentration of AO (about 2 μM, the extinction coefficient of 43,000 cm^{-1} M^{-1} at 490 nm) was constant in all samples. The inset shows the structure of the AO.

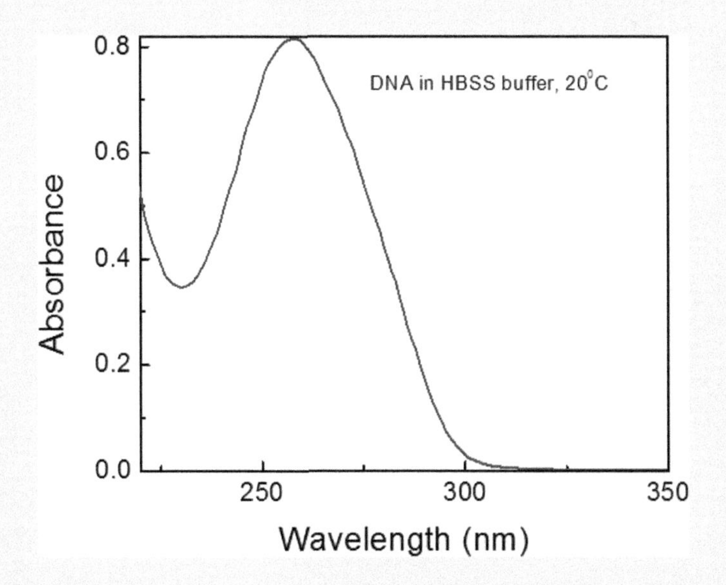

FIGURE 6-15.2 Absorption spectrum of DNA in HBSS buffer (stock solution diluted 10-fold). The calculated concentration of the DNA stock solution is 0.6 mM. We used the extinction coefficient by bp of 13,500 cm^{-1} M^{-1} at 259 nm). The DNA concentrations in the experiment varied from 0 mM to 0.31 mM.

(Continued)

EXPERIMENT 6-15 (CONTINUED) ANISOTROPY-BASED ASSAYS: ACRIDINE ORANGE BINDING TO DNA

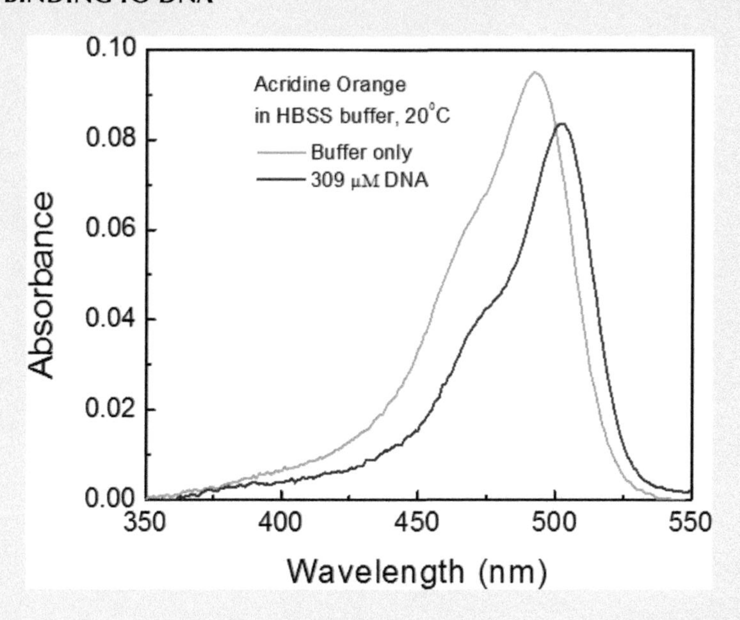

FIGURE 6-15.3 Absorption spectra of AO in the absence and presence of 309 μM of DNA.

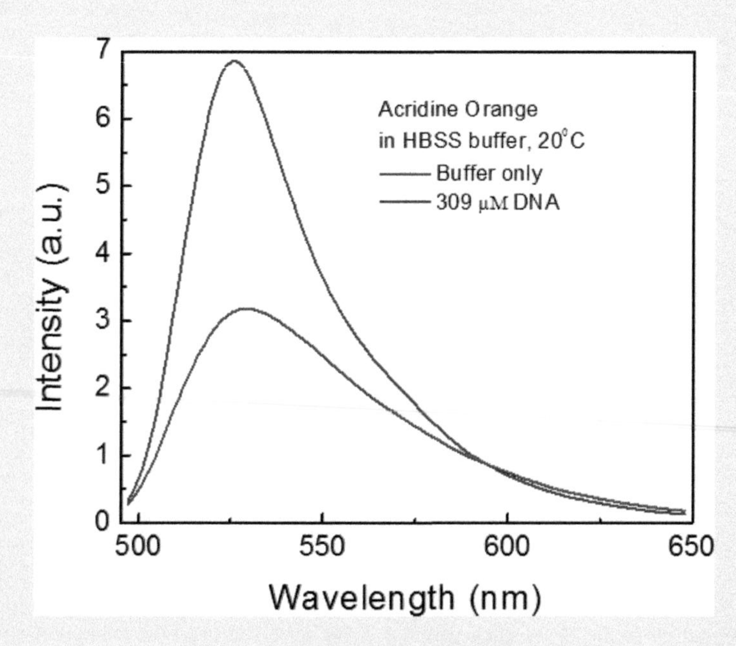

FIGURE 6-15.4 Fluorescence spectra of AO in the absence and presence of 309 μM of DNA. The spectra were corrected for the amount of absorbed light.

(Continued)

EXPERIMENT 6-15 (CONTINUED) ANISOTROPY-BASED ASSAYS: ACRIDINE ORANGE BINDING TO DNA

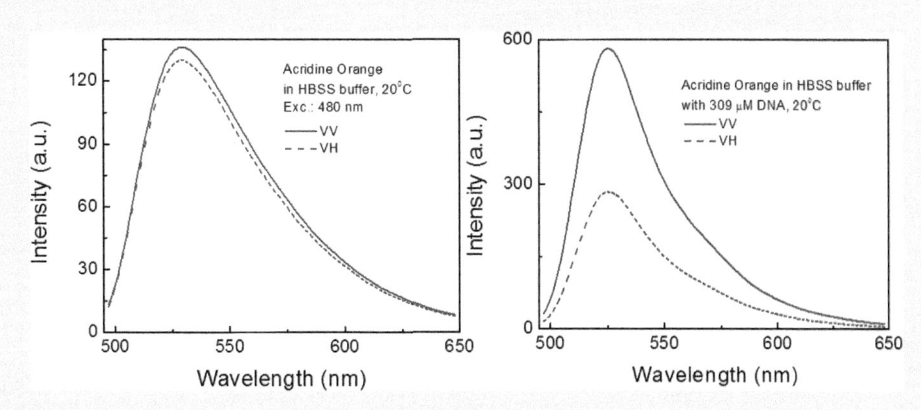

FIGURE 6-15.5 VV and VH components of AO fluorescence emission in the absence and presence of 309 μM of DNA. VH components were corrected for *G*-factor.

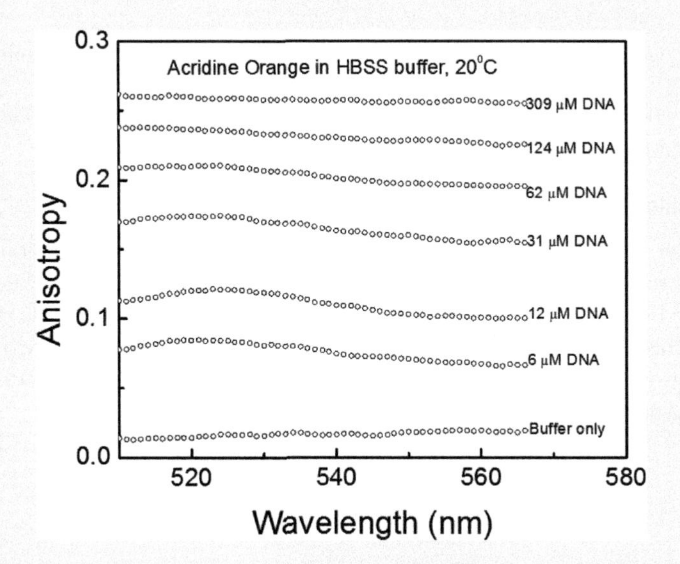

FIGURE 6-15.6 Fluorescence emission anisotropies of AO in the presence of various concentrations of DNA.

Next, we calculate and plot AO emission anisotropies for different concentrations of DNA, see Figure 6-15.6. Then, we read values of anisotropies at 525 nm observation, see inset in Figure 6-15.7.

Of course, it is also correct to measure the anisotropies at a single observation wavelength, usually at the maximum of the emission. Typically, we would select a small range of 10–20 nm from the anisotropy and use an average value. Also, in many cases, we could use just a bandpass filter covering a range over the emission spectrum and detect a signal for VV and VH components.

(Continued)

EXPERIMENT 6-15 (CONTINUED) ANISOTROPY-BASED ASSAYS: ACRIDINE ORANGE BINDING TO DNA

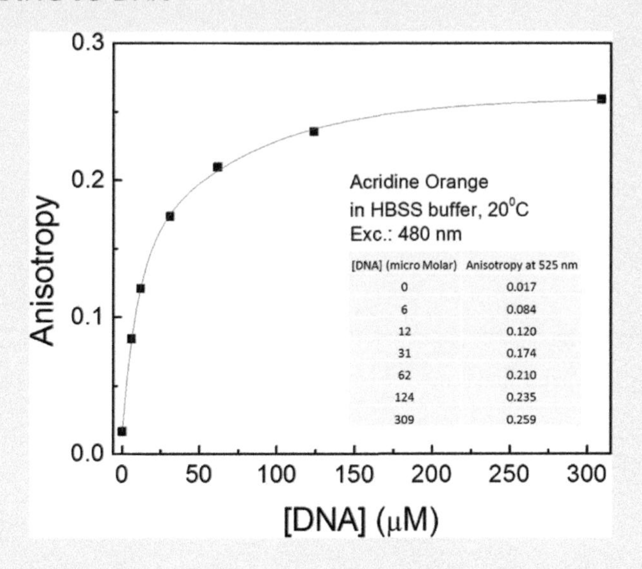

FIGURE 6-15.7 Dependence of the AO fluorescence anisotropy on the DNA concentration.

Finally, in Figure 6-15.7, we plot the dependence of fluorescence emission anisotropy as a function of DNA concentration.

CONCLUSIONS

Using the graph in Figure 6-15.7, one can easily find unknown concentrations of DNA by measuring the fluorescence anisotropy of AO. Anisotropy measurements are intrinsically ratio-metric and usually more accurate than simple intensity measurements. However, they are not applicable if the binding dye significantly increases its lifetime. A more detailed analysis of the anisotropies as a function of DNA concentration also allows for calculating the dye-DNA affinity (binding constant).

EXPERIMENT 6-16 DEPENDENCE OF FLUORESCENCE EMISSION SPECTRA ON EMISSION POLARIZER ORIENTATION IN VISCOUS SOLUTIONS

Keywords: fluorescence, polarizers, dansyl amide, polarization-dependent spectra.

Through all experiments, we emphasized that when measuring fluorescence spectra it is important to keep the magic angle (MA) conditions. In Experiment 4-3, we presented an experimental example of why this is important. The need to keep MA conditions is a direct consequence of photoselection (polarized excitation) and rotational diffusion (Brownian rotations) of excited molecules. The discrepancy between vertical (I_{VV}) and horizontal (I_{VH}) observed emission intensities depend on the ratio between the molecular speed of rotation and fluorescence lifetime.

(Continued)

EXPERIMENT 6-16 (CONTINUED) DEPENDENCE OF FLUORESCENCE EMISSION SPECTRA ON EMISSION POLARIZER ORIENTATION IN VISCOUS SOLUTIONS

Only in low viscosity solvents, the transition moment directions of excited molecules quickly randomize due to free Brownian rotations, and in such a case, we do not expect any dependence on the emission polarizer orientation. Similarly, for very viscous or rigid systems (e.g., polymer films), the distribution of excited molecules is fixed and we do not expect any changes. However, when a fluorophore lifetime is comparable to the solvent relaxation time, excited molecules will be able to reorient before emission. In this case, we can also observe the difference in the position of the emission spectra observed with vertical and horizontal polarization. The emission spectrum observed with horizontal polarizer (horizontal component) can be shifted toward longer wavelength as compared to vertically observed emission. Initially, after the vertically polarized excitation (V) more excited molecules have a V direction than horizontal (H) direction. During its lifetime, the excited molecules rotate and populate the (H) direction. The initial fluorescence is mostly from a not fully relaxed excited state and has preferential vertical polarization. The later emission is from the relaxed state contributing more to the horizontal emission component. In consequence, the vertically oriented observation will be more bias toward not fully relaxed molecules while horizontal observation will preferentially detect a more relaxed population. In effect observed emission spectra for vertical and horizontal observations could be different. We will illustrate this with dansyl amide (DA) fluorophore dissolved in propylene glycol (PG). This fluorophore has a large Stokes shift, unstructured fluorescence spectrum and fluorescence lifetime comparable to the solvent relaxation time in PG. A consequence of this process is emission anisotropy dropping toward the longer wavelength.

MATERIALS

- Dansyl amide (DA)
- Propylene glycol (PG)
- Glass or plastic vials, pipettes
- 1 cm × 1 cm cuvettes

EQUIPMENT

- Spectrophotometer
- Spectrofluorometer

METHODS

1. Make stock solutions of the studied compound (DA) in a few milliliters of PG. Use glass or plastic vials, sonicate.
2. Dilute small amounts of stock solutions in a few milliliters of PG to achieve a few micromolar concentrations. Check absorptions to be sure that the optical density of samples is low, below 0.1 (see Figure 6-16.1).
3. Measure the emission spectra at 350 nm excitation with magic angle (MA), vertical (VV) and horizontal (VH) observations and observe the maxima of emission (see Figure 6-16.2).
4. Calculate the isotropic spectrum composed from I_{VV} and I_{VH} as $(I_{VV} + 2I_{VH})/3$. Draw and compare with the spectrum measured for MA observation.

(Continued)

EXPERIMENT 6-16 (CONTINUED) DEPENDENCE OF FLUORESCENCE EMISSION SPECTRA ON EMISSION POLARIZER ORIENTATION IN VISCOUS SOLUTIONS

RESULTS

In Figure 6-16.1, we are presenting the absorption spectrum of DA in PG. The inset shows the chemical structure of DA.

Next, Figure 6-16.2 presents the emission spectra of DA measured at magic angle conditions, and polarized emission spectra measured with VV and VH polarizer configurations

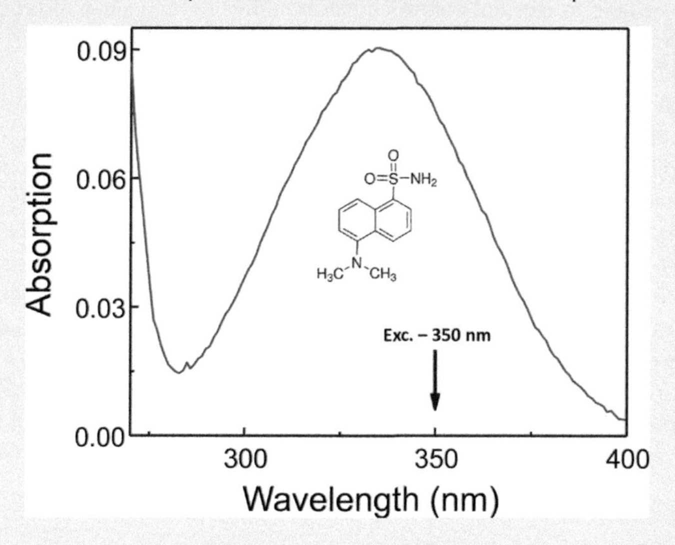

FIGURE 6-16.1 Absorption spectrum of dansyl amide (DA) in propylene glycol. The arrow indicates the wavelength used for fluorescence excitation. The inset shows the chemical structure of the DA.

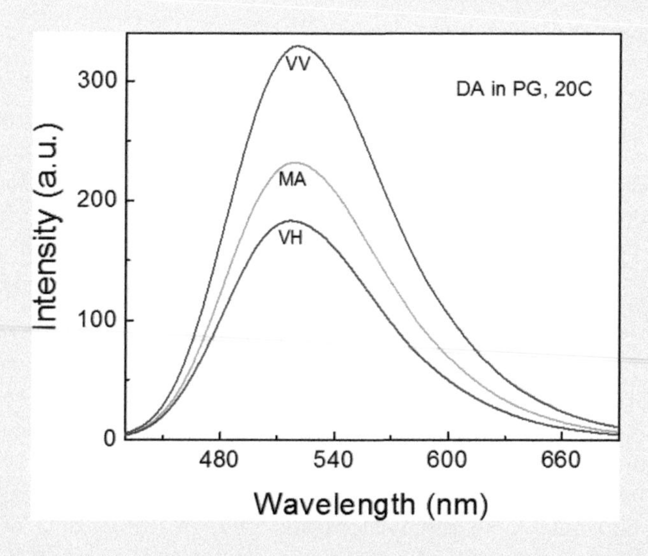

FIGURE 6-16.2 Fluorescence spectra of dansyl amide (DA) in propylene glycol (PG) observed at different orientations of the emission polarizer: 0° (VV, vertical), 54.7° (MA, magic angle) and 90° (VH, horizontal). The excitation at 350 nm was vertically (V) polarized.

(Continued)

EXPERIMENT 6-16 (CONTINUED) DEPENDENCE OF FLUORESCENCE EMISSION SPECTRA ON EMISSION POLARIZER ORIENTATION IN VISCOUS SOLUTIONS

excited with 350 nm. For comparison in Figure 6-16.3, we are presenting normalized I_{VV} and I_{VH} spectra. The presented spectra show a small shift. Next, in Figure 6-16.4, we present a comparison of the measured emission spectrum of DA solution at a magic angle and a spectrum calculated from I_{VV} and I_{VH} components $\left(\frac{1}{3}\left(I_{VV}+2I_{VH}\right)\right)$ that should correspond to the magic angle. Both spectra are overlapping quite well.

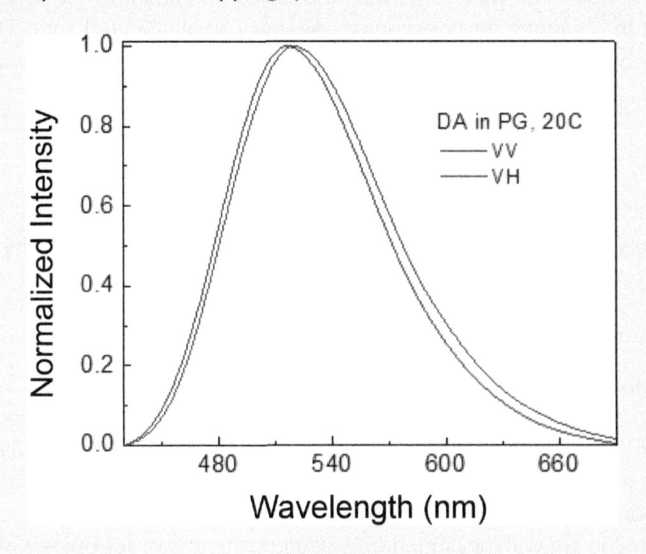

FIGURE 6-16.3 Normalized fluorescence spectra of dansyl amide (DA) in propylene glycol (PG) observed at vertical (VV) and horizontal (VH) orientation of the emission polarizer.

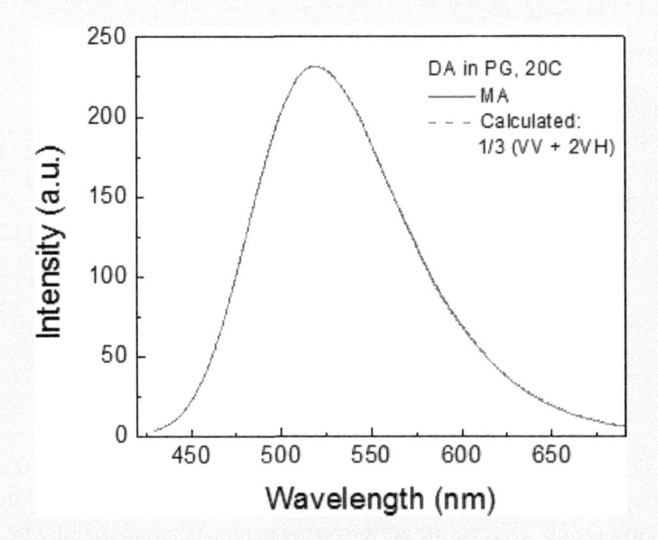

FIGURE 6-16.4 Comparison of fluorescence spectra of dansyl amide (DA) in propylene glycol (PG) measured at a magic angle (MA) and calculated from spectra measured at VV and VH observations.

(Continued)

EXPERIMENT 6-16 (CONTINUED) DEPENDENCE OF FLUORESCENCE EMISSION SPECTRA ON EMISSION POLARIZER ORIENTATION IN VISCOUS SOLUTIONS

CONCLUSIONS

In some cases (when fluorescence lifetime is comparable to molecular rotation), we can observe different (shifted) emission spectrum when measuring with the vertical and horizontal orientation of the observation polarizer. As a consequence, it is important to keep a magic angle (MA) condition to correctly measure the emission spectra. Not keeping MA conditions may result in obtaining improper intensity and a small spectral shift. The magic angle (MA) result can be achieved with the combination of vertical and horizontal observations $\left(\frac{1}{3}\left(I_{VV}+2I_{VH}\right)\right)$.

EXPERIMENT 6-17 PRECAUTIONS IN FLUORESCENCE ANISOTROPY MEASUREMENTS: TOO HIGH ABSORPTION

Keywords: fluorescence, anisotropy, fluorescence intensity.

Fluorescence anisotropy measurements require a low absorbance of the sample to avoid energy migration and reabsorption resulted in a secondary emission. *How much is too high?* This will depend on many factors like a cuvette that has been used for measurement or the dye itself. In this experiment and the following experiment, we will present examples of how sample absorption may affect measured anisotropy.

In this experiment, we will measure fluorescence emission anisotropies with different concentrations of the Rhodamine 110 dye. We will compare anisotropies measured for higher concentrations with the anisotropy measured for a low concentration (absorption below 0.05 in maximum) using a standard 1 cm × 1 cm cuvette.

MATERIALS

- Glycerol (G), ethanol (EtOH)
- Rhodamine 110 (R110)
- Glass filters (long pass 495 and 520 nm)
- Glass or quartz 1 cm × 1 cm cuvettes

EQUIPMENT

- Spectrophotometer
- Spectrofluorometer

METHODS

1. Prepare a stock-solution of R110 in ethanol. Prepare the samples of R110 in glycerol with three various optical densities, see Figure 6-17.1. Be sure that the solutions are homogeneous. Dissolution in glycerol requires long shaking. You may slightly heat the sample.
2. Measure the fluorescence emission spectra at the magic angle (MA) conditions, see Figure 6-17.2.

(Continued)

EXPERIMENT 6-17 (CONTINUED) PRECAUTIONS IN FLUORESCENCE ANISOTROPY MEASUREMENTS: TOO HIGH ABSORPTION

3. Measure the fluorescence emission anisotropies of R110 in glycerol with the excitation wavelength of 490 nm at 20°C. For this, you need to measure VV and VH components, see Figure 6-17.3. Apply *G*-factor for HV component correction (from measured HV and HH components) for each sample. For various concentrations, the *G*-factors can be slightly different, see Figure 6-17.4.
4. Plot and compare the emission anisotropies.

RESULTS

Figure 6-17.1 shows measured absorptions of three prepared samples using glycerol as a baseline. The highest concentration presents absorption of about 2 and the lowest is in the proper range of 0.1.

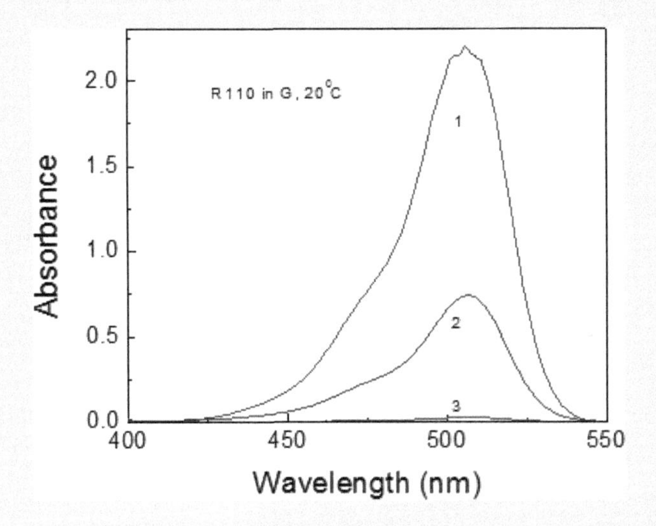

FIGURE 6-17.1 Absorption spectra of R110 in glycerol at 20°C.

For the fluorescence measurements, we will need to adjust the voltage on the detector. But do not change the size of the slit; keep the same geometrical conditions for all three samples.

Fluorescence spectra measured at the magic angle conditions are slightly shifted at higher concentrations, see Figure 6-17.2, because of the reabsorption.

Next, we measured the VV and VH components of the fluorescence emission presented in Figure 6-17.3. Be sure that the vertical intensity does not exceed the upper limit of the instrument range. Using horizontal excitation, we calculated the *G*-factor corrected horizontal component. *Note:* We recommend to separately measure *G*-factor for each concentration since due to the concentration we can have some small geometry change that will affect the *G*-factor. In Figure 6-17.4, we present the measured *G*-factor. As expected at higher concentrations, the fluorescence spot changes slightly its geometry, therefore each *G*-factor is slightly different.

(Continued)

EXPERIMENT 6-17 (CONTINUED) PRECAUTIONS IN FLUORESCENCE ANISOTROPY MEASUREMENTS: TOO HIGH ABSORPTION

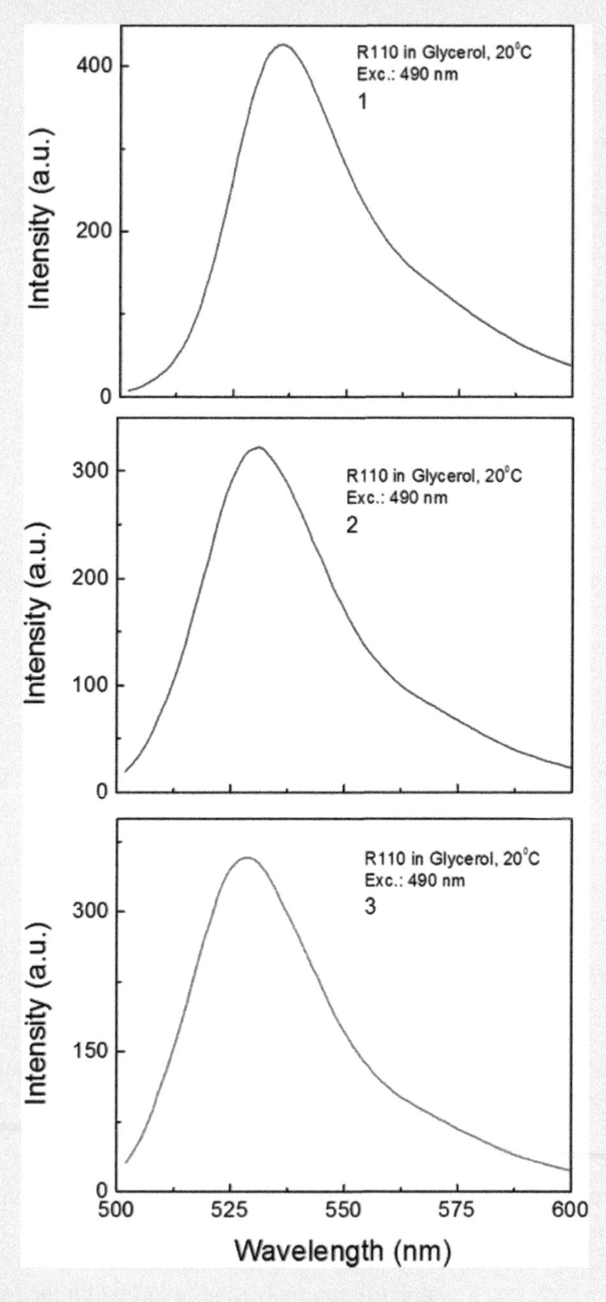

FIGURE 6-17.2 Fluorescence spectra of R110 in glycerol at 20°C measured at magic angle conditions.

(Continued)

EXPERIMENT 6-17 (CONTINUED) PRECAUTIONS IN FLUORESCENCE ANISOTROPY MEASUREMENTS: TOO HIGH ABSORPTION

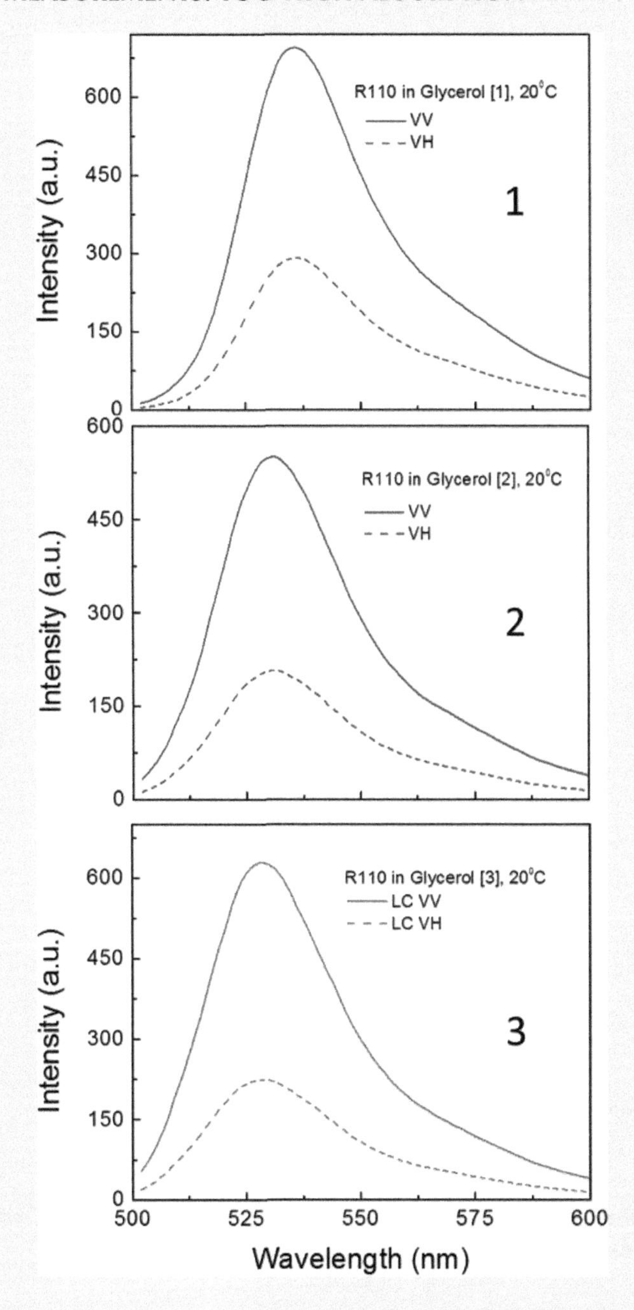

FIGURE 6-17.3 Corrected VV and VH components of R110 fluorescence emission in glycerol at 20°C.

(Continued)

EXPERIMENT 6-17 (CONTINUED) PRECAUTIONS IN FLUORESCENCE ANISOTROPY MEASUREMENTS: TOO HIGH ABSORPTION

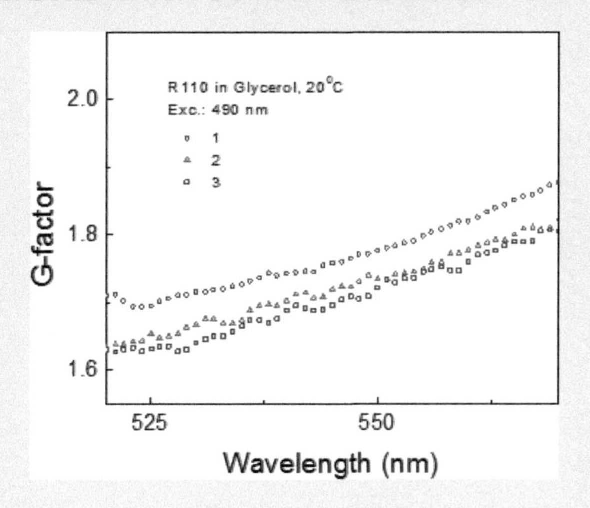

FIGURE 6-17.4 *G*-factors for individual samples.

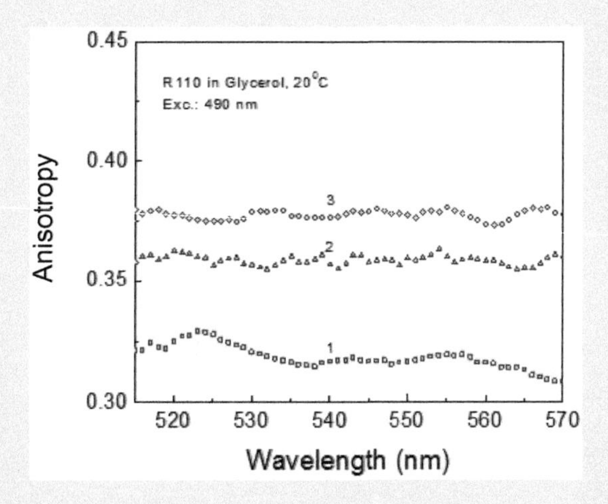

FIGURE 6-17.5 Fluorescence emission anisotropies of R110 in glycerol at 20°C.

Finally, in Figure 6-17.5, we plot measured emission anisotropies for each sample. As concentration increases, we observe a small drop in measured anisotropies.

CONCLUSIONS

Evidently, the measured anisotropy depends on the absorption (concentration) of the fluorophore. At higher concentrations, the anisotropy is artificially lowered. What is the reason for observed lower anisotropy? Even at the absorbance of 2, the concentration of R110 is about

(Continued)

EXPERIMENT 6-17 (CONTINUED) PRECAUTIONS IN FLUORESCENCE ANISOTROPY MEASUREMENTS: TOO HIGH ABSORPTION

2×10^{-5} M, which is too low for an efficient energy migration (homo-transfer). The reason for lower anisotropies is a secondary emission induced by the reabsorption. This secondary emission has a lower anisotropy than the original fluorescence. According to the additivity of the anisotropy, the total observed anisotropy is a sum of fractional anisotropies ($r = r_1 \times f_1 + r_2 \times f_2$, where r_i is the anisotropy of each fraction, f_i). In anisotropy measurements, it is important to keep the fluorophore concentration low. Unexperienced researchers often try to keep fluorescence signals as high as possible, which in the case of fluorescence anisotropy measurements, can lead to significant errors.

EXPERIMENT 6-18 PRECAUTIONS IN FLUORESCENCE ANISOTROPY MEASUREMENTS: SATURATED INTENSITY SIGNALS

Keywords: fluorescence, anisotropy, fluorescence intensity.

When measuring fluorescence anisotropy we need to remember that many factors may affect the results. One, it is a necessity of very low absorbance of the sample to avoid energy migration and reabsorption that would result in a secondary emission that will be significantly depolarized. Another precaution is the use of a proper cuvette. Often we use thin (1 mm × 10 mm) or micro cuvettes. In such a set-up, the wide side (short path) of the cuvette faces the excitation and narrow side faces the observation. *Note: When using a very thin cuvette (e.g., 0.5 mm) the emission from the center can reflect from the solution/glass interface before exiting the cuvette. This reflected emission will predominantly be vertically polarized (the reflectivity is greater for the vertical component) contributing to a total fluorescence signal. In effect, the fluorescence intensities measured along the vertical and horizontal axis will be altered, affecting the anisotropy.* However, the most often simple mistake is a saturated intensity signal. VV and VH components depend on the value of a G-factor. Usually, VV components show stronger fluorescence signals which may run over the range of the spectrofluorometer upper-limit (sometimes very close to the range limit the response is not fully linear). In this experiment, we will demonstrate how the saturated in maximum fluorescence signals affect the anisotropy measurements.

MATERIALS

- Glycerol (G), ethanol (EtOH)
- Acridine orange (AO)
- Glass filters (long pass 495 and 520 nm)
- 1 cm × 1 cm cuvette

EQUIPMENT

- Spectrophotometer
- Spectrofluorometer equipped with a temperature-controlled cuvette holder

(Continued)

EXPERIMENT 6-18 (CONTINUED) PRECAUTIONS IN FLUORESCENCE ANISOTROPY MEASUREMENTS: SATURATED INTENSITY SIGNALS

METHODS

1. Prepare stock-solutions of AO in ethanol. Prepare the samples of AO in glycerol with an optical density not exceeding 0.1, see Figure 6-18.1. Be sure that the solution is homogeneous. Dissolution in glycerol requires a lot of mixing (long shaking). You may slightly heat the sample.
2. Measure fluorescence emission and excitation spectra at magic angle (MA) conditions, see Figure 6-18.2.
3. Measure the fluorescence emission anisotropy of AO in glycerol with the excitation wavelength of 480 nm at 0°C. For this, you need to measure VV and VH intensity components, see Figure 6-18.3 (right). Set the voltage on the detector that the intensity of the VV component exceeds the upper limit (e.g., 1000 in Varian Eclipse). Apply *G*-factor for correcting the HV component (from measured HV and HH intensity components). This is an *incorrect* measurement which leads to false anisotropy spectra, see Figure 6-18.4.
4. Measure fluorescence excitation anisotropy with the observation at 525 nm, see Figure 6-18.3 (left) (use a long-pass 520 nm filter on observation). Use the *G*-factor for 525 nm in order to correct the VH component. These are *incorrect* measurements which lead to false anisotropy spectra, see Figure 6-18.4.
5. Repeat measurements of VV and VH components for both excitation and emission using a lower voltage on the detector, see Figure 6-18.5. Compare the correct and incorrect measurements, see Figure 6-18.6.

RESULTS

Figure 6-18.1 shows the measured absorption of the prepared sample. Be sure that solution is homogeneous. You can check this with a blue laser pointer. Next, we measured excitation and emission spectra of AO fluorescence at 0°C.

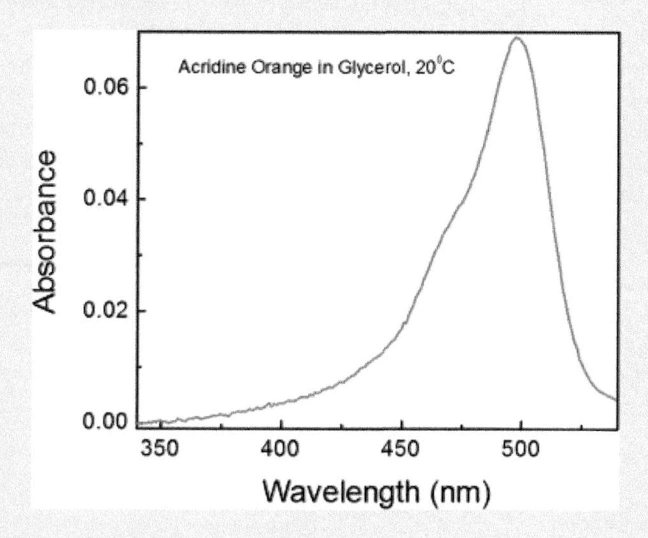

FIGURE 6-18.1 Absorption spectrum of AO in glycerol.

(Continued)

EXPERIMENT 6-18 (CONTINUED) PRECAUTIONS IN FLUORESCENCE ANISOTROPY MEASUREMENTS: SATURATED INTENSITY SIGNALS

Absorption and fluorescence spectra of AO suggest that within a wide range of wavelengths (350–520 nm), there is a single electronic transition ($S_0 \rightarrow S_1$) and we do not expect any interference from higher transitions. Expected anisotropy should be fairly constant.

Next, we measured VV and VH components of AO fluorescence in both excitation and emission regions. Set the voltage at such a level that the VV component will slightly exceed the upper limit of registered intensities. This is, of course, incorrect, but we want to see how this will affect measured anisotropies.

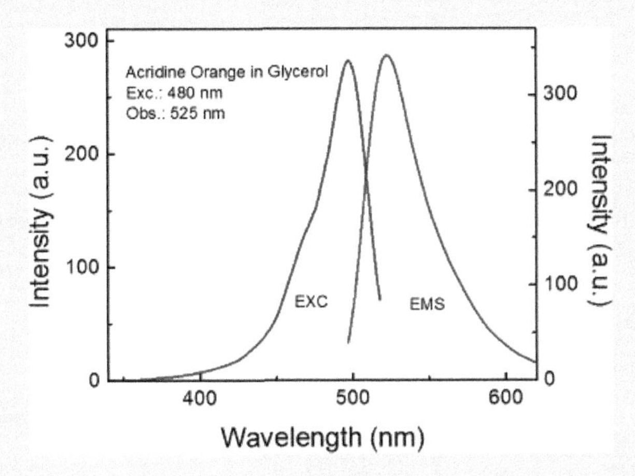

FIGURE 6-18.2 Fluorescence excitation and emission spectrum of AO in glycerol.

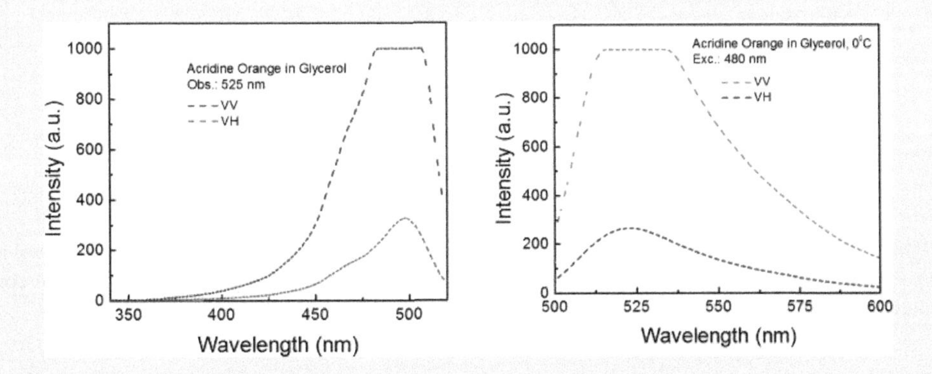

FIGURE 6-18.3 Incorrect measurements of VV and VH components of AO fluorescence in glycerol at 0°C. Left: In the excitation spectrum. Right: In the emission spectrum. The VH components were corrected for *G*-factor.

In Figure 6-18.4, we present calculate anisotropies. Clearly, in the range where we acceded the saturation range, we observe a bump in the measured anisotropy spectra.

To test that observed bump is an artifact, we lowered the voltage on the detector and repeat the measurements. In Figure 6-18.5, we present measured intensity components

(*Continued*)

EXPERIMENT 6-18 (CONTINUED) PRECAUTIONS IN FLUORESCENCE ANISOTROPY MEASUREMENTS: SATURATED INTENSITY SIGNALS

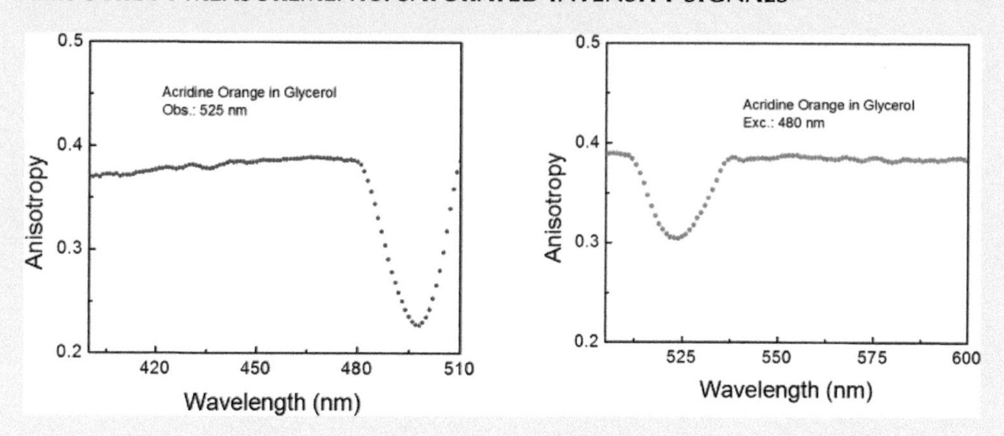

FIGURE 6-18.4 Incorrect measurements of fluorescence excitation (left) and emission (right) anisotropies of AO in glycerol at 0°C.

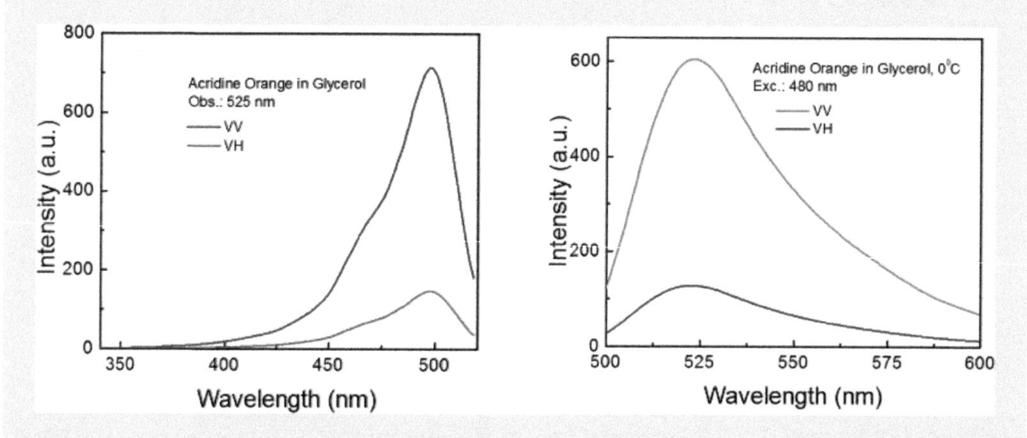

FIGURE 6-18.5 Correct measurements of VV and VH components of AO fluorescence in glycerol at 0°C. Left: In the excitation spectrum. Right: In the emission spectrum. The VH components were corrected for *G*-factor.

and in Figure 6-18.6, we present measured anisotropies (excitation and emission) as open circles. The previously measured anisotropies (filled circles) are also included for comparison.

(*Continued*)

EXPERIMENT 6-18 (CONTINUED) PRECAUTIONS IN FLUORESCENCE ANISOTROPY MEASUREMENTS: SATURATED INTENSITY SIGNALS

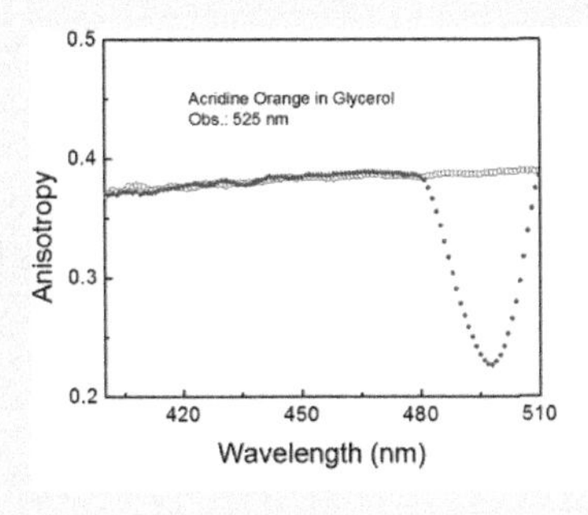

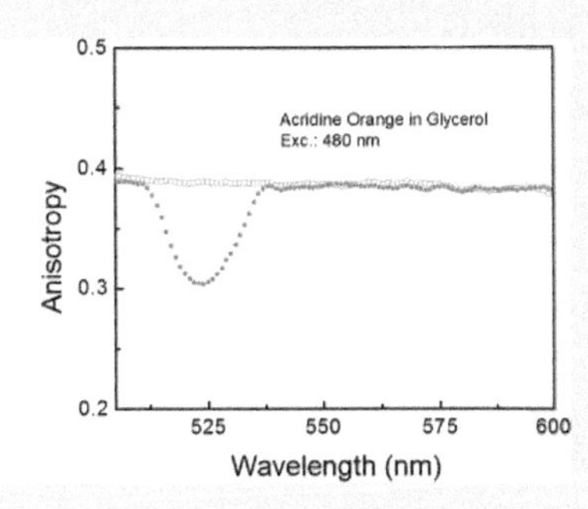

FIGURE 6-18.6 Corrected measurements of fluorescence excitation (left) and emission (right) anisotropies (open circles) of AO in glycerol at 0°C.

CONCLUSIONS

If the fluorescence intensity during the measurements even slightly exceeds the linearity limit, it will strongly affect the anisotropies. It is a very common mistake. If you set your instrument for good measurements at magic angle conditions, the VV component may easily exceed the scale or not be in the full linear range. This is always manifested by the sudden drop (bump) of measured anisotropy in the range of the peak.

EXPERIMENT 6-19 PRECAUTIONS IN FLUORESCENCE ANISOTROPY MEASUREMENTS: PRESENCE OF RAMAN SCATTERING AND SCATTERING

Keywords: fluorescence, anisotropy, fluorescence intensity, scattering, Raman scattering.

In previous experiments, we discussed artifacts that we can see when we reach the detector limit or we are using too high sample absorption. To measure fluorescence anisotropy, we would typically use diluted samples to minimize the reabsorption effect or energy migration. Also, occasionally, we will have to use very low sample concentrations for different reasons (like solubility, physiological conditions, etc.). In this case, we have to be alert about artifacts arising from the presence of light weak fluorescence signal)on and emission. The Raman scattering that is probably not convenient for you. en it would be gooscattering. The most important is the scattering of excitation light that leaks through the detection system (monochromator) and Raman scattering. The direct scattering of the excitation light can be limited by a proper setting of the detection monochromator and additional filters on excitation (e.g., short-pass filter) and emission (e.g., long-pass filter). The Raman scattering in many cases (e.g., weak fluorescence signal) is very difficult to avoid. For example, if the chromophore we are studying is excited at 350 nm or at longer wavelength range, the Raman scattering can be a significant problem and frequently researchers do not realize that. As we discuss in Chapter 1 and Experiments 4-9 through 4-13 for excitations longer than 350 nm the Raman peak will be shifted over 50 nm falling well into the emission spectrum. In many cases, it would be impossible to eliminate Raman scattering that will contribute with very high anisotropy (scattering is highly polarized) and even a relatively small contribution may significantly affect measured anisotropy.

(Continued)

EXPERIMENT 6-19 (CONTINUED) PRECAUTIONS IN FLUORESCENCE ANISOTROPY MEASUREMENTS: PRESENCE OF RAMAN SCATTERING AND SCATTERING

In this experiment, we will measure fluorescence anisotropy for diluted solutions of RhB. We want to demonstrate how the presence of Raman scattering impact measured anisotropy at any wavelength across the emission spectrum and how it would impact the average value when measuring the entire emission band using, for example, a cut-off filter (this is frequently done in a high throughput measurement).

MATERIALS

- Rhodamine B (RhB)
- Water
- Glass or quartz 1 cm × 1 cm cuvettes

EQUIPMENT

- Spectrophotometer
- Spectrofluorometer

METHODS

1. Prepare a stock-solution of RhB in water. Measure absorption and make appropriate dilution to have a stock solution concentration with an absorption in the good range (between 0.2 and 0.8). Measure absorption, see Figure 6-19.1.
2. Prepare the first sample of RhB in water with absorption in the range of 0.01. *Note: In practice, we would recommend preparing solution with the absorption of about 0.05 that can be precisely measured and dilute the sample appropriately to 0.01 optical density.* Check/measure the absorption.
3. Dilute the sample from #2 about 10 times with water. The absorption will be difficult to measure but still durable.
4. Dilute the sample from #3 another 10–20 times (in our experiment, we diluted ~20-fold). This sample will have absorption of about 0.0001 or less below the resolution of a standard spectrophotometer. *Note: Do not worry, the emission spectrum will still be easily measurable with only higher noise.*
5. Measure fluorescence emission spectra at the magic angle (MA) conditions. See Figure 6-19.2.
6. Measure *G*-factor using each solution. For that, rotate excitation polarization to horizontal and measure parallel (vertical) and perpendicular (horizontal) intensity components.
7. Rotate excitation polarizer to vertical and measure parallel (vertical) and perpendicular (horizontal) intensity components. Apply *G*-factor for HV component correction (calculated from measured HV and HH components). We used an average *G*-factor. *Note: Measured G-factor for each sample should be the same. However, for the lowest concentration, the signal is low and at long-wavelength emission tail, we may have significant noise. See Figure 6-19.3.*
8. Plot and compare the emission anisotropies.

(*Continued*)

EXPERIMENT 6-19 (CONTINUED) PRECAUTIONS IN FLUORESCENCE ANISOTROPY MEASUREMENTS: PRESENCE OF RAMAN SCATTERING AND SCATTERING

9. *Optional.* If you can bypass the monochromator in your spectrofluorometer (by using so-called zero mode or disconnecting the monochromator (not recommended for commercial instruments)) measure observed anisotropy using only a long-pass filter. *Note: This mode of observation is frequently used in high throughput screening approach.*
10. Using measured vertical and corrected horizontal emission components to calculate emission anisotropy that would be observed when we just use a cut-off filter (in our case 575 nm) and no monochromator.
11. Using wavelength-dependent anisotropies (Figure 6-19.3) and magic angle emission spectrum calculates the expected emission anisotropy that will be observed with the cut-off filter.
12. Compare anisotropies calculated and measured in points #9, #10, and #11.

RESULTS

Figure 6-19.1 presents normalized absorptions of RhB in water. In the figure, we marked the used excitation wavelength (535 nm from the laser diode). Figure 6-19.2 shows normalized emission spectra measured for all three solutions at magic angle conditions. Highest concentration shows clean RhB emission. A 10-fold dilution (dashed line) shows a small bump about 650 nm. Such a bump would be very difficult to identify without the emission spectrum measured with higher concentration. The highest dilution shows a very distinct peak at 650 nm. This is the Raman scattering from 535 nm excitation broadened by the emission monochromator.

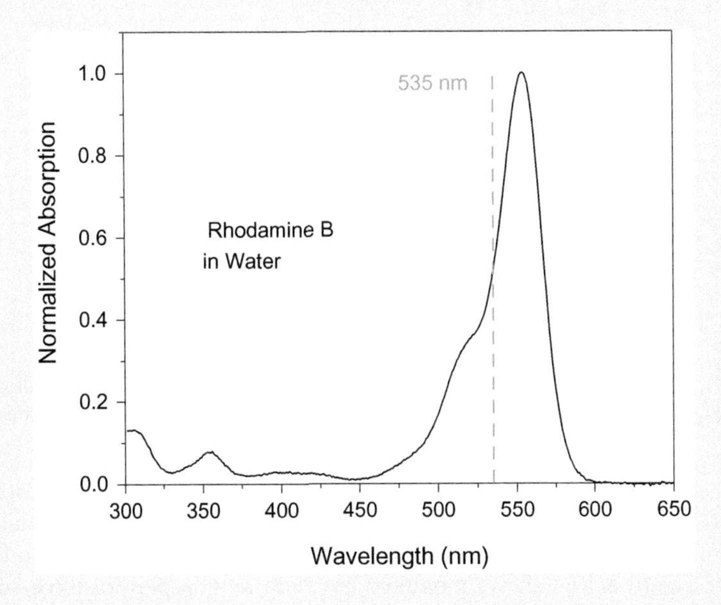

FIGURE 6-19.1 Normalized absorption spectrum of Rhodamine B (RhB) in water. Vertical dashed/green line indicates excitation wavelength.

(*Continued*)

EXPERIMENT 6-19 (CONTINUED) PRECAUTIONS IN FLUORESCENCE ANISOTROPY MEASUREMENTS: PRESENCE OF RAMAN SCATTERING AND SCATTERING

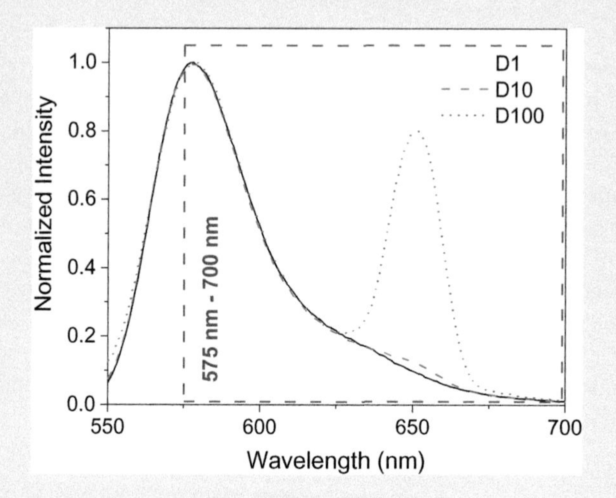

FIGURE 6-19.2 Normalized emission spectra of Rhodamine B (RhB) in water measured for three different dilutions. Dashed/red line indicates the range for calculating average anisotropy.

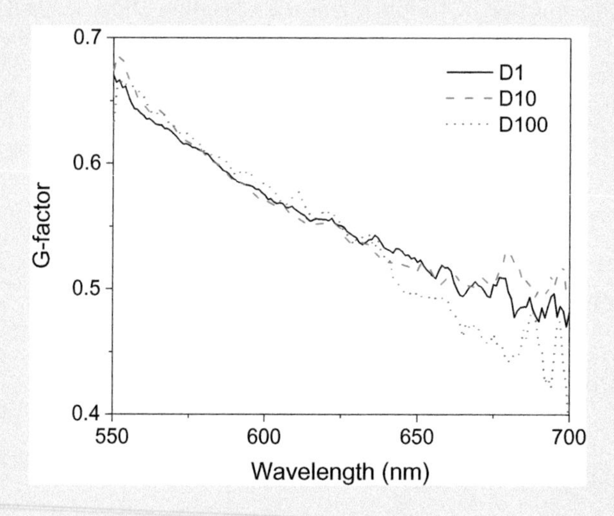

FIGURE 6-19.3 Calculated *G*-factors for three sample dilutions. The black thick line indicates *G*-factor measured for highest sample concentration that we later used for calculating emission anisotropies.

Next, we calculated the *G*-factors measured for each solution (Figure 6-19.3). It is clear that measured *G*-factors are very close and only the lowest concentration shows a small deviation. To calculate anisotropies we used the Raman scattering from *G*-factor measured for highest concentration. Figure 6-19.4 shows measured emission anisotropies for three solutions. As shown in the lower panel (expanded range) measured anisotropies in the range 570–670 nm are in very good agreement. However, at short wavelength the lowest concentration shows

(Continued)

EXPERIMENT 6-19 (CONTINUED) PRECAUTIONS IN FLUORESCENCE ANISOTROPY MEASUREMENTS: PRESENCE OF RAMAN SCATTERING AND SCATTERING

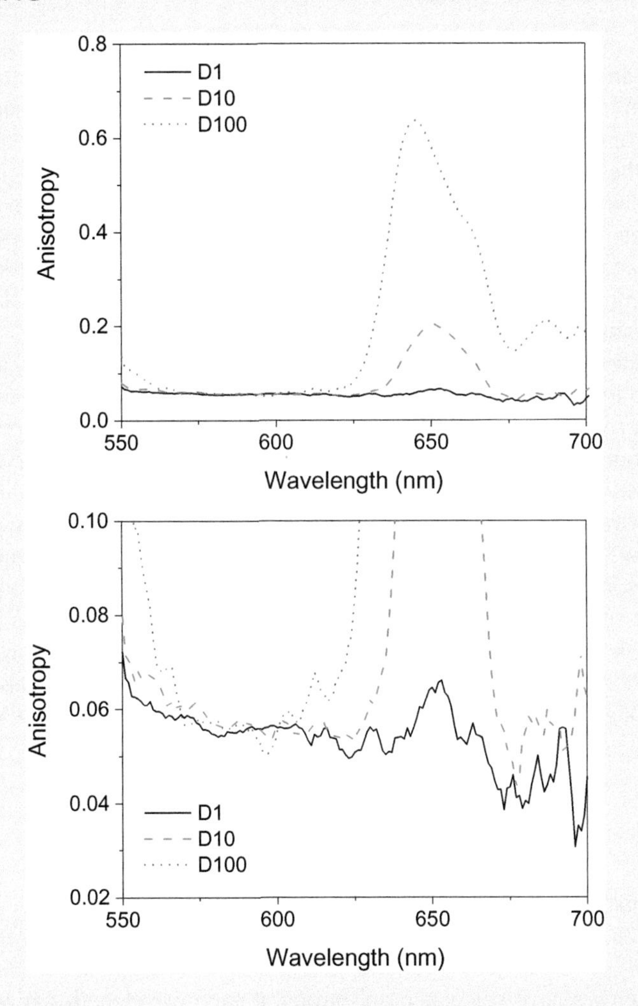

FIGURE 6-19.4 Measured anisotropies for three RhB dilutions. Lower panel presents expanded anisotropy view.

higher anisotropy. This is an indication of the presence of a small amount of scattered excitation light leaking through to the detector. It could be a surprise but measured anisotropy gets really high for longer 575–650 nm range where the lowest concentration presents anisotropy much over allowable value of 0.4. We could consider the signal to be too low but for wavelength longer than 650 nm (where the signal is really low), measured anisotropies are coming much closer (for intermediate concentration overlaps with values measured for high concentration).

It is evident that a small amount of scattered light leaking into the detector (at the short wavelengths emission and longer wavelengths Raman scattering) may significantly alter/increase observed anisotropy. For most samples that present good signal, the effect is negligible. However, when using very low concentrations of even a bright dye or when the probe

(Continued)

EXPERIMENT 6-19 (CONTINUED) PRECAUTIONS IN FLUORESCENCE ANISOTROPY MEASUREMENTS: PRESENCE OF RAMAN SCATTERING AND SCATTERING

has a low quantum yield, this can present a significant problem. In the case of excitation light scattering, we can use additional (or better) filters that will typically lower the effect. In the case of Raman finding, appropriate filters can be a problem since the Raman scattering falls into the middle of the emission spectrum. When measuring emission anisotropies in the full spectral range, the problem (distinct bump) is easy to identify. But in many cases in biomedical applications when the signal is low, we would collect an entire spectrum (just a long-pass cut-off filter to eliminate excitation light scattering). Anisotropy-based assays are very attractive for many practical applications since measured anisotropy is a ratiometric parameter that should not depend on sample signal. So, it is educational to consider if small Raman scattering would have any significant impact on overall measured anisotropy.

To demonstrate this effect we will do two things. We can measure overall anisotropy when observing the entire spectrum with only the cut-off filter on observation (no monochromator). For that, we just remove the monochromator from our ISS K2 system and measure emission anisotropies of our three samples using only 580 nm cut-off filter on observation. *Note: We do not recommend this for any inexperienced researcher. You will need a very high level of expertise to remove and put back monochromator without damaging your system.* In the second approach, we could calculate the expected overall anisotropy using measured vertical and horizontal components of emission spectra or just using already calculated anisotropy and an emission spectrum as measured with magic angle observation. This is easily durable since anisotropy is an additive parameter and an average value can be easily calculated (this is not a simple case for polarization). For direct calculation of emission anisotropy, r in any given spectral range between starting wavelength λ_s and final wavelength λ_f we have:

$$r = \frac{\sum_{\lambda_s}^{\lambda_f} I_{\parallel}(\lambda) - G(\lambda)\sum_{\lambda_s}^{\lambda_f} I_{\perp}(\lambda)}{\sum_{\lambda_s}^{\lambda_f} I_{\parallel}(\lambda) + 2G(\lambda)\sum_{\lambda_s}^{\lambda_f} I_{\perp}(\lambda)} \tag{6-19.1}$$

The sum is a simple sum of intensity readings at any wavelength from the starting value (in our case, we used 575 nm) and final value (700 nm in our case). $G(\lambda)$ is a wavelength-dependent G-factor.

Similarly, when we have calculated anisotropy at each wavelength $r(\lambda)$ we can calculate average anisotropy from any given spectral range, r as:

$$r = \sum \frac{I_{MA}(\lambda)}{\sum_{\lambda_s}^{\lambda_f} I_{MA}} r(\lambda) = I_{MA}^{N} r(\lambda) \tag{6-19.2}$$

where I_{MA}^{N} is a surface normalized intensity value measured at a magic angle (MA).

In Table 6-19.1, we are presenting values as calculated from the Equation (6.1), r_1 and (6.2), r_2, respectively and in the bottom line, we are presenting measured values r_M. As it is evident from the table, the agreement between all values is very good. Looking on the emission spectrum in Figure 6-19.2, we could anticipate that an effect of Raman scattering for the 10-fold dilution would be negligible since the Raman bump is barely visible. However, already here, the anisotropy increase is consistent.

(Continued)

EXPERIMENT 6-19 (CONTINUED) PRECAUTIONS IN FLUORESCENCE ANISOTROPY MEASUREMENTS: PRESENCE OF RAMAN SCATTERING AND SCATTERING

TABLE 6-19.1 Anisotropy Values as Calculated from the Equations (6.1), r_1 and (6.2), r_2, Respectively

	D1	D10	D100
r_1	0.052	0.068	0.233
r_2	0.052	0.067	0.224
r_M	0.054	0.067	0.240

Note: In the bottom line, we are presenting measured values r_M.

CONCLUSIONS

From presented measurements, we can clearly see that for low emission signals, the measured values of anisotropy can be significantly affected by excitation light scattering and Raman scattering. In the presented experiment, we purposely used 535 nm, significantly blue-shifted from absorption maximum to limit direct scattering of excitation light. But even in this case for low signals, this could be a problem. Eliminating Raman scattering in such cases is very difficult. One could suggest shifting excitation toward longer wavelength, e.g., 665 nm that would shift Raman beyond emission spectrum. This could help at the longer wavelength range but would dramatically increase the direct scattering of the excitation light.

In this experiment, we also demonstrate the additivity of the anisotropy and show that the average observed anisotropy (as seen from an entire spectrum or part of the spectrum) can be calculated from wavelength-dependent anisotropy and magic angle emission spectrum.

EXPERIMENT 6-20 EMISSION ANISOTROPY OF QUENCHED FLUOROPHORES

Keywords: fluorescence, fluorescence anisotropy, quenching.

In the previous chapter, we discussed how molecular quenching affects fluorescence intensity. A dynamic quenching decreases fluorescence intensity and fluorophore lifetime. Shorter fluorescence lifetime means that photons will be emitted within a shorter time while the rotational diffusion of the dye molecule will not be practically affected. As predicted by Perrin (Equation 6.24), a shorter fluorescence lifetime will result in a higher observed steady-state anisotropy. In this experiment, we will measure fluorescence emission anisotropies of progressively quenched Rhodamine 123 (R123) by KI.

(*Continued*)

EXPERIMENT 6-20 (CONTINUED) EMISSION ANISOTROPY OF QUENCHED FLUOROPHORES

MATERIALS

- Rhodamine 123 (R123)
- PBS buffer pH 7.4, ethanol (EtOH)
- Potassium iodide (KI)
- Glass or plastic vials
- 1 cm × 1 cm cuvettes

EQUIPMENT

- Spectrophotometer
- Spectrofluorometer

METHODS

1. Prepare stock solutions of R123 in ethanol.
2. Prepare a stock solution of KI in PBS buffer.
3. Prepare samples of fluorophores (R123) with various concentrations of KI up to 0.5 M in PBS buffer. Desired concentrations of KI can be achieved by mixing appropriate amounts of stock solutions of KI, sample, and PBS buffer, limit fluorophore (R123) concentration to have a low-optical density (OD < 0.1) but the same in all samples. For preparation details, see Experiments 5-11 through 5-13.
4. Measure fluorescence emission of each fluorophore at magic angle conditions.
5. Measure fluorescence emission anisotropies.

RESULTS

KI quenches R123 very efficiently (see Experiment 5-12) with Stern–Volmer constant, K_{SV} of about 40 M^{-1}. Taking into account, the initial lifetime of unquenched R123 is about 4 ns and the molecular rotation in the water environment is fast the anisotropy of unquenched R123 is expected to be very low (similar to R6G in the previous Experiments 6-5 and 6-6). The bimolecular constant, k_q of the quenching is 10^{-10} s^{-1} M^{-1}, which is close to the diffusion limit of small molecule like R123. As we add KI, the fluorescence intensity will dropdown.

In Figure 6-20.1, we are presenting measured fluorescence intensities (under MA conditions) of R123 for increasing KI concentrations.

In Figure 6-20.2, we are presenting quenching of R123 in the form of Stern–Volmer dependence. The estimated quenching constant is 40.5 M^{-1}.

Next, we measured polarized intensities (I_{VV} and I_{VH}) for all KI concentrations. Not only overall intensities decreased but relative intensities of I_{VV} and I_{VH} changed significantly. To better show how relative intensities for parallel and perpendicular polarization are changing in Figure 6-20.3, we present corrected I_{VV} and I_{VH} intensities for 0 M KI, 0.1 M KI, and 0.5 M KI.

Note: As the concentration of KI increases the fluorescence intensity significantly decreases. To better show the change, we were increasing voltage when adding KI to keep the total signal in the same range.

In Figure 6-20.4, we present anisotropies measured for all KI concentrations. As the concentration of KI increases the fluorescence anisotropy increases (Table 6-20.1).

(Continued)

EXPERIMENT 6-20 (CONTINUED) EMISSION ANISOTROPY OF QUENCHED FLUOROPHORES

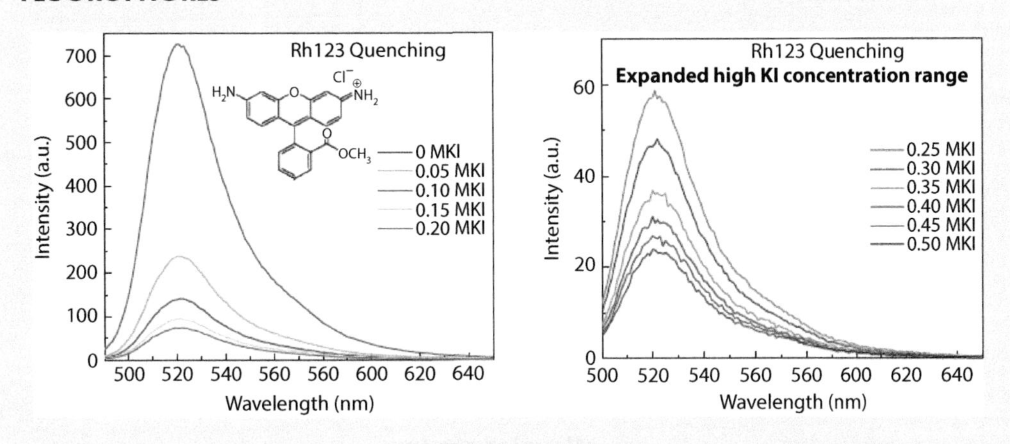

FIGURE 6-20.1 Measured fluorescence intensities of R123 as a function of KI concentration.

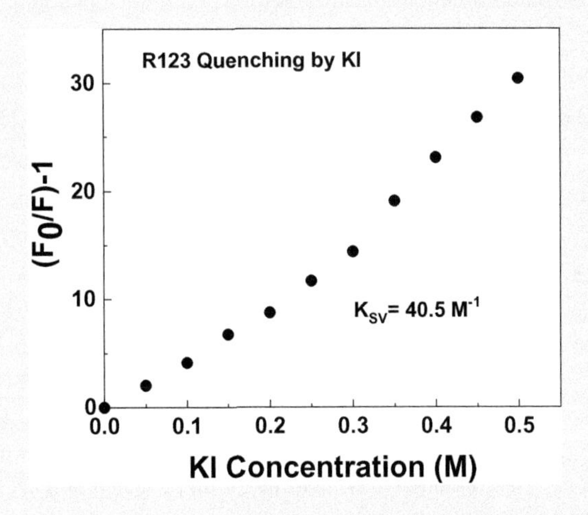

FIGURE 6-20.2 Stern–Volmer plots for KI quenching of R123 in PBS buffer at 20°C.

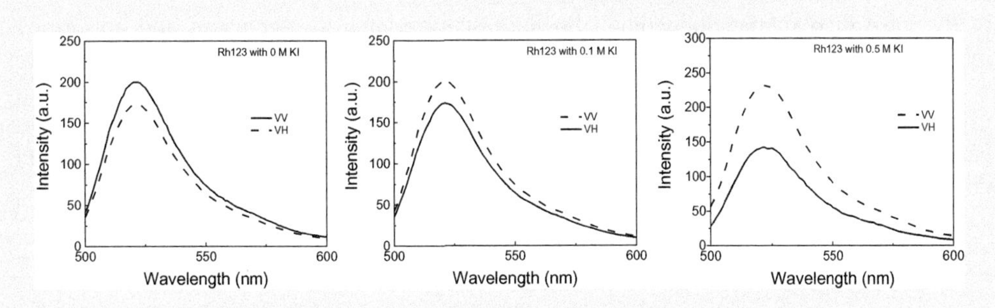

FIGURE 6-20.3 Measured fluorescence intensities of R123 for 0 M, 0.1 M, and 0.5 M KI concentrations. In the experiment, we increased the voltage to keep the signal in the same range.

(Continued)

EXPERIMENT 6-20 (CONTINUED) EMISSION ANISOTROPY OF QUENCHED FLUOROPHORES

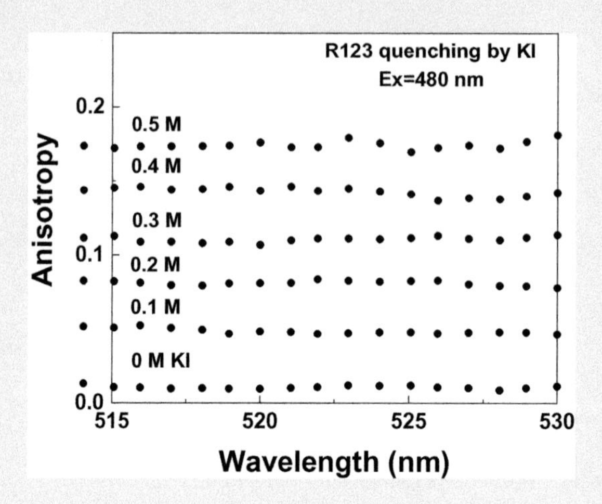

FIGURE 6-20.4 Measured fluorescence anisotropies for R123 in the function of increasing concentration of KI.

TABLE 6-20.1 Emission Anisotropies of Quenched Rh123

KI (M)	r	τ (ns)
0	0.013	4*
0.1	0.051	0.95
0.2	0.085	0.5
0.3	0.121	0.32
0.4	0.163	0.2
0.5	0.197	0.14

Note: Anisotropy values are taken at Maximum Fluorescence. Lifetime is calculated from the Perrin equation with assumption of 4 ns* lifetime for unquenched sample.

CONCLUSIONS

In the case of dynamic quenching, the fluorescence intensity decreases and anisotropy increases. This is general phenomena that is related to the shortening of fluorescence lifetime.

An important conclusion of this experiment is that we have to do a quenching experiment using magic angle conditions. Increase of polarization results in a different change in vertical and horizontal intensities (see Figure 6-20.3) and measurements done with VV or VH configurations will give different and incorrect results.

CHAPTER 7

Fluorescence

Time-Resolved Phenomena

INTRODUCTION

In the three previous sections, we were discussing steady-state fluorescence phenomena and some of its applications. As discussed in the first two chapters, an important characteristic of fluorescence is the so-called excited state decay called fluorescence lifetime. The definition of fluorescence lifetime arises from the fact that a transition of an excited molecule to the ground state is a purely statistical process that can be simply characterized with the rate constant (or probability for the transition). The transition to the ground state can be accompanied by the emission of a photon (radiative transition) or not result in photon emission (non-radiative transition). The fluorescence lifetime can then be calculated as a reciprocal of cumulative excited state decay rates:

$$\tau = \frac{1}{k} = \frac{1}{k_{nr} + \Gamma} \tag{7.1}$$

where Γ is the radiative decay rate (a rate of the excited state that results in photon emission, also sometimes called k_f) and k_{nr} is the sum of all other rates that result in the non-radiative decay of the excited state (e.g., heat or energy transfer), k is a total excited state decay rate, $k = \Gamma + k_{nr}$. Any modification of either rate results in an apparent change of observed fluorescence intensity and a change of fluorescence lifetime. Generally, time-resolved data are relatively complex and the interpretation of the fluorescence decays (lifetimes) often requires knowledge of the analytical expression for the decay function. Most common is an exponential decay model in which the time-dependent intensity decay after delta pulse excitation can be represented by an exponential function:

$$I(t) = I_0 e^{-kt} = I_0 e^{-t/\tau} \tag{7.2}$$

In a simple meaning, the fluorescence lifetime, τ represents the time after which observed fluorescence intensity decreases e-fold (the number of molecules in the excited state decreases e-fold where, $e = 2.72$ is the base of the natural logarithm).

The intensity decay can be multi-exponential in the form:

$$I(t) = \sum_i I_0 \alpha_i e^{-t/\tau_i} \sim \sum_i N_0 \alpha_i e^{-t/\tau_i} \tag{7.3}$$

where I_0 is the initial (zero time) intensity and N_0 is the initial number of molecule excited by the pulse. As we discussed in Chapter 1, the observed intensity (number of counts in TCSPC), I is proportional to the number of excited molecules, N. We will typically deal with large number of events that are statistically averaged (typically a very large molecules ensemble or in the case of single-molecule many photon emission events). Fluorescence intensity depends on many factors like sample concentration, intensity of excitation light, or other aspects resulting in signal attenuation. In contrast, fluorescence lifetime describes the observed intensity decay and will not depend on all these factors. These make measurements of fluorescence lifetimes most useful especially in experimental conditions where controlling many factors is difficult. Typical examples are measurements in cellular suspensions, tissue, or microscopy. Already for over 30 years in microscopy, it has been accepted that fluorescence lifetime imaging (FLIM) is the best approach to image molecular and cellular processes. The important advantage of intensity decay (fluorescence lifetimes) is that it is completely independent of excitation intensity, excitation wavelength, or overall fluorescence intensity, that could easily be affected by background absorption or any geometrical factors that would typically strongly affect the observed intensity. These being said we just want to indicate that intensity decay frequently will depend on the setting of polarizer on observation and it is crucial to keep magic angle conditions.

Experimental Factors Impacting Measured Intensity Decay—IRF

A typical measurement of fluorescence lifetime consists of two independent measurements. One measurement of observed fluorescence intensity decay and the second measurement of Instrument Response Function (IRF). Measured fluorescence intensity decay is always a combination/superposition of the true fluorophore intensity decay and inherent temporal response of the instrument (IRF). The IRF is a result of the delayed responses of the detector and electronics to the encounter of the photon (a process that in a very simplified way can be compared to the detection system inertia). Only by an appropriate subtraction of IRF from measured fluorescence intensity decay (a process called deconvolution), we obtain the clean/true fluorescence intensity decay of the studied fluorophore. This true intensity decay then can be fitted with a mathematical model. Later, we will discuss basics of data analysis and deconvolution. But the important question is: *How important is to obtain/measure and consider in calculation an appropriate IRF?* This will depend on the instrument we are using. Intuitively, we would expect that the temporal response (IRF) would depend on various factors like energy of the photon (wavelength) or temperature. Typically, for most of the instruments, the IRF will vary from a few tens of picoseconds to hundreds of picoseconds. As mentioned earlier, the IRF is a combination of the detector speed and electronic system response. If we are using a micro-channel plate photomultiplier tube (MCP-PMT) that has a very fast response time below 50 ps, the overall instrument response would be limited by the speed of electronics and will be in the range of 50–100 ps. However, this is a very expensive detector and most instruments will use just a photomultiplier (PMT) tube that has a response time of a few hundred picoseconds and depending on the system the IRF can be close to a nanosecond. Obviously, a proper subtraction (deconvolution) of IRF will allow for much better precision of obtained parameters. However, in reality, the precision of 10 ps even using MCP-PMT would be very demanding. So, when the measured (expected) lifetime components are below 100 ps, the experimentalist will have to pay special attention to the experimental conditions to have high confidence for obtained parameters.

Factors That May Affect IRF

There are many factors which would affect the temporal resolution of the measured decay. A trivial one is the speed of light. Even if it sounds incredibly fast, we need to remember that to travel 1 cm in air for light it takes about 33 ps and any difference in the physical distance that scattered light and fluorescence light travel before reaching the detector can be a factor. Such a difference can be a result of slightly different geometry (position) for fluorescence sample and scattering measurement. Another related factor is the different speed of light in dielectric materials. Frequently, we use filters to select a given wavelength or to attenuate the signal (typically reference/scattering). A filter thickness can easily reach 5 mm or more when multiple filters are used. Filters are made from glass or plastic that have a refractive index much higher than air (about 1.5). If we would use a 5 mm neutral density glass filter to attenuate the IRF, the additional pulse delay generated by the filter will be about 8 ps. Typically such simple delay is not a problem for time-domain where the pulse shift/delay is routinely built into data analysis but it could be a significant problem for frequency-domain since it is only effecting phase shift and not de-modulation.

Another more significant problem arises from the fact that fluorescence is typically observed at the optimal (maximum signal) wavelength but the IRF is measured at the excitation wavelength. Depending on the fluorophore, the wavelength difference can be at least 20 nm and easily up to 100 nm. For example, when using 480 nm excitation wavelength, we would observe fluorescence at 520 nm or sometimes even 600 nm wavelength. There are two problems with that. First is the different speed of light of different colors of light in dielectric materials (due to the wavelength-dependent refractive index). The difference in refractive index is typically small and it will result in a small (few picoseconds) relative shift of the IRF in respect to fluorescence observation wavelength. Much more significant is potentially different response time of the detector to a different color of light (a color effect of the temporal response of the detector). Photons of different wavelengths carry different energies that may affect temporal resolution. The fact that the detector has different sensitivities to different wavelengths (colors) is well recognized. But much less known is the wavelength-dependent temporal response of the detector. This is typically very difficult to measure in a common lab. The simplest would be to use very short pulses of different colors. This could be realized with ultrafast tunable lasers assuming that pulse width do not change with wavelength. An easier to realize approach would be to use Raman scattering. As we discussed in Chapter 1 and Experiments 4-9 and 4-10, Raman scattering of solvents (water) has a fixed energy shift as compared to the excitation wavelength. Scattering is practically instantaneous and is at a longer wavelength. For example, 405 nm excitation will show water Raman peak at about 470 nm. The Raman signal is typically very weak but with today's laser is easily detectable. As an example, we used 470 nm laser diode and measured the response function for (1) direct scattering from ludox solution observed at 470 nm and Raman scattering of just water observed at 556 nm. In Figure 7.1, we are presenting an example of measured responses using three different detectors. The top is the response measured for MCPPMT detector. IRF response measured with MCPPMT for direct scattering at 470 nm and Raman scattering at 556 nm are very close, just 102 and 92 ps, respectively. The MCPPMT detector has a minimal time-resolved color effect and both responses are very close. This is not the case for PMT and avalanche photodiode (APD) detectors. The wavelength difference is about 86 nm and for both detectors, the IRF measured for direct scattering and Raman scattering are significantly different. In many fluorescence experiments, the difference would be comparable or smaller and we should be alert about that when measuring and interpreting data. Especially, in cases like FRET where measured donor lifetimes are fairly short and multi-exponential.

We want to stress that presented differences will strongly depend on individual detectors and each system should be tested separately. But such a significant difference in the response function will definitely affect any recovered fluorescence lifetime, especially short-lived components.

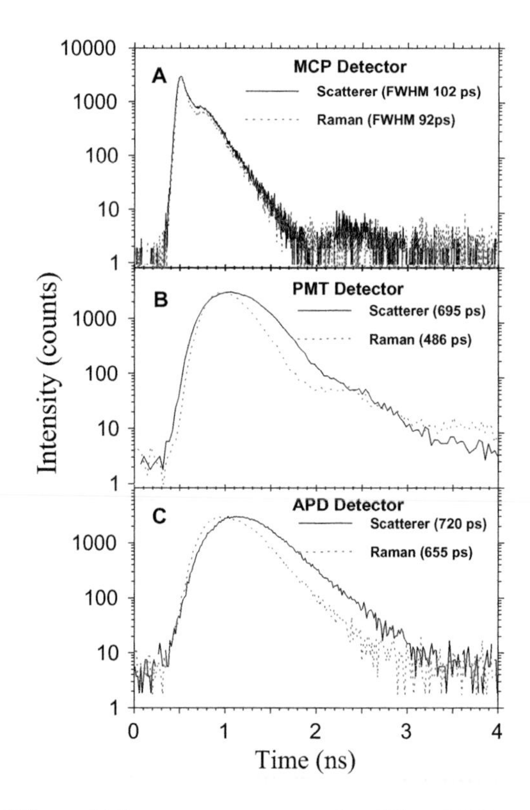

FIGURE 7.1 Measured different IRF responses measured at 470 and 556 nm (Raman) for MCPPMT, PMT, and APD.

Experimental Factors Impacting Measured Intensity Decay—Magic Angle Conditions

It should not be a surprise that the measured intensity decays observed in dynamic/mobile samples (like liquid solutions) would depend on polarization setting on observation. The typical excitation (even with non polarized light) will photoselect absorption transition moments resulting in an anisotropic population of excited molecules. If molecules may freely rotate (Brownian motion), the anisotropic distribution of excited molecules in time will relax to the isotropic distribution. Frequently, the correlation times (speed of rotation) are comparable to the fluorescence lifetimes. In this case, observed intensity decays will depend on the polarizer orientation. Imagine excitation with linearly polarized light (typical in most experiments). The initial population of excited molecules will be preferentially oriented along the excitation light polarization (for details see Chapter 1 and *Introduction to Chapter 6*). So, in an ideal case, the initial intensity observed with a polarizer oriented parallel to the excitation light polarization will be three times greater than the intensity observed through a polarizer oriented orthogonally. An average number of molecules, reorienting/rotating from parallel observation would be greater than from orthogonal orientation. We need to remember that if molecules oriented parallel to the polarizer reorient the apparent intensity observed through the parallel polarizer would artificially decrease. It is opposite for orthogonal observation since more molecules not seen before when rotating will be contributing to the orthogonal intensity components. In effect the population seen by the polarizer oriented parallel to the excitation light polarization will decay artificially faster than just expected from the decay of the excited-state population. The effect on intensity observed through orthogonally oriented polarizer will be opposite since molecules rotating out of parallel orientation will add to the population that contributes to the orthogonally polarized intensity. So, apparent intensity decay seen through the orthogonal polarizer will decay slower than expected from the decay of the excited state population.

In Figure 7.2, we schematically present intensity decays observed for parallel (red), orthogonal (blue), and 54.7° (green) orientation of observation polarizer. We assumed

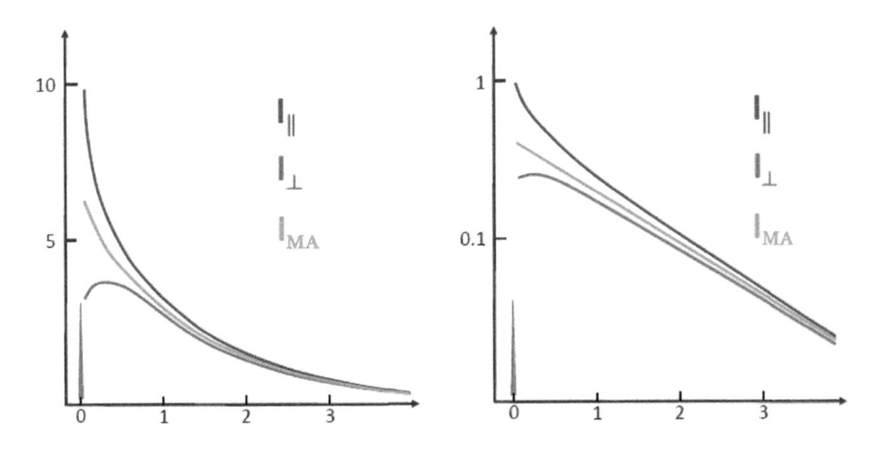

FIGURE 7.2 Schematic representation of fluorescence intensity decays observed through polarizer oriented parallel, orthogonal, and 54.7° to the polarization of excitation pulse. Left: Intensity decays in linear scale; Right: Intensity decays in logarithmic scale.

a single exponential intensity decay for fluorophore. After pulsed excitation, the initial intensity for parallel polarizer orientation is three times greater than that observed for orthogonal polarizer orientation. It is clear that all three intensities are different and decay in completely different form. On the right side, we present the intensity decays in logarithmic scale. The decay observed at 54.7° (MA) is clearly a straight line correctly reflecting single exponential decay. It is very important to keep MA conditions when measuring fluorescence lifetimes in liquid samples (samples where fluorophores have the freedom to reorient during the fluorescence lifetimes).

When using improper polarizer orientation (not MA condition) or no polarizer at all we can generate significant artifacts that will corrupt recovered parameters for intensity decays. In Experiments 7-10, we are presenting examples of potential mistakes.

Brief Concept of Data Analysis

When measuring time-dependent intensity response/decay, it is important to remember that observed intensity decay is a superposition of the excitation pulse and time-dependent intensity decay of the excited state (emission). The excitation pulse or more precisely an intensity response function (IRF) really contains the physical pulse width, the time-dependent response of the photomultiplier, and time-dependent response of the electronics which in overall can be comparable to the fluorescence lifetime. Even if we can have today rarely available systems with femtosecond pulses that are very close to δ-pulse, the overall instrument response function will typically be close to tens of picoseconds or more.

Today most time-resolved fluorescence measurements are made using time-domain technology (time-correlated single-photon counting—TCSPC). Occasionally, frequency-domain technology can also be used (we will not present experiments on frequency domain technology in this chapter). For time-domain experiments, the data set consists of time-dependent series of observations of the signal intensities (typically photon counts) that are convolutions of time-dependent fluorescence intensities and instrument response function (IRF). In fact, the instrument distorts the true intensity decay and we need to use analytical methods to separate the pure undistorted fluorescence intensity decay from the intensity decay convoluted with the excitation pulse (response function). Such deconvolution procedures of experimental data are complex and are frequently lengthy processes. Typical deconvolution methods are divided into two groups: those requiring an assumption of the functional form of the undistorted system response function (assumption of the decay law) and those that directly give the undistorted system decay law without any assumption about the decay law. In this chapter, we will only discuss the first group assuming an exponential decay form. But we want to stress that in the frequency domain physically the same deconvolution procedure is applied.

In a time-resolved experiment, we are measuring the quantity of time, t that is the value that is set by the instrumentation and we call it an *independent variable*. The actually measured quantities (number of counts, N, or corresponding intensity, I) by the experimental protocol, we call the *dependent variables*.

The goal for carrying out a time-resolved experiment and data analysis is to recover the actual undistorted intensity decay function, $I(t)$ separated from the instrument response function (IRF) that has to be evaluated independently. To recover the true fluorescence intensity decay, we assume physical models and examine it by mathematical analysis methods, typically least-square data analysis. All linear and nonlinear least-square data analyses are based on a series of assumptions that must be strictly obeyed in order to obtain valid results. Assuming the exponential model for intensity decay data analysis of the given arbitrary set of data, N_j and t_j, we find a set of parameters, $N_0\alpha_i = N_i$ and τ_i, of the fitting function, $I(N_0\alpha_j, t_j)$ that have the maximum likelihood (highest probability) of being correct. Furthermore, the obtained function for these optimal parameters must be a reasonable description of the actual experimental data in the absence of experimental uncertainties. Different regression algorithms have been developed to numerically extract the best satisfying parameter values. It can be proven mathematically that a typical least-square analysis correctly estimates the parameter values if the data (both the function $I(N_i, t_j)$ and the distribution of experimental uncertainties—errors) satisfy a series of assumptions. These assumptions are:

1. All of the uncertainties of the experimental data are only attributed to the dependent variables, N_i.
2. The experimental uncertainties of the dependent variable I_i obey a Gaussian distribution centered at the correct value of the dependent variable.
3. There are no unknown/uncontrollable systematic uncertainties in either the independent, t_i, or dependent, N_i, variables.
4. The assumed fitting function, $I(N_i, t_j)$ is the correct mathematical description of the experimental data.
5. There is a sufficient number of data points to yield a good random sampling of the random experimental uncertainty.
6. Each data point must be an independent observation.

The time-dependent analysis considers that the observed time-dependent signal is a convolution of undistorted fluorescence intensity decay with the excitation profile of IRF. The corresponding fitting function, $I(N_i, t_i)$, is a convolution integral of the true fundamental fluorescence response, $I(\alpha_i, t)$, and the instrument response function (IRF) also called lamp function, $L(t)$:

$$I(\alpha_i, \tau_i, t) = \int_0^{t_j} L(t)\, N(\alpha_i, t_j - t)\, dt \tag{7.4}$$

where the integration over time, t, is from time 0 to t_j corresponding to the time value of the particular data point. The vector of fitted parameters, α, consists of two values, the initial (time zero) number of counts, N_0, and fluorescence lifetime, τ. Consequently, one must solve the integral equation (7.4) that involves the undistorted luminescence system response function (undistorted intensity decay) and the independently measured time-dependent IRF. *Note:* Traditionally, IRF was referred to a pulse width but now we need to remember that with easily accessible ultrafast laser technology, the limitation is various instrumental factors like detector, electronics, or other optically active elements. The important typically untold requirements are the fact that IRF should be obtained in identical conditions as our original fluorescence intensity decay measurement. So, it is obvious that for example a 470 nm pulse from diode laser will be identical but detected response with the 470 nm observation could be very different than response detected during fluorescence intensity decay that is measured at 530 nm. The timing of the detector, electronics, and other elements may respond slightly different for different photon energy.

The calculated fluorescence intensity decay depends on fitted parameters, α_i, and τ_i. The difference between the measured number of counts (observed intensities), I_{ob}, and calculated number of counts (intensity), I_{cal} we call residual ρ_j for the j-th data point:

$$\rho_i = I_{ob}(t_{\sum j}) - I_{cal}(t_j) \tag{7.5}$$

The deconvolution problem reduces to obtaining the coefficients (fitted parameters) that minimize residuals ρ_i for all of the data points. The quality of the fit can then be judged by the magnitude of the sum of the weighted squares of the residuals as defined by:

$$\Phi = \sum_{j=1}^{n} w_j (I_{ob}(t_j) - I_{cal}(t_j))^2, \tag{7.6}$$

where t_j is the time at the j-th time interval for which a measurement or calculation is made (typically for photon-counting it corresponds to the channel number), and w_j is the weighting factor given to the square deviation between observed and calculated values. The weighting factor, w_i, generally will depend upon the experimental setup and is related to the corresponding data point variance, s_j^2:

$$s_j^2 = \left[\frac{I_{ob}(t_j) - I_{cal}(t_j)}{\sigma_j}\right]^2 \tag{7.7}$$

by the relation:

$$w_j = \frac{1/s_j^2}{(1/n)\sum_{j=1}^{n}(1/s_j^2)} \tag{7.8}$$

where σ_i is the standard deviation of the random experimental uncertainty for a particular data point. For photon-counting techniques, the statistical noise typically obeys Poisson statistics, approaching a Gaussian distribution at large numbers of counts. The standard deviation, σ_i, is proportional to the square root of the number of photons. The overall goodness-of-fit for all data points is the sum of the corresponding variances usually called chi-square (χ^2):

$$\chi^2 = \sum_{j=1}^{n} s_j^2 = \sum_{j=1}^{n} \left[\frac{I_{ob}(t_j) - I_{cal}(t_j)}{\sigma_j} \right]^2 \tag{7.9}$$

χ^2 depends on the number of data points and is not a convenient parameter to judge the goodness-of-fit. A much more convenient way is to use the value of so-called reduced chi-square, χ_R^2, which represents the standard weighted least-squares for the set of experimental data points and is given by:

$$\chi_R^2 = \frac{1}{NDF} \sum_{j=1}^{n} \chi_j^2 = \frac{1}{NDF} \sum_{j=1}^{n} \left[\frac{I_{ob}(t_j) - I_{cal}(t_j)}{\sigma_j} \right]^2 \tag{7.10}$$

where NDF is the number of degrees of freedom defined as the number of independent experimental observations (for time-domain the number of data points), n, minus the number of fitting parameters (number of floating parameters), p; NDF = $n-p$.

It is important to know how one can interpret the quality of curve-fitting (least-squares analysis) in order to judge the correctness of the assumed model. Various models (mathematical expressions) may yield reasonable fits to the set of experimental data points (reasonable value of χ_R^2), so the sum of the squares of the residual in eq. (7.6), Φ, is at a minimum with the magnitude determined by the intrinsic noise of the experiment.

There have been multiple approaches developed to test uncertainty of the curve fitting procedure in estimating the quality of data analysis and precision of fitted parameters. The expectation of experimental data analysis by a least-square procedure is not only to evaluate the parameters, α_i and τ_i, which have the maximum likelihood of being correct but also to assess a reasonable range of error for the estimated parameters, which will provide a measure of the overall precision of the estimated parameters. Unfortunately, there are no general approaches for estimating the reliability of the extracted values by nonlinear least-square procedures. Even if the uncertainties estimated from the least-square analysis are reported by most fitting programs, they are not always adequately considering parameter correlation and they are smaller than the actual uncertainties. We would warn the experimentalist to not completely trust to the degree of precision reported by some data analysis software.

One of the most common approaches to evaluate the range of estimated parameters is the procedure called support plane analysis. For regular analysis with all floating parameters yields, a minimum (smallest) value of χ_R^2 we call χ_R^2 (min). To examine the quality of estimated parameters, we would monitor changes in the value of χ_R^2 when one parameter is varied around its originally estimated value. In practice, the idea is to fix one parameter to a value different from its best estimate and then rerun the least-square analysis allowing the other parameters to adjust to a new minimum with the new value of χ_R^2. This new value obtained with one fixed parameter we call $\chi_R^2(fp)$. We increase the value of the fixed-parameter and repeat this routine each time allowing all other parameters to adjust to a new minimum. The procedure is repeated until the $\chi_R^2(fp)$ value for a given value of the parameter exceeds an acceptable value. Typically, the routine is repeated until the new value $\chi^2_R(fp)$ exceeds the value predicted by F-statistics for a given number of parameters (*p*), the number of degrees of freedom (NDF) and the chosen probability level, *P*. Probability, *P* describes the probability that the value of the *F*-statistic (ratio of chi-squares for two fits) is due to only random error in the data related to the standard deviation of the experiment. For a one-standard-deviation confidence interval, the value of *P* is approximately 67% and for two-standard deviation confidence intervals, the value of *P* increases to about 95%. We express *F* as:

$$F = \frac{\chi^2_R(fp)}{\chi^2_R(\min)} = 1 + \frac{p}{\text{NFD}} F(\alpha_i, \tau_i, \text{NFD}, P) \tag{7.11}$$

where *F*(*p*, NDF, *P*) is the *F*-statistic value with parameters α_i and τ_i, and NDF with a probability of *P*. The typical values for *F*-statistic can be found in the literature. The supporting plane analysis is probably the most convenient approach to graphically represent uncertainty in the values of determined parameters. In Experiments 7-8 and 7-11, we are presenting the detailed results of such analysis. As an example of support plane analysis, we present below the visualization of confidence intervals calculation. In Experiment 7-2, we measured a lifetime of an efficient fluorophore, Fluorol 7GA in ethanol. The recovered lifetime from the analysis with the program FluoFit4 (from PicoQuant, GmbH) revealed a lifetime of 9.05 ns with the accuracy of ±0.02 ns. *How was this accuracy estimated?* We fixed the lifetime at values near the lifetime (9.05 ns) and repeated the analysis. At each fixed value, we recorded χ^2_R. After normalization (calculating ratios of $\chi^2_R/\chi^2_{R\min}$) we plot the dependence of normalized χ^2_R as a function of a lifetime, see Figure 7.3.

In this case, and for most single-exponential decays, the dependence of normalized χ^2_R on the fixed lifetime is a symmetric parabola. Therefore, lower- and upper-confidence intervals are identical and we can estimate the accuracy with ±. In more complex decays, often the parabolic dependence is not symmetrical and the lower confidence can be different from the upper confidence interval.

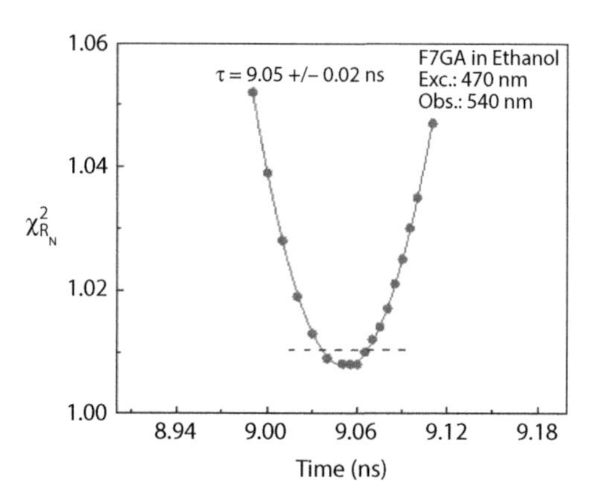

FIGURE 7.3 The support plane analysis (χ^2_R analysis) of Fluorol 7GA in ethanol lifetime. The dashed horizontal line corresponds to 67% of confidence. In TCSPC, this confidence corresponds to about 1.01 value of normalized χ^2_R.

EXPERIMENT 7-1 IMPULSE RESPONSE FUNCTION (IRF)

Keywords: fluorescence, fluorometer, lifetime reference.

As we discussed in the "Introduction" the important part of time-domain measurements is to obtain the best most accurate impulse Response function (IRF). The IRF contains multiple elements that include detector response time, electronic response time, wavelength (color effect), and geometrical factors. However, for a given instrument, we typically cannot change the detector or electronics that are in most cases fixed. But solid-state laser diodes and different cuvettes are now commonly available and very convenient to use. Since, in many instances, measured fluorescence lifetimes are short, the response to the excitation will always be convoluted with the excitation profile. In the case of time-domain measurements when we use fast detectors the excitation pulse-width and its shape could be a significant factor deciding about the appropriate resolution of short components. Unfortunately, the operation of the solid-state lasers depends on their power. In this experiment, we will demonstrate how the excitation pulse depends on the cuvette used and on the power applied to the laser diode. We will use Multi Channel Plate Photo-Multiplier (MCPPMT) detector that has very fast response much below 100 ps.

MATERIALS

- Water, silica nanospheres (ludox)
- Glass or quartz 1 cm × 1 cm cuvette, 1 mm × 10 mm cuvette, 2 mm × 10 mm cuvette, black tape, neutral density (ND) filters

EQUIPMENT

- Fluorometer FT200 (PicoQuant, GmbH, Germany) equipped with Hamamatsu MCPPMT detector
- Pulsed 470 nm laser diode

METHODS

1. Make a solution of the silica nanospheres (scatterer) by adding a drop of concentrated ludox to the cuvette filled with the water.
2. Measure time responses of the scattered light. Use a standard 1 cm × 1 cm cuvette. For the impulse function response (IRF) measurement, set the observation wavelength to the excitation wavelength (we would typically recommend being slightly of the excitation wavelength toward shorter wavelength; e.g., for 470 nm excitation we would set observation monochromator between 467 and 470 nm). *Note*, it is typically better to be on a shorter side to avoid potential leaking of the excitation pulse, sample emission, or Raman scattering. Use ND filter(s) on the excitation and detection paths. Optimize the laser operation by adjusting the power supply in order to obtain the shortest and "clean" pulses.
3. Repeat measurements with "underpowered" and "overpowered" laser diode. If you do not know the right power for the laser, adjust the power to a point where the laser starts operate and increase power in small increments each time measuring the response.
4. Repeat these measurements with 1 mm × 10 mm and 2 mm × 10 mm cuvettes placed in the holder that the excitation goes through the shorter path. Alternatively, you can place a slit on the observation front of the cuvette. Such slits can be constructed from a black electric tape (see Experiment 3-12).

(*Continued*)

EXPERIMENT 7-1 (CONTINUED) IMPULSE RESPONSE FUNCTION (IRF)

RESULTS

Most of the laser diodes available today will have regulated power output. This allows for easy and convenient regulation on the laser intensity but also it is affecting the pulse response function (shape of the pulse). It is important to carefully optimize the laser power supply in order to achieve the shortest IRF response. Check the cable connections and gently apply power on the laser. When changing the power (output of the laser diode), you will change peak power and pulse width but may also affect starting time (firing time for diode—the point where intensity starts rising up) and the slope of the rising edge of the pulse.

Next, we regulate the power and measure the pulse response. Repeating the measurements for different power settings and displaying measured pulses in one window allows for an easy way of determining the optimal pulse shape with desirable excitation power. As an example, in Figure 7-1.1, we are presenting a typical pulse response for the 470 nm laser diode as measured in a 10 mm × 10 mm fluorescence cuvette with MCPPMT detector. The upper graph shows the traditional logarithmic intensity scale and bottom linear intensity scale. The logarithmic scale clearly exposes pulse shape at the background level (pulse decaying edge).

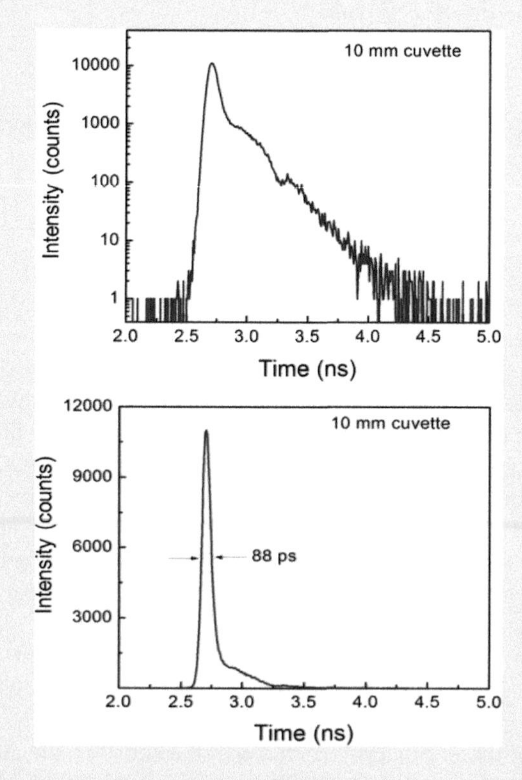

FIGURE 7-1.1 IRF measured for 470 nm laser diode in 1 cm × 1 cm cuvette. Top: logarithmic scale; Bottom: linear scale. The arrows point to the full width at the half of the maximum (FWHM).

(Continued)

EXPERIMENT 7-1 (CONTINUED) IMPULSE RESPONSE FUNCTION (IRF)

Laser Power Dependence. We want to test if the pulse shape depends on the laser power level. This will depend on the laser and the example we are presenting is for a typical laser diode. We will consider low power (underpowered diode) and high power (over-power diode) diode.

Under Power. What happens to the IRF if the power delivered is too low? Having a laser running with an average power adjust the power down. Remember, when adjusting the power down to a certain low power, the laser will shut down, so lower the laser power to a point where the laser is still working and then measure IRF. Repeat with slightly lower or higher power.

In Figure 7-1.2, we are presenting an example of a laser diode working at two power levels; low power (top) and low close to the threshold of shutting down (bottom).

It is clear that when lowering the power the pulse shape changes and the pulse width at half maximum (FWHM) increases. The increase can be very significant. Measuring the fluorescence intensity decay with higher power and IRF with lower it may result in shortening of the recovered (deconvoluted) intensity decay.

Over Power. What happens to the IRF if the laser power is too high? We can typically adjust the laser power up to the maximum allowed by the vendor (end of the scale).

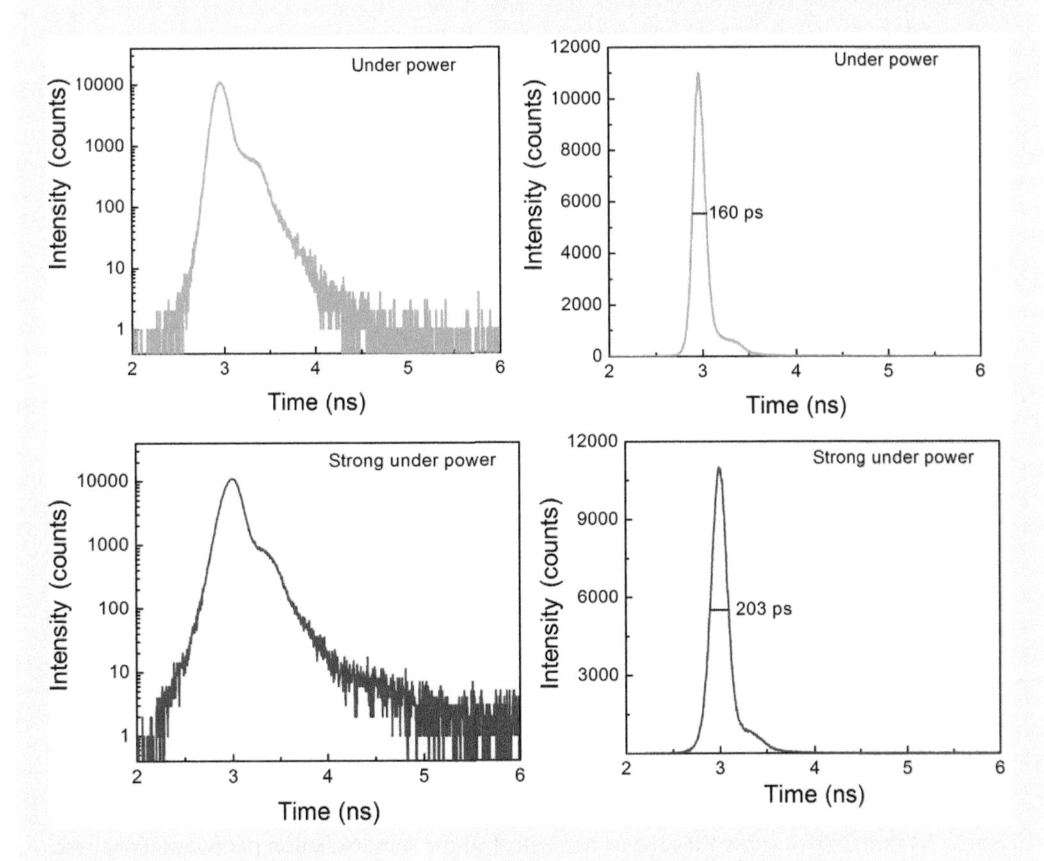

FIGURE 7-1.2 Effect of under-power on IRF. Left: logarithmic scale; right: linear scale.

(Continued)

EXPERIMENT 7-1 (CONTINUED) IMPULSE RESPONSE FUNCTION (IRF)

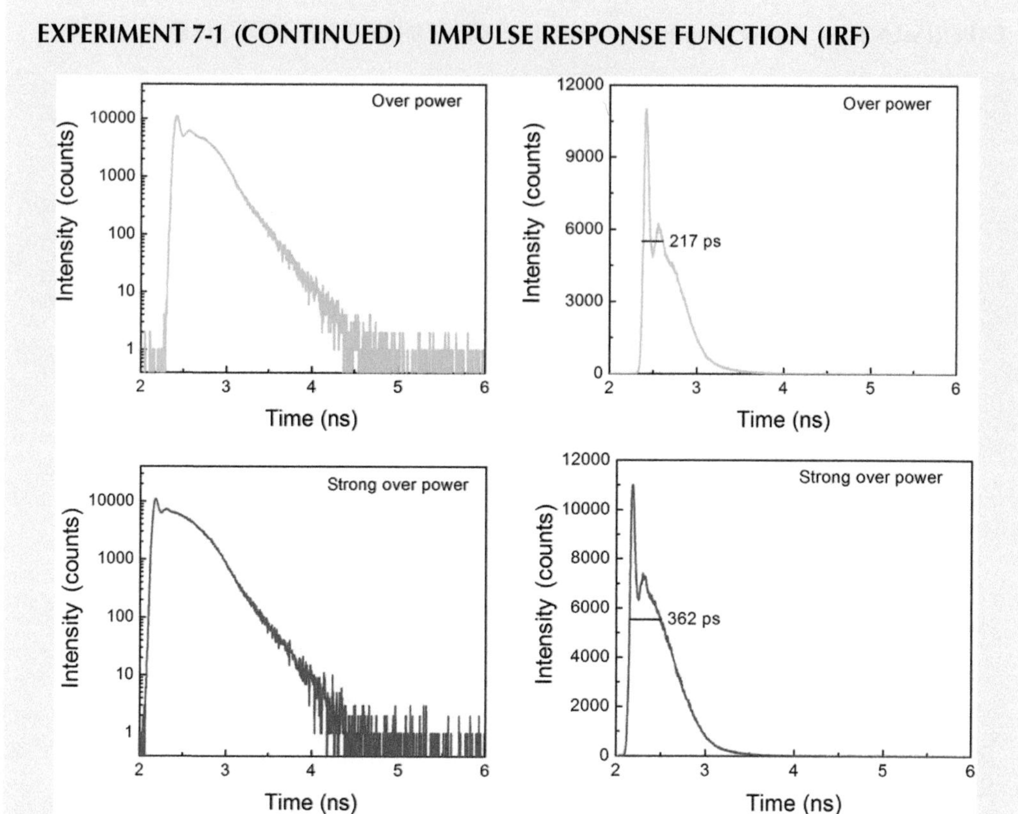

FIGURE 7-1.3 Effect of over-power on IRF. The same scatterer and experimental set-up were used as for "Under Power" experiment. Left: logarithmic scale; right: linear scale.

This typically may happen when the sample emission is low. In Figure 7-1.3, we are presenting measured IRF with the power higher than optimal (top) and power at the maximum value (bottom).

Higher power on the laser driver results not only with large FWHM of the IRF pulse but also in significant changes in the pulse shape.

Path Length Dependence. Next, we tested if the IRF response depends on the path length of the cuvette. For this experiment, we used 1 mm × 10 mm and 2 mm × 10 mm cuvettes and measure IRFs for 1 and 2 mm pathways. For most spectrofluorometers, the observation spot is at the center of the sample holder, typically 2–3 mm in size. As shown in the Experiment 4-5, the detector typically sees not more than this spot size. The reader may ask the question if a difference of a couple of millimeters can make a difference. But it is important to realize that light travels in 1 ns distance of about 30 cm and to travel the distance of 1 mm in water ($n = 1.33$) will take almost 5 ps. So, even a very small geometrical difference in the order of millimeter can make a significant change to the IRF.

Figures 7-1.4 and 7-1.5 show IRFs measured with 1 and 2 mm excitation pathways. For these measurements, we used the optimal setting for the excitation laser diode.

(*Continued*)

EXPERIMENT 7-1 (CONTINUED) IMPULSE RESPONSE FUNCTION (IRF)

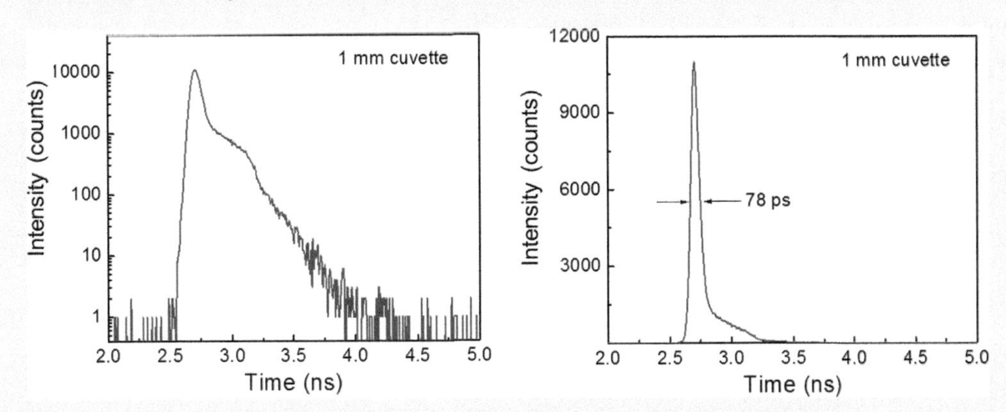

FIGURE 7-1.4 IRF observed with 1 mm × 10 mm cuvette. Left: in logarithmic scale; right: in linear scale. Arrows indicate FWHM.

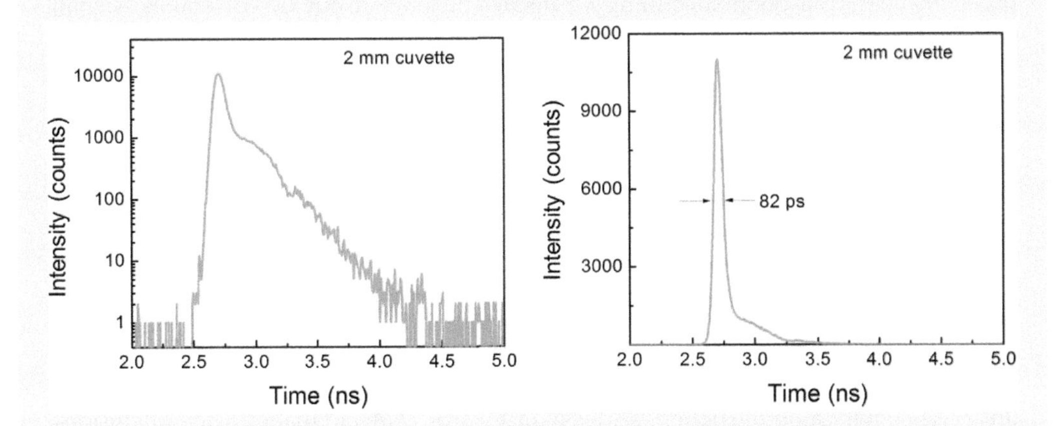

FIGURE 7-1.5 IRF observed with 2 mm × 10 mm cuvette. Left: logarithmic scale; right: linear scale.

It could be a surprise but FWHM depends on the size (width) of the cuvette and the difference between 1 and 2 mm path can be clearly detected. When comparing these results to a 1 cm cuvette (Figure 7-1.1), we can still see a small broadening. But it is important to remember that it is difficult to compare pulse shapes between different experiments.

CONCLUSIONS

These simple tests showed that measured pulse shape depends on laser power and cuvette size. The significance of these effects will depend on what lifetimes we want to measure. If our experiment is to measure lifetimes longer than 1 ns, the results will not depend on what size of the cuvette we are using, but if we are looking for very short components (below 100 ps), this could be a very significant factor. The broadening of the pulse due to the power

(Continued)

EXPERIMENT 7-1 (CONTINUED) IMPULSE RESPONSE FUNCTION (IRF)

level will have a similar effect. But, in principle, when we are able to keep a constant power through the experiment (sample and IRF measurements) the deconvolution routine will give proper results. However, the reader should remember that an underpowered or overpowered laser diode may frequently have an unstable output. So, even if we are not changing our power setting it would not be a surprise to see significant changes. So, it is recommended to properly adjust power on the laser driver and check the IRF response frequently.

EXPERIMENT 7-2 NUMBER OF COUNTS IN MAXIMUM

Keywords: fluorescence, fluorometer, lifetime, time-domain.

As in any measurement, the quality of recovered information (in our case fluorescence lifetime) will depend on a number of measured points. In steady-state experiments, in order to measure a good spectrum, we needed to have a number of counts (signal) significantly above the background and experimental noise. There are different ways of measuring intensity decays in the time-correlated single-photon approach. We could use the constant time for data collection, collect a dedicated total number of counts, or collect a dedicated number of counts at the peak. The most common approach is to use a set number of points at the peak. However, in anisotropy decay measurements, we have to use a constant time while collecting data for polarized components. Setting a time for collection requires prior knowledge of the decay to obtain a satisfactory (statistically meaningful) number of points. Setting a total number of points provides a comparison of lifetimes that are significantly different. In most of our experiments, we will use a number of points (so-called number of counts) in the peak. One of most often questions asked by students starting lifetime experiments is: "What should be a number of counts at maximum for good lifetime measurements?" First, we will check this for a simple, mono-exponential decay of Fluorol 7GA in ethanol. The conclusions of this experiment apply only to single, relatively long fluorescence lifetimes. For more complex decays (multiple fluorescence lifetimes), the required number of points will significantly increase.

MATERIALS

- Water, silica nanospheres (ludox)
- Neutral density (ND) filters
- Fluorol 7GA, ethanol
- Glass or quartz 1 cm × 1 cm cuvette

EQUIPMENT

- Fluorometer FT 200 (PicoQuant, GmbH, Germany) with 470 nm laser diode
- Spectrofluorometer (Varian Inc., Eclipse), Cary 50 Spectrophotometer (Varian Inc.)

(*Continued*)

EXPERIMENT 7-2 (CONTINUED) NUMBER OF COUNTS IN MAXIMUM

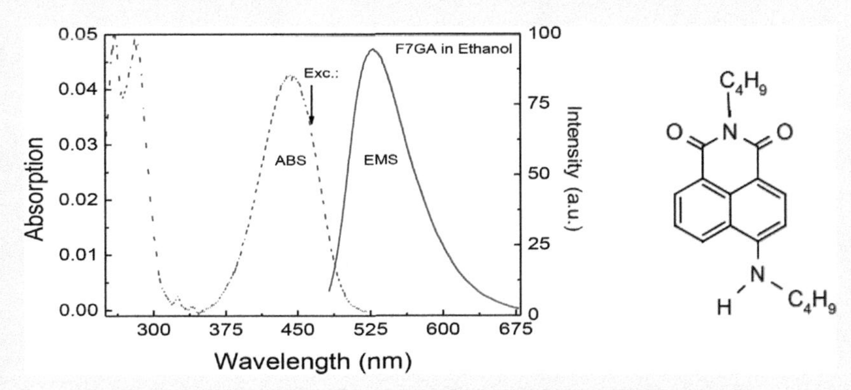

FIGURE 7-2.1 Absorption and fluorescence spectra of F7GA in ethanol and chemical structure of F7GA.

METHODS

1. Make a solution of the silica nanospheres (scatterer) by adding a drop of concentrated ludox to the cuvette filled with the water and measure IRF. *Note: Make sure the ludox is fresh and not contaminated. When stored for a long time it may grow bacteria preventing its use with UV excitation.*
2. Make a solution of Fluorol 7GA in ethanol with an optical density of about 0.05. Measure absorption and fluorescence spectra, see Figure 7-2.1.
3. Measure time responses of Fluorol 7GA for different counts in maximum.
4. Compare obtained lifetimes of Fluorol 7GA.

RESULTS

Fluorol 7GA (F7GA) has a strong green fluorescence emission with a maximum of about 530 nm, see Figure 7-2.1.

For our experiments, we set a number of counts collected for IRF at the maximum to 10,000 (see Experiment 7-1 for the IRF measurement). We will monitor a fit quality as we change the number of points collected at the maximum when measuring the sample (Fluorol 7GA, 530 nm observation). In Figure 7-2.2, we are presenting measured responses for IRF and for Fluorol 7GA when the maximum number of points collected in the maximum was set to 20,000 for Fluorol 7GA. The recovered fluorescence lifetime and error are presented in the figure. In Figure 7-2.3, we are presenting decays and results when we collected for the Fluorol 7GA 10,000 counts and following figures (Figures 7-2.4 through 7-2.7) show results for 5,000, 1,000, 500, and 100 counts at the maximum, respectively.

The accuracy of the measurement is given in the form of ± value following the value of the recovered lifetime. This ± number is estimated from the confidence intervals calculated for 67% of confidence. In practice, the confidence interval for a given parameter can be found by analyzing values of the goodness-of-fit $\left(\chi_R^2\right)$ at values near the value of the parameter (for more details see "*Introduction*" to this chapter). For example, fixing the lifetime of F7GA at values near 9 ns and running the analysis, again and again, we can collect values of χ_R^2 for different

(*Continued*)

EXPERIMENT 7-2 (CONTINUED) NUMBER OF COUNTS IN MAXIMUM

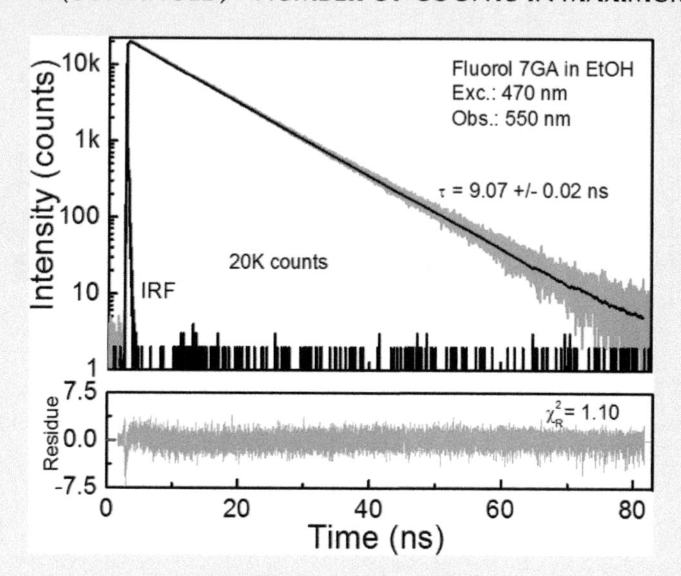

FIGURE 7-2.2 Fluorescence lifetime measurement of Fluorol 7GA in ethanol. The data collection has been stopped at 20,000 counts in maximum. The IRF is an impulse response from a ludox scatterer.

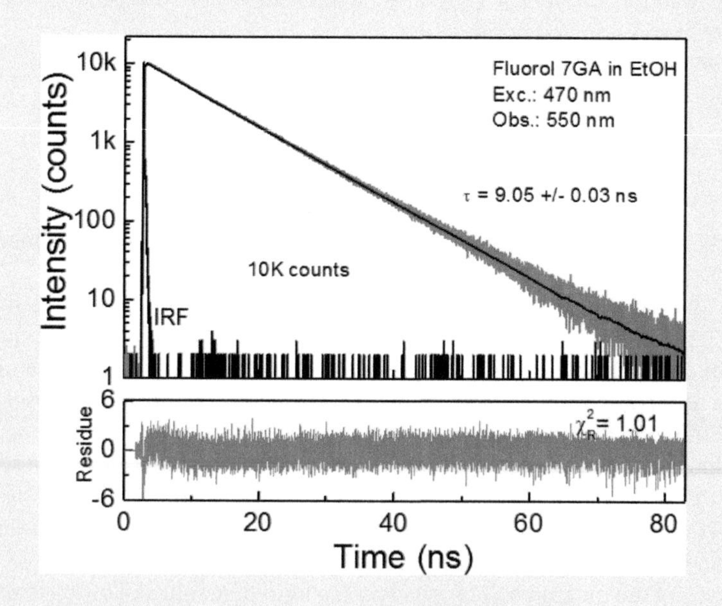

FIGURE 7-2.3 Fluorescence lifetime measurement of Fluorol 7GA in ethanol. The data collection has been stopped at 10,000 counts in the maximum.

(Continued)

EXPERIMENT 7-2 (CONTINUED) NUMBER OF COUNTS IN MAXIMUM

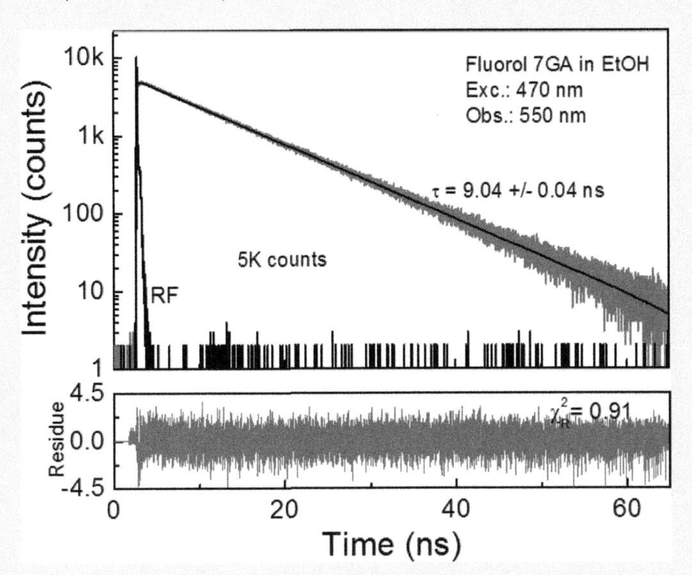

FIGURE 7-2.4 Fluorescence lifetime measurement of Fluorol 7GA in ethanol. The data collection has been stopped at 5,000 counts in the maximum.

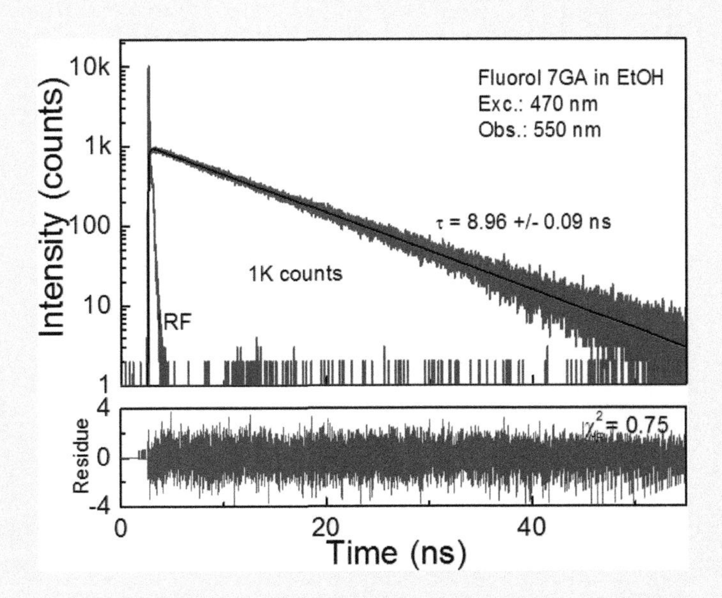

FIGURE 7-2.5 Fluorescence lifetime measurement of Fluorol 7GA in ethanol. The data collection has been stopped at 1,000 counts in the maximum.

(Continued)

EXPERIMENT 7-2 (CONTINUED) NUMBER OF COUNTS IN MAXIMUM

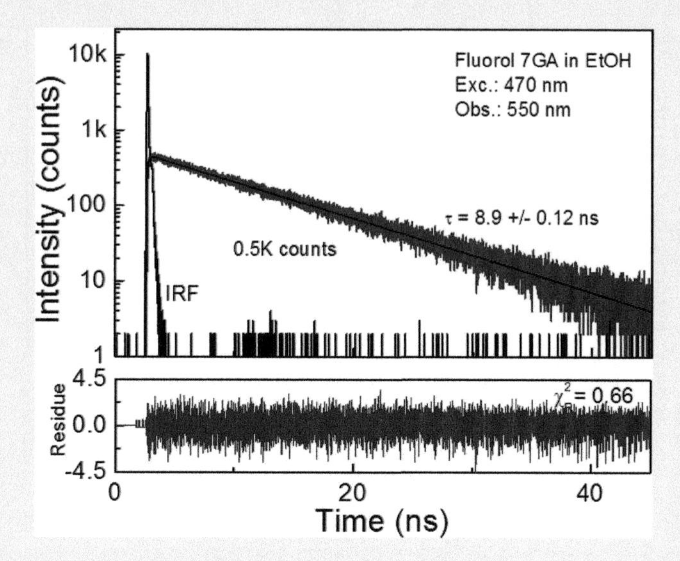

FIGURE 7-2.6 Fluorescence lifetime measurement of Fluorol 7GA in ethanol. The data collection has been stopped at 500 counts in the maximum.

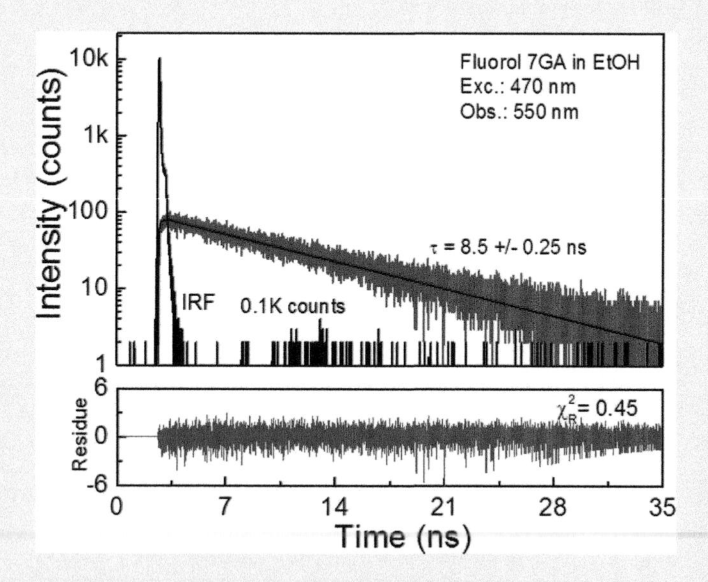

FIGURE 7-2.7 Fluorescence lifetime measurement of Fluorol 7GA in ethanol. The data collection has been stopped at 100 counts in the maximum.

fixed lifetimes. After normalizing the χ^2_R values, the plot of χ^2_R normalized values $\left(\chi^2_{RN}\right)$ will reveal a parabola-like dependence with a minimum at the studied parameter value (called also a χ^2_R − surface or support-plane analysis). The range of the confidence intervals is given by the normalized value of χ^2_{RN}. In TCSPC (time-domain), where the number of data points is large, this value is about 1.01. For details of the confidence intervals calculation, we refer to the

(Continued)

EXPERIMENT 7-2 (CONTINUED) NUMBER OF COUNTS IN MAXIMUM

Principles of Fluorescence Spectroscopy by J.R. Lakowicz and references there. We will use the visualization of confidence intervals (χ^2_R – surfaces) analyzing multi-exponential decays.

CONCLUSIONS

When analyzing Figures 7-2.1 through 7-2.6, we can safely conclude that down to 5,000 counts at the maximum the results are statistically equivalent. For 1,000 counts, the error is a little larger but not dramatically different. However, for 100 counts at the maximum results are not acceptable. A number of counts collected for the decay is an arbitrary number. From our experience, we consider 10,000 counts at the maximum sufficient for most experiments. A number of our colleagues consider it as not always sufficient, and the same number as too high for many experiments. Obviously, the obtained results would statistically improve as we increase the number of points but one has to remember that collecting more counts takes time and makes the experiment longer that may affect the sample. For example, the recovered difference between measurement with 20,000 counts and 10,000 counts has no statistical relevance but time is twice longer for an experiment where we collect 20,000 counts.

A standard of 10,000 counts at the maximum is an overkill for simple single exponential decays but for more complex multi-exponential decays that we will discuss later can be on the edge.

EXPERIMENT 7-3 CHOOSING A PROPER FITTING TIME AND REPETITION RATE OF THE EXCITATION PULSE

Keywords: fluorescence, fluorometer, lifetime, time-domain.

In time-correlated single-photon counting (TCSPC) measurements, a histogram of number of counts vs. channels is built over the time of data collection. Since the number of counts is proportional to the measured light intensity (number of photons emitted by the sample) and channel numbers are proportional to the time, usually TCSPC data are presented as the dependence of measured light intensity vs. time, also called a decay time. In the analysis of TCSPC data, a theoretical model is being assumed and the data points are fitted to this model. The most popular is a multi-exponent model and fitting based on least square minimization of error deviations. There are few important factors affecting the data analysis. The number of counts has been already discussed in the previous experiment. Here, we will focus on the repetition rate of the excitation source and on the time limits imposed on the analysis. It is important to set a proper repetition of the pulses in lifetime measurements. If the repetition is too low, the collection time becomes long, sometimes not acceptable because laser sources may have time drifts or the sample may change during the measurements. If the repetition rate is too high, the excited state decay is not completed and the detector sees also photons emitted from molecules excited by previous pulses. This will increase the base-line in the intensity decay. We will demonstrate this in the following experiment.

MATERIALS

- Water, silica nanospheres (ludox)
- Fluorol 7GA, ethanol (EtOH)
- Glass or quartz 1 cm × 1 cm cuvette, neutral density (ND) filters

(*Continued*)

EXPERIMENT 7-3 (CONTINUED) CHOOSING A PROPER FITTING TIME AND REPETITION RATE OF THE EXCITATION PULSE

EQUIPMENT

- Fluorometer FT 200 (PicoQuant, GmbH, Germany) with 470 nm laser diode

METHODS

1. Make a solution of the silica nanospheres (scatterer) by adding a drop of concentrated ludox to the cuvette filled with the water.
2. Measure time responses of the scattered light (IRFs). Set the repetition rate at 2.5 MHz. Use the ND filter(s) on the excitation and detection paths. Optimize the laser operation by adjusting the power supply in order to obtain the shortest and "clean" pulses.
3. Measure the time response for Fluorol 7GA in EtOH at the 2.5 MHz repetition rate.
4. Repeat these measurements with the repetition rate of 20 MHz.

RESULTS

Time Limits in TCSPC Measurements. First, we will analyze the low repetition rate (2.5 MHz) data with different time limits (time range in which we analyze the decay), see Figures 7-3.1 through 7-3.3.

Figure 7-3.1 shows the lifetime measurement for Fluorol 7G in EtOH fitted up to 100 ns.

This fitting is incorrect because the time limit of 100 ns is too high. The collected data above 75 ns are mining-less, contain mostly background noise. It results in a systematic squeeze of residuals and artificially lowers the goodness-of-fit.

Next, we analyze the data until 50 ns, see Figure 7-3.2. This analysis results in a fully random distribution of residuals but is also incorrect because it ends too early and we do

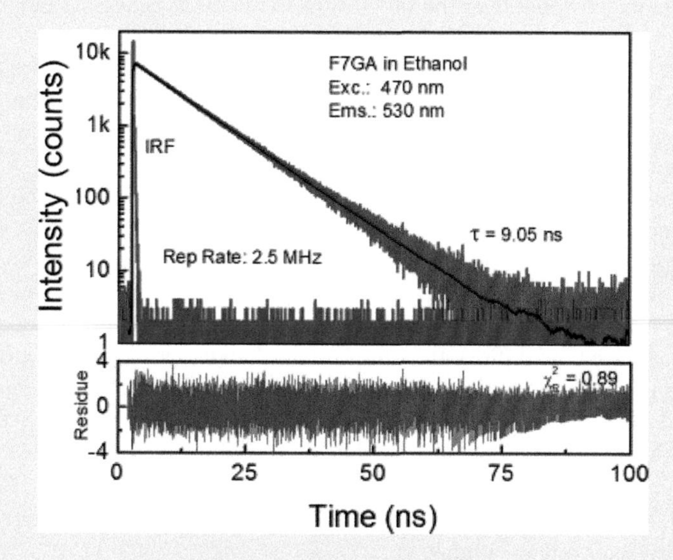

FIGURE 7-3.1 Fluorescence intensity decay of F7GA in ethanol analyzed up to 100 ns. The repetition rate of the 470 nm excitation laser is set at 2.5 MHz.

(Continued)

EXPERIMENT 7-3 (CONTINUED) CHOOSING A PROPER FITTING TIME AND REPETITION RATE OF THE EXCITATION PULSE

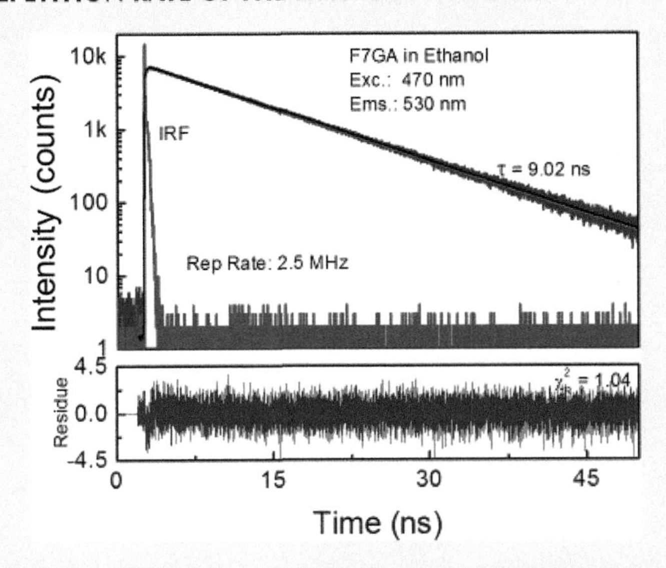

FIGURE 7-3.2 Fluorescence intensity decay of F7GA in ethanol analyzed up to 50 ns. The repetition rate of the 470 nm excitation laser is set at 2.5 MHz.

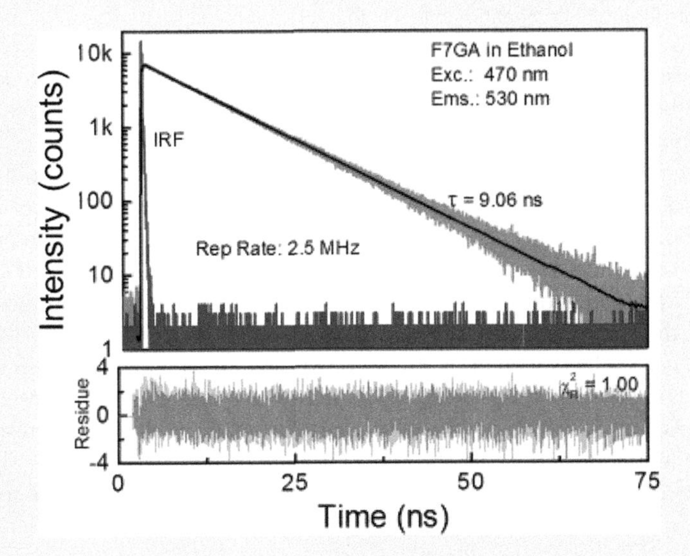

FIGURE 7-3.3 Fluorescence intensity decay of F7GA in ethanol analyzed up to 75 ns. The repetition rate of the 470 nm excitation laser is set at 2.5 MHz.

not know how the decay behaves after 50 ns. In the case of complex decay with a long-time component, the information about this long component is suppressed. Usually, a too-short upper time limit results in a slightly elevated goodness-of-fit.

Figure 7-3.3 presents the correct set-up of time limits in TCSPC measurements. The upper-time limit should only include a small amount of noise background counts to show the entire

(*Continued*)

EXPERIMENT 7-3 (CONTINUED) CHOOSING A PROPER FITTING TIME AND REPETITION RATE OF THE EXCITATION PULSE

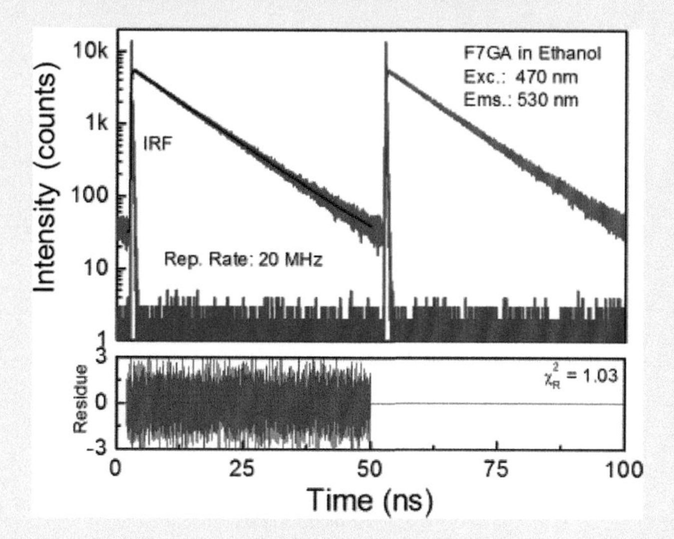

FIGURE 7-3.4 Lifetime measurement of Fluorol 7GA in EtOH with the 470 nm excitation pulses at 20 MHz repetition rate. The repetition rate is too high. The deactivation of excited molecules is not completed in the time between pulses. Although the software provided with a fluorometer can often compensate for too high repetition rate we do not recommend such measurements.

measured intensity. In other words, the upper time limit should end where the decay reaches the background counts. Including too much background at this side of the decay will artificially lower the chi-square value.

The lower-time limit is usually set before the excitation pulse. It should include the entire pulse and only a small trace of noise background before. Before the pulse arrival, the background should be low as it would be expected from the situation when a laser is off (it depends on collection time). Unexpected/unusual increase of background counts at this side is typically an indication that some lifetime component is too long for used repetition rate. Lowering the laser repetition rate will decrease the background as we discuss below.

Pulse Repetition. Selecting an appropriate repetition rate for a given fluorescence lifetime is crucial. *What happens at a higher repetition rate?* If the time between consecutive pulses from the excitation source is shorter than the decay time of the measured fluorescence decay, the analyzed intensity decay will be abruptly corrupted by the next pulse, see Figure 7-3.4.

The intensity decay still can be fitted between pulses but the longer time information can be lost.

CONCLUSIONS

It is important to set proper time limits in the TCSPC data analysis. On many occasions, we observed students "improving" the fit by extending time limits.

Although the software provided with the instrument (fluorometer) can often compensate for too high repetition rate (by including so-called "cyclic excitation"), we do not recommend such measurements. Try always set properly the time limits.

EXPERIMENT 7-4 CHOOSING A PROPER COUNTING RATE IN LIFETIME MEASUREMENTS

Keywords: fluorescence, fluorometer, lifetime, time-domain.

Students often have a tendency to collect lifetime data quickly, setting a too high counting rate (number of counts per second, CPS). But we need to remember that TCSPC electronics is designed to detect single photons for which the probability for photon arrival is equally spread along time. A high counting rate obviously speed-up detection (data collection) but frequently may result in a so-called pile-up effect which distorts the intensity decays and leads to a wrong interpretation of measured lifetimes. This we will demonstrate in the following experiment.

MATERIALS

- Water, silica nanospheres (ludox)
- Fluorol 7GA, ethanol
- Glass or quartz 1 cm × 1 cm cuvette, neutral density (ND) filters

EQUIPMENT

- Fluorometer FT 200 (PicoQuant, GmbH, Germany) with 470 nm laser diode

METHODS

1. Make a stock solution of Fluorol 7GA (see scheme in Figure 7-4.1) in ethanol.
2. Prepare a sample of Fluorol 7GA in ethanol with an optical density (absorption in 1 cm cuvette) below 0.05.
3. Collect time responses of Fluorol 7GA sample in ethanol with low (5×10^4) and high (1×10^6) CPS. Compare lifetime measurements with low and high counting rates.

RESULTS

Figure 7-4.1 presents lifetime measurement of Fluorol 7GA with a counting rate set at 5×10^4 CPS. This is a low/medium counting rate allowing the satisfactory collection of the data in a few minutes. The number of counts in the peak will depend on the lifetime of the fluorophore. 10,000 counts in maximum will be collected quicker for short-lived dyes. With the same CPS, long-lived dyes will require longer time. In order to keep a similar noise background, usually, we set the same time of the data collection for the IRF and fluorophore. However, for IRF the CPS need to be much lower, in range of 10^3 while for dyes usually 10^4–10^5. The decay can be very well approximated with a single lifetime of 9.07 ns.

Next, set the CPS rate at about 1×10^6 and repeat the lifetime measurement. Usually, you do this by removing neutral density (ND) filters on the excitation and/or observation paths. You cannot do this by increasing the gain on the laser/LED light source because it may change the IRF, see Experiment 7-1.

Figure 7-4.2 shows lifetime data collected with too high CPS. The decay cannot longer be approximated by the single-exponential model. The residuals show a systematic (not random)

(Continued)

EXPERIMENT 7-4 (CONTINUED) CHOOSING A PROPER COUNTING RATE IN LIFETIME MEASUREMENTS

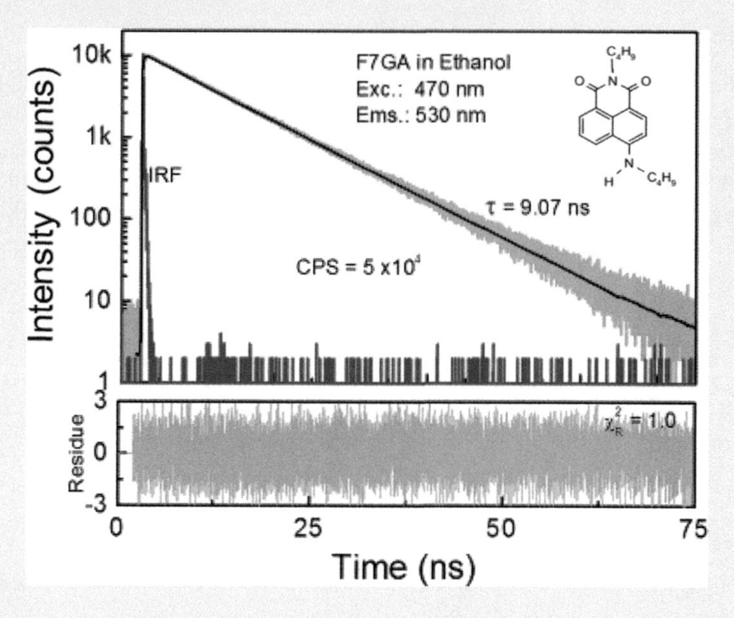

FIGURE 7-4.1 Lifetime measurement of Fluorol 7GA in EtOH with a low counting rate.

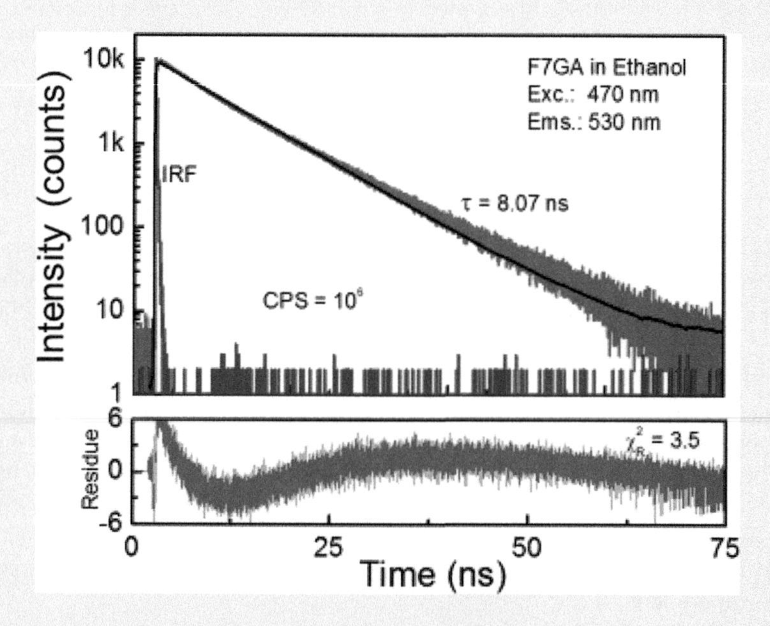

FIGURE 7-4.2 Lifetime measurement of Fluorol 7GA in EtOH with a high counting rate. Single-exponential fit.

(*Continued*)

EXPERIMENT 7-4 (CONTINUED) CHOOSING A PROPER COUNTING RATE IN LIFETIME MEASUREMENTS

deviations and the estimated lifetime is shortened by 1 ns. Of course, the collection time was much shorter with higher CPS.

What are the consequences of using too high CPS? First, the recovered lifetime is too short. Second, and more importantly, the observed fluorescence intensity decay becomes heterogeneous. In many (if not all) time-resolved fluorescence measurements, the information of interest is hidden in changes of the intensity decay profile. This information can be only recovered if there are no artificial changes in intensity decay.

The intensity decay measured with too high CPS, see Figure 7-4.2, clearly show that the decay is not single exponential. We will analyze this decay with the two-exponent model. If the lifetime of the fluorophore is significantly longer than excitation pulse often is enough to analyze just a fluorophore intensity decay, without the reconvolution, ignoring the IRF. Such fitting is called a "tail fitting."

Tail Fitting. In the case of F7GA, the lifetime is 100-fold longer than IRF. Entire IRF is concluded within the first 2 ns. Lets re-analyze the F7GA decay with tail fitting, see Figure 7-4.3.

Note the absence of IRF in Figure 7-4.3. The tail analysis takes into account only the dye intensity decay.

The recovered lifetime is almost exactly as from reconvolution analysis and residuals are randomly distributed.

The intensity decay of the data collected with high CPS requires two exponents to satisfactory fit the data, see Figure 7-4.4.

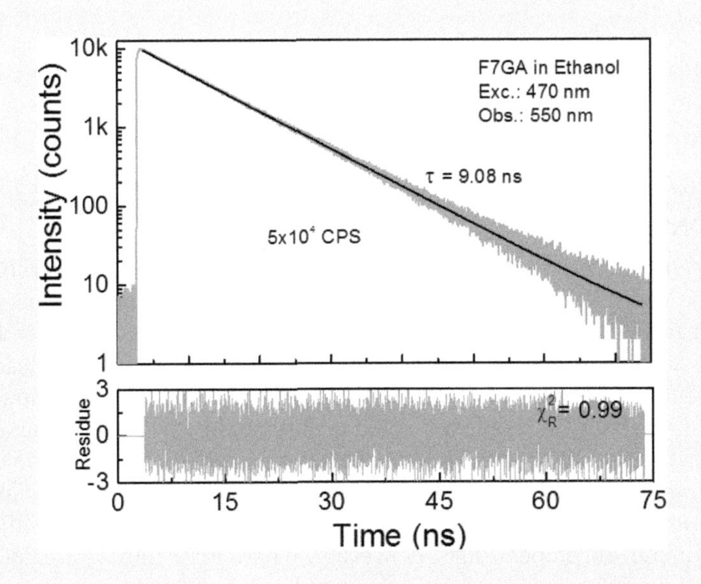

FIGURE 7-4.3 Lifetime measurement of Fluorol 7GA in EtOH with a low counting rate. Tail fitting analysis.

(Continued)

EXPERIMENT 7-4 (CONTINUED) CHOOSING A PROPER COUNTING RATE IN LIFETIME MEASUREMENTS

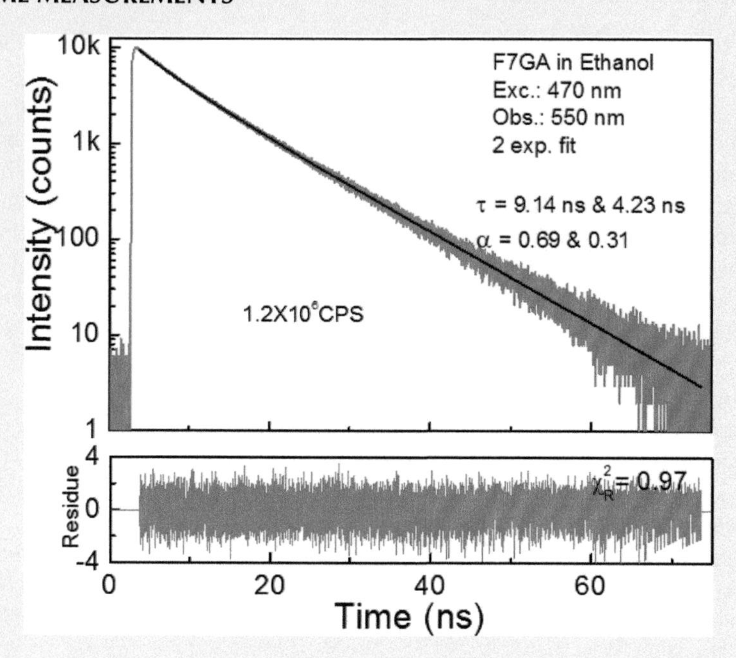

FIGURE 7-4.4 Lifetime measurement of Fluorol 7GA in EtOH with a high counting rate. Two-exponent fit.

It is important to note that proper pulse reconvolution analysis (we recommend for the reader to also reanalyze data when using pulse reconvolution and fitting with multiple exponentials) shows even more drastic differences.

CONCLUSIONS

Too high CPS rate may distort the collected data. The measurements with too high CPS rate can lead to the wrong conclusion about recovered lifetimes. Although the pile-up effect in some extend can be corrected with analyzing programs, we do not recommend using a too high CPS rate. Analysis of data collected with too high CPS rate will typically lead to artificial short components recovered from the decay. We can also ask the question "How to decide if the count rate is too high?" A simple test could be just measure intensity decay with one CPS rate and repeat the measurement with lower CPS (for example 50%). The recovered fluorescence lifetime should be practically the same. Any deviation indicates that higher CPS rate was too high. In the case of lifetimes much longer than the excitation pulse, the tail fitting, without IRF, can provide proper results. However, we have to be sure we are not missing any component.

EXPERIMENT 7-5 CHOOSING A PROPER RESOLUTION IN LIFETIME MEASUREMENTS

Keywords: fluorescence, fluorometer, lifetime, time-domain.

Most spectrofluorometers offer resolutions from a range of single picoseconds/channel to nanoseconds/channel. What is a proper resolution in my lifetime measurement? This is a question asked by almost everyone starting lifetime measurements. For lifetimes much longer than the excitation pulse this is not a big issue. However, for short lifetimes the resolutions must be as high as possible in order to recover a proper lifetime data. We will demonstrate this in the following experiment.

MATERIALS

- Water, silica nanospheres (ludox)
- 3, 3′-Diethyloxacarbocyanine Iodide (DOCI), Rhodamine 110 (R110)
- Ethanol (EtOH)
- Glass or quartz 1 cm × 1 cm cuvette, neutral density (ND) filters

EQUIPMENT

- Spectrofluorometer FT 300 (PicoQuant, GmbH, Germany) equipped with 470 nm laser diode and MCPPMT detector

METHODS

1. Make a solution of the silica nanospheres (scatterer) by adding a drop of concentrated ludox to the cuvette filled with water. Optimize and measure IRF.
2. Make a solution of DOCI and R110 in EtOH with optical densities of about 0.05. Make sure the fluorescence signal is good at the observation wavelength (e.g., 525 nm) and the count rate is appropriate.
3. Measure fluorescence lifetimes of DOCI and R110 in EtOH using the resolution settings of 4 ps/ch (high resolution), 16, 64, and 128 ps/ch. Compare intensity decays for each dye at a different resolution.
4. Make mixture solution of DOCI and R110 so the steady-state intensities will be equal. This is a 50:50 ratio. In practice, the easiest approach is to set laser excitation and measure the respective count rates for DOCI and R110. If the count rate for DOCI is A and R110 is B and you take volume V1 of DOCI you will need to add a volume of R110 $V_2 = V_1$ (A/B) to have identical steady-state intensities. *Note: In practice, we do not need to get exactly 50:50 and anything better than 46:54 or 54:46 will be good enough. Both dyes are in EtOH and mix well (frequently mixing different may result in interactions that will change spectra even if dyes are in the same solvent-see Experiment 3-23).*
5. Measure fluorescence lifetime of the mixture using different resolutions as in #3.

RESULTS

Figure 7-5.1 shows absorption spectra (dashed) and emission spectra (solid) for R110 (black), DOCI (red), and 50:50 mixture of DOCI and R110 (blue).

(Continued)

EXPERIMENT 7-5 (CONTINUED) CHOOSING A PROPER RESOLUTION IN LIFETIME MEASUREMENTS

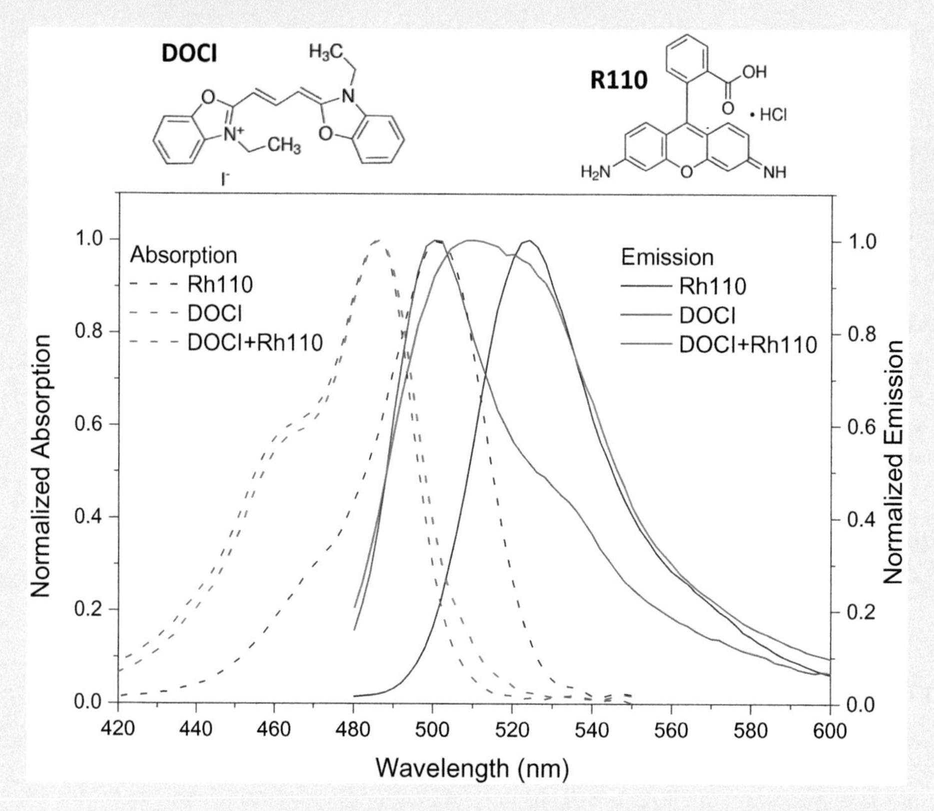

FIGURE 7-5.1 Normalized absorption and fluorescence spectra of DOCI, R110, and 50:50 mixture. The chemical structures are presented above.

DOCI dye, see Figure 7-5.1, is a saturable absorber for a Coumarin 102 dye laser but it is also an efficient laser dye. It has a greenish fluorescence and relatively short lifetime about 150 ps. The Rhodamine 110 (R110) is a popular dye with very high brightness and relatively long fluorescence lifetime of about 3.8 ns.

Both dyes are well soluble in ethanol and the mixture of these two dyes will present a system with two discrete fluorescence lifetimes (short and long lifetime components). The mixture we made is close to 50:50 steady-state intensities contributions. *Note: This is not 50:50 in concentration but 50:50 fluorescence intensity signal contributions as observed with the observation set for lifetime measurement with a given excitation. This is important to keep exactly the same conditions since the detected relative fractions depend on excitation wavelength and observation wavelength. The absorption of the mixture is only slightly different from the absorption of DOCI alone since only a small amount of R110 results in comparable emission signal at 525 nm observation.*

(Continued)

EXPERIMENT 7-5 (CONTINUED) CHOOSING A PROPER RESOLUTION IN LIFETIME MEASUREMENTS

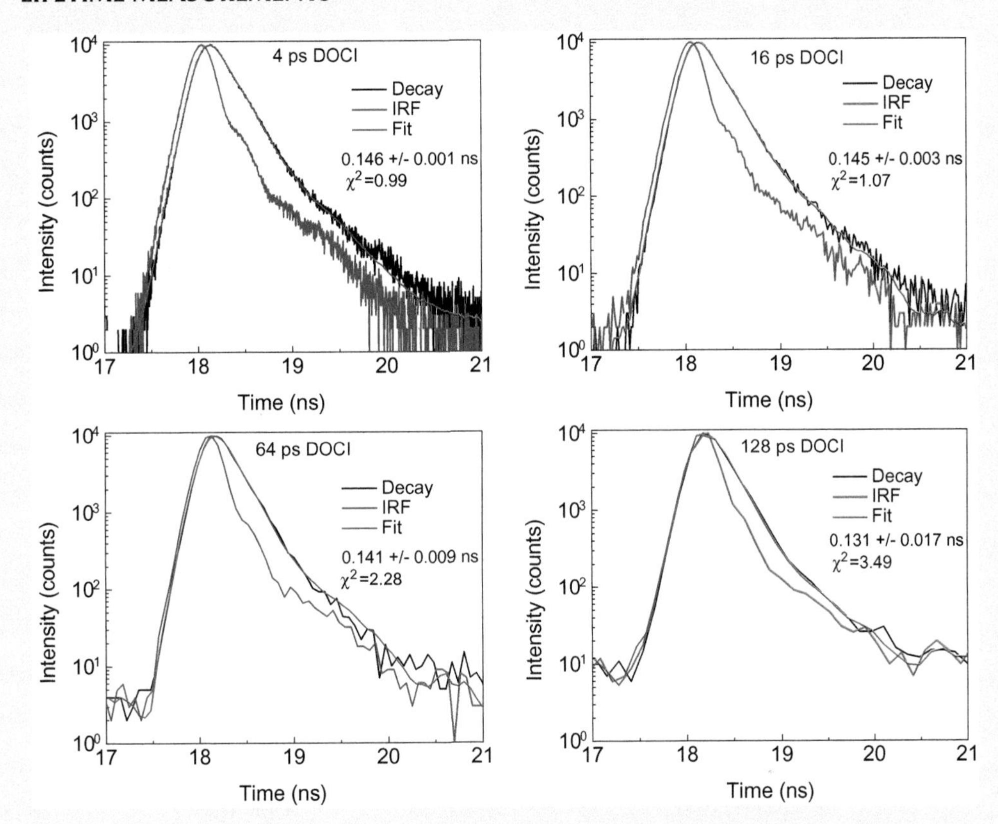

FIGURE 7-5.2 Intensity decays for DOCI in EtOH measured with different resolutions. IRF (blue), fluorescence intensity decays of DOCI in EtOH (black), and fits (red).

Next, we measure the fluorescence intensity decays for DOCI, R110, and 50:50 mixture in ethanol using different temporal resolution (4, 16, 64, and 128 ps/ch). In Figure 7-5.2, we show DOCI intensity decays measured with different resolutions. In the figure, we also present IRFs and resulting fits. The intensity decay of DOCI, although very short, is visibly distinguishable from the IRF. However, we need to remember that the MCPPMT detector is one of the fastest detectors with the response time below 100 ps. For 4 ps resolution, the decay can be satisfactorily fitted with a single exponent of about 0.146 ns. Intensity decay measured with 16 ps resolution still fits well with a single exponent. Only the error is slightly higher. For 64 ps resolution, the fit results in a shorter lifetime and significantly higher error. Finally, the measurement done with 128 ps resolution gives a value of recovered lifetime significantly different with much higher over 10% error.

(Continued)

EXPERIMENT 7-5 (CONTINUED) CHOOSING A PROPER RESOLUTION IN LIFETIME MEASUREMENTS

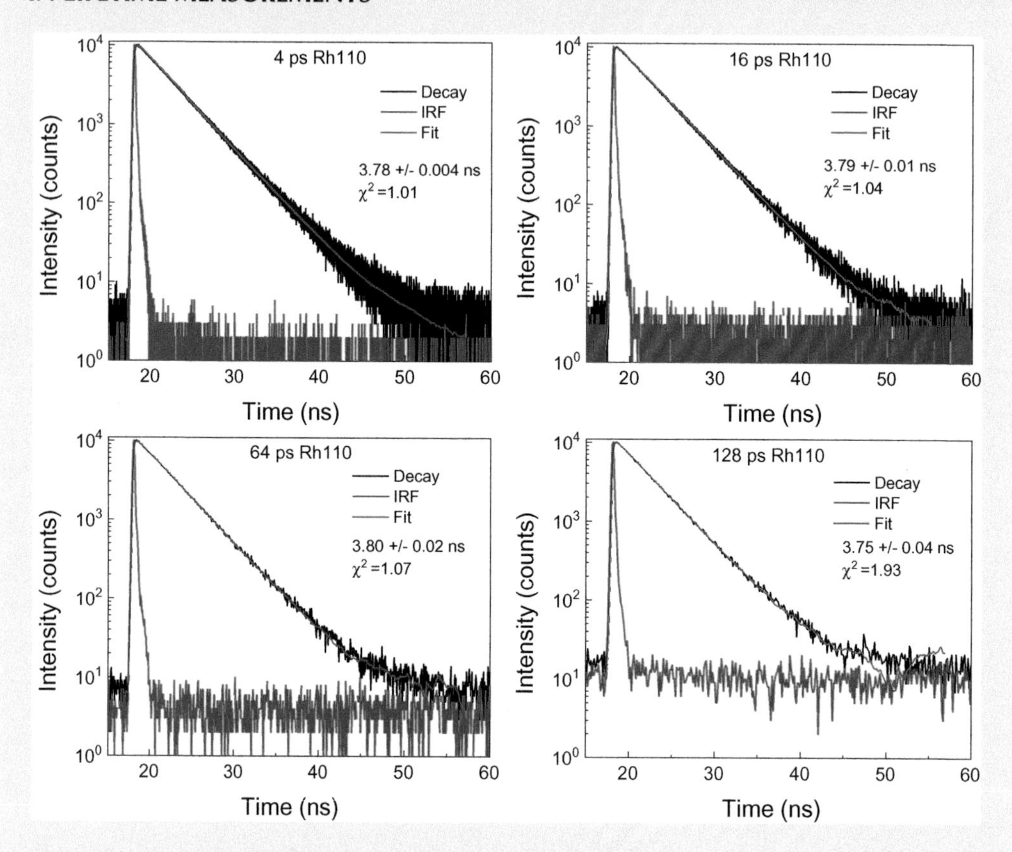

FIGURE 7-5.3 Intensity decays for R110 in ethanol measured with different resolutions. IRF (blue) and intensity decay of R110 in EtOH fluorescence (black), and fits (red).

Figure 7-5.3 shows intensity decays for R110 measured with different resolutions. Recovered values of fluorescence lifetimes are all very close to the expected value of 3.78 ns. Even the lowest resolution of 128 ps still gives within 1% error.

Finally, use the mixture solution and measure intensity decays with different resolution. In Figure 7-5.4, we show measured intensity decays with 4, 16, 64, and 128 ps resolution. The data were fitted with two exponential model and results are presented in Table 7-5.1. In Figure 7-5.4, we present recovered average fluorescence lifetimes. For the highest resolution (4 ps), recovered parameters are very close to that expected from 50:50 mixture. As the resolution decreases, we observe the change in recovered lifetimes (especially the short component) and a very significant modification in recovered fractions for both components. For the lowest resolutions (64 and 128 ps), the error in recovered fractions is significant.

(*Continued*)

EXPERIMENT 7-5 (CONTINUED) CHOOSING A PROPER RESOLUTION IN LIFETIME MEASUREMENTS

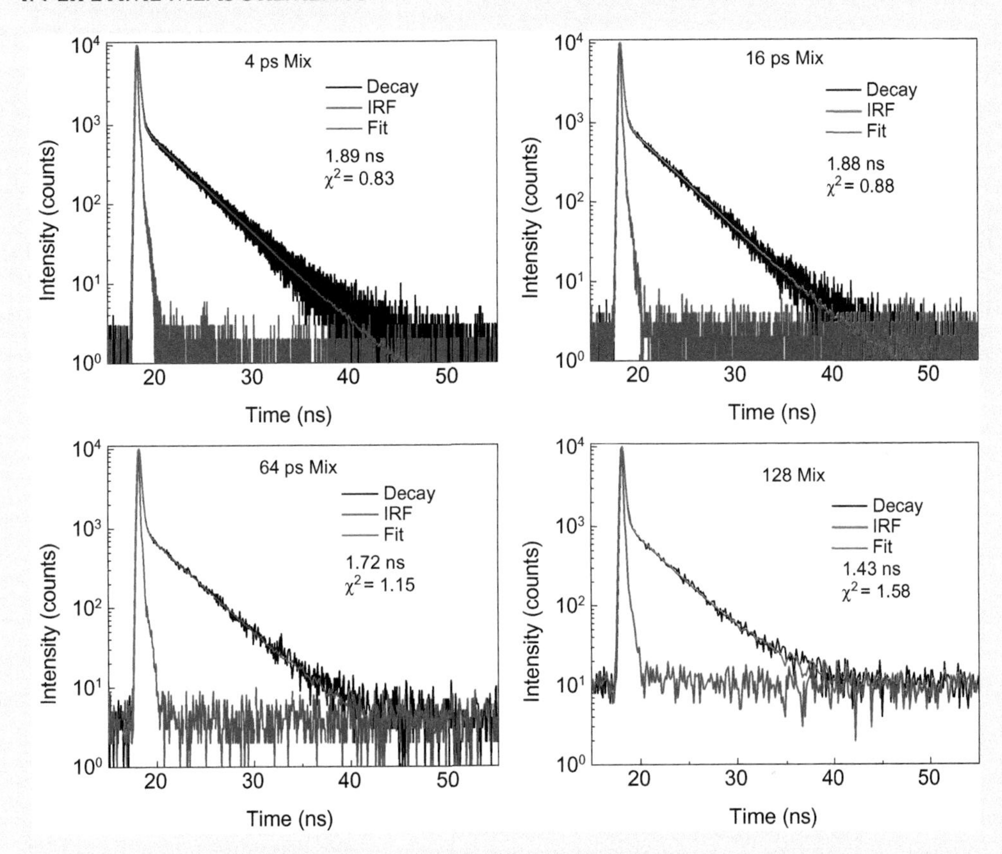

FIGURE 7-5.4 Intensity decays for DOCI:R110 mixture measured with different resolutions. IRF (blue) and intensity decay of R110 in EtOH fluorescence (black), and fits (red).

TABLE 7-5.1 Recovered Parameters for Measured Intensity Decays for DOCI:R110 Mixture Measured with Different Resolutions

Mix	Tau 1 (ns)	Fractional Intensity 1	Tau 2 (ns)	Fractional Intensity 2	Intensity Average Lifetime (ns)	X^2
4 ps	3.72	49%	0.144	51%	1.89	0.827
16 ps	3.74	48%	0.143	52%	1.87	0.88
64 ps	3.78	43%	0.141	57%	1.72	1.15
128 ps	3.67	37%	0.127	63%	1.42	1.58

(Continued)

EXPERIMENT 7-5 (CONTINUED) CHOOSING A PROPER RESOLUTION IN LIFETIME MEASUREMENTS

CONCLUSIONS

From the presented three examples, it is clear that when measuring lifetime the resolution for data collection should be significantly shorter than expected fluorescence lifetimes. In the case of short lifetime components, comparable to the IRF, the highest possible resolution should be used.

The selection of the resolution should take into account the time dedicated to the measurement. The data collection for measurement with the high resolution will take a longer time but the data will be more accurate. We recommend using common sense in the selection of the resolution. For long lifetimes, the resolution can be lower, speeding measurement, for short lifetimes and multi-exponential decays highest possible resolution, should be used.

EXPERIMENT 7-6 LIFETIMES OF SIMILAR COLOR FLUOROPHORES

Keywords: fluorescence, lifetime, fluorophores, fluorometer.

Fluorophores (dyes) may have different color, efficiencies of fluorescence emissions (quantum yield), and different lifetimes. Students frequently ask questions if these parameters are correlated. Other words can we predict the fluorescence lifetime of the dye from its color? In this experiment, we will show that fluorophores of similar color can have very distinct fluorescence lifetimes.

MATERIALS

- Rhodamine 110 (R110), acridine orange (AO), 3-aminofluoranten (3AF), 3,3-diethyloxacarbocyanine iodide (DOCI), Coumarin 153 (C153)
- Ethanol
- 1 cm × 1 cm disposable cuvettes

EQUIPMENT

- Spectrofluorometer (Varian Inc., Eclipse)
- Cary 50 Spectrophotometer (Varian Inc.)
- Fluorometer (FT 200, PicoQuant, GmbH, Germany) with 470 nm laser diode and MCPPMT

(*Continued*)

EXPERIMENT 7-6 (CONTINUED) LIFETIMES OF SIMILAR COLOR FLUOROPHORES

METHODS

1. Make stock solutions of the studied compounds (R110, AO, 3AF, DOCI, C153), each in few milliliters of ethanol; use a glass or plastic vials, see Figure 7-6.1. If a compound did not dissolve completely, add more ethanol, sonicate.
2. Prepare samples using stock solutions, see Figure 7-6.2. Measure absorptions. Adjust concentrations to achieve about 0.05 optical densities at maxima. Measure fluorescence spectra.
3. Measure fluorescence lifetimes of all five solutions.

RESULTS

Photographs of stock solutions and prepared samples are shown in Figures 7-6.1 and 7-6.2.

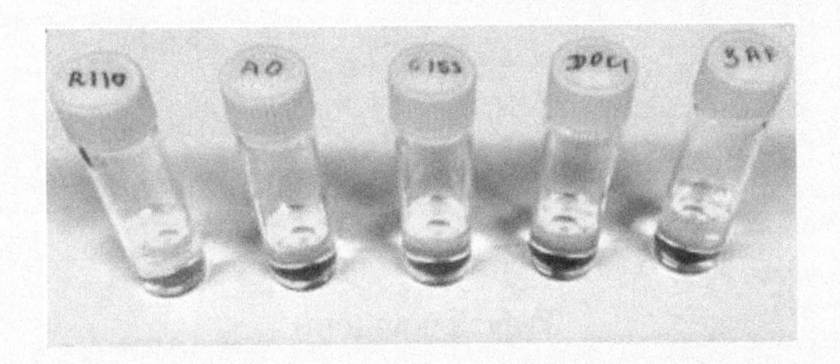

FIGURE 7-6.1 Stock solutions of various dyes.

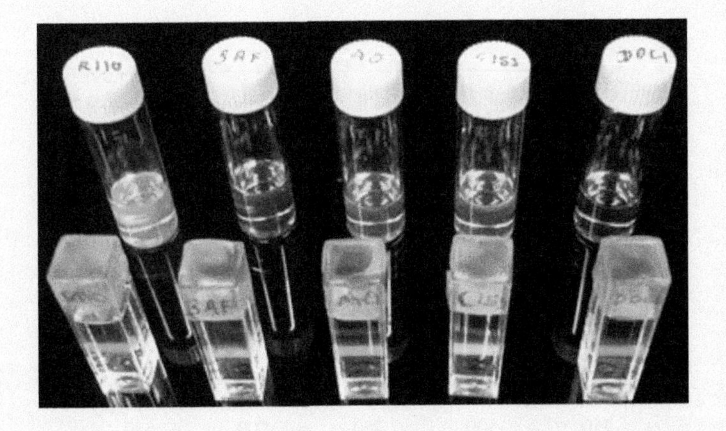

FIGURE 7-6.2 Samples of various fluorophores. All samples have similar fluorescence color, as seen from illumination with a blue laser pointer.

(Continued)

EXPERIMENT 7-6 (CONTINUED) LIFETIMES OF SIMILAR COLOR FLUOROPHORES

Absorption and fluorescence spectra of all samples are presented in Figures 7-6.3 through 7-6.7. All samples are easily excited at 470 nm and all fluorescence emissions are green, see Figure 7-6.2 (visually indistinguishable).

Figures 7-6.3 through 7-6.7 show absorptions and fluorescence spectra of dyes used in this experiment. For convenience, we also included inserts with chemical structures for each dye.

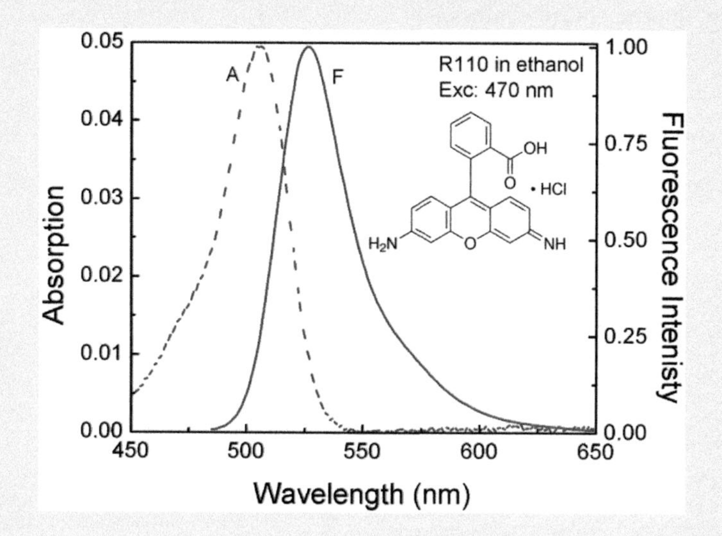

FIGURE 7-6.3 Absorption and emission spectra of R110. The chemical structure of R110 is shown as the insert.

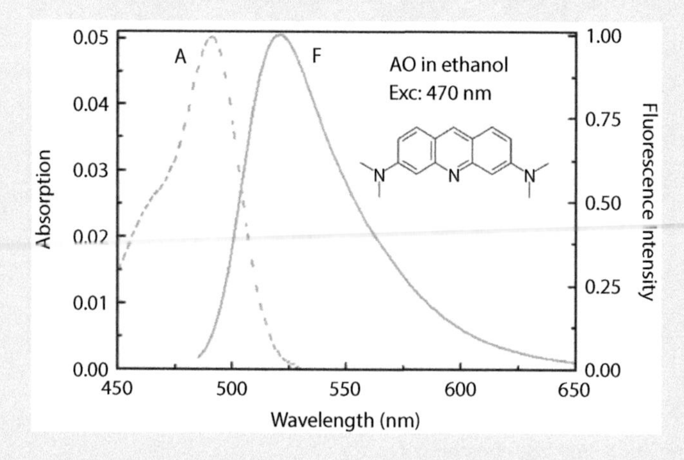

FIGURE 7-6.4 Absorption and emission spectra of AO. The chemical structure of AO is shown as the insert.

(Continued)

EXPERIMENT 7-6 (CONTINUED) LIFETIMES OF SIMILAR COLOR FLUOROPHORES

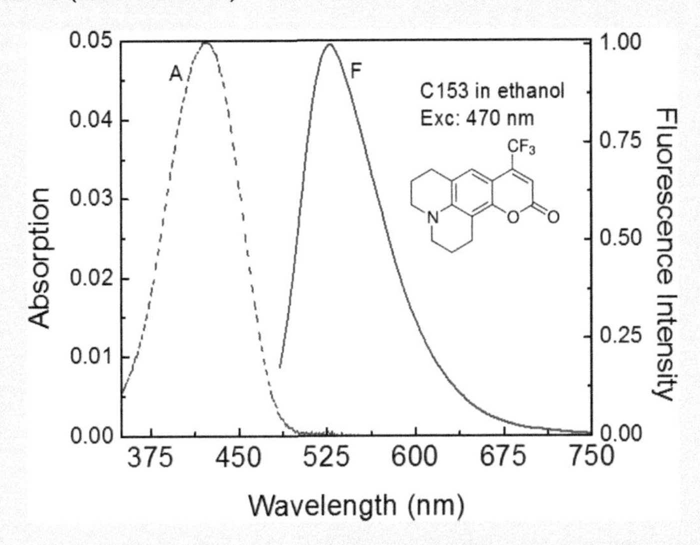

FIGURE 7-6.5 Absorption and emission spectra of C153. The chemical structure of C153 is shown as the insert.

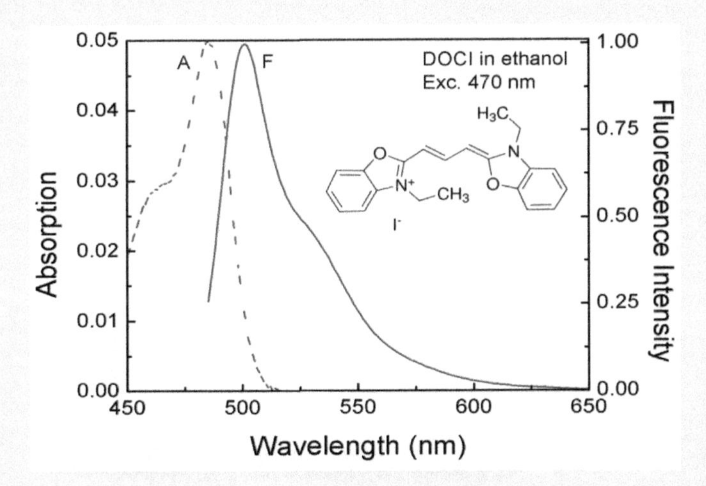

FIGURE 7-6.6 Absorption and emission spectra of DOCI. The chemical structure of DOCI is shown as the insert.

All samples can be efficiently excited by 470 nm laser diode. Next, we measured fluorescence lifetimes of each sample. Set the observation wavelengths near the maximum of the fluorescence. The fluorescence intensity decays are shown in Figures 7-6.8 through 7-6.12.

All the studied fluorophores can be satisfactorily fitted with a single exponent model. Because each individual fluorophore presents a single exponential decay, it can be used as a reference in time-resolved measurements.

(Continued)

EXPERIMENT 7-6 (CONTINUED) LIFETIMES OF SIMILAR COLOR FLUOROPHORES

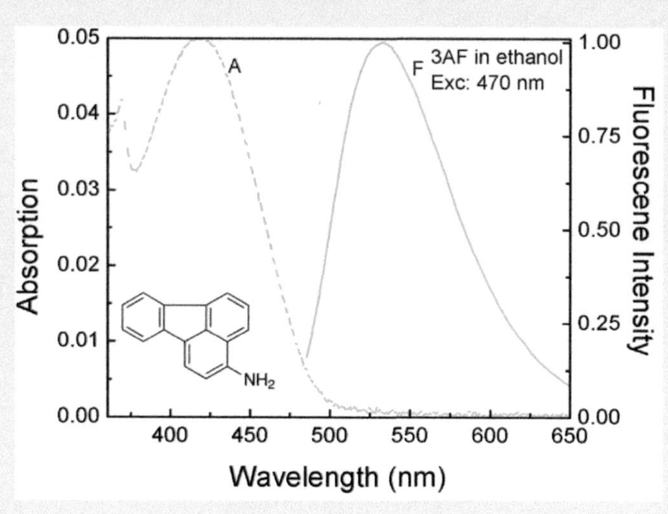

FIGURE 7-6.7 Absorption and emission spectra of 3AF. The chemical structure of 3AF is shown as the insert.

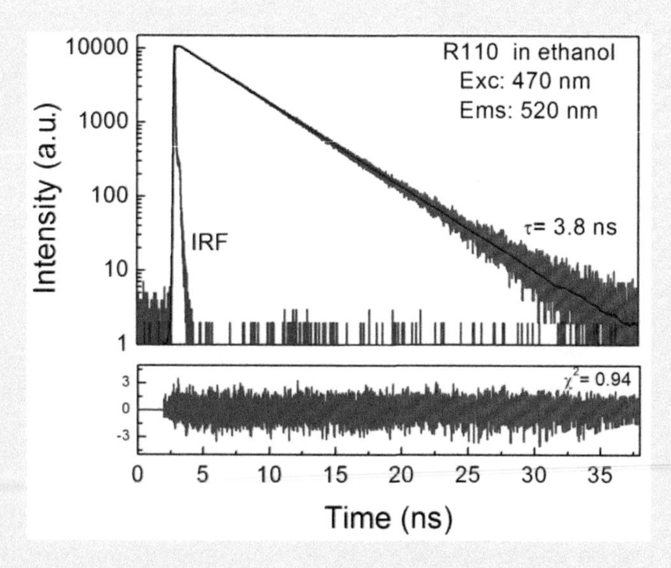

FIGURE 7-6.8 Fluorescence intensity decay of R110 fitted with a single exponent model.

(Continued)

EXPERIMENT 7-6 (CONTINUED) LIFETIMES OF SIMILAR COLOR FLUOROPHORES

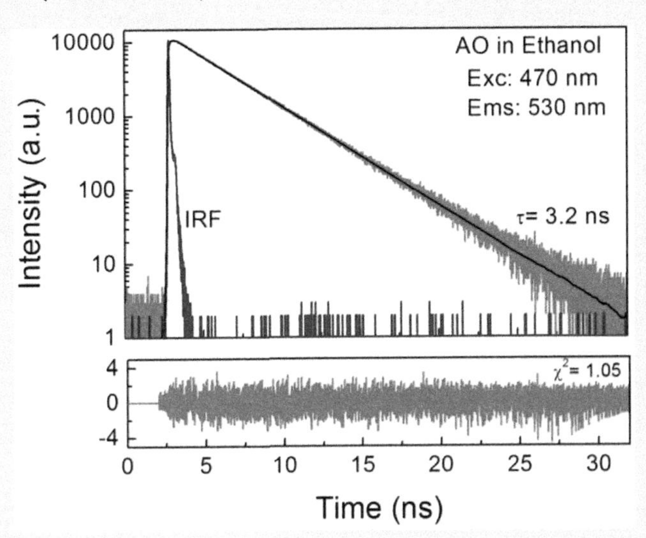

FIGURE 7-6.9 Fluorescence intensity decay of AO fitted with a single exponent model.

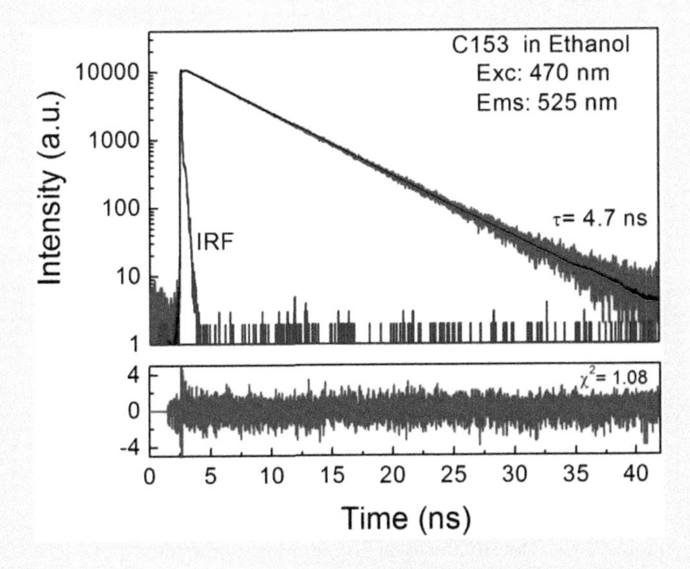

FIGURE 7-6.10 Fluorescence intensity decay of C153 fitted with a single exponent model.

(Continued)

EXPERIMENT 7-6 (CONTINUED) LIFETIMES OF SIMILAR COLOR FLUOROPHORES

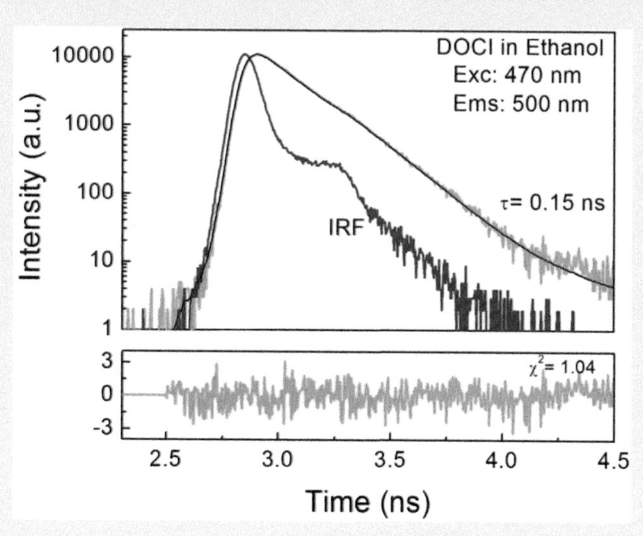

FIGURE 7-6.11 Fluorescence intensity decay of DOCI fitted with a single exponent model.

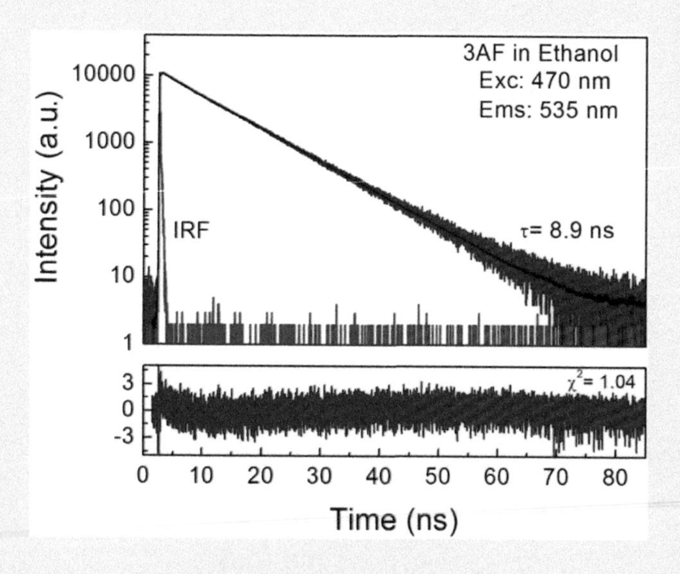

FIGURE 7-6.12 Fluorescence intensity decay of 3AF fitted with a single exponent model.

CONCLUSIONS

All five selected different fluorophores dissolved in ethanol have similar color and fluorescence spectra in a green spectral region, between 500 and 550 nm and very distinct lifetimes from almost 9 ns (3AF) to 0.15 ns (DOCI). It is not possible to predict lifetimes from the color of fluorescence. However, often lifetime depends proportionally to the brightness. Extremely short lifetimes usually are measured for low efficiency (low quantum yield) fluorophores.

EXPERIMENT 7-7 RECONVOLUTION VS TAIL FITTING: FLUORESCENCE INTENSITY DECAYS OF FLUOROPHORES

Keywords: fluorescence lifetime, Rhodamine 6G, erythrosine B, impulse response.

Fluorescence intensity decays are typically analyzed using the deconvolution procedure where the true decay of the fluorophore is deconvoluted from measured intensity decay and independently measured impulse response function (IRF). However, frequently, a simplified approach can be used which ignores the impulse response function (IRF) and analyzes only the fluorophore decay at the tail of the decay (analyzing intensity decay starting at time longer than expected IRF). Intuitively, it is obvious that the fluorophore that shows a single long-lived decay (much longer than IRF) can be fitted without the reconvolution (ignoring the excitation pulse) but short-lived fluorophores (with lifetimes shorter or comparable to IRF) or multi-exponential decays need to take into account the IRF. We will demonstrate both tail fitting and reconvolution for longer-lived Rhodamine 6G and short-lived Erythrosine B (chemical structures shown in Scheme 7-7.1). We will also apply both fitting procedures for the mixture of these dyes.

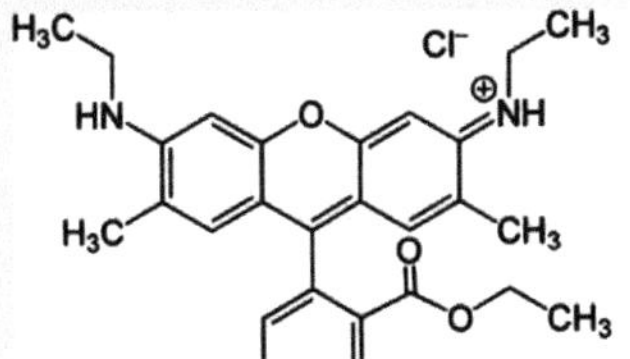

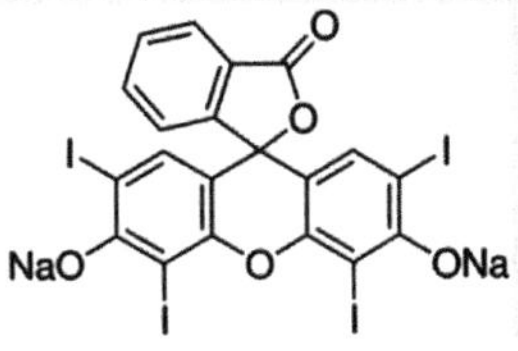

SCHEME 7-7.1 Chemical structures of R6G and ErB.

MATERIALS

- Rhodamine 6G (R6G), Erythrosin B (ErB)
- Water, ethanol
- 1 cm × 1 cm disposable cuvettes; vials; pipettes

EQUIPMENT

- Spectrofluorometer (Varian Inc., Eclipse)
- Cary 50 Spectrophotometer (Varian Inc.)
- Fluorometer FT 200 equipped with MCPPMT (PicoQuant, GmbH, Germany)

METHODS

1. Make stock solutions of studied compounds (R6G and ErB) in a few milliliters of ethanol. Use glass or plastic vials, sonicate.
2. Dilute small amounts of stock solutions in a few milliliters of water to achieve about 0.5 μM concentration of ErB and 0.02 μM of R6G. Measure absorption and fluorescence spectra of both fluorophores and their 1:1 mixture, see Figures 7-7.1 and 7-7.2.
3. Measure fluorescence lifetimes and analyze intensity decays with a multi-exponential model using IRF reconvolution and using tail fitting.

(Continued)

EXPERIMENT 7-7 (CONTINUED) RECONVOLUTION VS TAIL FITTING: FLUORESCENCE INTENSITY DECAYS OF FLUOROPHORES

RESULTS

Absorption Spectra Measurements. In order to achieve similar fluorescence signals, you need to use about 25-fold higher concentration of ErB than R6G, see Figure 7-7.1.

Fluorescence Spectra Measurements. At 470 nm excitation and 550 nm observation, fluorescence intensities of ErB and R6G are equal, see Figure 7-7.2 (left). The colors of ErB and R6G fluorescence emissions are difficult to distinguish as shown in Figure 7-7.2 (right).

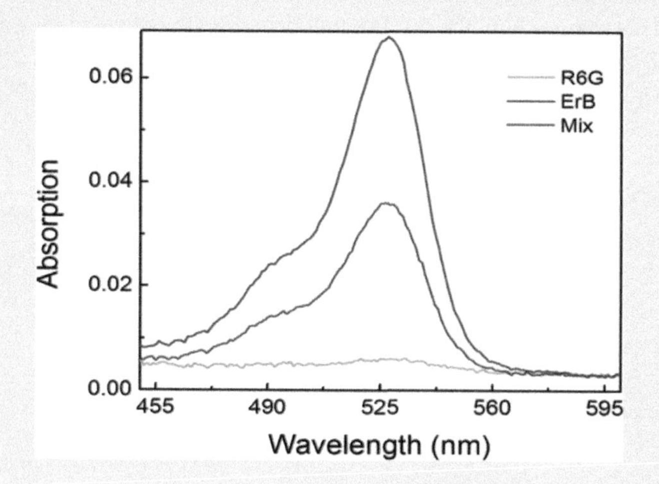

FIGURE 7-7.1 Absorption spectra of ErB, R6G, and 1:1 mixture.

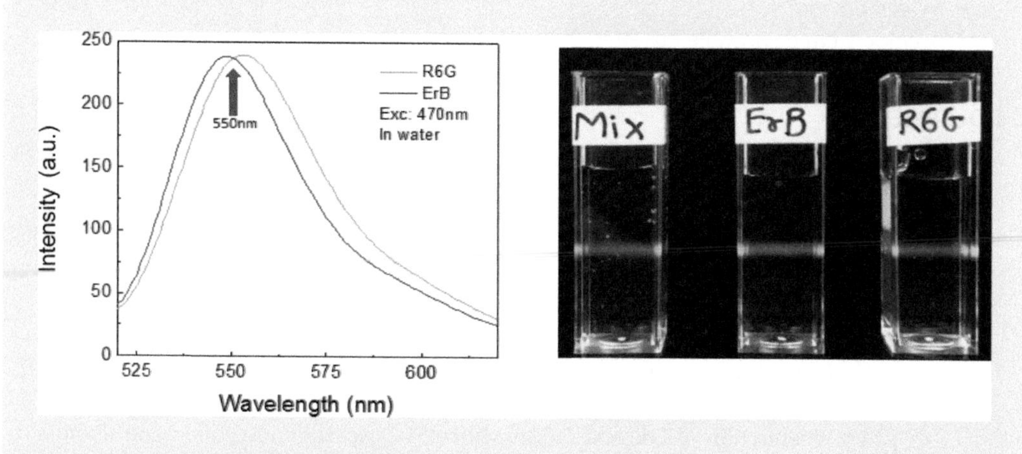

FIGURE 7-7.2 Left: Fluorescence spectra of ErB and R6G. The arrow indicates the observation wavelength in lifetime measurements. Right: Samples illuminated with a blue laser pointer.

(Continued)

EXPERIMENT 7-7 (CONTINUED) RECONVOLUTION VS TAIL FITTING: FLUORESCENCE INTENSITY DECAYS OF FLUOROPHORES

Next, measure IRF at 470 nm and time responses of individual fluorophores. *Note: For these measurements, we are using a system equipped with a multi-channel plate detector that has minimal color effect. When using a different detector that shows a significant color effect you cannot recover proper fluorescence lifetime for ErB.* Analyze decays with and without reconvolutions. Figures 7-7.3 and 7-7.4 show lifetimes of R6G and ErB analyzed with and without IRF. The arrow on the right intensity decay indicates a starting point for decay analysis. For R6G, the analysis starts at the point where IRF is practically ended and the number of meaningful points for analysis is large. For ErB intensity, decay is very short and analysis starts in the middle of IRF.

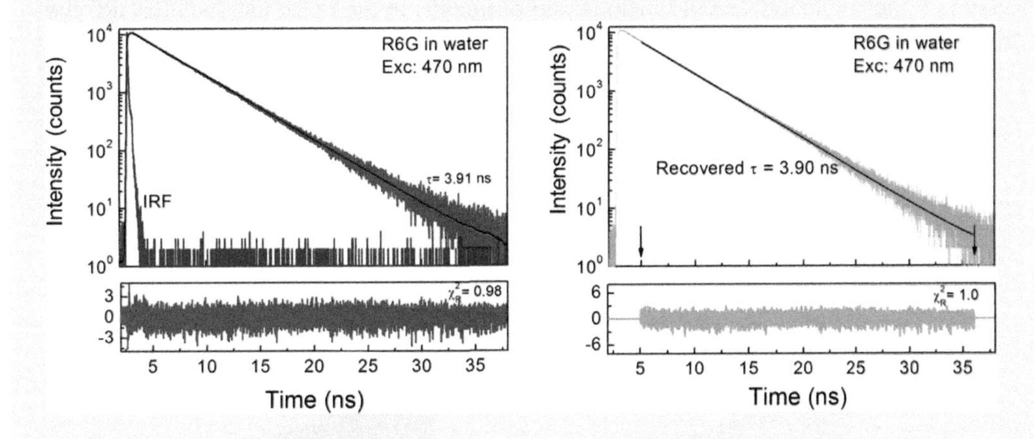

FIGURE 7-7.3 Fluorescence intensity decay of R6G in water. Left: Reconvolution with IRF. Right: Tail fit.

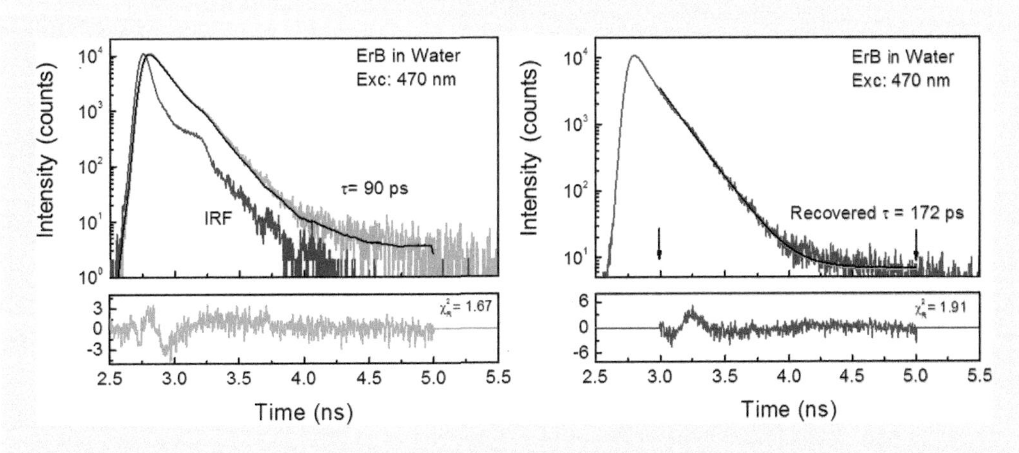

FIGURE 7-7.4 Fluorescence intensity decay of ErB in water. Left: Reconvolution with IRF. Right: Tail fitting.

(Continued)

EXPERIMENT 7-7 (CONTINUED) RECONVOLUTION VS TAIL FITTING: FLUORESCENCE INTENSITY DECAYS OF FLUOROPHORES

For R6G solution both methods of analysis give nearly the same result (same fluorescence lifetime). The intensity decay of R6G in water can be satisfactorily approximated with a single exponent of 3.9 ns. It is not always obvious where the analysis should start (lower time-limit) and where should end (upper time-limit), see arrows in Figure 7-7.4 (right). If the starting point includes IRF, the residuals will show systematic deviations at the beginning of the decay, we should avoid this by shifting the starting point. At the upper-time limit, we should avoid extending the limit too far that will include background noise that will artificially decrease the goodness-of-fit.

The lifetime of ErB is very short and the intensity decay should be analyzed with the IRF. The reconvolution analysis provides the correct fluorescence lifetime. Since the decay is very fast, the tail analysis would include part of the IRF. In the range that includes IRF, the intensity decay contain the IRF and analysis results with almost twice longer lifetime that is obviously incorrect.

Next, measure the lifetime of R6G and ErB mixture. Analyze the data with the two-exponential model, with and without reconvolution, see Figure 7-7.5.

The reconvolution method provides two lifetimes with equal intensity fractions. These are proper recovered parameters for the decay since the prepared mixture was intentionally

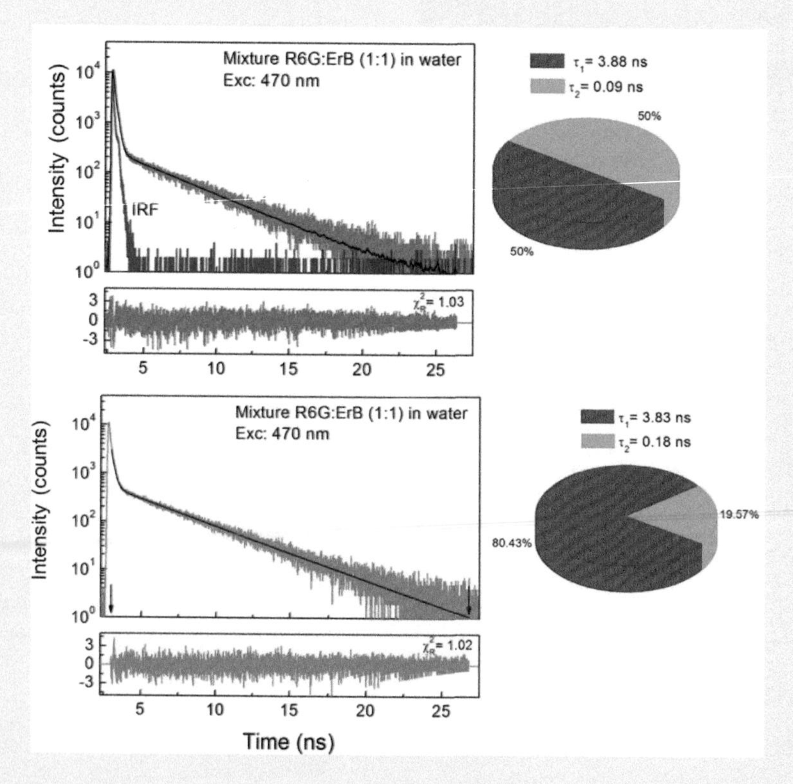

FIGURE 7-7.5 Fluorescence intensity decay of mixture R6G/ErB (1:1). Top: Reconvolution with IRF. Bottom: Tail fitting.

(Continued)

EXPERIMENT 7-7 (CONTINUED) RECONVOLUTION VS TAIL FITTING: FLUORESCENCE INTENSITY DECAYS OF FLUOROPHORES

made to contribute 50:50 intensities at 550 nm observation with 470 nm excitation. The tail fitting approach gives very different results. The recovered fluorescence lifetime for R6G is close to expected. However, the recovered fluorescence lifetime for the ErB is much longer than expected. But, more importantly, the recovered intensity fractions for each lifetime component are very different from expected values. Also recovered parameters strongly depend on the starting point.

CONCLUSIONS

Pulse reconvolution analysis gives correct results and is always preferred. However, for long lifetimes, tail fitting is faster and usually provides accurate results. However, when expecting more complex fluorescence intensity decays, like multi-exponential decay (e.g., a mixture of fluorophores), the analysis should include IRF with reconvolution. It is also important to use IRF as close as possible to true instrument response, as we will discuss many factors may influence the IRF.

EXPERIMENT 7-8 FLUORESCENCE INTENSITY DECAYS OF FLUOROPHORE MIXTURES

Keywords: fluorescence, fluorophore mixture, fluorometer.

Individual fluorophores in solutions may show mono-exponential intensity decays. But there are many reasons for the fluorescence decay not being mono-exponential and more than one exponent is needed to satisfactorily fit the intensity decay data. It could be a presence of different conformational states, fluorophore-solvent interactions, or just contamination by another fluorophore. It is instructive to show how the intensity decay of a mixture of two fluorophores with different fluorescence lifetimes differs from individual fluorophores. Two dyes have been selected: Rhodamine B and Rhodamine 101 (see structures below). Concentrations of these dyes are adjusted to yield the same fluorescence signals with 470 nm excitation and at 600 nm observation. The absorptions should be low, below 0.1. The goal of this experiment is to demonstrate how a multi-exponential decay would manifest in the observed intensity decay.

MATERIALS

- Rhodamine B (RhB), Rhodamine 101 (R101) (Scheme 7-8.1)
- Water, ethanol
- 1 cm × 1 cm disposable cuvettes, vials, pipettes

(Continued)

EXPERIMENT 7-8 (CONTINUED) FLUORESCENCE INTENSITY DECAYS OF FLUOROPHORE MIXTURES

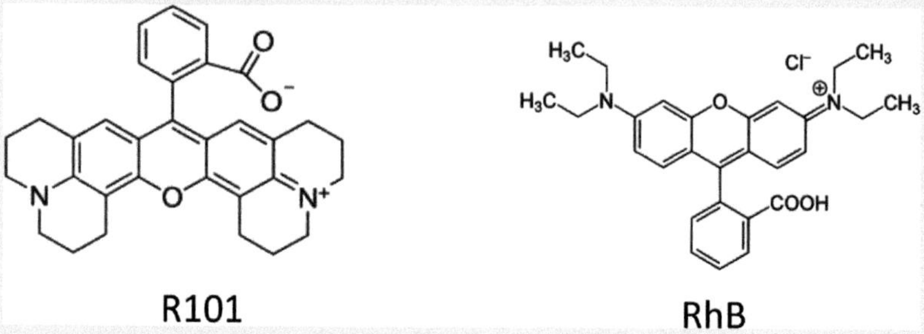

R101

RhB

SCHEME 7-8.1 Chemical structures of R101 and RhB.

EQUIPMENT

- Spectrofluorometer (Varian Inc., Eclipse)
- Carry 50 Spectrophotometer (Varian Inc.)
- Fluorometer FT 200 equipped with MCPPMT detector (PicoQuant, GmbH, Germany)

METHODS

1. Make stock solutions of the studied compounds (RhB and R101) in a few milliliters of ethanol. Use glass or plastic vials, sonicate.
2. Dilute small amounts of stock solutions in a few milliliters of water to achieve about 1 μM concentrations. Measure emission spectra and adjust concentrations until the fluorescence signal at 600 nm will be identical, see Figure 7-8.2. Remember the absorption should be below 0.1.
3. Mix 1 mL of RhB solution with 1 mL of R101 solution. Prepare also two times diluted (in water) solutions of RhB and R101 (use 1 mL dye solution and add 1 mL of water). With such preparation, the concentrations of dyes will be constant in all three samples. Measure absorption spectra of all three samples, see Figure 7-8.1.
4. Measure fluorescence lifetimes (the instrument response function (IRF) and samples intensity decay) and analyze intensity decays with a multi-exponential model, see Figures 7-8.3 and 7-8.4. Remember to use the magic angle (MA) conditions.

RESULTS

It is always recommended to keep sample absorption below 0.1 but in this experiment, we will use a higher absorption to show that the emission spectra can be affected by reabsorption while fluorescence lifetimes properly reflect sample conditions. When preparing samples for the experiment *adjust* the concentrations of RhB and R101 to obtain the same fluorescence signals at 600 nm observation with 470 nm excitation. Then make a 1:1 mixture of both dyes. When mixing two solutions 1:1 the relative concentrations of both dyes will decrease by half. To match individual signals from both dyes dilute original solutions of fluorophores by mixing each of the samples 1:1 with water.

(*Continued*)

EXPERIMENT 7-8 (CONTINUED) FLUORESCENCE INTENSITY DECAYS OF FLUOROPHORE MIXTURES

Absorption Measurements. It is important to measure the absorptions of the prepared samples and check if the spectrum of the mixture is a combination of RhB and R101. Figure 7-8.1 shows absorptions of the mixture and diluted individual fluorophores. It is clear that the absorption of the mixture is a simple sum of the absorption of individual components. No new absorption bands are noticed. If the absorption of the mixture of two fluorophores is not a simple combination of individual dye absorption it may indicate the interaction of these two substances. Often ground-state complexes are formed and they manifest itself with new absorption bands (for details of analyzing absorption spectra for a mixture see Experiment 3-23). The absorption of the mixture is significantly higher than 0.1. Too high absorption will affect the measured emission spectrum of the mixture.

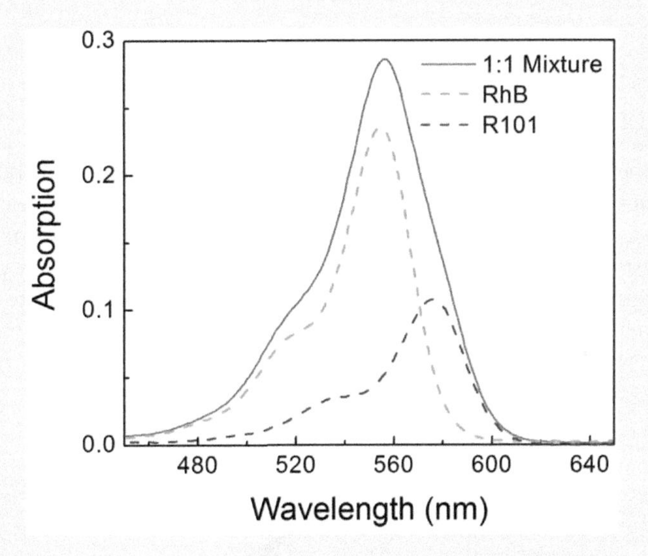

FIGURE 7-8.1 Absorption spectra of RhB, R101, and 1:1 mixture of both dyes.

Fluorescence Spectra Measurements. Next, using 470 nm excitation wavelength, we measured emission spectra of diluted samples of each dye and emission spectrum of 1:1 mixture, see Figure 7-8.2. It is clear that at short wavelength range, the emission spectrum of the mixture is not a simple sum of individual emissions of two components. This is a result of small sample reabsorption. However, above 580 nm where the absorption of the mixture drops below 0.1, the measured intensity of the mixture is a simple sum of the intensities contributed by RhB and R101. As it can be seen in Figure 7-8.1, the fluorescence intensities of RhB and R101 are the same at 600 nm when excited at 470 nm.

(Continued)

EXPERIMENT 7-8 (CONTINUED) FLUORESCENCE INTENSITY DECAYS OF FLUOROPHORE MIXTURES

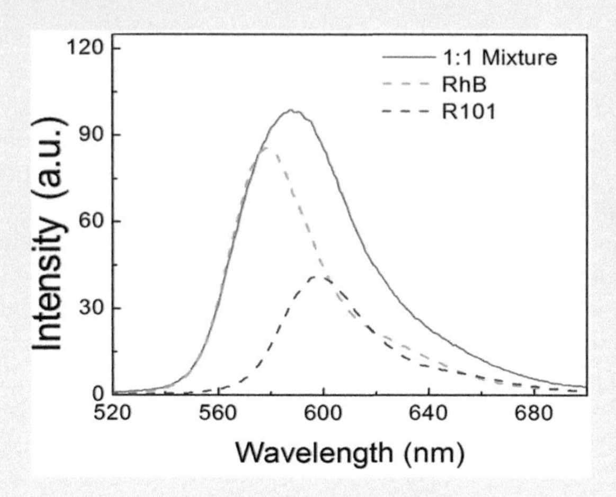

FIGURE 7-8.2 Fluorescence spectra of RhB, R101, and 1:1 mixture at 470 nm excitation. The fluorescence intensities at 600 nm are matched; fluorescent fractions of RhB and R101 are similar (at 600 nm observation).

Next, measure intensity decays of individual fluorophores and the mixture using 470 nm excitation and 600 nm observation. The individual samples will have the same intensities (count rate) and the mixture will present a twice higher signal.

Lifetime Measurements for Individual Fluorophores. In Figure 7-8.3, we present fluorescence intensity decays as measured for RhB and R101 solutions. The data were collected to achieve close to 10,000 counts in the decay peak.

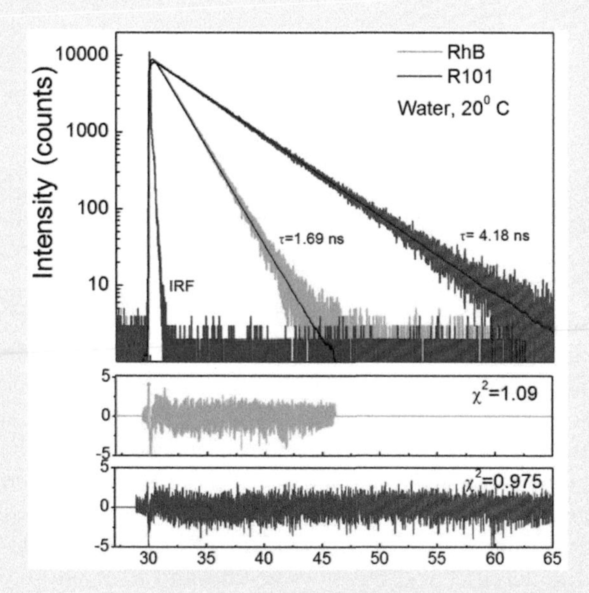

FIGURE 7-8.3 Fluorescence intensity decays of RhB and R101 at the excitation of 470 nm and observation at 600 nm. Both fluorophores can be fitted with a single exponential model.

(Continued)

EXPERIMENT 7-8 (CONTINUED) FLUORESCENCE INTENSITY DECAYS OF FLUOROPHORE MIXTURES

In a logarithmic scale, the single exponential decays appear as straight lines as seen in Figure 7-8.3. The lifetime of R101 is about 2.5 times longer than RhB. Both fluorophores can be satisfactorily fitted with a single-exponent model. Two exponential fits do not improve the goodness-of-fits.

Mixture of Fluorophores. Next, we measured the intensity decay of the mixture, see Figure 7-8.4. It is not possible to fit the intensity decay of the mixture with a single exponent with satisfactory error, see Figure 7-8.4 (left). The goodness of the chi-square is over 10 and standard deviations are clearly not random.

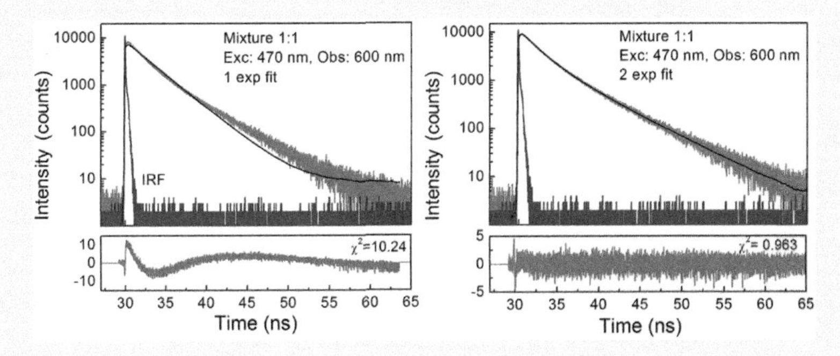

FIGURE 7-8.4 Fluorescence intensity decay of the mixture of RhB and R101 at the excitation of 470 nm and observation at 600 nm. Left: one-exponent fit; right: two-exponent fit.

However, the two-exponent fit shows an excellent approximation of the intensity decay with randomly distributed standard deviations and close to 1 goodness-of-fit, see Figure 7-8.4 (right). The recovered lifetimes correspond well to the lifetimes of individual fluorophores.

The concentrations of both fluorophores were adjusted to yield approximately the same emission intensities at 600 nm observation, and we expected to observe equal fractions of both decays. Recovered fractions are accurate within 5%, which is a good agreement with the expected results.

Recovered lifetimes using two-exponential fits are:

$$\tau_1 = 4.21 \pm 0.02(54.8\%),\ \tau_2 = 1.69 \pm 0.01(45.2\%)$$
$$\text{with goodness-of-fit close to 1 (0.963)}$$

Recovered lifetime using one-exponent fit:

$$\tau = 3.05(100\%),\ \text{but poor goodness-of-fit (10.24)}$$

As was discussed in the "Introduction" to this chapter, the confidence intervals for recovered lifetimes can be visualized with χ_R^2 surfaces presentation. The normalized values

(Continued)

EXPERIMENT 7-8 (CONTINUED) FLUORESCENCE INTENSITY DECAYS OF FLUOROPHORE MIXTURES

of χ_R^2 for both lifetimes are presented in Figure 7-8.5. The lifetime was fixed at a slightly different value at each step and the analysis has been run with all other floating parameters. The confidence at 67% corresponds to the normalized χ_R^2 of 1.01. From Figure 7-8.5, it is clear that both lifetimes are very well separated.

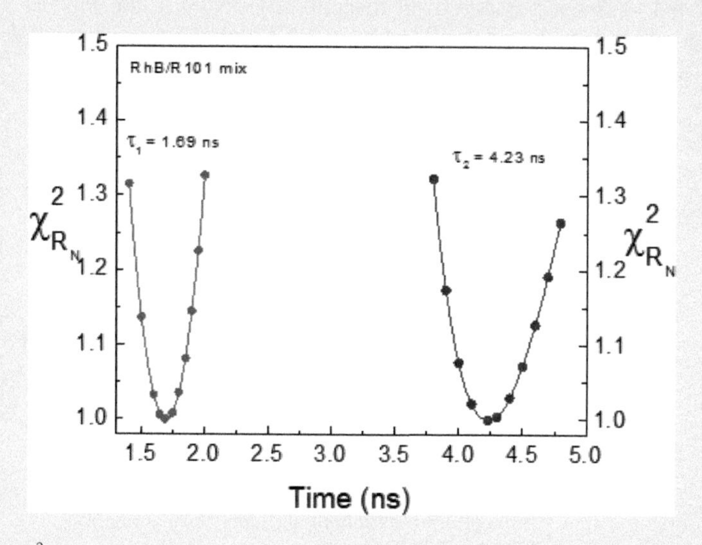

FIGURE 7-8.5 χ_R^2 surfaces for the recovered lifetimes from the mixture 1:1 of RhB and R101.

CONCLUSIONS

Mixtures of fluorophores with different fluorescence lifetimes show a complex fluorescence intensity decays. Such a mixture of two fluorophores with a distinct difference in fluorescence lifetimes can be adequately resolved with fluorescence intensity decay measurements. For this experiment, we used a mixture that had peak absorption much over 0.1. From emission spectra, we can clearly see the reabsorption effect at the short wavelength side of emission. However, when measuring intensity decay observed at a wavelength above 580 nm, the recovered lifetimes and fractions are unaffected by reabsorption.

EXPERIMENT 7-9 DEPENDENCE OF FLUORESCENCE INTENSITY DECAYS ON OBSERVATION WAVELENGTH

Keywords: fluorescence, lifetime, solvent relaxation.

It is important to realize that fluorophore transitions (relaxation) from the excited state to the ground state is a statistical process and the fluorophore decaying process spreads over time. So, we have fluorophores emitting photons immediately after excitation from a state un-relaxed and fluorophores decaying after significant time after excitation when the excited fluorophore reached full equilibrium with surrounding solvent molecules. Typically internal relaxation of the fluorophore itself after excitation is very fast (much shorter than typical fluorescence lifetimes). In contrast, solvent reorientation around the excited fluorophore is much slower and for viscous solvents frequently comparable to the excited state lifetime. In the case when the solvent relaxation time is comparable to the fluorophore lifetime, the observed intensity decay strongly depends on the observation wavelength. In this experiment, we will compare the intensity decays of the fluorophore in solvents with fast and slow relaxation times observed at different wavelengths. We selected Fluorol 7GA, a dye with a large Stokes shift, unstructured fluorescence spectrum, and relatively long fluorescence lifetime. In low viscous ethanol, the solvent relaxation, as well as Fluorol 7GA rotations, is expected to be very fast, much faster than the lifetime of the fluorophore. In very viscous glycerol, the solvent relaxation is slow and occurs during the lifetime of the fluorophore. Of course, the rotations of Fluorol 7GA molecules are also significantly hindered.

MATERIALS

- Fluorol 7GA (F7GA)
- Ethanol, glycerol
- 1 cm × 1 cm glass/quartz cuvettes, vials, pipettes

EQUIPMENT

- Spectrofluorometer (Varian Inc., Eclipse)
- Carry 50 Spectrophotometer (Varian Inc.)
- Fluorometer FT 200 (Picoquant, GmbH, Germany)

METHODS

1. Make stock solutions of studied compound (F7GA) in a few milliliters of ethanol. Use glass or plastic vials, sonicate.
2. Dilute small amounts of stock solutions in a few milliliters of ethanol and glycerol to achieve a few micromolar concentrations. Check absorptions to be sure that the optical density of samples is low, about 0.05. Measure emission spectra with 470 nm excitation in both solvents (Figure 7-9.1).
3. Measure fluorescence lifetimes with magic angle (MA) oriented emission polarizer at different observation wavelengths for both solvents.
4. Analyze lifetime data with a multi-exponential model (Scheme 7-9.1).

(Continued)

EXPERIMENT 7-9 (CONTINUED) DEPENDENCE OF FLUORESCENCE INTENSITY DECAYS ON OBSERVATION WAVELENGTH

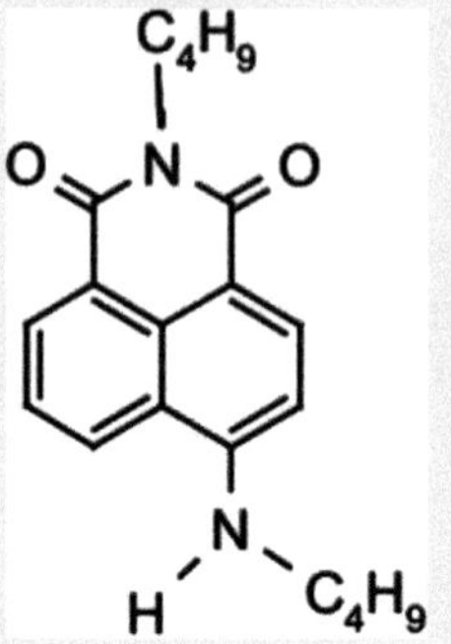

SCHEME 7-9.1 Chemical structure of Fluorol 7GA.

RESULTS

Fluorescence Spectra Measurements. Measured fluorescence spectra of F7GA in ethanol and glycerol are shown in Figure 7-9.1.

Fluorol 7GA is a bright dye with a green emission. The fluorescence spectrum of F7GA in ethanol has a maximum at 540 nm and in glycerol at 555 nm. We will measure intensity decays of F7GA in both solvents observing at the maxima, and at short (500 nm) and long (620 nm) edges of the emission spectra (wavelengths marked with arrows in Figure 7-9.1). The solvents, ethanol and glycerol, differ significantly with the viscosity.

While in ethanol a small F7GA molecule can rotate quickly, in glycerol the rotational diffusion is slower, in the nanosecond scale comparable to the lifetime of F7GA.

Lifetime Measurements. Next, we measured fluorescence intensity decays of F7GA in ethanol at different observation wavelength, at the maximum (540 nm), at the short edge of the emission spectrum (500 nm) and long edge (620 nm).

Figures 7-9.2 through 7-9.4 show lifetime measurements of Fluorol 7GA in ethanol at different observations with the fitting to a single exponential model.

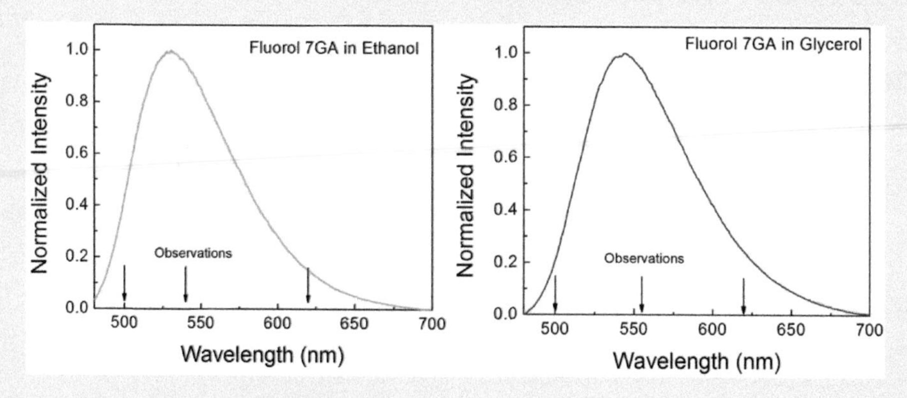

FIGURE 7-9.1 Fluorescence spectra of F7GA in ethanol (left) and glycerol (right) at 470 nm excitation. Arrows indicate observation wavelengths in lifetime measurements.

(Continued)

EXPERIMENT 7-9 (CONTINUED) DEPENDENCE OF FLUORESCENCE INTENSITY DECAYS ON OBSERVATION WAVELENGTH

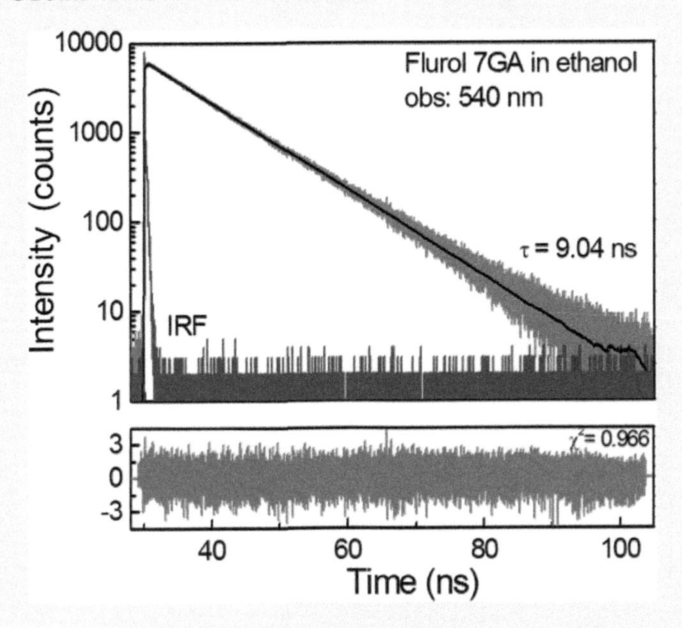

FIGURE 7-9.2 Fluorescence intensity decay of Fluorol 7GA in ethanol with the observation near maximum (540 nm).

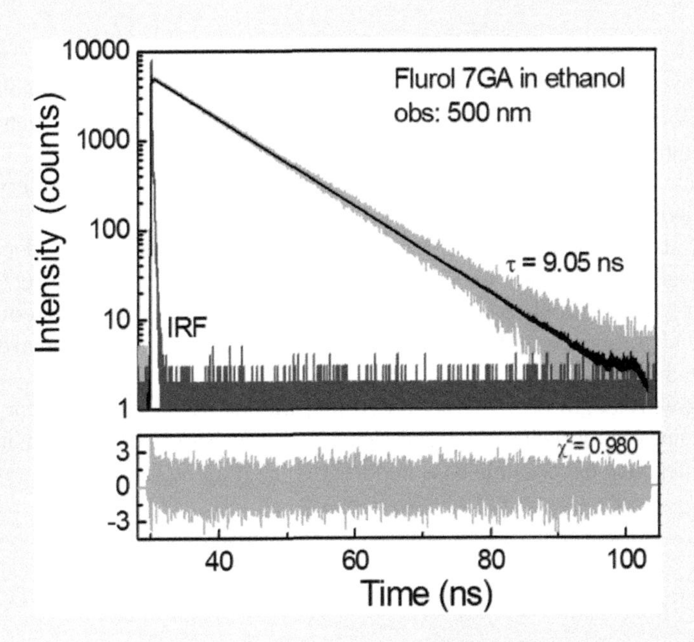

FIGURE 7-9.3 Fluorescence intensity decays of F7GA in ethanol with the short-wavelength observation at 500 nm.

(Continued)

EXPERIMENT 7-9 (CONTINUED) DEPENDENCE OF FLUORESCENCE INTENSITY DECAYS ON OBSERVATION WAVELENGTH

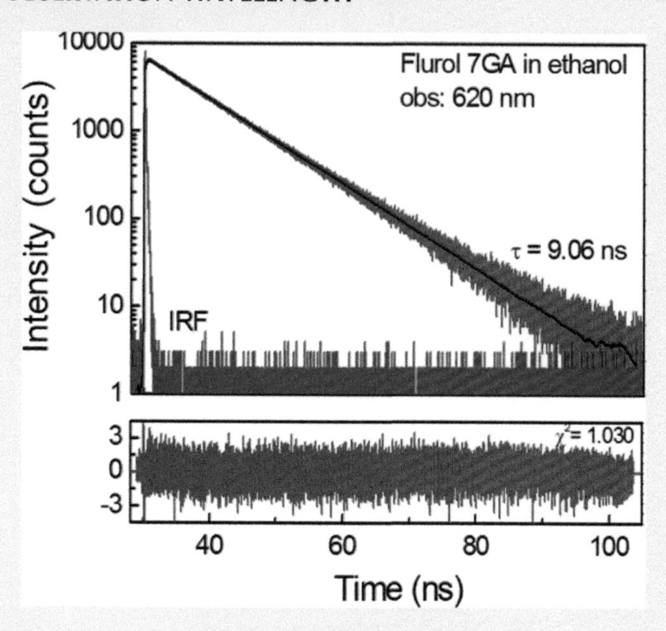

FIGURE 7-9.4 Fluorescence intensity decays of F7GA in ethanol with the long-wavelength observation at 620 nm.

At all three observations lifetimes of Fluorol 7GA in ethanol are practically the same. The single-exponent fits reveal about 9 ns lifetime with the accuracy better than ±0.01 ns, as estimated from the confidence intervals.

Now, we measure intensity decays of F7GA in glycerol at different observation wavelengths, see Figures 7-9.5 through 7-9.7.

In glycerol, at the observation near the maximum (555 nm), the intensity decay of Fluorol 7GA can be well approximated with a single exponent of 7.13 ns, see Figure 7-9.5. In contrast, the short-wavelength observation at 500 nm results in a very heterogeneous decay (see Figure 7-9.6) which requires three exponents to obtain an acceptable fit. The average lifetime shortened to 5.3 ns.

At the long-wavelength observation at 620 nm, the intensity decay also cannot be fitted well with the single-exponent model (see Figure 7-9.8). In order to obtain an acceptable fit, an additional negative exponent term is needed (see Figure 7-9.9).

(Continued)

EXPERIMENT 7-9 (CONTINUED) DEPENDENCE OF FLUORESCENCE INTENSITY DECAYS ON OBSERVATION WAVELENGTH

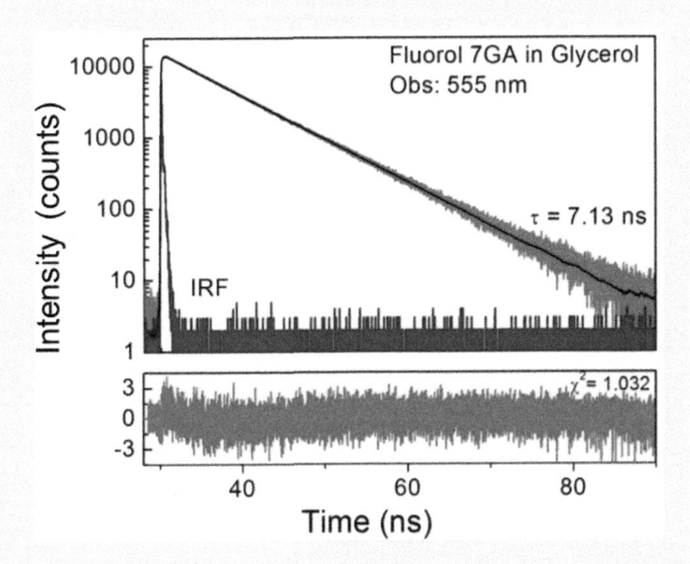

FIGURE 7-9.5 Fluorescence intensity decay of Fluorol 7GA in glycerol with the observation near emission maximum (555 nm).

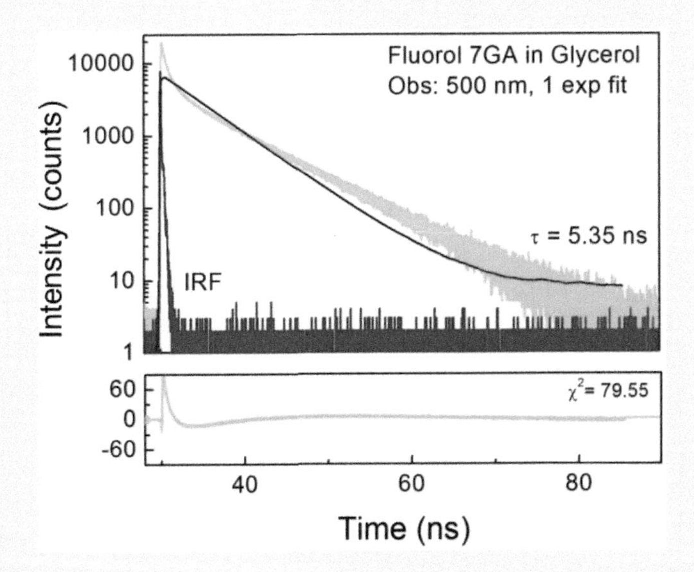

FIGURE 7-9.6 Fluorescence intensity decays of F7GA in glycerol with the short-wavelength observation at 500 nm. One-exponent fit.

(Continued)

EXPERIMENT 7-9 (CONTINUED) DEPENDENCE OF FLUORESCENCE INTENSITY DECAYS ON OBSERVATION WAVELENGTH

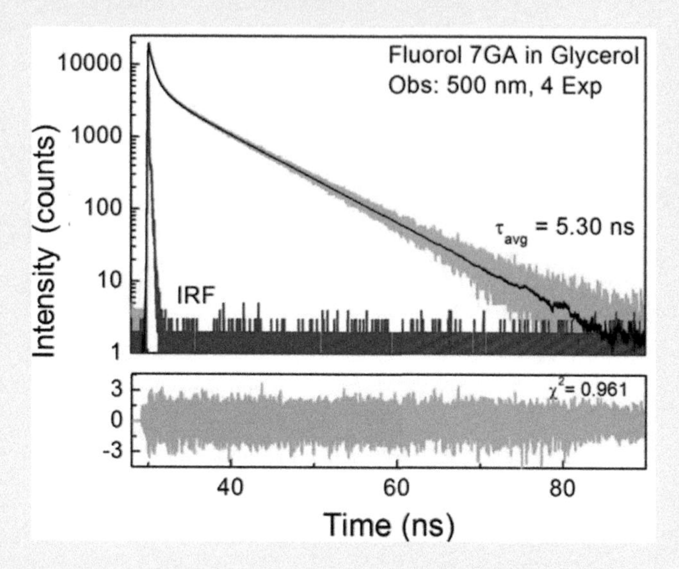

FIGURE 7-9.7 Fluorescence intensity decays of F7GA in glycerol with the short-wavelength observation at 500 nm. Four-exponents fit.

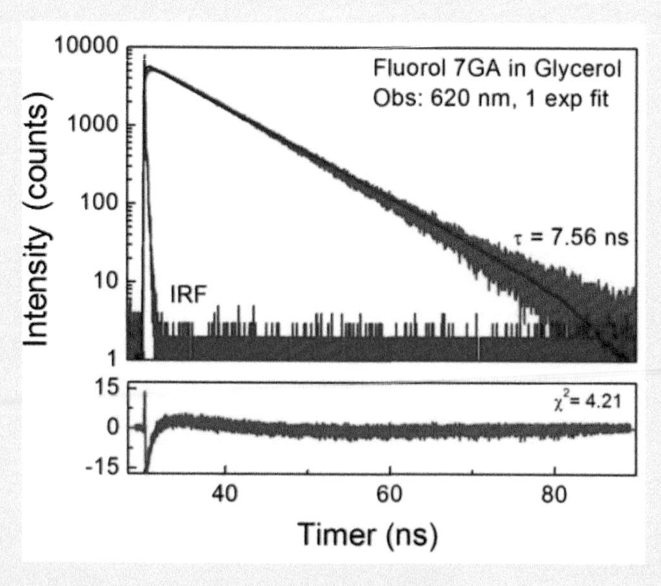

FIGURE 7-9.8 Fluorescence intensity decays of F7GA in glycerol with the long-wavelength observation at 620 nm. One-exponent fit.

(Continued)

EXPERIMENT 7-9 (CONTINUED) DEPENDENCE OF FLUORESCENCE INTENSITY DECAYS ON OBSERVATION WAVELENGTH

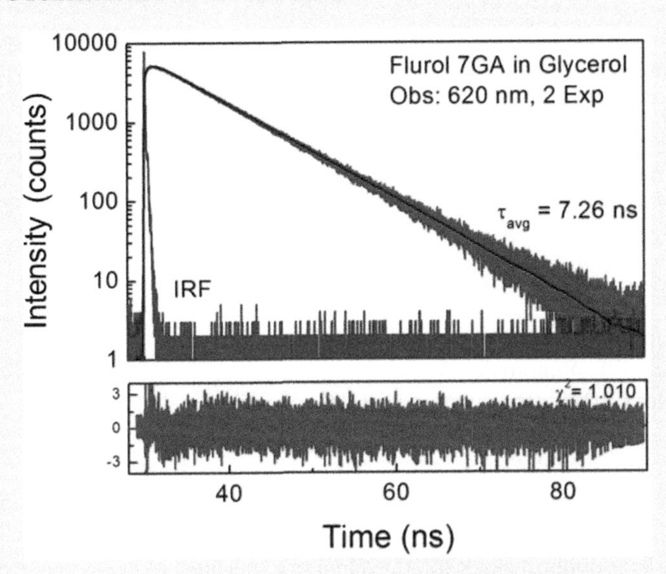

FIGURE 7-9.9 Fluorescence intensity decays of F7GA in glycerol with the long-wavelength observation at 620 nm. Two-exponent fit. The second component has a negative amplitude and the averaged lifetime is longer.

CONCLUSIONS

In the case of more viscous solutions, the relaxation time of solvent molecules and rotations of the dye molecule can be close to the lifetime of fluorophore. In such a case, the observation wavelength plays a very important role. A wrong observation wavelength may result in shorter and heterogeneous decay (at short-wavelength observation) or a longer lifetime with an artificial negative component (long-wavelength observation).

Warning: It is very easy to draw a false conclusion from measurements with an uncontrolled observation wavelength if relaxation processes are involved.

EXPERIMENT 7-10 DEPENDENCE OF FLUORESCENCE INTENSITY DECAYS ON OBSERVATION CONDITIONS: EMISSION POLARIZER ORIENTATION

Keywords: fluorescence, lifetime, dansyl amide, polarizers.

We earlier presented examples demonstrating that observation wavelength, the color effect of the detector, or other geometrical factors may sometimes significantly alter observed intensity decays. In some conditions, measured lifetimes depend on the detection pathway, especially on the observation polarizer orientation. We selected dansyl amide (DA), a dye with large Stokes shift, unstructured fluorescence spectrum, and relatively long fluorescence lifetime, see the structure and Figure 7-10.1. We will show that polarization conditions on the observation path

(*Continued*)

EXPERIMENT 7-10 (CONTINUED) DEPENDENCE OF FLUORESCENCE INTENSITY DECAYS ON OBSERVATION CONDITIONS: EMISSION POLARIZER ORIENTATION

play an important role in lifetime measurements. This experiment should be considered also as a warning for the necessity of using magic angle conditions in fluorescence lifetime measurements.

MATERIALS

- Dansyl amide (DA)
- Propylene glycol (PG)
- 1 cm × 1 cm glass/quartz cuvettes, vials, pipettes

EQUIPMENT

- Spectrofluorometer (Varian Inc., Eclipse)
- Carry 50 Spectrophotometer (Varian Inc.)
- Fluorometer FT 200 (Picoquant, GmbH, Germany)

METHODS

1. Make a stock solution of the studied compound (DA) in a few milliliters of PG. Use glass or plastic vials, sonicate.
2. Dilute small amounts of stock solutions in a few milliliters of PG to achieve a few micromolar concentration. Check the absorption to be sure that optical density of samples is low, below 0.1. Measure emission spectra with 375 nm excitation, see Figure 7-10.1.
3. Measure fluorescence lifetimes with magic angle (MA), vertical (VV), and horizontal (VH) observations (see Figures 7-10.2 through 7-10.6).
4. Analyze lifetime data with a multi-exponential model.

RESULTS

Absorption and Fluorescence Spectra Measurements. Figure 7-10.1 shows absorption and fluorescence spectra of DA in PG. Note the exceptionally large Stokes shift. Arrows indicate excitation and observation wavelengths.

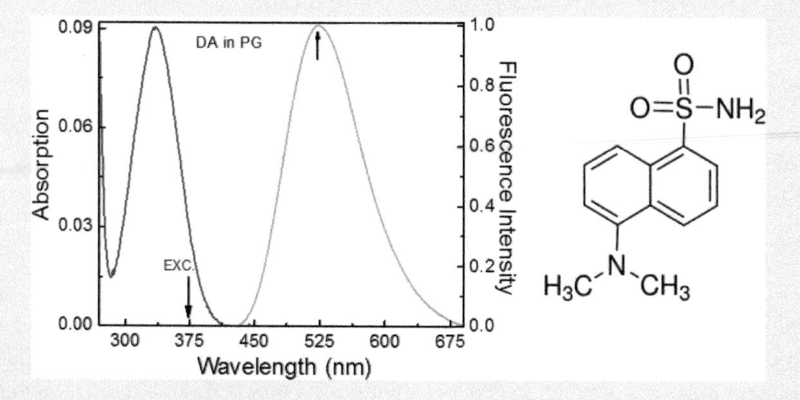

FIGURE 7-10.1 Absorption and fluorescence spectra of dansyl amide (DA) in propylene glycol (PG). Arrows indicate the excitation and observation wavelengths. The chemical structure of DA is on the right.

(Continued)

EXPERIMENT 7-10 (CONTINUED) DEPENDENCE OF FLUORESCENCE INTENSITY DECAYS ON OBSERVATION CONDITIONS: EMISSION POLARIZER ORIENTATION

Both absorption and fluorescence spectra of DA in PG are broad and unstructured. DA is a polar molecule, sensitive to the solvent. In polar solvents, DA has a very large Stokes shift. In less polar solvents like toluene, DA has a dark blue fluorescence, in more polar solvents, the DA fluorescence is light blue or green. PG is a polar solvent with a relatively high viscosity. DA fluorescence in PG has a green color.

Lifetime Measurements. We measured lifetimes of DA in PG with an excitation at 375 nm and observation 525 nm using different polarization conditions, see Figures 7-10.2 through 7-10.6. In the magic angle (MA) observation, the fluorescence decay intensity can be well approximated with a single exponential decay with a lifetime of 16.4 ± 0.02 ns (Figure 7-10.2). This is not the case with vertical (VV) and horizontal (VH) observations.

In the case of VV polarization conditions (vertical excitation and vertical observation), there is a significant deviation of residuals and the goodness-of-fit is very bad (22.3). A recovered lifetime from a single exponential fit is shorter, of about 12 ns. Three exponents are needed to satisfactorily fit this decay, see Figure 7-10.4. From recovered components, we can clearly see a significant contribution of shorter lifetime components that are not observed at MA conditions.

In the case of VH polarization conditions (vertically polarized excitation and horizontally polarized emission), the one-exponent fit is poor with a goodness-of-fit 4.7 with systematic, negative deviations of residuals. The recovered lifetime is longer of about 17 ns, see Figure 7-10.5. An attempt to fit these data with two components results in a good fit, but with a negative second component, see Figure 7-10.6.

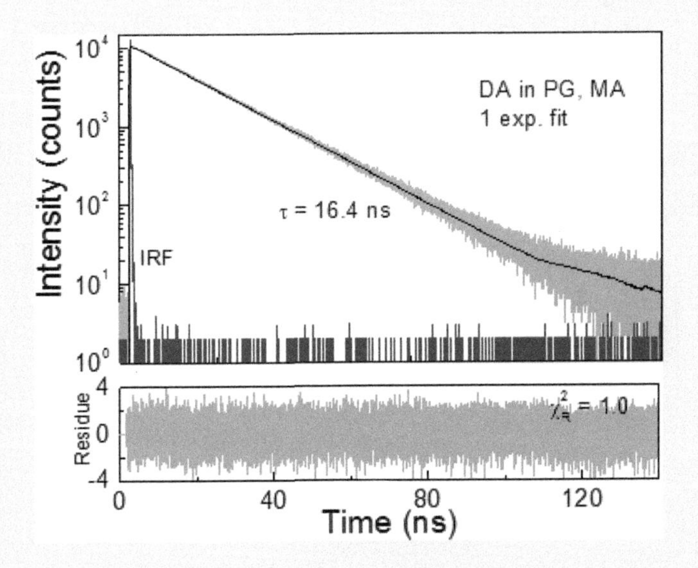

FIGURE 7-10.2 Fluorescence intensity decay of DA in PG. The emission polarizer was oriented at a magic angle (54.7°). The excitation was 375 nm, vertically polarized, and emission was observed at 525 nm.

(*Continued*)

EXPERIMENT 7-10 (CONTINUED) DEPENDENCE OF FLUORESCENCE INTENSITY DECAYS ON OBSERVATION CONDITIONS: EMISSION POLARIZER ORIENTATION

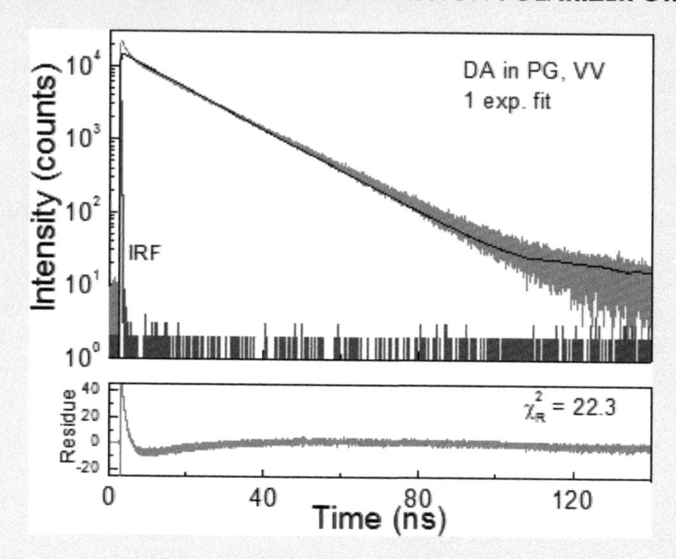

FIGURE 7-10.3 Fluorescence intensity decays of DA in PG at emission polarizer oriented vertically (VV). The excitation was at 375 nm and emission was observed at 525 nm. One-exponent model was used to fit the data.

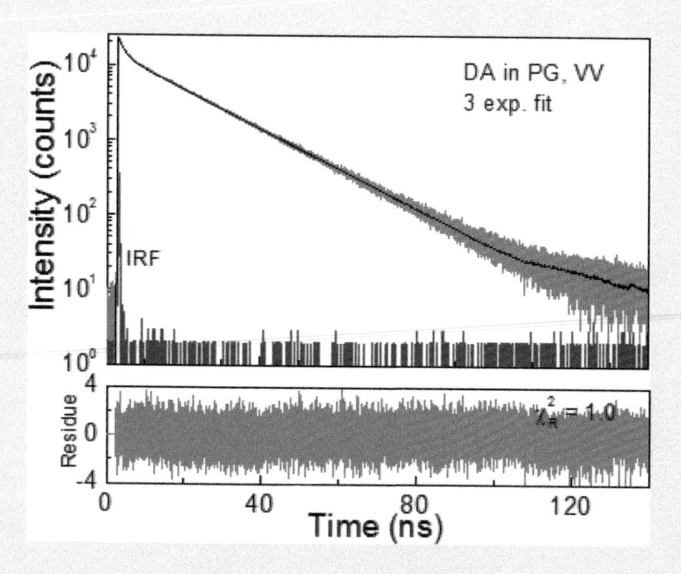

FIGURE 7-10.4 Fluorescence intensity decays of DA in PG at emission polarizer oriented vertically (VV). The excitation was at 375 nm and emission was observed at 525 nm. Three-exponent model was used to fit the data.

(Continued)

EXPERIMENT 7-10 (CONTINUED) DEPENDENCE OF FLUORESCENCE INTENSITY DECAYS ON OBSERVATION CONDITIONS: EMISSION POLARIZER ORIENTATION

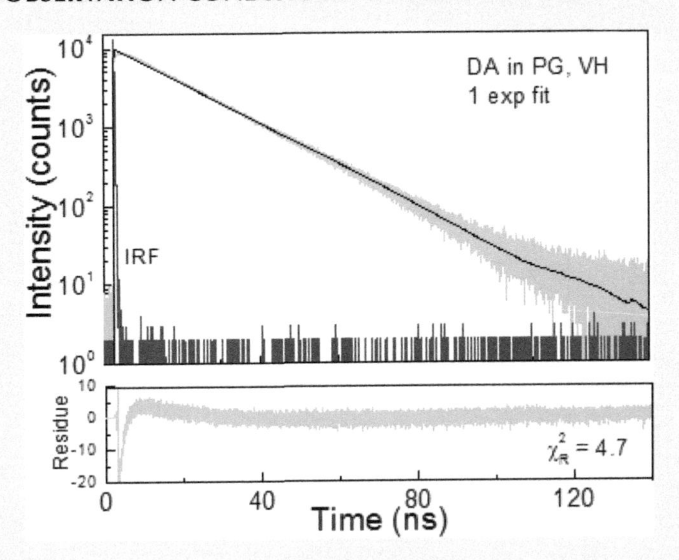

FIGURE 7-10.5 Fluorescence intensity decays of DA in PG at emission polarizer oriented horizontally (VH). The excitation was at 375 nm and emission was observed at 525 nm. One-exponent model was used to fit the data.

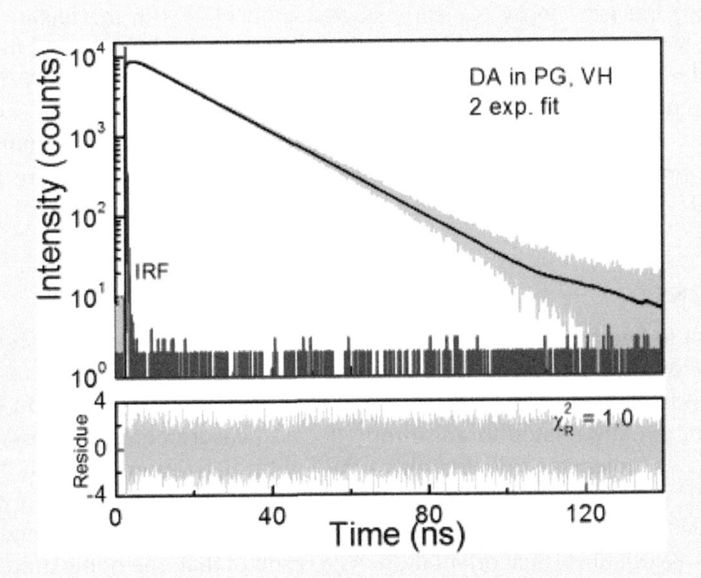

FIGURE 7-10.6 Fluorescence intensity decays of DA in PG at emission polarizer oriented horizontally (VH). The excitation was at 375 nm and emission was observed at 525 nm. Two-exponent model was used to fit the data with one negative component.

(Continued)

EXPERIMENT 7-10 (CONTINUED) DEPENDENCE OF FLUORESCENCE INTENSITY DECAYS ON OBSERVATION CONDITIONS: EMISSION POLARIZER ORIENTATION

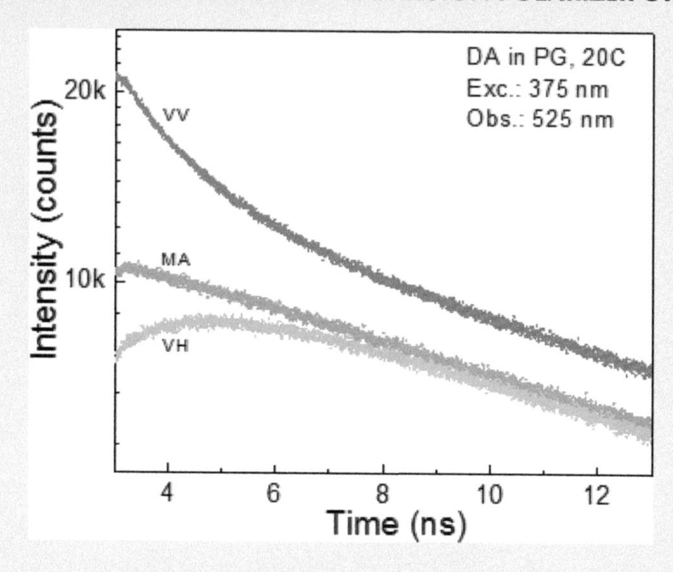

FIGURE 7-10.7 First 10 ns of the DA intensity decays in PG observed at different emission polarizer orientation.

Fluorescence lifetime strongly depends on the polarization conditions. At MA, the fluorescence intensity decay is clearly single-exponential with the lifetime of 16.4 ns. A vertical observation (VV) significantly alters (shorten) the lifetime and the decay can no longer be fitted with only one component. A horizontal observation (VH) results in a longer lifetime, and for an acceptable fit, a second (negative) component is needed.

The intensity decays at different polarization conditions differ mostly at the beginning of decays. Initial intensity decays for different polarization conditions are presented in Figure 7-10.7.

CONCLUSIONS

It is important to keep the MA (magic angle) conditions in lifetime measurements. Even a small deviation from 54.7° emission polarizer orientation can affect the recovered lifetime in viscous solutions. Described lifetime measurements with different polarization conditions are directly related to anisotropy decay measurements. Photoselection preferentially excites molecules oriented along the light polarization direction. So, intensity observed with vertically oriented emission polarizer is higher while intensity observed with horizontal orientation is much lower (see Figure 7-10.7). Due to molecular rotation, molecules rotate out of vertical orientation. As a result of that, the population of vertically oriented molecules decay artificially faster (both excited state decay and rotations contribute to the intensity decay). Population observed with a horizontally oriented polarizer will see an artificial increase in population of excited molecules yielding the negative component needed to fit intensity decay. See later experiments.

EXPERIMENT 7-11 FLUORESCENCE LIFETIME OF TRYPTOPHAN

Keywords: fluorescence, tryptophan, NATA, fluorometer.

Tryptophan is one of the three fluorescent amino acids that is mainly responsible for observed protein fluorescence. Other two, tyrosine and phenylalanine, have emission spectra weaker deeper in UV and their fluorescence is rarely studied. Because of its sensitivity to the environment, tryptophan can be considered as an intrinsic fluorescence probe. Tryptophan is among these fluorophores which show non-exponential intensity decay in solutions. We will demonstrate lifetime measurements of tryptophan in water and compare it with *N*-acetyl-tryptophan amide.

MATERIALS

- Tryptophan (TRP), and *N*-acetyl-tryptophan amide (NATA), both from Sigma-Aldrich
- Water, 1 cm × 1 cm quartz cuvettes

EQUIPMENT

- Spectrofluorometer (Varian Inc., Eclipse)
- Carry 50 Spectrophotometer (Varian Inc.)
- Fluorometer (FT 200, Picoquant, GmbH, Germany) equipped with UV LED

METHODS

1. Make stock solutions of TRP and NATA in a few milliliters of water, use glass or plastic vials, sonicate.
2. Prepare samples using stock solutions and measure absorptions. Adjust concentrations to achieve about 0.1 optical densities at maxima. Measure fluorescence spectra, see Figures 7-11.1 and 7-11.2.
3. Measure lifetimes of both TRP and NATA solutions using MA conditions.

RESULTS

In Figures 7-11.1 and 7-11.2, absorption and emission spectra for tryptophan and NATA are presented. For convenience, we also included chemical structures for TRP and NATA.

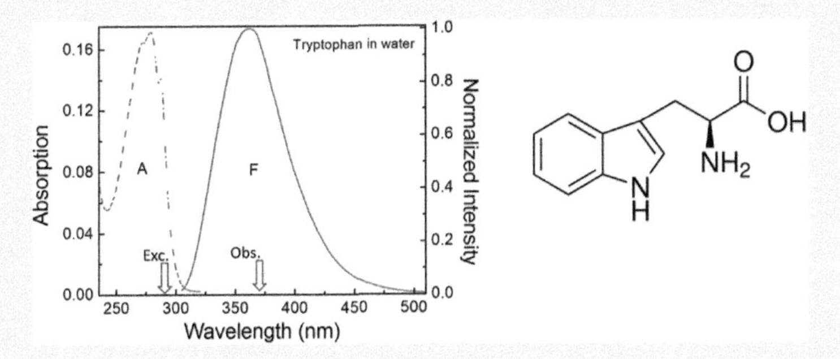

FIGURE 7-11.1 Absorption and emission spectra of TRP in the water at room temperature. Arrows indicate excitation and observation wavelengths in lifetime measurements. The chemical structure of TRP is on the right.

(Continued)

EXPERIMENT 7-11 (CONTINUED) FLUORESCENCE LIFETIME OF TRYPTOPHAN

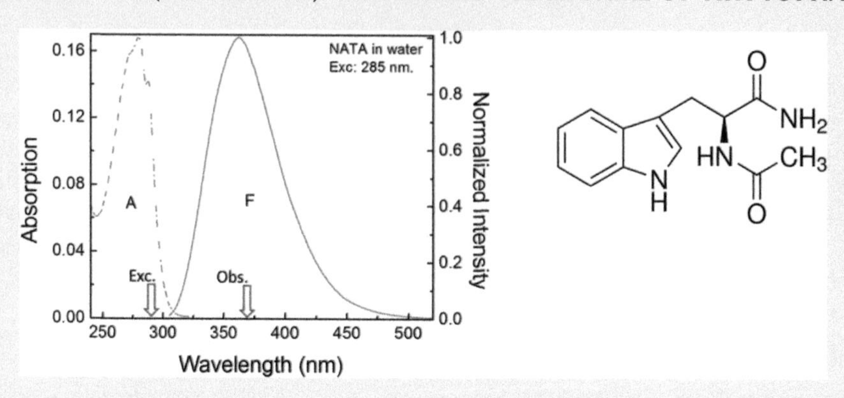

FIGURE 7-11.2 Absorption and emission spectra of NATA in the water at room temperature. Arrows indicate excitation and observation wavelengths in lifetime measurements. The chemical structure of NATA is on the right.

Absorption and fluorescence spectra of TRP and NATA are very similar. The molecular structures of TRP differ only slightly with a chemical group in NATA which blocks an amino group rotation in TRP.

Next, measure fluorescence intensity decays of TRP and NATA. Analyze decays with a multi-exponential model.

The intensity decay of TRP and one-exponent fit to the data are shown in Figure 7-11.3. This is not a fully satisfactory fit. The goodness-of-fit is elevated above 1.0 (1.4) and the residuals show a slight systematic deviation. Two-exponent fit is fully satisfactory, see Figure 7-11.4.

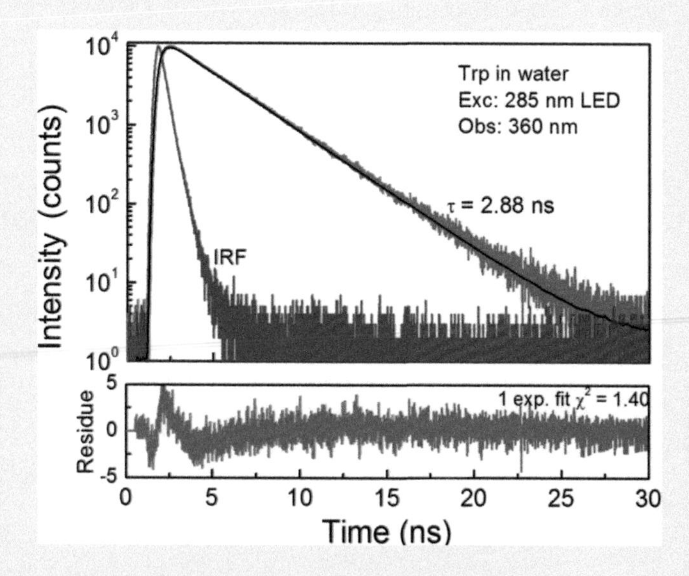

FIGURE 7-11.3 Fluorescence intensity decay of tryptophan in water. The intensity decay was fitted with a single-exponential model.

(Continued)

EXPERIMENT 7-11 (CONTINUED) FLUORESCENCE LIFETIME OF TRYPTOPHAN

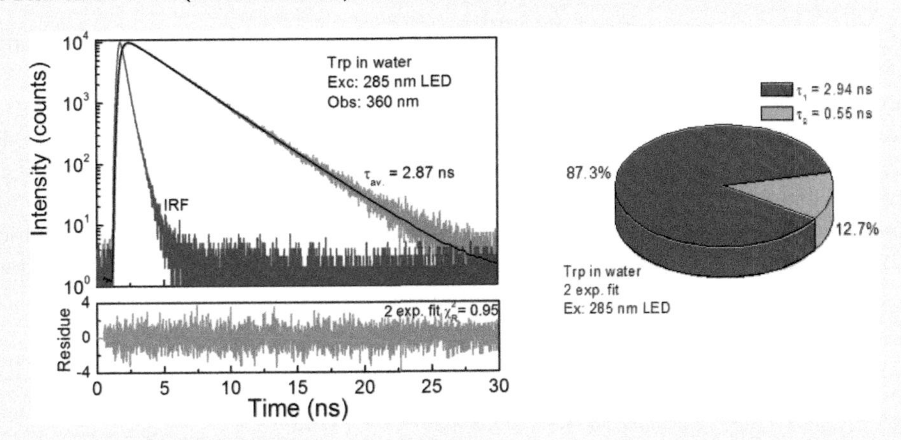

FIGURE 7-11.4 Tryptophan fluorescence intensity decay fitted with a two-exponential model.

The second, sub-nanosecond component with an amplitude of about 13% is a result of possible different conformations of TRP molecule (rotamers). In the TRP derivative NATA, the rotations of the amino group are blocked and different rotamers are not present (Figure 7-11.5).

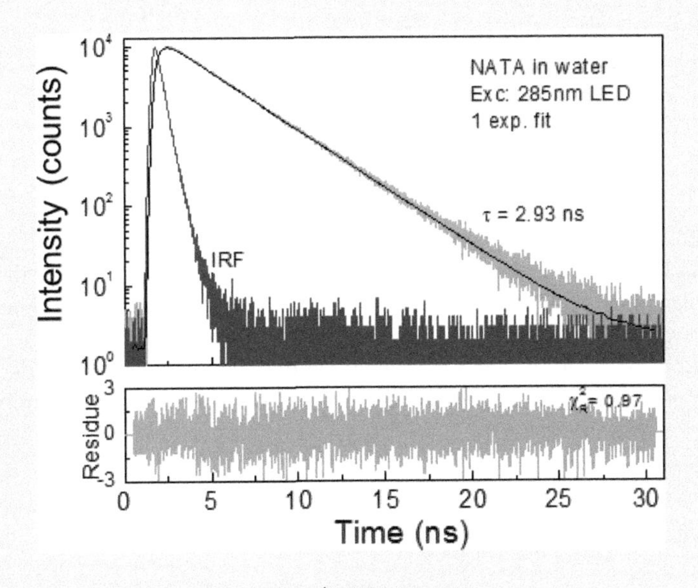

FIGURE 7-11.5 Fluorescence intensity decay of NATA in water.

(Continued)

EXPERIMENT 7-11 (CONTINUED) FLUORESCENCE LIFETIME OF TRYPTOPHAN

In fact, fluorescence intensity decay of NATA in water can be well fitted with a single exponent with a lifetime of 2.93 ± 0.01 ns.

Are we confident that TRP intensity decay is really two-exponential?

The confidence intervals analyses provide accuracies for recovered lifetimes: (1) 2.94 ± 0.01 ns, and (2) 0.55 ± 0.07 ns. In order to visualize the resolution, we plotted the χ_R^2 surfaces, see Figure 7-11.6. It looks that the resolution of 2.94 ns lifetime is much better than for the second component of 0.55 ns. Please note that the χ_R^2 surfaces are not symmetrical parabolas, which means that lower limit confidence can be different than the upper limit.

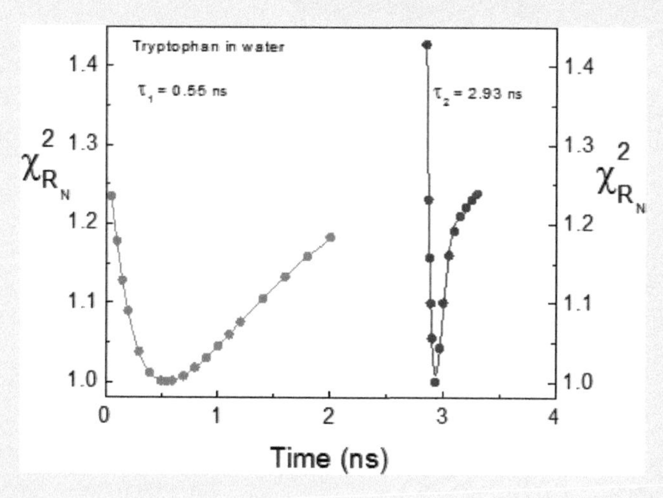

FIGURE 7-11.6 χ_R^2 surfaces for recovered lifetimes of tryptophan in water.

CONCLUSIONS

Whereas NATA fluorescence intensity decay in water is well approximated with a single component the TRP intensity decay requires two exponents for an acceptable fit. In this case, a possible explanation is a hindrance of rotation of amino group limiting number of possible rotamers. We want to stress that this is one of possible explanations.

The excitation source in this (and following) experiment is not a laser diode but UV LED (light-emitting diode). The IRF is not as short as from the laser and reconvolution analysis is required.

EXPERIMENT 7-12 FLUORESCENCE LIFETIMES OF PROTEINS

Keywords: fluorescence, lifetime, proteins.

Fluorescence of proteins is dominated by tryptophan. In the case of absence of tryptophan, proteins may show fluorescence emission of tyrosine upon UV excitation. In many cases, tryptophan is shielded from water when buried in the protein structure. Both emission spectrum and lifetime are affected by a change of the TRP moiety environment. We will demonstrate lifetime measurements of two proteins, human serum albumin (HSA, one tryptophan) and bovine serum albumin (BSA, two tryptophans). We will also demonstrate that fluorescence intensity decay of NATA in glycerol may mimic proteins intensity decays.

MATERIALS

- Human serum albumin (HSA), bovine serum albumin (BSA), *N*-acetyl-tryptophan amide (NATA) from Sigma-Aldrich.
- Water, glycerol
- 1 cm × 1 cm cuvettes quartz cuvettes

EQUIPMENT

- Spectrofluorometer (Varian Inc., Eclipse)
- Carry 50 Spectrophotometer (Varian Inc.)
- Fluorometer (FT 200, Picoquant, GmbH, Germany) equipped with UV LED

METHODS

1. Make stock solutions of HSA and BSA in a few milliliters of water, use a glass or plastic vial, sonicate.
2. Prepare samples using stock solutions and measure absorptions. Adjust concentrations to achieve about 0.1 optical densities at maxima. Measure fluorescence spectra, see Figures 7-12.1 and 7-12.2.
3. Measure lifetimes of all samples, NATA, HSA, and BSA solutions using MA conditions.
4. Make a solution of NATA in glycerol with an absorption below 0.1 in 1 cm cuvette. Measure absorption and fluorescence spectra.
5. Measure the lifetime of NATA in glycerol using MA conditions.

RESULTS

Absorption and fluorescence spectra of both proteins, HSA and BSA, are shown in Figures 7-12.1 and 7-12.2. The emission spectra are short-wavelength shifted comparing to tryptophan and NATA, see previous Experiment 7-11. BSA has one TRP moiety (134) exposed to the aqueous phase. Therefore, the BSA spectrum is at slightly longer wavelengths.

(*Continued*)

EXPERIMENT 7-12 (CONTINUED) FLUORESCENCE LIFETIMES OF PROTEINS

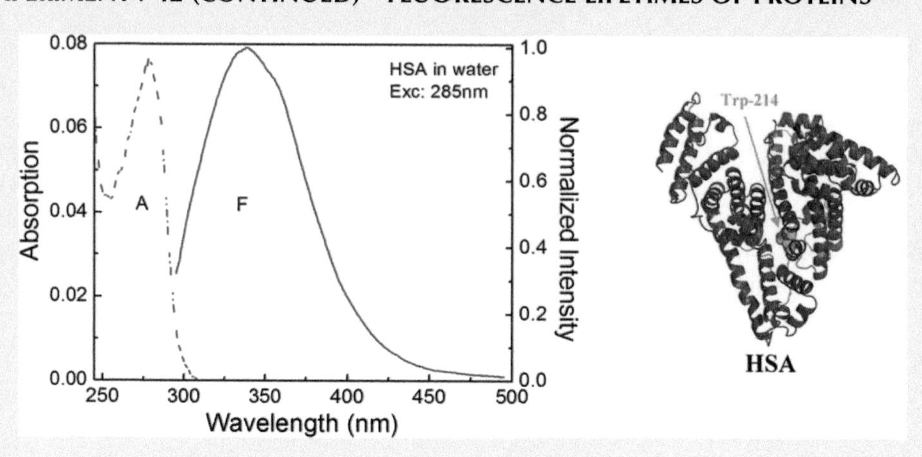

FIGURE 7-12.1 Absorption and emission spectra of HSA in the water at room temperature. The HSA structure is shown on the right. TRP moiety is marked green.

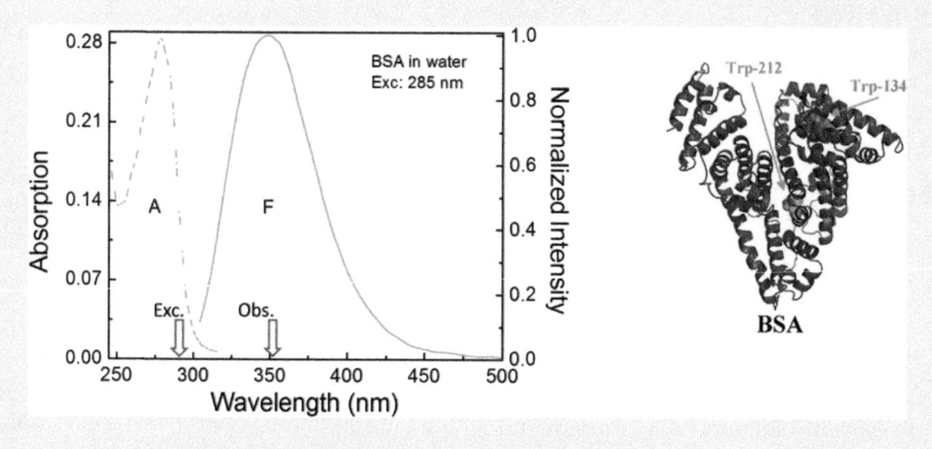

FIGURE 7-12.2 Absorption and emission spectra of BSA in the water at room temperature. The BSA structure is shown on the right. TRP moieties are marked green.

For the lifetime measurements, we used for both solutions, the excitation of 285 nm and the observations 345 nm for HSA and 355 nm for BSA, respectively. Figures 7-12.3 and 7-12.4 show the lifetime measurements for HSA and BSA.

Please note that the average lifetimes of intrinsic fluorescence of both proteins is almost twice longer than TRP or NATA in water. Both intensity decays, HSA and BSA can not be fitted satisfactorily with a single exponential model, two or three exponents are needed. This is a common feature of proteins intrinsic fluorescence. It is very difficult to predict efficiency and lifetimes of individual proteins. However, TRP moieties are well immobilized in a protein matrix. Lets compare fluorescence of proteins with NATA immobilized in glycerol. In fact,

(*Continued*)

EXPERIMENT 7-12 (CONTINUED) FLUORESCENCE LIFETIMES OF PROTEINS

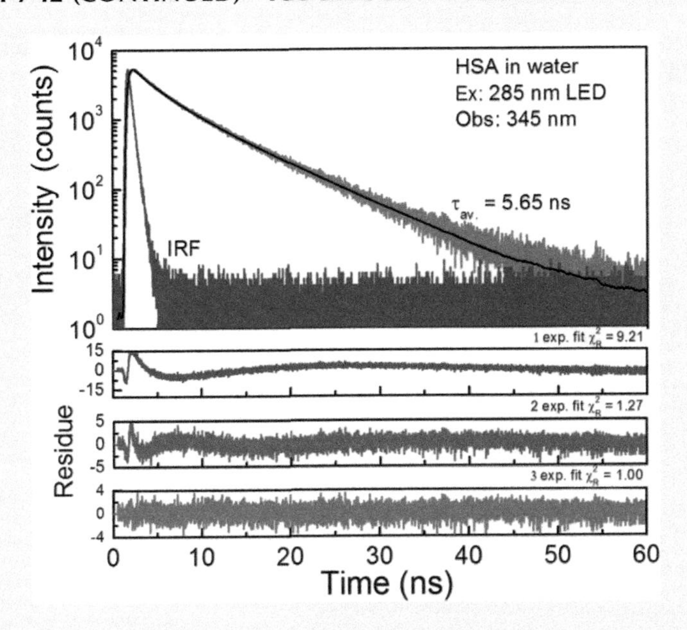

FIGURE 7-12.3 Fluorescence intensity decay of HSA in the water at room temperature. Recovered lifetimes are: 0.48 ± 0.05 ns, 2.79 ± 0.06 ns, and 7.22 ± 0.05 ns, with amplitudes 0.305, 0.370, and 0.325, respectively.

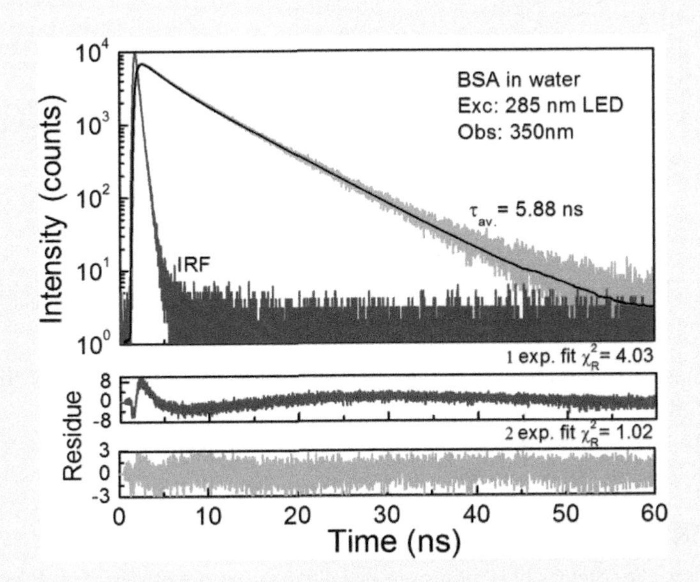

FIGURE 7-12.4 Fluorescence intensity decay of BSA in the water at room temperature. Recovered lifetimes are: 2.38 ± 0.08 ns and 6.55 ± 0.03 ns, with amplitudes 0.344 and 0.656, respectively.

(Continued)

EXPERIMENT 7-12 (CONTINUED) FLUORESCENCE LIFETIMES OF PROTEINS

the fluorescence spectrum of NATA in glycerol is slightly shifted toward shorter wavelengths comparing with aqueous solution, see Figure 7-12-5.

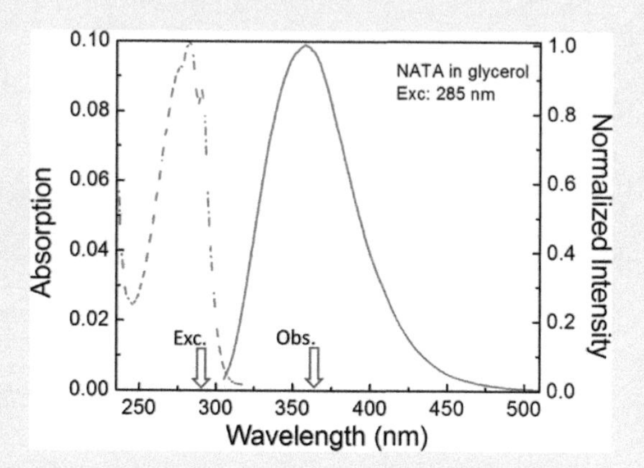

FIGURE 7-12.5 Absorption and emission spectra of NATA in glycerol at room temperature. Arrows indicate excitation and emission wavelengths.

Next, we measure the lifetimes of NATA in glycerol (Figure 7-12.6).

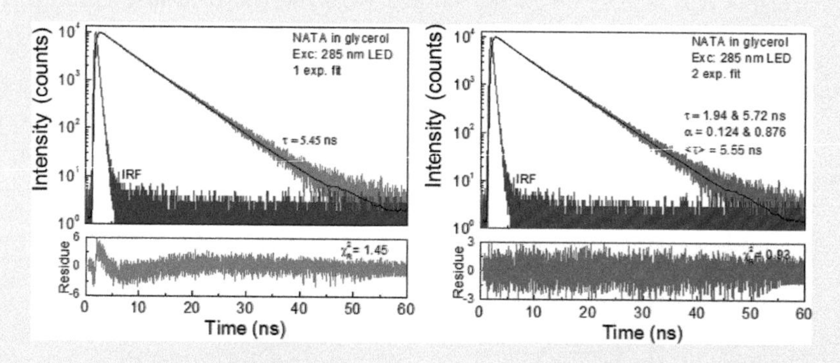

FIGURE 7-12.6 NATA in glycerol intensity decay fitted with one- (left) and two-exponential (right) models.

As in proteins case, the lifetime of NATA in glycerol is about twice longer than in water. Also, the intensity decay cannot be well fitted to a single-exponent model.

Two exponents are needed to fit the data. Please, note that NATA in water has a single exponential decay, but not in glycerol.

CONCLUSIONS

Fluorescence intensity decays of proteins are usually non-exponential and require two or even three exponents for acceptable fits. Lifetimes of fluorophores in viscous solutions depend on the observation wavelength because of the solvent relaxation. TRP residue in the protein matrix is well packed in the structure and face relaxation processes with neighboring amino acids as well. A viscous solvent like glycerol mimics roughly the protein matrix environment. NATA in glycerol shows two exponential decay when observing a maximum of fluorescence emission.

EXPERIMENT 7-13 LIFETIMES OF SINGLE FLUOROPHORES: EFFECT OF INTRINSIC HEAVY ATOMS

Keywords: fluorescence, lifetime, heavy atoms.

Fluorophores may have different efficiencies of fluorescence emissions (for intensity see Experiment 5-9). The most common reason is the presence of heavy metal atoms in the molecule structure. The presence of heavy atoms in the molecule structure enhances non-radiative transitions from the excited state of the molecule and one significantly increased is an intersystem crossing transition to the triplet state. We will show how lifetime depends on the presence of heavy atoms in the molecular structure.

MATERIALS

- Rhodamine 123 (R123)
- Eosin Y
- Erythrosine B (ErB)
- Water
- 1 cm × 1 cm cuvettes

EQUIPMENT

- Spectrofluorometer (Varian Inc., Eclipse)
- Cary 50 Spectrophotometer (Varian Inc.)
- Fluorometer (FT 200, Picoquant, GmbH)

METHODS

1. Make a stock solution of the studied compound in a few milliliters of water, use a glass or plastic vial. If the compound did not dissolve completely, add more water, see Figure 7-13.1 (left).
2. Prepare samples using stock solutions, see Figure 7-13.1 (right).
3. Measure absorption and fluorescence spectra of all three samples, see Figures 7-13.2 through 7-13.4.
4. Measure lifetimes of R123, Eosin Y, and ErB using MA conditions, see Figures 7-13.5 through 7-13.7.

RESULTS

Fluorescence emission of all three samples is green and difficult to distinguish, except ErB which is much less efficient, see Figure 7-13.1.

(Continued)

EXPERIMENT 7-13 (CONTINUED) LIFETIMES OF SINGLE FLUOROPHORES: EFFECT OF INTRINSIC HEAVY ATOMS

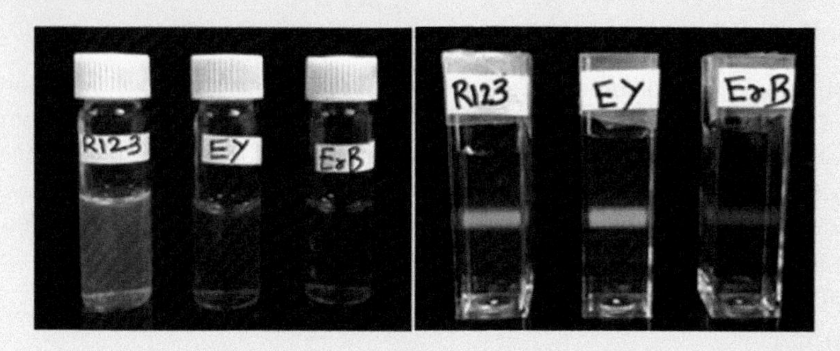

FIGURE 7-13.1 Photographs of stock solutions and samples. Left: Stock solutions of R123, Eosin Y, and ErB; Right: Samples used for lifetime measurements illuminated with 532 nm laser pointer. Absorption of each sample was adjusted to about 0.05 at maximum.

First, we measure absorption and emission spectra of all three samples, see Figures 7-13.2 through 7-13.4.

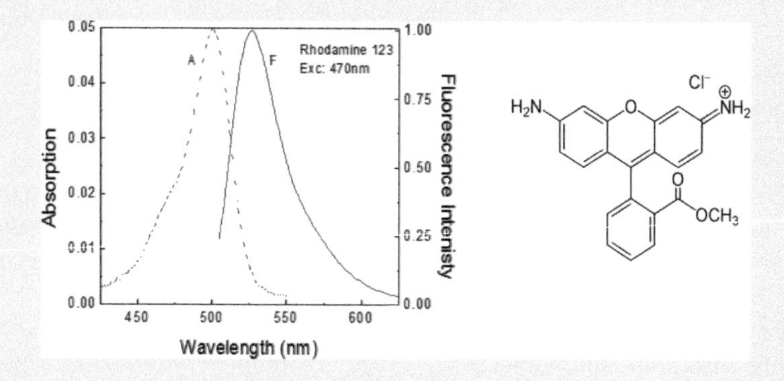

FIGURE 7-13.2 Absorption and emission spectra of R123 in water. The chemical structure of R123 is on the right.

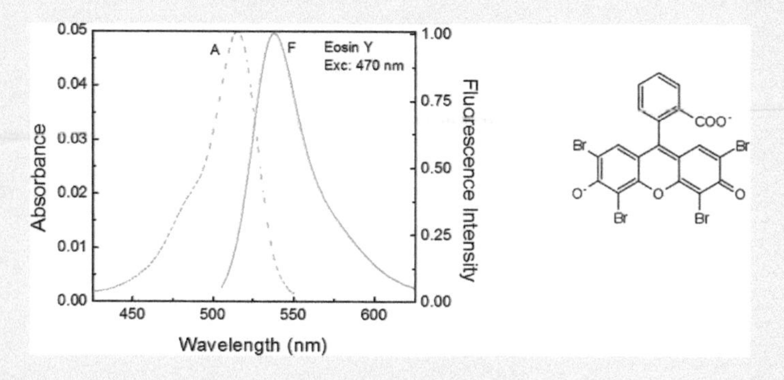

FIGURE 7-13.3 Absorption and emission spectra of Eosin Y in water. The chemical structure of Eosin Y is on the right.

(*Continued*)

EXPERIMENT 7-13 (CONTINUED) LIFETIMES OF SINGLE FLUOROPHORES: EFFECT OF INTRINSIC HEAVY ATOMS

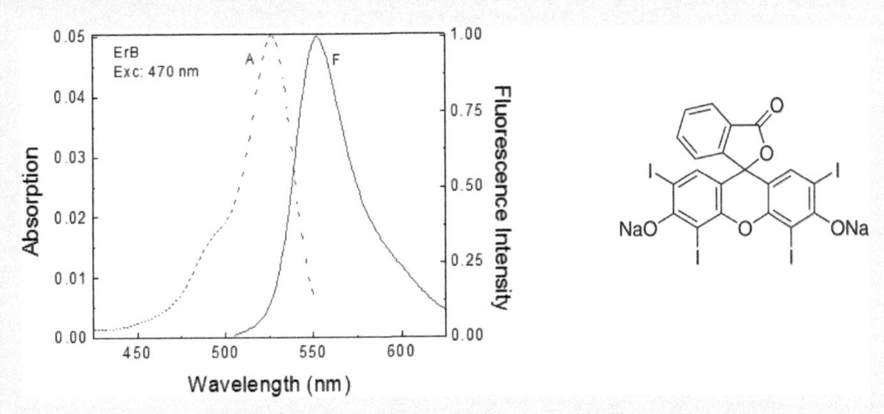

FIGURE 7-13.4 Absorption and emission spectra of ErB in water. The chemical structure of ErB is on the right.

The fluorescence maxima are about 530 nm for R123, 545 nm for Eosin Y, and 555 nm for ErB. Lifetime measurements will be done with these observations.

Next, we measure fluorescence lifetimes of all three compounds (remember to keep MA conditions), see Figures 7-13.5 through 7-13.7.

The fluorescence intensities of all three compounds can be reasonably fit with a single exponential model. However, the lifetimes are dramatically different.

The accuracies of the estimated lifetimes are better than ±0.01 ns for all three studied compounds.

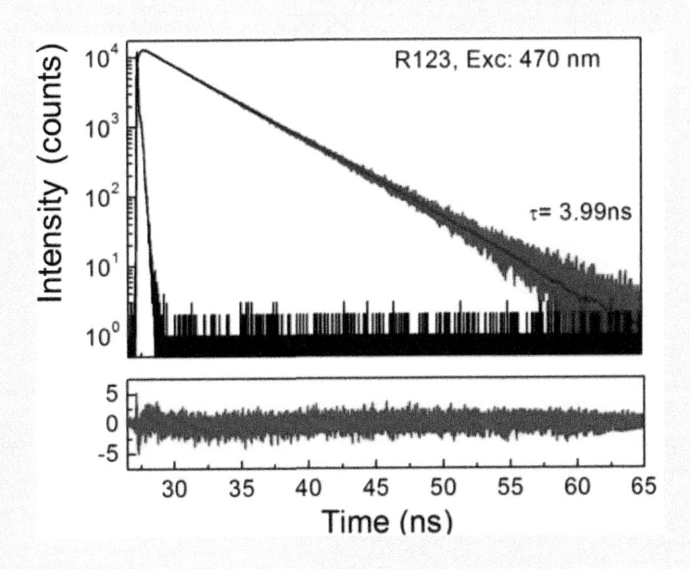

FIGURE 7-13.5 Fluorescence intensity decay of R123 in water.

(Continued)

EXPERIMENT 7-13 (CONTINUED) LIFETIMES OF SINGLE FLUOROPHORES: EFFECT OF INTRINSIC HEAVY ATOMS

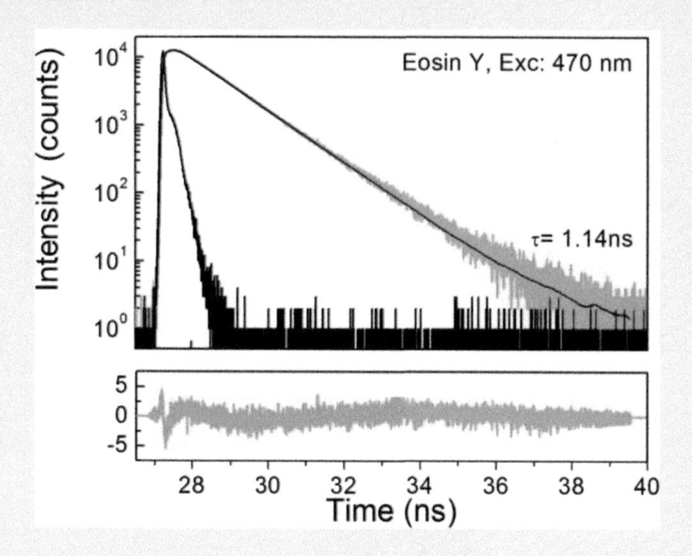

FIGURE 7-13.6 Fluorescence intensity decay of Eosin Y in water.

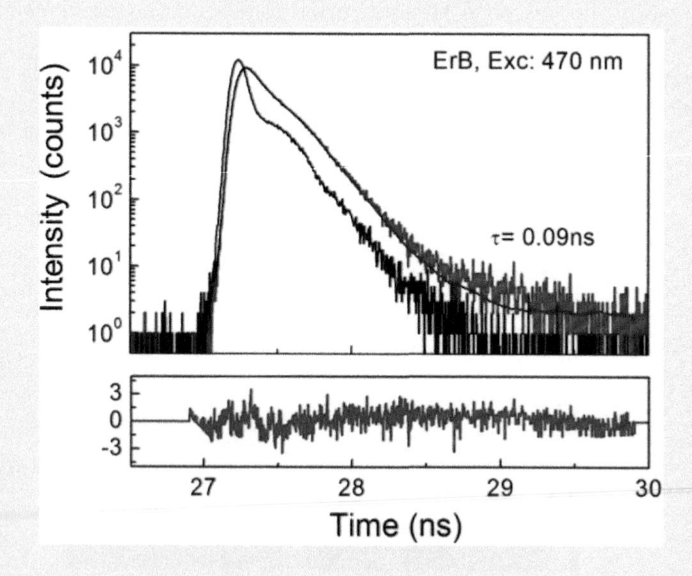

FIGURE 7-13.7 Fluorescence intensity decay of ErB in water.

The values of R123, Eosin Y, and ErB lifetimes follow a relation 1:3.3:44, which roughly corresponds to the relation of fluorescence intensities, see Experiment 5-9.

CONCLUSIONS

In many cases, the fluorescence efficiency can be predicted from the chemical structure of fluorophore. Heavy atoms, Cl, Br, and I usually internally quench the fluorescence.

EXPERIMENT 7-14 ANISOTROPY DECAYS

Keywords: fluorescence, anisotropy decay, polarization rotation.

Fluorophores in solutions move freely with the speed depending on the solvent viscosity. A translational displacement (diffusion) does not change the orientation of the direction of the molecule transition moment. However, a rotational diffusion (rotation) of the fluorophore usually changes its transition moment orientation. Therefore, rotational diffusion is a major factor affecting fluorescence anisotropy. As Perrin's equation describes, the fluorescence anisotropy depends on the ratio of lifetime and correlation time. The rotational correlation time describes molecular rotations and depends on the solvent viscosity. In general, in low viscous solutions, fluorophores rotate faster and the initial anisotropy induced by the excitation pulse decreases quickly in time. In this case, the rotational correlation time will be short. If the solvent is viscous, the same fluorophore will rotate slower and the correlation time will be longer. The change of the fluorescence anisotropy in time is called anisotropy decay. We will demonstrate the anisotropy decays of the fluorophore Rhodamine 700 in propylene glycol solutions and describe the analysis.

MATERIALS

- Rhodamine 700 (R700)
- Propylene glycol (PG)
- 1 cm × 1 cm fluorometric cuvettes, vials, plastic polarizers

EQUIPMENT

- Spectrofluorometer (Varian Inc., Eclipse)
- Cary 50 Spectrophotometer (Varian Inc.)
- Fluorometer (FT200, Picoquant, GmbH, Germany)
- Camera or cell phone

METHODS

1. Make a stock solution of R700 in PG, sonicate.
2. Prepare the sample of R700 in PG using the stock solution and measure its absorption. Adjust concentration to achieve about 0.05 optical density at maximum. Measure absorption and fluorescence spectra, see Figure 7-14.1.
3. Measure G-factor of the fluorometer (measure HV and HH intensities).
   a. Rotate polarization on excitation light from vertical (V) to horizontal (H). This can be done either by rotating the laser head, by using polarization rotator (expensive option) or by using two plastic polarizers (cheap option). In the last case, when the excitation polarization is fixed to the vertical orientation, place the horizontally oriented polarizer in the front, there will not be any light going through, and next insert the second polarizer oriented at 45° between the laser and the first (H) polarizer (see Figure 7-14.2). You may calculate the excitation intensity from Malus's law of transmission. In the ideal case, we can obtain 25% of the original laser intensity with horizontal polarization.
   b. Collect the signals for the same time for HV and HH configurations. In an ideal case, the signals should be the same. However, the collection optics (especially

(*Continued*)

EXPERIMENT 7-14 (CONTINUED) ANISOTROPY DECAYS

monochromators) transmits not equally V and H light components and a correction should be done in order to make them equal. Usually, the V component gives a larger signal than H. Therefore, one component must be corrected (V component must be divided by HV/HH ratio or H component multiplied by this ratio (see G-factor definition in Chapter 6)).

4. Measure intensity decays (using a 635 nm pulsed laser) at the magic angle, VV, and VH conditions, and observation at the maximum of fluorescence (about 690 nm).

The intensity decays will be fitted to a multi-exponential model:

$$I(t) = \sum_i \alpha_i \exp(-t/\tau_i)$$

where τ_i are lifetimes and α_i are amplitudes associated with the lifetimes.

The anisotropy decay will be fitted to an exponential anisotropy decay model:

$$r(t) = r_0 \exp(-t/\theta)$$

where r_0 is the initial anisotropy and θ is the correlation time.

RESULTS

First, we measured absorption and fluorescence spectra of R700, see Figure 7-14.1. Anisotropy measurements always require a correction of one component of the emission (usually VH) for the geometry of the instrument (including the uneven transmissions of the

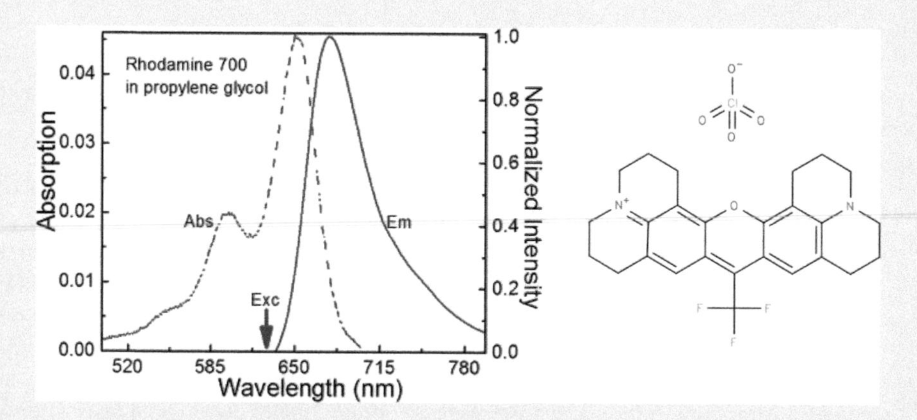

FIGURE 7-14.1 Absorption and emission spectra of R700 in PG. The arrow indicates a 635 nm excitation wavelength from a pulsed laser diode. The observation wavelength was 690 nm. The chemical structure of R700 is on the right.

(Continued)

EXPERIMENT 7-14 (CONTINUED) ANISOTROPY DECAYS

VV and VH components through a monochromator), called the correction factor or G-factor (see Chapter 6). Not always it is easy to obtain H polarization of the excitation beam, especially with fixed laser sources. The easiest way to rotate polarization plane from vertical to horizontal orientation is to use a combination of two polarizers with the second oriented at H direction, see Figure 7-14.2.

FIGURE 7-14.2 Top: Window views through two parallel polarizers (left) and crossed (middle). On the right, a third polarizer (45°) was inserted between crossed polarizers allowing a partial view. Bottom: Rotation of the laser light for the horizontal excitation. Crossed polarizers block completely the excitation (middle).

Rotate the excitation polarization to the horizontal orientation (H). Figure 7-14.2, bottom shows the polarization rotation when using two polarizers. Measure HV and HH components of the R700 emission at 690 nm observation. Calculate G-factor.

Next, we measure fluorescence intensity decays with MA (Figure 7-14.3). This measurement will reveal the proper fluorescence lifetime of R700 in PG.

(Continued)

EXPERIMENT 7-14 (CONTINUED) ANISOTROPY DECAYS

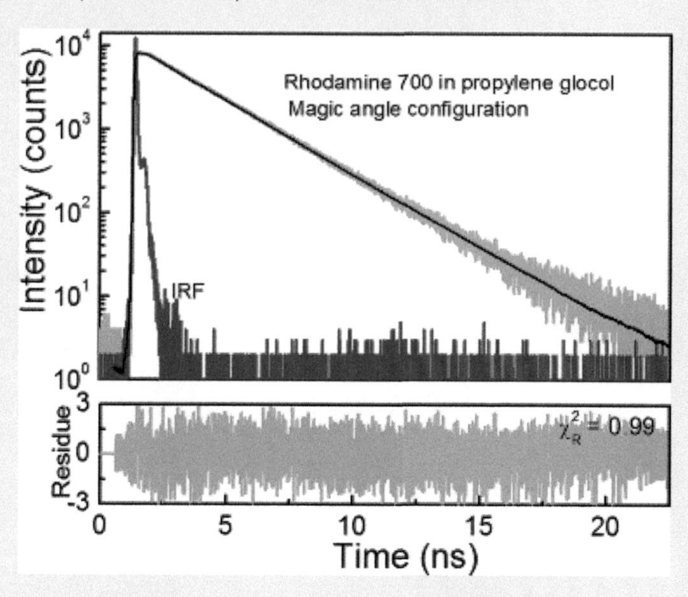

FIGURE 7-14.3 Lifetime measurements of R700 in PG. The excitation was from a pulsed 635 nm laser diode and the observation was at 690 nm, near the maximum of fluorescence. The decay is well approximated with a single exponent model with a lifetime of 2.44 ± 0.01 ns.

The lifetime of R700 in PG is about 2.5 ns which is close to the time of solvent relaxation and molecular rotation. It is good to collect anisotropy data over the entire lifetime of the fluorophore. In the case of a very long lifetime, molecule rotations and solvent relaxations will be completed in a fraction of the lifetime, the rest of the data will be meaningless. On the other hand, in the case of a very short lifetime of the fluorophore, there will not be fluorescence photons left in the measured VV and VH component after a quick fluorescence decay. The meaningful data will be only from the time of the decay (typically less than 10 fluorescence lifetimes).

The uncorrected VV and VH components of R700 in PG fluorescence intensity decays are shown in Figure 7-14.4. Note the slight upper-curvature of the VH component profile. Such a curvature can be fitted only with a second, negative exponent. It is typically an indication of an excited-state process (process occurring in the excited state of the molecule). During the lifetime, some excited R700 molecules will rotate to H polarization. This is similar to the "pumping" of the excited state (really population of excited molecules with the horizontal component of emission) during the fluorescence lifetime.

(Continued)

EXPERIMENT 7-14 (CONTINUED) ANISOTROPY DECAYS

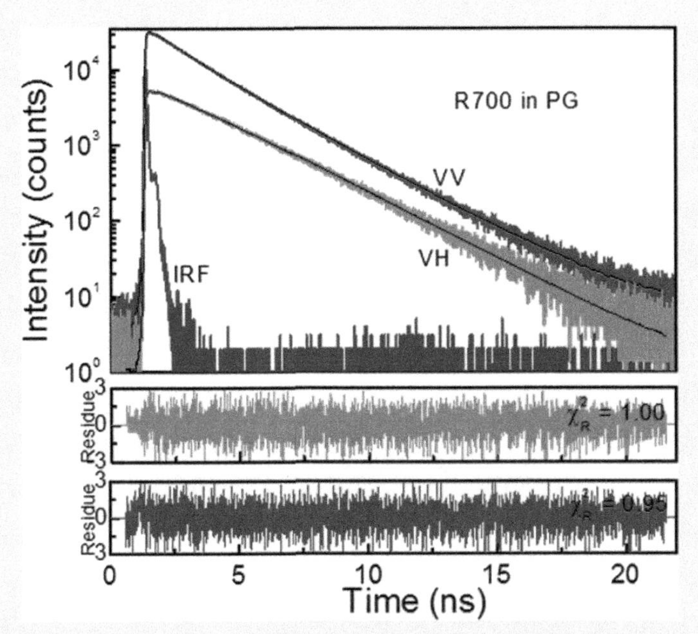

FIGURE 7-14.4 Fluorescence intensity decays of VV and VH components of R700 in PG. Not corrected for G-factor. The excitation was from a pulsed 635 nm laser diode and the observation was at 690 nm, near the maximum of fluorescence. The VV component can be well approximated with three exponent model with lifetimes 0.27, 1.42, 2.36 ns and amplitudes 0.048, 0.358, 0.594, respectively. The VH component can be well fitted with the two exponent model with lifetimes 1.75 ns, 2.33 ns, and amplitudes −0.31, 0.69, respectively. Note, the first amplitude is negative.

The implementation of G-factor (multiplication of VH component by G-factor) compensates for an uneven detection of polarized components. The intensity decays of VV and VH components after correction for G-factor are presented in Figure 7-14.5. After just 8 ns (about three lifetimes) after the excitation, the decays of VV and VH components are identical. And, after 8 ns, the anisotropy must be close to zero. This happens only for freely rotating molecules. In the case of immobilized molecules or when rotations are restricted/hindered, the VV and VH components may not merge.

Now, we are ready to calculate (fit) anisotropy decay. How does one set the lower and upper time limits in the fitting procedure? This is not obvious in the case of fast correlation times because of IRF interference. In the case of longer correlation times, it is good to first plot the anisotropy decay profile in a wide span of the time, see Figure 7-14.6. In this case, the fitting was done from 0 to 20 ns. Before the excitation pulse (0–2 ns) and after time longer than 15 ns there is only a noise because of the absence of fluorescence photons. The meaningful data is only between times indicated by arrows.

(*Continued*)

EXPERIMENT 7-14 (CONTINUED) ANISOTROPY DECAYS

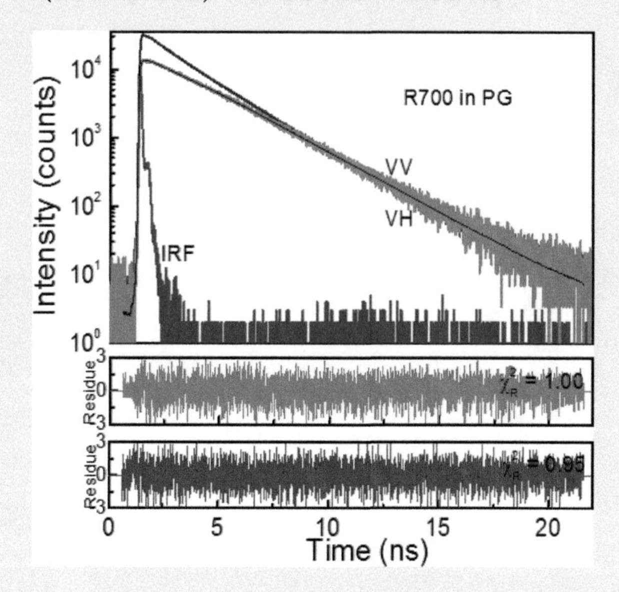

FIGURE 7-14.5 Fluorescence intensity decays of VV and VH components of R700 in PG. The VH component was corrected for G-factor (3.46).

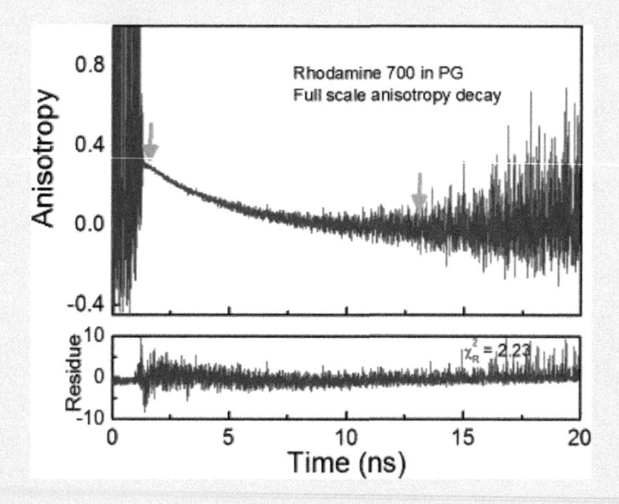

FIGURE 7-14.6 Anisotropy decay of R700 in PG calculated from zero to 20 ns. Arrows point the "reasonable" time which should be used for the calculation.

The fitting of the meaningful data is presented in Figure 7-14.7. The anisotropy decay of R700 in PG can be satisfactorily fitted with one correlation time of 3.43 ns. The residuals are randomly distributed. In the analysis of anisotropy decays the goodness-of-fit is rarely close to 1. This is because the calculation of anisotropy values require the ratio of intensities and individual errors propagate. The judgment should be rather based on the distribution of residuals.

(*Continued*)

EXPERIMENT 7-14 (CONTINUED) ANISOTROPY DECAYS

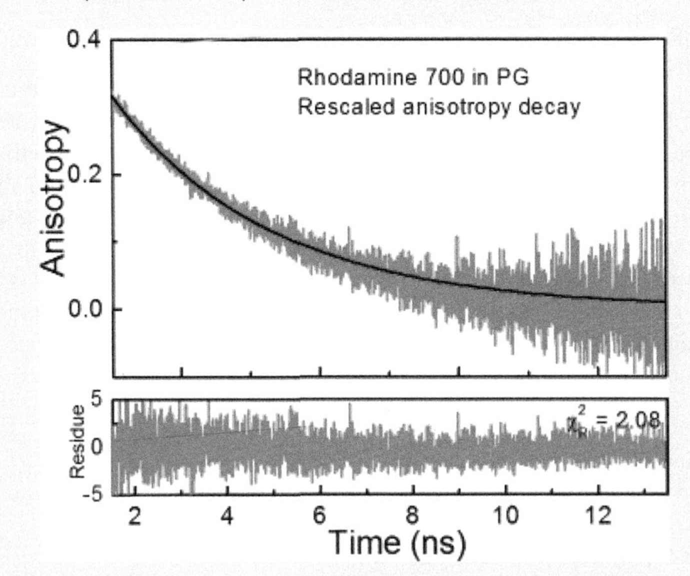

FIGURE 7-14.7 Anisotropy decay of R700 in PG. Recovered parameters are: initial anisotropy $r = 0.32 \pm 0.002$ and correlation time $\theta = 3.43 \pm 0.01$ ns, as estimated from confidence intervals.

CONCLUSIONS

Intensity decays of VV and VH components differ significantly at shorter time collection, see Figure 7-14.5. The VH component carries a negative amplitude of one of the exponents in the multi-exponent model used in the analysis. This is characteristic of the processes occurring in the excited state. The excitation process with V polarized light promotes V-oriented transition moments of molecules, which is called a photoselection. During the lifetime of the molecules, their transition moments randomize as result of Brownian rotations; the V component of the fluorescence decreases and the H component "builds up" in time. At a longer time, the VH and HH components are similar. As seen in Figure 7-14.6, the anisotropy values within the excitation pulse and a longer time, above 15 ns, are very scattered and not well defined. Therefore, it is necessary to set the time limits in the anisotropy analysis, see arrows in Figure 7-14.6. This is an arbitrary procedure which may affect recovered parameters, initial anisotropy, and correlation time. Typically, we would not recommend analysis above 6–7 fluorescence lifetimes. The recovered initial anisotropy is always lower than a fundamental anisotropy (0.4). The reason for this is fast depolarizing movements of molecules occurring within the excitation pulse. It should be noted that in ordered systems (like stretched polymer films) or in two or more photon excitations the initial anisotropies can be higher than 0.4.

Note: Anisotropy decays are usually presented in a linear scale, in contrast to the intensity decays.

EXPERIMENT 7-15 ANISOTROPY DECAYS OF QUENCHED FLUOROPHORES

Keywords: fluorescence, anisotropy decay, Rhodamine 6G, fluorescence quenching.

The main factor responsible for anisotropy decay is a rotational diffusion of the fluorophore which changes the direction/orientation of the molecule transition moment in space. In a completely rigid environment, the fluorescence anisotropy will not change with time. In low viscous solution, the fluorescence anisotropy decay depends on the size and shape of fluorophore and on solvent viscosity. Intuitively, the rotational diffusion should not depend on the lifetime of the fluorophore. In this experiment, we will measure anisotropy decays of a quenched and unquenched fluorophore and compare the decay parameters. Quenching does not lead to formation of molecular complex and fluorophore rotation (rotational correlation time) is unchanged but fluorescence lifetime is significantly modified.

MATERIALS

- Rhodamine 6G (R6G)
- Potassium iodide (KI)
- Water
- 1 cm × 1 cm fluorometric cuvettes, vials

EQUIPMENT

- Spectrofluorometer (Varian Inc., Eclipse)
- Cary 50 Spectrophotometer (Varian Inc.)
- Fluorometer (FT 200, Picoquant, GmbH, Germany)

METHODS

1. Make a stock solution of R6G in water with the absorption of about 0.1 in maximum.
2. Make a 0.4 M stock solution of KI in water.
3. Prepare an R6G sample by mixing 2 mL of R6G stock solution and 2 mL water. This will be an unquenched sample. The quenched sample you will obtain by mixing 2 mL of R6G stock solution and 0.4 M KI stock solution. Both unquenched and quenched samples will have the same R6G concentration. The quenched sample will contain 0.2 M KI.
4. Measure absorption and fluorescence emission of both samples, see Figures 7-15.1 and 7-15.2.
5. Measure fluorescence intensity decays of both samples with vertically polarized 470 nm excitation from the laser diode and observation polarizer oriented at the magic angle (54.7°), see Figure 7-15.3.
6. Measure anisotropy decays of both samples, see Figures 7-15.4 and 7-15.5. A G-factor value for the observation wavelength (553 nm) is needed to fit the anisotropy decays. You can find the G-factor value from HV and HH components. Alternatively, you can fit the decay to a longer time scale (10–20 ns) and reveal the G-factor value, assuming that free rotations of a small R6G molecule will quickly randomize excited dipole moments. At a longer time, the anisotropy of freely rotated molecules should approach the zero value (randomized distribution of transition moments orientation).

(*Continued*)

EXPERIMENT 7-15 (CONTINUED) ANISOTROPY DECAYS OF QUENCHED FLUOROPHORES

The intensity decays will be fitted to a multi-exponential model:

$$I(t) = \sum_i \alpha_i \exp(-t / \tau_i)$$

where τ_i are lifetimes and α_i are amplitudes associated with the lifetimes.

The anisotropy decay will be fitted to an exponential model:

$$r(t) = r_0 \exp(-t/\theta)$$

where r_0 is the initial anisotropy and θ is the correlation time.

RESULTS

We selected R6G, a very bright fluorophore, sensitive to iodide (Scheme 7-15.1).

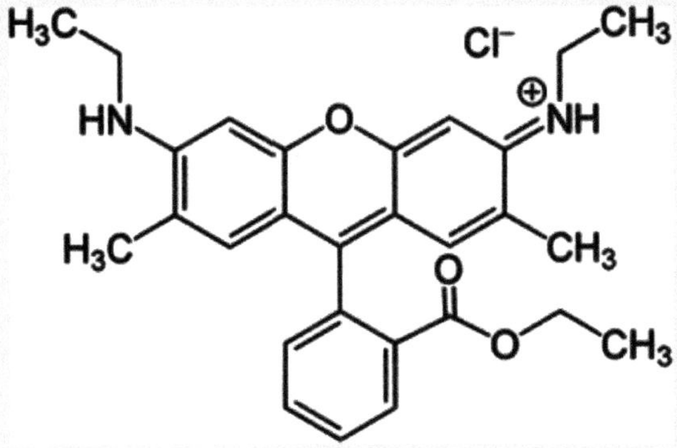

SCHEME 7-15.1 Chemical structure of R6G.

It is important to compare the absorption of fluorophore in absence and presence of the quencher. Often, the fluorophore and quencher form complexes in the ground state which may result in a static quenching.

Absorption spectra of unquenched and quenched samples are shown in Figure 7-15.1. Absorptions of R6G in presence and absence of 0.2 M KI are identical, no complexes are being formed.

First, we measure the fluorescence emission spectra of unquenched and quenched R6G, see Figure 7-15.2.

The fluorescence of R6G in the presence of 0.2 M KI is quenched about nine-fold. This indicates a Stern–Volmer constant of about 40 LM^{-1}, which is expected for a dynamic quenching.

(Continued)

EXPERIMENT 7-15 (CONTINUED) ANISOTROPY DECAYS OF QUENCHED FLUOROPHORES

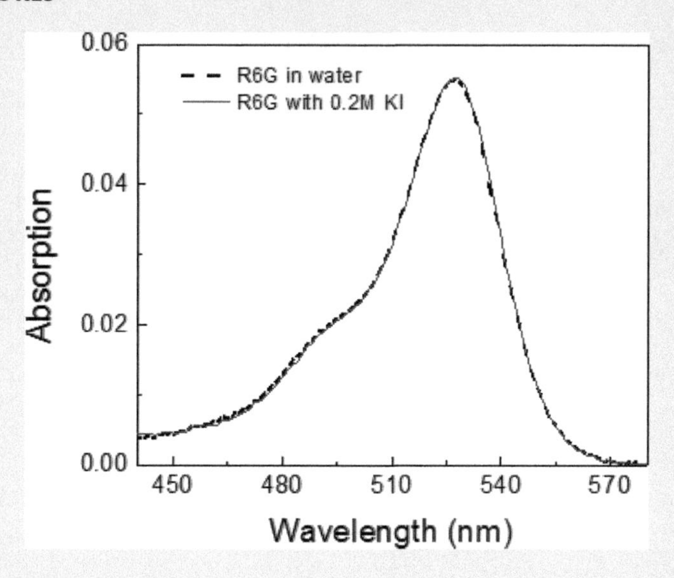

FIGURE 7-15.1 Absorption spectra of unquenched (dashed line) and quenched (solid line) samples. There are no new absorption bands.

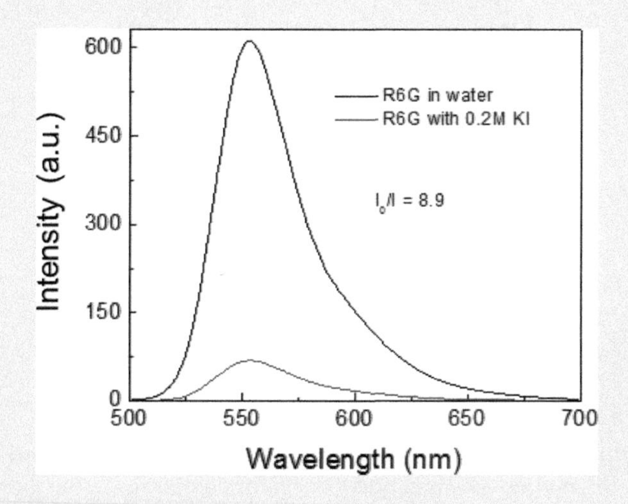

FIGURE 7-15.2 Fluorescence spectra of unquenched (black) and quenched (red) samples.

Next, we measure fluorescence intensity decays, see Figure 7-15.3. Both decays can be well fitted with a single-exponent model and lifetimes of 3.93 ns (unquenched sample) and 0.44 (quenched sample).

The lifetimes also indicate a Stern–Volmer constant of about 40 LM^{-1}, with a good agreement with the intensity data.

Next, we measure fluorescence anisotropy decays for both unquenched and quenched samples. The measurements consist of VV and VH intensity decay components corrected for G-factor.

(Continued)

EXPERIMENT 7-15 (CONTINUED) ANISOTROPY DECAYS OF QUENCHED FLUOROPHORES

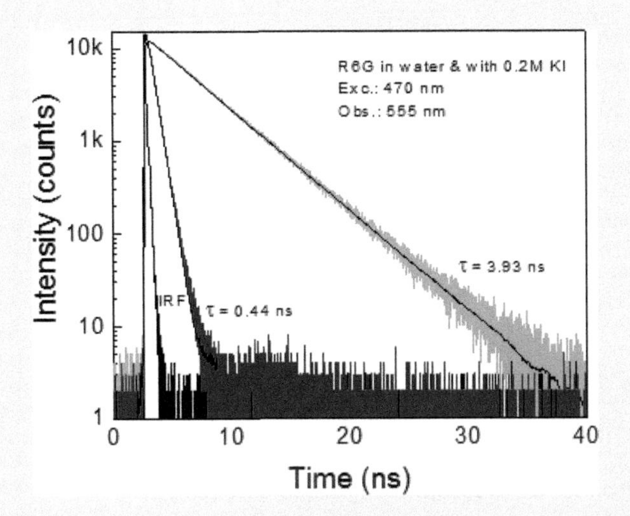

FIGURE 7-15.3 Fluorescence intensity decays of unquenched (green) and quenched (burgundy) samples.

The anisotropy decays of unquenched and quenched samples reveal the same correlation times of about 0.27 ns and the same initial anisotropies of 0.385, see Figures 7-15.4 and 7-15.5.

In both cases, unquenched and quenched sample, the accuracies in recovered correlation times were better than ±0.01 ns, and in initial anisotropies better than ±0.001, as estimated from confidence intervals.

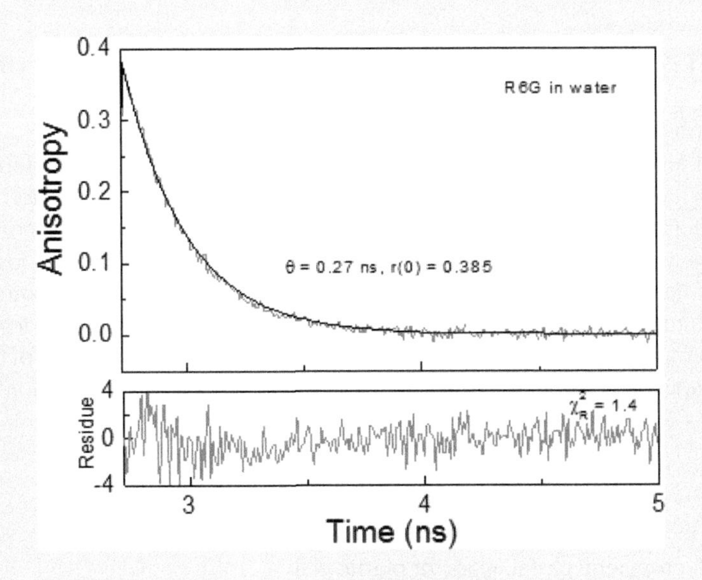

FIGURE 7-15.4 Fluorescence anisotropy decay of unquenched sample.

(Continued)

EXPERIMENT 7-15 (CONTINUED) ANISOTROPY DECAYS OF QUENCHED FLUOROPHORES

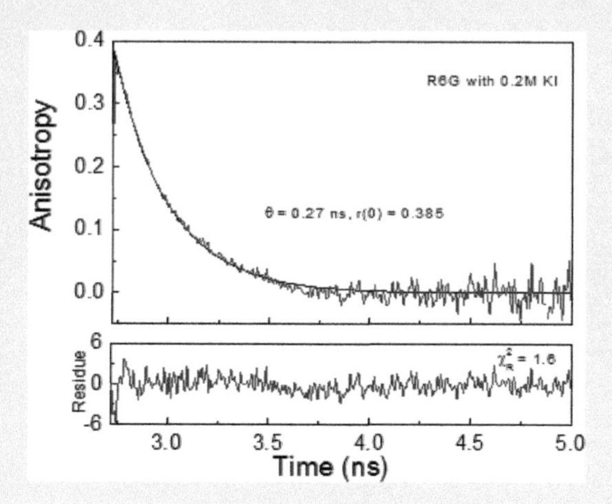

FIGURE 7-15.5 Fluorescence anisotropy decay of quenched sample.

CONCLUSIONS

Fluorescence anisotropy decays do not depend on the lifetime of the fluorophore. The revealed anisotropy decay parameters are the same for both samples.

Why this is important? The meaningful anisotropy data are these measured within the lifetime of the fluorophore. If the fluorophore has a much longer lifetime than investigated correlation times, it is reasonable shortening this lifetime to the scale of studied relaxation processes.

EXPERIMENT 7-16 SOLVENT VISCOSITY-DEPENDENT ANISOTROPY DECAYS

Keywords: fluorescence, anisotropy decay, viscosity.

In the case of an ideal fluorophore in a rigid isotropic medium, it is expected to see fluorescence anisotropy close to a limiting value of 0.4 resulting from photo selection. In any time during the fluorescence intensity decay (lifetime), the value of anisotropy is constant. In the case of a low viscosity medium, the observed fluorescence anisotropy is lower because during the lifetime excited molecules can rotate and change a transition moment direction/orientation. Immediately after a short-pulse excitation, the anisotropy is high (close to 0.4 or limiting anisotropy) but decreasing in time of the observation to zero. The rate of this decrease is viscosity dependent and is governed by the Perrin equation. We will demonstrate anisotropy decays of Coumarin 152 in low and high viscosity solvents.

MATERIALS

- 7-Dimethylamino-4-trifluoromethylcoumarin (C152)
- Ethanol, propylene glycol, glass, or plastic vials
- 1 cm × 1 cm disposable fluorometric cuvettes

(*Continued*)

EXPERIMENT 7-16 (CONTINUED) SOLVENT VISCOSITY-DEPENDENT ANISOTROPY DECAYS

EQUIPMENT

- Spectrofluorometer (Varian Inc., Eclipse)
- Carry 50 Spectrophotometer (Varian Inc.)
- Fluorometer (FT 200, PicoQuant, GmbH, Germany)

METHODS

1. Make about 1 mL stock solution of C152 (see Scheme 7-16.1) in ethanol with a concentration of about 1 mM (use extinction coefficient 18,000 L/mol × cm), sonicate the solution. Prepare vials with 5 mL of ethanol and PG and add 10 μL of the C152 stock solution to each vial. Optical densities of prepared samples should not exceed 0.05.
2. Measure absorption and fluorescence spectra, see Figure 7-16.1.
3. Measure lifetimes and anisotropy decays of both ethanol and PG solutions of C152, see Figures 7-16.2 and 7-16.3.

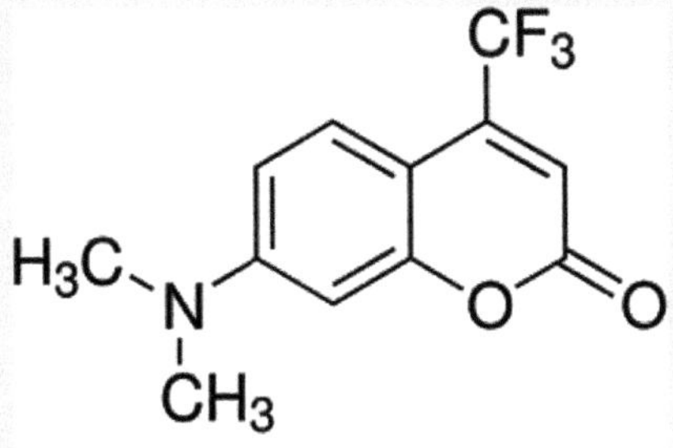

C152

SCHEME 7-16.1 Chemical structure of C152.

RESULTS

First, measure absorption and fluorescence spectra of C152 in both solvents (Figure 7-16.1). The absorption and fluorescence spectra of C152 are very similar in both solvents. Greenish emissions have a maxima at about 510 and 520 nm in ethanol and PG, respectively. We will observe fluorescence emissions at these wavelengths. Next, measure the lifetimes of C152 in both solvents (Figure 7-16.2).

(*Continued*)

EXPERIMENT 7-16 (CONTINUED) SOLVENT VISCOSITY-DEPENDENT ANISOTROPY DECAYS

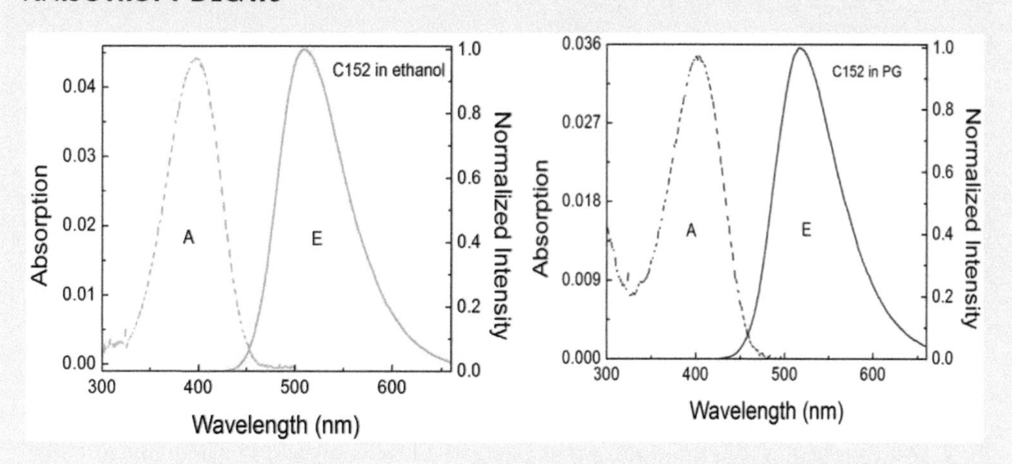

FIGURE 7-16.1 Absorption and emission spectra of C152 in ethanol (left) and PG (right) at room temperature.

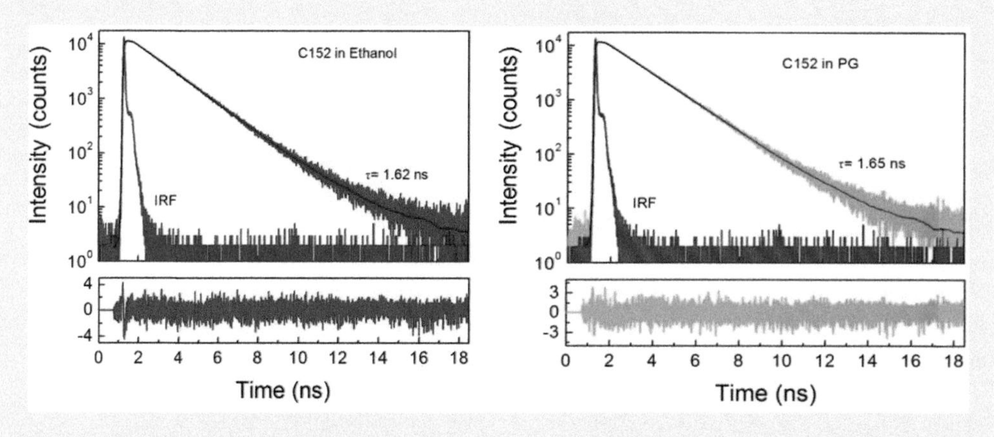

FIGURE 7-16.2 Fluorescence intensity decays of C152 in ethanol (left) and PG (right). Note, both lifetimes are very close.

Lifetimes of C152 in ethanol and PG are similar (almost identical) and both are well approximated by a single exponential model.

Next, measure fluorescence anisotropy decays. These measurements involve VV and VH intensity decays and G-factor correction.

Both anisotropy decays can be satisfactorily fitted with the one correlation time model, see Figure 7-16.3.

Recovered correlation times are very different. In ethanol, the rotations of C152 molecules are very fast, only 0.14 ± 0.01 ns. The recovered zero time anisotropy is slightly below

(Continued)

EXPERIMENT 7-16 (CONTINUED) SOLVENT VISCOSITY-DEPENDENT ANISOTROPY DECAYS

0.3 because of IRF interference, see Figure 7-16.3 (left). In the case of C152 in PG, the IRF interference is not as severe and recovered zero time anisotropy is about 0.35 ± 0.002. The correlation time of C152 in PG is 2.22 ± 0.02 ns.

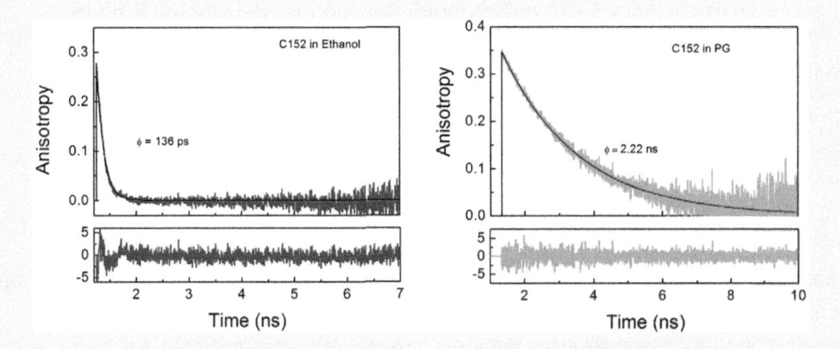

FIGURE 7-16.3 Fluorescence anisotropy decays of C152 in ethanol (left) and PG (right).

CONCLUSIONS

The anisotropy decays depend strongly on the solvent viscosity. The correlation time is a good measure of a local micro-viscosity. In the case when the measurement conditions cannot be kept steady, the time-resolved measurements are preferable.

EXPERIMENT 7-17 DEPENDENCE OF ANISOTROPY DECAYS ON THE MOLECULE SIZE

Keywords: fluorescence, fluorometer, lifetime, anisotropy.

Rotational Brownian motions are a major factor for fluorescence depolarization in diluted solutions. Of course, larger objects, like proteins, rotate slower than free fluorophore molecules. However, it is not obvious if small differences in fluorophores sizes can be observed in anisotropy decays. We selected two molecules, Coumarin 152 and Rhodamine 700. Core aromatic rings are the same, but the Rhodamine 700 is clearly a bigger molecule (see Figure 7-17.1). We will compare anisotropy decays of both molecules.

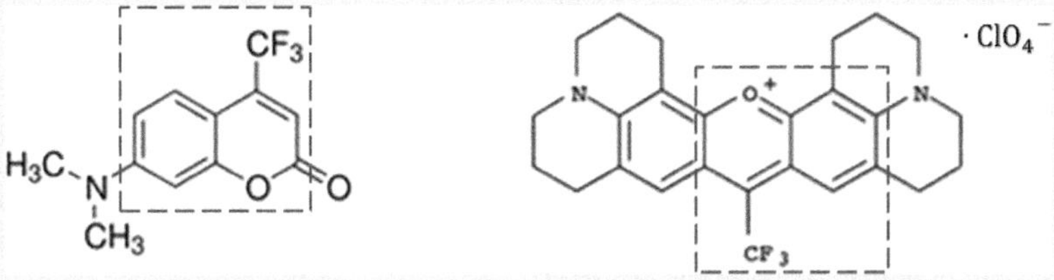

FIGURE 7-17.1 Chemical structures of C152 and R700. Core aromatic rings are the same.

(Continued)

EXPERIMENT 7-17 (CONTINUED) DEPENDENCE OF ANISOTROPY DECAYS ON THE MOLECULE SIZE

MATERIALS

- Coumarin 152 (C152), Rhodamine 700 (R700)
- Dimethyl sulfoxide (DMSO)
- Glass or quartz 1 cm × 1 cm cuvette, pipettes, neutral density (ND) filters

EQUIPMENT

- Spectrofluorometer (Varian Inc., Eclipse)
- Carry 50 Spectrophotometer (Varian Inc.)
- Fluorometer (FT200, PicoQuant, GmbH, Germany)

METHODS

1. Make solutions of C152 and R700 (Figure 7-17.1) in DMSO, adjust concentrations to achieve absorptions of about 0.1 (Figure 7-17.2).
2. Measure fluorescence spectra with excitations 405 and 635 nm for C152 and R700, respectively (Figure 7-17.3).
3. Measure lifetimes and anisotropy decay of C152 and R700 in DMSO (Figures 7-17.4 through 7-17.7).

RESULTS

First, we measure the absorption spectra of both fluorophores in DMSO, see Figure 7-17.2.

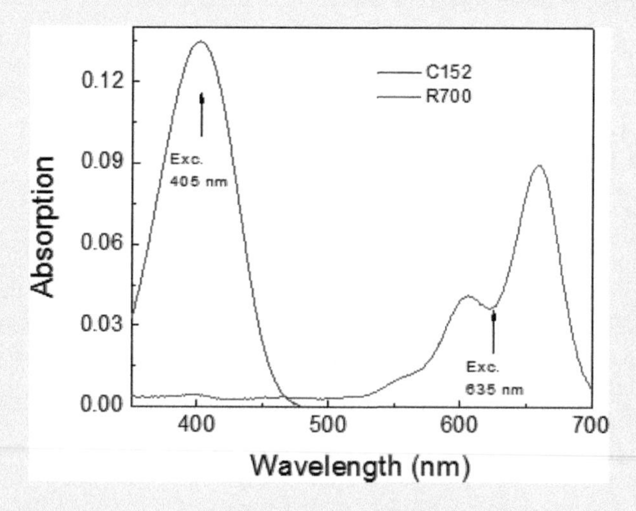

FIGURE 7-17.2 Absorption spectra of C152 and R700 in DMSO. Arrows indicate excitation used in fluorescence measurements.

Next, we measure the fluorescence spectra of both samples (Figure 7-17.3) and indicate observation wavelengths. Lifetimes and anisotropy decays will be measured at 515 nm observation for C152 and 690 nm for R700. Both dyes are efficient fluorophores.

(*Continued*)

EXPERIMENT 7-17 (CONTINUED) DEPENDENCE OF ANISOTROPY DECAYS ON THE MOLECULE SIZE

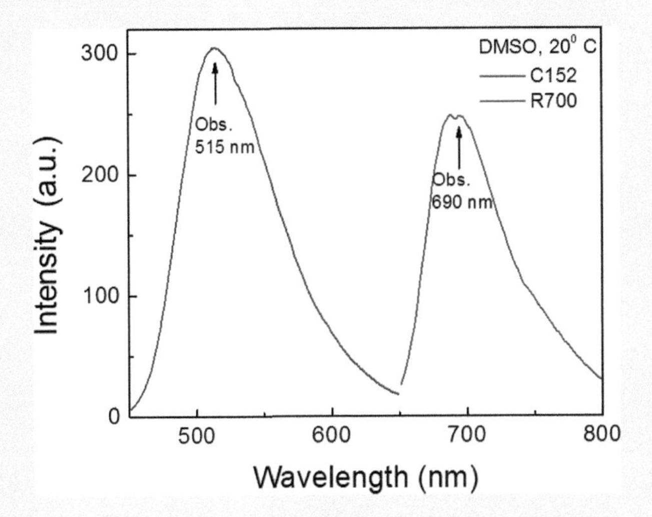

FIGURE 7-17.3 Fluorescence emission spectra of C152 and R700 in DMSO. Arrows indicate the observations used in time-resolved measurements.

Next, measure lifetimes of both fluorophores, see Figures 7-17.4 and 7-17.5. Both intensity decays can be well fitted with a single exponential model.

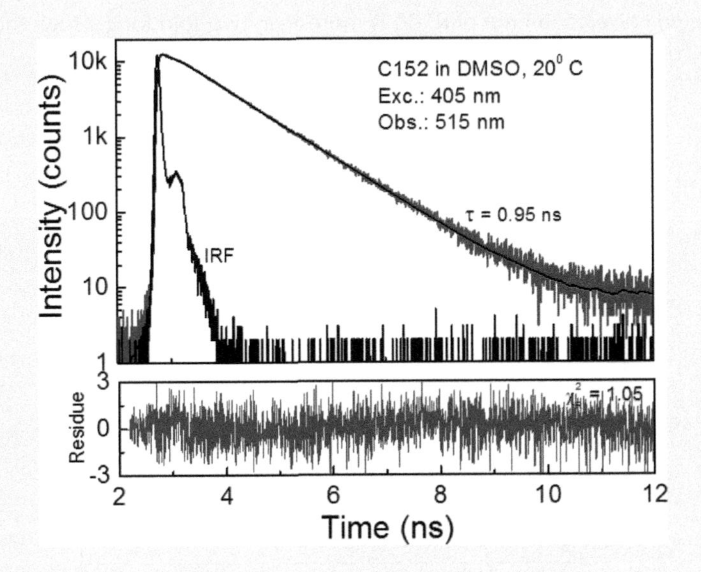

FIGURE 7-17.4 Fluorescence intensity decay of C152 in DMSO.

(Continued)

EXPERIMENT 7-17 (CONTINUED) DEPENDENCE OF ANISOTROPY DECAYS ON THE MOLECULE SIZE

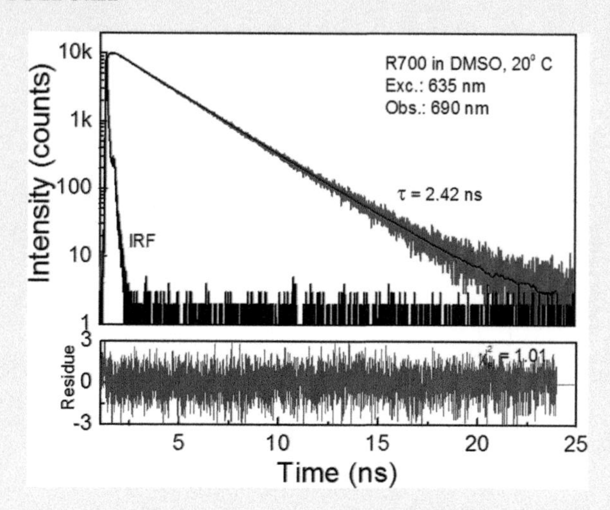

FIGURE 7-17.5 Fluorescence intensity decay of R700 in DMSO.

Finally, we measure anisotropy decays of both fluorophores in DMSO, see Figures 7-17.6 and 7-17.7. These measurements require G-factor measurement and intensity decays of polarized components VV and VH.

The recovered correlation time of R700 is more than two-fold longer than the correlation time of C152, which is visualized in Figure 7-17.8.

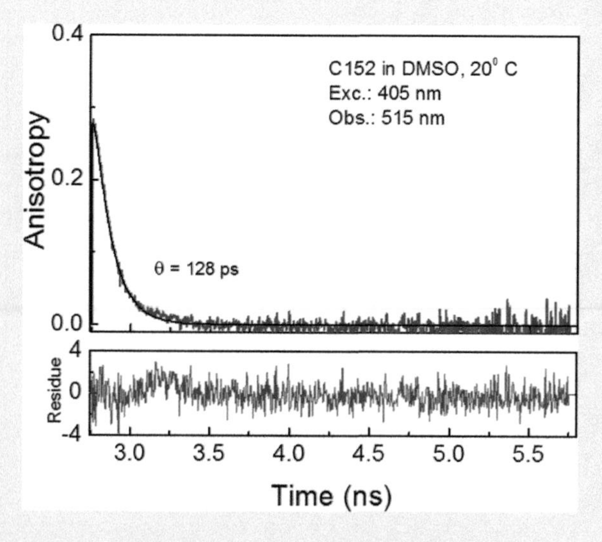

FIGURE 7-17.6 Fluorescence anisotropy decay of C152 in DMSO. The accuracy of the estimated correlation time is ±0.004, as estimated from confidence intervals.

(Continued)

EXPERIMENT 7-17 (CONTINUED) DEPENDENCE OF ANISOTROPY DECAYS ON THE MOLECULE SIZE

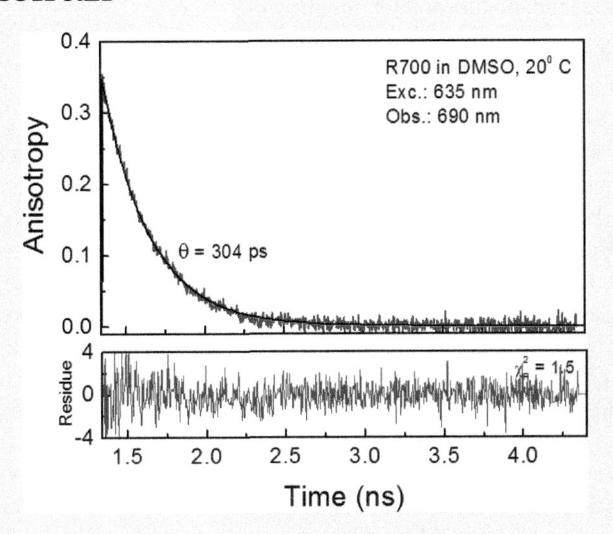

FIGURE 7-17.7 Fluorescence anisotropy decay of R700 in DMSO. The accuracy of the recovered correlation time is ±0.003, as estimated from confidence intervals.

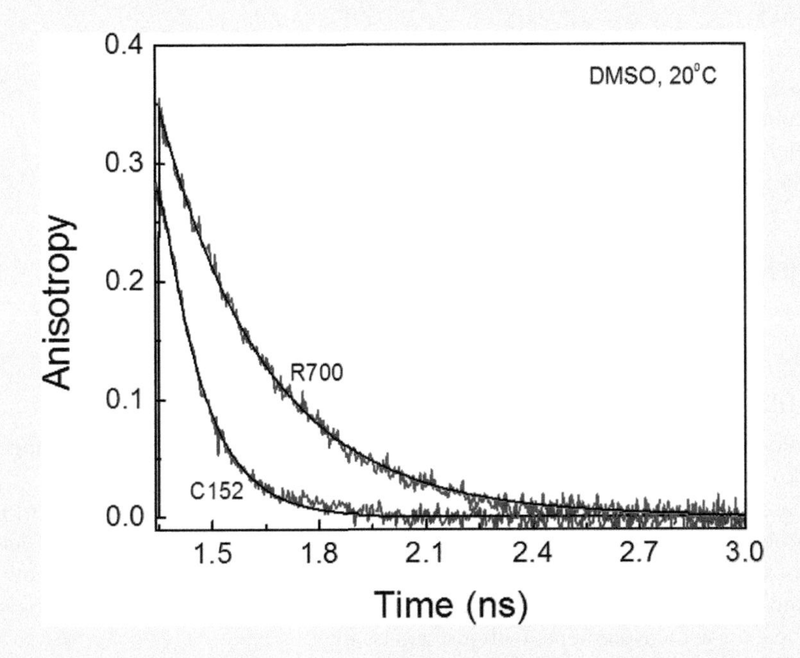

FIGURE 7-17.8 Comparison of fluorescence anisotropy decays of C152 and R700 in DMSO.

CONCLUSIONS

The correlation time (rotation of molecule) depends on the size of molecule. The anisotropy decay measurements can distinguish differences in sizes of small molecules. Of course, this effect is stronger when fluorophore binds to a macromolecule that is much larger.

EXPERIMENT 7-18 OPTICAL DELAYS MEASURED WITH IRF; SPEED OF LIGHT

Keywords: fluorescence, fluorometer, lifetime reference.

Light travels through the medium slower than through a vacuum. The velocity of light in the medium with a refractive index n is equal c/n where c is the speed of light in the vacuum (299,792 km/s or 29.98 cm/ns). The air has the refractive index very low, of about 1.0003, and practically the speed of light in the air is equal to the speed of light in vacuum. If in the way of the incoming light we place any transparent medium like glass or water, the light will be delayed, and the delay will depend on the thickness of the medium and its refractive index. Imagine that we record an incoming train of short pulses. If we now place in the path of the light a transparent object with the thickness of l and refractive index of n, the record will show the train of pulses shifted in time because instead in the air the light goes through the medium a distance l. The shift will be equal $t_m - t_a$, where t_m is the time needed to travel distance l in the medium and t_a is the time that light travels distance l in the air. We can also write that $t_m - t_a = l/v - l/c$, where v and c are speeds of light in the medium and air, respectively. Taking into account that $v = c/n$, we can find that $t_m - t_a = (n-1)l/c$. If we know the temporal shift of the pulse and refractive index of the medium, we can estimate the speed of light. In this experiment, we will measure optical delays of pulses from the picosecond laser diode cost by different thicknesses of the sapphire plate and a common glass filter.

MATERIALS

- Water, silica nanospheres (ludox)
- Glass or quartz 1 cm × 1 cm cuvette
- Neutral density (ND) filters
- Sapphire or glass thick plate
- 5 mm thick glass filter

EQUIPMENT

- Fluorometer FT 200 (Picoquant, GmbH, Germany) with 470 nm laser diode

METHODS

1. Make a solution of the silica nanospheres (scatterer) by adding a drop of concentrated ludox to the cuvette filled with the water.
2. Measure time responses of the scattered light. Use a standard 1 cm × 1 cm cuvette. For IRF measurement, set the observation wavelength to the excitation wavelength.
3. Repeat measurements with a sapphire plate placed on the way of the excitation laser beam. If the plate has all clear sides, repeat measurements through different thicknesses.
4. Repeat these measurements with just a glass filter.
5. Measure time shifts between IRFs without and with sapphire/glass.
6. Estimate speed of light in the air.

(*Continued*)

EXPERIMENT 7-18 (CONTINUED) OPTICAL DELAYS MEASURED WITH IRF; SPEED OF LIGHT

RESULTS

First, we adjusted and optimized IRF with a ludox scatterer (see Experiment 7-1). *Note: Remember we want to have as narrow pulse as possible. For that, we need to optimize the power of the diode (see Experiment 7-1).* We used a repetition rate of 5 MHz, but any other will work as well. In Figure 7-18.1, we present the IRF in logarithmic (left) and linear (right) scales. Both representations are equivalent. For the analysis, the more convenient is the logarithmic scale because of an exponential nature of fluorescence decays. For the estimation of a full-width half maximum (FWHM), the more convenient is the linear scale.

For maintaining optical delays we will use an all-clear sapphire plate (Figure 7-18.2) with dimensions 1 cm × 3 cm × 4 cm. Using different configurations, we can achieve 1, 3, and 4 cm thicknesses of sapphire. We will also check what optical delay will be achieved with a common 5 mm thick glass filter.

Measured IRFs responses without and with the sapphire plate of different thickness are presented in Figure 7-18.3 (note linear scale for intensity counts).

Figure 7-18.4 presents a snap-shot from the instrument monitor during the IRF measurement with 4 cm of sapphire. With the resolution of 4 ps/ch, the shift of IRFs without and with the sapphire is easily measurable.

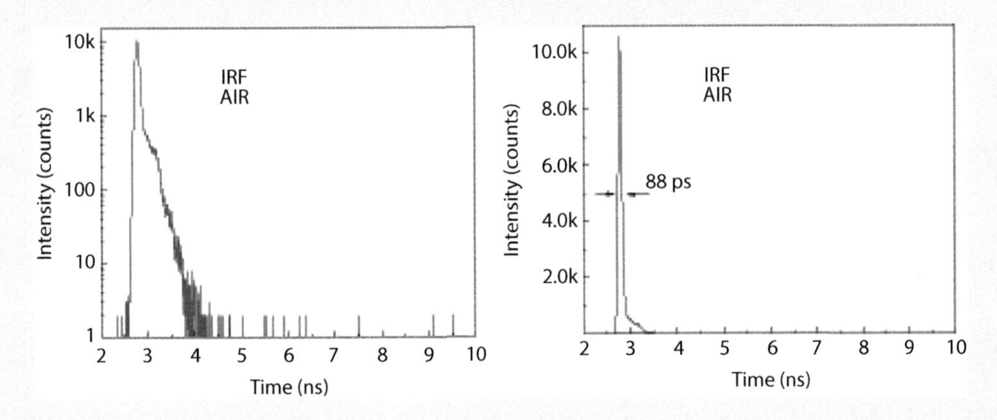

FIGURE 7-18.1 Impulse response function of 470 nm laser diode measured with the scatterer (ludox). Left: In a logarithmic scale; Right: In linear scale.

FIGURE 7-18.2 Sapphire plate 1 × 3 × 4 cm and 4-96 glass filter 0.5 cm thick used in the experiment.

(Continued)

EXPERIMENT 7-18 (CONTINUED) OPTICAL DELAYS MEASURED WITH IRF; SPEED OF LIGHT

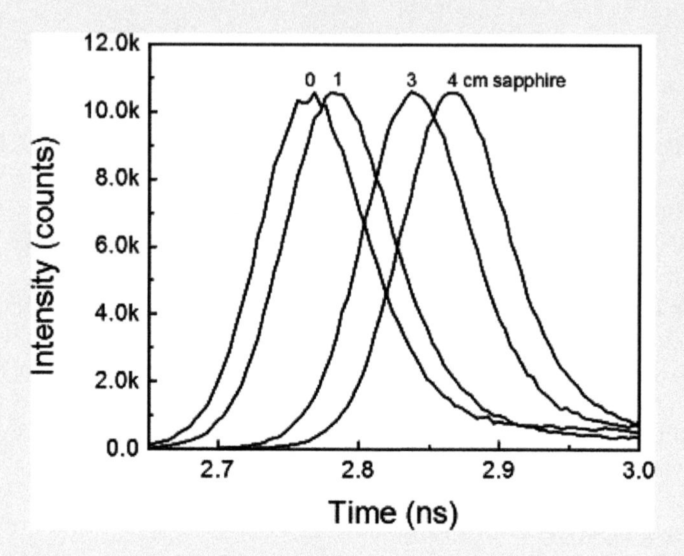

FIGURE 7-18.3 IRFs measured without and with sapphire plate. The light traveled through 1, 3, or 4 cm of sapphire. The temporal shift is responsible for a lower velocity of light in the sapphire.

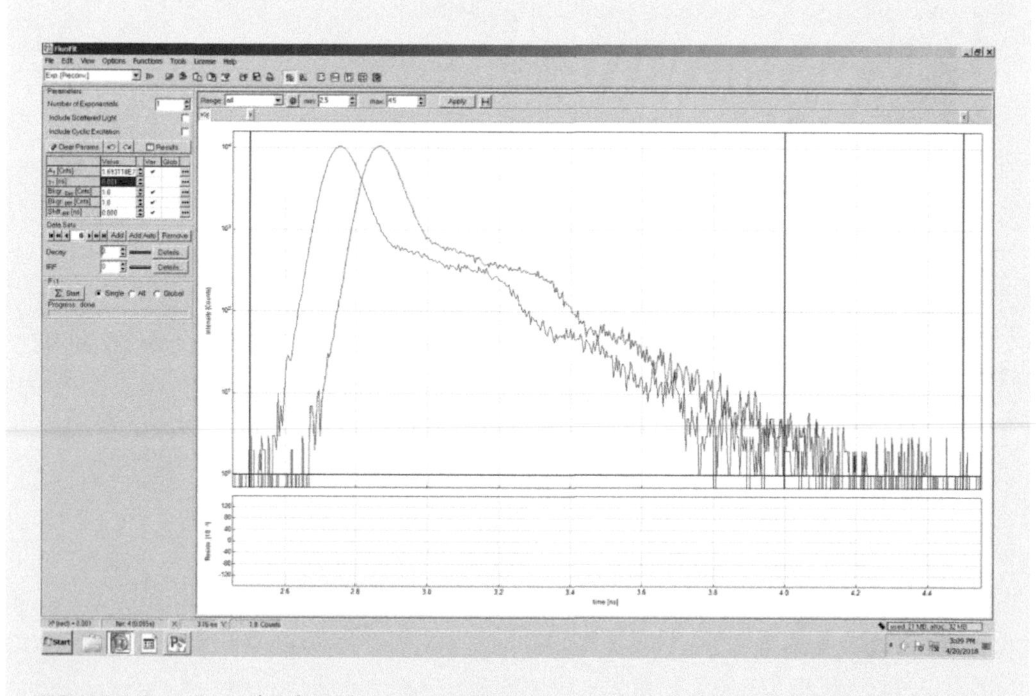

FIGURE 7-18.4 Snap-shot from the measurement of the optical delay with 4 cm sapphire.

(Continued)

EXPERIMENT 7-18 (CONTINUED) OPTICAL DELAYS MEASURED WITH IRF; SPEED OF LIGHT

How to measure the shift?

You can read the positions of maxima while plotting decays. Alternatively, we can analyze decays by fitting the decay of the sample (sapphire) using IRF (air) with a fixed very short lifetime. The program includes a parameter "shift" and properly adjusts the decay and IRF, see Figure 7-18.5.

The shifts for 1 cm sapphire and 3 cm sapphire are 26 ps and 80 ps, respectively. The dependence of the temporal shift on the thickness of the sapphire is presented in Figure 7-18.6.

The dependence of the shift on the optical delay, $t_m - t_a = (n-1)/c \times l$ is represented in Figure 7-18.6. The slope of the linear fit is equal $(n-1)/c$. Taking the refractive index of sapphire $n = 1.78$ and the slope 26.4, we can calculate the parameter $c = 0.0295$ cm/ps, or 29.5 cm/ns, or 295,000 km/s which is in a very good agreement with the speed of light.

Next, we check how the optical delay can be created by a single glass filter. Figure 7-18.7 shows a small but measurable shift. The fit (Figure 7-18.8) reveals delay of 9 ps. Using the

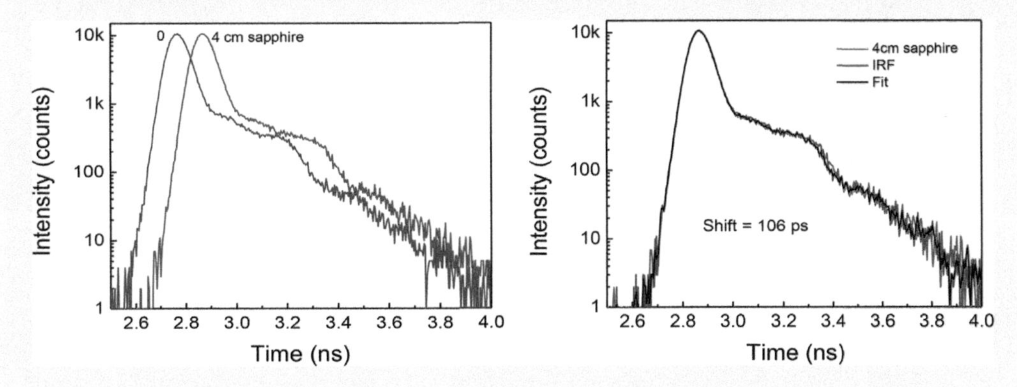

FIGURE 7-18.5 IRFs responses in air and with 4 cm sapphire (left). Fit to a fixed very small lifetime reveals the temporal shift between two IRFs (right).

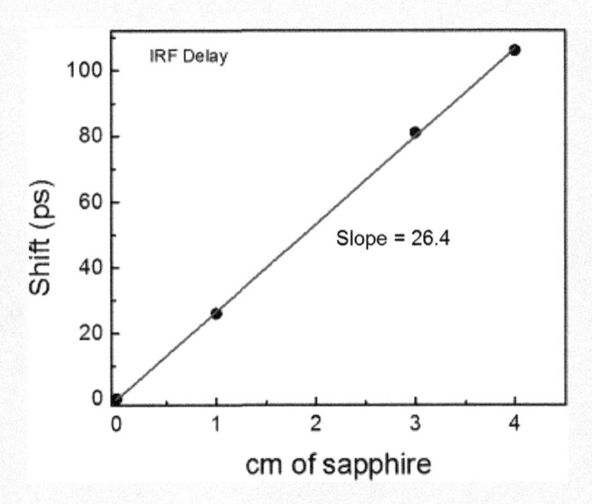

FIGURE 7-18.6 The dependence of the temporal shift on the thickness of the sapphire.

(Continued)

EXPERIMENT 7-18 (CONTINUED) OPTICAL DELAYS MEASURED WITH IRF; SPEED OF LIGHT

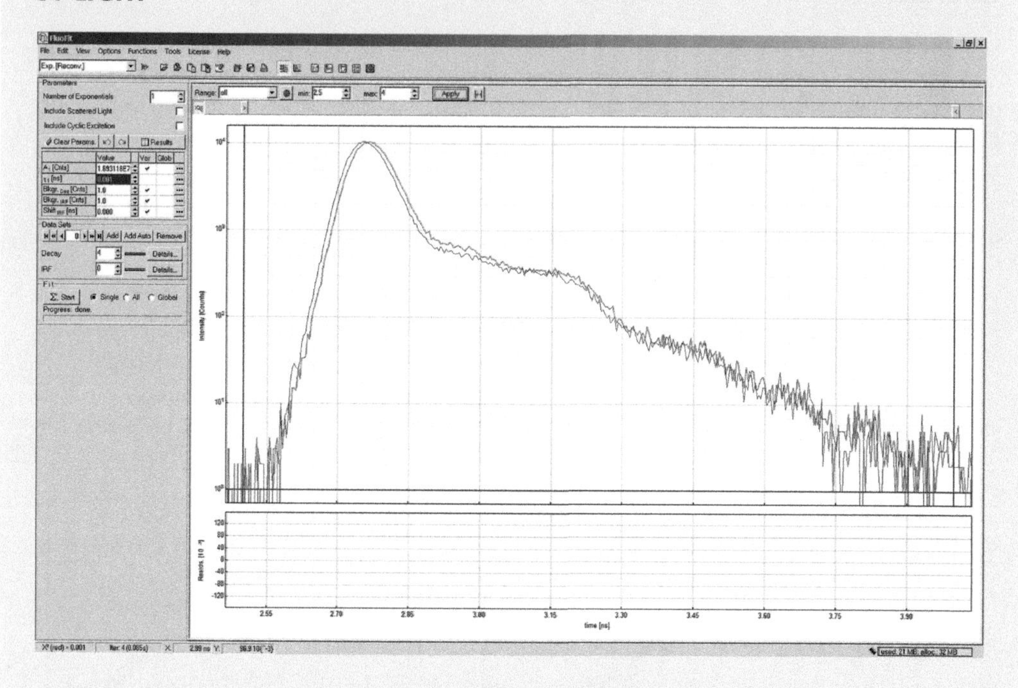

FIGURE 7-18.7 Snap-shot from the measurement of the optical delay created by a single 5 mm thick glass filter.

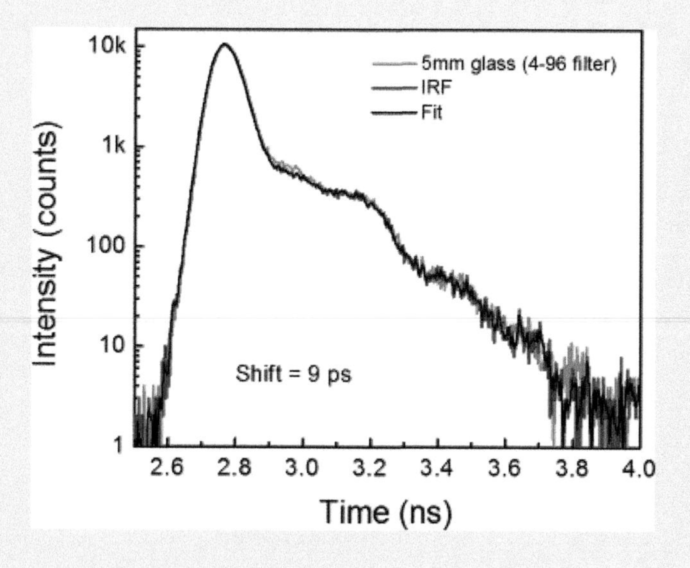

FIGURE 7-18.8 The fit of the delayed pulse reveals the time shift of 9 ps.

(Continued)

EXPERIMENT 7-18 (CONTINUED) OPTICAL DELAYS MEASURED WITH IRF; SPEED OF LIGHT

same dependence for the temporal shift we can rearrange the equation to $n = 1 + (t_m - t_a)c/l$. Using the 9 ps delay (0.009 ns), 30 cm/ns speed of light and $l = 0.5$ cm, we calculate $n = 1.54$, which is a very reasonable value for the glass.

CONCLUSIONS

It is not necessary to have a sapphire plate, any glass or transparent liquid will work as well.

Of course, a longer optical delay will result in a more precise speed of light measurement. But it is important to realize that even a single glass filter costs a measurable light delay. It is not important where you place the filters (maintain an optical delay), on the excitation or observation path.

It is important to realize that in routine lifetime measurements, we are using multiple filters (frequently different for reference and sample measurement) it may play a significant role for recovered lifetimes. Most fitting routines for TCSPC has now a shift parameter included in the fit that accounts for this effect, however, in the frequency domain it is very difficult to account for that. This can be a significant problem for measuring very short fluorescence lifetime (<100 ps).

EXPERIMENT 7-19 PRESENCE OF SCATTERING IN LIFETIME MEASUREMENTS

Keywords: fluorescence, fluorometer, lifetime, time-domain, scattering.

In earlier chapters, we discussed the presence of the Rayleigh and Raman scatterings in absorption and fluorescence spectra measurements. Here, we will discuss the presence of sample scattering in lifetime measurements. Biological samples that are macromolecules (e.g. proteins) or cells are expected to be strongly scattering. This heavily affects measured absorption (see Experiments 3-17 through 3-20) and emission spectra but also may affect measurements of fluorescence lifetimes. We will demonstrate the effect of the presence of scattering using silica nanoparticles (ludox) which we will add to a solution of Rhodamine 110.

MATERIALS

- Water, silica nanospheres (ludox)
- Rhodamine 110 (R110)
- Glass or quartz 1 cm × 1 cm cuvette, neutral density (ND) filters

EQUIPMENT

- Glass or quartz 1 cm × 1 cm cuvette Fluorometer FT200 (PicoQuant, GmbH, Germany) with 470 nm laser diode

(Continued)

EXPERIMENT 7-19 (CONTINUED) PRESENCE OF SCATTERING IN LIFETIME MEASUREMENTS

METHODS

1. Make a solution of the silica nanospheres (ludox scatterer) by adding a drop of concentrated ludox to the cuvette filled with water. Optimize and measure IRF.
2. Make about 5 mL solution of R110 in water with an optical density of about 0.1. Split the solution into two samples. Add a drop of ludox to one sample and a drop of water to another sample.
3. Measure absorption and fluorescence spectra of both samples, see Figures 7-19.1 and 7-19.2.
4. Measure lifetime of R110 without and with Ludox with the resolution set at 4 ps/ch (high resolution). Analyze lifetimes without and with scattering parameter.

RESULTS

First, we measured absorptions of both samples, with and without ludox scatterer using water as a baseline, see Figure 7-19.1. The presence of the scattering is obvious (see Experiments 3-18 through 3-20). How does this strong scattering will affect fluorescence spectra and lifetimes?

Fluorescence spectra of R110 in absence and presence of ludox are shown in Figure 7-19.2. The observed spectra are fairly similar.

Next, we measured the intensity decays of R110 in the absence and presence of the ludox scatterer, see Figures 7-19.3 and 7-19.4. The intensity decay of R110 in water can be very well approximated by a single exponent, revealing the lifetime of 3.94 ns. However, the intensity decay of R110 in the presence of Ludox scattering cannot be simply fitted to a single exponent model, see Figure 7-19.4.

This problem can be resolved in two ways. If there is no need to resolve ultra-short components in the lifetimes, one can use a tail-fitting and analyzing a longer part of the decay profile. Alternatively, an extra function in software analysis, called "scattering" can be used, see Figure 7-19.5.

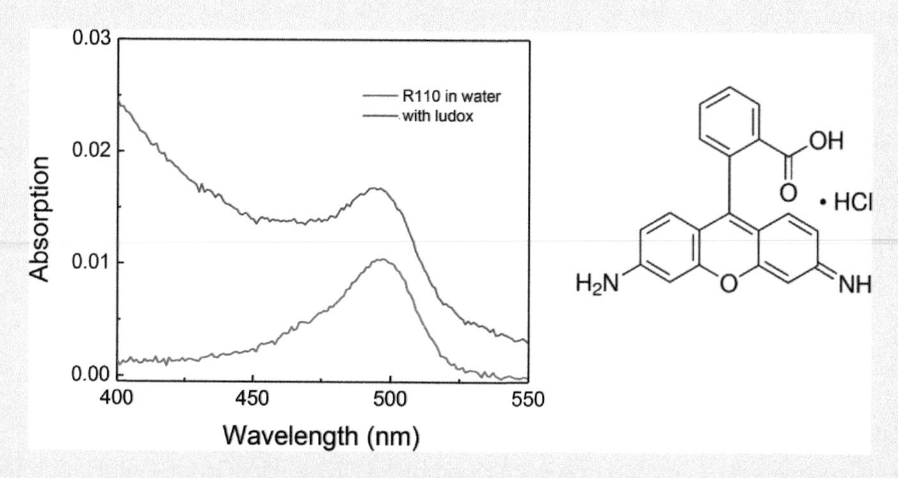

FIGURE 7-19.1 Absorption spectra of R110 in absence and presence of scattering. The R110 chemical structure is on the right.

(Continued)

EXPERIMENT 7-19 (CONTINUED) PRESENCE OF SCATTERING IN LIFETIME MEASUREMENTS

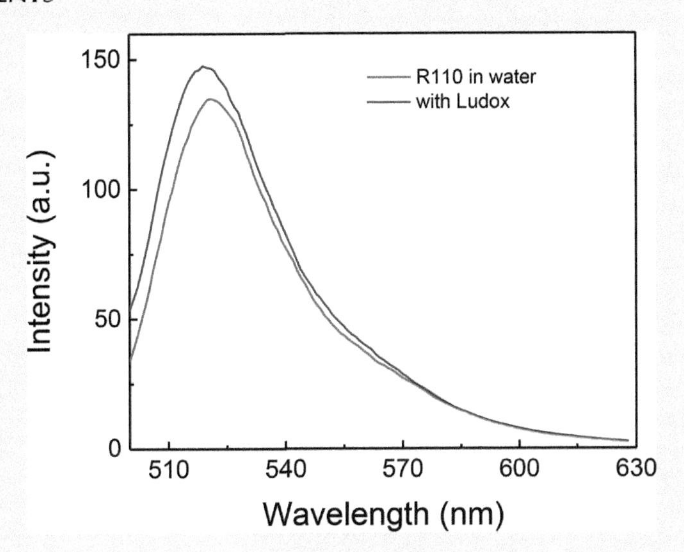

FIGURE 7-19.2 Fluorescence spectra of R110 in water in the absence and presence of scattering.

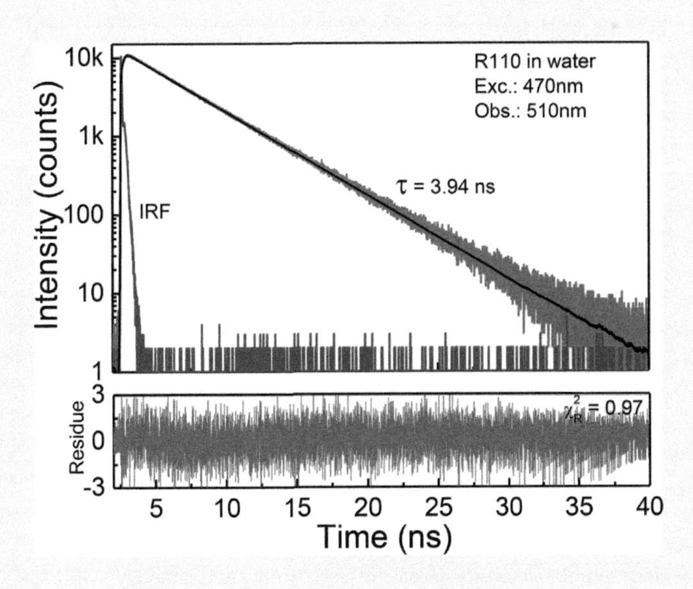

FIGURE 7-19.3 Fluorescence intensity decay of R110 in water.

(Continued)

EXPERIMENT 7-19 (CONTINUED) PRESENCE OF SCATTERING IN LIFETIME MEASUREMENTS

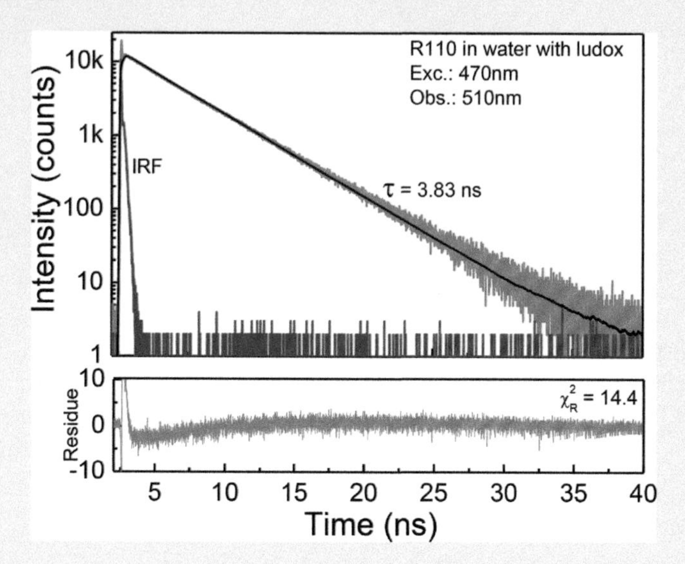

FIGURE 7-19.4 Fluorescence intensity decay of R110 in water in the presence of scattering. One-exponent fit without correction for the scattering.

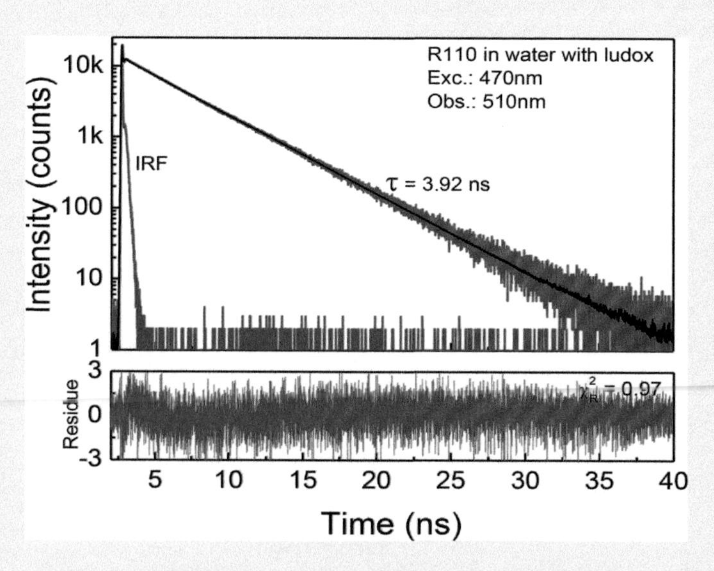

FIGURE 7-19.5 Fluorescence intensity decay of R110 in water analyzed with the correction for scattering.

(*Continued*)

EXPERIMENT 7-19 (CONTINUED) PRESENCE OF SCATTERING IN LIFETIME MEASUREMENTS

CONCLUSIONS

In this experiment, we used a so-called monochromatic laser diode excitation and monochromator that is a standard in FT200 system. *Why we see scattering?* There are two reasons for that. The main one there is no perfect monochromators and every monochromator would transmit a small amount of light at wavelength longer (or shorter) than the wavelength selected. The amount of transmitted light depends how far the wavelength from the set wavelength is. This is clearly seen in Figure 7-19.2 where the spectrum is the most disturbed at shorter wavelength than at longer wavelength. When measuring spectra of highly scattering systems we always recommend additional support with appropriate filters as we discussed in Chapters 4 and 5. Because scattering has a lifetime of zero (it is an instantaneous process), it can be resolved easier in time-resolved than in steady-state measurements. It will typically affect the few first bins in the collected decay. The correction can be done by using the software with "scattering" parameter or by a tail analysis, ignoring the initial time (about a pulse duration) of the fluorescence decay.

EXPERIMENT 7-20 PRESENCE OF LONG LIVED LUMINESCENCE IN FLUORESCENCE LIFETIME MEASUREMENTS

Keywords: fluorescence, fluorometer, lifetime, luminescence, delayed fluorescence, phosphorescence.

In the previous Experiment 7-19, we discussed the presence of scattering in fluorescence lifetime measurements. Here, we will discuss the presence of long components in lifetime measurements. In some samples with weak fluorescence emission, other luminescence radiations may be present. The small energy gap between the first excited singlet state and the triplet state enables thermal activation of the fluorescence state from the triplet state. This photo-process (reverse to intersystem crossing) is governed by a Boltzmann distribution law. Of course, the emissive transition to the ground state (fluorescence) will be in this case delayed (excited molecules "visit" the triplet state for a while and then return to the S_1 state). This fluorescence emission with the triplet state involvement is called *delayed fluorescence*. We will demonstrate the presence of delayed fluorescence and phosphorescence in fluorescence lifetime decays of Erythrosin B embedded in PVA film.

MATERIALS

- Water, silica nanospheres (ludox)
- Erythrosin B (ErB), Rhodamine B (RhB)
- Poly(vinyl alcohol) (PVA)

(Continued)

EXPERIMENT 7-20 (CONTINUED) PRESENCE OF LONG LIVED LUMINESCENCE IN FLUORESCENCE LIFETIME MEASUREMENTS

EQUIPMENT

- Spectrophotometer
- Spectrofluorometer with a front-face attachment
- Fluorometer FT200 (PicoQuant, GmbH, Germany) with 485 nm laser diode and front-face attachment

METHODS

1. Prepare PVA films doped with ErB and RhB. Remember to make also a PVA film only for the baseline and background reference. See Experiment 3-24.
2. Measure absorption and fluorescence spectra of ErB and RhB in PVA films.
3. Measure lifetimes of ErB and RhB in PVA films at 485 nm excitation and 560 nm observation. Be sure that the collection time is the same in each measurement, which assures similar dark current-related backgrounds.
4. Measure lifetimes of ErB and RhB in PVA films at 485 nm excitation and 660 nm observation.

RESULTS

Using a reference PVA film as a baseline, measure the absorption spectrum of ErB in PVA film, see Figure 7-20.1. This spectrum has a slight short shift to its wavelength compared to ErB absorption in water, see Experiment 7-13.

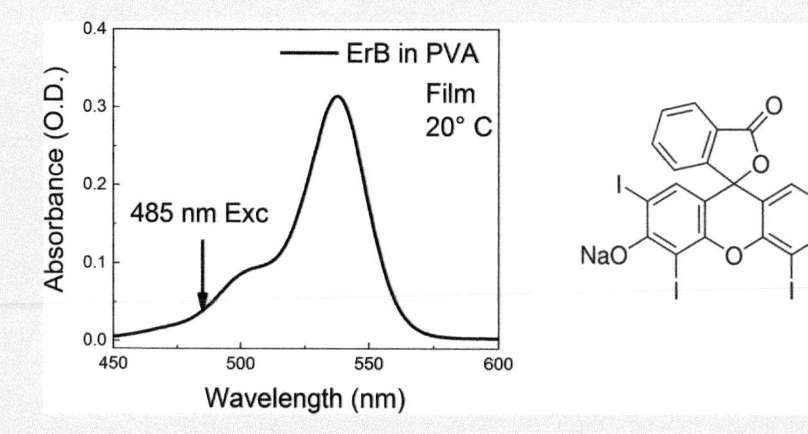

FIGURE 7-20.1 Absorption spectrum of ErB in PVA film. The arrow indicates the excitation used in fluorescence measurements. The chemical structure is on the right.

(Continued)

EXPERIMENT 7-20 (CONTINUED) PRESENCE OF LONG LIVED LUMINESCENCE IN FLUORESCENCE LIFETIME MEASUREMENTS

Next, measure the fluorescence spectrum of ErB in PVA film, see Figure 7-20.2. This spectrum slightly differs from the fluorescence spectrum of ErB in water. The difference is in the long-wavelength part of the spectrum, above 650 nm, a small "bump" which is invisible in the water. Responsible for this is a phosphorescence of ErB.

Next, we will measure the fluorescence intensity decays of ErB in PVA film at two observation wavelengths, near the main peak (560 nm) and near the phosphorescence peak (660 nm). We will compare these decays with RhB decays (also in PVA film). In the case of 560 nm observation, the ErB decay shows a much higher background than observed for IRF and for RhB (Figure 7-20.3). This elevated background indicates a very long component in the fluorescence intensity decay of ErB. Delayed fluorescence of ErB is responsible for this. The RhB has a very efficient and prompt fluorescence and minimal delayed fluorescence.

The observation at 660 nm provides a more dramatic effect of the elevated background, see Figure 7-20.4 (top). In this case, the phosphorescence of ErB also participates in the observed intensity decay. This effect is not visible in RhB which has minimal phosphorescence and the background matches IRF background.

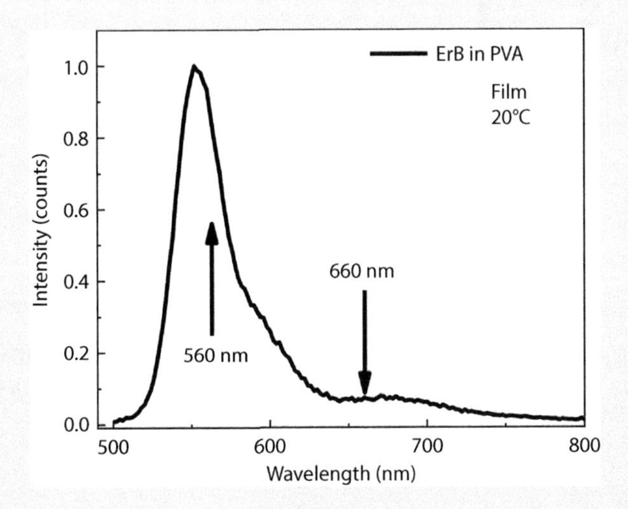

FIGURE 7-20.2 Fluorescence spectrum of ErB in PVA film. The arrows indicate observation wavelengths used in fluorescence lifetime measurements.

(Continued)

EXPERIMENT 7-20 (CONTINUED) PRESENCE OF LONG LIVED LUMINESCENCE IN FLUORESCENCE LIFETIME MEASUREMENTS

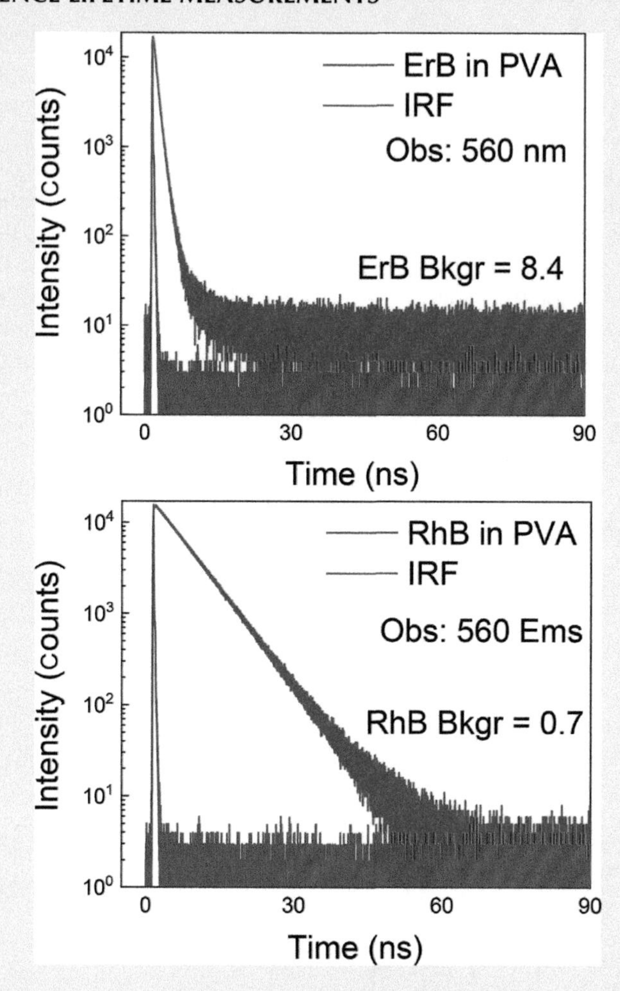

FIGURE 7-20.3 Fluorescence intensity decays of ErB (top) and RhB (bottom) in PVA films. The time collection of both decays and IRF was the same (200 s). While RhB decays to the noise background (0.6), ErB background is elevated to 8.4. Responsible for this is the presence of delayed ErB fluorescence.

(Continued)

EXPERIMENT 7-20 (CONTINUED) PRESENCE OF LONG LIVED LUMINESCENCE IN FLUORESCENCE LIFETIME MEASUREMENTS

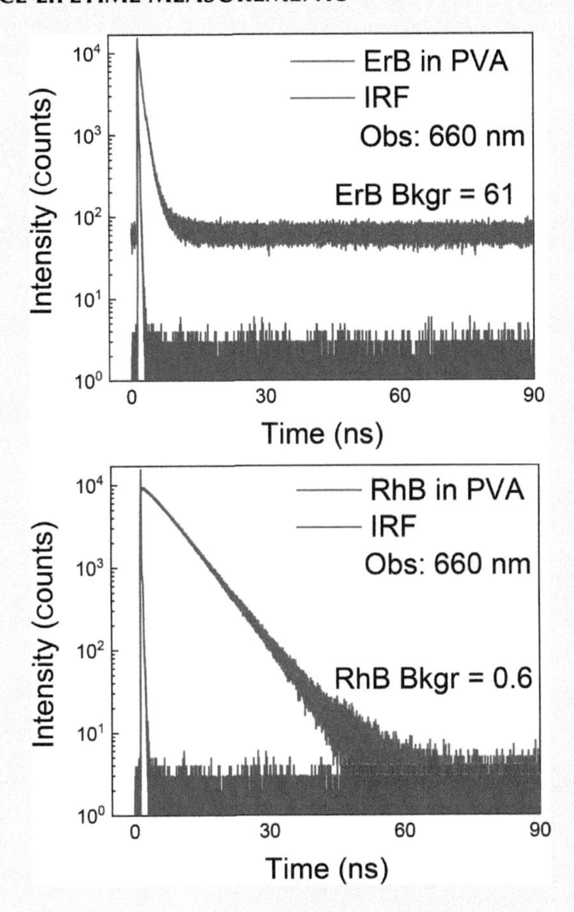

FIGURE 7-20.4 Fluorescence intensity decays of ErB (top) and RhB (bottom) in PVA films observed at 660 nm. The time collection of both decays and IRF was the same (200 s). While RhB decays to the noise background (0.6), ErB background is elevated to 61. Responsible for this is the presence of ErB phosphorescence.

CONCLUSIONS

This experiment is a warning of sometimes unexpected background elevation in lifetime measurements. In rigid samples, delayed fluorescence often is present. Of course, the observation within the phosphorescence peak will also result in a strong elevation of the background. It should be noted that the presence of delayed fluorescence or phosphorescence, in most cases, does not affect the fluorescence decay measurements because in the fitting programs the decay, as well as IRF backgrounds, are taken into account.

CHAPTER 8

Advanced Experiments

INTRODUCTION

In this chapter, more advanced experiments on fluorescence will be presented. In most cases, these experiments require both steady-state and time-resolved measurements. We will use these experiments to introduce readers to more advanced fluorescence measurements like polarization measurements in thin films, complex anisotropy decays, Forster resonance energy transfer (FRET), excimer and exciplex fluorescence emissions, delayed fluorescence and phosphorescence, solvatochromism, and solvent effects.

Fluorescence Anisotropy

Complex Anisotropy Decays. In a series of experiments, we will demonstrate complex anisotropy decays. Experiments 8-8 and 8-9 present anisotropy decays of dyes labeled on DNA and lipid membranes. In the case of labeled DNA, the anisotropy decay reflects complex motions of the DNA strands. At least three correlation times are needed to satisfactorily fit the data. The multi-exponential model is usually used to fit the anisotropy decay:

$$r(t)=\sum_i r_i \exp(-t/\theta_i) \tag{8.1}$$

In lipid membranes, an elongated fluorophore cannot rotate freely but only "wobble" within a narrow-angle. In this case, a "hindered" rotator model is more appropriate:

$$r(t)=r(\text{inf})+[r_0-r(\text{inf})]\exp(-t/\theta) \tag{8.2}$$

The term $r(\text{inf})$ represents a constant value of anisotropy.

In most cases, the anisotropy decay of the free molecule can be described with a single correlation time. However, in a specific case of a planar molecule in non-polar solvents, fluorophore depolarizing rotations can occur along two or three axes. These rotations will independently contribute to the anisotropy decay. Experiment 8-13 demonstrates two-exponential anisotropy decay as a result of anisotropic rotations of a rigid perylene molecule.

Rotations of the same fluorophore may depend on the solvent polarity. In Experiment 8-8, we will demonstrate this using a polar fluorophore in solvents with similar viscosities but different polarities.

In the case of fluorophores with extremely short lifetimes, measured anisotropies are high. Usually, in water and low viscosity solvents, these dyes have also very low quantum yields. In higher viscosity solvents, these fluorophores are brighter, but most fluorophores in viscous solvents have also relatively long lifetimes. In Experiment 8-17, we will demonstrate extremely high anisotropy of free fluorophore in propylene glycol.

Perrin Equation. The excitation of fluorophores in isotropic solutions results in a partial polarization of observed fluorescence. This process, called photo-selection, gives at time zero a *fundamental anisotropy* with a theoretical maximum value of 0.4. This is a maximum of possible fluorescence anisotropy observed for the isotropic solution with normal, one-photon excitation. In the excited state of molecules during their lifetime, some depolarization processes always occur. The most significant are: rotational motions (changing directions of molecules transition moments), energy migration, called also homo-transfer (emission originates from different molecule) and torsional motions in the excited state. These motions are responsible for observing less than 0.4 anisotropy even in frozen solutions. The maximal anisotropy observed in absence of rotational motions and energy migration is called *limiting anisotropy*. The limiting anisotropy can be close, but never reaches the *fundamental* value of 0.4. In liquid and diluted solutions, the dominant depolarization factor is rotational motions of molecules. Of course, molecules with a longer lifetime will make more rotations during their lifetime than molecules with a short lifetime. Intuitively, the change of the initial (zero time) limiting anisotropy, observed in steady-state experiments, should depend on the ratio of molecule lifetime to the time needed for the rotation of its transition moment. This time is called a *correlation time* and is inversely related to a rotational diffusion coefficient of the molecule. First who derived a dependence of polarization change on the rotational diffusion coefficient (or correlation time) and molecule lifetime was Francis Perrin. With the introduction of fluorescence anisotropy by Alexander Jablonski, the equation can be written in a shorter form, similar to Stern–Volmer equation (for fluorescence quenching):

$$r_0/r = 1 + \tau/\theta \tag{8.3}$$

where r_0 is the limiting anisotropy, r is measured fluorescence anisotropy, τ is the lifetime, and θ is the correlation time. In the case of isotropic rotations, $\theta = 1/6D$, where D is the rotational diffusion coefficient. This coefficient depends on viscosity and temperature of the solution as well as on the volume of rotating object, and is given by the Stokes-Einstein equation: $D = RT/6V\eta$, where V is the hydrodynamic volume, η is the viscosity of the medium, T is absolute temperature (in Kelvin degrees), and R is a gas constant. For more details, we refer readers to the recommended books, see Preface.

Associated Anisotropies. Extremely complex anisotropy decay is demonstrated with using incomplete labeling of the protein by the dye, see Experiment 8-12. It is called an associated anisotropy decay because lifetimes and correlation times of the dye are associated with the bound/unbound fractions of the dye-macromolecule system. The associated decay can be described with an appropriate mathematical model, which can be found in recommended textbooks. Experiment 8-12 will only demonstrate this unusual decay. The concept of the associated anisotropy decay clearly demonstrates the advantage of *anisotropy* notation over *polarization.*

Forster Resonance Energy Transfer (FRET)

One donor-one acceptor systems. The FRET concept has been introduced already in Chapter 1. This through-space molecular interaction has found multiple applications in biology, especially in co-localization of proteins and interactions with ligands studies. Therefore, the most investigated FRET case is a one donor-one acceptor system. Here, we will discuss this with an example of dually labeled peptide selectively cleaved by a metalloproteinase enzyme MMP-9. As a donor, we used a rhodamine derivative TAMRA, see Figure 8-1.

Note: Some results were published in Methods and Applications in Fluorescence Vol 4 (2016) 047002. Figures 8-2–8-6 are used with permission from IoP.

The donor-acceptor system is presented in Figure 8-2.

As an acceptor reference, a free HiLyte647 is used. All three compounds (D, DA, and A) at identical molar concentrations were dissolved in a liquid 10% (w/w) PVA and left for drying in Petri dishes (see Experiment 3-24 for PVA films preparation). After drying, the films were peeled from the dishes. The thicknesses of all films were the same at about 0.2 mm (see absorption spectra in Figure 8-3).

Emission spectra measured for D, DA, and A show clearly the transfer of the excitation energy from TAMRA to HiLyte647 dye. In the presence of the HiLyte647, the TAMRA donor fluorescence is strongly quenched and the fluorescence of the acceptor is increased, see Figure 8-4.

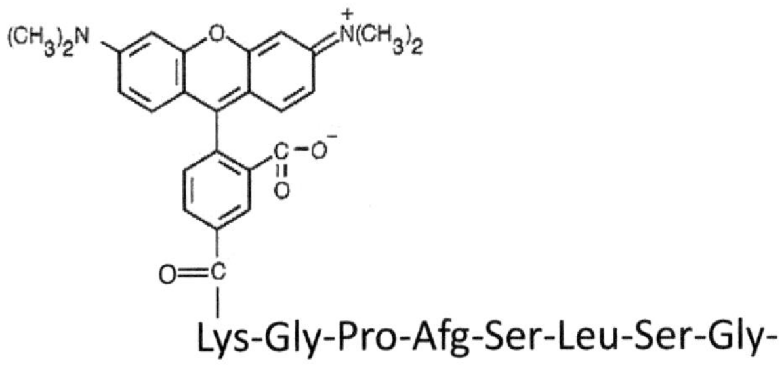

FIGURE 8-1 Structure of a TAMRA labeled on the peptide. It will be used as a donor reference (D).

FIGURE 8-2 TAMRA-peptide-HiLyte donor-acceptor system (DA).

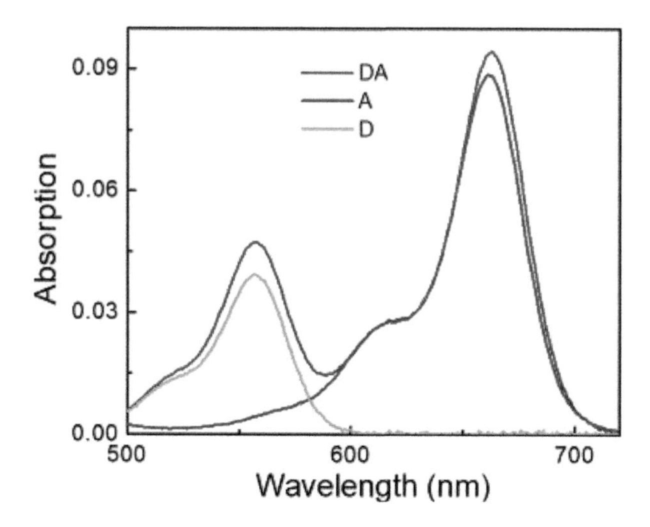

FIGURE 8-3 Absorption spectra of the donor (D, 5,6-TAMRA), acceptor (A, free HiLyte647), and donor-acceptor system (DA) in PVA films. Concentrations were the same (about 20 μM) for all compounds. The DA absorption is well-approximated by combining both D and A spectra.

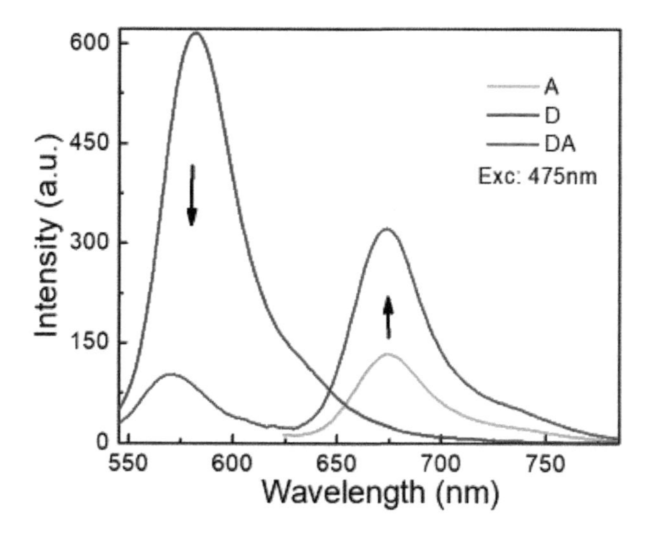

FIGURE 8-4 Fluorescence emission spectra of the donor (D, 5,6-TAMRA), acceptor (A, HiLyte647), and donor-acceptor linked system (DA). Whereas fluorescence of the donor in the DA system decreased about six-fold, the emission of the acceptor increased 2.4 times which indicates a very efficient FRET.

A very rigorous test for the FRET presence is the measurement of the excitation spectrum of the DA system. In the excitation spectrum of TAMRA-peptide-HiLyte647 observed at the HiLyte647 emission at 675 nm, one can clearly see the presence of TAMRA absorption, see Figure 8-5. This could happen only if the excitation energy goes from TAMRA to HiLyte647.

The donor emission at 675 nm (green) is minimal. However, in the DA system, the emission at 675 nm (red) is strongly enhanced in the region of TAMRA absorption.

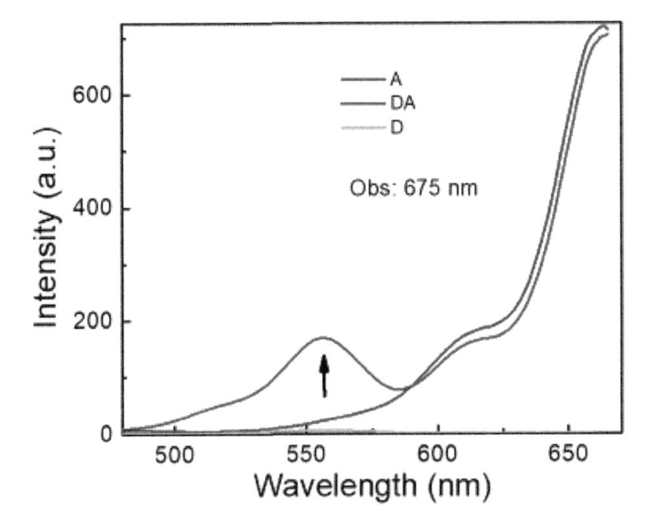

FIGURE 8-5 Excitation spectra of the donor (D, 5,6-TAMRA), acceptor (A, free HiLyte647), and donor-acceptor system (DA) observed at 675 nm (the acceptor emission wavelength).

This indicates that the acceptor excited state is strongly populated as a result of FRET from the donor (5,6-TAMRA).

Lifetime measurements also indicate a strong quenching of the TAMRA donor emission in the presence of HiLyte647 acceptor, see Figure 8-6.

The donor (D) intensity decay is homogeneous and can be satisfactorily approximated with a single exponential fit. However, the donor-acceptor system (DA) intensity decay is

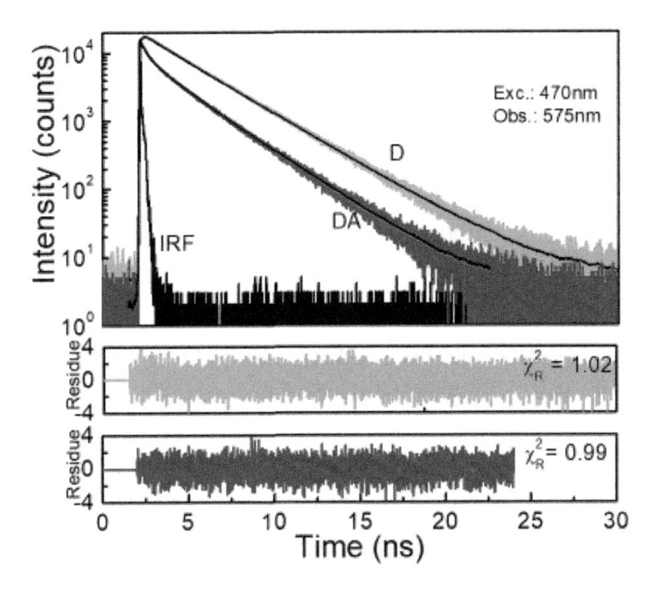

FIGURE 8-6 Fluorescence intensity decay of the donor (D, 5,6-TAMRA) and donor-acceptor system (DA). The donor intensity decay can be well-approximated by a single exponential fit with a lifetime of 3.78 ns. In contrast, the emission of the donor (at 575 nm) in the DA system is very heterogeneous and can be approximated with a three exponent fit yielding an amplitude-averaged lifetime of 0.76 ns.

TABLE 8-1 Multi-exponential Analyses of D and DA Intensity Decays

Compound	n	α_1	α_2	α_3	τ_1	τ_2	τ_3	$\bar{\tau}$	$\langle\tau\rangle$	χ^2_R
5,6-TAMRA-peptide	1	1	—	—	3.78	—	—	3.78	3.78	1.02
5,6-TAMRA-pep-HiLyte647	1	1	—	—	2.80	—	—	2.80	2.80	31.4
	2	0.567	0.433	—	0.33	3.16	—	2.82	1.56	2.61
	3	0.613	0.185	0.202	0.03	0.53	3.25	2.82	0.76	0.99

N: Number of exponents.

$\bar{\tau} = \sum_i f_i \tau_i$ where $f_i = \frac{\alpha_i \tau_i}{\sum_i \alpha_i \tau_i}$

$\langle\tau\rangle = \sum_i \alpha_i \tau_i$

very heterogeneous and a satisfactorily fit requires a more complex fit. The multi-exponential analysis of the donor emission in the DA system is presented in Table 8-1.

Lifetime Analysis. The donor (D) intensity decay is homogeneous and can be satisfactorily approximated with a single exponential fit. However, the donor-acceptor system (DA) intensity decay is very heterogeneous and a satisfactory fit requires a more complex fit. The multi-exponential analysis of the donor emission in the DA system is presented in Table 8-1.

Transfer Efficiency. Energy transfer (FRET) efficiency was estimated from:

$$E = 1 - \frac{I_{DA}}{I_D} \tag{8.3}$$

An I_{DA}/I_D ratio of about six results in a transfer efficiency of 0.84.

The critical distance at which the energy transfer efficiency is 0.5 (Forster distance) was estimated at 5.2 nm (51.9 Å) according to this equation:

$$R_0^6 = \alpha_i \tau_i = \frac{9000\ (\ln 10) k^2 Q_D^0}{128 \pi^5 N_A n^4} \int_0^\infty F_D(\lambda) \varepsilon_A(\lambda) \lambda^4 d\lambda \tag{8.4}$$

where Q_D is the quantum yield of the donor in the absence of acceptor, n is the solvent refractive index, and N_A is the Avogadro's number. κ^2 is the orientation factor, usually 0.476 for the rigid isotropic PVA system with no mobility. The normalized donor fluorescence intensity is represented by F_D. The ε is the acceptor's extinction coefficient.

The normalized spectra for R_0 calculation are shown in Figure 8-7.

An apparent donor-to-acceptor distance can be calculated from:

$$r = R_0 \sqrt[6]{\frac{1-E}{E}} \tag{8.5}$$

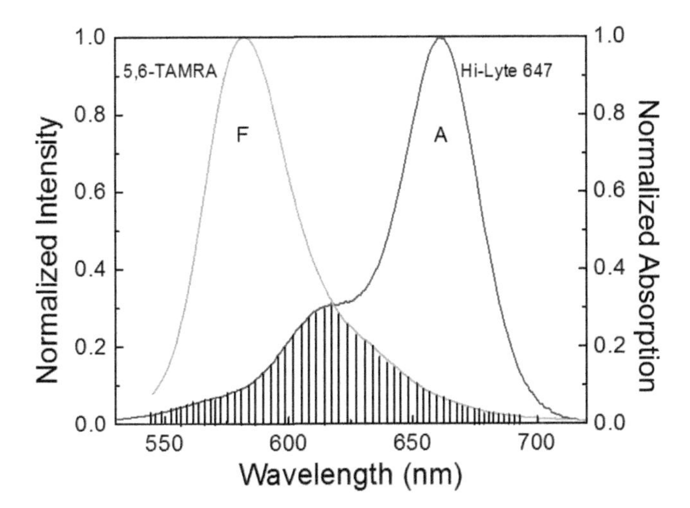

FIGURE 8-7 Absorption spectrum of HiLyte647 (A) and fluorescence spectrum of 5, 6-TAMRA-peptide (D). The shadowed area represents the spectral overlap.

For $E = 0.84$ and $R_0 = 5.2$ nm, the apparent DA distance is about 3.9 nm.

It should be noted that the orientation factor κ^2 may play here an important role because both donor and acceptor molecules can have different orientations in each DA pair.

FRET in Solutions (3D). FRET in solutions (3D) is demonstrated in Experiments 8-10 and 8-11. In both cases, rigid polymer and viscous liquid, very thin layers of the samples are needed and the measurements and should be done in a front-face configuration. This is because the acceptor concentrations must be very high to assure short average donor-acceptor distances.

FRET in One Dimension. High concentrations of dyes are not required for dyes located on DNA, see Experiment 8-20. Both donor and acceptor molecules are bound/intercalated to the double DNA strand forming a linear (1D) configuration for FRET.

Energy Migration. FRET occurring between the same dye molecules is called *Energy Migration* (EM) or Homo-FRET. In solutions, EM requires very high concentrations of the dye, similar to the concentrations of acceptor in the 3D FRET. Homo-FRET results in self-quenching and fluorescence depolarization. These effects will be presented in the Experiment 8-7.

Excimers and Exciplexes

Excimers. Dissolved fluorophores can diffuse freely and interact with other molecules. In some cases, molecules can attract each other and form complexes. This requires usually high concentrations of fluorophores. Such complexes formed in the ground state of molecules are called *dimers.* Dimers have different properties than monomers, different absorptions and fluorescence. Often dimers are not fluorescent and are responsible for the self-quenching occurring at high concentrations of dyes (see Experiment 8-7). The complex formed between an excited molecule and not excited molecule is called *excimer.*

The excimer fluorescence is shifted toward the longer wavelengths (because some energy is utilized to form the excimer). First observation of excimer fluorescence was reported in 1950 by Forster and co-workers. We will demonstrate the excimer fluorescence of pyrene in the Experiment 8-8.

Exciplexes. The excimers form between the same fluorophore molecules. A complex of one excited fluorophore and another (different, not excited) fluorophore is called *exciplex.* The formation of the exciplex is usually accompanied with a strong charge-transfer process. We will describe the exciplex formation between anthracene molecules and diethylaniline molecules in Experiment 8-19. It should be noted that both excimers and exciplexes do not have ground states, and their absorptions cannot be measured.

Solvent Effects

Solvatochromism. Solvatochromism is the phenomenon of color change for the dye dissolved in solvents of different polarities. It is the ability of a chromophore (https://en.wikipedia.org/wiki/Chemical_substance) to change its color toward the red when dissolved in a more polar solvent. Negative solvatochromism corresponds to hypsochromic shift (or blue shift) with increasing solvent polarity (Figures 8-8 and 8-9).

A positive solvatochromism corresponds to bathochromic shift (redshift).

Polarities of the ground and excited state of a chromophore are different, therefore a change in the solvent polarity will lead to different stabilization of the ground and excited states, and thus, a change in the energy gap between these electronic states.

Solvent effect and dipole moments. Dye molecules are built of atoms and possess positive and negative electric charges. In effect, any molecule can be represented by an electric dipole with a dipole moment $\boldsymbol{p}$. In solutions, solvent molecules surround each dye molecule to create a cavity with an electric field $\boldsymbol{E}$. In such cavity (called Onsager cavity), the dye

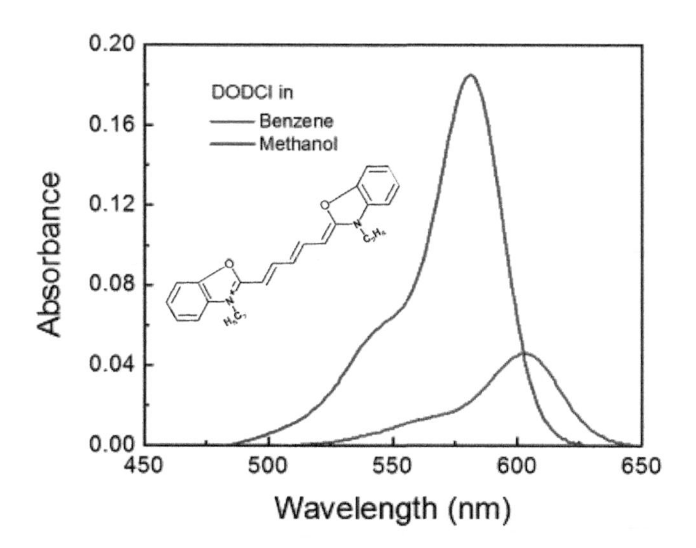

FIGURE 8-8 Example of a hypsochromic shift in the absorption of 3,3'-diethyloxadicarbocyanine iodide (DODCI).

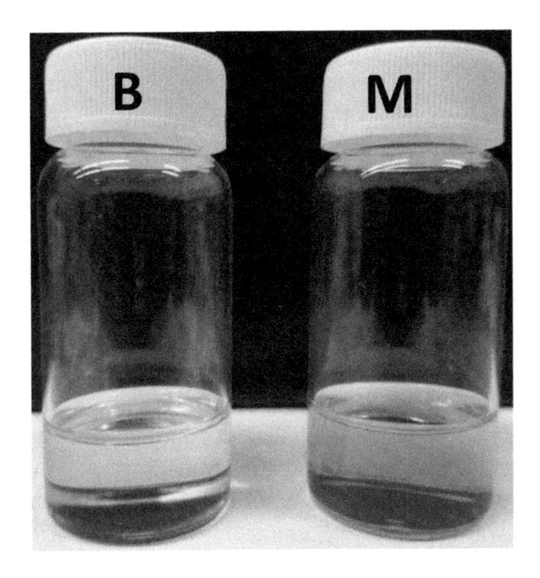

FIGURE 8-9 Photograph of DODCI solutions in benzene (B) and methanol (M).

molecule needs to allocate itself according to the force induced by electric field. Because the radius of the cavity (Onsager radius) is very small, in a range of few Angstroms, the intensity of the electric field within the cavity is colossal. The potential energy of the molecule dipole in the electric field (energy of dipole-field interaction) is a dot product of the dipole moment and field vectors:

$$U = p \cdot E \tag{8.6}$$

This energy depends on the molecule electric dipole and the magnitude of the electric field. Of course, in polar solvents, the electric field in the Onsager cavity is much stronger than in non-polar. As a consequence, the energy levels of the molecule electronic states will change. This will result in spectral shifts. In the excited state of the molecule, the distribution of electric charges (electrons) is different than in the ground state. Usually, the fluorophore dipole moment in the excited state is larger, which means a stronger interaction with the electric field and a stronger shift of fluorescence spectra.

A phenomenological theory of the solvent effect treats the solvent as a continuous dielectric characterized by a dielectric constant ε and refractive index n. Such assumption allows estimating a fluorophore dipole moment change upon the excitation. With an independent measurement of the fluorophore ground state dipole moment, solvatochromic measurements give the unique opportunity to estimate electric dipole moments in the excited state. It is only required to measure absorption and fluorescence spectra in solvents of different polarities. We will present such measurements in Experiment 8-17.

More often the positive solvatochromism is observed. The *solvatochromic effect* (*solvatochromic shift*) describes the dependence of absorption and emission spectra with the solvent polarity. Since polarities of the ground and excited state of a chromophore are different, a change in the solvent polarity will lead to different stabilization of the ground and excited states, and thus, a change in the energy gap between these electronic states. Consequently, variations in the position, intensity, and shape of the absorption spectra can be direct measures of the specific interactions between the solute and solvent molecules. We will demonstrate the solvatochromic effect using Fluorol 7GA dye in benzene-methanol mixtures in Experiment 8-21.

EXPERIMENT 8-1 EMISSION FILTER SELECTION FOR FLUORESCENCE MEASUREMENTS IN IN-LINE GEOMETRY

Keywords: fluorescence, filters, background.

A filter selection on the observation path often can decide about the proper results of fluorescence measurements. An observed emission depends on the strength of fluorescence signal, excitation light, and configuration of the instrument used. If the fluorescence is weak and the excitation is broadband from the lamp, it is possible that a long-wavelength part of the excitation beam will leak into the emission channel. For a clear, weakly scattered sample and square-geometry configuration, the leak of the excitation is often negligible. In this case, long-pass glass filters are fully sufficient. However, in the case of highly scattered samples, front-face, and/or in-line geometry, these filters may introduce significant artifacts in fluorescence measurements. There are few reasons for that. Typical glass cut-off filters (color filters) are highly fluorescent when excited with shorter wavelength. This could be direct excitation light, out-scattered excitation light, or even short wavelength emission. Emission of the filters is mixed with the sample emission resulting in a distorted spectrum.

In this experiment, we will explore how to avoid an unwanted secondary emission of fluorescence filters.

MATERIALS

- Water, NaOH
- 1 cm × 1 cm quartz cuvette
- A set of Schott glass filters
- Rhodamine 6G (R6G)
- Potassium dichromate, $K_2Cr_2O_7$
- Blue laser pointer, 473 nm

EQUIPMENT

- Spectrophotometer (Cary 50, Varian, Inc.)
- Ocean Optics USB 4000 spectrometer equipped with 1 mm fiber
- Cell phone or camera

METHODS

1. Make a concentrated (saturated) water solution of potassium dichromate. Pour it in two 1 cm × 1 cm cuvettes, 2 mL in each cuvette.
2. Add to one of the cuvettes one pastille of NaOH, observe a change in the color.

(Continued)

EXPERIMENT 8-1 (CONTINUED) EMISSION FILTER SELECTION FOR FLUORESCENCE MEASUREMENTS IN IN-LINE GEOMETRY

3. Measure transmissions of the two potassium dichromate solutions, see Figure 8-1.1.
4. Set an in-line configuration for fluorescence measurements with the Ocean Optics spectrometer using fiber optics, see Figure 8-1.2 (this type configuration is frequently used in sensing).
5. Make stock solutions of R6G in EtOH (use a glass vial). Prepare a sample of R6G in water (using EtOH stock solution) with a low optical density of about 0.01.
6. Measure fluorescence using a cuvette with high pH solution of potassium dichromate as an emission filter (in front of the end of optical fiber), see Figure 8-1.3 (left).
7. Repeat these measurements with various long-pass glass filters (495–545 nm).
8. Illuminate different filters with a 473 nm laser pointer and make photographs of observed fluorescence spots (under the excitation), see Figure 8-1.4.
9. Measure fluorescence of individual filters illuminated with 473 nm laser, see Figure 8-1.5.

RESULTS

The solution of potassium dichromate has an orange color and transmits light above 600 nm and blocks the light with wavelengths below 550 nm. The transmission of potassium dichromate depends on pH and can be tuned within a wide range of wavelengths. Addition of NaOH (about 1 M) results in a short wavelength shift of the transmission (color of the solution becomes yellow).

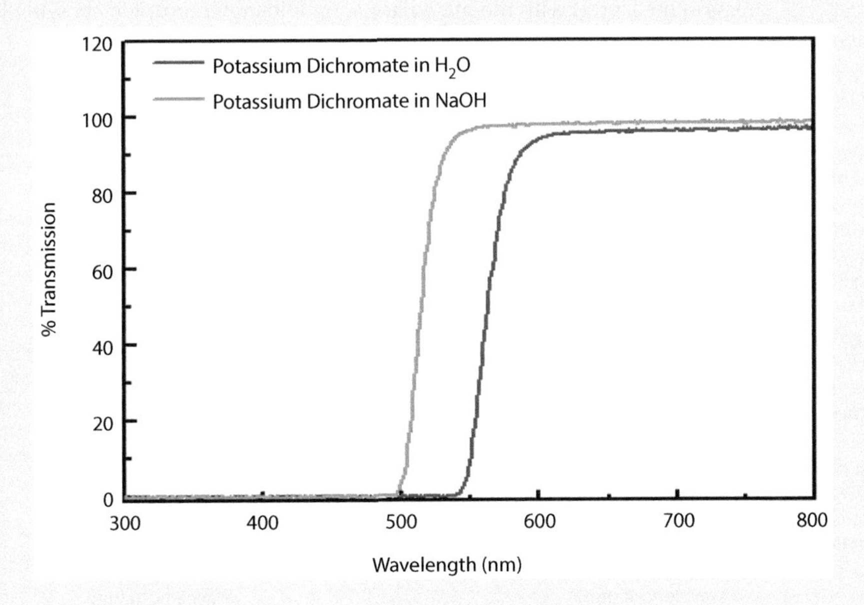

FIGURE 8-1.1 Transmissions of liquid potassium dichromate filters.

(Continued)

EXPERIMENT 8-1 (CONTINUED) EMISSION FILTER SELECTION FOR FLUORESCENCE MEASUREMENTS IN IN-LINE GEOMETRY

To test how to filter emission can distort measured sample emission, we will measure Rhodamine 6G (R6G) solution using in-line geometry as shown in Figure 8-1.2. This configuration is used in microscopy and can also be used for sensing devices. In this configuration, the excitation is blocked by the barrier filter (long wavelength cut-off).

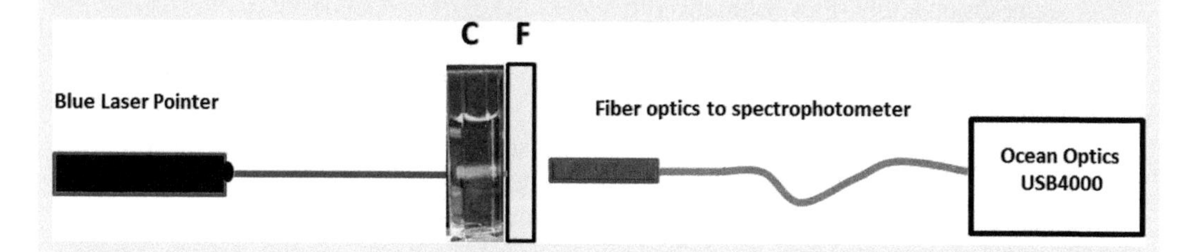

FIGURE 8-1.2 In-line configuration for fluorescence measurements. Cuvette (C) is a photograph of the sample used in the experiment. The green fluorescence emission of R6G is observed through either potassium dichromate liquid filter or through 495 nm glass long-pass filter. The excitation is from the 473 nm laser pointer.

The spectra of R6G solution measured with liquid potassium dichromate filter (left) and glass filter (right) are shown in Figure 8-1.3. The measured emission spectra are drastically different. The emission measured with the glass filter is significantly corrupted showing long-wavelength dominant peak.

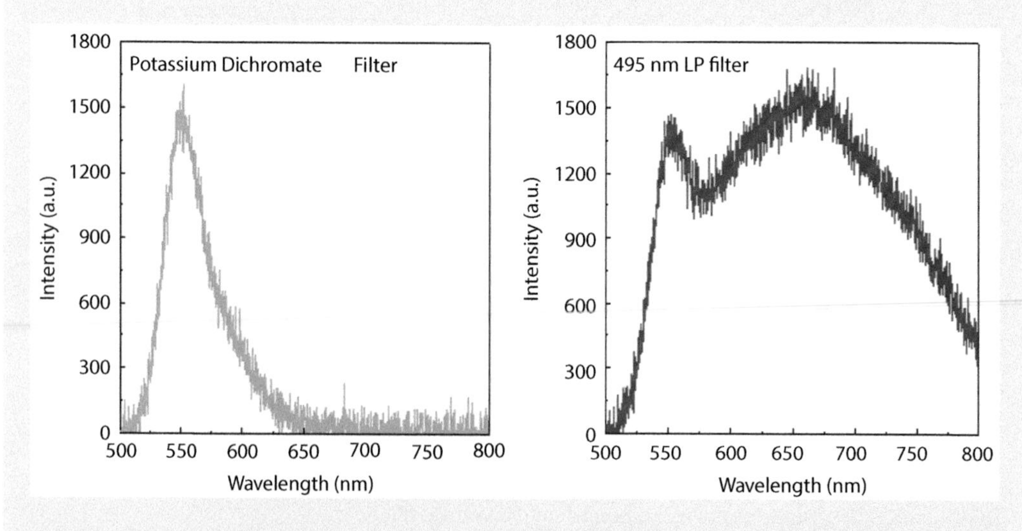

FIGURE 8-1.3 Fluorescence of an Rh6G solution with the liquid potassium dichromate filter (left) and a 495 nm long-pass glass filter on the observation (right). The measurements were done using the in-line configuration with optic fiber detection.

(Continued)

EXPERIMENT 8-1 (CONTINUED) EMISSION FILTER SELECTION FOR FLUORESCENCE MEASUREMENTS IN IN-LINE GEOMETRY

To explain the origin of long-wavelength emission, we selected few filters from our set of filters and exposed them to blue laser excitation (see photographs in Figure 8-1.4). A clear yellowish spot is visible on each filter but solution of potassium dichromate show nothing.

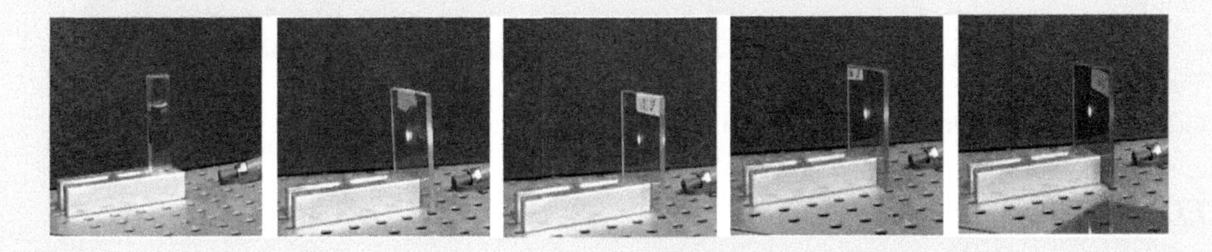

FIGURE 8-1.4 Photographs of various filters illuminated with 473 nm laser pointer. Only the potassium dichromate liquid solution does not show any emission.

Next, we measured emission spectra of each illuminated filter and solution of potassium dichromate (Figure 8-1.5). Long-pass glass filters have relatively strong emission covering wide range of the spectrum. In contrast, the liquid potassium dichromate filter shows no detectable emission upon illumination with the 473 nm laser pointer. On many occasions, it is difficult to avoid this unwanted emission.

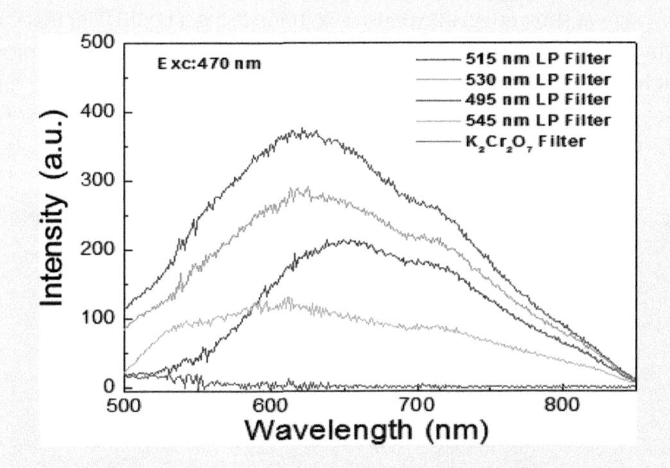

FIGURE 8-1.5 Emission of glass and potassium dichromate filters illuminated with a 473 nm laser pointer.

It is easy to see luminescence emissions of glass filters upon UV illumination, see Figure 8-1.6.

If someone plans to use a short wavelength excitation and expects a weak fluorescence should first check for potential emission of used filters.

(Continued)

EXPERIMENT 8-1 (CONTINUED) EMISSION FILTER SELECTION FOR FLUORESCENCE MEASUREMENTS IN IN-LINE GEOMETRY

FIGURE 8-1.6 Emission of glass filters on the UV illuminator.

CONCLUSIONS

In most fluorescence measurements, the unwanted emission of filters can be neglected. However, when the observed fluorescence is weak and the configuration is front-face or in-line, the filter emission may create problems. Especially, one should consider this problem when constructing a sensing device where in-line geometry is preferable as well as laser/LED excitation. In such a case, we would typically recommend to avoid glass filters and use interference filters or liquid filters. Depending on wavelength and filters availability, we may need to use glass filters. In this case, the best results are achieved when combinations of interference (liquid) and glass filter are used. It is important to remember that in such case, interference/liquid (non-fluorescent) filter should be placed before the glass filter. This blocks excitation preventing (limiting) glass filter excitation. Placing the glass filter first (on the side of the sample) will excite the filter and resulting emission will freely pass through the interference (liquid) filter.

In this experiment, we showed an inexpensive solution using a liquid filter that can be inexpensively made in any chemistry laboratory. Good interference filters are typically more expensive.

Potassium dichromate filter does not have to be only in a liquid form. It is easy to prepare the PVA film (see Experiment 3-24) containing $K_2Cr_2O_7$, see Figure 8-1.7. To have polymer-doped with $K_2Cr_2O_7$ is very convenient for the construction of sensing devices.

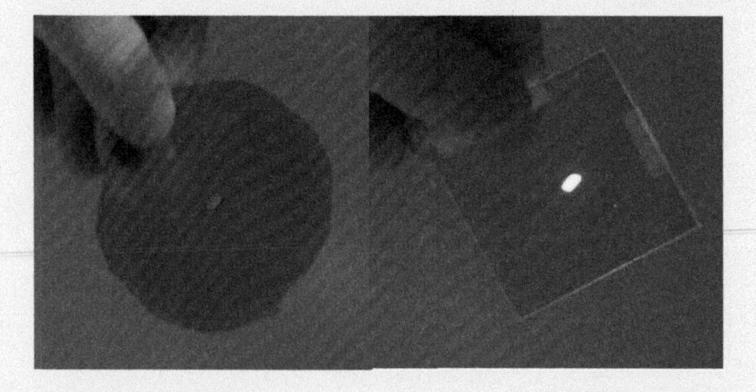

FIGURE 8-1.7 K2Cr2O7-doped PVA film (left) and Corning 495 nm long-pass glass filter illuminated with a blue laser pointer.

Note: Some results from this experiment were published in "*Methods and Applications in Fluorescence*." vol 7, (2019) 037001. Figures 8-1.1 through 8-1.3 and 8-1.6 have been used with permission from IOP.

EXPERIMENT 8-2 RAMAN SCATTERING OF DIFFERENT SOLVENTS

Keywords: Raman scattering, water, ethanol, cyclohexane.

It is expected that the Raman signal from different solvents will be different. In fluorescence measurements, especially for low emission intensities, the Raman scattering is an unwanted background which can corrupt your measurements. We already presented Raman for different solvents in Experiments 4-9 and 4-10. It is important to recognize the magnitude and wavelength location of the Raman signal. We will present Raman signals for several popular solvents measured at the same conditions. We extended these measurements for more solvents and vary the excitation wavelengths.

MATERIALS

- 1 cm x 1 cm quartz cuvette
- Water
- Methanol, water, ethanol, ethyl acetate, cyclohexane, *n*-heptane, toluene

EQUIPMENT

- Spectrofluorometer (Ocean Optics USB 4000 spectrometer equipped with 1 mm fiber)
- 473 nm laser pointer, 520 nm long-pass filter, camera or cell phone

METHODS

1. Measure the Raman spectra of various solvents using 473 nm excitation. Raman signal is low and you will probably need to use maximum allowed voltage on spectrofluorometer but narrow excitation and observation slits.
2. Compare Raman signals of water and ethanol and prepare a mixture of 50/50 water-ethanol and measure the Raman spectrum of the mixture.
3. Measure Raman signals for cyclohexane and water at different excitation wavelengths.
4. Expose the cuvette with any solvent to 473 nm laser pointer beam (suggested: cyclohexane). Make a photograph without and with 520 nm cut-off filter.

RESULTS

First, we measure Raman signals for various solvents, see Figure 8-2.1. In these measurements, we did not average signals (as it was done in Experiments 4-9 and 4-10) and used slits

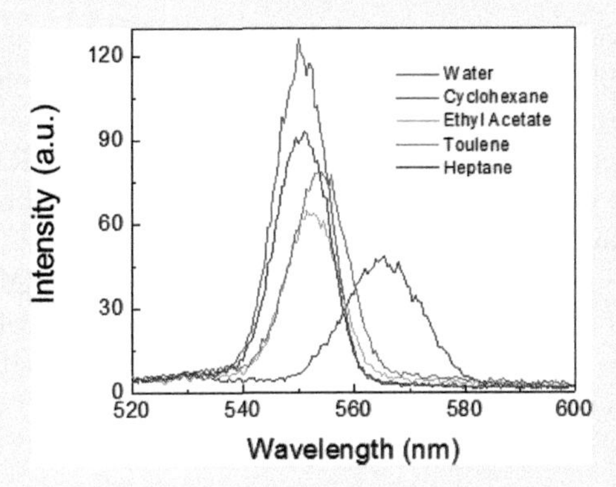

FIGURE 8-2.1 Raman signals observed for 473 nm excitation. The shift depends on the solvent.

(Continued)

EXPERIMENT 8-2 (CONTINUED) RAMAN SCATTERING OF DIFFERENT SOLVENTS

5 nm for both excitation and observation. In organic solvents, the Raman signals are different in the intensity but close in the wavelengths. In water, the signal is slightly weaker and significantly shifted toward longer wavelengths. This gives the opportunity for simply estimation of water contamination in mixtures with other solvents.

Next, compare Raman shifts for water and ethanol.

The data presented in Figure 8-2.2 uses 95% pure ethanol, thus a small Raman signal of water is present in the ethanol spectrum (a small bump at a longer wavelength of ethanol spectrum). The purple curve is the mixture of ethanol and water. It can be observed that the ethanol intensity is higher. This is because at the wavelength used the Raman cross-section for ethanol is greater than that for water.

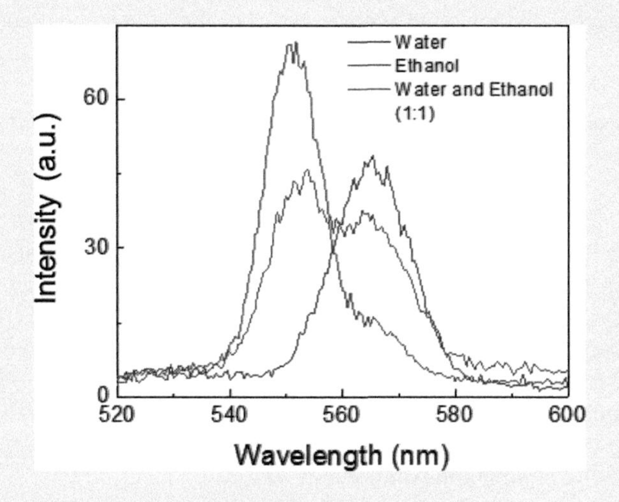

FIGURE 8-2.2 Raman spectra of water, ethanol, and 50/50 water-ethanol mixture with the excitation of 473 nm. The comparison of Raman signal for water/ethanol was already shown in Experiment 4-10.

In fluorescence experiments, various excitation wavelengths are used. Proteins, nucleotide analogs and many scintillators require ultraviolet excitation. Coumarins, fluoresceins as well as short-wavelength Alexa-type dyes excitations are in the blue region. Rhodamins require a green/orange excitations and CY5-type dyes are excited in red. How Raman signals will affect fluorescence of different dyes? We will check the Raman spectra for cyclohexane and water at different excitation wavelengths, see Figures 8-2.3 and 8-2.4.

The Raman signal clearly depends on the excitation wavelength. In this experiment, we did not measure the intensity of the excitation for individual wavelengths, therefore this dependence is valid only for the instrument used (Varian). However, the output of a xenon flash lamp in the 400–500 nm region is stronger than in 300–400 nm (see *Principles of Fluorescence Spectroscopy,* J.R. Lakowicz and/or a spectrofluorometer instruction).

(*Continued*)

EXPERIMENT 8-2 (CONTINUED) RAMAN SCATTERING OF DIFFERENT SOLVENTS

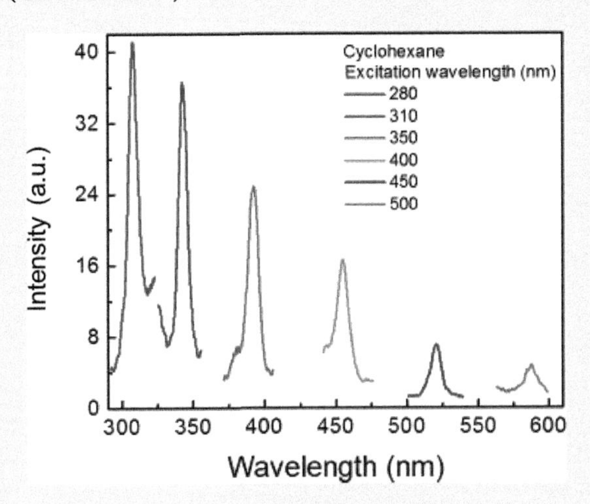

FIGURE 8-2.3 Raman spectra of cyclohexane at different excitation wavelengths.

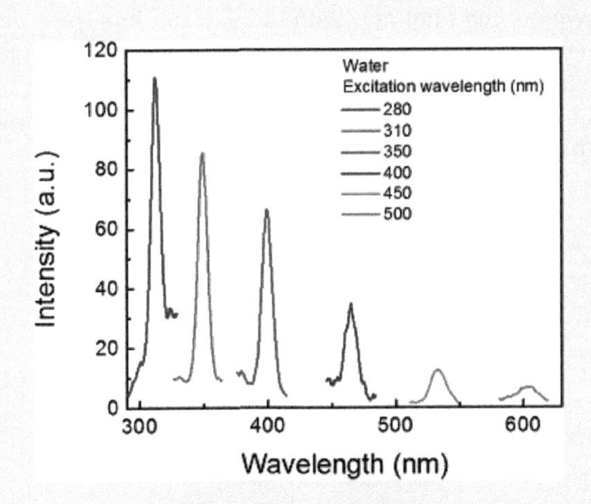

FIGURE 8-2.4 Raman spectra of water at different excitation wavelengths.

Next, we repeated this experiment with water (HPLC grade) using exactly the same excitation wavelengths and experimental condition.

Raman shifts for cyclohexane and water are presented in Figure 8-2.5.

In order to observe Raman signals in commercial spectrofluorometers, a high sensitivity (high voltage on the detectors) is needed. The Raman will interfere only with weak fluorescence signals, like in extreme quenching measurements or with much diluted solutions. However, at the possibility of illumination with laser light source, the Raman signal can be easily visualized.

(*Continued*)

EXPERIMENT 8-2 (CONTINUED) RAMAN SCATTERING OF DIFFERENT SOLVENTS

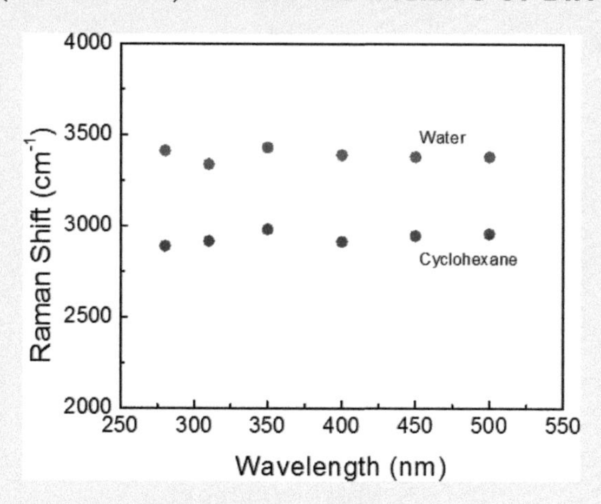

FIGURE 8-2.5 Raman shifts for cyclohexane and water at different excitation wavelengths.

Next, we illuminated the cuvette with cyclohexane outside the spectrofluorometer and observed the signal without the filter and with the 520 nm long-pass filter. The photograph of Raman signal from cyclohexane illuminated with 473 nm laser pointer, see Figure 8-2.6. Without the filter, the 473 nm laser beam is clearly visible. But this wavelength is not transmitted by 520 nm long-pass filter. However, we can clearly see the green laser beam pass what is just signal of Raman scattering.

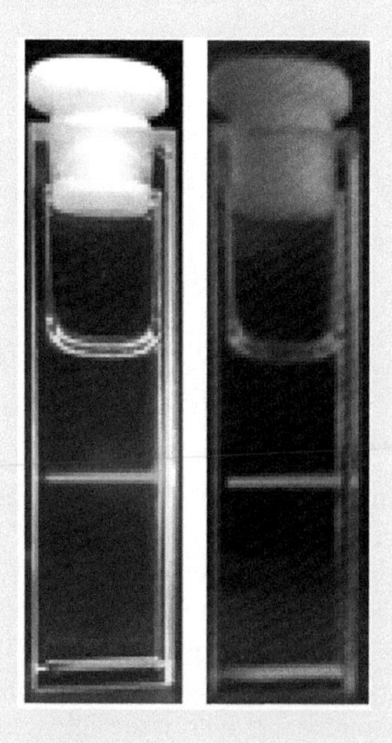

FIGURE 8-2.6 Photographs of the Raman signals from cyclohexane illuminated with 473 nm laser. Left: Without any filter; Right: With 520 nm long-pass filter.

(*Continued*)

EXPERIMENT 8-2 (CONTINUED) RAMAN SCATTERING OF DIFFERENT SOLVENTS

CONCLUSIONS

Each solvent shows different intensity and shift of the Raman signal. The strongest Raman signal among used solvents was observed for cyclohexane. The different Raman shifts and intensities allow the estimation of the amount of solvent components in mixtures; for example, the amount of ethanol in the mixture ethanol/water. It is important to consider that there may be multiple Raman active species in a sample and that the signal from all of them will combine, much in the same way as emission. The Raman signal depends on the excitation wavelength; at the long wavelengths excitations it is much weaker than in UV. The position of the Raman peak can be precisely predicted.

EXPERIMENT 8-3 THERMALLY INDUCED SPECTRAL PROPERTIES

Keywords: absorption, fluorescence, PVA films, cis-trans isomerization, 1,8-diphenyloctatetraene.

Diphenyl polyenes can exist in two forms; trans- and cis-form. In organic solvents, they exist dominantly in a trans-form. However, it is known that a cis-form can exist, depending heavily on the solvent. In particular, water strongly promotes an isomerization into the cis-form, which is better soluble. In the cis-form spectral properties of the compound are dramatically different than in the trans-form. We will demonstrate this with the 1,8-diphenyloctatetraene, a bright fluorophore in organic solvents but almost not fluorescent in water. We will also show that it is possible to "freeze" cis-forms of this dye in poly(vinyl) alcohol films, which are prepared in water and later dried. After exposure of these films to the heat (above 100°C) the absorption and strong fluorescence appear.

MATERIALS

- 1 cm × 1 cm fluorometric cuvette
- 1,8-Diphenyloctatetraene (DPO)
- Ethanol, water, poly(vinyl) alcohol (PVA, 130,000 MW)
- Glass or plastic vials, Petri dishes

EQUIPMENT

- Spectrofluorometer (Varian, Inc., Eclipse) with a front-face attachment
- Carry 50 Spectrophotometer (Varian, Inc.)
- Cell phone or camera, UV illuminator

METHODS

1. Make stock solutions of DPO in ethanol.
2. Prepare 10% (w/w) PVA solution, see Experiment 3-24.

(Continued)

EXPERIMENT 8-3 (CONTINUED) THERMALLY INDUCED SPECTRAL PROPERTIES

3. Add the same amount of the DPO stock solution to 20 mL of ethanol and 20 mL of PVA solution in vials. Shake well both bottles until solutions are homogeneous. Wait a little for air bubbles to vanish and put the vials on a UV illuminator to make a photo, see Figure 8-3.1.
4. Measure the absorptions of both samples, see Figure 8-3.2.
5. Measure fluorescence spectra of DPO in ethanol and in 10% PVA solution, see Figure 8-3.3.
6. Pour about 20 mL of PVA solution (baseline reference) and the same amount of the DPO solution in PVA into Petri dishes and dry them (this can take a few days; see Experiment 3-24 for preparation details). When the films are completely dry, pile them from the dishes. Cut the films into two pieces each. One-half of the sample will serve as a reference, another half will be exposed to heating, see Figure 8-3.4.
7. Measure absorption and fluorescence spectra of both films, heated and not heated, see Figures 8-3.5 and 8-3.6.

RESULTS

In ethanol, DPO has a strong blue-green fluorescence as seen in Figure 8-3.1. In contrast, in 10% PVA water solution, DPO is almost not fluorescent. Absorption measurements reveal that with the same amount of DPO contained, the PVA solution has an absorbance more than 10-fold lower and different in shape than DPO in an ethanol solution, see Figure 8-3.2.

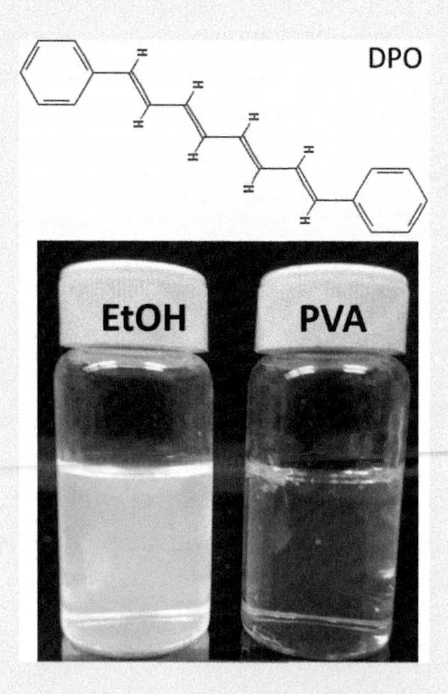

FIGURE 8-3.1 Photograph of DPO in ethanol (left) and in 10% PVA solution (right) positioned on a UV illuminator. Top: chemical structure of DPO as a trans-form. The solutions contain the same amount of DPO.

(*Continued*)

EXPERIMENT 8-3 (CONTINUED) THERMALLY INDUCED SPECTRAL PROPERTIES

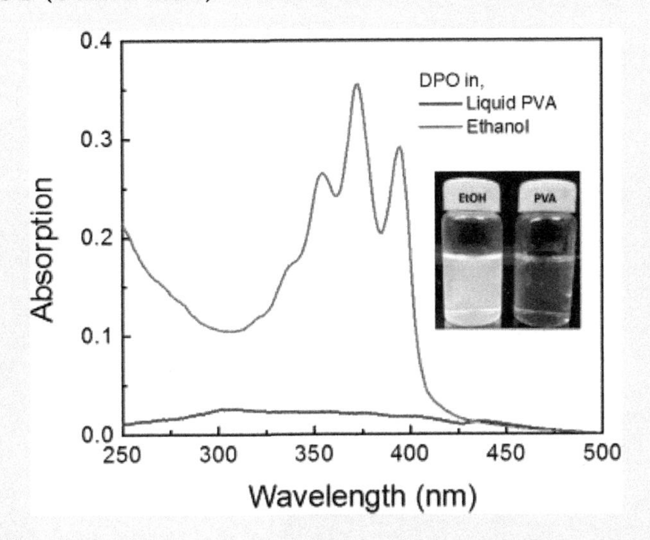

FIGURE 8-3.2 Absorption spectra of DPO in ethanol and in 10% PVA solution. The solutions contain the same amount of DPO.

We measured fluorescence with 360 nm excitation using a regular cuvette (we recommend to use plastic fluorometric cuvette since washing PVA solution can be difficult). The fluorescence of DPO in PVA solution is negligible compared to the emission of DPO in ethanol.

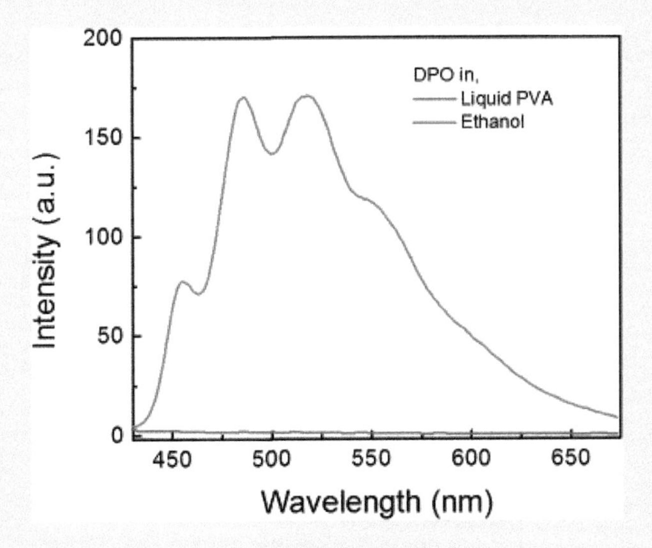

FIGURE 8-3.3 Fluorescence spectra of DPO in ethanol and in 10% PVA solution. Both solutions contain the same amount of DPO.

PVA films doped with the DPO are transparent and not distinguishable from not doped (reference) PVA film. We cut the DPO-doped PVA film into two parts. One part was heated on the heat plate for 10 minutes at about 130°C, another part of the film was not heated.

(Continued)

EXPERIMENT 8-3 (CONTINUED) THERMALLY INDUCED SPECTRAL PROPERTIES

The comparison of heated and not heated films on the UV illuminator is presented in Figure 8-3.4. We recommend doing the same with reference PVA film. The heated and non-heated film will have an identical color.

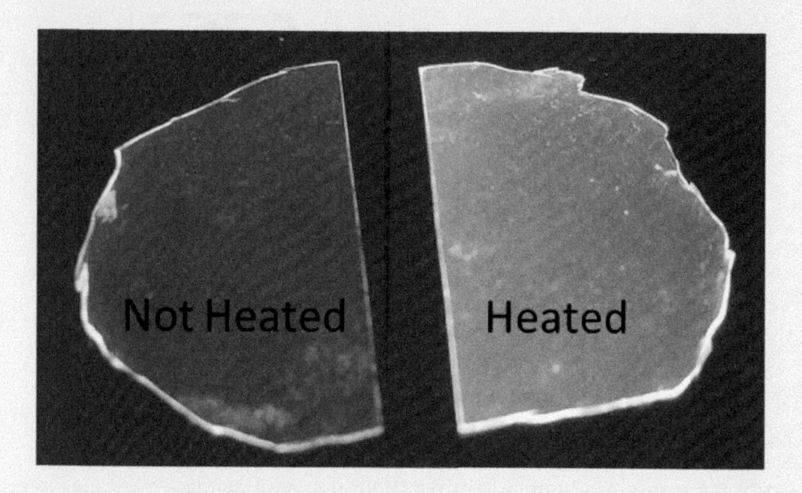

FIGURE 8-3.4 Photograph of dried PVA films doped with DPO. Left: Half of the film without heating; Right: Second half of the film heated at 130°C for about 10 min.

Next, we measured absorptions of both parts of the DPO-doped films, see Figure 8-3.5. The absorbance of the heated film is significantly stronger, and the shape is similar to the absorption in an organic solvent, compare with Figure 8-3.2.

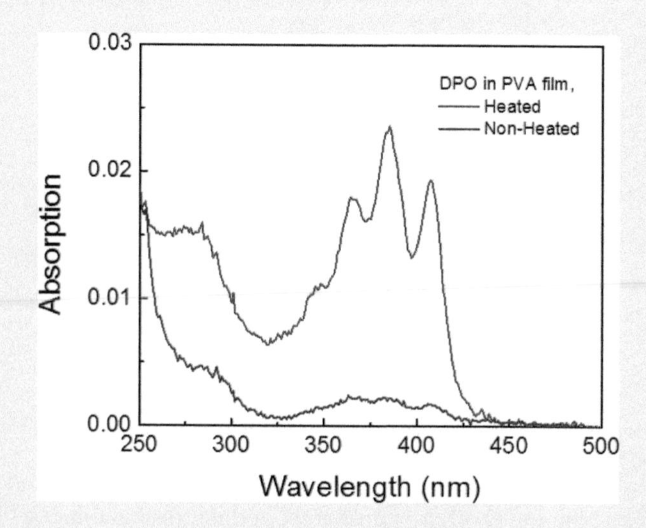

FIGURE 8-3.5 Absorption spectra of heated and not heated PVA films doped with DPO. Baselines were taken from heated and not heated PVA films without DPO.

(Continued)

EXPERIMENT 8-3 (CONTINUED) THERMALLY INDUCED SPECTRAL PROPERTIES

Similarly to absorption, the fluorescence emission of the heated DPO-doped PVA film is significantly stronger, see Figure 8-3.6.

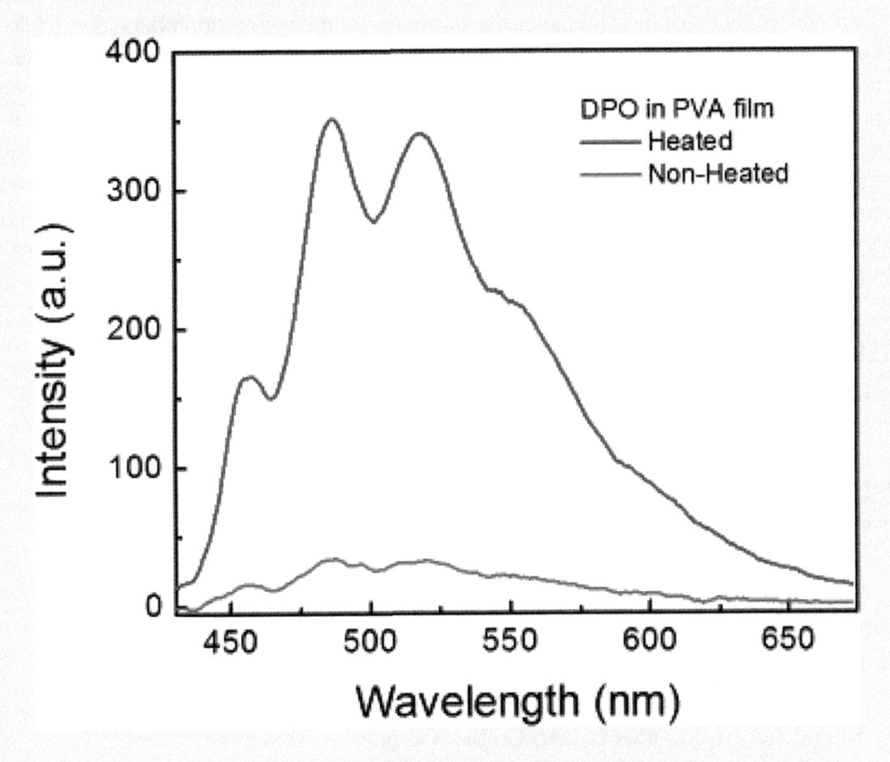

FIGURE 8-3.6 Fluorescence spectra of heated and not heated PVA films doped with DPO. Fluorescence emissions from heated and not heated reference PVA films without DPO are negligible.

CONCLUSIONS

Heating of the DPO-doped PVA film-induced the absorbance and fluorescence of the DPO. The DPO in the heated film behaves like in organic solvents. In liquid PVA water solution, the DPO molecules exist in cis-form(s) because trans-DPO is very weakly soluble in a polar solvent. During a drying process, the viscosity increases, and even if there are no more water molecules in the film, the DPO molecules are trapped in the cis-form(s). The heating to the temperature close to a melting point of the PVA polymer lowers the viscosity. Without the water molecules, DPO relaxes to its natural trans-form. There are several potential applications of such a phenomenon like secret writing or memory storage.

EXPERIMENT 8-4 EMISSION ANISOTROPY AND CHROMOPHORE LIFETIME

Keywords: fluorescence, anisotropy, lifetime.

As we have shown in Chapter 6, fluorescence anisotropy depends on a number of factors. This includes excitation wavelength and solvent viscosity. The change in measured anisotropy value is a consequence of the change of the transition moment direction/orientation while the molecule remains in the excited state. For long-wavelength excitation, if the transition moment direction is not changed, the maximum observed anisotropy is 0.4, as given by the photo-selection. However if the molecule can reorient (rotate) before emission of photon its anisotropy will be lower. In effect, the rotational motion of the molecule is the dominant reason for the loss of the emission anisotropy. The viscosity of the solvent directly impacts the rotational freedom and anisotropy. In liquid samples where molecule can rotate freely the anisotropy loss will depend on the lifetime of the fluorophore, as given by the Perrin equation. In this next experiment, we will observe how the anisotropies of similar size/shape molecules depend on their lifetimes.

MATERIALS

- Rhodamine 123 (R123)
- Eosin Y (Eos)
- Erythrosin B (ErB)
- Ethanol
- Ethyl acetate
- 1 cm × 1 cm fluorometric cuvettes

EQUIPMENT

- Spectrophotometer (Cary 50, Varian, Inc.)
- Spectrofluorometer (Eclipse, Varian, Inc.)
- Fluorometer FT 200 (PicoQuant, GmbH, Germany) with 470 nm laser diode

METHODS

1. Prepare stock-solutions of R123, Eos, and ErB in EtOH.
2. Make individual water solutions of R123, Eos, and ErB with optical densities not exceeding 0.1.
3. Measure fluorescence spectra of individual fluorophores at VV and VH polarization conditions. Apply the correction for G-factor, see Figure 8-4.2. The spectra of these fluorophores at MA conditions were presented in Experiment 7-13.
4. Review Experiment 7-13 and note lifetimes of studied fluorophores, see Figure 8-4.1.
5. Calculate and draw fluorescence emission anisotropies of individual fluorophores, see Figure 8-4.3.

RESULTS

We selected the same fluorophores as in Experiment 5-9. These dyes have a very similar size and shape (see chemical structures shown in Scheme 8-4.1) and are expected to rotate in solution with a similar rate (correlation times). However, these fluorophores differ significantly in fluorescence lifetimes, from 3.99 ns for R123, 1.14 ns for Eos to 0.09 ns for ErB, see

(Continued)

EXPERIMENT 8-4 (CONTINUED) EMISSION ANISOTROPY AND CHROMOPHORE LIFETIME

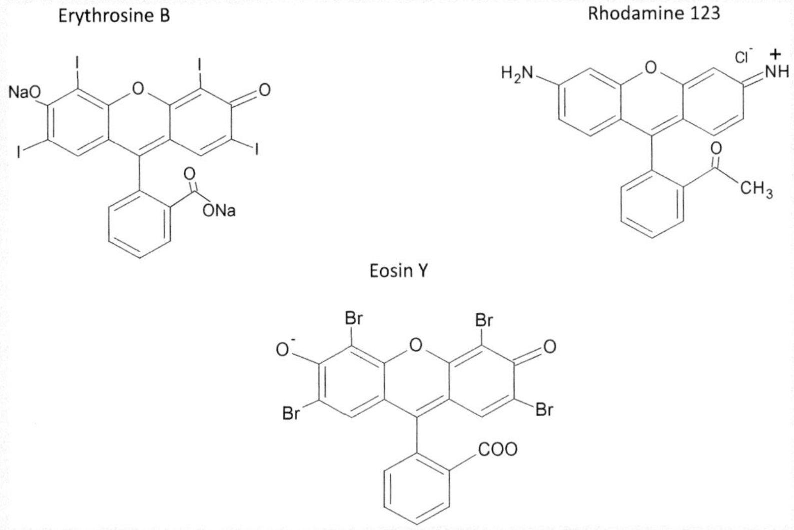

Erythrosine B

Rhodamine 123

Eosin Y

SCHEME 8-4.1 Chemical structures of used fluorophores. The sizes of all three fluorophores are similar.

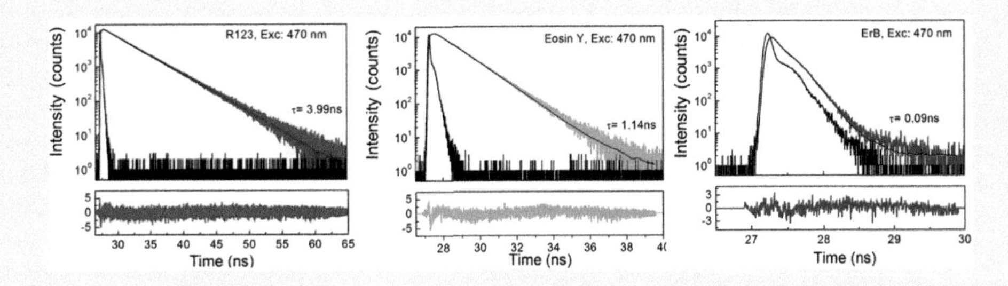

FIGURE 8-4.1 Lifetimes of studied fluorophores (from Experiment 7-13).

Figure 8-4.1 and Experiment 7-13. When the fluorophore lifetime is very short, the molecule will not move significantly during the lifetime and we expect higher anisotropy.

As shown in Figure 8-4.1 selected fluorophores differ significantly in lifetimes (see Experiment 7-13).

Next, measure fluorescence spectra at different polarization polarizer orientations (VV and VH) at 480 nm excitation, see Figure 8-4.2. The corrections for G-factor were applied as described in Experiment 5-15.

Next, we calculated emission anisotropies, see Figure 8-4.3.

The observed anisotropies depend on the lifetimes of the fluorophores. Very high fluorescence anisotropy of ErB in water is a result of its ultra-short lifetime. This lifetime was first estimated by F. Perrin on the base of the anisotropy data collected by W.L. Levshin almost a century ago.

(Continued)

EXPERIMENT 8-4 (CONTINUED) EMISSION ANISOTROPY AND CHROMOPHORE LIFETIME

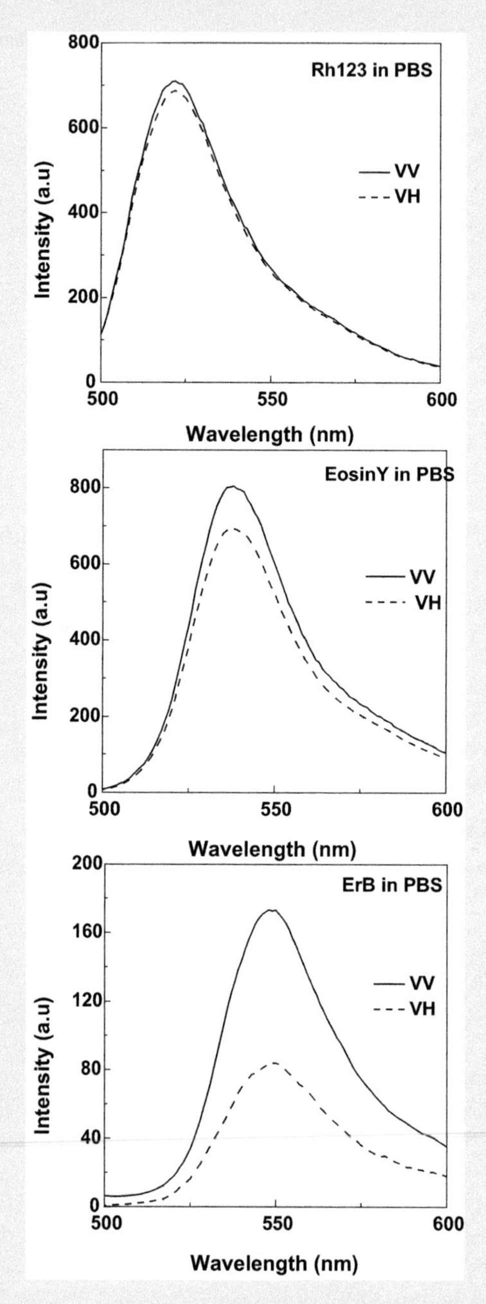

FIGURE 8-4.2 Polarized components of fluorescence spectra of R123 (top), Eos (middle), and ErB (bottom) in PBS buffer. The spectra measured at VH polarization were corrected with G-factor.

(*Continued*)

EXPERIMENT 8-4 (CONTINUED) EMISSION ANISOTROPY AND CHROMOPHORE LIFETIME

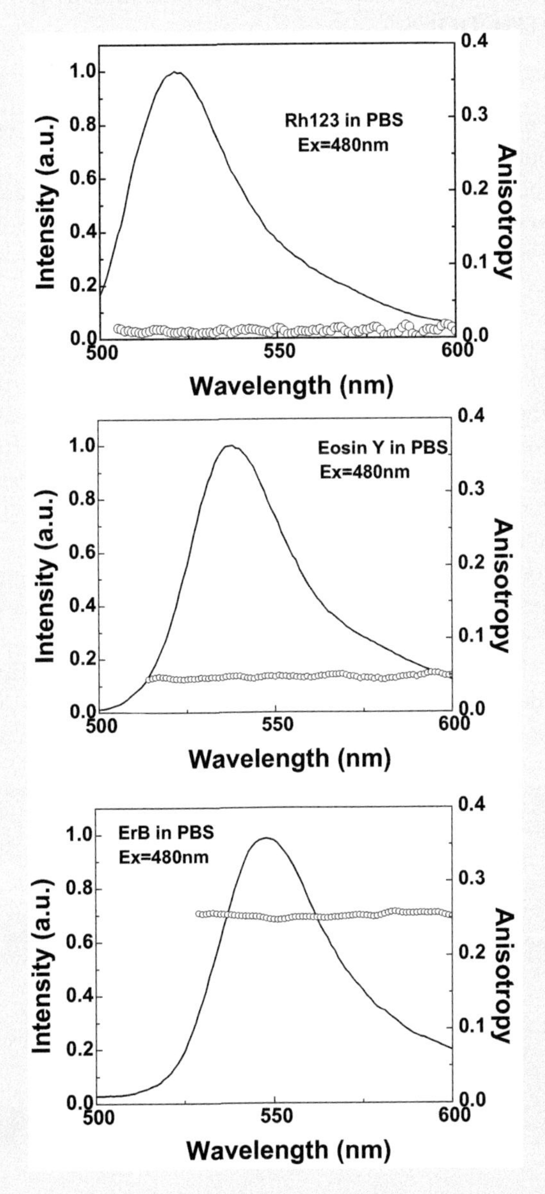

FIGURE 8-4.3 Fluorescence emission anisotropy measurements of R123 (top), Eos (middle), and ErB (bottom) in PBS buffer.

CONCLUSIONS

A lifetime of the fluorophore and solvent viscosity (correlation time of fluorophore) decide in most cases about observed fluorescence anisotropy. Both parameters are linked in the Perrin equation.

EXPERIMENT 8-5 PRECAUTIONS IN FLUORESCENCE ANISOTROPY MEASUREMENTS: DEPOLARIZATION THROUGH REABSORPTION AND INTERNAL REFLECTIONS

Keywords: fluorescence, anisotropy, fluorescence intensity, excitation energy migration.

There are many processes that may lead to lower anisotropy. Typical effects like rotational depolarization in solutions, less common Raman scattering, and too high chromophore/dye concentration where excitation energy migration (homo-FRET—radiationless excitation energy transfer between same chromophores) leads to fluorescence depolarization have been already discussed in Chapter 6. In many practical applications like sensing or diagnostics, we will frequently use thin layer samples to measure emission and/or polarization. In this experiment, we will show three examples of how such measurements can be done and what mistakes one should avoid. In these examples, we will present how geometry of the experiment/sample and emission light reflections or secondary emission due to reabsorption may affect the outcome of measured emission spectra and emission anisotropies. In Experiment 6-11, we discussed artifacts due to too high sample absorption. Too high sample absorption is straightforward to identify and the problem can easily be avoided. However, in some cases, measured sample absorption can be in the right range but geometry of the sample can impose unexpected problems.

As an example, we will use solid samples in the form of polymer films. We will use films made out of PVA as described in the Experiment 3-20. In Figure 8-5.1, we present examples of PVA films doped with different dyes. The films are positioned on a UV illumination stage and fluorescence can be clearly seen. These films are typically thin 50–300 μm and measured absorption of embedded chromophores is apparently very low. Interestingly, looking on the films in Figure 8-5.1 one would notice a stronger emission glowing from the edges of the film.

FIGURE 8-5.1 Examples of PVA films with different dyes on the UV illumination stage.

For the presented three examples, we will use a PVA film with Rhodamine 6G (R6G). An example of absorption of the film doped with R6G is presented in Figure 8-5.2. The measured absorption is well within the "good" fluorescence range (below 0.05). Since the film is about 70 μm thick the concentration of R6G is relatively high (absorbance of the 1 cm thick film would be above 3). However, this high concentration is still much below the range where energy migration would play any significant role.

(*Continued*)

EXPERIMENT 8-5 (CONTINUED) PRECAUTIONS IN FLUORESCENCE ANISOTROPY MEASUREMENTS: DEPOLARIZATION THROUGH REABSORPTION AND INTERNAL REFLECTIONS

The goal of this exercise is to show how such films can be measured properly and with what kind of problems we could expect and avoid. We will discuss three approaches:

1. Using standard square geometry configuration and a regular 1 cm cuvette. This appears to be a simple straightforward approach but it is very prompt to multiple artifacts.
2. Using standard square geometry configuration and 1 mm × 10 mm cuvette, and observing fluorescence from the film edge. This approach is easy to realize but dye reabsorption could be a problem.
3. Finally, we will show measurements in the front-face configuration. In such a configuration, when done properly we can obtain the most adequate results. When measuring thin samples or dyes deposited on the surface this is the configuration that we would typically recommend.

MATERIALS

- PVA, ethanol (EtOH)
- Rhodamine 6G (R6G)
- Mineral oil
- Bandpass interference filter (for the range 470–520 nm) and a long-pass filter (LP 530 nm)
- Glass or quartz 1 mm × 10 mm flat cuvette and 10 mm × 10 mm square cuvette (12 mm × 12 mm external dimensions), see Figure 8-5.3

EQUIPMENT

- Spectrophotometer
- Spectrofluorometer
- Square geometry cuvette mount allowing position change (cuvette displacement orthogonal to the excitation line). See Figures 8-5.3 and 8-5.4 for example.
- Front-face adapter

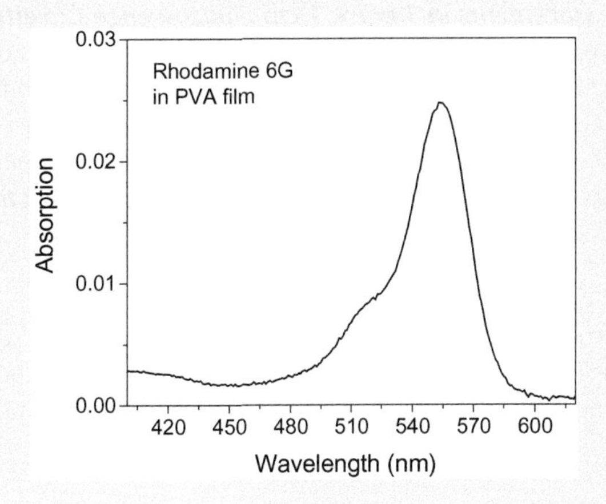

FIGURE 8-5.2 Corrected absorption spectrum of Rhodamine 6G in the PVA film.

(Continued)

EXPERIMENT 8-5 (CONTINUED) PRECAUTIONS IN FLUORESCENCE ANISOTROPY MEASUREMENTS: DEPOLARIZATION THROUGH REABSORPTION AND INTERNAL REFLECTIONS

Figure 8-5.2 shows background-corrected absorption spectrum of R6G in the PVA film. To properly correct absorption for PVA background, see Experiment 3-22. We used film with relatively high concentration of R6G. The film thickness was ~70 μm. The absorption spectrum well corresponds to R6G absorption in EtOH.

Note: We will describe three experimental configurations used for measurements before proceeding to a comparative discussion of measured spectra and polarization. For excitation, we will use 515 nm wavelength supported by the BP filter. As presented in the Experiment 6-4, the excitation anisotropy of R6G is high and constant within the first absorption band (above 480 nm). Using shorter excitation allows us to limit leaking of the excitation light into the emission channel. Solid samples like films or dye deposited on the surfaces produce much more scattering than typical solutions measured in a square configuration. It is important to realize that standard monochromators are not perfect. In reality, an excitation monochromator transmits a very small amount of light of wavelength greater (and smaller) than the selected wavelength. For example, the monochromator set to 515 nm will transmit a small amount of light at shorter and longer wavelengths, so call transmission wings. The intensity of light in the wings is negligible as compared to intensity at the peak (about 515 nm) but when compared to a much weaker fluorescence signal it can become significant. Especially when we use weak fluorophores at very low concentration. This small amount of light off the main transmitted peak when overlapping with emission is scattered and is seen/transmitted by the observation monochromator. In addition, the observation monochromator will transmit minimal amount of main scattered excitation light. Because of that, we purposely shift excitation to a shorter wavelength. But we need to remember that too large of a shift may affect our measured polarization. Best solution is to use additional band-pass interference filers on the excitation that transmit excitation light (in our case, we used 470–520 nm BP filter) and a long-pass filter on emission (in our case, 530 LP filter). The proper filter composition is the best approach to minimize (eliminate) residual light intensity leaking through monochromators.

Square Geometry Measurements in 1 cm × 1 cm Fluorescence Cuvette. Frequently experimentalists will not have a front-face adapter and will do measurements of thin samples in the square geometry using a standard 1 cm × 1 cm cuvette. Typically, the film will be mounted diagonally across the cuvette so the film surface will form about 45° to excitation and emission lines, see Figure 8-5.3. This is very easy to do without any adjustments to the spectrofluorometer sample compartment, but we do not recommend measuring thin samples like that.

METHODS

1. Prepare a stock-solution of R6G in ethanol or water. Prepare PVA solution and PVA films as described in Experiment 3-24. Typically, we would recommend preparing a few films with varying dye (R6G) concentrations. Drying films will take a couple of days.

(*Continued*)

EXPERIMENT 8-5 (CONTINUED) PRECAUTIONS IN FLUORESCENCE ANISOTROPY MEASUREMENTS: DEPOLARIZATION THROUGH REABSORPTION AND INTERNAL REFLECTIONS

To be able to correct for PVA film absorbance and reflection on-air/PVA interfaces, we always recommend preparing a couple of blank PVA films to make sure that dye-doped films and blank films are identical.

2. Measure absorptions of prepared films and select films with absorption between 0.01–0.05 per 100 μm thickness (measured absorption divided by thickness in microns times 100) so, the total absorption of the film below 0.1. Remember to properly correct for PVA film absorbance (see Experiment 3-25).
3. We will use a square configuration/geometry and a standard 1 cm × 1 cm fluorescence cuvette. You can use a standard cuvette holder as is provided with your system. To better present potential problems, we will use a custom assembled holder that allows mounting cuvette at the bottom making it completely visible (see Figure 8-5.3). Cut a strip of PVA film about 1.4 cm wide and 3–4 cm long. Insert the strip of PVA film diagonally into 10 mm × 10 mm square cuvette mounted on the stage (see Figure 8-5.3). The surface of the film forms about 45° as compare to excitation and emission lines. *Note that film is oriented in a way to reflect excitation beam out of the detector. Orienting the film at a complementary 45° angle will reflect excitation toward the detector. We definitely want to avoid that!*
4. Measure the emission spectrum using vertical excitation and magic angle observation. *Note: When measuring solid samples with no rotational diffusion we do not have to use magic angle observation. However, some limited mobility can be present even in a solid matrix. In addition, 54.7° angle on observation significantly reduces highly polarized scattering. Therefore, we recommend MA conditions for emission spectra measurements jus as a precaution.*
5. To measure polarization, we will use vertical excitation and measure emission with vertical and horizontal observations.
6. *Measure G-factor*: To measure G-factor, we used the film and 45° oriented polarizer on the observation. We can also use the solution in a cuvette and horizontal excitation as typically recommended for square geometry. *Note: We do NOT recommend to measure G-factor using horizontal excitation with the film for two reasons: (1) The excitation light when entering the film will change direction of propagation (refracts) and changes the direction for horizontal light polarization; (2) if the film is not perfectly centrally mounted, the G-factor can be slightly different.*
7. Correct emission and calculate anisotropies.
   a. Use the same 10 mm × 10 mm cuvette but now fill it with mineral oil. Figure 8-5.3 (right). The refractive index of oil is close to glass and PVA film. Repeat #4 and #5. *Note: The oil will not dissolve PVA.* It is important to use a solvent that will not interact with the film. For example, water will dissolve the film quickly.
8. Measure G-factor as in #6. *Note: This time if the refractive index is well matched we can also measure G-factor using horizontal excitation.*

(Continued)

EXPERIMENT 8-5 (CONTINUED) PRECAUTIONS IN FLUORESCENCE ANISOTROPY MEASUREMENTS: DEPOLARIZATION THROUGH REABSORPTION AND INTERNAL REFLECTIONS

RESULTS

Figure 8-5.3 shows mounted cuvette without oil (left) and filled with oil (right). Note that without the mineral oil reflections of the excitation light can be clearly visible. Also, the film is very clearly visible. Filling the cuvette with oil (right), we notice that film practically disappeared and you cannot clearly distinguish film edges. In addition, the excitation spot looks much dimmer and shows proper yellowish color (emission) since we eliminated almost completely light reflections.

FIGURE 8-5.3 Photograph of the cuvette with the PVA film in standard 1 cm × 1 cm cuvette mounted diagonally. Left: Empty cuvette. Right: Cuvette filled with mineral oil.

In Figure 8-5.4 are presented normalized emission spectra and measured emission anisotropies for R6G in PVA film using square geometry configuration. The black solid line represents the film in an empty cuvette and dashed orange line cuvette filled with oil. The emission spectrum measured with oil shows maximum about 570 nm that well corresponds to the expected emission maximum for R6G in PVA film. However, the emission spectrum measured in empty cuvette (black solid line) is distinctly shifted toward the red with maximum much over 565 nm. Such a distinct shift suggests that we have significant reabsorption of emission. This is a clear indication that significant part of emitted light travels through a significant distance in the film. This is due to multiple reflections of excitation and emission light inside the film before exiting the film. In fact, a PVA film works partially as a waveguide. This makes the emission light path much longer than expected from the film thickness making reabsorption a significant problem.

(*Continued*)

EXPERIMENT 8-5 (CONTINUED) PRECAUTIONS IN FLUORESCENCE ANISOTROPY MEASUREMENTS: DEPOLARIZATION THROUGH REABSORPTION AND INTERNAL REFLECTIONS

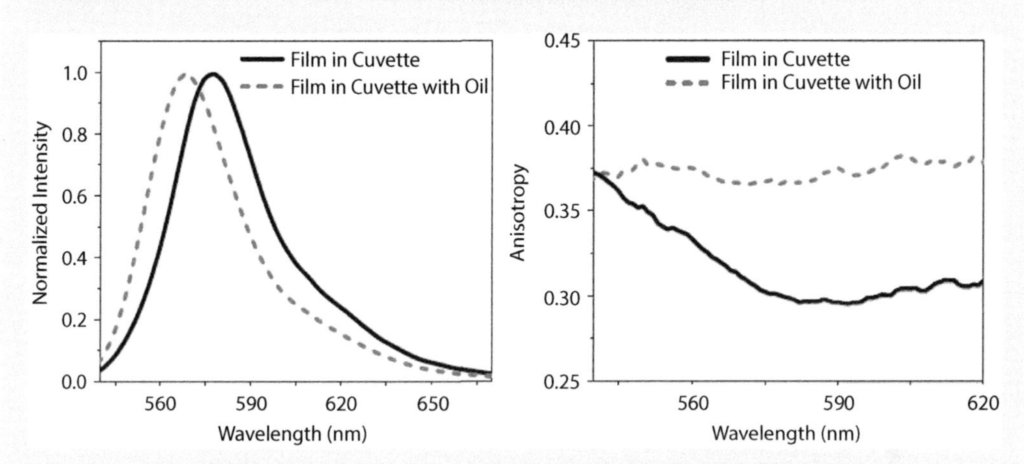

FIGURE 8-5.4 Emission spectra (left) and emission anisotropies of R6G in PVA film (right) measured in a square geometry using standard 1 cm × 1 cm cuvette. Black line represents the film in empty cuvette (black) and dashed orange line cuvette filled with mineral oil.

In Figure 8-5.4 (right) are shown measured emission anisotropies. The anisotropy measured for the film in a cuvette filled with oil pretty well correspond to anisotropy expected from immobilized molecules of R6G. Contrary, the anisotropy measured with empty cuvette is completely different. The emission anisotropy at short wavelength starts with values above 0.36 (anisotropy expected for the film) and quickly drops with observation wavelength to about 0.3. Such red edge depolarization confirms reabsorption and reemission processes in the film.

This results clearly demonstrate that we cannot use film mounted in an empty 1 cm × 1 cm fluorescence cuvette for measuring emission spectra or polarizations. The error can be significant. However, when filling the cuvette with appropriate index matching solution the measured spectra and polarizations are correct.

Square Geometry Measurements in 1 mm × 10 mm Fluorescence Cuvette. Similar to the previous example with standard fluorescence cuvette frequently experimentalists may try to use more specialized 1 mm × 10 mm fluorescence cuvette with standard outside dimensions. Such cuvette can be easily used with a standard sample holder, see Figure 8-5.5. This could be attractive since measurement will not require any modifications to the system/sample compartment and at first glance, the conditions for proper square geometry should apply (satisfied conditions for magic angle or G-factor measurement). In addition, as seen in Figure 8-5.1, films edges show bright fluorescence (glowing), typically much greater than fluorescence from the film surface. This can make measurements of weak fluorescence signal easier.

(Continued)

EXPERIMENT 8-5 (CONTINUED) PRECAUTIONS IN FLUORESCENCE ANISOTROPY MEASUREMENTS: DEPOLARIZATION THROUGH REABSORPTION AND INTERNAL REFLECTIONS

METHODS

9. Use a square geometry configuration as in previous example (just your standard cuvette holder). To demonstrate potential problems, we will use as previously custom assembled holder with cuvette mount in such a way that allows moving cuvette (with PVA film) orthogonal to the excitation line (along the observation line). Insert the strip of PVA film into 1 mm × 10 mm square cuvette (12 mm × 12 mm external dimensions) and mount on the stage (see Figure 8-5.5). The surface of the film is orthogonal to the excitation and emission is observed from the edge. We will excite the film in the center (about 3–5 mm from the edge) and close to the edge on the side of observation. For these measurements, we will use film with low R6G concentration (~0.02 absorbance per 100 μm).
10. Use vertical excitation and measure emission spectra using the magic angle observation (54.7°).
11. Use vertical excitation and measure vertical and horizontal emission components as observed from the film edge.
12. *Measure G-factor.* To measure G-factor, we can use R6G solution in a 1 mm cuvette and horizontal excitation or just use the R6G film and a 45° oriented polarizer on the observation (see Experiment 6-7). *Note: Even if we have perfect square geometry we do NOT recommend to use the R6G film emission excited with horizontal polarization (approach typically recommended for square geometry). As we will discuss in the conclusions multiple reflections in the film polarize/depolarize light and may lead to the wrong result. However, the 45° oriented polarizer will give proper G-factor.*
13. Correct emission spectra and calculate anisotropies.
14. Use the same 1 mm × 10 mm cuvette but now fill it with mineral oil. The refractive index of oil is close to glass and PVA film. Repeat # 11, #12, and # 13.

In Figure 8-5.5, we are presenting square geometry with a 1 mm × 10 mm fluorescence cuvette with standard external dimensions 12 mm × 12 mm. The view of the cuvette setup with the R6G containing PVA film inside is shown in Figure 8-5.5 (top). Such cuvettes are frequently used to measure fluorescence spectra and emission polarization of liquid samples when only limited volumes of the sample are available. We are not using standard cuvette holder but a home built adapter to make possible to move/displace the cuvette along the observation line (orthogonal to the excitation line) to enable excitation at the film edge (from the observation side (Figure 8-5.5, bottom right)) or in the center (in fact any position within the film width (Figure 8-5.5, bottom left)). It is important to move cuvette orthogonal to the excitation line since this will not change the position of the emission spot in relation to the observation. But will only change the distance light is traveling in the film before escaping at the edge (moving the excitation beam will change the emission spot in relation to the observation lens). It appears very convenient to use such 1 mm or even 0.5 mm cuvette to be able to measure emission from thin polymer films. When using just standard cuvette holder and 0.5 mm (or 1 mm) thin fluorescence cuvette, we do not need to build any film holder or arrange front-face excitation configuration. In addition, the direction of excitation and observation lines are oriented exactly at 90° and measuring a

(*Continued*)

EXPERIMENT 8-5 (CONTINUED) PRECAUTIONS IN FLUORESCENCE ANISOTROPY MEASUREMENTS: DEPOLARIZATION THROUGH REABSORPTION AND INTERNAL REFLECTIONS

G-factor appears to be straight forward. Also, as seen from the photograph in Figure 8-5.1, we can see significantly stronger emission from the film edge. However, we do NOT recommend doing such measurements.

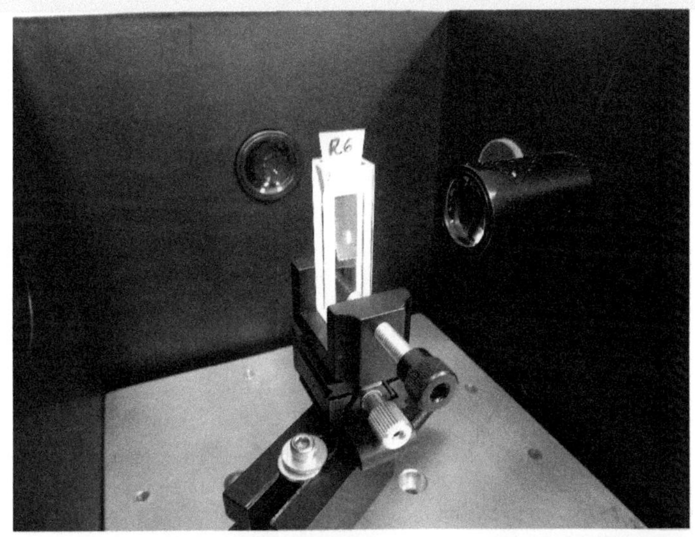

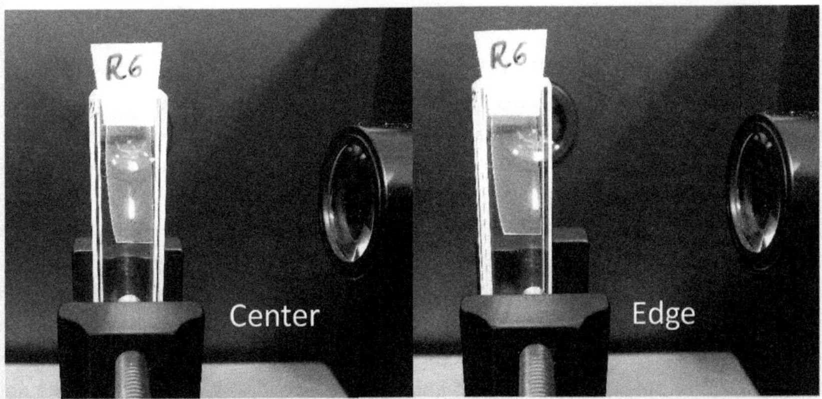

FIGURE 8-5.5 Schematic representation of emission from the thin layer film when observed from the edge.

In Figure 8-5.6, we present normalized emission spectra of the R6G film measured from the edge and center of the film. For comparison, we also included the emission spectrum as measured in the square geometry in 1 cm × 1 cm cuvette with oil (Figure 8-5.4, left: orange dashed line). The emission spectra measured from the film edge do not significantly depend on the presence/absence of the oil in the cuvette.

(*Continued*)

EXPERIMENT 8-5 (CONTINUED) PRECAUTIONS IN FLUORESCENCE ANISOTROPY MEASUREMENTS: DEPOLARIZATION THROUGH REABSORPTION AND INTERNAL REFLECTIONS

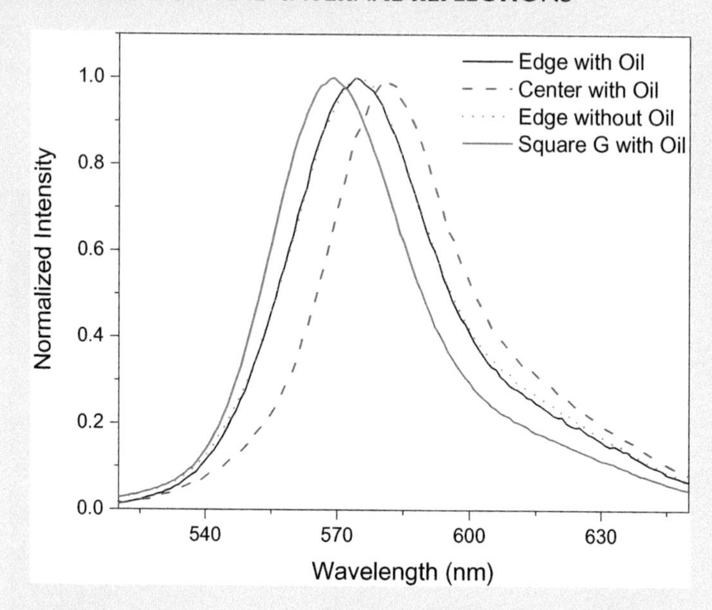

FIGURE 8-5.6 Emission spectra of R6G-doped PVA film measured from the film center or edge in an empty cuvette and in cuvette filled with mineral oil. For comparison, we also present the emission spectrum of this film measured in a front-face configuration.

The emission observed from the edge when excited at the edge of the film is shifted toward the red (orange dotted line) and the emission excited in the center is shifted even more (blue dashed line). This is a very significant spectral shift even for excitation on the edge. However, looking on the bottom-right photograph in Figure 8-5.5, one can clearly see that the center of emission spot is over a millimeter from the edge. *Note: When using the lamp excitation, it is very difficult to collimate the beam to a spot smaller than a millimeter.* Also, to avoid scattering on the edge you want to be a little shifted from the edge. Therefore, even for the excitation at the edge, the emission light will travel through 1–2 mm of the film increasing the apparent absorption about 10-fold as compared to the measured film absorption (Figure 8-5.2). Obviously, when exciting in the center, the light traveling toward the detector has to pass through about 5 mm of the film and the reabsorption will be very high, yielding a very significant shift of the emission spectrum toward the red. *Note: We do not recommend measurements from the film edge.*

Front-face Configuration. In Figure 8-5.7, we present a proper front-face configuration that can be simply realized with practically any sample compartment. On the left, we are presenting a case where we use a proper sample holder that holds polymer film perfectly flat (slightly tensioned) and orthogonal to the direction of the observation. *This is a proper way of measuring thin samples* (polymer films, sample deposited on a glass slide, etc.). Frequently an experimentalist will not have easy access to such a sample holder and can

(*Continued*)

EXPERIMENT 8-5 (CONTINUED) PRECAUTIONS IN FLUORESCENCE ANISOTROPY MEASUREMENTS: DEPOLARIZATION THROUGH REABSORPTION AND INTERNAL REFLECTIONS

use a simpler approach using thin 1 mm cuvette in place of the film holder as shown in Figure 8-5.7 (right). The PVA film containing R6G is inserted in the 1 mm cuvette. The greenish emission spot can be seen in the film. It is important to align the position of the spot exactly at the focal point of the observation lens (*Note: We do not use focusing lens typically placed in the excitation light entry. Instead, the first mirror is a focusing cylindrical mirror to ensure a small well-collimated spot on the film*). This spot should be very close to (exactly at) the position of emission spot in the square geometry (center of the square-geometry cuvette holder). For the discussion on how the position of an emitting spot affects fluorescence signal, visit Experiment 4-16. This position should correspond to the maximum of the signal that can be observed from the film. The first reflective mirror could be equipped with the micro-positioner screws to adjust position of the spot. To measure emission spectra, we should use magic angle conditions (vertical excitation and 54.7° observation). To measure polarization, we use vertical excitation and vertical and horizontal observations. To measure G-factor, we can use zero-anisotropy standard or 45° polarizer on observation (for details see Experiments 6-1 and 6-2).

METHODS

15. We use the front-face configuration to measure fluorescence emission and emission anisotropies of R6G. The example of the front-face configuration used by us in this experiment is presented in Figure 8-5.7.
    a. First, we will use a proper film holder as discussed in previous chapters. *Note: Remember in such a set-up it is crucial to properly align the excitation spot in reference to the observation line. The detection system is very sensitive to the position of the emission spot (see Experiments 4-4 and 4-5).*
    b. Next, we will use simple modification with a thin 1 mm × 10 mm cuvette to hold the film. This set-up is easier to assemble. To measure emission spectra, remember to use magic angle conditions (for front-face configuration as shown in Figure 8-5.7 (vertical excitation) and 54.7 (observation)). To measure anisotropy, we use vertical excitation and measure VV and VH emission components. We will repeat the measurement with the film that contains low R6G concentration (~0.01 absorbance per 100 μm) and film containing higher concentration of R6G (>0.05 absorbance per 100 μm) *Note: It is important to make sure that film surface from which fluorescence is observed is not folded and well flat, and orthogonal to the direction of the emission observation.*
    c. Next, fill the cuvette containing film with mineral oil. Make sure there are no air bubbles. Repeat measurement of emission anisotropy in the front-face configuration. The refractive index of oil is close to glass and PVA film what will limit unwanted reflections at the film surface (between film surfaces) and between film surfaces and cuvette walls. *Note: The oil will not dissolve PVA.* It is important to use a solvent that will not interact with the film. For example, water will dissolve film quickly.

(Continued)

EXPERIMENT 8-5 (CONTINUED) PRECAUTIONS IN FLUORESCENCE ANISOTROPY MEASUREMENTS: DEPOLARIZATION THROUGH REABSORPTION AND INTERNAL REFLECTIONS

16. Measure G-factor. To measure G-factor, use a solution of long-lived dye in low viscosity solvent in the same (identical) cuvette, or use 45° polarizer on observation and use the film directly. For evaluating G-factor, see Experiments 6-1 and 6-2. *Note: In the front-face configuration, we* ***cannot*** *use horizontal excitation as it can be done in square geometry.*
17. Correct emission intensities and calculate measured anisotropies as measured for the film in film holder, cuvette, and cuvette filled with oil.

In Figure 8-5.7, we present a photograph of front-face configurations which use proper sample holder (left) and 1 mm cuvette (right).

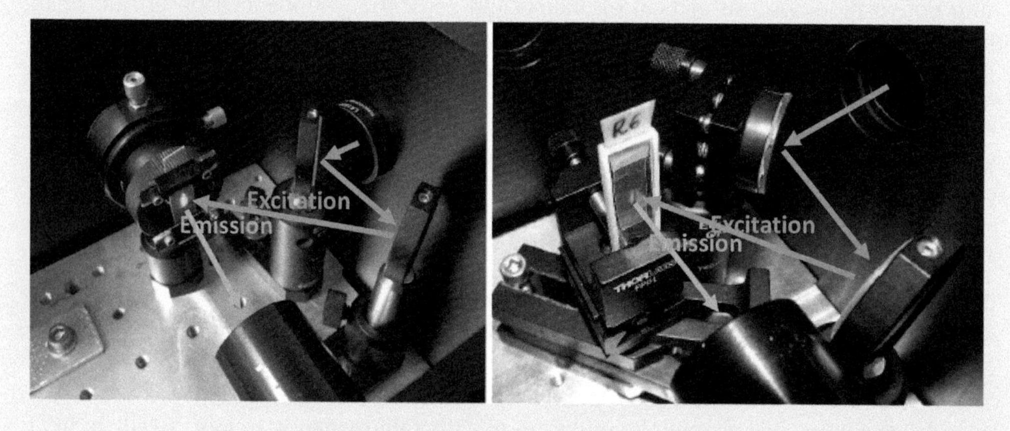

FIGURE 8-5.7 Front-face configuration for measuring emission of the film. Left: Special adapter for holding film well straighten. Right: Film inserted to the empty 1 mm × 10 mm cuvette.

Figure 8-5.8 (left) shows measured emission spectra for film mounted in holder, cuvette, and cuvette filled with oil. *Note: Not shown, but when filling cuvette with oil film is not visible and only fluorescence spot can be seen (see Figure 8-5.3 (right)).* All three spectra are practically identical showing emission maximum about 565 nm. Figure 8-5.8 (right) shows measured emission anisotropies for all three cases. The anisotropies for just film in the cuvette is clearly lower than in the case of the film in the holder and cuvette filled with oil. Both anisotropies measured with film holder and in the cuvette filled with oil are close and about 0.37.

(Continued)

EXPERIMENT 8-5 (CONTINUED) PRECAUTIONS IN FLUORESCENCE ANISOTROPY MEASUREMENTS: DEPOLARIZATION THROUGH REABSORPTION AND INTERNAL REFLECTIONS

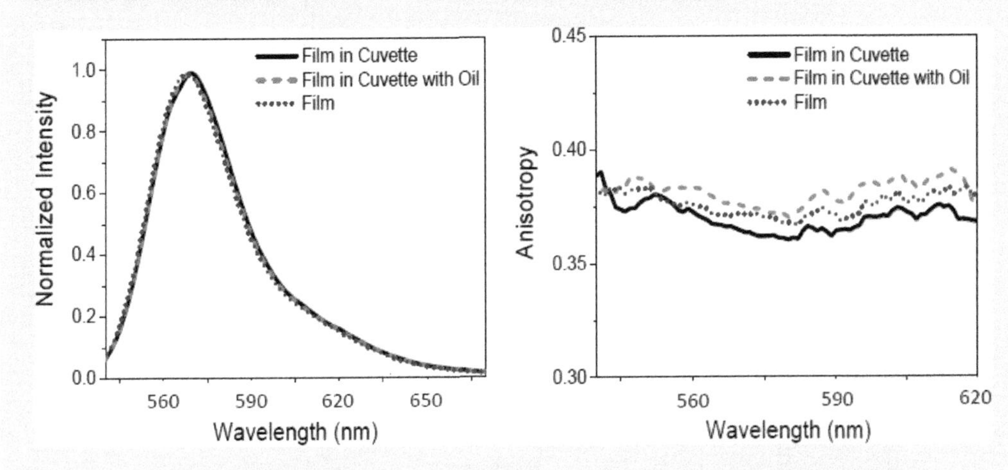

FIGURE 8-5.8 Emission spectra (left) and measured emission anisotropies of R6G in PVA film measured in front-face configuration using film holder (blue), the film in empty 1 mm × 10 mm cuvette (black), and film in the cuvette filled with the mineral oil.

In Figure 8-5.9 are compared anisotropies as measured in all configurations. But for the front-face measurement, we also use film with higher R6G concentration (absorbance about 0.03 in a 70 µm thick film) inserted into the 1 mm × 10 mm cuvette. The measured anisotropy (orange stars) is high at the blue edge (about 0.38) that is dropping to about 0.33 at the red edge of the spectrum. Such drop of anisotropy at red edge of the spectrum is not expected for R6G and is due to reabsorption and reemission (we have very thin film but due to high refractive index of PVA emitted light can be significantly reflected at PVA/air interface that will significantly affect apparent path length for a small part of observed emitted light). Also, we may have very limited excitation energy migration lowering the anisotropy at the red emission edge. To test this, we next used a lower concentration of R6G in the film (absorbance about 0.02 in 110 µm thick film—about one-third of high concentration). Measured anisotropy shown as green squares starts at about 0.38 and drops much less at read spectral edge to about 0.35. This anisotropy drop is more reasonable. However, to eliminate potential reflections (both the excitation and emission light), we filled cuvette with mineral oil. *Note: A good matching of refractive indexes is demonstrated by almost complete disappearance of film edges.* Emission anisotropy observed from the film with low concentration (LC) filled with mineral oil is represented by green filled triangles. Anisotropy starts from 0.38 and only drops to 0.37 showing only small lowering anisotropy toward the longer wavelength. This is a very large anisotropy and observed anisotropy drop is minimal (expected at the red emission range).

(Continued)

EXPERIMENT 8-5 (CONTINUED) PRECAUTIONS IN FLUORESCENCE ANISOTROPY MEASUREMENTS: DEPOLARIZATION THROUGH REABSORPTION AND INTERNAL REFLECTIONS

Measured anisotropies from the film edge are more problematic (shown as open triangle and square). The red-edge drop is much more pronounce and in the case without oil measured anisotropy stars from 0.45 that is an impossible value for an isotropic system (maximum anisotropy is 0.4). The starting anisotropy may vary but frequently it will be between 0.4 and 0.5 what is very problematic (impossible for an isotropic system).

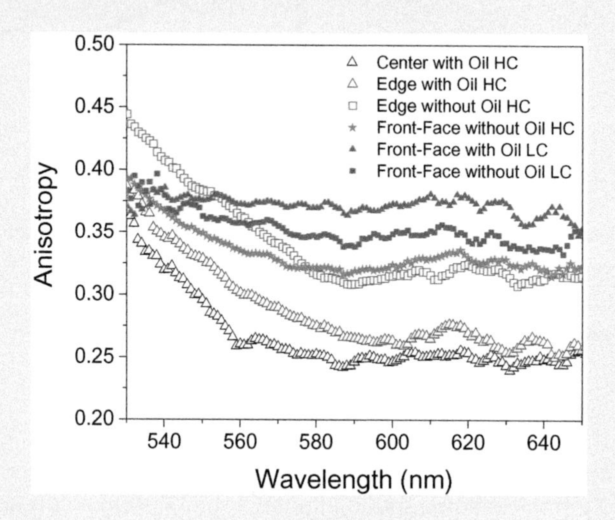

FIGURE 8-5.9 Measured emission anisotropies of R6G in the PVA film in front-face configuration and from the film edge in empty cuvette and cuvette filled with mineral oil.

Why measured values of anisotropy are exceeding the theoretical value? Obviously, this is impossible and it is a result of a mistake or some artifact. First, let's consider edge emission of a dye embedded to a thin film as schematically shown in Figure 8-5.10. We are only considering films that are over a micron thick (typically up to few hundred microns), a thickness that is much larger than the wavelength of light and we can ignore any near-field effects erasing from dipoles emitting at the interface. The emitting point radiates/emits uniformly in all directions. The emitted light will reach the surface and can be partially reflected back and partially refracted and escape polymer/dielectric. The refractive index of the film is about 1.5, much larger than refractive index of surrounding air. In such interface of two dielectrics (polymer and air), the light transmittance and reflection will strongly depend on the incident angle. For angles greater than critical angle for total internal reflection (above 60° for PVA/air interface), the light will not escape the polymer and will be going through multiple reflections as it reaches the film edge (similar to the light-guiding in the fiber optics). In addition, for angles smaller than critical angle the reflection will strongly depend on light polarization yielding to artificial light polarization/depolarization by reflection. There are two important observations: (1) light that is going toward the direction of the detector travels in the polymer film a distance much greater than film thickness, and (2) significant portion of observed fluorescence has gone through multiple reflections and refraction that change original light polarization.

(Continued)

EXPERIMENT 8-5 (CONTINUED) PRECAUTIONS IN FLUORESCENCE ANISOTROPY MEASUREMENTS: DEPOLARIZATION THROUGH REABSORPTION AND INTERNAL REFLECTIONS

So, measured absorbance below 0.1 in a 100 μm thick film should principally be correct for emission measurement but when observed from the edge the light may easily travel in the polymer through a distance of a few millimeters yielding an apparent absorption much over 1. This, for dyes that exhibit significant spectral overlap, will result in large apparent emission shifted toward the red spectral edge (the blue edge will be reabsorbed). Also multiple reabsorptions and reemission processes will strongly depolarize the observed emission light. This depolarization effect will be strongly manifested at the red spectral edge.

As discussed earlier, the process that may increase apparent emission polarization is light reflected at the PVA/air interference. As schematically shown in Figure 8-5.10, a significant portion of the light that goes toward the detector is after reflections at the PVA/air interface. Such a reflection, when are not total internal reflections, will strongly affect light polarization. An s-polarized light component will be reflected more (with better efficiency) on interface than a p-polarized component. This will be manifested by much higher apparent polarization. A simple way to limit this is to fill cuvette with the mineral oil to match refractive index of the film.

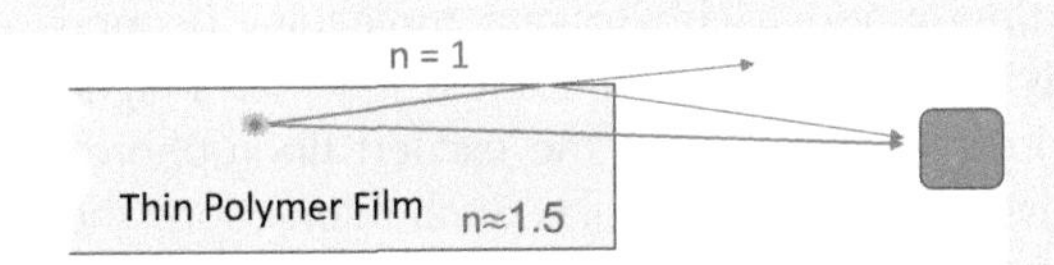

FIGURE 8-5.10 Top: The view of the cuvette with the PVA film mounted in the spectrofluorometer chamber. Bottom shows two positions for excitation; center (left) and edge (right).

CONCLUSIONS

There are many important conclusions from these experiments. First, using square geometry, we have to be aware of sample geometry. As shown for the film without oil, we can easily measure very wrong anisotropy. Even if the conditions are proper for G-factor evaluation (square geometry), we still can make a significant error. *Note: Doing a G-factor directly from the film when using horizontal light polarization for excitation does not correct for polarization problems. In fact, the G-factor can be erratic leading sometimes too much greater discrepancies.*

High anisotropy values in the PVA film are expected. This is a rigid matrix strongly limiting any mobility of the chromophore. From our experience among many polymer matrixes and various glasses, the PVA is the most effective in immobilizing chromophores. *Note: It may sound surprising since PVA polymer at room temperature is fairly elastic and at about 70°C becomes very flexible.* But its internal structure with very long polymer chains and a limited amount of crystallites make this microscopically a very dense matrix. In effect, measured anisotropies in the PVA films are very high even in higher temperatures (up to 60°C).

(Continued)

EXPERIMENT 8-5 (CONTINUED) PRECAUTIONS IN FLUORESCENCE ANISOTROPY MEASUREMENTS: DEPOLARIZATION THROUGH REABSORPTION AND INTERNAL REFLECTIONS

It is clear that reabsorption and reemission lead to two effects. The apparent shift of emission spectrum toward the red and significant anisotropy lowering at a longer wavelength (red emission edge). Importantly, it may appear very convenient to measure emission and anisotropies in the film using 1 mm (0.5 mm) cuvettes or just sandwich the film between two slides but this may lead to different problems that are very difficult to control. Also measurements from the film edge are easy to do in square configuration and offer more light but multiple reflections in the thin film are very likely to corrupt such measurements.

EXPERIMENT 8-6 DEPENDENCE OF ANISOTROPY DECAYS ON THE SOLVENT POLARITY

Keywords: fluorescence, fluorometer, lifetime, anisotropy.

The interaction of fluorophores with solvent molecules depends on the polarity of the solvent used. Usually, the molecule in the excited state has a higher electric dipole moment. The dipole–dipole interaction between the excited fluorophore and solvent molecules is stronger in polar solvents than in non-polar. Such interactions should slow down Brownian motions and can potentially affect the fluorophore's correlation times. Are these differences detectable? We will measure anisotropy decays of a polar molecule Coumarin 153 in polar and non-polar solvents with similar viscosities.

MATERIALS

- Coumarin 153 (C153)
- Methanol (MeOH), benzene (B), cyclohexane (C), water (W), ethanol (EtOH)
- Glass or quartz 1 cm × 1 cm cuvette, pipettes, neutral density (ND) filters

EQUIPMENT

- Spectrofluorometer (Varian, Inc., Eclipse)
- Cary 50 Spectrophotometer (Varian, Inc.)
- Fluorometer (FT 200, PicoQuant, GmbH, Germany)

METHODS

1. Make solutions of C153 in MeOH, benzene, cyclohexane, and water using a stock solution of C153 in EtOH. Adjust concentrations for absorptions below 0.1. Measure absorption spectra (Figures 8-6.1 and 8-6.2).
2. Measure fluorescence spectra with excitations 405 nm (Figures 8-6.3 and 8-6.4).
3. Measure lifetimes and anisotropy decay of C153 in all four solvents (Figures 8-6.5 through 8-6.10).

(Continued)

EXPERIMENT 8-6 (CONTINUED) DEPENDENCE OF ANISOTROPY DECAYS ON THE SOLVENT POLARITY

RESULTS

Both methanol and benzene have almost identical viscosities, as well as water and cyclohexane 0.6 and 1.0 cP, respectively. A polar C153 molecule interacts stronger with polar solvents molecules (methanol, water) than with non-polar (benzene, cyclohexane). The result of stronger dipole–dipole interactions with solvent molecules is a bathochromic (red) shift of the absorption and emission spectra. The even stronger shift in the emission indicates a larger electric dipole moment of C153 in the excited state.

Figures 8-6.1 and 8-6.2 show absorption spectra of C153 in various solvents.

Measure fluorescence spectra of C153 in all four solvents, see Figures 8-6.3 and 8-6.4.

Measure lifetimes of C153 in all four solvents, see Figures 8-6.5 through 8-6.8.

Next, measure anisotropy decays of C153 in all four solvents, see Figures 8-6.9 through 8-6.12.

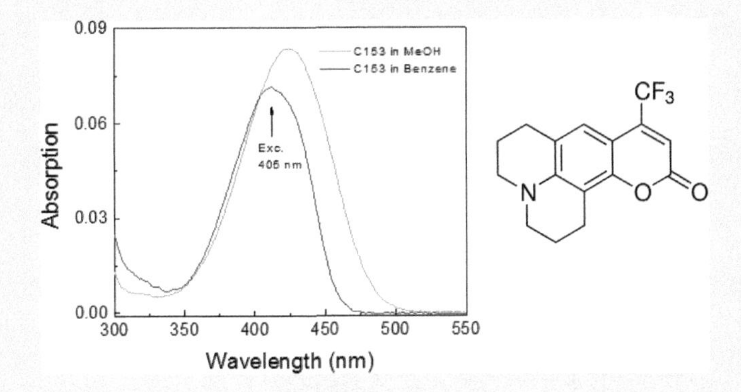

FIGURE 8-6.1 Absorption spectra of C153 in MeOH and benzene. Chemical structure of C153 is on the right.

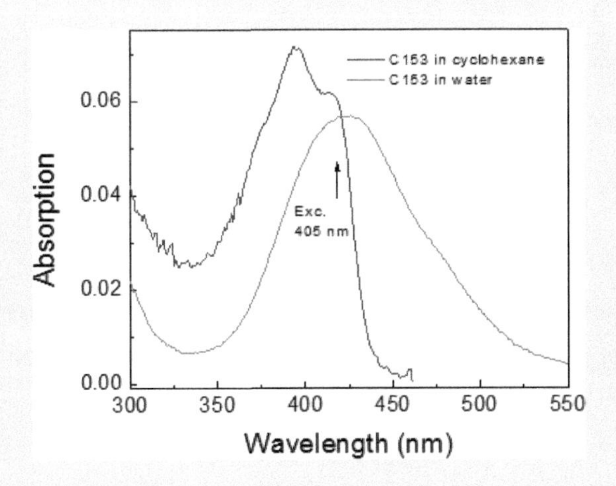

FIGURE 8-6.2 Absorption spectra of C153 in water and cyclohexane.

(Continued)

EXPERIMENT 8-6 (CONTINUED) DEPENDENCE OF ANISOTROPY DECAYS ON THE SOLVENT POLARITY

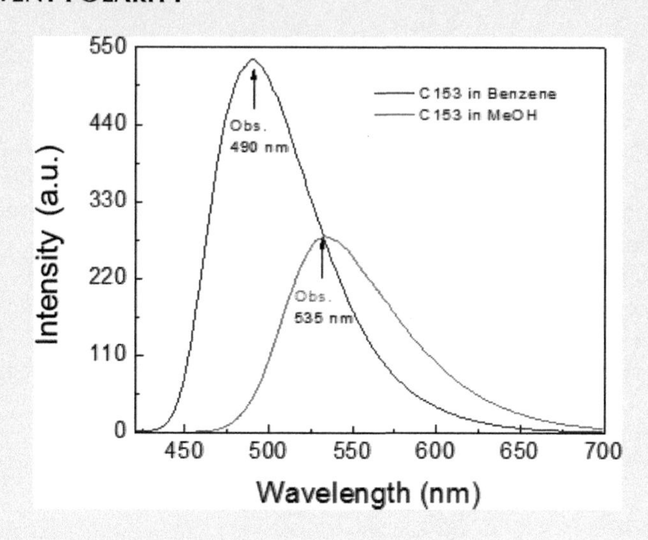

FIGURE 8-6.3 Fluorescence spectra of C153 in MeOH and benzene. Arrows indicate observation wavelengths in time-resolved measurements.

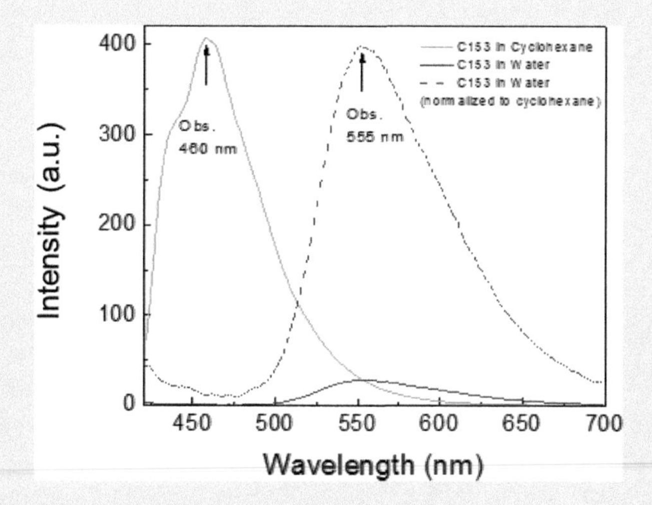

FIGURE 8-6.4 Fluorescence spectra of C153 in water and cyclohexane. Arrows indicate observation wavelengths in time-resolved measurements.

(Continued)

EXPERIMENT 8-6 (CONTINUED) DEPENDENCE OF ANISOTROPY DECAYS ON THE SOLVENT POLARITY

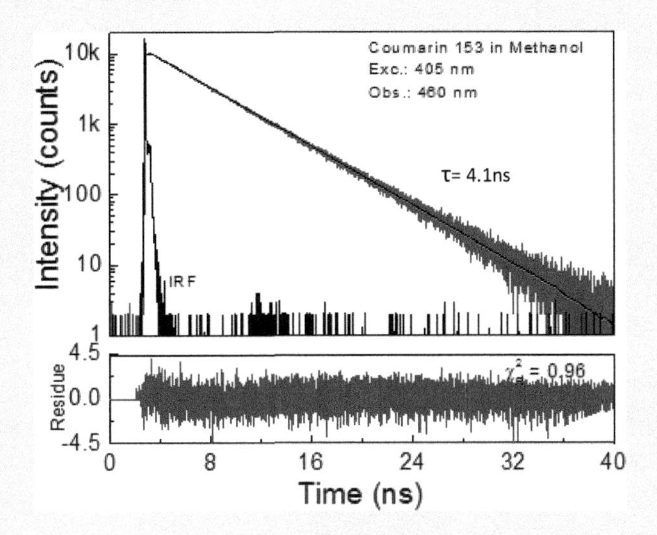

FIGURE 8-6.5 Fluorescence intensity decay of C153 in MeOH.

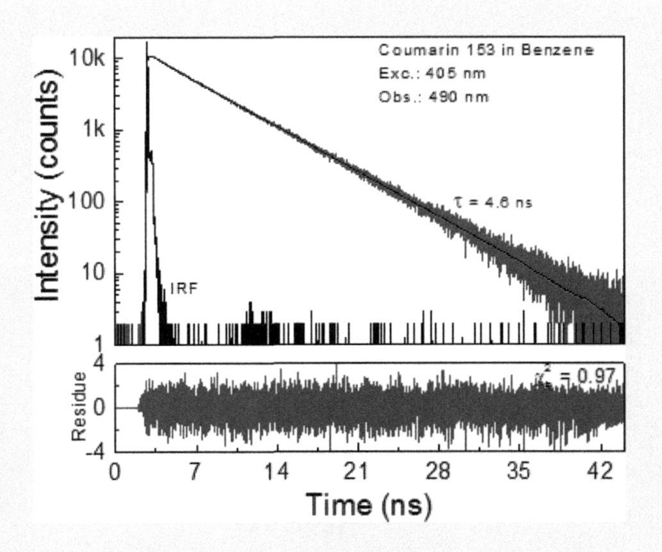

FIGURE 8-6.6 Fluorescence intensity decay of C153 in benzene.

(*Continued*)

EXPERIMENT 8-6 (CONTINUED) DEPENDENCE OF ANISOTROPY DECAYS ON THE SOLVENT POLARITY

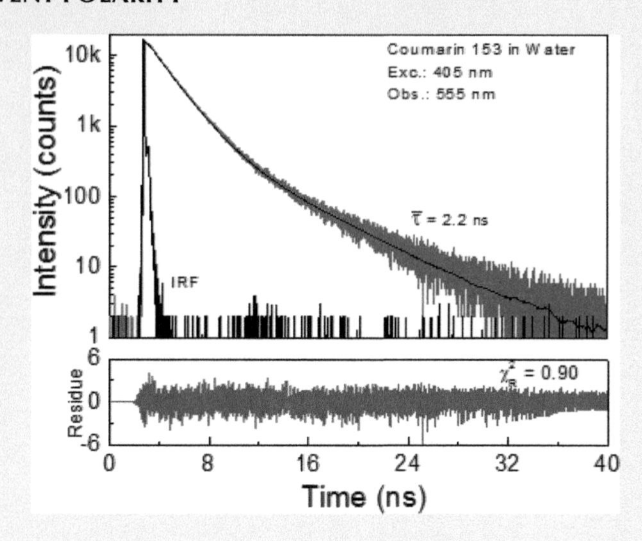

FIGURE 8-6.7 Fluorescence intensity decay of C153 in water.

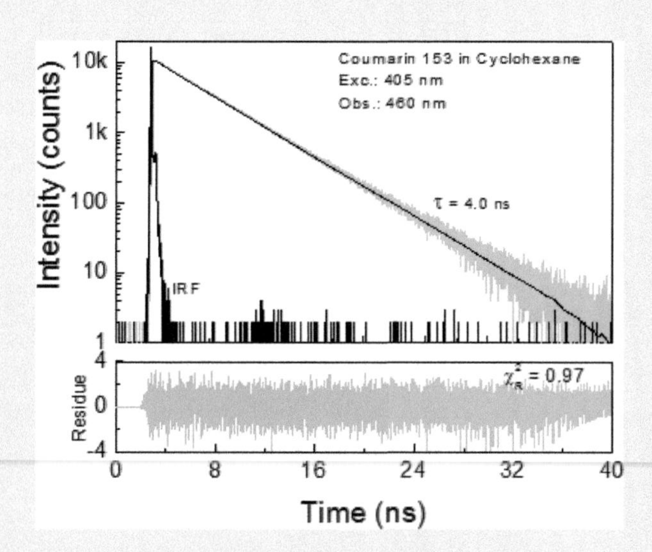

FIGURE 8-6.8 Fluorescence intensity decay of C153 in cyclohexane.

(Continued)

EXPERIMENT 8-6 (CONTINUED) DEPENDENCE OF ANISOTROPY DECAYS ON THE SOLVENT POLARITY

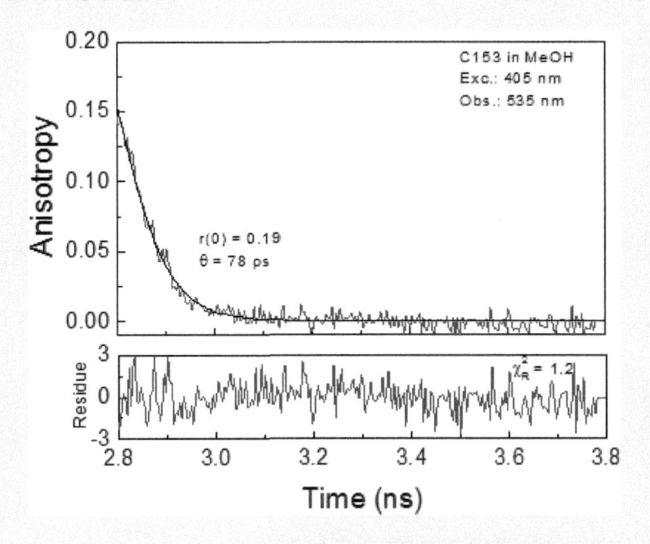

FIGURE 8-6.9 Fluorescence anisotropy decay of C153 in MeOH.

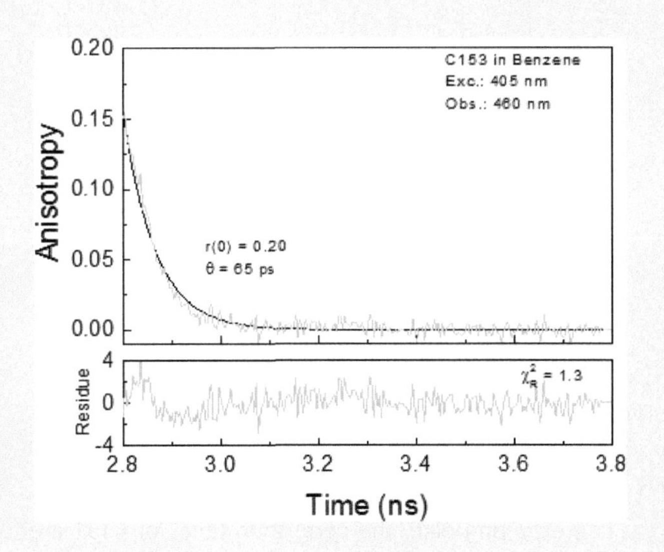

FIGURE 8-6.10 Fluorescence anisotropy decay of C153 in benzene.

(Continued)

EXPERIMENT 8-6 (CONTINUED) DEPENDENCE OF ANISOTROPY DECAYS ON THE SOLVENT POLARITY

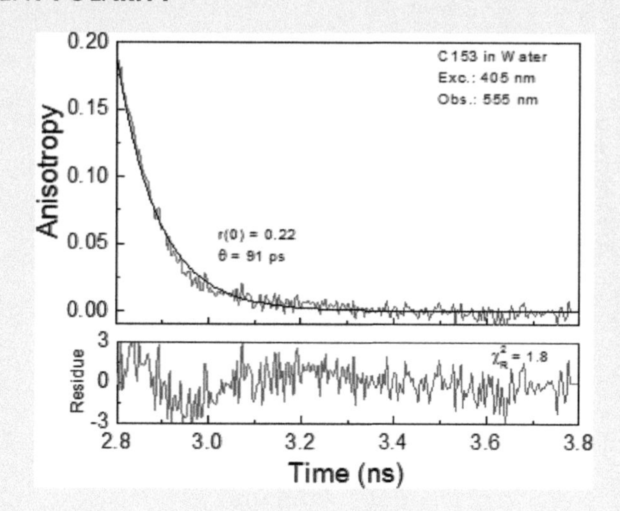

FIGURE 8-6.11 Fluorescence anisotropy decay of C153 in water.

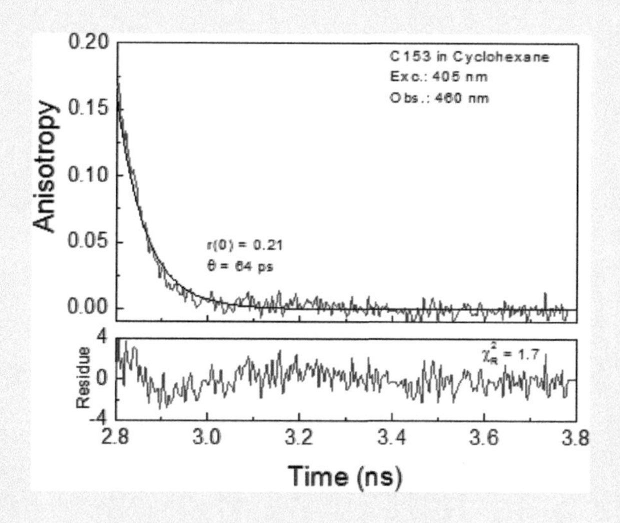

FIGURE 8-6.12 Fluorescence anisotropy decay of C153 in cyclohexane.

CONCLUSIONS

In both solvent pairs (polar/non-polar), the correlation times of C153 are longer for polar solvent (water, methanol) than for non-polar (cyclohexane, benzene). Viscosities of solvents in each pair were similar and one could expect similar values of correlation times. However, C153 is a polar molecule (see large spectral shifts in Figures 8-6.1 through 8-6.4) and electric dipole–dipole interactions with solvent molecules depend on the solvents polarities. These interactions are responsible for the slow-down rotations of C153 molecules. It can be said that in polar solvents, polar fluorophores rotate in "stick" conditions whereas in non-polar solvents in "slip" conditions.

EXPERIMENT 8-7 ENERGY MIGRATION, SELF-QUENCHING, AND DEPOLARIZATION OF FLUORESCENCE

Keywords: fluorescence, self-quenching, energy migration.

At low concentrations of fluorophores, where distances between dye molecules are long, fluorescence emission is not perturbed and is independent on variations of the dye concentration. However, at higher dye concentrations, the separations between fluorophores become shorter and intermolecular interactions becomes more probable. The FRET occurs through space over the distance that depends on the characteristic Forster distance (R_0, see Chapter 1). In the case of the same dye (one component solution), FRET is called either *homo-FRET* or *energy migration*. In solutions, molecular diffusion during the lifetime of the excited state adds to a higher probability for molecular interactions. Also, high concentrations of the dye can result in aggregates, dimers, or higher-order ground state complexes. They will be seen in the absorption spectrum as new bands or a broadening of the monomer spectrum. In effect, at high concentrations, the fluorescence emission becomes weaker, a phenomenon called *self-quenching*. The mechanism of the self-quenching, although the phenomenon is known for a hundred years, is not fully explained. It is commonly accepted that energy migration may result in the excitation of dimers or aggregates that dissipate excitation energy in a non-radiative way and are non- or weakly fluorescent. In general, energy migration means that other than originally excited fluorophores can emit fluorescence. Since fluorophores are randomly oriented this will result in depolarization that will lower measured fluorescence anisotropy, a phenomenon called *concentration depolarization* of fluorescence. In this experiment, we will demonstrate mentioned above phenomena. To limit diffusion and rotational depolarization, we will use a solid PVA film/matrix in which fluorophores are well immobilized.

MATERIALS

- Rhodamine 6G (R6G)
- Poly (vinyl alcohol), MW: 130,000, Aldrich
- Ethanol, water
- Glass vials
- Microscope slides, pipettes

EQUIPMENT

- Spectrophotometer (Cary 50, Varian, Inc.)
- Spectrofluorometer (Eclipse, Varian, Inc.)
- Fluorometer FT200 (PicoQuant, GmbH, Germany) for time-resolved measurements
- Camera or cell phone, UV illuminator

METHODS

1. Prepare 10% and 1% (w/w) solutions of PVA. Pour 5 mL of each PVA solution into two vials. Two vials will be for samples and two for background controls.
2. Make a stock solution of R6G with a concentration of about 1 mM in a few milliliters of ethanol. Use glass or plastic vials, sonicate.

(*Continued*)

EXPERIMENT 8-7 (CONTINUED) ENERGY MIGRATION, SELF-QUENCHING, AND DEPOLARIZATION OF FLUORESCENCE

3. Prepare samples for the demonstration using the stock solutions: add 50 μL of the stock solution to about 5 mL of PVA solutions (10% and 1%) to achieve about 0.01 mM concentrations of R6G in the solution. As PVA solution will dry the volumes of PVA films will be about 10 times smaller resulting in a different concentration of R6G in the PVA matrix. Add 50 μL of just ethanol to the remaining vials with PVA solutions. These will be reference/control films for 10% and 1% samples of R6G in PVA. Shake well the vials to make sure they are well mixed and the solution is homogenous.
4. Deposit about 1 mL of each solution and PVA controls on the microscope slides as described in Experiment 8-10 *"Demonstration of FRET in PVA Films"* and leave them for drying. The drying will take few hours (alternatively leave slides over the night).
5. Measure absorption spectra of dried samples using PVA only slides as baselines. The absorption will depend on the place on the slide because the thickness of the sample layer will vary along the slide. Choose places on the slides where absorptions are about 0.1, mark the positions.
6. Measure fluorescence spectra at magic angle conditions for both slides (be sure that the excitation spots are at the marked places on the slides) using front-face format and 470 nm excitation wavelength.
7. Measure fluorescence anisotropies of both slides. Use a correction for the G-factor.
8. Measure fluorescence lifetimes of both slides.

RESULTS

Figure 8-7.1 shows photographs of slides coated with PVA films doped with R6G. The low concentration (LC) and high concentration (HC) of R6G were achieved with 10% and 1% PVA solutions, respectively. The low concentration sample made out of 10% PVA solution is about 10-fold thicker. Note that visually colors of both films are similar. The film on right is thinner but the R6G concentration is higher. However, UV illuminated films show very different fluorescence even if absorptions (Figure 8-7.2) are almost identical.

It should be noted that energy migration is NOT equivalent to the self-quenching. It is only the process which enables/facilitates the quenching. For many fluorophores, including fluoresceins and rhodamines, the concentration-dependent quenching is very efficient.

Figures 8-7.2 presents the absorption spectrum measured for both concentrations. We selected PVA regions in such a way that measured absorption is close to 0.1 (a thinner film for a slide with 1% PVA). The absorption spectra are very similar and a small distortion can be due to some minimal aggregations and not a perfectly matched reference slides (films).

In contrast, emission spectra presented in Figure 8-7.3 are very different. The emission spectra for each film were measured in a front-face configuration from the same spot that absorption was measured. Even if the sample does not have rotational diffusion (molecular rotations are blocked by the film matrix), we used the magic angle condition to compare sample intensities. A much weaker emission observed from the 1% PVA film is a clear demonstration of *self-quenching* of R6G. With about the same absorption, the fluorescence from the high concentration sample is much weaker.

(*Continued*)

EXPERIMENT 8-7 (CONTINUED) ENERGY MIGRATION, SELF-QUENCHING, AND DEPOLARIZATION OF FLUORESCENCE

FIGURE 8-7.1 Slides with low (left) and high (right) concentrations of R6G in PVA films. Top: In daylight, bottom: on the UV illuminator. Photographs were taken through a 520 nm long-wavelength pass filter. Upon the same illumination, the low concentration sample (bottom, left) is significantly brighter than the high concentration sample (bottom, right). The amount of R6G molecules is the same in both samples.

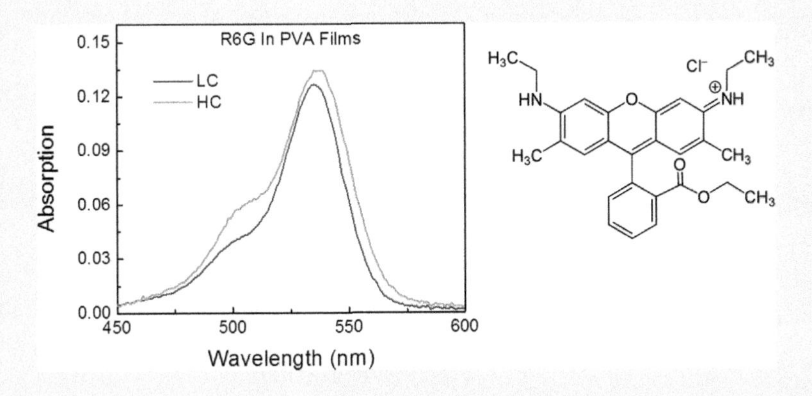

FIGURE 8-7.2 Absorption spectra of R6G in PVA films with low and high concentrations of the dye.

(Continued)

EXPERIMENT 8-7 (CONTINUED) ENERGY MIGRATION, SELF-QUENCHING, AND DEPOLARIZATION OF FLUORESCENCE

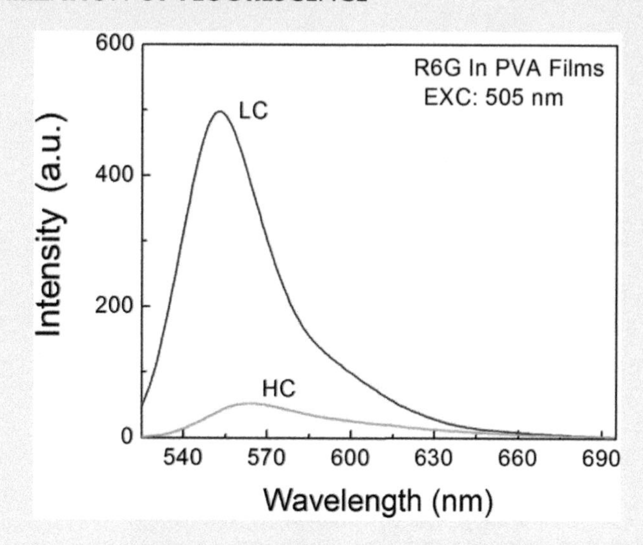

FIGURE 8-7.3 Fluorescence emission spectra of R6G in PVA films with low and high concentrations of the dyes. The excitation was 505 nm.

Figures 8-7.4 and 8-7.5 show corrected for G-factor polarized emission components of R6G fluorescence. Again, these measurements were done in the front-face configuration. Special care needs to be taken for G-factor measurement (see Experiments 6-6 and 6-7). A smaller difference between VV and VH components observed for high concentration sample (1% PVA) indicate smaller anisotropy of that sample.

Calculated from polarized components emission anisotropies are shown in Figure 8-7.6.

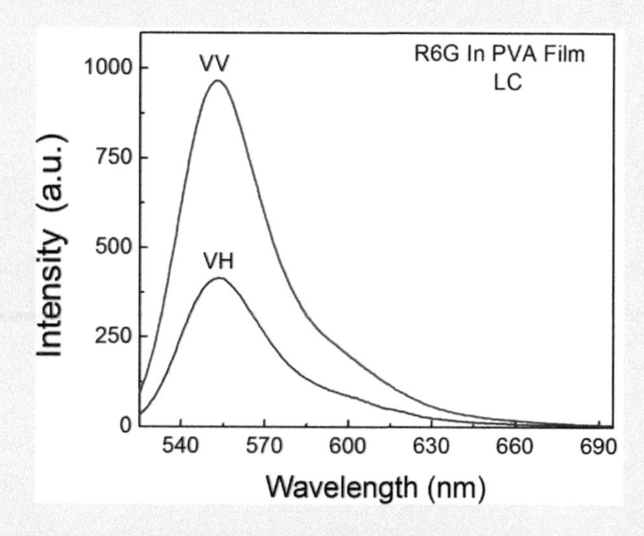

FIGURE 8-7.4 Vertical and horizontal components of the fluorescence of R6G in the PVA film with a low concentration of the dye. The excitation wavelength was 470 nm.

(Continued)

EXPERIMENT 8-7 (CONTINUED) ENERGY MIGRATION, SELF-QUENCHING, AND DEPOLARIZATION OF FLUORESCENCE

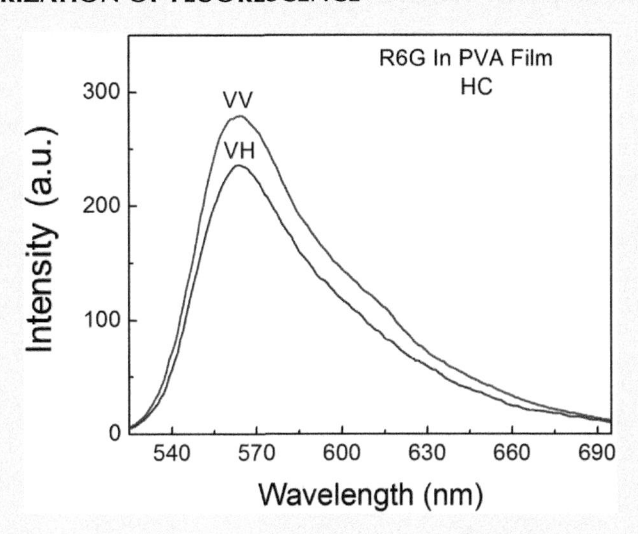

FIGURE 8-7.5 Vertical and horizontal components of the fluorescence of R6G in the PVA film with a high concentration of the dye. The excitation wavelength was 470 nm.

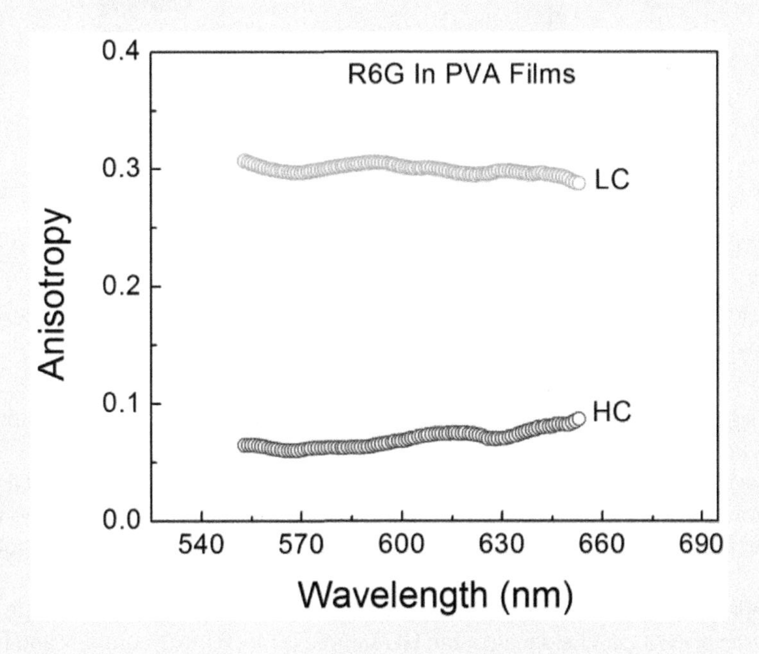

FIGURE 8-7.6 Fluorescence emission anisotropies of R6G in PVA films with low and high concentrations of the dye.

(Continued)

EXPERIMENT 8-7 (CONTINUED) ENERGY MIGRATION, SELF-QUENCHING, AND DEPOLARIZATION OF FLUORESCENCE

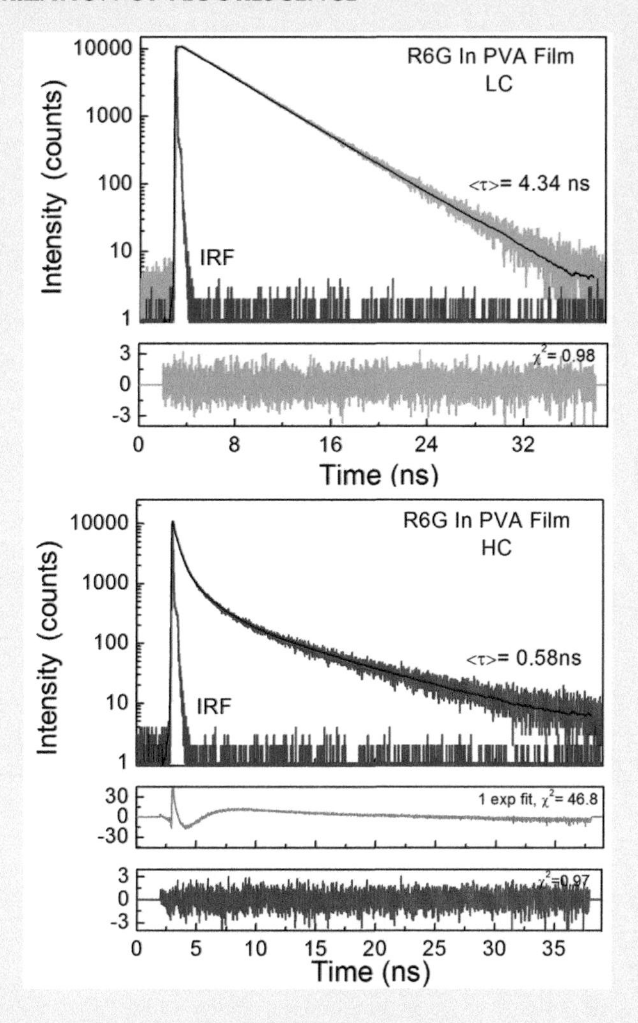

FIGURE 8-7.7 Fluorescence intensity decays of R6G in PVA films with low (top) and high (bottom) concentrations of the dye.

A very significant decrease of anisotropy observed for high concentration sample is the effect of *concentration depolarization*.

Next, we measured fluorescence lifetimes for both samples. For that, we used a front-face configuration, vertical excitation, and magic angle observation. Figure 8-7.7 shows measured intensity decays for low sample concentration—LC (top), and high sample concentration—HC (bottom).

The intensity decay of the LC sample (top) can be satisfactorily fitted with a single exponent with a lifetime of 4.34 ns. The decay of the HC sample is much more complex and four exponents are needed to satisfactory fit the data. An amplitude averaged lifetime decreases about eight-fold.

(Continued)

EXPERIMENT 8-7 (CONTINUED) ENERGY MIGRATION, SELF-QUENCHING, AND DEPOLARIZATION OF FLUORESCENCE

CONCLUSIONS

Energy migration results in a dramatic decrease of fluorescence anisotropy as shown in Figure 8-7.6. It can be shown that single step FRET in isotropic solution (random molecular orientation) lowers the emission anisotropy almost 10-fold. In effect energy migration is a process that very effectively lower observed anisotropy. The concentration-dependent depolarization is a first sign of the energy migration (homo-transfer) phenomenon. The observed decrease in fluorescence lifetime concomitant with very heterogenic behavior (multi-exponential decay) is typically observed but its origin is not well understood yet. The observed shortening of fluorescence lifetime can be a result of dimers or aggregates present in a solution that very dramatically quench the excited state.

EXPERIMENT 8-8 DEMONSTRATION OF EXCIMERS FORMATION IN SOLUTIONS

Keywords: fluorescence, excimers, lifetimes.

After excitation fluorophores in the excited state may form complexes with molecules/ fluorophore in the ground state. Such a complex of two fluorophores consistent of excited/not excited molecules is capable to emit fluorescence and is called an excimer. Excimer fluorescence is long-wavelength shifted comparing to normal monomer fluorescence because some energy is needed to form the complex with the not excited molecule. The excimer does not have its own ground state, therefore, it is not possible to measure its absorption spectrum. In this experiment, we will demonstrate excimer formation of pyrene in triacetin. We selected this solvent because of its higher viscosity which is convenient for the sample measurements in a front-face configuration. This configuration is needed because of high concentrations (absorptions) of samples used.

MATERIALS

- Pyrene, triacetin, dimethyl sulfoxide (DMSO)
- Disposable cuvettes, microscopy slides

EQUIPMENT

- Spectrofluorometer (Varian, Inc., Eclipse)
- Carry 50 Spectrophotometer (Varian, Inc.)
- Fluorometer (FT200, PicoQuant, GmbH, Germany)

METHODS

1. Make a stock solution of 0.4 M pyrene in 1 mL of DMSO, use a glass or plastic vial, sonicate.
2. Prepare sample solutions using stock solutions; to 0.9 mL of triacetin add 0.1 mL of stock solution. This will result in about 10 times dilution to about 0.04 M pyrene. This sample we will call "high concentration." For a second sample, "low concentration" use

(*Continued*)

EXPERIMENT 8-8 (CONTINUED) DEMONSTRATION OF EXCIMERS FORMATION IN SOLUTIONS

0.99 mL of triacetin and 0.01 mL of stock solution which will result in 0.004 M pyrene which we will call "low concentration."

3. Prepare samples for front-face measurements. Deposit 10 μL of sample solution on the microscope slide and cover it with another slide, wait about 5 minutes, see Figure 8-8.1. As a baseline reference use 10 μL of triacetin.
4. Measure absorption, fluorescence, and fluorescence excitation spectra of assembled samples.
5. Measure the lifetimes of assembled samples.

RESULTS

FIGURE 8-8.1 Left: Low concentration of pyrene (0.004 M); right: high concentration of pyrene (0.04 M). Assembled (sandwiched) slides are placed on a UV illuminator. Samples were prepared from 10 μL of solutions disperses homogeneously between two microscope slides, 2.5 cm × 7.5 cm. The estimated apparent thickness of the sample is 5.3 μm.

Figure 8-8.2 shows absorption spectra of samples with high and low pyrene concentrations. The spectra differ only with intensity, not the shape.

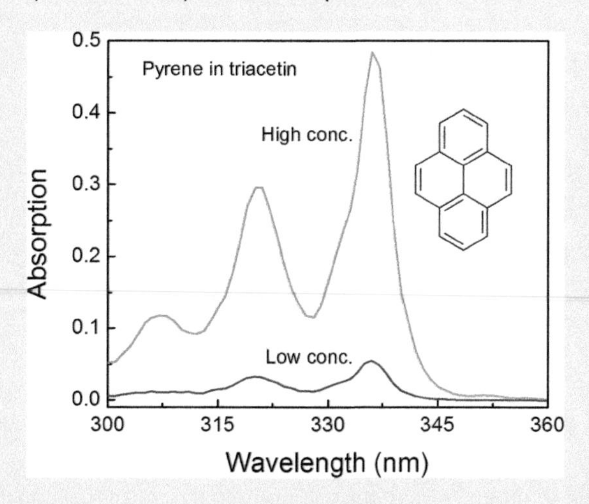

FIGURE 8-8.2 Absorption spectra of pyrene in triacetin. Solutions (10 μL) were placed between two microscope slides.

(Continued)

EXPERIMENT 8-8 (CONTINUED) DEMONSTRATION OF EXCIMERS FORMATION IN SOLUTIONS

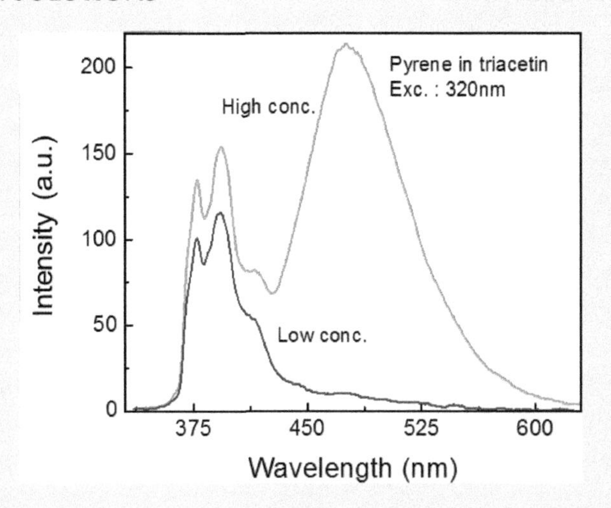

FIGURE 8-8.3 Fluorescence spectra of pyrene in triacetin at 320 nm excitation. The measurements were done in front-face geometry.

Figure 8-8.3 shows measured emission spectra for low and high concentration in the front-face configuration. The "high concentration" sample shows a long wavelength band—an *Excimer* fluorescence.

Next, measure excitation spectra of both samples. For "high concentration" sample, use two observation wavelengths 393 nm for monomer and 480 nm for excimer emissions, see Figure 8-8.4. For both observation wavelengths, the excitation spectra are identical.

Measure the lifetimes of both samples, see Figures 8-8.5 and 8-8.6. The lifetime of "low concentration" sample can be approximated with a single exponent. "High concentration" sample shows a complex decay with a negative component characteristic for the excited state process (pumping).

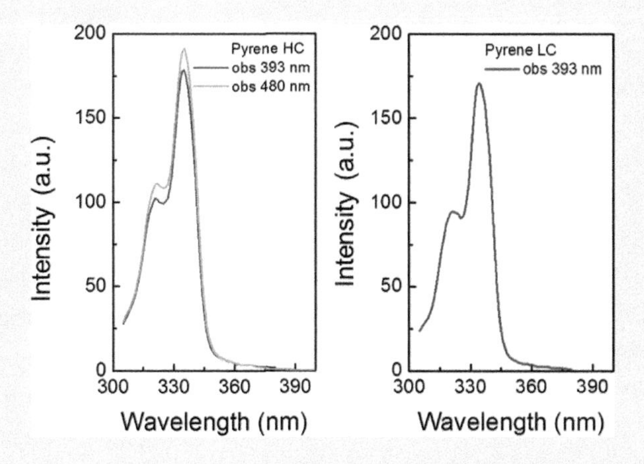

FIGURE 8-8.4 Fluorescence excitation spectra of pyrene in triacetin. Left: High concentration with two observations, at 393 and 480 nm. Right: Low concentration with 393 nm observation.

(Continued)

EXPERIMENT 8-8 (CONTINUED) DEMONSTRATION OF EXCIMERS FORMATION IN SOLUTIONS

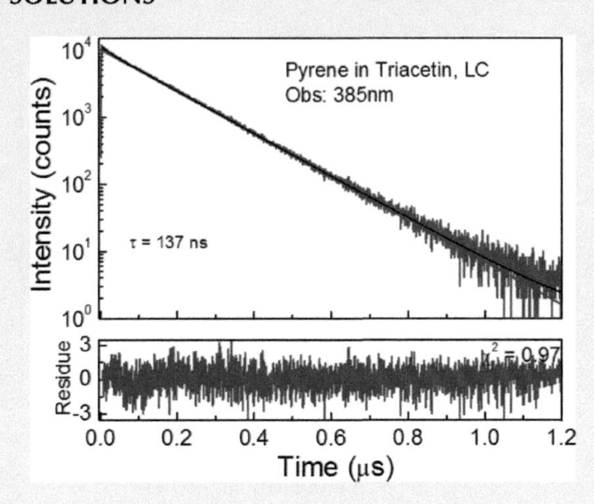

FIGURE 8-8.5 Fluorescence intensity decay of pyrene in triacetin (low concentration). Excitation was from 340 nm pulsed LED with repetition rate of 100 kHz.

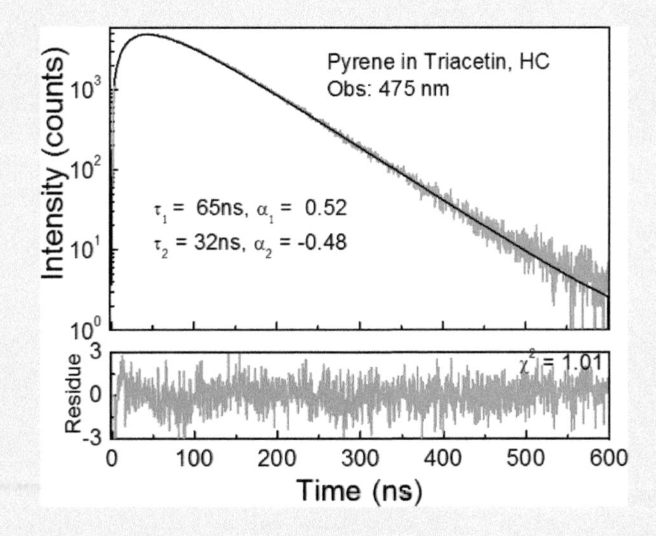

FIGURE 8-8.6 Fluorescence intensity decay of pyrene in triacetin (high concentration). Excitation was from 340 nm pulsed LED with repetition rate of 100 kHz.

(Continued)

EXPERIMENT 8-8 (CONTINUED) DEMONSTRATION OF EXCIMERS FORMATION IN SOLUTIONS

CONCLUSIONS

How does excimer fluorescence differs from monomer? First, its spectrum is long-wavelength shifted and unstructured. Second, the excitation spectrum of excimers is identical to the excitation spectrum of monomers. It means that there is no ground state of excimers that are formed before the excitation. Third, the intensity decay of excimer fluorescence shows very clearly the initial rise which can be fitted only with a negative exponent, the evidence that the process of excimer formation occurs in the excited state. The formation of excimers is diffusion dependent and more efficient in low viscosity solvents. For the observation of excimer fluorescence, it is not necessary to use front-face configuration. In right-angle (cuvette) experiments, it is necessary to use a red edge excitation of high concentration sample. Front-face configuration enables correct measurements of the excitation spectrum.

EXPERIMENT 8-9 ANISOTROPY DECAYS IN LIPID MEMBRANES

Keywords: fluorescence, anisotropy decay, DPPC lipid.

Fluorophores in lipid membranes cannot move freely. Rotations of fluorophores are hindered and only restricted motions are possible. This, of course, will affect both steady-state and time-resolved fluorescence anisotropy. We will compare anisotropy decays of DPH fluorophore in solvent (Triacetin) where fluorophore molecules can move freely and DPH in DPPC lipid where rotations of fluorophores are hindered (see below chemical structure of DPPC and Figure 8-9.1).

MATERIALS

- Diphenyl-hexatriene (DPH)
- Triacetin, chloroform, dimethyl sulfoxide (DMSO) PBS buffer, DPPC lipid (1,2-dipalmitoyl-sn-glycero-3-phosphocholine)
- 1 cm × 1 cm fluorometric cuvettes, vials, plastic polarizers, 0.02 μm syringe filters

EQUIPMENT

- Spectrofluorometer (Varian, Inc., Eclipse)
- Carry 50 Spectrophotometer (Varian, Inc.)
- Fluorometer (FT200, PicoQuant, GmbH, Germany)

METHODS

1. Make a stock solution of DPH in DMSO, sonicate.
2. Prepare samples of DPH in triacetin using the stock solution and measure absorptions. Adjust concentration to achieve about 0.1 optical density at maximum. Measure fluorescence spectrum, see Figures 8-9.2 (left).

(Continued)

EXPERIMENT 8-9 (CONTINUED) ANISOTROPY DECAYS IN LIPID MEMBRANES

3. Prepare DPPC lipid vesicle with and without DPH. Follow this procedure:
   Unilamellar lipid vesicles were prepared using DPPC (1,2-dipalmitoyl-sn-glycero-3-phosphocholine) purchased from Avanti Polar Lipids Inc., USA. Appropriate amount of DPPC lipid and the DPH were dissolved in chloroform (the lipid:dye ratio was 800:1) in a glass bottle. The solvent was evaporated under an oxygen-free nitrogen stream and left overnight to remove any traces of organic solvents. Next, about 1 mL 10 mM PBS (phosphate buffer saline) pH 7.4 was added followed by strong sonication at about 40°C to get giant multilamellar vesicles. Moreover, in order to obtain unilamellar vesicles, multilamellar vesicles were passed through 100 and 0.02 μm membrane syringe filters. As obtained lipid vesicles with appropriate dilution were used for fluorescence measurements.

4. Measure absorption and fluorescence spectra.
5. Measure fluorescence intensity decays (lifetimes) of DPH in triacetin and in DPPC (remember to use magic angle conditions), see Figure 8-9.3.
6. Measure anisotropy decays of DPH in triacetin and in DPPC lipid.

RESULTS

Anisotropy decay usually is fitted to an exponential model:

$$r(t) = r_0 \exp(-t/\theta)$$

In the case of restricted molecular motions (called also "hinder rotator"), the molecule cannot completely rotate and will wobble in a restricted angle. So, the observed anisotropy cannot be completely rotated out and an additional term r (inf) applies:

$$r(t) = r(\text{inf}) + [r_0 - r(\text{inf})]\exp(-t/\theta)$$

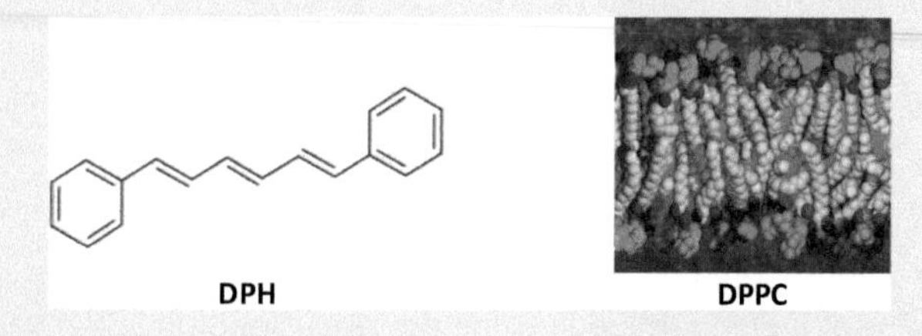

Figure 8-9.1 DPH (left) is a non-polar compound, not soluble in water. Added to the DPPC (right), solution localizes inside the membrane apart from polar groups.

(Continued)

EXPERIMENT 8-9 (CONTINUED) ANISOTROPY DECAYS IN LIPID MEMBRANES

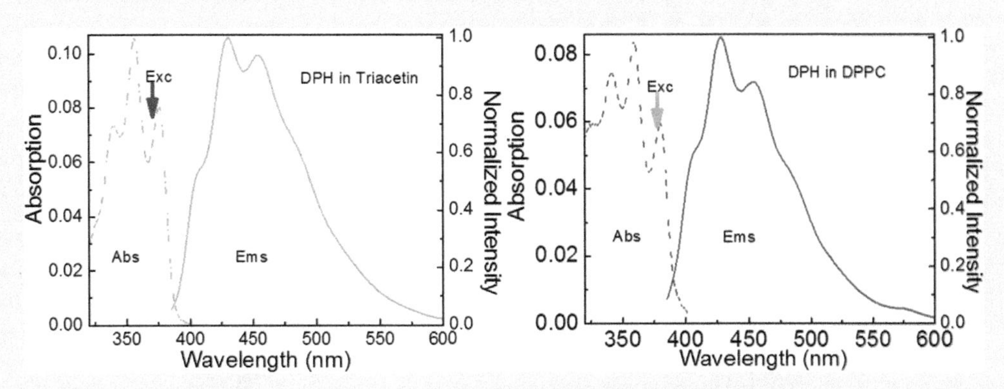

FIGURE 8-9.2 Absorption and emission spectra of DPH in triacetin (left) and in DPPC lipid (right). Arrows indicate 375 nm excitation wavelength used in time-resolved measurements.

Measure absorption and fluorescence spectra of free DPH molecules in triacetin (in this solvent, DPH molecules move and rotate freely) and in DPPC, see Figure 8-9.2. In both media, spectra are similar with slightly more visible oscillation structure in a less polar DPPC environment.

Next, measure DPH lifetimes in triacetin and DPPC, see Figure 8-9.3.

Both intensity decays can be well approximated with a single exponent fit with lifetimes 7.9 ns and 9.6 ns, respectively.

Next, measure polarized components of intensity decay (VV and VH) and correct them for G-factor. Corrected for G-factor polarized components of DPH intensity decays in triacetin are shown in Figure 8-9.4. Please note that after few nanoseconds from the excitation the VV and VH components are not distinguishable in contrast to DPH in DPPC, see Figure 8-9.5. In the case of DPPC, VV and VH components do not merge. This means that anisotropy does not reach zero at any time of fluorescence.

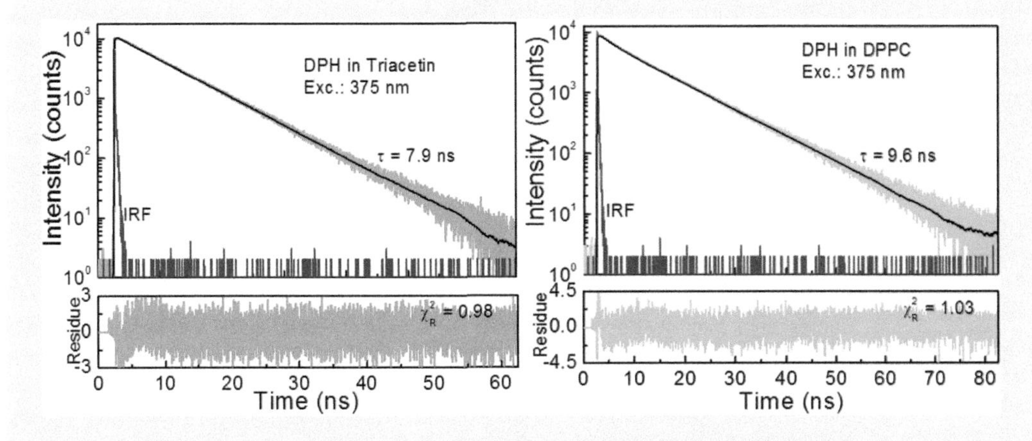

FIGURE 8-9.3 Fluorescence intensity decays of DPH in triacetin (left) and in DPPC lipid (right).

(Continued)

EXPERIMENT 8-9 (CONTINUED) ANISOTROPY DECAYS IN LIPID MEMBRANES

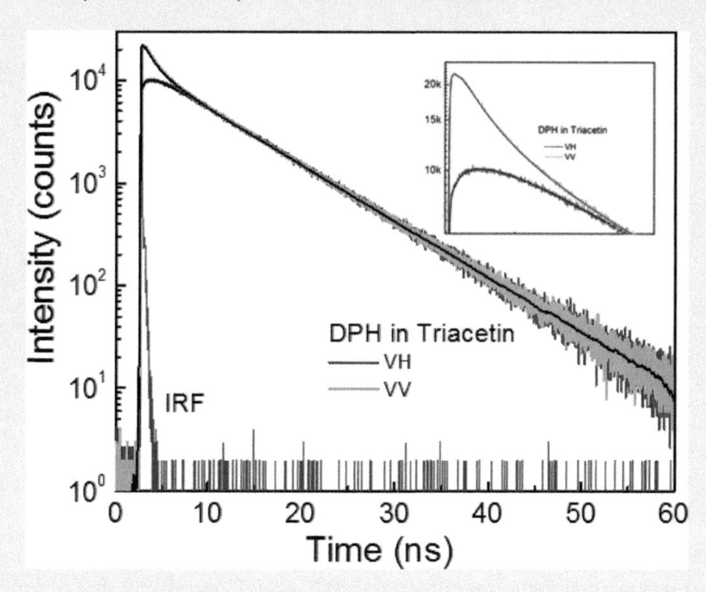

FIGURE 8-9.4 Fluorescence intensity decays of DPH in triacetin observed with VV and VH polarization conditions. VH component was corrected for G-factor.

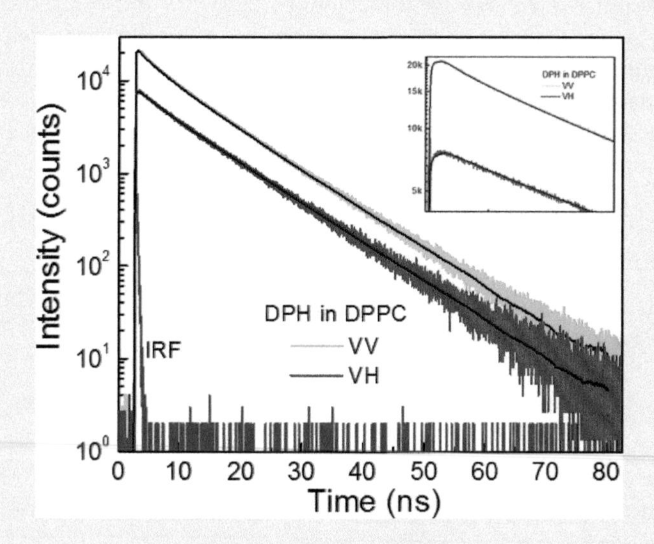

FIGURE 8-9.5 Fluorescence intensity decays of DPH in DPPC lipid observed in VV and VH polarization conditions. VH component was corrected for G-factor.

(Continued)

EXPERIMENT 8-9 (CONTINUED) ANISOTROPY DECAYS IN LIPID MEMBRANES

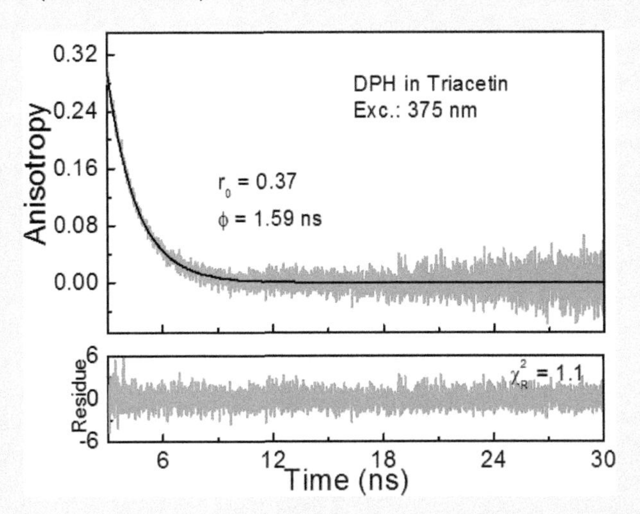

FIGURE 8-9.6 Fluorescence anisotropy decay of DPH in triacetin. One correlation time fit.

Next, calculate anisotropy decays of DPH in triacetin and DPPC lipid membrane, see Figures 8-9.6 through 8-9.8.

The anisotropy decay of DPH in DPPC lipid membrane cannot be fitted with a single correlation time, see Figure 8-9.7. But can be easily fitted with a constant anisotropy value *r* (inf), see Figure 8-9.8.

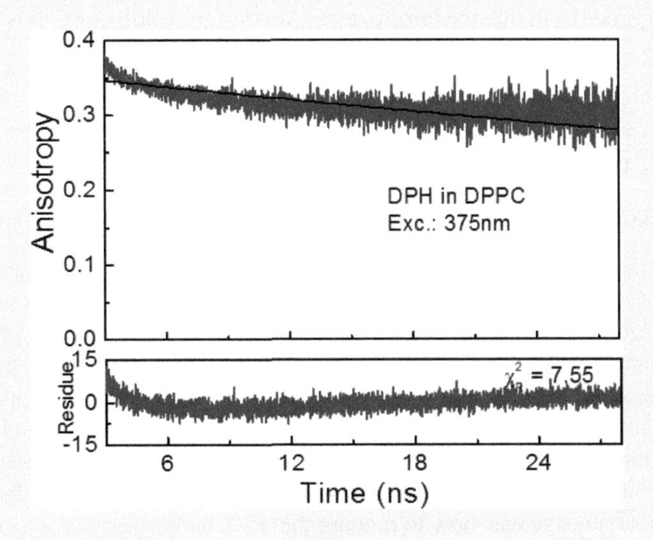

FIGURE 8-9.7 Fluorescence anisotropy decay of DPH in DPPC lipid. One correlation time fit.

(Continued)

EXPERIMENT 8-9 (CONTINUED) ANISOTROPY DECAYS IN LIPID MEMBRANES

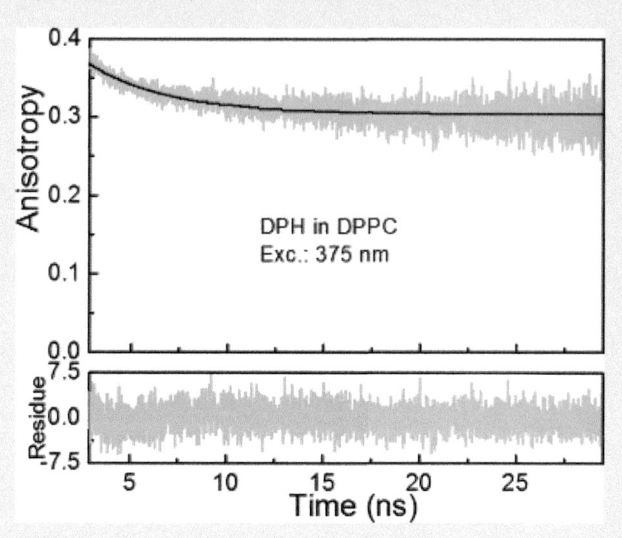

FIGURE 8-9.8 Fluorescence anisotropy decay of DPH in DPPC lipid. One correlation time fit with infinity anisotropy.

CONCLUSIONS

Although spectra and lifetimes of DPH in triacetin and DPPC lipid membrane are similar, the anisotropy decays are drastically different. In DPPC, the time-dependent anisotropy never reaches zero. In order to fit the anisotropy decay, a constant value of *r* (inf) is needed.

EXPERIMENT 8-10 DEMONSTRATION OF FRET IN PVA FILMS

Keywords: fluorescence, lifetime, FRET, PVA film.

In previous experiments, we discussed a convenient approach to immobilize fluorophores in the polymer matrix. One of the easiest polymers for the preparation of a film is poly (vinyl alcohol) (PVA). It is also a good way to obtain high concentrations of dyes because the polymer solution decreases its volume while drying. For the FRET demonstration, it is important to keep a low optical density of studied samples. It means that the thickness of the sample (polymer) should be small. For example, if the extinction coefficient of the dye is 100,000 L/mol × cm, and the desired concentration is 1 mM, the thickness of the polymer layer should be below 10 μm to achieve absorption less than 0.1. The preparation of PVA films is described in Experiment 3-24. Here, we will show an easy way how to prepare thin PVA films.

MATERIALS

- Rhodamine 110 (R110) donor, Rhodamine 101 (R101) acceptor
- Poly (vinyl alcohol), MW: 130,000, Aldrich, ethanol

(Continued)

EXPERIMENT 8-10 (CONTINUED) DEMONSTRATION OF FRET IN PVA FILMS

- Petri dishes, glass vials, water, ethanol
- Microscope slides (25 mm × 75 mm), pipettes
- Camera or cell phone

EQUIPMENT

- Spectrofluorometer (Varian, Inc., Eclipse) with a front-face attachment
- Carry 50 Spectrophotometer (Varian, Inc.)
- Fluorometer (FT200, PicoQuant, GmbH, Germany) for time-resolved measurements

METHODS

1. In order to estimate dye concentrations in PVA, it is convenient to know how the volumes change upon drying. Prepare 10% (w/w) solution of PVA. Pour 20 mL of PVA solution into an 8 cm diameter petri dish and leave for drying. It will take few days to get a completely dry film. Peel off the PVA film, cut it to small pieces and put it to the 8 mL vial. Add calibrated volume of ethanol to the vial and measure its volume. *Note: Do not use water that quickly dissolves PVA. Ethanol dissolving PVA much slower.* Estimate the volume of dried PVA. Calculate the ratio of liquid and dried PVA. For 130,000 MW PVA, this ratio is about 12.
2. Make stock solutions of studied compounds (about 4 mM of R110 donor, and 40 mM of an R101 acceptor) in few milliliters of ethanol. Use glass or plastic vials, sonicate.
3. Prepare samples for FRET demonstration using the stock solutions. For donor control, acceptor control and donor-acceptor mixture use 10 mL of PVA in vials. Add 50 µL of R110 (D) stock solution to donor control and donor-acceptor mixture, and 50 µL of R101 (A) stock solution to the mixture and acceptor control. With such preparation, the concentrations of the donor in the donor control and DA system will be the same, as well as concentrations of the acceptor in DA system and acceptor control. Mix, shake, and sonicate prepared samples until dyes will be homogeneously dissolved.
4. Deposit about 1 mL of each solution and PVA control (no dye) on the microscope slides as shown in Figure 8-10.1. Adjust the slope of slides to about 10°. PVA solution should very slowly flow on the slides. Leave the slides for drying, it will take a few hours.
5. Measure absorption spectra of dried samples using PVA only slide as a baseline, see Figure 8-10.2.

   The absorption will depend on the place on the slide because the thickness of the sample layer will vary along the slide. Choose places on the slides where absorptions are lower than 0.1, mark the positions.
6. Measure fluorescence spectra at magic angle conditions for all slides, D, A, and DA (be sure that the excitation spots are at the marked places on the slides) using front-face format. For a description of a proper front-face configuration, see Chapter 5. Assemble together D and A slides and measure fluorescence spectrum.
7. Measure lifetimes of all slides at 470 nm excitation and 520 nm observation. Measure lifetimes of A and at 595 nm observation for acceptor only and DA system.

(Continued)

EXPERIMENT 8-10 (CONTINUED) DEMONSTRATION OF FRET IN PVA FILMS

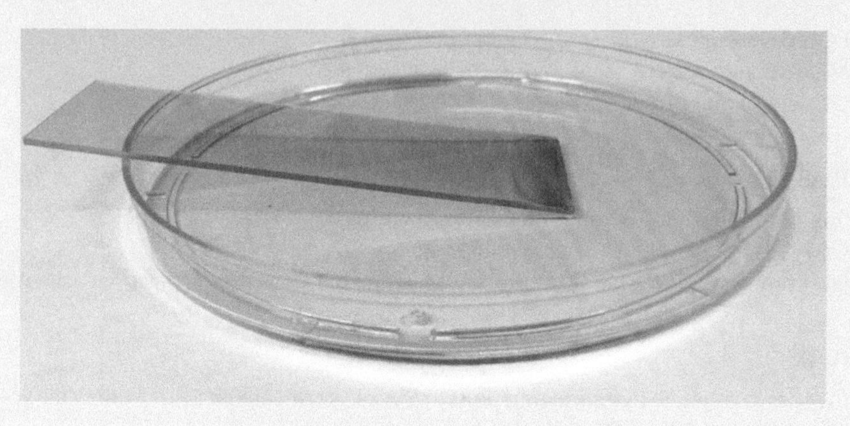

FIGURE 8-10.1 Photograph of the slide preparation.

RESULTS

Figure 8-10.1 shows the simple set up for sample preparation for FRET measurements in thin PVA films.

Absorption Measurements. It is important to measure the absorptions of the prepared samples. First, check if the spectrum of DA system is a combination of donor and acceptor spectra and no extra absorption bands are present. For details, see Experiment 3-23. Second, be sure that absorptions are below 0.1, see Figure 8-10.2.

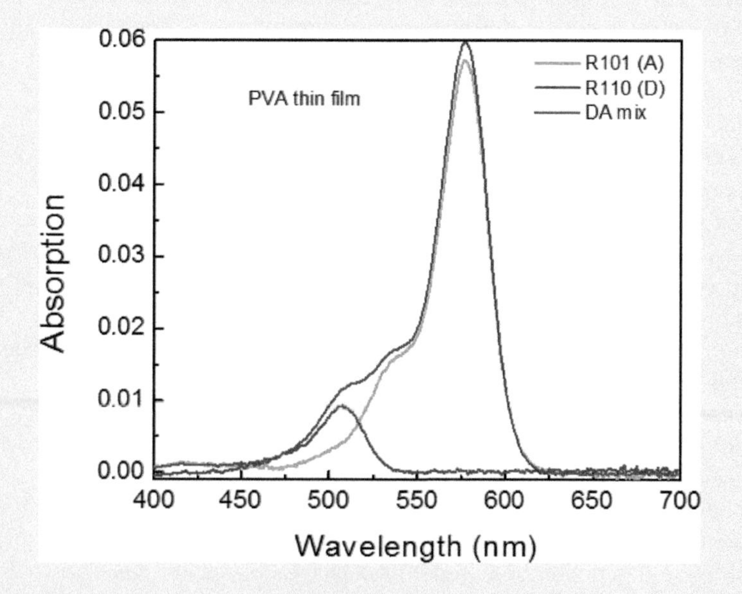

FIGURE 8-10.2 Absorption spectra of R110 donor (blue), R101 acceptor (green), and DA system (red). The spectrum of the mixture of donor-acceptor (green) is a combination of the donor an acceptor absorptions. No new absorption band is being observed.

(Continued)

EXPERIMENT 8-10 (CONTINUED) DEMONSTRATION OF FRET IN PVA FILMS

Use PVA only control on the microscope slides as the background for the baseline (remember reflectivity and absorption of just glass slide and slide with a thin layer of deposited PVA are slightly different). For details, see Experiment 3-25.

Why we selected this donor-acceptor pair for FRET demonstration?

First, the donor has a high quantum yield and second, at the absorption region of the donor, the acceptor absorption is low which will result with a weak direct excitation of the acceptor, see Figure 8-10.2. Also, R101 acceptor has high extinction coefficient resulting in larger R_0.

The fluorescence of the donor highly overlaps with the absorption of the acceptor (Figure 8-10.3, shadowed area). These are favorable for FRET conditions with Forster distances of about 65 Å.

Fluorescence Measurements of the FRET. The fluorescence measurements were done in a front-face geometry with a vertically polarized excitation and the observation polarizer oriented at the magic angle.

In the DA mixture, the fluorescence intensity of the donor decreases while the acceptor intensity increases. This is a direct indication of excitation energy transfer. Notice, that absorptions of the samples are very low which makes a reabsorption process negligible.

A very different situation is observed with the combination of R110 donor and R101 acceptor assembled samples (this is a sandwich assembly where we squeezed together with a slide

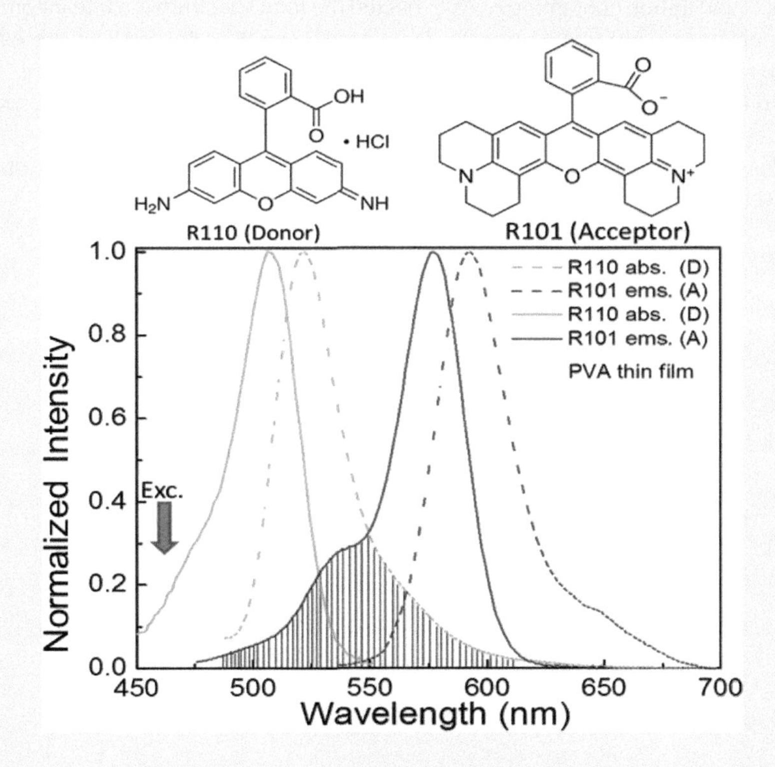

FIGURE 8-10.3 Normalized absorption (solid lines) and emission (dashed lines) spectra of R110 donor (green) and R101 acceptor (red). A shadowed area represents spectral overlap.

(Continued)

EXPERIMENT 8-10 (CONTINUED) DEMONSTRATION OF FRET IN PVA FILMS

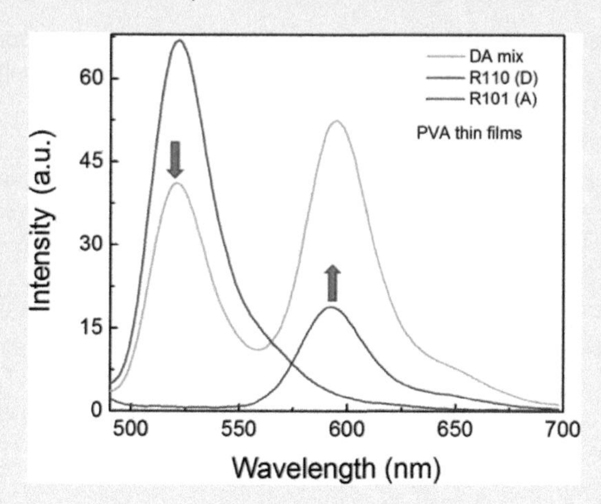

FIGURE 8-10.4 Fluorescence emission spectra of the donor (blue), acceptor (red), and mixture of donor-acceptor (green). A decrease in the donor emission is accompanied by a strong increase of acceptor fluorescence.

containing D and a slide containing A. As expected the total spectrum is a linear combination of individual spectra. No energy transfer is being observed. In the path of the excitation, there is the same amount of donor and acceptor molecules in D-A mixture (Figures 8-10.4, green) as in two assembled slides (Figure 8-10.5). However, measured emission spectra are different (compare green line in Figure 8-10.4 and Figure 8-10.5).

Time-Resolved Measurements of FRET. Below, we show the demonstration of FRET in time-resolved (lifetime) measurements.

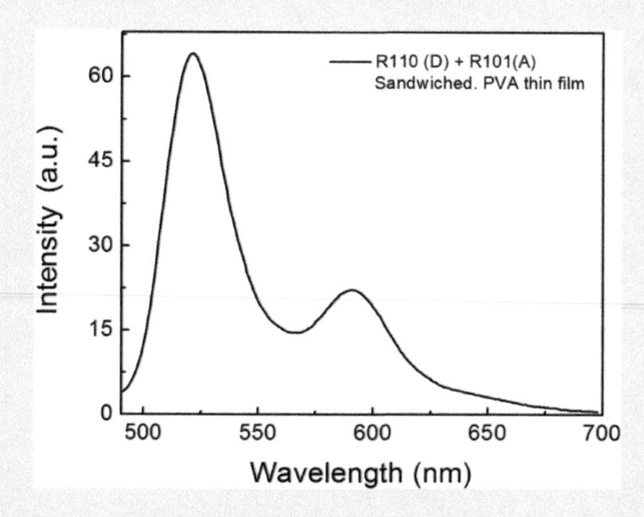

FIGURE 8-10.5 Fluorescence spectrum of a combination of both R110 (D) and R101 (A) assembled together. A total spectrum is a sum of individual spectra, no FRET is observed.

(Continued)

EXPERIMENT 8-10 (CONTINUED) DEMONSTRATION OF FRET IN PVA FILMS

FRET process additionally depopulates the excited state of the donor which results in the decrease of intensity and lifetime. In solution, the acceptors are randomly distributed around donors. Some acceptor molecules are closer; some are more distant from the donor. Therefore, some donors are quenched faster which demonstrates in a non-homogeneous donor decay in the presence of an acceptor. The lifetime of the donor becomes heterogeneous and can be correctly fitted with a proper decay low which includes energy transfer. In a low viscous solution, the diffusion of the molecules should be also included. The heterogeneity of the lifetimes is immediately seen if the decay and requires more terms in the multi-exponential approximation.

The donor (R110) decay can be satisfactorily fitted with a single exponent of 3.65 ns, see Figure 8-10.6. However, the intensity decay of the DA mixture cannot be satisfactorily fitted with the one exponent model, see Figure 8-10.7. The satisfactory three exponent fit reveals lifetimes: 0.28 ns, 1.18 ns, and 2.87 ns with amplitudes: 0.102, 0.213, and 0.685, respectively. The amplitude averaged lifetime for this decay is 2.31 ns. The transfer efficiency in DA mix is about 37%.

Next, we observe the intensity decays of A and DA mixture at the acceptor emission (595 nm) (Figure 8-10.8).

The intensity decay of acceptor A (R101) can be satisfactorily fit with single exponent model with a lifetime of 4.9 ns. The intensity decay of DA mixture is complex and show initial increase (pumping) characteristic for excited-state processes. This decay can be satisfactorily fitted only with a negative amplitude for one component or an appropriate energy transfer model.

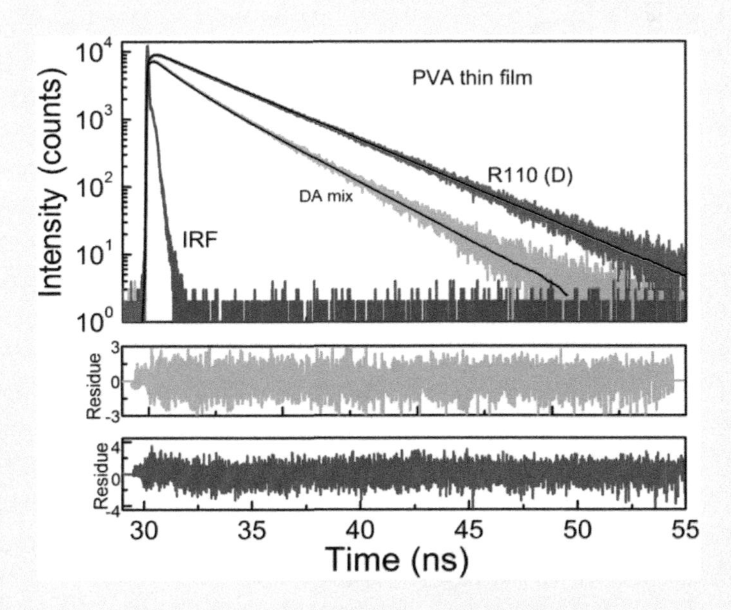

FIGURE 8-10.6 Fluorescence intensity decays of R110 donor alone and in mixture with R101 acceptor. The excitation was 470 nm and the observation was at 520 nm.

(Continued)

EXPERIMENT 8-10 (CONTINUED) DEMONSTRATION OF FRET IN PVA FILMS

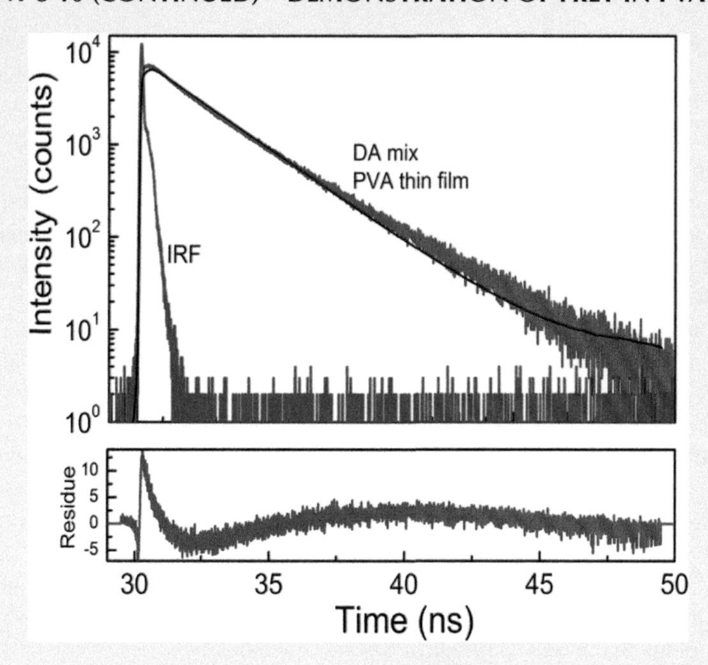

FIGURE 8-10.7 DA mixture cannot be satisfactorily fitted with a single exponent fit.

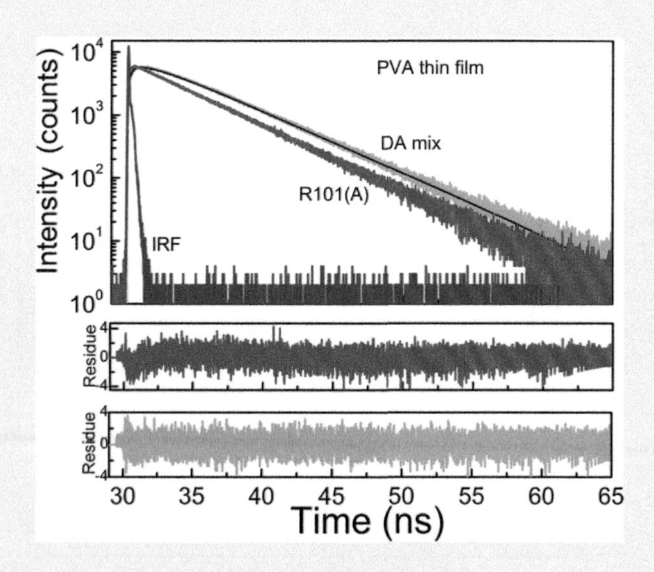

FIGURE 8-10.8 Fluorescence intensity decays of R101 acceptor alone and in the presence of the donor. The excitation was 470 nm and the observation was at 590 nm.

(Continued)

EXPERIMENT 8-10 (CONTINUED) DEMONSTRATION OF FRET IN PVA FILMS

CONCLUSIONS

FRET measurements require very short distances between donor and acceptor molecules. In free solutions, it means very high concentrations. These FRET measurements are very difficult in the square geometry but in a front-face set-up are readily possible. The goal of this experiment was to demonstrate 3D FRET. For theoretical descriptions and applications, we refer to the recommended textbooks.

EXPERIMENT 8-11 DEMONSTRATION OF FRET IN SOLUTIONS

Keywords: fluorescence, FRET, lifetime.

The resonance energy transfer requires a close distance between the donor and acceptor molecules. The efficiency of the transfer depends also on several other factors like the quantum yield of the donor, refractive index of the solution, spectral overlap between emission of the donor and absorption of the acceptor and relative orientation of interacting molecules. Usually, to observe the FRET, the average donor-acceptor distance should be well below 10 nm (100 Å). In order to achieve such short distances, the concentration of the acceptor must be very high, in the range of millimolar or higher. Solutions of such high concentrations cannot be reliably measured in regular cuvettes. First, the excitation beam cannot penetrate the cuvette, and second, the fluorescence emission is strongly reabsorbed. In general, to avoid these effects there should be a low absorption of the sample, below 0.1. In order to have low absorptions of highly concentrated solutions, the thickness of the samples must be a few microns short. We will use a viscous solvent propylene glycol for samples solutions and place a small amount of the studied sample between microscope slides. Such "sandwiched" samples will be measured in front-face geometry. A donor-acceptor (DA) system of free donor and acceptor molecules (U-R101) will be used to demonstrate FRET in solutions.

MATERIALS

- Uranin (U) donor, Rhodamine 101 (R101) acceptor, propylene glycol (PG)
- 1 cm × 1 cm disposable cuvettes, microscope slides (25 mm × 75 mm), pipettes
- Camera or cell phone

EQUIPMENT

- Spectrofluorometer (Varian, Inc., Eclipse) with a front-face attachment
- Carry 50 Spectrophotometer (Varian, Inc.)
- Fluorometer (FT200, PicoQuant, GmbH, Germany)

METHODS

1. Make stock solutions of studied compounds (about 0.2 mM of U donor, and 2 mM of an R101 acceptor) in a few milliliters of PG. Use glass or plastic vials, sonicate.

(Continued)

EXPERIMENT 8-11 (CONTINUED) DEMONSTRATION OF FRET IN SOLUTIONS

2. Dilute small amounts of stock solutions in PG to achieve about 1 μM concentrations. Check absorptions in 1 cm × 1 cm cuvettes; be sure that optical densities are below 0.1. Measure absorption and emission spectra, see Figure 8-11.1.
3. Prepare samples for FRET demonstration using the stock solutions. For donor controls, use 50% of stock solutions, for example 0.5 mL of the stock solution and 0.5 mL of PG which will result in about 0.1 mM concentration of the donor. For the acceptor controls, use 0.5 mL of the stock solution and 0.5 mL of PG which will result in 1 mM concentration of the acceptor. For the donor-acceptor system, use 0.5 mL of donor and acceptor stock solutions, which will result in 0.1 and 1 mM of the donor and acceptor concentrations, respectively. With such preparation, the concentrations of the donor in the donor control and DA system will be the same, as well as concentrations of the acceptor in DA system and acceptor control.
4. Place 10 μL of the samples between microscope slides, as shown in Figure 8-11.2.
5. Measure absorption spectra using a slide "sandwich" with PG only for the baseline (see Figure 8-11.3).
6. Measure fluorescence spectra of U, R101, and U-R101 mixture, see Figure 8-11.4.
7. Measure the fluorescence spectrum of a combination of both U and R101 "sandwiched" samples. Use a drop of PG to stick them together, see Figure 8-11.5.
8. Measure fluorescence excitation and emission anisotropies, see Figure 8-11.6.
9. Measure lifetimes of donor (U) and DA samples at 470 nm excitation and 520 nm observation, see Figures 8-11.7 and 8-11.8.
10. Measure lifetimes of DA and acceptor (R101) samples at 470 nm excitation and 595 nm observation, see Figures 8-11.9 and 8-11.10.

RESULTS

Why we selected this donor-acceptor pair for FRET demonstration?

First, the donor has a high quantum yield and second, at the absorption region of the donor, the acceptor absorption is low which will result with a weak direct excitation of the acceptor, see Figure 8-11.1. Also, R101 acceptor has a high extinction coefficient. The fluorescence of the donor highly overlaps absorption of the acceptors (Figure 8-11.1, shadowed area). These are favorable for FRET conditions with Forster distances of about 60–70 Å.

Sample Preparation for FRET Measurements in Solutions. Place 10 μL of the sample in the middle of a microscope slide, as shown in Figure 8-11.2. Cover it with another microscope slide. Wait about 10 minutes.

Why Use PG as a Solvent? PG has relatively high viscosity of about 60 cP (viscosity of water is 1 cP) and high boiling temperature, therefore, it not vaporize quickly. Also, microscopy slides stick nicely together with PG between them.

It is easy to estimate the apparent thickness of the sample from the volume (10 μL) and the area of the slide (25 mm × 75 mm). It is about 5 μm, which is confirmed by absorption measurements with known extinction coefficients and concentrations.

Absorption Measurements. It is important to measure the absorptions of the prepared samples. First, check if the spectrum of DA system is a combination of donor and acceptor

(*Continued*)

EXPERIMENT 8-11 (CONTINUED) DEMONSTRATION OF FRET IN SOLUTIONS

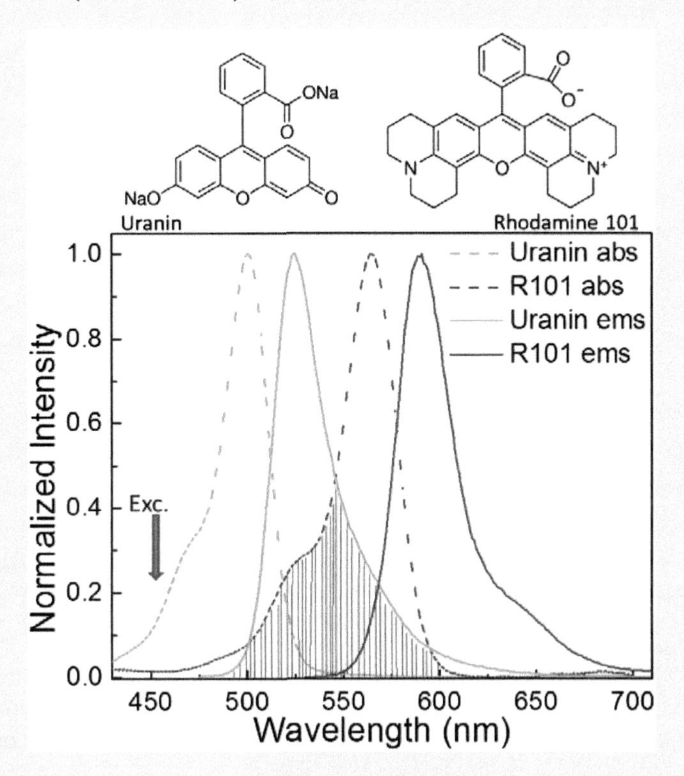

FIGURE 8-11.1 Normalized absorption (dashed lines) and emission (solid lines) spectra of U donor (green) and R101 acceptor (red). A shadowed area represents spectral overlap.

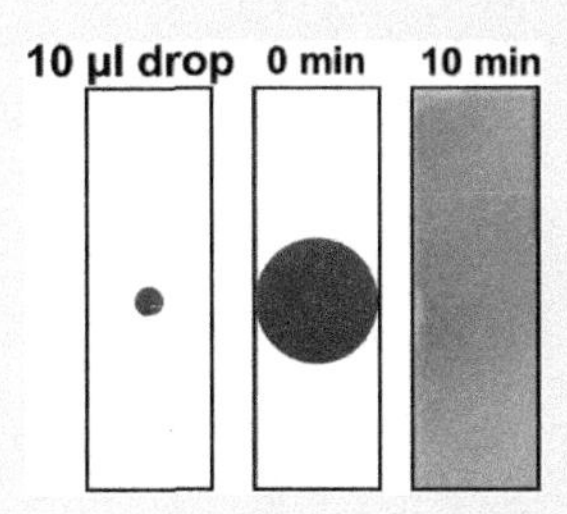

FIGURE 8-11.2 Sample preparation for FRET measurements in solutions. Left: 10 µL drop of DA solution; middle: immediately after covering with another slide; right: sample after 10 minutes. Photographs were done with a cell phone.

spectra and no extra absorption band is present. Second, be sure that absorptions are below 0.1, see Figure 8-11.3.

Use the 10 µL drop of PG (between slides) to prepare a reference background for absorption and fluorescence measurements.

(Continued)

EXPERIMENT 8-11 (CONTINUED) DEMONSTRATION OF FRET IN SOLUTIONS

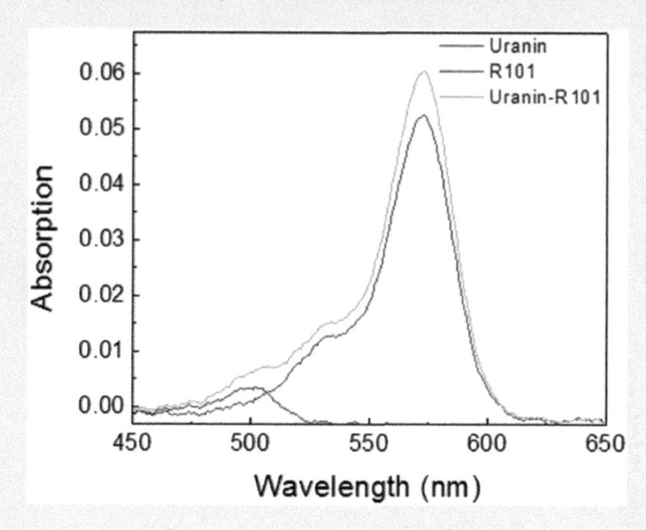

FIGURE 8-11.3 Absorption spectra of the donor (blue), acceptor (red) and DA system (green). The samples were placed between microscopes slides, as described above. The spectrum of the mixture of donor-acceptor (green) is a combination of the donor an acceptor absorptions. No new absorption band is being observed.

Fluorescence Measurements. The fluorescence measurements should be done in front-face geometry with a vertically polarized excitation and the observation polarizer oriented at the magic angle.

A strong FRET effect is observed for the mixture donor (U) and acceptor (R101), see Figure 8-11.4.

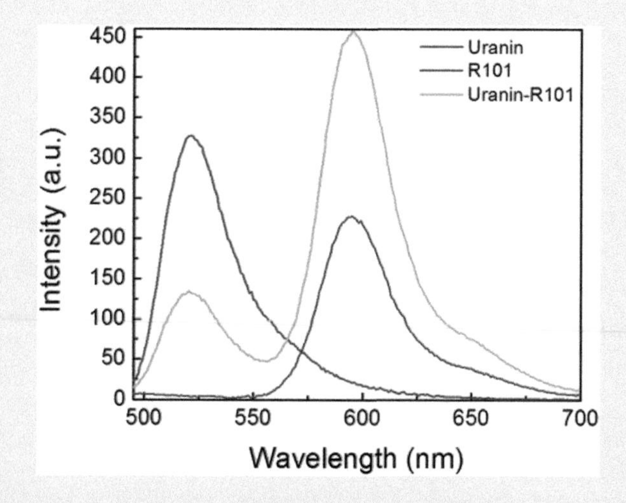

FIGURE 8-11.4 Fluorescence emission spectra of the donor (blue) acceptor (red) and a mixture of donor-acceptor (green) with the excitation at 470 nm. A significant decrease in the donor emission is accompanied by a strong increase of acceptor fluorescence.

(Continued)

EXPERIMENT 8-11 (CONTINUED) DEMONSTRATION OF FRET IN SOLUTIONS

A very different situation is observed with the combination of U donor and R101 acceptor assembled samples. No energy transfer is being observed. In the path of the excitation, there is the same amount of donor and acceptor in Figures 8-11.4 and 8-11.5.

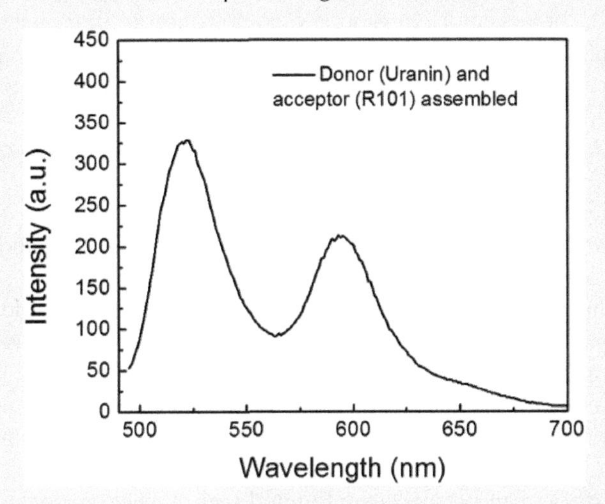

FIGURE 8-11.5 Fluorescence spectrum of a combination of both U donor and R101 acceptor samples assembled together. The donor and acceptor are not mixed, no FRET is observed.

As expected, the total spectrum is a linear combination of individual spectra in the mixture and assembled individual samples.

Anisotropy Measurements. For the correction of anisotropy values, measure G-factor as described in Experiments 6-6 and 6-7.

Measure VV and VH components for each sample; be sure that the surface of the sample is orthogonal to the observation axis. Calculate fluorescence emission anisotropies and use corrections for G-factor (see Figure 8–11.6).

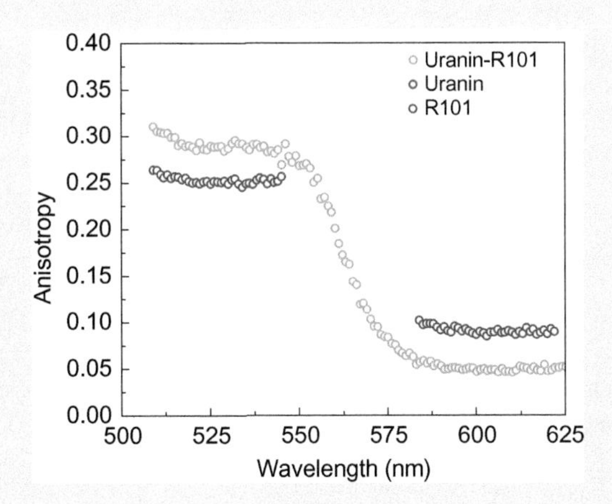

FIGURE 8-11.6 Fluorescence emission anisotropies of the donor (blue), acceptor (red), and mixture of donor-acceptor (green).

(Continued)

EXPERIMENT 8-11 (CONTINUED) DEMONSTRATION OF FRET IN SOLUTIONS

Why fluorescence emission anisotropies of the donor and acceptor are affected by FRET? FRET decreases the fluorescence lifetime of the donor, therefore, the anisotropy of the donor in the presence of the acceptor is higher (less time for rotational depolarization). The observed fluorescence within the acceptor emission is a combination of three fractions: (1) Directly excited acceptor; (2) donor fluorescence; and (3) acceptor fluorescence after the FRET. In the case of significant FRET, the third fraction (not polarized emission) is significant, therefore acceptor anisotropy decreases.

Time-resolved Measurements. Below, we show the demonstration of FRET in time-resolved (lifetime) measurements.

FRET process depopulates the excited state of the donor which results in the decrease of intensity and lifetime. In solution, the acceptors are randomly distributed around donors. Some acceptor molecules are closer; some are father from the donor. Therefore, some donors are quenched faster which demonstrates in non-homogeneous donor decay in the presence of acceptor. The lifetime of the donor becomes heterogeneous and can be correctly fitted with a proper decay low which includes energy transfer. In a low viscous solution, the diffusion of the molecules should be also included. The heterogeneity of the lifetimes is immediately seen if the decay requires more terms in the multi-exponential approximation.

The donor (U) decay can be satisfactorily fitted with a single exponent of 3.65 ns, see Figure 8-11.7.

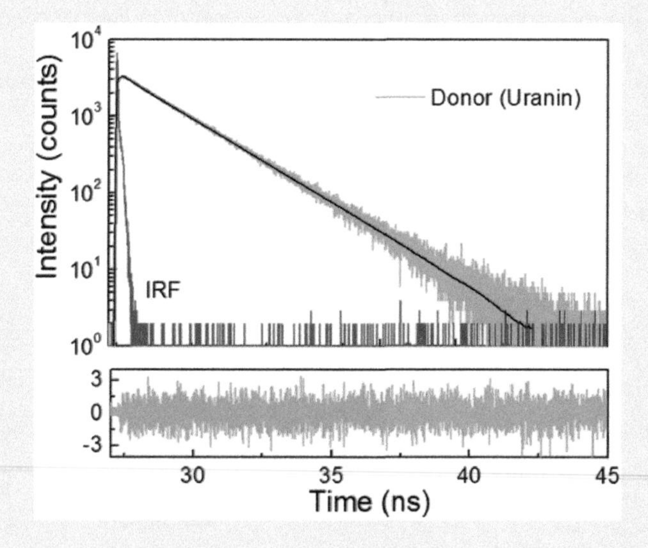

FIGURE 8-11.7 The intensity decay of the donor (U) in propylene glycol. The excitation was 470 nm and observation 520 nm.

(*Continued*)

EXPERIMENT 8-11 (CONTINUED) DEMONSTRATION OF FRET IN SOLUTIONS

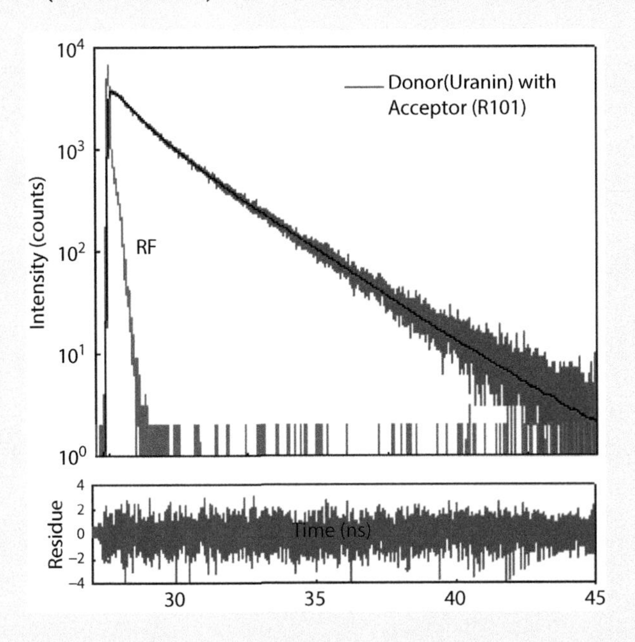

FIGURE 8-11.8 The intensity decay of the donor (U) in the presence of 1 mM of the acceptor (R101). The excitation was 470 nm and observation 520 nm.

In contrast, the intensity decay of the donor (U) in the presence of an acceptor (R101) is strongly heterogeneous and cannot be approximated with a single exponent decay model, see Figure 8-11.8. In order to satisfactorily fit this decay, three exponents are needed. The average lifetime of this decay is 1.96 ns (intensity weighted) and 1.34 ns (amplitude weighted).

Which average lifetime should be used for calculation of the transfer efficiency and the comparison with spectral (steady-state) data? The total amount of emitted photons is proportional to the area under the decay curve. Therefore, the correct average for comparison with spectral data and calculation of the transfer efficiency is the amplitude weighted lifetime, in this case, 1.34 ns. The ratio of the average lifetime of the donor in the absence and in the presence of acceptor is 3.65/1.34, about 2.7. The ratio of the fluorescence intensities (see Figure 8-11.4) is about 2.6, which is in excellent agreement with the lifetime's data.

It is interesting to compare the lifetimes of the acceptor alone and acceptor in the presence of the donor. In the absence of the donor, the acceptor decay (Figure 8-11.9) can be approximated with two positive exponents (4.16 ns and 0.49 ns) with amplitude weighted average of 3.38 ns.

(Continued)

EXPERIMENT 8-11 (CONTINUED) DEMONSTRATION OF FRET IN SOLUTIONS

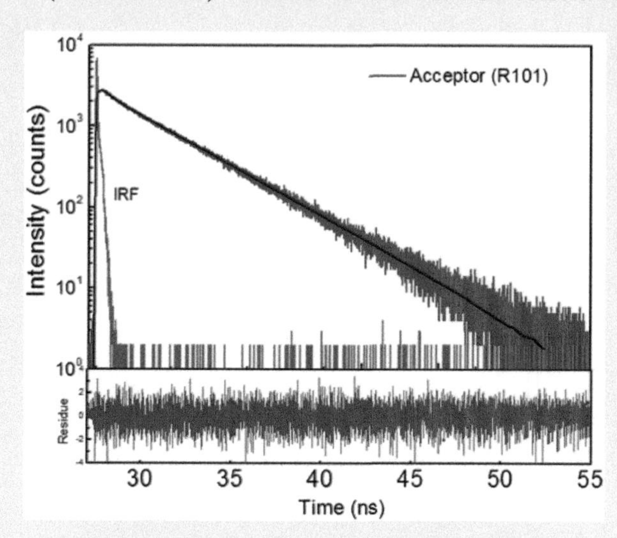

FIGURE 8-11.9 The intensity decay of the acceptor (R101) alone in PG with the excitation at 470 nm and observation at 595 nm.

In the presence of the donor, the acceptor decay cannot be longer fitted with positive exponents. A significant part of the acceptor excitation is due to FRET. It means that over time, the population of excited acceptor molecules is reinforced. We observe a "pumping" of the acceptor excited state. This effect manifest itself with an upper-curvature of the decay profile, see Figure 8-11.10.

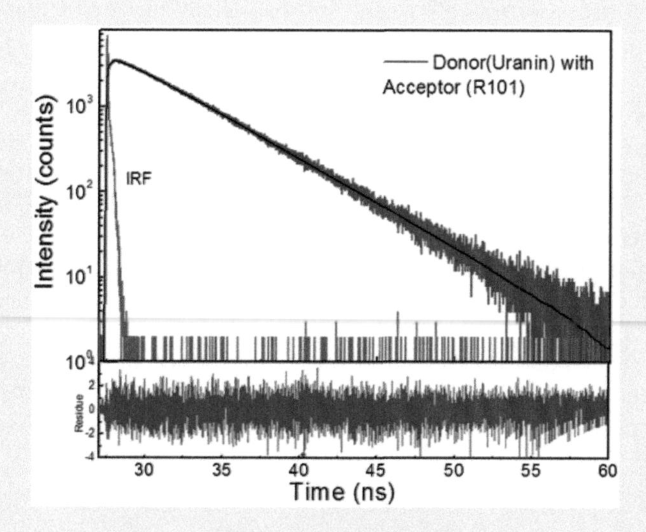

FIGURE 8-11.10 The intensity decay of the acceptor (R101) in the presence of the donor (U) in PG. PG with the excitation at 470 nm and observation at 595 nm.

(Continued)

EXPERIMENT 8-11 (CONTINUED) DEMONSTRATION OF FRET IN SOLUTIONS

In order to satisfactorily fit this decay, a negative component is required. This decay can be fitted with two exponents, one, 4.2 ns with a positive amplitude of 0.81, and second, 1.3 ns with a negative amplitude of 0.19.

The upper-curvature of an intensity decay indicates that an excited state process is involved in the photophysics. In this case, it is a FRET from U donor to R101 acceptor. The energy transfer process occurs during the lifetime of the donor; over all of this time, the R101 molecules are accepting the excitation energy from U donor. Other processes include an electron or proton transfer in the excited state. The upper-curvature is also observed in anisotropy measurements in solutions when observing the perpendicular (horizontal) component. *Note:* Absorptions of the "sandwiched" samples may change in time. We recommend checking the absorption before and after fluorescence/anisotropy measurements.

Note: Results from this experiment were published in *"Methods and Applications in Fluorescence. Vol. 4, (2016) 015001."* Figures have been used with permission from IOP.

EXPERIMENT 8-12 ASSOCIATED AND NOT ASSOCIATED ANISOTROPY DECAYS

Keywords: fluorescence, anisotropy decay, BSA, ANS.

Individual fluorescent species, fluorophores, or labeled macromolecules usually display anisotropy decays which can be satisfactorily approximated with single exponential or two/three exponential models. An interesting case is when fluorescence emission occurs from two fluorescent species with distinct lifetimes and correlation times. At any time, the fluorescence anisotropy decay of two emitting species is given by: $r(t) = r_1 f_1 + r_2 f_2$, where r_i are anisotropies of individual species and f_i are normalized fractional intensities of these species. If the considered fluorophores have different lifetimes, the relative fractions f_i will change during the emission because the short-lived fluorophores will depopulate faster from the excited state and the observed emission will be dominated by the longer-lived fluorophore. This may happen when monitoring fluorophore binding to a larger macromolecule (e.g., protein or DNA). We will demonstrate such a case for mixtures of ANS fluorophore and BSA protein. The free (unbound) ANS has a short lifetime and correlation times and ANS bound to BSA protein has a long lifetime and long correlation time. Of course, the effect depends on the ratio of ANS/BSA molecules.

MATERIALS

- Spectrophotometer (Cary 50, Varian, Inc.)
- 8-Anilino-1-naphthalenesulfonic acid (ANS), bovine serum albumin (BSA)
- Water, glass or plastic vials
- 1 cm × 1 cm disposable fluorometric cuvettes

(Continued)

EXPERIMENT 8-12 (CONTINUED) ASSOCIATED AND NOT ASSOCIATED ANISOTROPY DECAYS

EQUIPMENT

- Spectrophotometer (Cary 50, Varian, Inc.)
- Spectrofluorometer (Eclipse, Varian, Inc.)
- Fluorometer (FT200, PicoQuant, GmbH)

METHODS

1. Make stock solutions of ANS and BSA in a few milliliters of water; use glass or plastic vials. Be sure that both compounds dissolved completely.
2. Using stock solutions prepare samples: ANS in water (Figure 8-12.1), BSA + ANS with the ratio 20:1 (Figure 8-12.2, left) and BSA + ANS with the ratio 1:20 (Figure 8-12.2, right) keeping the same concentration of ANS of about 20 μM in all three samples.
3. Measure fluorescence spectra, see Figures 8-12.1 and 8-12.2.
4. Measure the lifetimes of all three samples (Figures 8-12.3 and 8-12.4).
5. Measure anisotropy decays of all three samples (Figures 8-12.5 and 8-12.6).

RESULTS

Greenish fluorescence of ANS in water is weak. In the presence of BSA, however, it becomes bluish and much stronger with the maximum shifted to about 480 nm, see Figures 8-12.1 and 8-12.2.

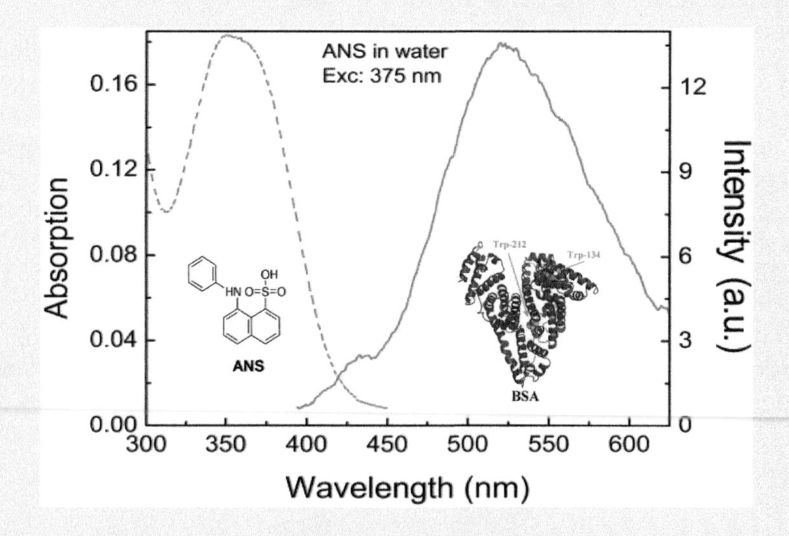

FIGURE 8-12.1 Absorption and emission spectra of ANS in the water at room temperature.

First, we measure absorption and emission spectra of samples with different ratios of BSA/ANS, see Figure 8-12.2.

(Continued)

EXPERIMENT 8-12 (CONTINUED) ASSOCIATED AND NOT ASSOCIATED ANISOTROPY DECAYS

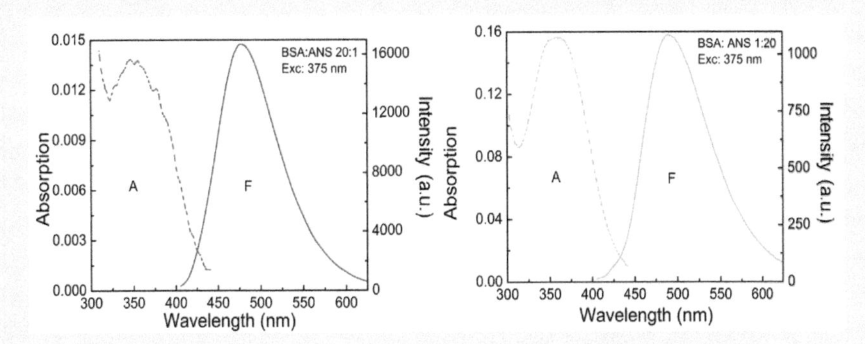

FIGURE 8-12.2 Left: Absorption and emission spectra of BSA-ANS (20:1) mixture in the water at room temperature. Right: Absorption and emission spectra of BSA-ANS (1:20) mixture in water at room temperature.

With the excess of BSA (Figure 8-12.2, left) the emission spectrum is short wavelength shifted. At the ratio of 20:1 of BSA:ANS, it is safe to assume that almost all ANS molecules are bounded to BSA protein. At the reversed ratio of BSA:ANS (1:20), Figure 8-12.2 (right), the maximum of emission is slightly shifted to longer wavelengths. Some of ANS molecules are not bounded to BSA and remain free in the solution.

Next, we measure the lifetimes of free ANS in water (Figure 8-12.3, left) and bound to BSA (Figure 8-12.3, right). There is a drastic difference in the intensity decays. The free ANS has a very short lifetime, well below 1 ns, whereas ANS bound to BSA has a relatively very long lifetime, above 16 ns. Both intensity decays can be satisfactorily fitted with a single exponent model.

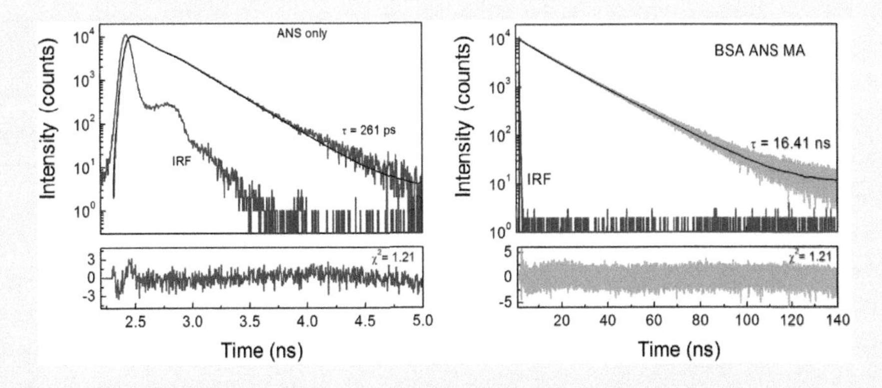

FIGURE 8-12.3 Left: Fluorescence intensity decay of ANS in water. Right: Fluorescence intensity decay of BSA-ANS (20:1) mixture in water.

In contrast, the intensity decay of the sample with the excess of ANS (1:20) cannot be fitted with a single exponent model. The fit to two exponential model reveals lifetimes corresponding to free and bound ANS (0.3 ns and 16.5 ns), see Figure 8-12.4.

(Continued)

EXPERIMENT 8-12 (CONTINUED) ASSOCIATED AND NOT ASSOCIATED ANISOTROPY DECAYS

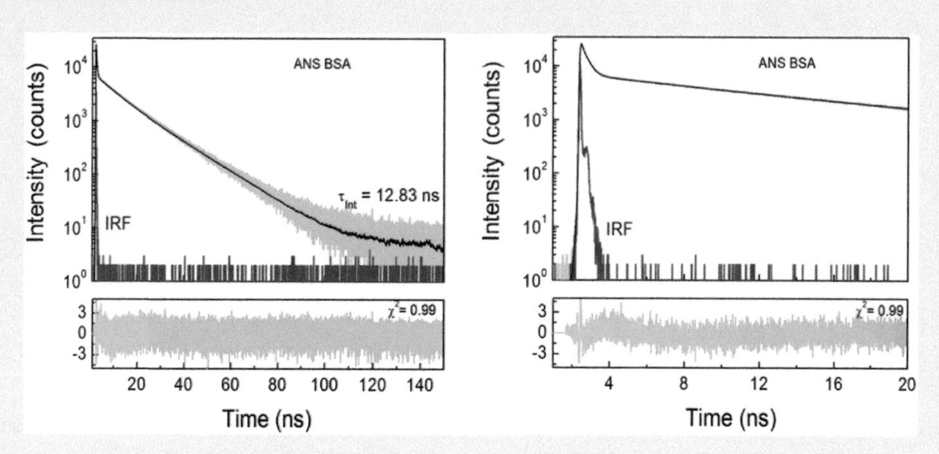

FIGURE 8-12.4 Fluorescence intensity decay of mixture BSA-ANS (1:20) in water. Two exponent fit. Left: full-time scale; right: first 20 ns of fluorescence intensity decay.

The anisotropy decays of free and bound ANS (Figure 8-12.5) can be fitted with single correlation times of 0.2 ns and 40 ns, respectively.

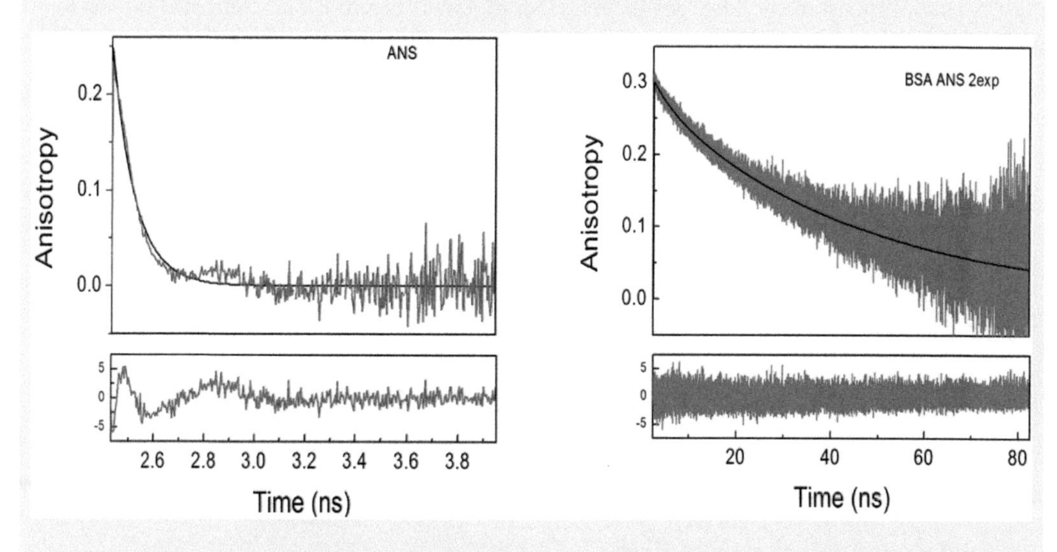

FIGURE 8-12.5 Left: Fluorescence anisotropy decay of ANS in water. Right: Fluorescence anisotropy decay of BSA-ANS (20:1) mixture in water.

The anisotropy decay of the sample with the excess of ANS (1:20) has an unusual profile, see Figure 8-12.6.

After an initial decrease, it rises again and then slowly decays.

(Continued)

EXPERIMENT 8-12 (CONTINUED) ASSOCIATED AND NOT ASSOCIATED ANISOTROPY DECAYS

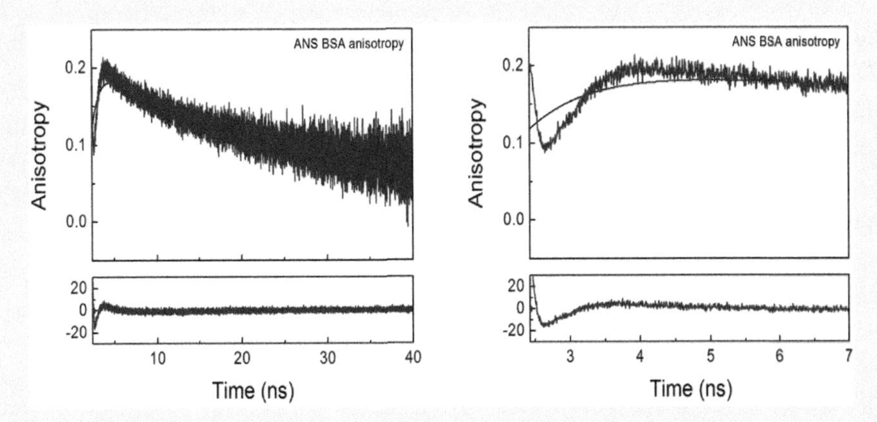

FIGURE 8-12.6 Fluorescence anisotropy decay of BSA-ANS (1:20) mixture in water. Left: full-time scale; right: first few ns of the anisotropy decay.

CONCLUSIONS

In the presence of an excess of ANS, we could clearly observe the fast anisotropy decay corresponding to quickly rotating free ANS. Because the lifetime of free ANS is much shorter than bound ANS, its fraction quickly decreases and after about 1 ns a small fraction of bound ANS starts dominate the signal. However, bound ANS rotates much slower and contributes to higher anisotropy. This is the reason why after 1 ns, we observe an anisotropy increase. Such behavior can be observed when relative fractions of two species are changing during the excited state lifetime. It was originally observed for systems that undergo reaction (is changing) during the excited state lifetime. Associated anisotropy decays illustrate nicely the advantage of anisotropy notation over polarization. The additive properties of anisotropy allow predicting anisotropy values at any time of the decay. The fraction associated with the free ANS molecule decreases quickly and after about 1 ns the decay is dominated by the bound ANS. Such measurements can be used to determine presence and amounts of unbound fluorophores. This anomalous anisotropy decay cannot be fitted with a simple exponential model unless a negative pre-exponential factor is used. However, the negative amplitude is not realistic and cannot be used for quantitative interpretation. But using the proper model of an excited state reaction will yield, a good description (fit) of the observed process.

EXPERIMENT 8-13 NON-EXPONENTIAL ANISOTROPY DECAY OF PERYLENE

Keywords: fluorescence, anisotropy decay, perylene.

If a planar molecule is not interacting with solvent molecules, there are possible various rotations. Rotations in-plane do not require significant solvent molecules displacement and are faster. Rotation out-of-plane will have to displace solvent molecules and are slower. If both

(*Continued*)

EXPERIMENT 8-13 (CONTINUED) NON EXPONENTIAL ANISOTROPY DECAY OF PERYLENE

rotations (fast frequently called slippery) and slow are leading to the emission transition moment displacement, both rotations will lead to different anisotropy decays. This results in a non-exponential anisotropy decay. We will demonstrate a non-exponential decay of symmetric non-polar molecule perylene in a medium viscosity solvent-triacetin. For comparison, we will demonstrate a single exponential anisotropy decay of a polar molecule—Coumarin 152.

MATERIALS

- Perylene (Per), Coumarin 152 (C152)
- Triacetin, glass or plastic vials
- 1 cm × 1 cm fluorometric cuvettes

EQUIPMENT

- Spectrophotometer (Cary 50, Varian, Inc.)
- Spectrofluorometer (Eclipse, Varian, Inc.)
- Fluorometer (FT200, PicoQuant, GmbH, Germany)

METHODS

1. Make stock solutions of C152 and perylene in triacetin.
2. Prepare samples using stock solutions and measure absorptions. Adjust concentrations to achieve about 0.1 optical densities at maxima. Measure absorption and fluorescence spectra, see Figure 8-13.1.
3. Measure lifetimes of both C152 and perylene samples, see Figure 8-13.2.
4. Measure fluorescence anisotropy decays of both samples.

Intensity and anisotropy decays will be fitted to multi-exponential models:

$$I(t) = \sum_i \alpha_i \exp(-t/t_i)$$

and

$$r(t) = \sum_i r_i \exp(-t/\theta_i)$$

RESULTS

First, measure absorption and fluorescence spectra, see Figure 8-13.1. The fluorescence spectra, lifetimes, and anisotropy decays were measured at 405 nm excitation from a pulsed laser diode and the observation wavelength was set at 475 nm for both molecules.

Intensity decays of both compounds can be well fitted with the single exponents with 4.69 ns for perylene and 5.02 ns for C152, see Figure 8-13.2.

Next, measure anisotropy decays of both compounds. Remember about correction for the G-factor. Fit both anisotropic decays with one correlation time model, see Figure 8-13.3.

(*Continued*)

EXPERIMENT 8-13 (CONTINUED) NON EXPONENTIAL ANISOTROPY DECAY OF PERYLENE

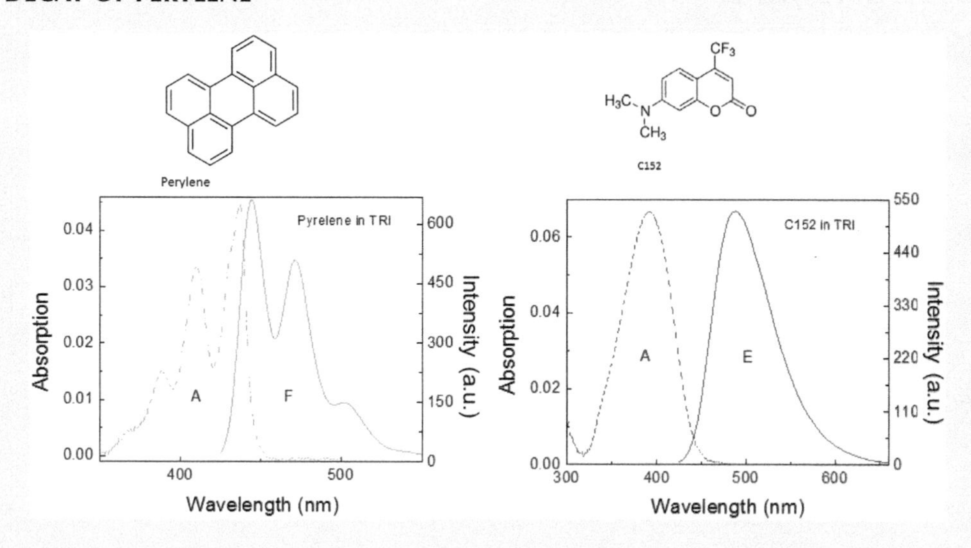

FIGURE 8-13.1 Absorption and emission spectra of perylene (left) and C152 (right) in triacetin at room temperature. Top: Chemical structures of perylene and C152.

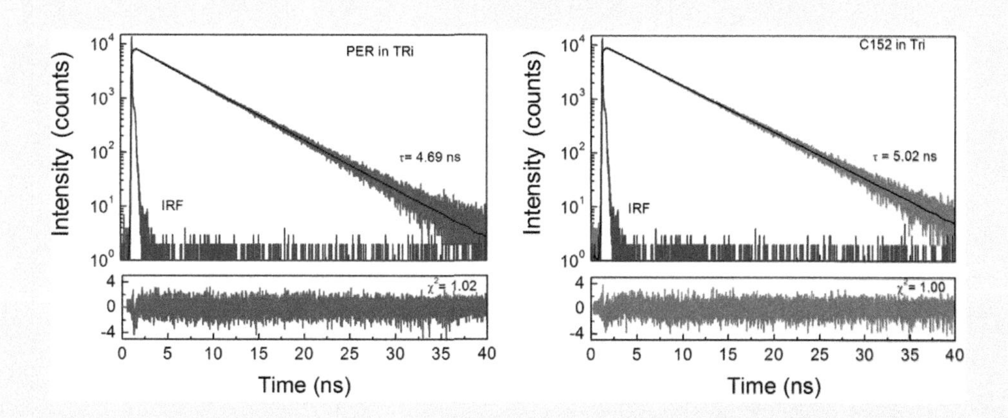

FIGURE 8-13.2 Fluorescence intensity decays of perylene (left) and C152 (right) in triacetin at room temperature. Recovered lifetimes are: perylene: 4.69 ± 0.02 ns; C152: 5.02 ± 0.02 ns.

While the anisotropy decay of C152 can be well fitted with one correlation time 0.84 ± 0.01 ns (Figure 8-13.3, right), the fit for perylene is not satisfactory. It shows a systematic deviation of residuals and rather large goodness of fit (6.67).

Next, we apply two correlation times fit, see Figure 8-13.4. Now the fit is good with two components of 0.166 ns and 0.663 ns. These two correlation times are well separated as demonstrated by χ_R^2 surface analysis; see Figure 8-13.5.

The χ_R^2 surface analyses illustrate the resolutions in correlation times, see Figure 8-13.5.

(Continued)

EXPERIMENT 8-13 (CONTINUED) NON EXPONENTIAL ANISOTROPY DECAY OF PERYLENE

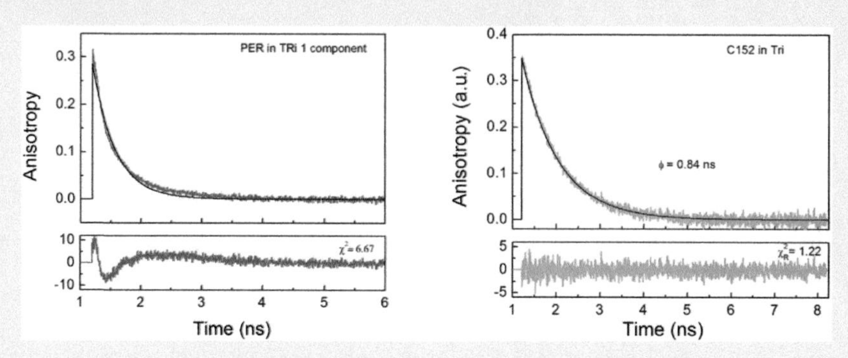

FIGURE 8-13.3 Fluorescence anisotropy decays of perylene (left) and C152 (right) in triacetin at room temperature. Single-exponential model was used for the analysis.

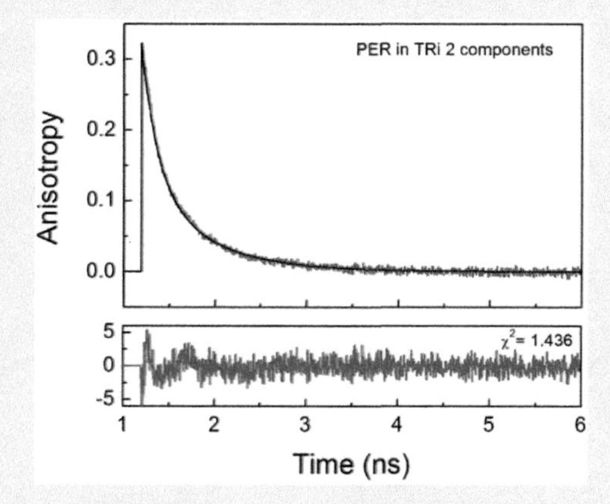

FIGURE 8-13.4 Anisotropy decay of perylene in triacetin analyzed with a two-exponential model. Recovered correlation times are: 0.17 ± 0.01 ns and 0.66 ± 0.01 ns, with anisotropies 0.19 and 0.13, respectively.

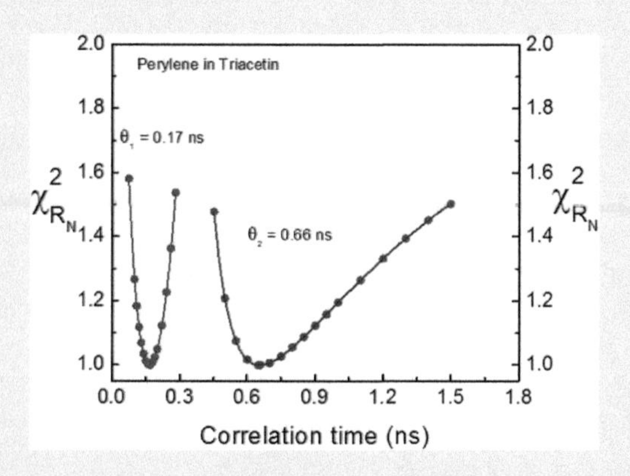

FIGURE 8-13.5 χ_R^2 surface analyses for correlation times of perylene in triacetin.

(*Continued*)

EXPERIMENT 8-13 (CONTINUED) NON EXPONENTIAL ANISOTROPY DECAY OF PERYLENE

CONCLUSIONS

Why C152 anisotropy decay can be fitted with one correlation time and perylene needs two correlation times? The C152 molecule is relatively polar and has a structure which is close to a spherical shape (note the side groups on the structure). In contrast, perylene is a planar non-polar molecule with a disk-like shape. The rotations in-plane are faster than rotations out of-plane. The ratio of correlations times is about 4, which can be interpreted that shorter correlation time (0.17 ns) corresponds to slippery rotations in-plane.

EXPERIMENT 8-14 ROOM TEMPERATURE PHOSPHORESCENCE

Keywords: fluorescence, phosphorescence, 5,6-benzoquinoline, PVA films.

A radiative transition from the lowest excited triplet state (T_1) of the molecule to its ground state (S_0) is responsible for phosphorescence emission. By the symmetry rule, this is a forbidden transition with a very low probability. Therefore, very low efficiencies and very long lifetimes are expected for the phosphorescence. In addition, the long lifetime enables an efficient quenching of molecules by oxygen. In effect, the phosphorescence is practically not observed in solutions. Of course, such measurements are possible but require a careful oxygen removal from the solution, low temperature, and special precautions. However, when using very viscous solutions or polymer matrices that has low permeability for oxygen, phosphorescence can be easily detected. In this experiment, we will demonstrate the phosphorescence of 5,6-benzoquinoline embedded in the PVA film.

MATERIALS

- 5, 6-Benzoquinoline (BQ), poly(vinyl) alcohol (PVA, 130,000 MW)
- Glass vials, pipettes, Petri dishes, water, ethanol

EQUIPMENT

- Spectrophotometer (Cary 50, Varian, Inc.)
- Spectrofluorometer (Eclipse, Varian, Inc.)
- UV illuminator, camera or cell phone

METHODS

1. Prepare about 100 mL solution of 10% (w/w) PVA, see Experiment 3-24.
2. Make a stock solution of BQ in ethanol.
3. Pour about 20 mL of liquid PVA into few vials. One vial with PVA only will serve as a reference. To other vials, add different amounts of the BQ stock solution, for example, 0.1 mL, 0.2 mL, 0.3 mL,…, and shake. Be sure that solutions are homogeneous. With such preparation, the PVA films will contain progressively higher concentrations of BQ.
4. Pour solutions to Petri dishes and leave for drying. It could take few days until films will be completely dry.

(Continued)

EXPERIMENT 8-14 (CONTINUED) ROOM TEMPERATURE PHOSPHORESCENCE

5. Peel the dry films from Petri dishes. Measure their absorptions using PVA only film as a baseline. Select the film with absorption 0.5–1 in the maximum, see Figure 8-14.1. For a precise absorption measurement in PVA film, see Experiment 3-25.
6. Put the selected film on the UV illuminator. Observe fluorescence (Figure 8-14.2) and phosphorescence (Figure 8-14.3). *Note: Phosphorescence can be seen after switching off illuminator for a few seconds.*
7. Measure the fluorescence spectrum of the BQ-doped PVA film using a front-face geometry, see Figure 8-14.4.
8. Measure the phosphorescence spectrum of the film. For this, you need to set the spectrofluorometer in a *phosphorescence* mode, see Figure 8-14.5. *Note: This is possible with Varian Eclipse but not all spectrofluorometers offer such option. However, it is relatively easy to adopt a typical spectrofluorometer to be able to measure such phosphorescence.*
9. Measure the phosphorescence spectra at 20, 40, and 60°C.

RESULTS

From prepared dried films, select one with relatively high absorption of about 1. The goal of this experiment is to observe a phosphorescence which will appear at longer wavelengths far from BQ absorption and will not be reabsorbed by BQ.

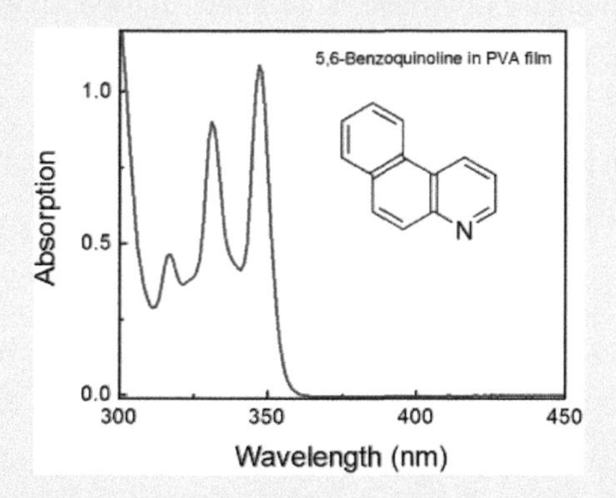

FIGURE 8-14.1 Absorption spectrum of BQ-doped PVA film.

Placed on the UV illuminator, the BQ-doped film has a dark blue emission, brightest on the edges of the film, see Figure 8-14.2.

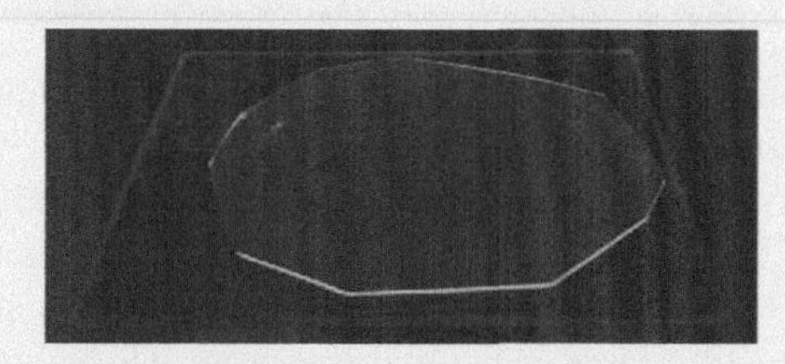

FIGURE 8-14.2 BQ-doped PVA film on UV illuminator. Dark blue fluorescence emission is brightest on the edges.

(*Continued*)

EXPERIMENT 8-14 (CONTINUED) ROOM TEMPERATURE PHOSPHORESCENCE

However, after switching-off the illuminator, the film is still glowing for about 2–3 seconds in a blue-greenish color, see Figure 8-14.3. Of course, this cannot be fluorescence emission, which is dark blue, mostly below 400 nm, see Figures 8-14.2 and 8-14.4.

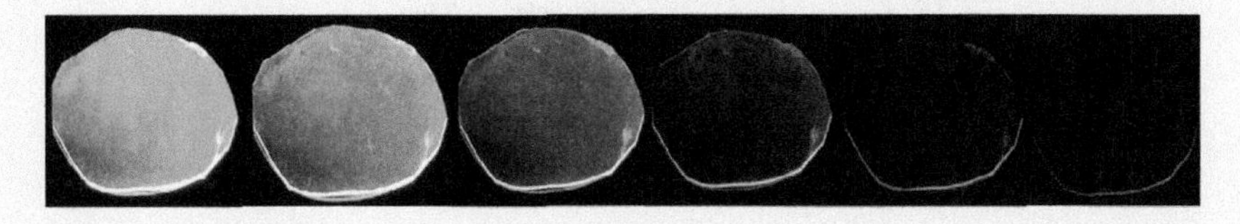

FIGURE 8-14.3 Snap-shots photographs taken immediately after switching the illuminator. Entire series of photographs was taken within about 2 seconds.

Figure 8-14.4 shows a fluorescence spectrum of BQ in PVA film. Looking closer to the long-wavelength part of this spectrum, we see traces of structured emission near 500 and 550 nm. At 640 nm, it is seen a harmonic from the 320 nm excitation.

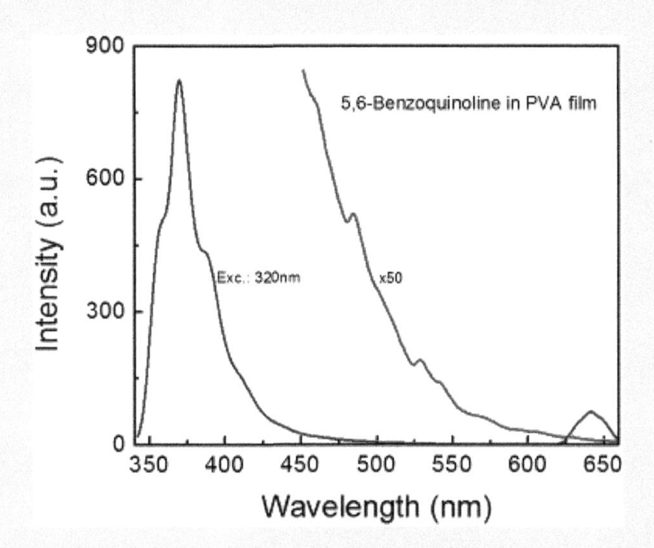

FIGURE 8-14.4 Fluorescence spectrum of BQ-doped PVA film.

Next, turn the spectrofluorometer to the phosphorescence mode and measure the phosphorescence spectrum, see Figure 8-14.5. In this figure are included parameters required for the phosphorescence measurement when using Varian Eclipse. You may try to change these parameters and optimize detection.

The BQ-doped PVA film has a structured phosphorescence centered at 500 nm with peaks at 450, 500, and 550 nm.

We measured BQ phosphorescence at a few higher temperatures, see Figure 8-14.6. The phosphorescence strongly depends on temperature, at 60°C, the intensity drops to only 10% compared to room temperature.

(*Continued*)

EXPERIMENT 8-14 (CONTINUED) ROOM TEMPERATURE PHOSPHORESCENCE

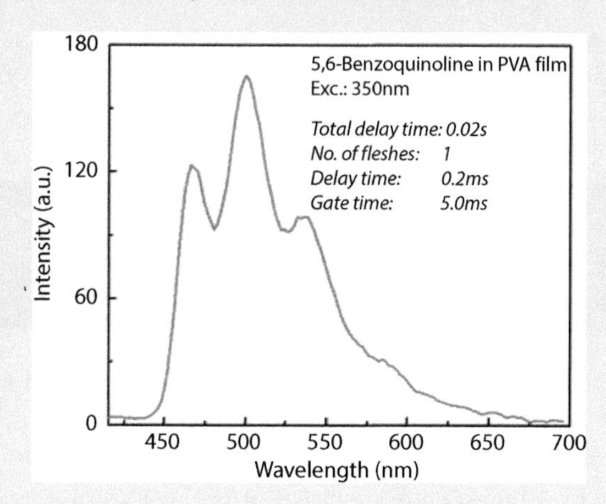

FIGURE 8-14.5 Phosphorescence spectrum of BQ-doped PVA film. In italics are parameters used in the measurement.

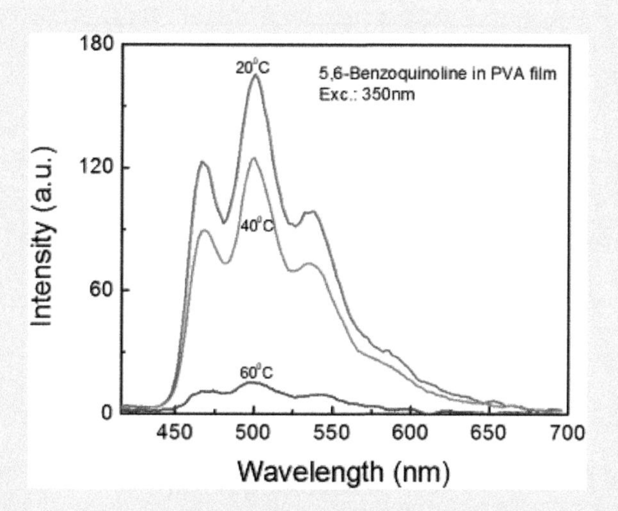

FIGURE 8-14.6 Temperature dependence of BQ-doped PVA film phosphorescence.

CONCLUSIONS

It is possible to measure phosphorescence emission in rigid polymer films. In the case of BQ-doped PVA film, the phosphorescence lifetime is close to 1 s, which is easily detectable by human eyes. Are any potential applications of BQ-doped PVA films? One would be a temperature sensor. With a proper calibration, the phosphorescence intensity and lifetime will indicate temperatures with a high precision. Another potential application will be the detection of the fluorescence/phosphorescence ratio which dramatically changes in time. This unique feature can be applied in counterfeiting and marking valuable documents.

EXPERIMENT 8-15 DELAYED FLUORESCENCE

Keywords: fluorescence, delayed fluorescence, phosphorescence, Eosin Y, PVA films.

Delayed fluorescence (DF) belongs to the long-lived luminescence phenomena. The transition responsible for DF is still $S_1 \rightarrow S_0$, but with the involvement of the triplet state T_1. After non-radiative intersystem transition from S_1 to T_1, part of the molecules can be activated thermally back into the S_1 state. This can happen only if there is not a big difference between energies of S_1 and T_1 states. Naturally, the DF emission spectrum should be no different from the normal fluorescence spectrum. However, the lifetime of DF fluorescence is much longer because of the triplet state involvement. It is also expected that the DF intensity will increase with the temperature (see Boltzmann factor) in contrast to the phosphorescence. We will demonstrate these features with an example of Eosin Y embedded in a poly(vinyl) alcohol matrix.

MATERIALS

- Eosin Y, poly(vinyl) alcohol (PVA, 130,000 MW)
- Glass vials, pipettes, Petri dishes, water, ethanol

EQUIPMENT

- Spectrophotometer (Cary 50, Varian, Inc.)
- Spectrofluorometer (Eclipse, Varian, Inc.)
- UV illuminator, camera or cell phone

METHODS

1. Prepare about a 100 mL solution of 10% (w/w) PVA, see Experiment 3-24.
2. Make a stock solution of Eosin Y in ethanol.
3. Pour about 20 mL of liquid PVA into a few vials. One vial with PVA only will serve as a reference. To other vials add different amounts of the eosin Y stock solution, for example, 0.1 mL, 0.2 mL, 0.3 mL,..., and shake. Be sure that solutions are homogeneous. With such preparation, the PVA films will contain progressively higher concentrations of Eosin Y. The solutions should have a light peach color.
4. Pour solutions to Petri dishes and leave for drying. It could take a few days until the films will be completely dry.
5. Peel the dry films from Petri dishes. Measure their absorptions using PVA only film as a baseline. Select the film with absorption 0.05–0.1 in the maximum, see Figure 8-15.1.
6. Measure the fluorescence spectrum of the Eosin Y-doped PVA film using a front-face geometry. Use the excitation of 500 nm and scan emission wavelengths from 520 nm to 820 nm, see Figure 8-15.3.
7. Measure luminescence spectrum of the film. For this, you need to set the spectrofluorometer in a *phosphorescence* mode, see Figure 8-15.4.
8. Measure the luminescence spectra in room temperature and slightly heated, see Figures 8-15.5.

(*Continued*)

EXPERIMENT 8-15 (CONTINUED) DELAYED FLUORESCENCE

RESULTS

Absorption of Eosin Y occurs in a blue/green region of the spectrum and the color of a low concentration solution/film is a light peach. The fluorescence emission is green-to-yellow; see Figures 8-15.1 and 8-15.2.

Figure 8-15.2. Eosin Y-doped PVA film on UV illuminator.

The fluorescence spectrum of Eosin Y-doped PVA film in room temperature has a maximum near 550 nm. In a long-wavelength tail of the fluorescence spectrum, at about 675 nm, a small irregularity can be recognized, see Figure 8-15.3.

The measurements in a phosphorescence mode with a time delay reveal long-lived phenomena—delayed fluorescence (DF) and phosphorescence, see Figure 8-15.4. DF emission overlaps precisely the fluorescence spectrum of Eosin Y-doped PVA film (compare with Figure 8-15.3). The phosphorescence is centered at 675 nm and extends up to 800 nm.

How to confirm that measured DF is not a leak of normal fluorescence? We selected a Rhodamine 6G (R6G)-doped PVA film with similar fluorescence intensity as Eosin Y-doped PVA film and measured it in a phosphorescence mode. There is no detected DF or

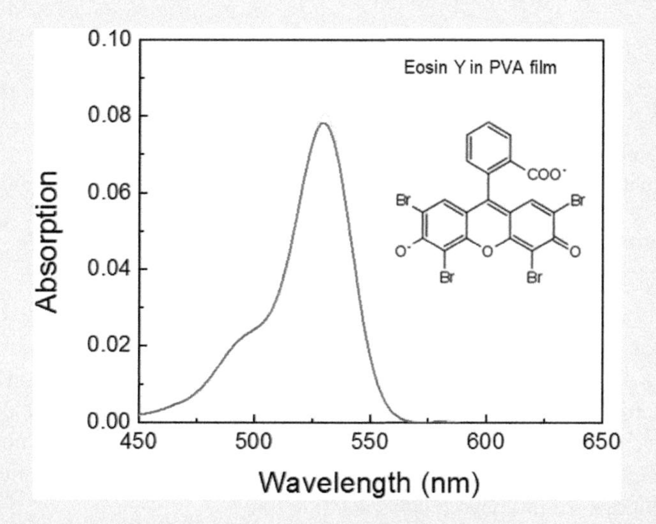

FIGURE 8-15.1 Absorption spectrum of Eosin Y-doped PVA film.

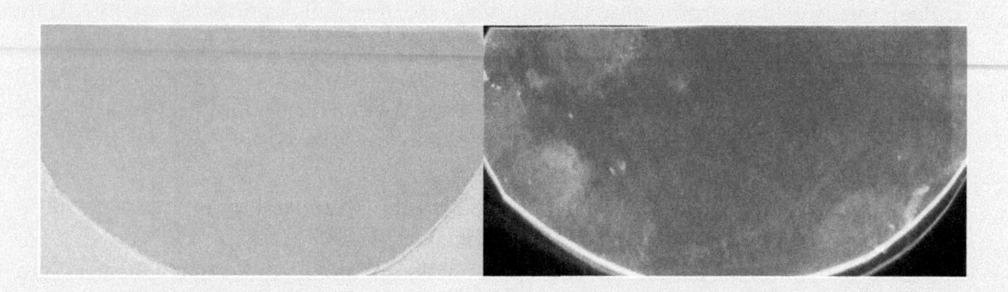

FIGURE 8-15.2 Eosin Y-doped PVA film. Left: In the room light. Right: On the UV illuminator.

(Continued)

EXPERIMENT 8-15 (CONTINUED) DELAYED FLUORESCENCE

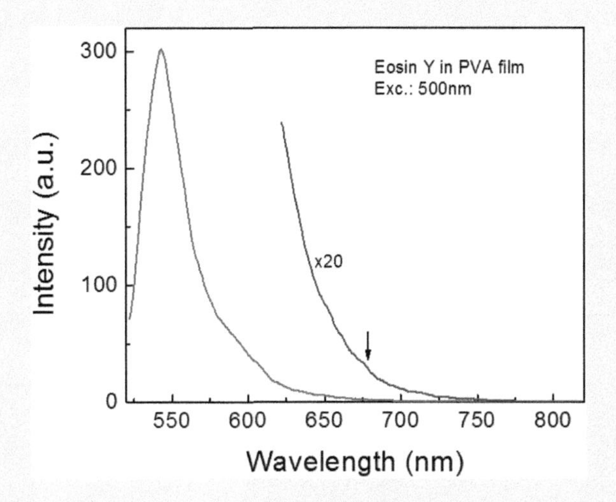

FIGURE 8-15.3 Fluorescence spectrum of Eosin Y-doped PVA film. The arrow indicates presence of phosphorescence.

phosphorescence from R6G, see Figure 8-15.4. R6G has fluorescence with almost absolute quantum yield and in room temperature, DF and phosphorescence are not detectable.

Next, we measured Eosin Y-doped PVA film in phosphorescence mode at higher temperature, see Figure 8-15.5. While the phosphorescence intensity slightly decreases, the DF intensity significantly increases. This is additional proof that observed short-wavelength emission is in fact DF. Normal fluorescence usually decreases with temperature.

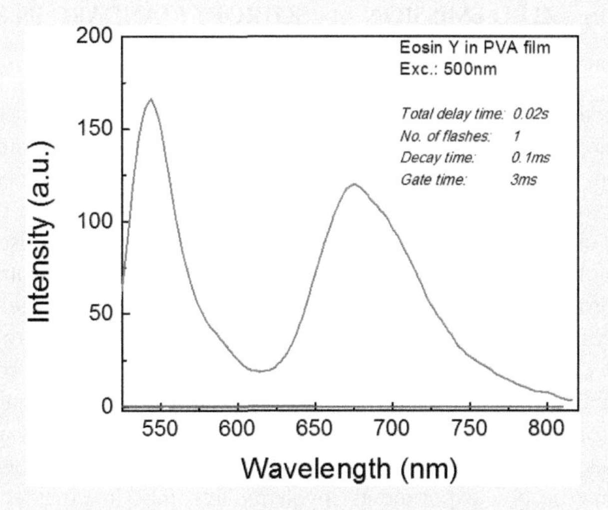

FIGURE 8-15.4 Luminescence spectrum of Eosin Y-doped PVA film. In italics are parameters used in the measurement. The green line is a signal from R6G-doped PVA film.

(Continued)

EXPERIMENT 8-15 (CONTINUED) DELAYED FLUORESCENCE

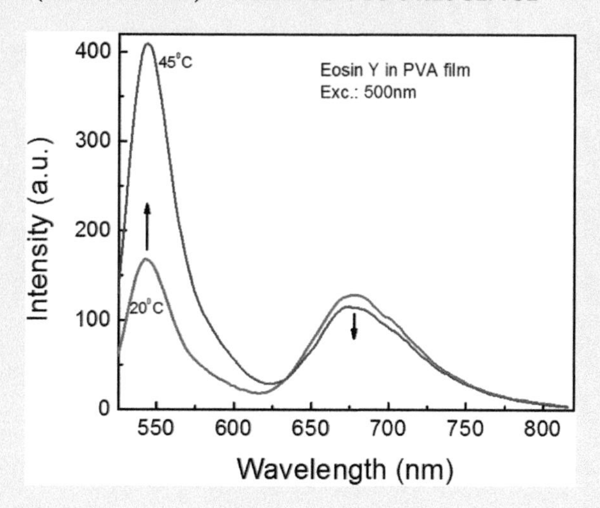

FIGURE 8-15.5 Temperature dependence of Eosin Y-doped PVA film luminescence.

CONCLUSIONS

Thermally activated DF is also called delayed fluorescence type E because first was observed for Eosin. Luminescence spectra of Eosin Y offer simultaneous observation of DF and phosphorescence. The ratio of intensities at 550/675 nm is very sensitive to the temperature and can be used as a ratiometric sensor.

EXPERIMENT 8-16 ZERO EMISSION ANISOTROPY STANDARD IN SOLUTIONS

Keywords: fluorescence, anisotropy, anisotropy standard.

Due to photoselection in the absorption process, fluorescence emission is partially polarized. Maximum anisotropy for an isotropic dye solution is 0.4 (fundamental anisotropy). However, even in frozen solutions lower than 0.4 values of fluorescence anisotropies are observed because of limited mobility and torsional motions of dye molecules. These anisotropies, measured for immobilized molecules, are called limiting anisotropies. In liquid solutions where molecules rotate more freely observed, steady-state anisotropies depend on the ratio of a fluorophore fluorescence lifetime and correlation time (see Perrin equation). In most solvents with moderate viscosities, fluorescence anisotropies are low (below 0.1) because dye molecules rotate quickly during their lifetime. Dyes with rotational correlation times comparable to their lifetimes will show anisotropies of about half of their limiting anisotropies. Fluorophores with very long fluorescence lifetimes will show anisotropies close to zero. We use this fact in Experiment 6-6 for G-factor evaluation using low viscous solvent. In spectroscopy/microscopy experiments involving free (not in cuvette) samples, sometimes more time is needed for alignments and adjustments. In such case, the low viscous solvents, usually quickly vaporizing, cannot be used. Here, we will present a long-lived dye,

(Continued)

EXPERIMENT 8-16 (CONTINUED) ZERO EMISSION ANISOTROPY STANDARD IN SOLUTIONS

ruthenium-based metal-ligand complex, which shows near-zero emission anisotropy in more viscous and less vaporizing solvents at room temperature. We suggest the $[Ru(bpy)_3]^{2+}$ solution in propylene glycol as a potential standard for alignments of high-throughput screening (HTS) instruments and microscopes for polarization measurements.

MATERIALS

- Ruthenium bipyridyl, tris(2,2′-bipyridyl) ruthenium(II) chloride hexahydrate, $[Ru(bpy)_3]^{2+}$ (Ru)
- Propylene glycol (PG)
- 1 cm × 1 cm cuvettes

EQUIPMENT

- Spectrophotometer (Cary 50, Varian, Inc.)
- Spectrofluorometer (Eclipse, Varian, Inc.)
- Fluorometer (PicoQuant FT200)

METHODS

1. Prepare a stock-solution of Ru in water, sonicate.
2. Using the stock-solution make the solution of Ru in PG with an optical density of about 0.1 in maximum. Measure absorption spectra.
3. Measure fluorescence emission spectra of the Ru sample at magic angle (MA) conditions, see Figure 8-16.1.
4. Measure fluorescence emission spectra of the Ru sample at HH and HV polarizers' orientations. Calculate G-factors across fluorescence spectra, see Figures 8-16.2 and 8-16.3.
5. Measure fluorescence emission spectra of the Ru samples at VV and VH polarization conditions, see Figure 8-16.4. Correct the VH component for G-factor, see Figure 8-16.5.
6. Plot the emission anisotropy of Ru in PG, see Figure 8-16.6.

RESULTS

Absorption of Ru covers a visible spectral region from 350 nm to 550 nm and a broad fluorescence emission lies between 525 nm and 750 nm, which are the most desired regions in fluorescence spectroscopy/microscopy. Because of its very long emission lifetime, the anisotropy of Ru is expected to be low, close to zero.

First, we will measure fluorescence spectra and anisotropies of Ru in propylene glycol. For the excitation, we will use 440 nm, as shown in Figure 8-16.1.

First, we need to determine the G-factor for the spectrofluorometer. For this, we will measure polarized intensities when using horizontally oriented excitation polarizer (H). In an ideal instrumental system both V and H components of emission should be equal. In the real instruments, mostly because of a dispersing element-grating, H and V polarized light is transmitted differently, see Figure 8-16.2. In order to be equal to HV, HH-component must be adjusted (first index stays for the excitation, second for the emission). The adjustment factor can be calculated as a ratio of I_{HV}/I_{HH} intensity at each emission wavelength. For details on how to determine G-factor, see Chapter 6.

(*Continued*)

EXPERIMENT 8-16 (CONTINUED) ZERO EMISSION ANISOTROPY STANDARD IN SOLUTIONS

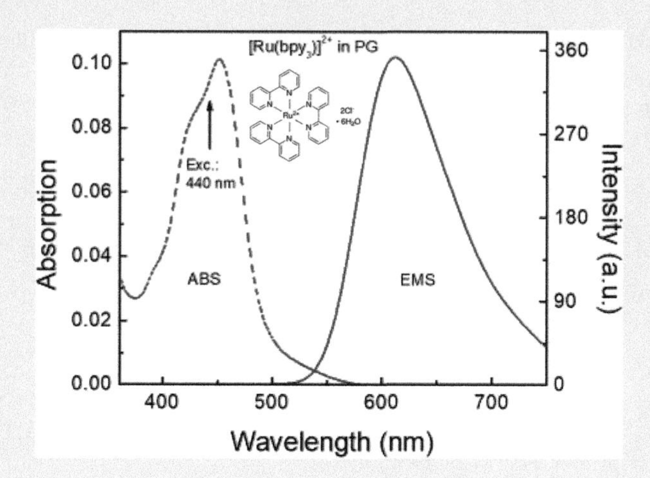

FIGURE 8-16.1 Absorption and emission spectra of $[Ru(bpy)_3]^{2+}$ in PG. Top: Chemical structure of $[Ru(bpy)_3]^{2+}$.

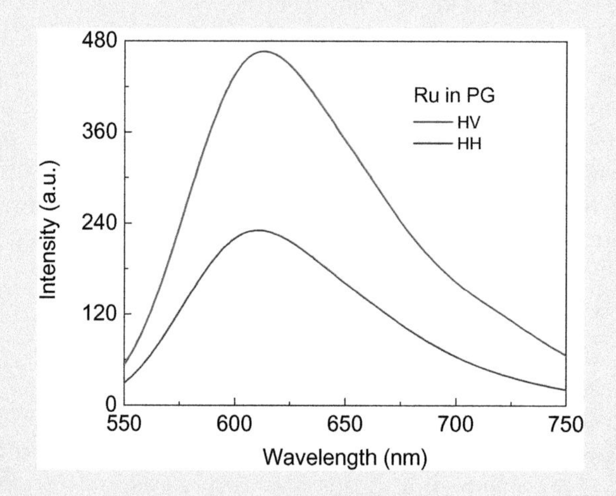

FIGURE 8-16.2 HV and HH components of Ru emission in PG.

The ratios of I_{HV}/I_{HH} are plotted in Figure 8-16.3.

In all future measurements, the measured I_{VH} intensities, see Figure 8-16.4, should be multiplied by the G-factor. The multiplication of I_{VH} by G-factor results in the correct polarized emission components, see Figure 8-16.5.

Corrected for G-factor intensities are used for anisotropy calculation according to:

$$r = \frac{I_{VV} - I_{VH}}{I_{VV} + 2I_{VH}}$$

(Continued)

EXPERIMENT 8-16 (CONTINUED) ZERO EMISSION ANISOTROPY STANDARD IN SOLUTIONS

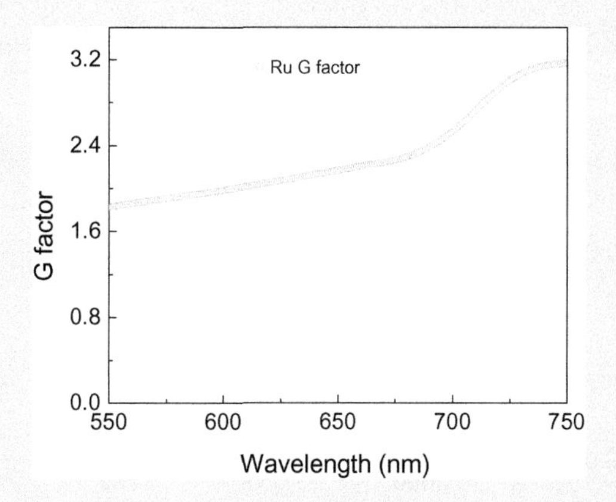

FIGURE 8-16.3 G-factor calculated from HV and HH components of Ru emission in PG.

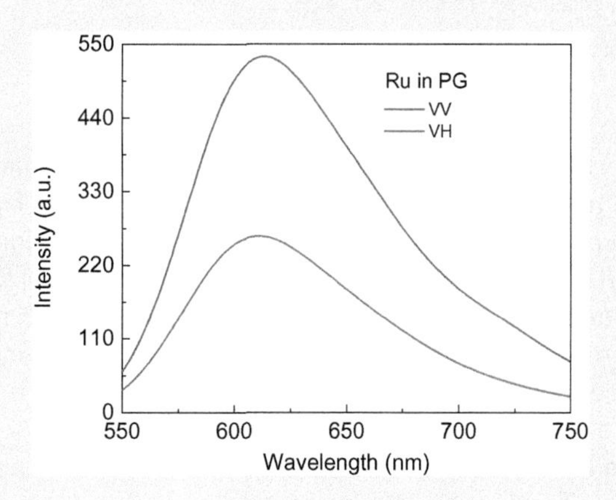

FIGURE 8-16.4 VV and VH components of Ru emission in PG. NOT corrected for G-factor.

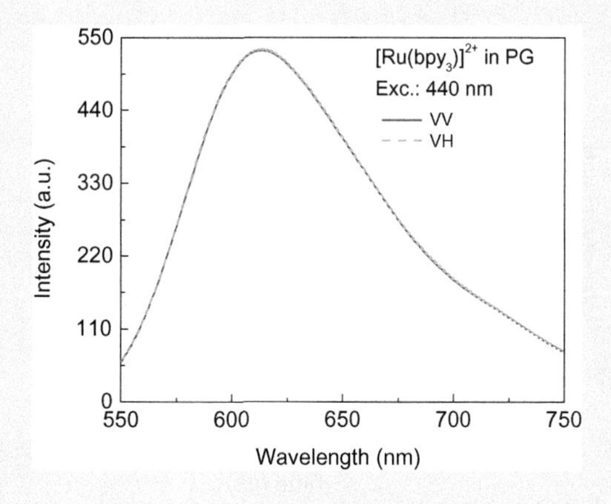

FIGURE 8-16.5 VV and VH components of Ru emission in PG. The VH component was corrected for G-factor.

(Continued)

EXPERIMENT 8-16 (CONTINUED) ZERO EMISSION ANISOTROPY STANDARD IN SOLUTIONS

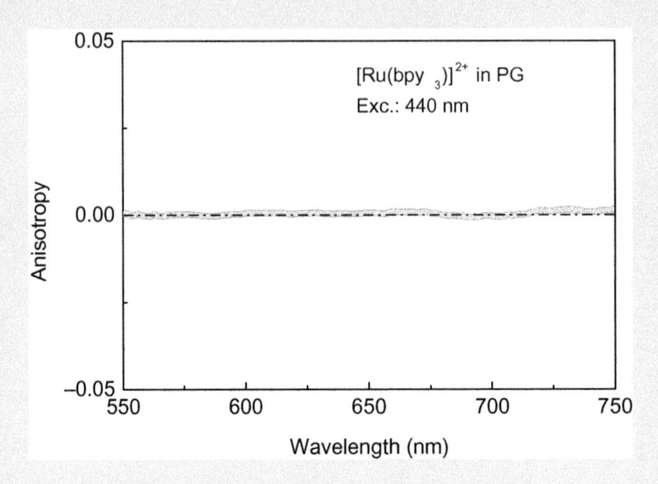

FIGURE 8-16.6 Emission anisotropy of Ru in PG.

Emission anisotropies of Ru in PG are presented in Figure 8-16.6. In the entire range 550–750 nm, these anisotropies are near-zero, which intuitively was obvious from Figure 8-16.5 because VV and VH emission components are identical after G-factor correction.

Below are presented lifetime and anisotropy decay measurements of Ru in PG. The lifetime can be satisfactorily fitted by a single exponent with a lifetime of 638 ns, see Figure 8-16.7. The anisotropy decay reveals the correlation time of 2.11 ns, see Figure 8-16.8.

The relation between anisotropy, lifetime, and correlation time is given by the Perrin equation. Please note a relatively low value of limiting anisotropy of Ru.

The Perrin equation allows to predict fluorescence anisotropy:

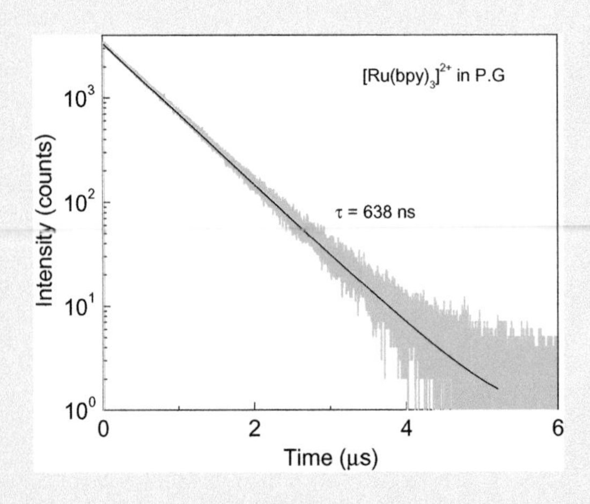

FIGURE 8-16.7 Lifetime measurements of Ru in PG.

(Continued)

EXPERIMENT 8-16 (CONTINUED) ZERO EMISSION ANISOTROPY STANDARD IN SOLUTIONS

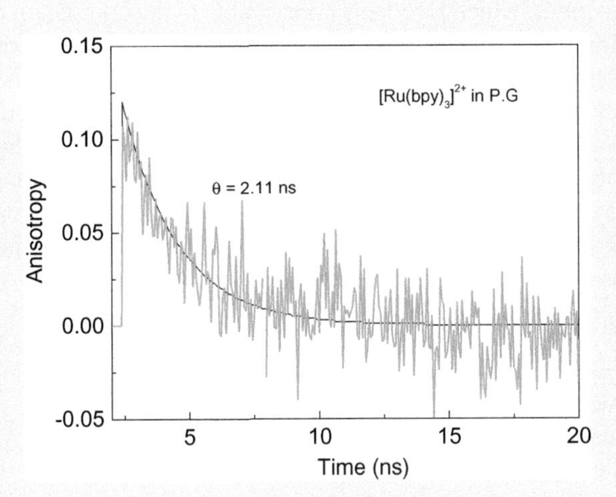

FIGURE 8-16.8 Anisotropy decay measurements of Ru in PG.

$$r = \frac{r_0}{\left(1 + \frac{\tau}{\theta}\right)}$$

where r_0 is the limiting anisotropy.

Taking lifetime of 638 ns, correlation time of 2.11 ns, and limiting anisotropy of 0.12, one can predict emission steady-state anisotropy to be lower than 0.0004.

CONCLUSIONS

$[Ru(bpy)_3]^{2+}$ is easily soluble in water and in most organic solvents. It is also very photo stable. The convenient spectral properties and a remarkably long lifetime (above 350 ns in water and 600 ns in PG) make Ru a preferred standard for anisotropies measurements in the red spectral range (550–750 nm). It is very convenient to have a reference in liquid, not quickly vaporizing solution, with zero anisotropy for polarization-based assays and other studies involving anisotropy measurements. This standard is especially useful for microscopes calibration for anisotropy-based imaging measurements.

EXPERIMENT 8-17 EXTREME ANISOTROPIES IN SOLUTIONS

Keywords: fluorescence, anisotropy, anisotropy standard.

Observed fluorescence anisotropies in solutions depend on the ratio of a fluorophore lifetime and correlation time (see Perrin equation). Fluorophores with very long lifetimes will show anisotropies close to zero even in viscous solvents. However, most dyes have lifetimes in the

(Continued)

EXPERIMENT 8-17 (CONTINUED) EXTREME ANISOTROPIES IN SOLUTIONS

range of few nanoseconds, which is roughly comparable to the correlation times in a medium viscosity solvent. Therefore, observed fluorescence anisotropies in solutions of medium viscosities (1–50 cP) are in the range of 0–0.1. Fluorophores with extremely short lifetimes will show extremely high anisotropies. In this experiment, we will measure extremely high anisotropies of the Allura Red dye in propylene glycol. With its short lifetime, Allura Red could be a very convenient fluorescence anisotropy standard in wide range of wavelength. Moreover, the propylene glycol is a relatively viscous solvent in room temperature, transparent in UV-VIS-RED spectral region, and not vaporizing during measurements which is very important in microscopy and HTS applications.

MATERIALS

- Allura Red (AR)
- Propylene glycol (PG)
- 1 cm × 1 cm cuvettes

EQUIPMENT

- Spectrophotometer (Cary 50, Varian, Inc.)
- Spectrofluorometer (Eclipse, Varian, Inc.)
- Fluorometer (PicoQuant FT200)

METHODS

1. Prepare a stock-solution of AR in water, sonicate.
2. Using the stock-solution, make the solution of AR in PG with an optical density of about 0.2 at maximum. Measure absorption spectra.
3. Measure fluorescence emission spectra of the AR sample at magic angle (MA) conditions, see Figure 8-17.1.
4. Measure fluorescence emission spectra of the AR samples at HH and HV polarizers' orientations. Calculate G-factors across fluorescence spectra, see Figures 8-17.2 and 8-17.3.
5. Measure fluorescence emission spectra of the AR samples at VV and VH polarization conditions, see Figure 8-17.4. Correct VH component for G-factor, see Figure 8-17.5.
6. Plot the emission anisotropy of AR in PG, see Figure 8-17.6.
7. Measure excitation fluorescence anisotropy of AR in PG at 620 nm observation, see Figure 8-17.7.

RESULTS

Absorption of AR covers a visible spectral region from 375 nm to 575 nm and a broad fluorescence emission lies between 525 nm and 750 nm, which are the desired regions in fluorescence spectroscopy/microscopy. Because of its very short fluorescence lifetime, the anisotropy of AR is expected to be high, close to the limiting value of 0.4. First, we will measure the fluorescence spectra and anisotropies of AR in propylene glycol.

(Continued)

EXPERIMENT 8-17 (CONTINUED) EXTREME ANISOTROPIES IN SOLUTIONS

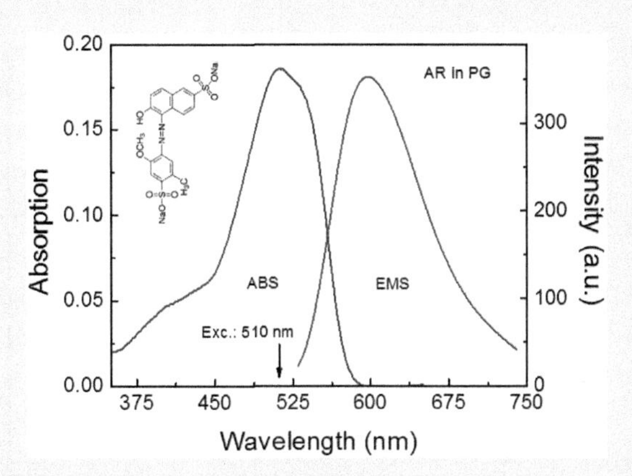

FIGURE 8-17.1 Absorption and emission spectra of AR in PG. Chemical structure of AR is in the insert.

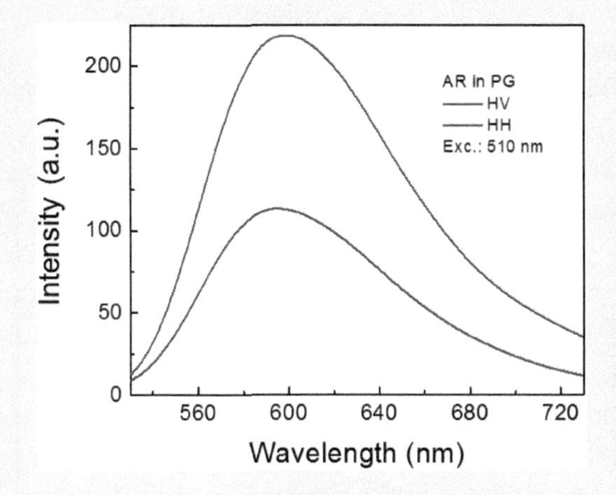

FIGURE 8-17.2 HV and HH components of AR emission in PG.

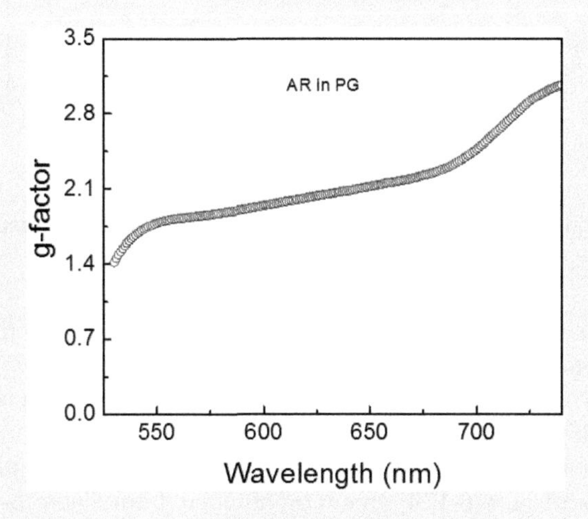

FIGURE 8-17.3 G-factor calculated from HV and HH components of AR emission in PG.

(Continued)

EXPERIMENT 8-17 (CONTINUED) EXTREME ANISOTROPIES IN SOLUTIONS

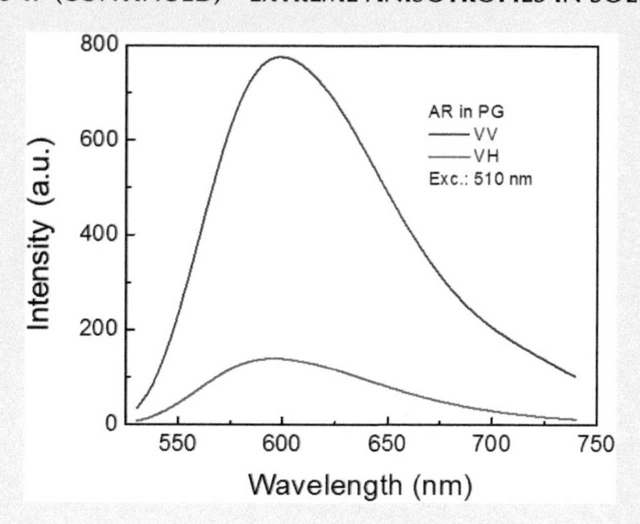

FIGURE 8-17.4 VV and VH components of AR emission in PG. The VV component was NOT corrected for G-factor.

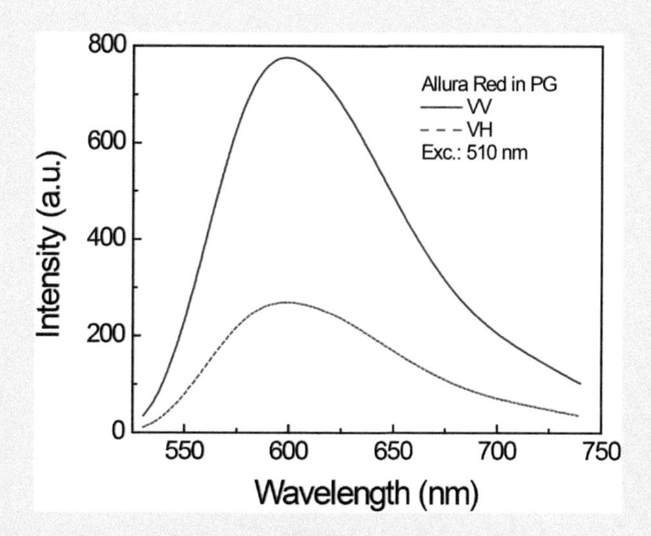

FIGURE 8-17.5 VV and VH components of AR emission in PG. The VV component was corrected for G-factor.

For anisotropy measurements in square geometry, first, you need to measure the HV and HH components (Figure 8-17.2) and calculate G-factor (Figure 8-17.3).

Next, measure VV and VH components, correct the VH component for G-factor and plot emission anisotropy (Figures 8-17.4 and 8-17.5).

Correction of the HH component for G-factor requires multiplication of HH fluorescence intensities (from Figure 8-17.4, green) by numbers from Figure 8-17.3. The polarized fluorescence components of AR in PG emission, corrected for G-factor, are presented in

(Continued)

EXPERIMENT 8-17 (CONTINUED) EXTREME ANISOTROPIES IN SOLUTIONS

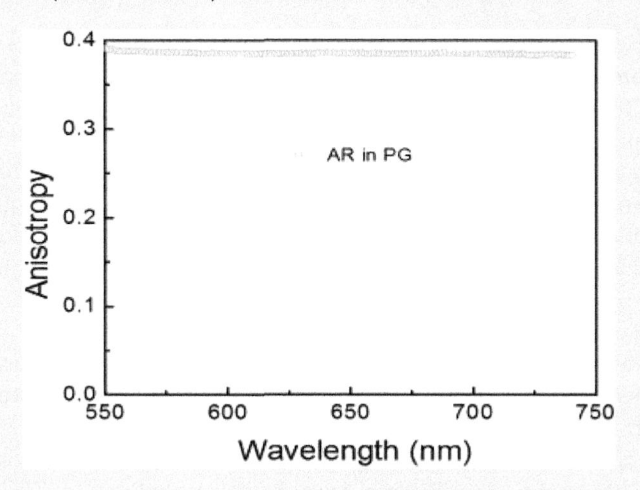

FIGURE 8-17.6 Emission anisotropy of AR in PG.

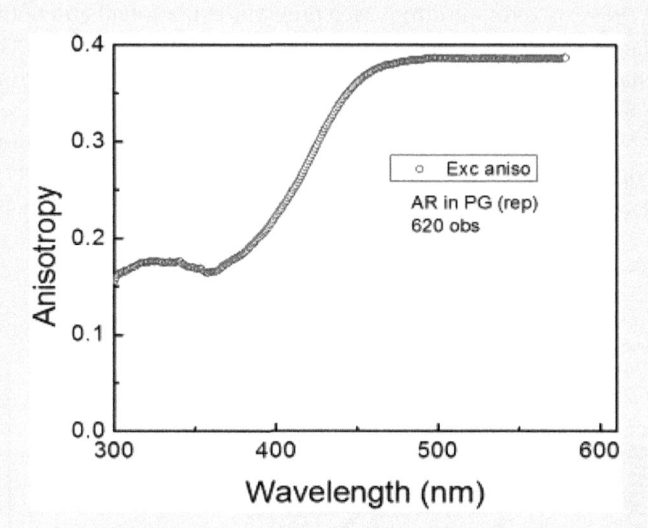

FIGURE 8-17.7 Excitation fluorescence anisotropy of AR in PG.

Figure 8-17.5. Now, the emission fluorescence anisotropy can be calculated at each wavelength across emission wavelengths according to:

$$r = \frac{I_{VV} - I_{VH}}{I_{VV} + 2I_{VH}}$$

Sometimes, it is convenient to use the ratio of $I_{VV} / I_{VH} = \Lambda$. With this notation, the anisotropy values are given by:

$$r = (\Lambda - 1)/(\Lambda + 2)$$

(Continued)

EXPERIMENT 8-17 (CONTINUED) EXTREME ANISOTROPIES IN SOLUTIONS

And all that needs to be done is to calculate ratios of polarized components Λ. For the maximum (fundamental) anisotropy, 0.4, this ratio is 3. The fluorescence emission anisotropy of AR in PG (0.385) is very close to the fundamental value and practically constant from 550 nm to 740 nm, see Figure 8-16.6.

Next, we measure excitation fluorescence anisotropy at 620 nm observation (Figure 8-17.7). For this, we need to measure polarized components of excitation spectra at 620 nm observation and multiply the HH component by the value of G-factor at 620 nm (see Figure 8-17.3).

In the range 475 nm and 575 nm, the observed excitation anisotropy is constant at the value of 0.385.

For comparison in Figure 8-17.8, we are presenting the intensity decay of AR in PG. Fluorescence intensity decay of AR in PG can be satisfactorily approximated with a single exponent of 58 ps, see Figure 8-17.8.

Interestingly, high steady-state anisotropy can be easily visualized. For this, we used solutions of AR and $[Ru(bpy)_3]^{2+}$ in PG. Both dyes can be excited with a 475 nm laser diode and emission can be observed through 495 nm long-pass filter to eliminate excitation light. The Ru complex presents zero anisotropy (see previous experiment) and both vertically and horizontally polarized intensity should be equal. AR presents high polarization and vertically polarized intensity should be about three times stronger than the horizontally polarized intensity. Figure 8-17.9 shows photography of two adjunct cuvettes filled with AR and Ru solutions, respectively. Photography on the left was made through a vertically oriented polarizer and on the right through a horizontally oriented polarizer. The intensity of Ru solution does not change while AR significantly changed.

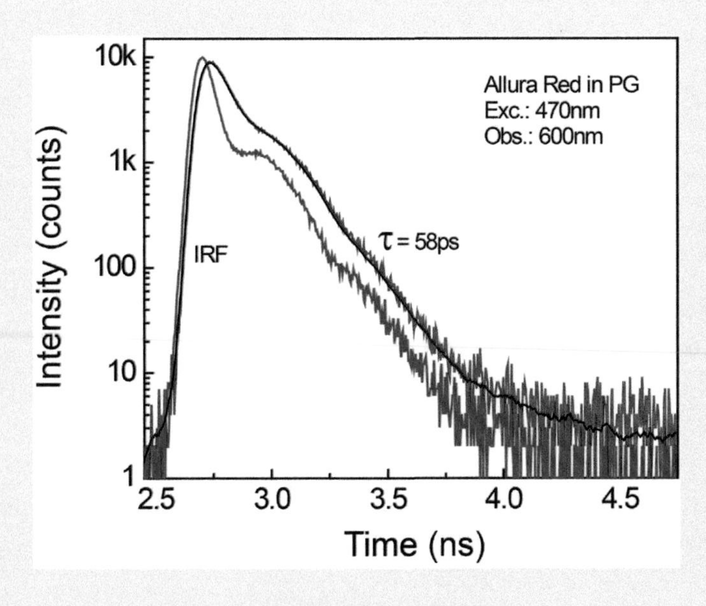

FIGURE 8-17.8 Lifetime measurements of AR in PG.

(Continued)

EXPERIMENT 8-17 (CONTINUED) EXTREME ANISOTROPIES IN SOLUTIONS

FIGURE 8-17.9 Photographs of AR in PG with vertically polarized excitation from a blue laser pointer (473 nm). Left: Observation through a vertically oriented polarizer (VV component). Right: Observation through horizontally oriented polarizer (VH component). For comparison, the second cuvette contains long-lived (zero anisotropy) $[Ru(bpy)_3]^{2+}$ solution in PG. Photographs were taken through 495 nm long-pass filter.

CONCLUSIONS

Allura Red is a known food colorant used in the food industry. It is easily soluble in water and in most organic solvents. It is also very photo stable. The convenient spectral properties and a remarkably short lifetime (about 10 ps in water and 58 ps in PG) make AR a preferred standard for lifetimes and anisotropies measurements. For the excitations wavelengths between 480 nm and 550 nm, the anisotropy is constant (0.385) and extinction coefficient is high. It is very convenient to have a reference in liquid, not quickly vaporizing solution, with extreme anisotropy for polarization-based assays and other studies involving anisotropy measurements. Another important reference for anisotropy is $[Ru(bpy)_3]^{2+}$ with near-zero anisotropy, discussed in previous experiments.

EXPERIMENT 8-18 SOLVENT EFFECTS ON ELECTRONIC SPECTRA

Keywords: absorption, fluorescence, solvent polarity.

In diluted solutions, dye molecules are surrounded by solvent molecules. Due to small sizes of molecules and small distances between them, the electric interactions between the fluorophore and solvent molecules might be very strong. If the dye molecule is polar (possesses an electric dipole moment), it creates a strong electric field around itself. The solvent molecules are interacting with this field. If the solvent molecules are also polar, the interaction is similar to the Stark effect (interaction of dipole moment with an electric field). This dye-solvent interaction changes electronic states of the fluorophore. In different solvents, the energy of the fluorophore electronic states will be different. This will manifest with the changes (shifts)

(Continued)

EXPERIMENT 8-18 (CONTINUED) SOLVENT EFFECTS ON ELECTRONIC SPECTRA

in absorption and fluorescence spectra. There are many theories describing fluorophore-solvent interaction. It is very difficult to describe precisely this interaction because there is no existing satisfactory theory of liquids. One possible approximation is to treat the solvent as a continuous dielectric characterized by the dielectric constant (ε) and refractive index (n). Higher dielectric constants describe polar solvents, for non-polar solvents, the dielectric constant will be close to the square of refractive index (n^2). A solvent polarity function $f(\varepsilon,n)$:

$$f(\varepsilon,n)=\frac{2n^2+1}{n^2+2}\left(\frac{\varepsilon-1}{\varepsilon+2}-\frac{n^2-1}{n^2+2}\right)$$

provides dielectric properties of the solvents. Stoke's shifts of fluorophore spectra are different in different solvents, and are described by the equation:

$$\tilde{\nu}_a-\tilde{\nu}_f=m_1 f(\varepsilon,n)+const.$$

where m_1 is equal:

$$m_1=\frac{2(\mu_e-\mu_g)^2}{hca^3}$$

where μ_g and μ_e are electric dipole moments of the fluorophore in the ground and excited state, h-Planck constant, c-velocity of the light, and a is a cavity (Onsager) radius. The plot of Stoke's shifts vs. polarity function enables the determination of the m_1 parameter and estimation of the change of the dye's dipole moment upon the excitation.

MATERIALS

- Coumarin 153 (C153)
- *n*-Heptane (H), cyclohexane, benzene, toluene (T), butyl acetate (BA), ethyl acetate (EA), dichloromethane, butanol, ethanol, methanol, dimethylformamide (DMF), water (W)
- Camera or cell phone
- 1 cm x 1 cm glass/quartz cuvettes

EQUIPMENT

- Spectrophotometer (Cary 50, Varian, Inc.)
- Spectrofluorometer (Eclipse, Varian, Inc.)

METHODS

1. Prepare a few milliliters stock-solutions of C153 in all used solvents.
2. Prepare C153 samples in all solvents, keep optical density below 0.1.
3. Make a table with dielectric properties of all solvents.
4. Measure absorption and fluorescence spectra of C153 in all solvents. For each sample, use a proper solvent for baseline/background reference.

(Continued)

EXPERIMENT 8-18 (CONTINUED) SOLVENT EFFECTS ON ELECTRONIC SPECTRA

5. Plot absorption and fluorescence spectra measured for all samples.
6. Normalize all absorption and fluorescence spectra. Plot all spectra as a function of wavenumbers. Estimate maxima. Make a table with C153 spectral properties for all solvents.
7. Plot the difference of wavenumbers for maximum absorption and maximum emission (Stoke's shifts) vs. the polarity function $f(\varepsilon,n)$.
8. Estimate the change of C153 dipole moment upon the excitation.

RESULTS

Dielectric properties of selected solvents are presented in Table 8-18.1. The polarity function changes from zero (highly non-polar solvents) to over 0.8 (highly polar solvents).

C153 fluorophore shown in Scheme 8-18.1 is very sensitive to the solvent polarity. It significantly changes color, from blue to green-yellow, going from non-polar *n*-heptane to polar ethanol and water, see Figure 8-18.1.

First, we measure absorption spectra of all 12 solutions with appropriate baselines, keeping optical densities about 0.05, see Figure 8-18.2. In *n*-heptane and cyclohexane, the spectra show structures, in other solvents they are unstructured.

Next, using the same solutions, we measure fluorescence spectra, see Figure 8-18.3.

Next, we normalized and redraw both absorption and fluorescence spectra in the function of wavenumbers (cm^{-1}), see Figures 8-18.4 and 8-18.5 (Table 8-18.2).

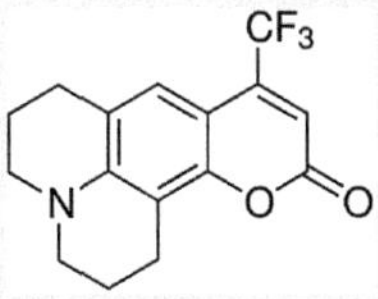

SCHEME 8-18.1 Chemical structure of C153.

TABLE 8-18.1 Dielectric Properties of Used Solvents

Solvent	*n*	*ε*	*f(ε,n)*
n-Heptane	1.388	1.922	−0.001
Cyclohexane	1.426	2.022	−0.002
Benzene	1.501	2.3	0.01
Toulene	1.496	2.39	0.032
Butyl acetate	1.394	5.1	0.419
Ethyl acetate	1.372	6.09	0.493
Dichloromethane	1.424	9.1	0.596
Butanol	1.399	17.8	0.753
Ethanol	1.361	25	0.815
Methanol	1.328	33.52	0.857
DMF	1.43	38.3	0.839
Water	1.333	80.36	0.914

(Continued)

EXPERIMENT 8-18 (CONTINUED) SOLVENT EFFECTS ON ELECTRONIC SPECTRA

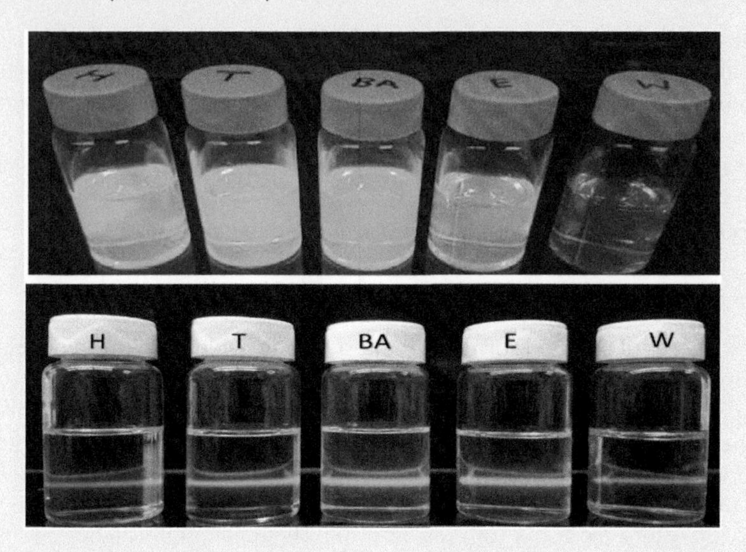

FIGURE 8-18.1 Photographs of C153 solutions in different polarity solvents: *n*-heptane (H), toluene (T), butyl acetate (BA), ethanol (E), and water (W). Top: On UV illumination plate; bottom: illuminated with blue laser pointer.

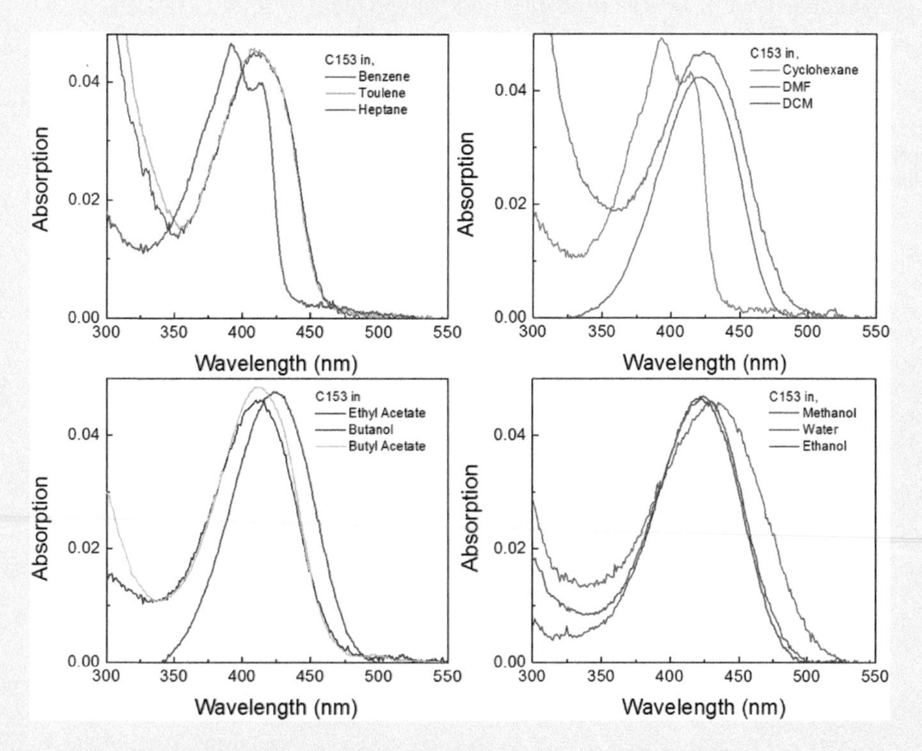

FIGURE 8-18.2 Absorption spectra of C153 in different solvents. The spectra are presented on a wavelength scale.

(Continued)

EXPERIMENT 8-18 (CONTINUED) SOLVENT EFFECTS ON ELECTRONIC SPECTRA

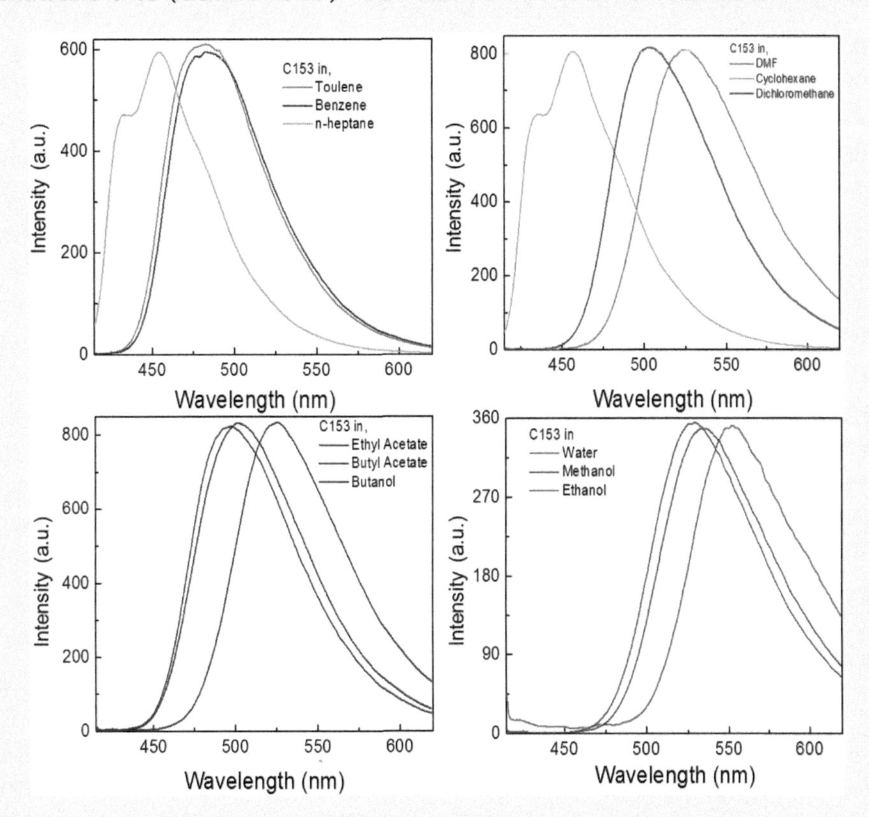

FIGURE 8-18.3 Fluorescence spectra of C153 in different solvents. The spectra are presented on a wavelength scale.

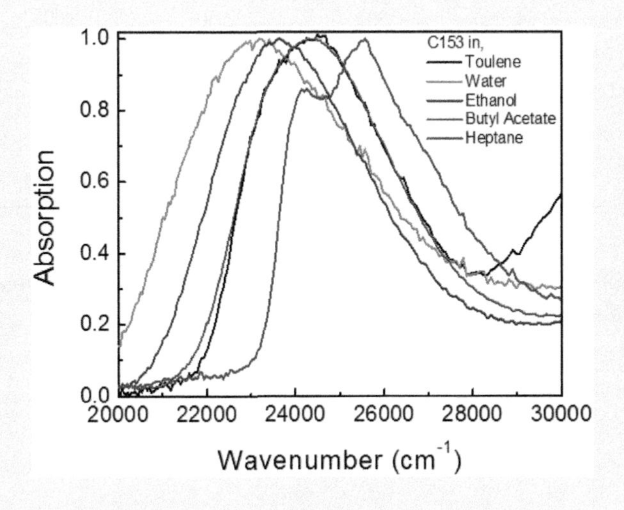

FIGURE 8-18.4 Normalized absorption spectra of C153 in different solvents.

(Continued)

EXPERIMENT 8-18 (CONTINUED) SOLVENT EFFECTS ON ELECTRONIC SPECTRA

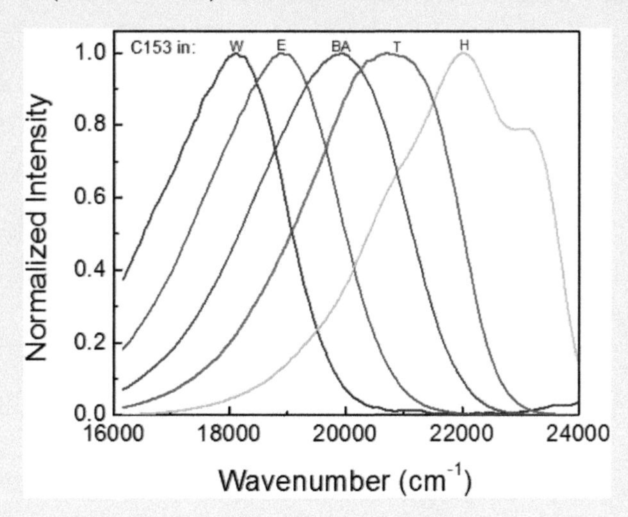

FIGURE 8-18.5 Normalized fluorescence spectra of C153 in different solvents.

TABLE 8-18.2 Absorption and Emission Maxima of C153 Spectra in Different Solvents

Solvent	λ_a (nm)	λ_f(nm)	$\tilde{\nu}_a$ (cm^{-1})	$\tilde{\nu}_f$ (cm^{-1})	$\tilde{\nu}_a - \tilde{\nu}_f$ (cm^{-1})
n-Heptane	391	454	25,600	22,050	3550
Cyclohexane	393	456	25,500	21,950	3550
Benzene	411	483	24,350	20,700	3650
Toulene	411	482	24,350	20,750	3700
Butyl acetate	411	502	24,350	19,950	4400
Ethyl acetate	413	497	24,250	20,100	4150
Dichloromethane	420	503	23,850	19,900	4150
Butanol	424	525	23,600	19,050	4550
Ethanol	422	529	23,700	18,900	4800
Methanol	424	534	23,600	18,700	4800
DMF	423	525	23,650	19,050	4600
Water	433	551	23,100	18,150	4950

Finally, we plot the Stoke's shift values (–) in function of solvent polarity $f(\varepsilon, n)$, see Figure 8-18.6. The slope reveals parameter m_1.

(*Continued*)

EXPERIMENT 8-18 (CONTINUED) SOLVENT EFFECTS ON ELECTRONIC SPECTRA

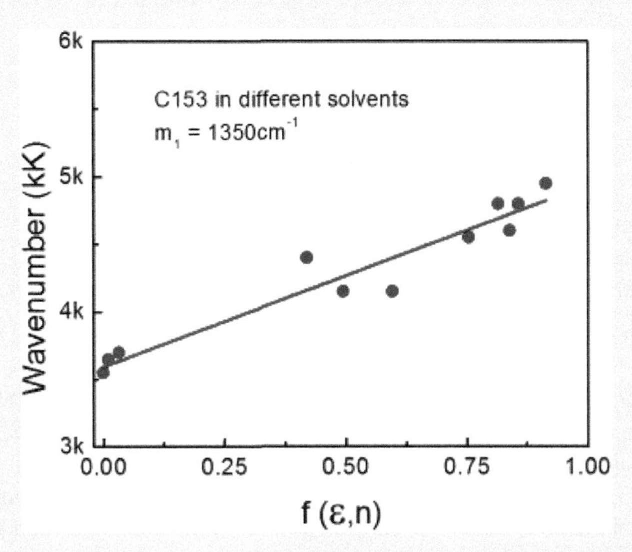

FIGURE 8-18.6 Stokes shifts as a function of solvent polarity function.

CONCLUSIONS

Spectral shifts in solvents of different polarities give the chance to evaluate electric dipole moments of studied molecules and their changes upon the excitation. The use of the spectral shifts and macroscopic solvent parameters can be used to determine microscopic molecular parameters. Estimate the change of C153 electric dipole moment upon the excitation.

EXPERIMENT 8-19 EXCIPLEX FORMATION IN SOLUTIONS

Keywords: fluorescence, anthracene, diethylaniline, exciplex, lifetimes.

In Experiment 8-8, we demonstrated an excimer formation between pyrene molecules. An excited pyrene molecule formed a complex with another (not excited) pyrene molecule. The term "excimer" stays for an excited dimer. The complex formation is also possible between different molecules and is based on a charge-transfer process. These complexes are called exciplexes, which stays for excited complexes. A most studied exciplex is a complex of excited anthracene and not excited diethylaniline. Similarly to excimers, exciplex fluorescence is long-wavelength shifted. The exciplex also does not have its ground state, therefore, there is not possible to measure its absorption spectrum. In this experiment, we will demonstrate exciplex formation between anthracene and diethylaniline molecules in benzene. In contrast to excimer studies, the concentration of the fluorophore can be low and measurements can be done in square geometry.

(*Continued*)

EXPERIMENT 8-19 (CONTINUED) EXCIPLEX FORMATION IN SOLUTIONS

MATERIALS

- Anthracene (A), diethylaniline (DEA), benzene
- Quartz cuvette, 1 cm × 1 cm, glass vials

EQUIPMENT

- Spectrophotometer (Cary 50, Varian, Inc.)
- Spectrofluorometer (Eclipse, Varian, Inc.)
- Fluorometer (FT200, PicoQuant, GmbH, Germany) with 370 nm laser diode

METHODS

1. Make a stock solution of anthracene in benzene. Prepare about 10 mL of anthracene in benzene with an absorbance of about 0.1 in a long-wavelength part of absorption spectrum (about 360 nm), see Figure 8-19.1.
2. Divide the prepared solution to two parts, 5 mL each. Add to the one part, 0.16 mL of diethylaniline and to another part, 0.16 mL of benzene. With such preparation, the anthracene concentrations are the same in both samples. The concentration of diethylaniline should be about 0.2 M.
3. Measure fluorescence emission of both samples, anthracene alone and with diethylaniline, see Figures 8-19.2 and 8-19.3.
4. Measure lifetimes of both samples with 370 nm excitation and 405 nm observation (anthracene monomer), see Figures 8-19.4 and 8-19.5.
5. Measure the lifetime of the exciplex emission at 505 nm, see Figure 8-19.6.

RESULTS

The absorption spectrum of anthracene in benzene is nicely structured and located at the violet/UV spectral region, see Figure 8-19.1. We selected a 356 nm (major peak of the spectrum) for the excitation of the fluorescence.

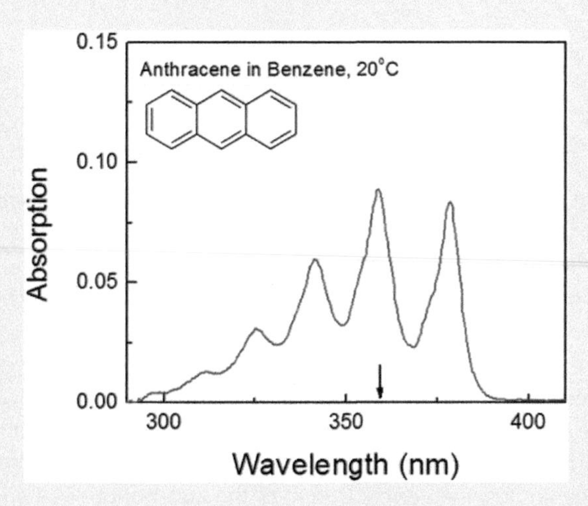

FIGURE 8-19.1 Absorption spectrum of anthracene in benzene.

(Continued)

EXPERIMENT 8-19 (CONTINUED) EXCIPLEX FORMATION IN SOLUTIONS

The fluorescence of anthracene in benzene has a dark blue color and covers wavelengths from 375 nm to 500 nm, see Figure 8-19.2.

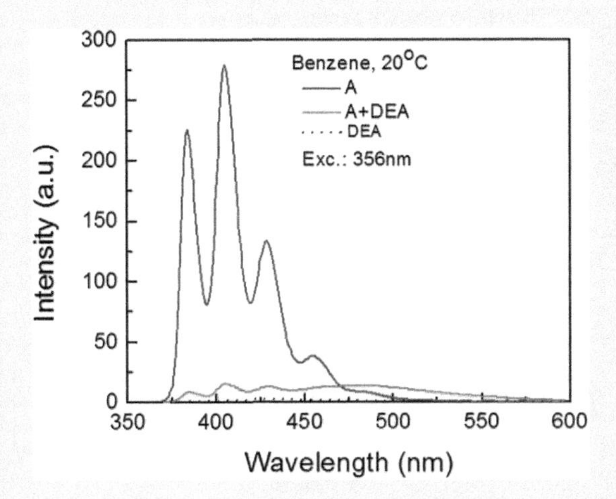

FIGURE 8-19.2 Fluorescence spectra of anthracene in benzene in absence and presence of diethylaniline as measured.

The fluorescence of anthracene in benzene in the presence of diethylaniline is strongly quenched and displays a long-wavelength unstructured band (exciplex emission) up to 600 nm.

Please note that 0.2 M diethylaniline in benzene has no fluorescence emission (dotted line in Figure 8-19.2) in the experimental conditions.

The exciplex emission is clearly visualized with a closer evaluation of the fluorescence spectrum, see Figure 8-19.3. This spectrum is a combination of a monomer anthracene fluorescence (short wavelengths and structured) and unstructured long-wavelength band (exciplex fluorescence).

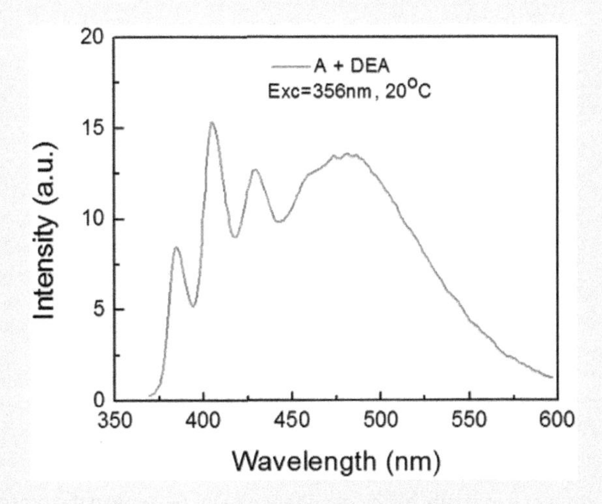

FIGURE 8-19.3 Fluorescence spectrum of anthracene in benzene in the presence of 0.2 M diethylaniline.

(Continued)

EXPERIMENT 8-19 (CONTINUED) EXCIPLEX FORMATION IN SOLUTIONS

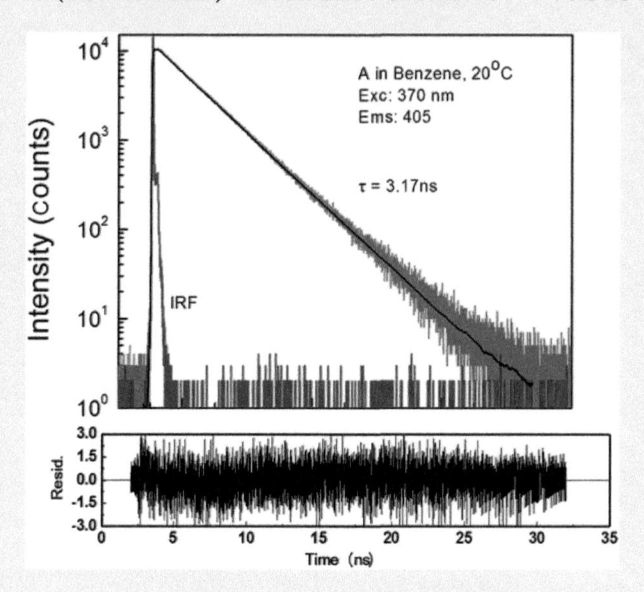

FIGURE 8-19.4 Fluorescence intensity decay of anthracene in benzene.

The fluorescence intensity decay of monomer anthracene in benzene can be well fitted to a single exponent with the lifetime of 3.17 ns, see Figure 8-19.4.

However, in the presence of 0.2 M diethylaniline, the intensity decay is significantly shorter and heterogeneous. Two exponents are needed to achieve a satisfactory fit, see Figure 8-19.5. This is a direct proof that the quenching of anthracene fluorescence by diethylaniline is a dynamic process occurring in the lifetime of anthracene fluorescence.

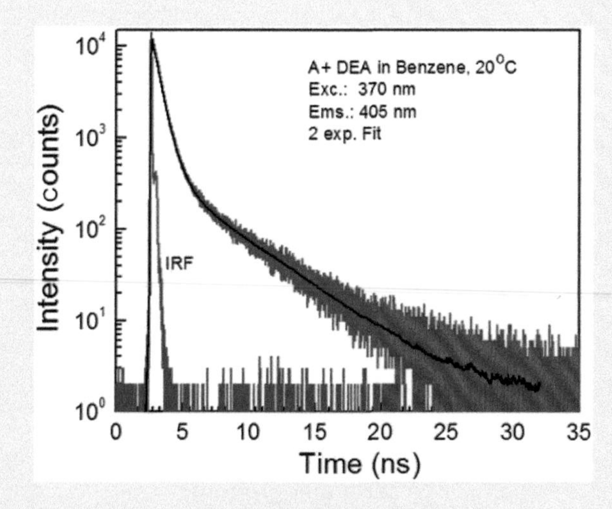

FIGURE 8-19.5 Fluorescence intensity decay of anthracene in benzene in the presence of 0.2 M diethylaniline observed at 405 nm (monomer fluorescence).

(Continued)

EXPERIMENT 8-19 (CONTINUED) EXCIPLEX FORMATION IN SOLUTIONS

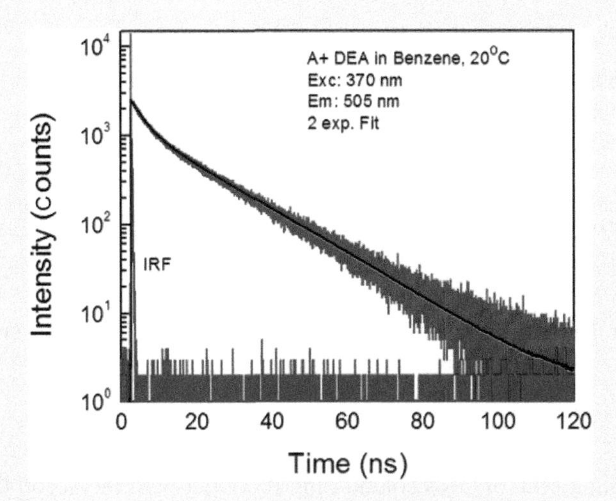

FIGURE 8-19.6 Fluorescence intensity decay of anthracene in benzene in the presence of 0.2 M diethylaniline observed at 505 nm (exciplex fluorescence).

The fluorescence decay of the exciplex is less heterogeneous than the quenched anthracene monomer but significantly longer, see Figure 8-19.6.

CONCLUSIONS

The exciplex formation between excited anthracene and diethylaniline is accompanied by the change of fluorescence color from dark blue to light celestial. Diethylaniline is a colorless liquid but in time became lightly yellow. This change of color has no influence on the experiment as long as diethylaniline does not have an intrinsic fluorescence.

EXPERIMENT 8-20 FRET IN ONE DIMENSION

Keywords: fluorescence, FRET, lifetime, DNA, ethidium bromide, acridine orange.

The Förster resonance energy transfer (FRET) requires a close distance between donor and acceptor molecules. Although it is possible to make FRET measurements in solutions (see Experiments 8-10 and 8-11), it is much easier to use a "vehicle" for donors and acceptor molecules. There are molecules with high affinities to the DNA macromolecule which attach or intercalate to specific places and stay there. We will demonstrate FRET phenomenon using two different dyes attached to the DNA. One dye has a short wavelength fluorescence emission in the region of a second dye absorption. These two dyes, acridine orange and ethidium bromide, form a donor-acceptor system located on the DNA strand. Instead of huge dyes concentrations needed for the FRET in three dimensions, we will use minute amounts of both donor and acceptor fluorophores.

(*Continued*)

EXPERIMENT 8-20 (CONTINUED) FRET IN ONE DIMENSION

MATERIALS

- Acridine orange (AO), ethidium bromide (EB), DNA (calf thyme)
- PBS buffer, vials
- 1 cm × 1 cm fluorometric cuvettes

EQUIPMENT

- Spectrophotometer (Cary 50, Varian, Inc.)
- Spectrofluorometer (Eclipse, Varian, Inc.)
- Fluorometer (FT200, PicoQuant, GmbH, Germany) with 370 nm laser diode
- Camera or cell phone, UV illuminator

METHODS

1. Prepare about 10 mL of DNA solution in PBS buffer with the absorption of about 0.5 at 259 nm, see Figure 8-20.1. The apparent concentration of the DNA (per base-pair) will be about 40–50 µmol. Use for the estimation of DNA concentration, an extinction coefficient of 13,500 M^{-1} cm^{-1} per base pair. Split the DNA solution into four parts (about 2.5 mL each).
2. Make stock solutions of AO and EB in PBS buffer with concentrations of about 0.4 mM. Use 27,000 M^{-1} cm^{-1} and 5,800 M^{-1} cm^{-1} as extinction coefficients for AO and EB, respectively.
3. Prepare 2.5 mL samples of AO, EB, and mixture (1:1) of AO + EB in PBS buffer with and without DNA: add 25 µL of each dye (AO, EB) to 2.5 mL of DNA solution and to PBS buffer only. The concentrations of AO and EB in samples will be about 4 µmol. At the end of the preparation, it should be eight solutions in vials: three samples (AO, EB, and AO + EB) in PBS buffer without DNA, and similar three samples with the DNA; plus PBS buffer and DNA only as references/baselines. At such preparation, there will be about 10–12 base-pairs DNA per one AO and EB molecules.
4. Measure absorption spectra of the samples in the buffer, see Figure 8-20.2.
5. Measure fluorescence spectra of both sample series, without and with DNA; see Figures 8-20.3 and 8-20.4.
6. Measure lifetimes of AO and AO/EB in presence of DNA, see Figures 8-20.6 and 8-20.7.

RESULTS

Why we selected this donor-acceptor pair for FRET demonstration?

First, the fluorescence of AO fairly overlaps the absorption of EB. Second, the emission of EB is red-shifted allowing a clean observation of AO fluorescence at 530 nm without EB emission interference. Third and most important, both dyes bind efficiently to the DNA macromolecule.

The concentrations of DNA and dyes were adjusted to assure at least 10 base-pairs per one dye molecule. In such conditions, a minimal amount of free dyes (unbound) will be present in the sample.

(*Continued*)

EXPERIMENT 8-20 (CONTINUED) FRET IN ONE DIMENSION

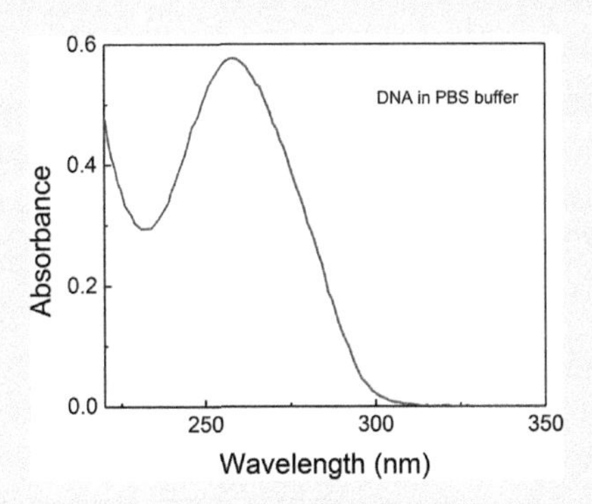

FIGURE 8-20.1 Absorption spectrum of DNA in PBS buffer.

Absorptions of AO and EB are additive in PBS solution with and without DNA. At the same concentrations (about 4 μmol), the absorption of AO is about five times stronger than EB absorption, which precisely corresponds to relative extinction coefficients of both dyes, see Figure 8-20.2.

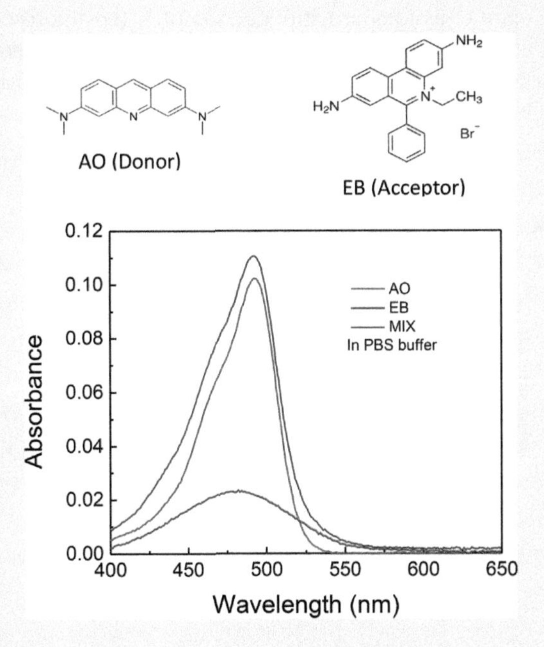

FIGURE 8-20.2 Absorption spectra of AO, EB, and mixture of AO + EB (1:1) in PBS buffer. Top: Structures of AO and EB.

Next, we measured fluorescence spectra of AO, EB, and mixture AO + EB in PBS buffer. As seen in Figure 8-20.3, the fluorescence of AO is significantly stronger (at the excitation at 470 nm) than EB fluorescence. Also, the spectrum of the AO + EB mixture is clearly a sum of individual dyes spectra.

(Continued)

EXPERIMENT 8-20 (CONTINUED) FRET IN ONE DIMENSION

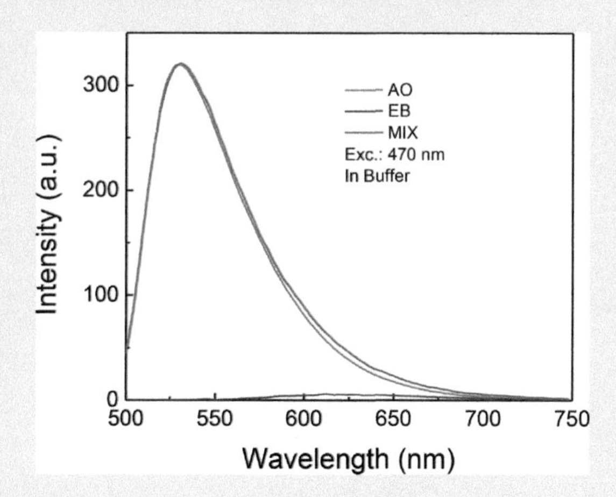

FIGURE 8-20.3 Fluorescence spectra of AO, EB, and mixture of AO + EB (1:1) in PBS buffer.

A dramatically different situation is in the presence of DNA (Figure 8-20.4). First, both (AO and EB) fluorescence emissions are stronger. Second, the fluorescence of the mixture AO + EB in the presence of DNA shows a significant decrease of AO emission with simultaneous increase of EB emission fluorescence. This is consistent with a significant FRET from AO to EB molecules.

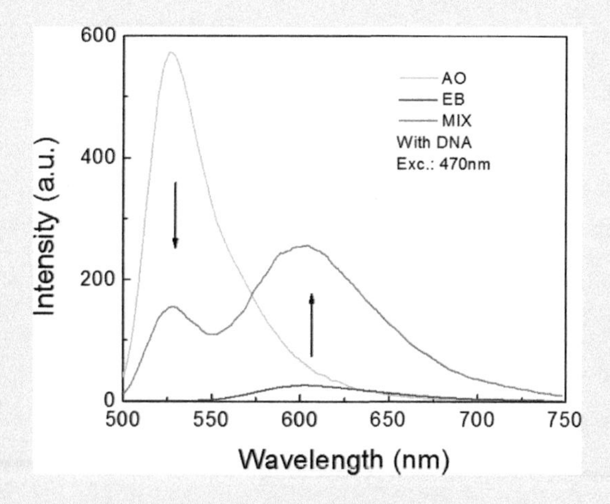

FIGURE 8-20.4 Fluorescence spectra of AO, EB, and mixture of AO + EB (1:1) in presence of DNA.

A significant change in fluorescence emissions of AO and EB in the presence of DNA is visualized in Figure 8-20.5, which shows a photograph of solutions on UV illuminator.

(Continued)

EXPERIMENT 8-20 (CONTINUED) FRET IN ONE DIMENSION

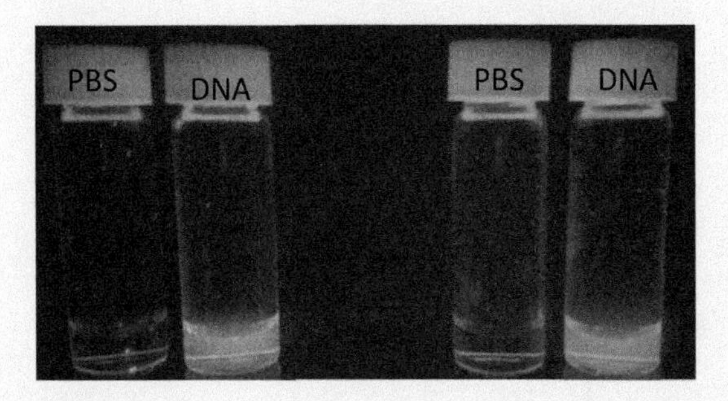

FIGURE 8-20.5 Photographs of AO (left) and EB (right) on UV illuminator in the absence and presence of DNA.

Next, we measured lifetimes of AO and mixture AO + EB in the presence of DNA, see Figures 8-20.6 and 8-20.7.

Fluorescence intensity decay of AO-labeled DNA in PBS buffer is clearly single-exponential with the lifetime of about 5 ns, see Figure 8-20.6. This is an additional proof that all AO molecules are bound to the DNA (free AO in PBS buffer has twice shorter lifetime).

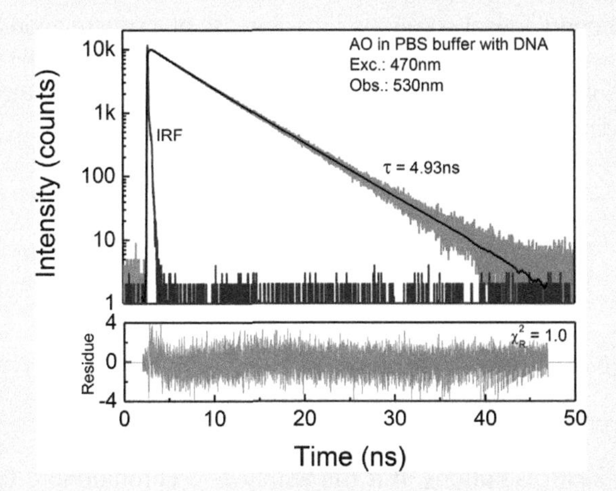

FIGURE 8-20.6 Fluorescence intensity decays of AO in PBS buffer in the presence of DNA.

There is a dramatic change of AO-labeled DNA lifetime in the presence of EB, see Figure 8-20.7. In order to reasonably fit this, intensity decay at least three exponents are needed. The amplitude averaged lifetime is only 1.57 ns, more than three-fold shorter than in the absence of EB. This change in lifetime roughly corresponds to the change in AO fluorescence intensity, compared with Figure 8-20.4.

(Continued)

EXPERIMENT 8-20 (CONTINUED) FRET IN ONE DIMENSION

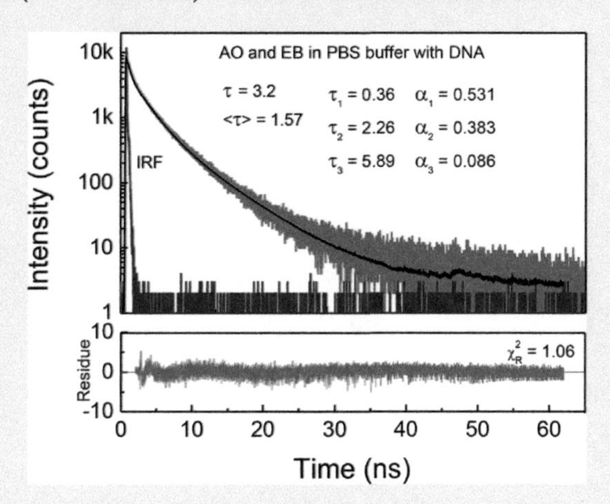

FIGURE 8-20.7 Fluorescence intensity decays of the mixture of AO + EB (1:1) in presence of DNA.

CONCLUSIONS

Fluorophores bound to the DNA are forced to be at a close distance. This enables an efficient FRET process. Measurements of such FRET have significant advantages. First, the amounts of dyes are minimal. Second, low absorptions allow the use of a square geometry and measurements are free of inner filter effects. The goal of this experiment was to demonstrate the FRET by using DNA as a linker between donor and acceptor molecules. For a more detailed theory of the FRET in one dimension see the recommended textbooks.

EXPERIMENT 8-21 SOLVATOCHROMISM

Keywords: absorption, fluorescence, solvent polarity.

Solvatochromism refers to solutions. It is the ability of a chromophore to change its color when dissolved in a more polar solvent. Negative solvatochromism corresponds to hypsochromic shift (or blue shift) with increasing solvent polarity. Respectively, positive solvatochromism corresponds to bathochromic shift (or red shift). More often the positive solvatochromism is observed. The *solvatochromic effect* (*solvatochromic shift*) describes the dependence of absorption and emission spectra with the solvent polarity. Since polarities of the ground and excited state of a chromophore are different, a change in the solvent polarity will lead to a different stabilization of the ground and excited states, and thus, a change in the energy gap between these electronic states. Consequently, variations in the position, intensity, and shape of the absorption spectra can be direct measures of the specific interactions between the solute and solvent molecules. We will demonstrate the solvatochromic effect using Fluorol 7GA dye in benzene–methanol mixtures.

(*Continued*)

EXPERIMENT 8-21 (CONTINUED) SOLVATOCHROMISM

MATERIALS

- Fluorol 7GA (F7GA)
- Benzene (B), methanol (M)
- 1 cm ×1 cm glass/quartz cuvettes

EQUIPMENT

- Spectrophotometer (Cary 50, Varian, Inc.)
- Spectrofluorometer (Eclipse, Varian, Inc.)
- Camera or cell phone

METHODS

1. Prepare a few milliliters of F7GA stock-solutions.
2. Prepare about 20 mL of both solvents, B and M. Add a few microliters of F7GA from stock solutions to achieve concentrations in both samples of 10–20 μmol (use F7GA extinction coefficient of 14,000 L $(mol\ cm)^{-1}$.
3. Make mixtures of B and M samples containing 0%, 0.2%, 0.5%, 1%, 2.5%, 5%, 10%, 20%, 50%, 75%, and 100% of M.
4. Measure absorption and fluorescence spectra of F7GA in all mixtures, see Figures 8-21.1 through 8-21.4.
5. Estimate maxima. Make a table with F7GA spectral properties for all mixtures.
6. Plot the spectral shifts for absorption and fluorescence as a function of methanol molar fraction, see Figures 8-21.5 and 8-21.6.

RESULTS

F7GA dye (Scheme 8-21.1) shows clearly a positive solvatochromic effect. The same amount of the dye added to B and M solvents results in a significant change of color, see Figure 8-21.1.

The change of color is a result of more than a 20 nm shift in absorption spectra, see Figure 8-21.1, right. In order to visualize more this shift, we plotted just maxima of absorption spectra in extended scale, see Figure 8-21.2.

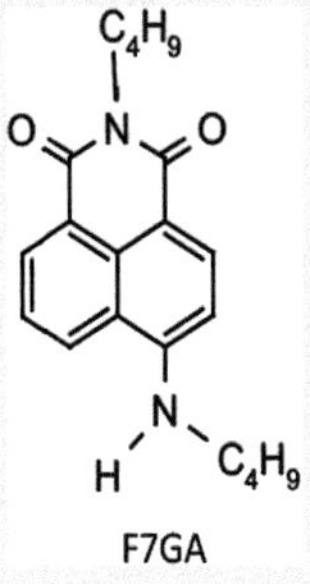

F7GA

SCHEME 8-21.1 Chemical structure of Fluorol 7GA.

(Continued)

EXPERIMENT 8-21 (CONTINUED) SOLVATOCHROMISM

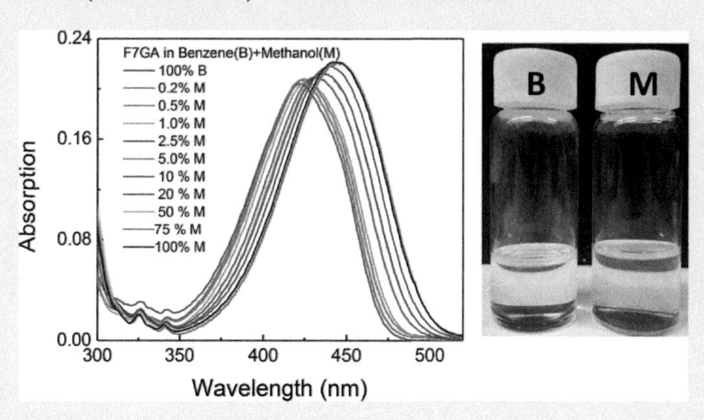

FIGURE 8-21.1 Absorption spectra of F7GA in mixtures benzene-methanol. The spectra are presented in a wavelength scale. On right is a photograph of F7GA in benzene and methanol in room light.

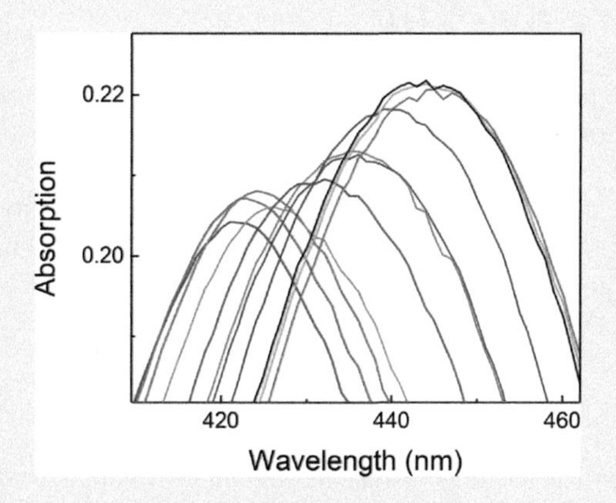

FIGURE 8-21.2 Absorptions maxima of F7GA in mixtures benzene–methanol (expanded).

Even more dramatic is the solvatochromic effect in fluorescence, see Figure 8-21.3.

In the case of F7GA fluorescence, the solvatochromic shift is almost 50 nm, see Figure 8-21.4 and Table 8-21.1.

Next, we estimated the maxima of absorption and fluorescence spectra and calculated spectral shifts, see Table 8-21.1.

Relative absorption and fluorescence shifts of F7GA in B-M mixtures are plotted in Figure 8-21.5 as a function of methanol molar fraction in mixtures.

At a methanol molar fraction of 0.1 (which corresponds to 5% of methanol in the mixture), more than 50% of total shifts are completed.

(*Continued*)

EXPERIMENT 8-21 (CONTINUED) SOLVATOCHROMISM

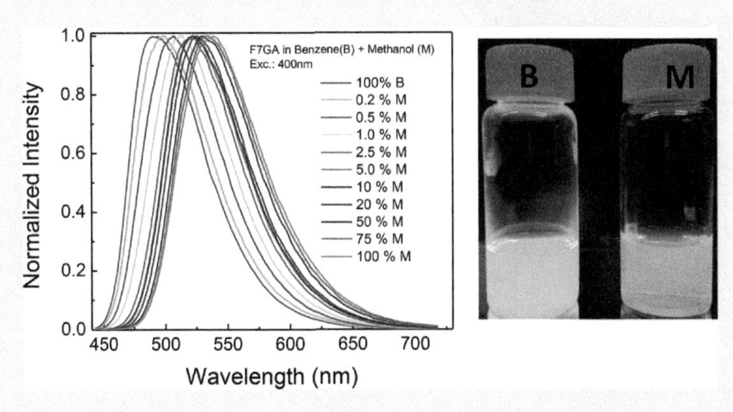

FIGURE 8-21.3 Normalized fluorescence spectra of F7GA in mixtures benzene–methanol. The spectra are presented in a wavelength scale. On right is a photograph of F7GA in benzene and methanol on a UV illuminator.

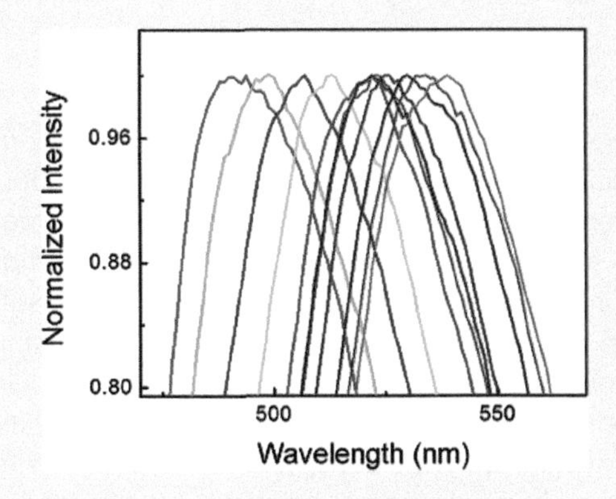

FIGURE 8-21.4 Fluorescence maxima of F7GA in mixtures benzene–methanol expanded from Figure 8-21.3. The spectra are presented in a wavelength scale.

TABLE 8-21.1 Absorption/Fluorescence Maxima and Spectral Shifts of F7GA in B-M Mixtures

M:B (VV) Mix	M Mol. Fraction	λ_A (nm)	λ_F (nm)	$\tilde{\nu}_a$ (cm^{-1})	$\tilde{\nu}_f$ (cm^{-1})	$\Delta\tilde{\nu}_a$ (cm^{-1})	$\Delta\tilde{\nu}_f$ (cm^{-1})
0	0	421	490	23,750	20,400	0	0
0.2	0.004	422	498	23,700	20,100	50	300
0.5	0.011	424	506	23,600	19,750	150	650
1	0.022	425	512	23,550	19,550	200	850
2.5	0.053	431	521	23,200	19,200	550	1200
5	0.104	435	523	23,000	19,100	750	1300
10	0.196	437	524	22,900	19,050	850	1350
20	0.354	439	526	22,800	19,000	950	1400
50	0.687	443	530	22,600	18,850	1150	1550
75	0.868	444	532	22,500	18,800	1250	1600
100	1	444	535	22,500	18,700	1250	1700

(Continued)

EXPERIMENT 8-21 (CONTINUED) SOLVATOCHROMISM

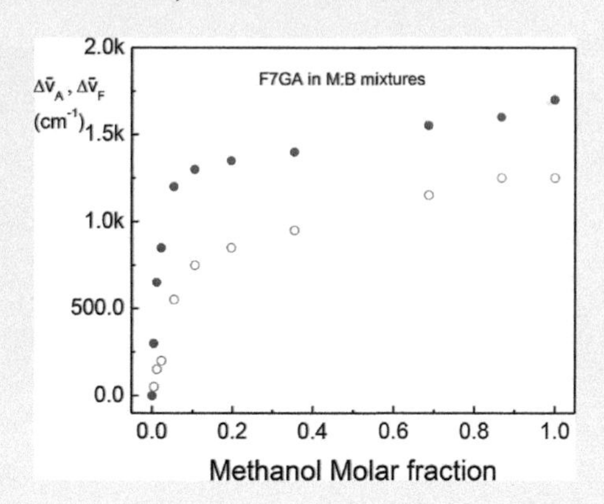

FIGURE 8-21.5 Absorption (blue open circles) and fluorescence (filled-red circles) plotted as a function of methanol molar fraction in mixtures.

CONCLUSIONS

Why we plotted spectral shifts as a function of a polar solvent molar fraction?

The solute molecule (F7GA) is surrounded by solvent molecules. But, how many molecules of B and M are in a given mixture? These solvents have different molar weights (MW) and different densities. Therefore, we recalculated % of M and B into molarities and calculated the molar fraction of M in each mixture. Next question is: why the solvatochromic shifts are almost completed with 0.2 M molar fraction? The polar fluorophore molecule preferentially attracts polar solvent molecules which form around the dye molecule a solvation shell. Of course, Brownian motions and Boltzmann factor determine the solvation statistics. Although nonpolar–polar solvent mixtures provide continuous changes in the dielectric constant, the mixtures should not be used in dipole moments estimation. A local dielectric constant in the solvent shell is different than in a bulk solution.

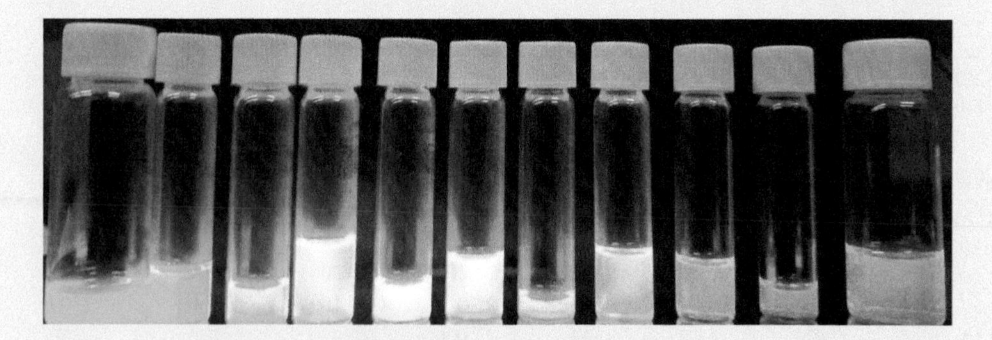

FIGURE 8-21.6 Photograph of F7GA in all benzene methanol used mixtures on a UV illuminator.

EXPERIMENT 8-22 FRET IN DONOR-ACCEPTOR PAIRS

Keywords: fluorescence, FRET, lifetime, TAMRA, AMCA.

FRET concept has been introduced already in Chapter 1. This through-space molecular interaction has found multiple applications in biology, especially in co-localization of proteins and interactions with ligands studies. Therefore, the most investigated FRET case is one donor-one acceptor system. Here, we will discuss this with an example of dually labeled peptide selectively cleaved by a metalloproteinase enzyme MMP-9. Typically, specific peptide sequence can now be routinely ordered from a commercial source. In our case, we used a Lys(AMCA)-Gly-Pro-Arg-Ser-Leu-Ser-Gly-Lys(TAMRA)-NH2 peptide that was synthesized by AnaSpec (AnaSpec, Freemont, CA, USA). As a donor, we used a coumarin derivative AMCA and as an acceptor a rhodamine derivative TAMRA, see Figure 8-22.1.

MATERIALS

- AMCA-peptide (donor, D)
- TAMRA-peptide (acceptor, A)
- AMCA-peptide-TAMRA (donor-acceptor, DA)
- Glycerol, ethanol, vials
- 1 cm × 1 cm fluorometric cuvettes

EQUIPMENT

- Spectrofluorometer (Varian, Inc., Eclipse)
- Carry 50 Spectrophotometer (Varian, Inc.)
- Fluorometer (FT200, PicoQuant, GmbH, Germany)

METHODS

1. Prepare stock solutions of D, A, and DA in ethanol. Shake solutions until compounds dissolve, but do not sonicate. Estimate concentrations of stock solutions using extinction coefficients for AMCA 19,000 cm^{-1} M^{-1} and 90,000 cm^{-1} M^{-1} for TAMRA.
2. Using stock solutions make equimolar concentrations (about 1 μM) of D, A, and DA in few milliliters of glycerol. Shake gently until solutions are homogeneous (check with a laser pointer).

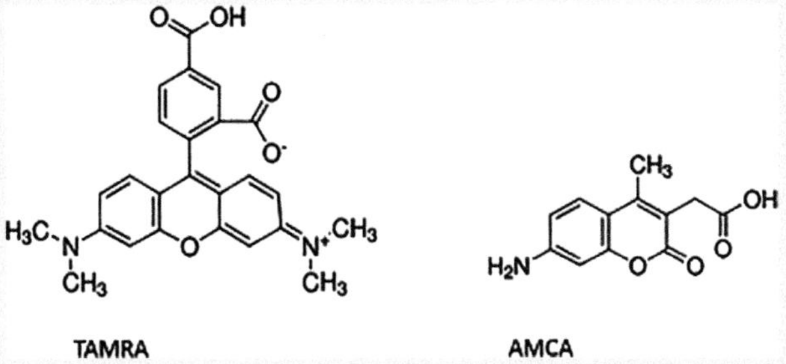

TAMRA AMCA

FIGURE 8-22.1 Structures of a donor AMCA and acceptor TAMRA used for the peptide labeling.

(Continued)

EXPERIMENT 8-22 (CONTINUED) FRET IN DONOR-ACCEPTOR PAIRS

3. Measure fluorescence spectra of all three solutions at 370 nm excitation. Maintain MA conditions, see Figure 8-22.3.
4. Measure lifetimes of D and DA fluorescence at 370 nm excitation and 450 nm observation. Maintain MA conditions, see Figures 8-22.4 through 8-22.6.
5. Estimate the energy transfer efficiency in the DA system from spectra and from lifetimes.

RESULTS

The donor-acceptor system, as well as donor and acceptor controls, is schematically presented in Figure 8-22.2. On top of the figure, we present model of one possible conformation. As for the extended peptide conformation, the separation between donor and acceptor

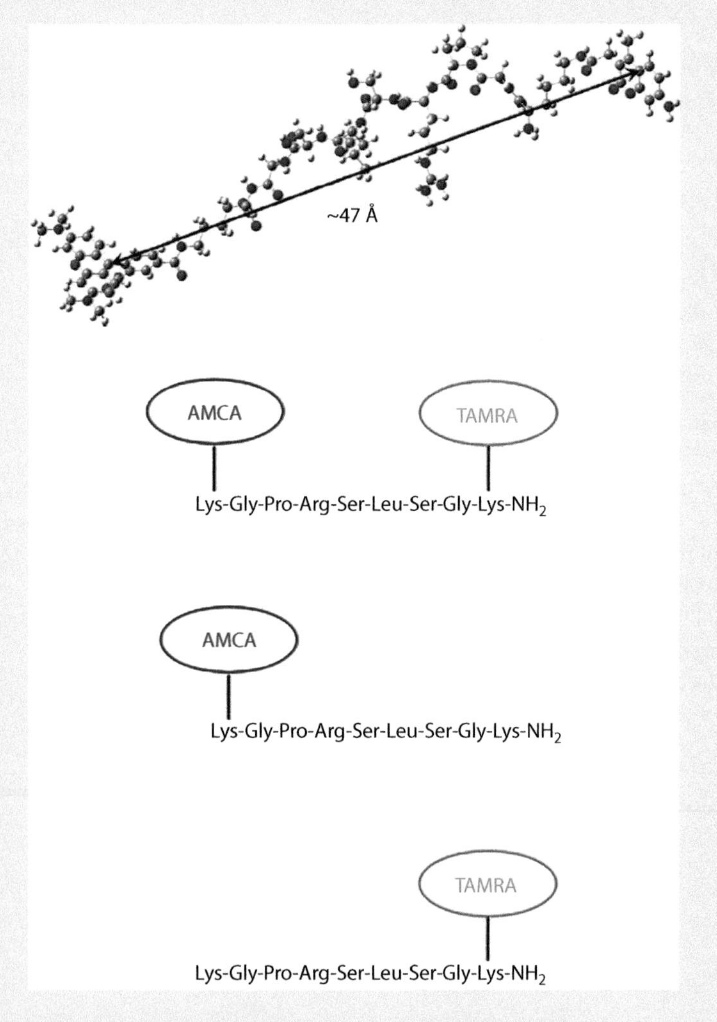

FIGURE 8-22.2 AMCA-peptide-TAMRA donor-acceptor system (top), AMCA-peptide donor (middle) and peptide-TAMRA acceptor (bottom).

(Continued)

EXPERIMENT 8-22 (CONTINUED) FRET IN DONOR-ACCEPTOR PAIRS

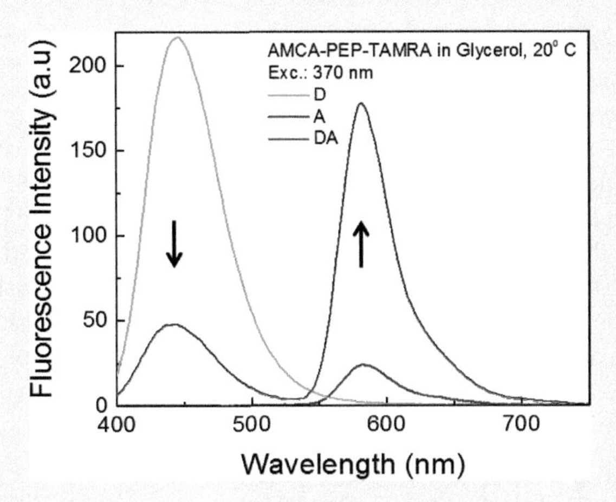

FIGURE 8-22.3 Fluorescence spectra of the AMCA-peptide donor (green), AMCA-peptide-TAMRA donor-acceptor system (red), and peptide-TAMRA acceptor (magenta).

is about 47 Å. However, we need to remember that for many possible conformations this distance will be much shorter. Estimated Förster distance for the selected donor-acceptor pair is about 39 Å.

First, we measure fluorescence spectra of D, A, and DA, see Figure 8-22.3. In DA system, the fluorescence of AMCA is strongly quenched while Fluorescence of TAMRA is significantly enhanced. This is consistent with FRET process from AMCA to TAMRA.

Next, we measured the fluorescence lifetime of the AMCA-donor in the absence and presence of TAMRA-acceptor. The AMCA donor shows a single-exponential intensity decay with a lifetime about 4 ns, see Figure 8-22.4.

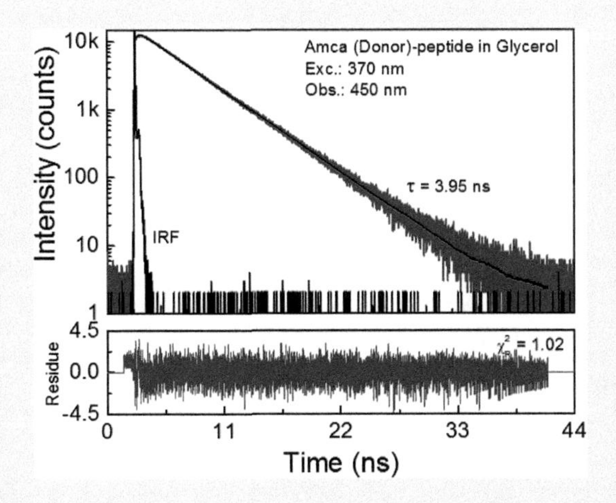

FIGURE 8-22.4 Fluorescence intensity decay of AMCA-donor in glycerol. The decay can be satisfactorily fitted with a single exponent of 3.95 ns.

(Continued)

EXPERIMENT 8-22 (CONTINUED) FRET IN DONOR-ACCEPTOR PAIRS

The fluorescence of AMCA is strongly quenched in the presence of TAMRA and is also very heterogeneous, see Figure 8-22.5. This intensity decay cannot be satisfactorily fitted with a single exponent.

The fluorescence intensity decay of AMCA-donor in the presence of TAMRA-acceptor requires three exponents to satisfactorily fit the data, see Figure 8-22.6. *Reminder.* For the calculation of FRET efficiency, an amplitude averaged lifetime should be used.

Heterogeneity of the AMCA-Peptide-TAMRA intensity decay observed at the donor emission (Figure 8-22.6) is a result of a distribution of D-A distances. The labeled peptide system exists in different conformations in the solution. The donor-acceptor distance will be different for each conformation. This means that for the observed FRET multiple different constant

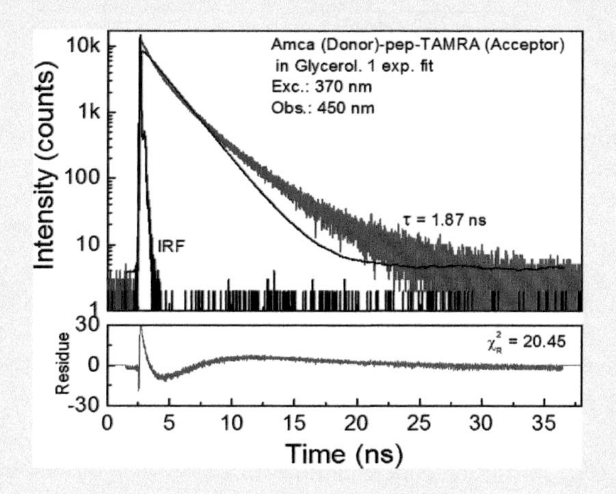

FIGURE 8-22.5 Fluorescence intensity decay of the AMCA-peptide-TAMRA donor-acceptor system observed at the donor emission (450 nm). A single-exponent fit.

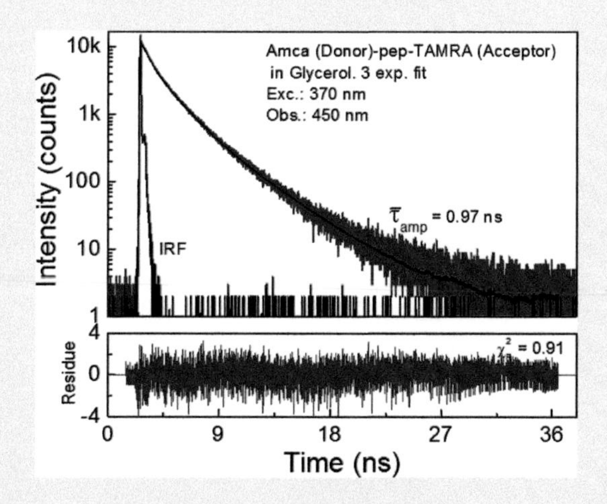

FIGURE 8-22.6 Fluorescence intensity decay of the AMCA-peptide-TAMRA donor-acceptor system observed at the donor emission (450 nm). A three-exponent fit.

(*Continued*)

EXPERIMENT 8-22 (CONTINUED) FRET IN DONOR-ACCEPTOR PAIRS

rates k_T are responsible, which results in a non-exponential (multi-exponential) intensity decay of the donor.

In summary, the fluorescence intensity of the AMCA-donor decreases in the close proximity of TAMRA-acceptor (in the AMCA-peptide-TAMRA) while TAMRA-acceptor fluorescence increases (Figure 8-22.3). The lifetime of the AMCA-donor also decreases in the presence of TAMRA-acceptor. The average transfer efficiency of about 60–70% can be estimated from both steady-state spectra data and from the lifetime change according to:

$$E = 1 - \frac{I_{DA}}{I_D} = 1 - \tau_{DA}/\tau_D$$

Note: We selected glycerol solutions to avoid a diffusion. How diffusion influences the FRET? If during its lifetime, the donor moves closer to the acceptor, the FRET process will be more probable and the donor will be quenched more efficiently. Below are presented measurements of the same AMCA-peptide-TAMRA system in PBS buffer. Of course, the water-based buffer has a significantly lower viscosity than glycerol and we expect stronger FRET.

Already from the fluorescence spectra, an extremely strong FRET is observed, see Figure 8-22.7.

The lifetime of the AMCA-peptide donor is longer in the buffer than in glycerol, see Figure 8-22.8.

Longer lifetime of the donor enhances a diffusion effect.

The quenching of the donor by FRET is very strong and reduces the donor lifetime to 0.3 ns, see Figure 8-22.9.

Diffusion significantly enhances FRET. The application of FRET is presented below. Addition of an MMP-9 enzyme (metalloproteinase-9) cleaves the peptide (separates donor

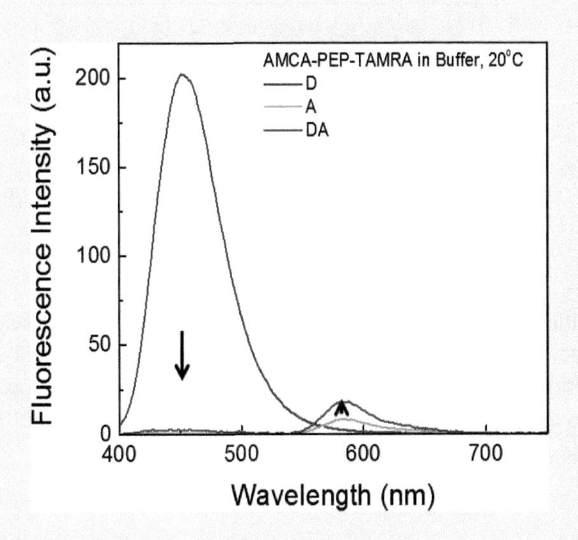

FIGURE 8-22.7 Fluorescence spectra of the AMCA-peptide-TAMRA donor-acceptor system in PBS buffer at the excitation of 370 nm.

(Continued)

EXPERIMENT 8-22 (CONTINUED) FRET IN DONOR-ACCEPTOR PAIRS

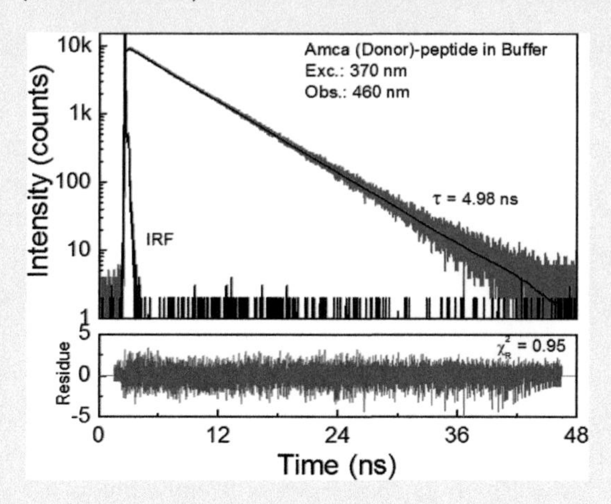

FIGURE 8-22.8 Fluorescence intensity decay of AMCA-donor in PBS buffer.

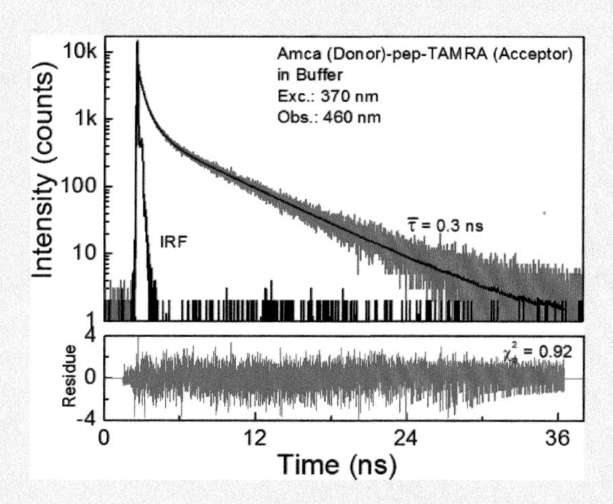

FIGURE 8-22.9 Fluorescence intensity decay of the AMCA-peptide-TAMRA donor-acceptor system observed at the donor emission (460 nm) in PBS buffer. A three-exponent fit.

and acceptor) resulting in a significant change of FRET. After the cleavage of the peptide linker by the enzyme, the donor and acceptor are separated and no FRET for this D-A pair is observed. A calibration with different concentrations of the enzyme allows for detecting the presence of the enzyme in the sample (effectively the MMP-9 sensing method). The observed intensity changes are presented in Figure 8-22.10.

(Continued)

EXPERIMENT 8-22 (CONTINUED) FRET IN DONOR-ACCEPTOR PAIRS

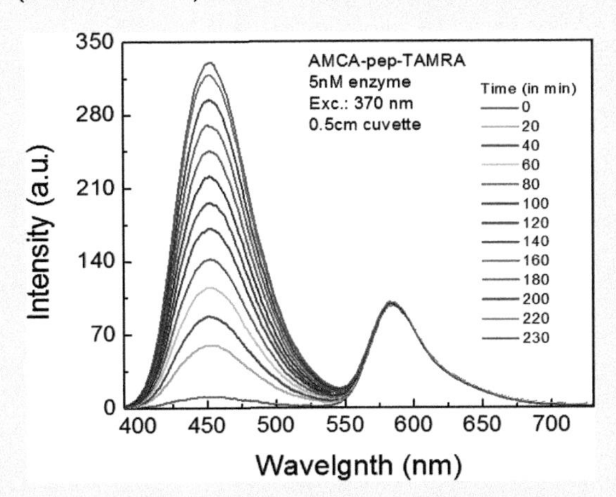

FIGURE 8-22.10 Cleavage of peptide in DA system by MMP9 enzyme results in the release of FRET. AMCA-peptide donor fluorescence increases. The cleavage process depends on time and on enzyme concentration/activity.

CONCLUSIONS

The linked donor-acceptor system requires low concentration of the FRET pair and can be easily measured in cuvette. Various applications of such FRET pairs are possible. Detection of active enzyme or viscosity measurements are some of them. If embedded in a polymer, the FRET pair emission will be sensitive to the polymer stretching/orientation. Small changes in the distance can be detected with a ratiometric (donor/acceptor) fluorescence measurements.

Note: Some results from this experiment were published in *"Dyes and Pigments 158 (2018), 60-64."* Figure 8-22.3 has been used with permission from Elsevire.

Index

C

D

E

I

J

K

L

M

N

R

S

9781032337371